PROCEEDINGS SERIES

NUCLEAR SAFEGUARDS TECHNOLOGY 1978

PROCEEDINGS OF A SYMPOSIUM
ON NUCLEAR MATERIAL SAFEGUARDS
ORGANIZED BY THE
INTERNATIONAL ATOMIC ENERGY AGENCY
AND HELD IN VIENNA, 2–6 OCTOBER 1978

In two volumes

VOL.I

INTERNATIONAL ATOMIC ENERGY AGENCY
VIENNA, 1979

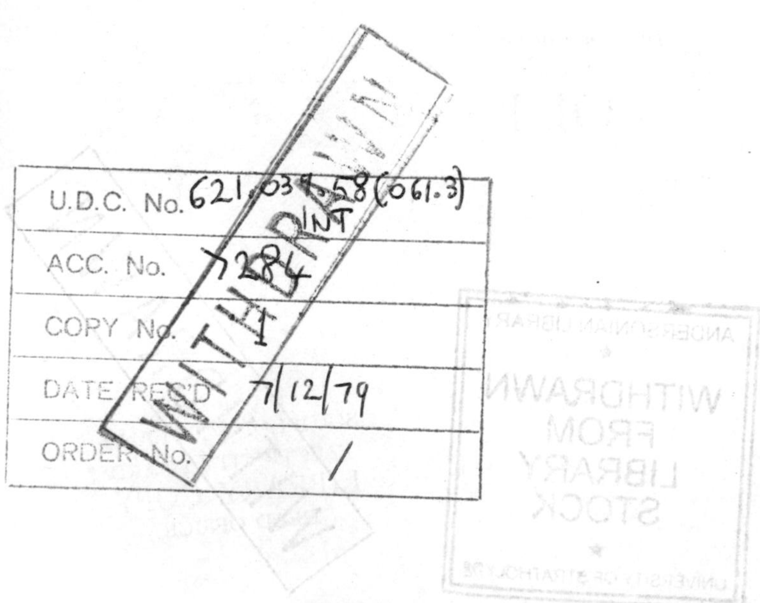

NUCLEAR SAFEGUARDS TECHNOLOGY 1978
IAEA, VIENNA, 1979
STI/PUB/497
ISBN 92–0–070079–9

FOREWORD

This was the fourth time that the IAEA had held an international symposium on the general theme of nuclear material safeguards, and the timeliness of this symposium, held in Vienna from 2 to 6 October 1978, was evident to all. World interest in international safeguards, noticeably increased by the ongoing International Nuclear Fuel Cycle Evaluation (INFCE), was high. So the topics of research and development in the field of safeguards, which had anyway been growing in importance, were significantly expanded. There was a clear need for researchers to come together in a common technical forum to discuss and evaluate their results and their expectations, and the date of the symposium had the combined advantages of being late enough to allow time for significant accomplishments to be reported and yet early enough to produce a meaningful input to INFCE.

The organization of a symposium such as this one always presents serious problems. International safeguards is itself a broad subject, extending from fundamental questions of how to design facilities to make them more easily safeguardable or how to design safeguards approaches to optimize effectiveness while minimizing such factors as intrusiveness or cost, to the more detail questions of how to measure specific material quantities or how to analyse statistically the resulting materials accountancy data. All these topics are important to international safeguards, and all have a rightful claim to a place in any broad-based symposium on nuclear material safeguards.

With all signs indicating that the symposium would be a large one, nobody should have been surprised that there were over 275 participants. Only rarely were there fewer than 100 listeners in the meeting room, and lively discussions were held throughout. The full texts of the papers, together with the discussions, are published here in two volumes.

The foreword of the proceedings of the last symposium in 1975 argued that "it cannot be claimed that all the problems of international safeguards have been solved". It may be disappointing to some to learn that, three years later, some of those problems still have not been solved. Technical problems, however, are noted for their capability of self-proliferation. Some problems have been solved, and significant progress has been achieved on a number of others. Those that still remain will be tackled with determination and confidence.

EDITORIAL NOTE

The papers and discussions have been edited by the editorial staff of the International Atomic Energy Agency to the extent considered necessary for the reader's assistance. The views expressed and the general style adopted remain, however, the responsibility of the named authors or participants. In addition, the views are not necessarily those of the governments of the nominating Member States or of the nominating organizations.

Where papers have been incorporated into these Proceedings without resetting by the Agency, this has been done with the knowledge of the authors and their government authorities, and their cooperation is gratefully acknowledged. The Proceedings have been printed by composition typing and photo-offset lithography. Within the limitations imposed by this method, every effort has been made to maintain a high editorial standard, in particular to achieve, wherever practicable, consistency of units and symbols and conformity to the standards recommended by competent international bodies.

The use in these Proceedings of particular designations of countries or territories does not imply any judgement by the publisher, the IAEA, as to the legal status of such countries or territories, of their authorities and institutions or of the delimitation of their boundaries.

The mention of specific companies or of their products or brand names does not imply any endorsement or recommendation on the part of the IAEA.

Authors are themselves responsible for obtaining the necessary permission to reproduce copyright material from other sources.

CONTENTS OF VOLUME I

GENERAL PAPERS (Session I)

FACILITY DESIGN CRITERIA FOR INTERNATIONAL SAFEGUARDS (Session II, Part 1)

ELECTRONIC PROCESSING OF SAFEGUARDS INFORMATION (Session II, Part 2)

SAFEGUARDS TECHNOLOGY FOR URANIUM ENRICHMENT FACILITY (Session III, Part 1)

SAFEGUARDS TECHNOLOGY FOR FUEL FABRICATION FACILITIES (Session III, Part 2)

SAFEGUARDS FOR NUCLEAR POWER REACTORS (Session III, Part 3, and Session IV)

CONTAINMENT AND SURVEILLANCE (Session V, Part 1)

DESTRUCTIVE AND NON-DESTRUCTIVE MEASUREMENT TECHNOLOGY (Session V, Part 2)

Session I

GENERAL PAPERS

Chairman: H.W. SCHLEICHER (CEC)

EXPERIENCE IN THE APPLICATION OF IAEA INSPECTION PRACTICES

T. HAGINOYA, M. FERRARIS, W. FRENZEL,
I. KISS, F. KLIK, V. POROYKOV, L. THORNE,
S. THORSTENSEN
Department of Safeguards,
International Atomic Energy Agency, Vienna

Abstract

EXPERIENCE IN THE APPLICATION OF IAEA INSPECTION PRACTICES.

In 1977, more than 700 times Agency inspectors visited nuclear installations in 45 States throughout the world to verify nuclear material through record audits and material characterization or measurements. The experience during the three years since the last symposium proved that there are only a few types of facilities at which the Agency can fully base its statements on quantitative verification of mateiral, while there are many areas where the inspector must rely on semi-quantitative information like γ-spectroscopy, or qualitative information like containment and surveillance measures. Techniques have been improved, but a number of procedures to be further developed were identified by the inspectorate.

1. INTRODUCTION

Papers presented at previous safeguards symposia have described the inspectorate's approach to the safeguarding of facilities such as reactors, research establishments and fuel fabrication plants. At the time of the 1970 Karlsruhe symposium, descriptions of field work were confined to isolated examples of safeguards exercises aimed at establishing the feasibility for the future of safeguards at basic plants such as LWR reactors. Routine inspection activities, such as those carried out at research reactors and laboratories, did not warrant description at the Karlsruhe symposium.

By the time of the 1975 symposium in Vienna[1] the scope and scale of the inspectorate's activities had significantly changed. At that symposium in papers describing the Operations Division's activities, accounts were given of work at power reactors, low-enriched fabrication plants and large research centres. At that time the main effort was being devoted to establishing meaningful procedures at fabrication plants and the paper reflected this.

[1] Safeguarding Nuclear Materials (Proc. Symp. Vienna, 1975) **1** and **2**, IAEA, Vienna (1976).

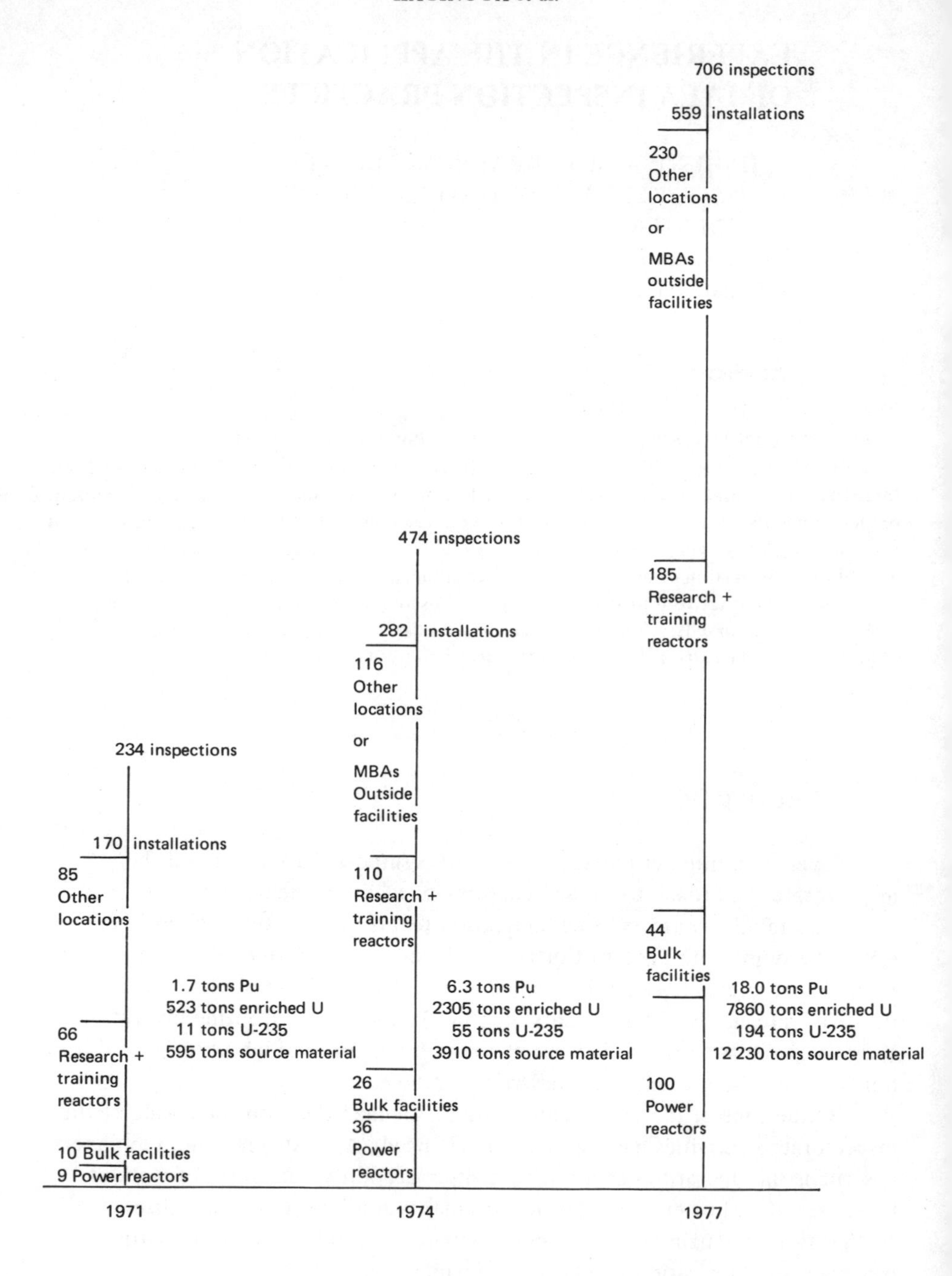

FIG.1. Scope of IAEA safeguards operations.

Now, three years later, the picture has again changed, reflecting an acceleration in the problems confronting the Agency's field inspectorate. Figure 1 shows the scope and scale of IAEA safeguards operations from 1971 to 1978. Since 1975 the number of facilities in all categories has more than doubled, the amount of Pu and source materials under safeguards has increased threefold, and the number of inspections has been increasing correspondingly. Last year (1977) saw the start of continuous inspection at reprocessing plants, the implementation of inspection at fast reactors, which are in operation, and the successful inventory checking of fast critical assemblies and of large mixed-oxide fuel fabrication plants. In short, safeguards should also be applied to enrichment plants and thus the stage will be reached where the inspectorate's daily work covers the entire fuel cycle in its most advanced form.

2. REACTORS

The application of safeguards to reactors continues to be based upon the verification and counting of items supported by containment and surveillance measures.

The most extensive experience has been accumulated in connection with the application of safeguards to light-water reactors. At present the Agency's two Operations Divisions apply safeguards to more than 100 power reactors and large research reactors.

Light-water reactors have their own intrinsic containment, the pressure vessel, for the inaccessible part of the material.

Regarding certain types of on-load refuelled reactors, considerable problems are caused by the large number of items; this applies in particular to the irradiated fuel storage. The on-load refuelling methods do not provide intrinsic containment. Consequently attempts are being made to verify fuel movements in and out of the reactor core. The development and application of satisfactory camera systems – TV or optical – is one part of the problem. The other part is devising means of reviewing the record of such devices. In this respect the very success of a device can produce its own problems. The first successful surveillance devices were simple movie cameras operating in the 'single-shot' mode. Individual photographs were taken at semi-random intervals. The total time period which could be covered between servicings, was a function of film length and the allowable interval between shots. This service interval was about two months for light-water reactors. The introduction of TV and video recorders brought about a marked improvement in the surveillance capabilities. The tape length became orders of magnitude longer than the photographic films, thus permitting very short intervals to be set between the shots and also for the total time between services to be extended. Events are registered even in the almost total absence of discernable light. At some stage, however, the video tapes have to be reviewed by an inspector, usually on site, which presents problems in maintaining an alertness over long periods to detect a significant event in a background of trivia. Similar problems have occurred in astronomy and nuclear physics, where large numbers of photographic records must be scanned to detect extremely low frequency events. Some of the successful techniques from this field can also doubtless be applied in evaluating the recordings of surveillance caneras.

Techniques for the non-destructive testing of irradiated fuel have not yet reached the stage of development where they can be applied routinely by non-specialists. To date the most intensive use of such techniques has been in connection with the verification of fuel from natural uranium reactors. These verification operations require considerable planning and the engagement of specialists. The procedures are based upon gamma-measurements of certain fission products which lead to an estimate of the burnup and the cooling time of fuel assemblies.

3. FUEL FABRICATION PLANTS

The field activities in applying safeguards to low-enriched fuel fabrication plants have been described elsewhere. The SAM-II instrument remains the inspector's fundamental tool for the physical verification of nuclear material.

However, in high-enriched and plutonium fuel fabrication plants great progress has been made in non-destructive assay techniques. A high-level neutron coincidence counter has been used successfully for measuring mixed-oxide and fast-reactor fuel assemblies for verifying plutonium. (In at least one major facility it can be described as almost routine.) At the present state of development the instrument works well but requires semi-permanent installation and cannot be readily calibrated in the field. At present analysis and evaluation of the measurements taken in the field require several man-weeks of effort though considerable progress has been made in automating the data reduction with a mini-computer. Further development of implementation procedures is continuing in order to make the system a tool which will fulfil the required qualifications.

Since a large part of this symposium is devoted to accounts of instrument development throughout the worked, it seems to be worthwhile pausing to consider which of an instrument's attributes are virtues in the eyes of the user, the field inspector.

First and above all is its reliability under extreme conditions – the instruments are used worldwide; the components in the measurement chain must be assembled by non-specialists; and little or no maintenance is possible under field conditions. No matter how clever or sophisticated the instrument might be, it is useless for safeguards purposes if it cannot function reliably. Next in importance to reliability is simplicity of operation, followed by ruggedness and ease of transport. During a large number of inspections the SAM-II has shown that it has these attributes to a very high degree. Successors have been put forward but to date none of them has proved to be as suitable for routine inspection as the SAM-II.

This lesson should be applied elsewhere, for example in the seals area. Many sophisticated types of seal have been developed but so far none has managed to win acceptance by field staff, the reason being its complications without compensating benefits. The original simple metal seal is still the type most used.

Regarding finished fuel assemblies, there is a demand for an NDA method which permits, even with a relatively large limit of error, the composition of their nuclear material contents to be measured.

4. CRITICAL ASSEMBLIES

For some time the inspectorate has faced problems in safeguarding critical assemblies which often contain material of high strategic value – particularly important examples are fast critical assemblies. A characteristic of such a facility is that the fuel is in the form of small plates or rods, which can be easily handled and easily slipped into a pocket. Surveillance and monitoring devices have been developed to detect the passage – for example, in a coat pocket – of such fuel. Until recently the problem has been that of adequately verifying the inventory of such a facility as a preliminary to installing surveillance devices. A complete verification by item counting and non-destructive measurement has now been carried out in such a facility. The data reduction problem involved in the measurements has proved, however, to be immense. Although, clearly this aspect of the problem will be eased with subsequent verifications, alternative approaches are desirable. One, which is currently receiving attention, is the possibility of monitoring the performance of the assembly when criticality is reached to ensure that the characteristics are typical of a device with the stated composition.

5. REPROCESSING PLANTS

The first reprocessing plants to come under Agency safeguards, other than as a demonstration exercise, have now been operating for one year. During this time they have been under continuous inspection. Inspectors have been continuously present in the plant or in its vicinity, ready to witness important operations. Much has been learned of the effectiveness of the applied techniques but areas of weakness have also become clear.

One of the biggest problems in these plants has been the difficulty of dealing with samples of the input and output solutions of the chemical treatment sections. The actual taking of the samples has proved no problem thanks to the co-operation between the inspectors and the plant operators. Problems remain in the subsequent treatment and transport of the samples. There is a great need for further development of techniques for material verification, which do not necessitate the skilled labour of conventional methods. In addition, some problems related to national regulations for domestic and international transport of highly active solutions require attention.

A further aspect of these plants which presents problems is the general design of the facility in relation to safeguards. In many areas the safeguarding problems would have been much easier had the importance of safeguards been realized by the designer. This is relevant to all plants – fabrication plants, reactors or otherwise. Some of the prime factors of importance are:

Ease of surveillance of key points such as plutonium off-loading stations; and
Ease of measuring material quantities (simple, direct, easily verified procedures) and avoidance of design features which introduce confusion in accountancy verification.

The study of safeguards features and the respective improvement in plant design may reward all involved.

6. SUMMARY AND CONLUSIONS

Experience in the application of inspection practices shows progress in techniques in all areas of inspection activities:

TV cameras have been introduced as surveillance tools. This has proved to be a significant improvement in comparison with simple movie cameras, permitting a much longer period and high frequency of shots. Still the reliability of the equipment and the records evaluation in the field remain problems to be solved.

In the use of non-destructive assay techniques the greatest progress has been made in the area of highly enriched and plutonium fuel.

The first year of continuous inspection at reprocessing plants has been completed and much has been learned.

Experience confirms that benefits for the inspector very often mean benefits for the operator.

DISCUSSION

D. GUPTA: In your presentation you mentioned the problem of treatment and transport of analysis samples from a reprocessing facility. Have you considered the possibility of getting such samples analysed at the facility's laboratory? By verifying the performance of the operator's measurement system, the Agency inspectors could then get very fast results and avoid possible effects on the results due to treatment and transport conditions.

T. HAGINOYA: Yes. We are considering this possibility at the IAEA. In fact, we are using a facility laboratory to reduce the volume of samples by evaporation.

EXPERIENCE GAINED WITH EURATOM'S NUCLEAR MATERIALS ACCOUNTING AND REPORTING SYSTEM

M. SCHMITT, H. KSCHWENDT, A.G. MAXWELL,
M. LITTLEJOHN
Euratom Safeguards Directorate,
Luxembourg

Abstract

EXPERIENCE GAINED WITH EURATOM'S NUCLEAR MATERIALS ACCOUNTING AND REPORTING SYSTEM.

The entry into force of the Verification Agreement in early 1977, linked to the wish to update the old Euratom System created in 1959, required that a new Euratom system (Community Regulation) be established. The main aspects of this new system, together with the practical experience gained in one and a half years operation, are presented. Certain basic accounting principles incorporated in the Euratom system, which are somewhat different from IAEA principles, are discussed in detail. This includes the notion of accounting date, some correction procedure aspects as well as the continuous updating of the book inventory to the physical reality in form of inventory changes. The effect of these differences when comparing IAEA and Euratom data is also mentioned. Furthermore, certain of the verifications carried out routinely on the operator's reports as well as on the reports submitted by Euratom to IAEA, are described and quantifications are given. Some mention is also made of areas where Euratom's role goes beyond that of the IAEA, i.e. the reporting implications of accounting for material by origin and control of particular use of the materials as well as verification of ore production and processing activities. Finally, improvements and simplifications concerning reports to the IAEA are proposed.

1. INTRODUCTION

In 1973, the IAEA, the European Atomic Energy Community (hereafter called Euratom) and seven Member States signed an Agreement in implementation of the Non-Proliferation Treaty, which entered into force on 21 February 1977 [1]. In 1976 and 1978, the two remaining Member States, together with Euratom and IAEA, signed similar Agreements, one of which entered into force on 14 August 1978. Euratom itself has possessed a supranational safeguards system since 1959. It was implemented via the European Commission Regulations 7 and 8.

The rapid development of nuclear industry, together with a parallel development of safeguards techniques, created within Euratom already in the late 'sixties the wish to update this Euratom system. The 18 years experience gained with this system, together with the above-mentioned Agreements and the Euratom Treaty [2], formed the basis for the new European Commission Regulation No.3227/76 [3], which replaced and repealed old Regulations 7 and 8 and which entered into force on 15 January 1977.

Any nuclear material control system makes use of: Measurement; Accountancy; Inspection; and Evaluation. This paper deals exclusively with accountancy. It reports on the practical accounting experience gained in the past one and a half years with this new Euratom System and compares it with the IAEA system. Such a comparison was already discussed on a theoretical basis three years ago [4]. We are now in a position to evaluate those differences which have practical consequences either on the accounting/reporting level or during inspection.

The Euratom accounting system requires among other things that operators of nuclear installations keep an internal accountancy (records) and submit accounting reports to Euratom at regular time intervals. Euratom in turn is required under the Agreement to submit reports to the IAEA.

When the seven Member States Agreement entered into force, no Facility Attachments were finalized. This situation created special problems during the initial phase since the operators could not be advised on the particular requirements imposed on them.

Hence, reports could not be sent in the form and format of the Agreement at the outset. To minimize time taken in changing, Euratom undertook, with co-operation from the operators, an extensive educational campaign including seminars and test reporting old and new systems in parallel. In close collaboration with the IAEA, reports were progressively converted to the IAEA format. Now, more than 90% of the installations (over 200) situated in the seven non-nuclear weapon States of the European Community have been converted to the new accounting system.

2. ACCOUNTING PRINCIPLES

In creating its new accounting system, Euratom has remained faithful to its basic accounting principles, which are compatible with world-wide accounting practice and which have proved their effectiveness during the 18 years of existence of Euratom safeguards. The accounting system of Euratom is, therefore, in essence, in no way different from any other type of modern accountancy. The accounting principles apply both to operators and to Euratom's nuclear material accountancy. The reports to Euratom must be justified by the operator's accountancy.

To appreciate the working of the Euratom system, the most important accounting principles upon which this system is based are now enumerated; their application is, in some cases, the key to the differences between the Euratom and IAEA systems.

The first accounting principle is the *accuracy* of data which are input into the accountancy system; the accountancy input must be well founded, e.g. the figures must be based on the best available information.

Another principle is the *completeness* and *continuity* of the accountancy, i.e. all operations must be recorded chronologically, continuously and in the right sequence; consequently there must be no gaps.

A further principle is that for every operation a *supporting document* must exist; no accounting entry can be accepted without adequate record justifying it.

The next principle is *self-consistency* of the accountancy. This means that the various types of accounting document, i.e. delivery notes, transfer documents, analysis reports, general and subsidiary ledgers etc., must constitute a coherent system.

Another principle is the *timeliness* of the accountancy. This is only possible if internal communications function and exact internal instructions are given which stipulate the maximum recording delay permissible for any entry. This delay should be kept as short as possible to guarantee an accountancy as up to date as possible.

Custodianship is the last principle mentioned in this connection as it means that, within the installation's organization, responsibility is clearly and officially defined and assigned. For each operation the person responsible is designated so that, for example, no material transfer can take place without those responsible for both the receiving and the shipping having their signature on the transfer note.

If all the above criteria are fully met the accounting system can be considered as trustworthy.

The above principles ensure that the installation accountancy is in good order and that it is faithfully reproduced in operators' reports to Euratom. The reports submitted to Euratom are, therefore, a precise mirror of the internal accountancy.

3. COMPARISON WITH THE IAEA SYSTEM

In this section we discuss in more detail certain differences between the Euratom and the IAEA accounting systems. The implementation of the accounting principles described above, and in particular the continuous updating of the accountancy and subsequent reporting, provides in the Euratom system the possibility of timely reproduction of the accounting data of the operator. This is particularly important when inspectors in the field can be supplied with information which corresponds exactly to the accounting in the installations. The IAEA system on the other hand requires less data to be reported or certain data to be reported at a later date.

The additional data of the Euratom system concern mainly the inventory change report. Euratom, therefore, for accounting purposes, considers the inventory change report as *the* basic source of information since all events affecting the accountancy must be reported therein.

The main differences between the two accounting systems are now enumerated and the consequences are discussed.

3.1. Accounting date

The basic difference between the two systems is the fact that the Euratom system requires two dates to be reported in the inventory change report: the *accounting date* and the *original date,* whereas the IAEA system requires only the original date.

The accounting date is the date at which an inventory change is accounted for, whereas the original date is the date on which a physical event occurred.

Normally, these two dates are equal, for example in the case of transfer of nuclear material. There are, however, situations where these two dates are different. Let us suppose that the operator has shipped on 28 January a quantity of 100 to another installation, but, by mistake, he has accounted for and reported 200. On 9 March, the mistake is detected. The operator therefore corrects his accountancy as well as the report by entry lines which show as accounting date 9 March, whereas the original date is 28 January. The same procedure applies in the case of a physical inventory taking where the measurement results – and consequently the material unaccounted for (MUF) – are sometimes known only several weeks after the inventory taking. The entry lines adjusting the book inventory to the physical reality must in the records and reports show as accounting date the date when the measurement results became known and as original date the day of the physical event, i.e. the inventory taking.

A direct consequence of the concept of the accounting date is its influence on the book inventory. In accordance with the accounting principle of continuity, operators are obliged to account, in a sequential manner, for all inventory changes. The book inventory at the end of a day is then the algebraic sum of the book inventory at the end of the previous day and of all inventory changes accounted for during that day. Since corrections and other entries bearing a prior original date are treated in the same way, the book inventory at a given date *does not* change any more at a later date. This has practical advantages because a book inventory, once verified, need not be verified systematically again at later time.

In the IAEA system, on the other hand, the book inventory is based on the original date only. Any quantitative correction to a previous entry will therefore change *retroactively* the book inventory. For example, this can mean that on 26 June, the book inventory of 31 January is 100, whereas on 27 June, after processing of a correction, the book inventory of 31 January became 200.

This fact has practical consequences on IAEA book balances. The first balance will start with a figure A and end with a figure B. The next balance will then *not* start with figure B, but with another figure C, because B has been modified in the meantime by corrections. This obviously renders any book verification difficult. The IAEA inspector arrives on inspection with book inventories which may month by month all be different from those of the Euratom inspector and the operator. Consequently, time is spent in reconciling the figures to the detriment of other verification and auditing activities. Periods which had already been verified during previous inspections must sometimes be re-verified by the IAEA. Euratom does not have this problem because of its accounting principle of continuity.

We fully recognize the need for the original date for evaluation purposes. However, the two objectives – accounting verification and evaluation – can only be achieved if both the accounting date (for accounting purposes) and the original date (for evaluation purposes) are implemented. The transformation of the data from the one purpose to the other is then a pure computer matter.

3.2. Corrections

Many circumstances require that records and reports are corrected. There might be errors at the installation or the operators' record and report provisional data, which have to be corrected later when precise measurement results are known. This calls for the provision of a correction procedure.

The correction procedure built into the Euratom system consists of two types of entry, where one is a cancellation entry (i.e. removal of the incorrect data, called "deletion"), followed by an insertion entry (addition of the correct data) [5]. Correction entry lines always bear both dates, the accounting date and the original date.

In the Euratom system, because of the principle of continuity, corrections affect the book inventory on the accounting date only and leave previous book inventories unchanged. In the IAEA system, corrections can only be performed under the original date since it is the only date required by the IAEA system and, therefore, they affect the book inventory retroactively.

3.3. New measurement

Operators often perform new measurements of quantities of nuclear material for internal pruposes. These new measurements are either performed for single batches or for whole production campaigns, i.e. for partial inventory taking. Following the accounting principles of continuity and timeliness, the results of such new measurements must, in the Euratom system, be reflected in the accountancy in terms of book inventory changes. This procedure guarantees that the operator's as well as Euratom's accountancy correspond as much as possible to the physical reality. In the case of physical inventory taking, as required by the Agreements, the books are adjusted in the same manner.

The differences accounted for as new measurements are MUF components. The IAEA system, however, requires that such differences are reported only after a physical inventory taking as one consolidated entry in the material balance report. MUF is therefore known to IAEA only after the end of a material balance period, i.e. after a relatively long time, thus affecting the principle of timeliness.

This situation leads to further discrepancies between IAEA's and Euratom's book inventories with the consequences already described under 3.1. The same is true for rounding adjustments, which are also not required in the IAEA inventory change report.

3.4. Reported book inventory

To be able to verify if the accounting principle of self-consistency has been properly applied, the Euratom system requires that operators report monthly in the inventory change report the ending book inventory per category of nuclear material. Even if there has been no inventory change during a month, the operator is obliged to report thus. This obliges the operator to verify his accountancy before dispatching the reports since a great number of clerical mistakes can be detected in this way – reports inadvertently not dispatched by operators, reports mislaid or lost in the post, information drop-out during computer processing, or even simple writing mistakes.

The IAEA system does not require a monthly reported book inventory. Therefore, no systematic check on the timely receipt of all reports is possible and a cross-check between the operator's and Euratom's accountancy on the one hand, and the IAEA accountancy on the other, is not possible before the next inspection takes place. The tracing and identification of eventual differences during inspection for periods which last up to six or twelve months could, however, be complex and time-consuming.

3.5. Minor differences

There are further systematic differences between the two accounting systems. The most significant one is the fact that Euratom has subdivided the enriched uranium into two categories, with the boundary at 20% enrichment. This has been done in order to allow the attribution of MUF to a smaller enrichment band than IAEA does.

Other differences are the inventory change code "nuclear transformation" which covers both nuclear production (positive figures) and nuclear loss (negative figures). Transfers between the nuclear weapon and non-nuclear weapon States of the Community are, according to the Euratom Treaty, considered as domestic transfers whereas, for IAEA, they are foreign transfers. There are similar differences for the starting point of safeguards which do not exist at all in the Euratom system.

In its computer system the IAEA has implemented a very rigid sequential numbering system for entry lines. Euratom has a similar system. In both accounting systems, this numbering is used to check the clerical completeness of the reports.

The Euratom numbering system cannot be used for the reports dispatched by Euratom to IAEA, because the number of entry lines is different in both systems. Certain entries are not reported at all to IAEA (for example, entries concerning uniquely those areas beyond IAEA safeguards) whereas others are split into two or more entry lines (for example, inventory changes concerning rebatching).

3.6. Material balance report

The Euratom accounting principles also affect the material balance report. The Euratom material balance report is different from that of the IAEA because of the concept of accounting date, the new measurement entries, the different book inventory and consequently MUF. In addition, certain transfers such as those between the nuclear weapon and non-nuclear weapon States of the Community, which, according to the Euratom Treaty, are domestic transfers, whereas for the IAEA they are foreign transfers, have to be treated in a different way. The Euratom computer system is able to reproduce both types of material balance report.

The main role of the Euratom material balance report is to allow additional verifications of the accuracy, continuity, self-consistency, completeness and timeliness of the information reported to Euratom. It furthermore conveys an overall picture over a certain period in time.

On the basis of all this information, Euratom is in a position to calculate both the Euratom and the IAEA MUF in two different ways based on two different sets of data. One way is according to the usual material balance scheme. The other way uses the inventory change report alone. MUF, as defined by IAEA, is the sum of all new measurement inventory changes, whereas MUF, as defined by Euratom, is the sum of all inventory changes with original date before and accounting date after a physical inventory taking.

3.7. Areas outside IAEA safeguards

Euratom's role goes beyond that of IAEA in certain areas. According to Article 2(e) of the Euratom Treaty, Euratom shall "make certain, by appropriate supervision, that nuclear materials are not diverted to purposes other than those for which they are intended". This goes beyond the objective of IAEA Safeguards, which, according to Article 28 of the Agreement, "is the timely detection of diversion of significant quantities of nuclear material from peaceful nuclear activities ...". This means that in the Euratom system *all* nuclear materials are under Safeguards and that there is *no* termination of Safeguards.

Consequently, ore movements and ore processing activities have to be reported to Euratom, but are not reported to the IAEA. For the same reason, the Euratom system does not allow exemption from safeguards, whether for nuclear or for non-nuclear activities, but only exemption from declaration.

Another important area where Euratom goes beyond IAEA safeguards are the "particular safeguard obligations" assumed by the Community in agreements concluded with non-Member States. This means in practice that the Euratom system must separate accounting-wise material under such obligations and establish separate balances per obligation.

Finally, we mention that the Euratom system includes reporting on the "use" of nuclear material.

3.8. Transformation from the Euratom to the IAEA system

Because of the systematic differences between the two accounting systems, as already explained, the Euratom accounting data have to be transformed in such a way that they comply with the formal reporting requirements of the IAEA. The Euratom system is designed in such a way that the necessary transformations can be performed completely by computer.

In this context, we would like to emphasize that the Euratom system can accept a maximum of data, even if an entry line contains errors. We see no reason, for example, to exclude an entry line from a book inventory calculation if a key measurement code is missing or a material description code is wrong. Such errors, however, are *flagged* and the flag is reproduced on the listings until the errors are removed. Reports as a whole are never rejected. This ensures that maximum information is contained in the Euratom accountancy.

Each month, two runs on the computer are performed. The first run comprises the processing of the data as reported by the operator and the second run includes all the corrections resulting from the verifications described in Section 4 and others. The updated data then form the basis for the reports to the IAEA.

The transformation from the Euratom to the IAEA system is already in itself a considerable task. Each entry line of the inventory change reports and of the physical inventory listings undergoes individually the necessary transformations. For material balance reports the situation is more complex. The transformation may consist of deletion, modification or generation of information.

The following information is deleted: accounting date, the inventory changes concerning new measurement and rounding adjustment (deleted from inventory change reports but included later in material balance reports), category changes between low- and high-enriched uranium and all information concerning areas where Euratom's role goes beyond IAEA safeguards (ores, particular safeguards obligation, use, etc.).

Modification of information mainly concerns certain inventory changes such as nuclear transformation, which becomes either nuclear production or nuclear loss, transfers between nuclear weapon and non-nuclear weapon States of Euratom, which are domestic transfers for Euratom and foreign transfers for the IAEA, and transfers which cross the IAEA starting point of safeguards, which are again domestic transfers in the Euratom system. Further examples are splitting of entry lines, either because the IAEA system requires two entries whereas Euratom requires only one (for example, rebatching), or because figures are too large (Euratom has 10 digit quantity fields whereas IAEA has 8). Modification of information already requires more complex transformatlion procedures than deletion of information.

The most complex part of the transformation is the generation of information. This concerns three items – the correction by replacement, the sequential numbering of the entry lines, and the calculation of the material balance report.

In the IAEA system, corrections are performed under the original date as replacements of the wrong entry by the correct one. These replacements are nothing more than a combination of Euratom's deletions and additions at the same logical place in the natural sequence of the entry lines.

After transformation, the entry lines are numbered sequentially as required by the IAEA system. Special care must be given to correction entries because they bear, in addition to their own serial number, a reference to the entry to be corrected.

The material balance reports cannot be directly transformed from that reported to Euratom. Let us take an example. An installation in a non-nuclear weapon State reports two shipments, one to an installation in another non-nuclear weapon State and the other one to a nuclear weapon State. In the Euratom material balance report there will be only one domestic shipment entry covering both transfers. In the IAEA material balance report, however, two entries are required since one is a domestic shipment whereas the other is a foreign shipment. This example shows that one always needs the already transformed inventory change report for the establishment of the IAEA material balance report. The situation becomes even more complicated when corrections to an already reported material balance report have to be made. Special procedures based on comparison of the data before and after correction had to be developed to cater for this problem.

4. ACCOUNTING VERIFICATIONS

In this section we describe the major accounting verifications carried out routinely by the Euratom Accounting Department. In this paper, by accounting verifications we understand all those activities which ensure the accuracy, continuity, self-consistency, completeness and timeliness of the accounting system. The other control aspects such as measurement, inspection and evaluation, are not dealt with in this paper.

Many of these verifications seem to be obvious. But, when one appreciates the large amount of data reported every month to Euratom, one will see that these verifications have to be carefully organized and clear internal instructions and responsibilities must exist to ensure that they are carried out properly.

In past years much has been said and written about evaluation techniques like measurement precision and MUF evaluation. We fully recognize the importance of this evaluation, but no meaningful evaluation can be carried out if it is not ensured that the data forming the basis of the evaluation are complete and correct. This statement seems to be self-evident, but our experience shows that generally the importance of accounting verifications is underestimated. In this context, we would like to refer to an affair concerning the disappearance of a large amount of nuclear material about 10 years ago, which had received extensive coverage in the press. This disappearance had not been detected by, for example, MUF evaluation, but by simple comparison of shipper's and receiver's declarations, the latter having been absent in this particular situation. This example shows the fundamental importance of systematic and efficient accounting verifications.

Euratom has over 400 nuclear and non-nuclear installations under its control. Under the old Euratom system these installations reported approximately 20000 entry lines per month to Euratom. The corresponding figure in the new system cannot be accurately estimated because in some cases the facility attachments are not yet finalized (or yet relevant – nuclear weapon States), and consequently the reporting obligations are not yet fully defined.

The first and most basic accounting verification is to check that all reports are received and on time. This is possible because a monthly report is normally required even if there were no inventory changes during the month. Routine telexes are sent to the operators if their report has not been received within the reporting deadline.

During computer processing, arithmetic checks are carried out. The reported monthly book inventory is compared automatically with the book inventory calculated by the computer. Every inconsistency is signalled and immediately followed up. This check ensures that each month the book inventory as kept by Euratom agrees with the book inventory kept by the operator. Experience has shown that in the past about 4% of monthly balances contain this type of error.

Another fundamental verification is to check whether all material movements within Euratom tally, i.e. whether the receiver reports the same quantity, quality etc., as the shipper. Previous experience shows that each month about 100 differences, both large and small, were found in the course of this type of check, or it was discovered that one of the installations involved forgot to report at all. These inconsistencies lead to routine follow-up procedures. This could be avoided altogether if the operators would simply respect the rule that the receiver always reports the shipper's data precisely. This in no way precludes the possibility of the receiver carrying out subsequent modifications if he does not agree, for one reason or another, with the shipper's quantities. This is certainly an area where Safeguards Authorities have still some educational work to do with the operators.

In this context, another problem is encountered, namely the coding in the shipper's reports of the receiving material balance area in the case of shipment. Complex installations are subdivided into a number of material balance areas. Thus, a shipper often does not know into which material balance area nuclear material goes in the receiver's installation. In such a situation we require that the shipper ask the receiver for the correct code and, in the meantime, to submit to Euratom the best information available at the moment of reporting, which is normally the installation code. Any correction should then appear in the reports of the following months.

To simplify this problem, reporting of the corresponding installation code instead of the material balance area code could be envisaged. Both codes convey practically the same information in this particular situation but the error rate is considerably lower. Discussions between the two Safeguards Authorities have already been initiated on this subject.

A further check, which is also extremely important, is the check on all transfers which result in material either entering Euratom's control or leaving it — in other words, imports, exports, etc. There must be a guarantee that in such situations exact quantities are established for material entering or leaving our controls. To facilitate verifications in the case of imports and exports, the operator must submit to Euratom advanced notifications which ensure that appropriate checks can be conducted on site before material enters or leaves our control. Procedures with Supplying States have been established in order to obtain systematic notification of imports and exports, or negotiations on this subject are in hand.

When a physical inventory has been taken, the operator is obliged to report a physical inventory listing together with material balance reports. The physical inventory listing is reviewed by the inspector who has carried out the physical inventory verification. Any discrepancies are then followed up and might require the installation to correct some entry lines. The category totals of the physical inventory listings are furthermore compared with the accountancy (inventory change report) and the material balance report and any inconsistency is detected and followed up.

IAEA is in the fortunate situation that the above-mentioned verifications are already carried out by Euratom and, consequently, the follow-up work need not be repeated by IAEA. The quality of the reports submitted to IAEA is therefore higher than the quality of the reports submitted originally to Euratom before the corrections have been carried out.

5. CONCLUSIONS

This paper deals exclusively with accountancy and not with other control aspects such as measurement, inspection and evaluation. The new Euratom accounting system is, in its basic principles, in no way different from any other type of modern accountancy where the highest standards of controls are sought. The reports received by Euratom are a precise mirror of the internal accountancy of the operators of nuclear installations. This, however, creates certain differences between the Euratom and IAEA accounting systems, although the Euratom system is capable of transforming automatically the accounting data to the form and format required by the IAEA. Nevertheless, the question arises if a harmonization between the two accounting systems could be envisaged, at least in those areas where the removal of the differences between the two systems would lead to more accurate, complete and timely sets of accounting data. In addition, this would simplify the verification activities as well as the use of the accountancy, particularly during inspection, as has been shown by practical experience.

REFERENCES

[1] INTERNATIONAL ATOMIC ENERGY AGENCY, INFCIRC-193, IAEA, Vienna (1973).

[2] EURATOM TREATY, Treaty Establishing the European Atomic Energy Community (Euratom), Rome (1957).

[3] EURATOM, Commission Regulation (Euratom) No.3227/76, Official Journal of the European Communities **19** L363 (1976) 1.

[4] SCHMITT, M., KSCHWENDT, H., "Some aspects of IAEA materials accountancy in relation to the future Euratom system", Safeguarding Nuclear Materials (Proc. Symp. Vienna, 1975) **1**, IAEA, Vienna (1976) 269.

[5] LOVETT, J.E., Nuclear Materials: Accountability, Management and Safeguards, Am. Nucl. Soc. (1974).

DISCUSSION

R. PARSICK: Book inventory is usually taken to mean the amount of material to be accounted for. If the Euratom system permits adjustments in the book inventory value, e.g. new measurements, how does it compute the amount of material that the operator should be held accountable for at the time of the physical inventory?

H. KSCHWENDT: Accounting for new measurement results during a material balance period ensures that the Euratom book inventory corresponds as closely as possible to the physical reality. The total MUF (in IAEA terms) for the period is then composed of two parts: the part which is detected by the physical inventory taking, and the part which had already been accounted for during the period as new measurement results.

D. GUPTA: In Section 3.3 of your paper you say that the Euratom Accounting System provides earlier knowledge of the MUF components than the Agency's system thanks to the new measurement results with which the book inventory can be updated between two inventory takings. Since MUF is defined as the difference between book and physical inventories, no conclusion can be reached regarding MUF (and hence possible diversion) until a physical inventory has been taken. What then is the advantage in having an incomplete knowledge of certain components of MUF before a physical inventory has been taken?

H. KSCHWENDT: A conclusion regarding total MUF or possible diversion falls under the heading "evaluation" and is outside the scope of this paper. The purpose of accounting for MUF components during a period is simply to have the best available information — in this particular case new measurement results — reflected in the accountancy so that the records and reports correspond as closely as possible to the physical reality.

W. GMELIN: I should like to mention a few points in regard to your paper:

(1) The main difference between the IAEA and the Euratom systems is that the IAEA system is an information system, of which accountancy forms just a part, whereas the Euratom system seems to be a pure accountancy system.

(2) The term "new measurement" is misleading and is a relic of the past. A better term would be "interim book inventory adjustment".

(3) The breakdown into enrichment classes is not necessary (at the input level) in a computerized information system, since the computer is able to divide a fissile component by a total quantity, allowing any sub-division of classification which may be required.

(4) It is clear that any system must store calendar dates of various types in addition to the original date, and the IAEA system does this. However, a book inventory, or any other material balance component which has been reported wrongly, must be capable of correction as of *any* date required, not only as of the date of the finding. Similarly, an information system must be able to handle different types of correction (retroactive for certain canned outputs or, if so required, corrections of transactions), since obvious diversion strategies are possible when wrong accounts cannot be corrected or reverified.

H. KSCHWENDT: I will answer your questions in the order in which you put them.

(1) I spoke only on the pure accounting aspects of the Euratom system. Obviously, the system has other features (e.g. evaluation), but they are not covered by the scope of my present paper.

(2) I agree that the term "new measurement" is not very satisfactory. However, I think we should confine our discussions to the concepts involved rather than the nomenclature.

(3) Your statement is correct if you consider only transfers or similar types of inventory change, but it is wrong if you are dealing with blending or similar operations, because these operations change the degree of enrichment, and consequently the effective kilogrammes available, in some cases without even the operation having been reported. Euratom has subdivided the enriched uranium into two categories in order to allow the attribution of MUF to a smaller enrichment band than in the IAEA system.

(4) I think there is a misunderstanding. I did not say that inventory changes cannot be corrected or re-verified. I only said that in the Euratom system there is no need for *systematic* re-verification.

The purpose of an invariant book inventory is simply to be able to reproduce the operator's book inventory, which is also invariant. If you want to know the retroactive effect of a correction, the Euratom system can easily accomplish this using a computer program based on the original date only. This is always done when we prepare reports for the IAEA.

P.O. FREDERIKSEN: I should like to draw attention to another difference between the IAEA system and the Euratom system. The printout of the IAEA computer has a different layout from that of the Euratom computer. This complicates the inspections and means that they take more time than if the printouts were of the same layout. It would be useful if uniform layouts could be used. However, maximum efficiency can be achieved only if the IAEA accepts "new measurement" reports so as to have the same book inventory as the operator and Euratom.

H. KSCHWENDT: I fully agree with your remarks and I hope the time will come when the IAEA inspector arrives at an installation with a book inventory which is the same as those held by Euratom and the operator.

IAEA-SM-231/112

TECHNICAL CRITERIA FOR THE APPLICATION OF IAEA SAFEGUARDS

G. HOUGH, T. SHEA, D. TOLCHENKOV
Department of Safeguards,
International Atomic Energy Agency, Vienna

Abstract

TECHNICAL CRITERIA FOR THE APPLICATION OF IAEA SAFEGUARDS.

A system of technical criteria to be used in the design, operation and evaluation phases of safeguards implementation is under development. The framework for these criteria and preliminary values of external criteria are presented.

1. Introduction

IAEA safeguards are expected to serve a major role with respect to the realization of non-proliferation of nuclear weapons. Towards that purpose, IAEA safeguards are designed to detect diversion of materials committed to peaceful uses and the misuse of equipment or facilities subject to IAEA safeguards, and to deter such undertakings through the risk of early detection.

The Agency implements its safeguards based on the IAEA Statute, in accordance with two main types of Agreements. As described in documents INFCIRC/153 [1] and INFCIRC/66/Rev. 2 [2], these Agreements are concluded with States party to the Non-Proliferation Treaty (NPT) and non-NPT-States, respectively. The implementation of safeguards is intended to be uniform to the extent possible, except as required in specific provisions of different Agreements.

In response to requests from the Board of Governors of the IAEA, the Secretariat is endeavouring to improve the extent to which Agency safeguards effectiveness is assessed on the basis of quantitative as opposed to qualitative factors. A necessary first step towards that objective is acceptance of a system of technical criteria to be used in the design, operation and evaluation phases of safeguards implementation. These criteria are viewed as guides for Agency planning and or potential goals to be achieved, rather than requirements. The system of technical criteria is intended to contribute to safeguards effectiveness and effeciency, by:

- establishing specific definitions of adequate safeguards;
- providing a quantitative basis for system design, operation and evaluation; and
- enabling facility and national authorities to better anticipate the impact of IAEA safeguards.

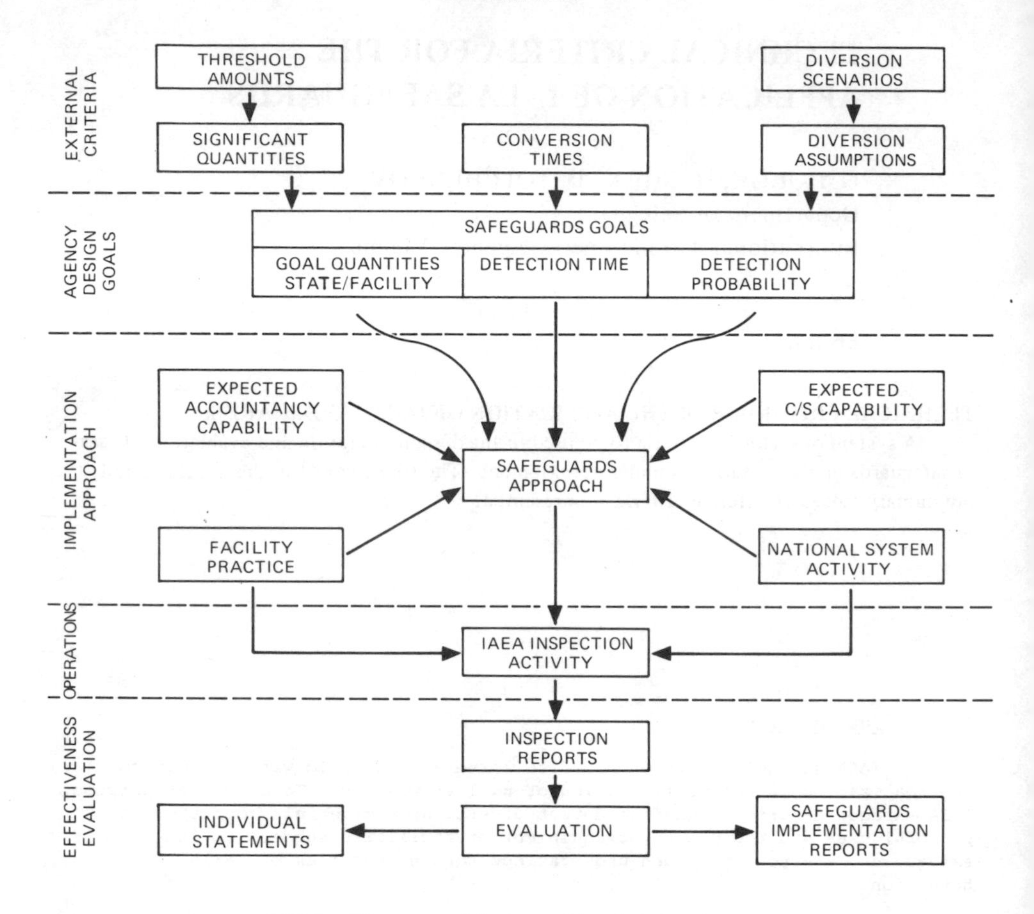

FIG.1. Elements of design and implementation.

The work on technical criteria has included and will be continued with input from the Standing Advisory Group on Safeguards Implementation (SAGSI), which provides recommendations to the Director General on technical aspects of safeguards implementation. The criteria developed for "timely detection" and "significant quantities" have been authorized by the Director General for use by the Agency in its safeguards system.

Adjustments are being made to the Agency safeguards system to incorporate these quantitative criteria. The system of technical criteria is not complete, as yet. Our intent in this paper is to describe the system as currently perceived, incorporating the criteria adopted to date.

2. SYSTEM FRAMEWORK

Agency safeguards center on four task elements : setting performance goals, selecting procedures, applying safeguards and evaluating results. Figure 1 illustrates the interactions and considerations essential in proceeding through these elements.

The approach is : (1) define external criteria, (2) determine desired performance goals on the basis of credible diversion possibilities, taking into consideration technical capabilities and the burden on facility operators, (3) design the approach for each State and facility in accordance with those goals, (4) implement the safeguards activities, and then (5) evaluate the results of those efforts vis à vis the performance goals. The approach calls for determining the system structure and criteria following a top-down approach, then evaluating performance in a bottom-up approach.

3. EXTERNAL SAFEGUARDS TECHNICAL CRITERIA

In order to provide a technical basis for establishing safeguards quantitative criteria, the Agency has carried out a study [3] of nuclear material quantities required for the manufacture of a nuclear explosive, and of the minimum time required for the conversion of different forms of safeguarded material into metallic components of a nuclear explosive device.

3.1. Significant Quantities

The definition of the central criterion "significant quantity" is related to "threshold amount", which is understood to be the approximate quantity of special fissionable material required for a single nuclear explosive device. The values for "threshold amounts" have been chosen on the basis of data provided in a UN report on this matter [4] and recent publications indicating somewhat lower values [5].

A "significant quantity" is understood to be the approximate quantity of nuclear material with respect to which - taking into account any conversion processes involved - the possibility of manufacturing a nuclear explosive device cannot be excluded.

For convenience, materials which can be converted to nuclear explosives without transmutation or further enrichment are defined as "direct-use" materials, while materials which must be enriched or transmuted to form direct-use materials are defined as "indirect-use" materials. Table 1 gives the significant quantity values currently in use by the Agency. These significant quantity values reflect estimates of the amount of material which must be included in a nuclear explosive, plus an additional allowance for process losses in the conversion operation.

Four factors determine the amount of a particular nuclear material which must be included in a nuclear explosive : design yield, design sophistication or skill, component composition and material isotopic composition.

A useful reference point in consideration of minimum amount of material required for a nuclear explosive is the minimum reflected sphere critical mass value for each direct-use material. The amounts actually required by a State in a first effort are expected to be larger owing to a lack of skill, as gained through experience, especially through testing. The amount of feed material necessary to produce one explosive depends on the chemical form selected. It may be possible to construct nuclear explosives with other than metallic nuclear material components. But it would appear that the likelihood that non-metallic devices would fail to explode is greater than for a metallic device, especially in a first attempt. Thus, the significant quantities indicated in Table 1 assume that metal components would be used. Note that larger amounts would appear to be necessary for any other chemical form. Isotopic composition also appears to influence the amount of material required to realize a given explosive yield. Our studies suggest that the use of "reactor grade" plutonium in nuclear explosives may indeed be credible, but that the design complexity, amounts required and handling problems increase as the content of Pu-238, 240 and 242 increase due to spontaneous fission neutrons, dilution, and decay heat.

TABLE 1. QUANTITIES OF SAFEGUARDS SIGNIFICANCE

	Material	Quantity of Safeguards Significance (SQ)	SQ applies to:
"Direct-use" material	Pu	8 kg	Total element
	^{233}U	8 kg	Total isotope
	U($^{235}U \geq 20\%$)	25 kg	^{235}U
	Plus rules for mixtures where appropriate		
"Indirect-use" material	U($^{235}U < 20\%$) [a]	75 kg	^{235}U
	Th	20 t	Total element
	Plus rules for mixtures where appropriate		

[a] Including natural and depleted uranium.

Our studies also suggest that an explosive based on less than 20% enriched uranium would require extraordinary design sophistication. Thus, a safeguards cut-off for practical direct use of enriched uranium at 20% enrichment has been adopted.

It would appear that nuclear explosives can also be constructed using different direct-use materials, possibly mixed intrinsically to form alloys, or fabricated into separate parts, each part fabricated using a single type of direct-use material. Formulas for computing significant quantities of mixtures have been developed for use as appropriate.

Significant quantities for low enrichment uranium and thorium were determined by considering the requirements for the amounts of direct-use material to be produced and the production yield of the intermediate processes.

3.2 Conversion Time

Criteria for timely detection were recently discussed at a series of meetings of SAGSI, which made provisional recommendations to the Director General that "detection time" be used as a parameter for timeliness and that it should correspond in order of magnitude to the "conversion time". "Conversion time" has been defined as the minimum time required to convert different forms of nuclear material to the metallic components of a nuclear explosive device. "Detection time" is defined as the maximum time which may elapse between an assumed diversion and its detection by Agency safeguards. The values of "conversion time" recommended by SAGSI for different material categories are given in Table 2.

TABLE 2. ESTIMATED MATERIAL CONVERSION TIMES

MATERIAL CLASSIFICATION	BEGINNING MATERIAL FORM	END PROCESS FORM	ESTIMATED CONVERSION TIME
1.	Pu, HEU[a], or U-233 Metal	Finished Plutonium or Uranium Metal Components	Order of days (7-10)
2.	PuO_2, $Pu(NO_3)_4$ or other pure compounds. HEU or U-233 oxide or other pure compounds.	"	Order of weeks [b] (1-3)
	MOX or other non-irradiated pure mixtures of Pu or U [(U-233 + U-235) $\geq$ 20%]. Pu, HEU and/or U-233 in scrap or other miscellaneous impure compounds.	"	"
3.	Plutonium, HEU or U-233 in irradiated fuels ($\geq 10^5$ Ci/kg HEU or U-233 or plutonium)	"	Order of months (1-3)
4.	Uranium containing $\leq$ 20% U-235 and U-233; thorium	"	Order of one year

[a] Uranium enriched to ⩾ 20% in the isotope ^{235}U.

[b] While no single factor is completely responsible for the indicated range of 1–3 weeks for conversion of these plutonium and uranium compounds, it was noted in the discussion that the pure compounds will tend to be at the lower end of the range and the mixtures and scrap at the higher end.

Conversion time includes that period of time commencing with receipt of the diverted material at the conversion facility and ending with the manufacture of the components required for a single device. It does not include the time required to transport the diverted material to the conversion facility, nor the time to assemble the device for use, or any subsequent time period.

In estimating the conversion times, the diversion activity is assumed to be part of a planned sequence of actions chosen to give a high probability of success in manufacturing one or more nuclear explosives with minimal risk of discovery, until at least one such device is manufactured. It is therefore assumed that before a diversion would take place, the explosive device has been designed with considerable care; non-nuclear components of the device have been manufactured, assembled, and tested; and engineering drawings for the nuclear components have been approved, including dimensions, metallurgical requirements and tolerances for all manufacturing parameters.

In estimating the conversion times, conversion facilities are assumed to exist, tailored to convert the chemical form of the diverted plutonium, high enrichment uranium or uranium-233 to the corresponding metal. It is assumed that this conversion capability has been tested using appropriate surrogate materials and is available, complete with operators, on receipt of the diverted material. Finally, it is assumed that the required shop capabilities exist to machine the metal into finished components for the explosive in accordance with the engineering drawings, and that these manufacturing processes have been tested by manufacturing dummy components using appropriate surrogate materials.

3.3. Diversion Assumptions

Agency safeguards are implemented at the request of a Member State on behalf of all Member States. Verification activities are undertaken in partnership with each Member State, to demonstrate that the conditions of Agreements with the Agency are being honoured. In order to provide credible assurance of non-diversion, it is assumed in the Agency system that a State may divert nuclear materials for the production of nuclear explosives. For this reason, the Agency must take into account the possibility that the State will engage the resources necessary to achieve its objectives with minimum risk of premature discovery. Accordingly, two main diversion assumptions are considered by the Agency:

At one extreme, concern is focussed on the abrupt diversion and prompt conversion of material for conversion from safeguarded forms to at least one nuclear explosive. Countering this possibility requires that the Agency conduct inspection activities at a frequency which corresponds to the estimated conversion time for the safeguarded materials.

At the other extreme, concern is focussed on the minimum credible diversion rate. In facilities holding bulk nuclear materials, small quantities of nuclear material may be systematically diverted over long periods of time, to avoid detection indefinitely. Countering this protracted diversion strategy requires defining a credible minimum diversion rate and implementing high sensitivity tests to detect diversion at or above that minimum rate. It is assumed in the Agency system that a situation of proliferation would exist if a single nuclear explosive were acquired. Thus, the focus of Agency safeguards is to detect, with high confidence, the diversion of a minimum amount of material at a rate necessary for the production of a single nuclear explosive per year.

3.4. Diversion Modes or Scenarios

A state with a peaceful nuclear power programme will operate a number of facilities for fuel enrichment, manufacturing and reprocessing, for power production, waste treatment and research and development. These facilities may serve as diversion sources or as collusion centers to mask diversion from source facilities.

There are many diversion paths which may be open depending on the facilities which exist within the State. If multiple facilities of the same type exist, the number of paths increases accordingly. It should be noted that since there are no nuclear explosives production facilities under safeguards, at some step in the manufacturing process, non-safeguarded facilities must be employed. Thus, if any diversion scheme is to be considered credible, the existence of the necessary process capabilities must also be assumed. The Agency's safeguards system, including any diversion analysis, is limited to the detection of diversion of materials committed to peaceful uses and the misuse of equipment of facilities subject to IAEA safeguards.

Three diversion modes become possible as a State broadens its peaceful nuclear activities:

Single Source : diversion from a single facility with concealment restricted to that facility;

Parallel Source : diversion from two or more facilities holding the same material with cross concealment possibilities; and
Series Source : diversion from facilities interrelated in a fuel cycle with concealment extending through the fuel cycle.

4. SYSTEM PERFORMANCE GOALS

These external criteria, i.e., significant quantities, conversion times and the diversion assumptions, define the problem to be addressed by the IAEA safeguards system. For the Agency safeguards system to be effective, it must counter both abrupt and protracted diversion strategies at the MBA level, and also at the State level.

The system must be able to detect anomalies in material control which would accompany these diversion strategies and scenarios. Furthermore, such anomalies must be detected and resolved in a time frame which is adequate to allow diplomatic initiatives to stop further movement towards proliferation. Resolution in this case must be sufficiently clear to warrant returning the system to its normal status, or to initiate the notification provision set forth in the Agency's Statute. Through implementing such a capability, it is hoped that a State considering diversion would be deterred by the risk of early detection.

At this time, the system of technical criteria is not complete to the point where the linkage between external criteria and system performance goals is resolved. In the following section, an implementation structure is described which represents current Secretariat views.

5. IMPLEMENTATION OF AGENCY SAFEGUARDS

For Agency safeguards to be technically credible as an instrument of non-proliferation, its implementation must be planned in a manner which is logically consistent, applied according to that plan, and evaluated in respect to the underlying objectives.

As noted in the preceding parts of this paper, initial focus for planning is at the State level. First consideration is given to characterizing the nuclear industry within the State, obtaining information on the flow of materials. From this information, State level safeguards considerations are identified, including:

- Possible diversion sources for each material type.
- Possible cross concealment activities involving declared facilities; and
- State level verification goal quantities.

One test under consideration to determine the likelihood that interrelated fuel cycle facilities may have been involved in a series collusion scheme to conceal diversion is a consistency check of fuel flows through the fuel cycle, in comparison to similar fuel cycle activities in other States. Of importance are power production per kilogram of fuel produced, nuclear production and nuclear loss, and plant efficiency, expressed in amount of product per kilogram of feed material. Isotope correlation techniques and input-output analysis methods are under examination for these purposes.

The following checks are under consideration to determine the effectiveness of the safeguards system against credible diversion schemes involving parallel collusion.

- consistency of interfacility and import/export data, including shipper/receiver data;

- acceptability of combined accountancy values;
- detection of undeclared inventory shuffling; and
- review of material-in-transit.

Following this State-level consideration, the focus shifts to planning safeguards at the facility/MBA level. At the MBA level, safeguards performance goals are determined on the basis of the material types and quantities, while the approach is determined by the performance goals, the material forms, and ambient characteristics where the safeguards are to be applied.

Basic concepts for achieving timely detection of abrupt and protracted diversion of significant quantities of nuclear material at the MBA level are:

a) Verification of information declared in records and reports by the facility operator by internal records audits to verify that the declared information is correct, and external audits in conjunction with shipper/receiver documents to verify that the declared information is complete.

b) Effective safeguarding of nuclear material by the use of instruments and other techniques at strategic points established to maintain continuity of knowledge of the flow and inventory of nuclear material, taking into account corresponding detection time requirements. Adequate access to nuclear material is essential for Agency verification purposes, especially taking into account the short detection time requirements for "direct-use" materials such as plutonium and high enriched uranium. Analysis shows the necessity for high frequency or continuous inspections for sensitive facilities for the Agency to reach its objectives.

c) The periodic closing of the material balance by the operator, including the measurement of receipts, shipments and physical inventories, and their verification by the Agency.

The frequencies for physical inventory taking depend on the types of nuclear material used at the facility, and its throughput and inventory. In this respect, physical inventories should be taken twice a year at low-enriched uranium fabrication and conversion facilities and four times a year at facilities which process highly enriched uranium or plutonium. The actual frequency for physical inventory taking, to be agreed between the Agency and the State, is set forth in the particular facility attachment. This frequency is chosen to reflect the effectiveness and accuracy of the operator's flow control system and specific conditions for safeguarding at the facility. In this connection, adequate accounting procedures, evaluating the accuracy of operator measurement, and measurement of discarded and process hold-up are essential for the effectiveness of the Agency material balance verifications.

To facilitate effective safeguards, the minimum accuracy of material balances which to be achieved by facility operators should correspond to international standards. International standards of measurement accuracy are expressed as the expected standard deviation of a material balance closing (σ_{MUF}) for facilities where mass accountancy is relevant, as shown in Table 3. Multiples of these standard deviations are used for inspection planning, depending upon statistical sampling considerations to achieve the desired probability of detection (β) and false alarm probability (α). These measurement accountancy goals have been used for several years and serve the very limited purpose of calculating statistical sample sizes for independent measurements that are not excessive, yet provide adequate assurance of detecting significant measurement biases in the material balance, and deterrence of data falsification to conceal diversion.

TABLE 3. EXPECTED ACCURACY (STANDARD DEVIATION) OF A MATERIAL BALANCE AND VERIFICATION ACCURACY GOALS EXPRESSED AS PER CENT OF INVENTORY OF THROUGHPUT

Facility Type	Expected Accuracy of σ_{MUF}	Facility Detection Goal $\alpha = \beta = 0.05$	Inspection Detection Goal[a] $\alpha = \beta = 0.05$
Uranium Enrichment	0.2	0.7	1.0
Uranium Fabrication	0.3	1.0	1.5
Plutonium Fabrication	0.5	1.7	2.5
Reprocessing, Uranium	0.8	2.6	4.0
Reprocessing, Plutonium	1.0	3.3	5.0

[a] Includes inspector measurement errors.

d) Independent verification by the Agency of the entire material balance for nuclear material subject to safeguards, by item counting and identification to establish the total population of items to be verified, and by measurement to verify the declared amounts of material, using chemical analysis and non-destructive measurements. Agency measurements are based upon statistical sampling procedures to provide the capability to detect, with a high degree of confidence (e.g. 90 - 95%), the diversion of a "significant quantity" of nuclear material.

e) Containment and surveillance measures applied to complement material accountancy, as discussed in a separate paper [6]. These measures may include, for example, the observation (human surveillance) of on-going operational activities related to the control of nuclear material for the bulk facilities having significant quantities of "direct-use" materials. The analysis of process data relevant to safeguards is also meaningful in such cases.

In the practical implementation of Agency safeguards, the above mentioned concepts are incorporated in the safeguards approaches for different types of facilities. The particular safeguards approach will depend on many factors such as design of the facility, organization of material accounting, capability of the operator's measurement system, specific operational conditions, types, forms and quantities of nuclear material, and an assessment of credible diversion possibilities.

Table 4 summarizes general safeguards criteria for the different types of facilities which use significant quantities of nuclear material.

4. OPERATION AND EVALUATION

Once the requirements are defined and the facility attachments are negotiated, safeguards are applied in a manner which becomes increasingly routine. Inspectors are recruited, trained, designated and sent to the facilties in accordance with the prescribed purposes. Logistical support, instruments, seals, calibrations, etc. are provided as required.

TABLE 4. SAFEGUARDS CRITERIA FOR DIFFERENT TYPES OF FACILITIES WHICH USE SIGNIFICANT QUANTITIES OF NUCLEAR MATERIAL

Types of Facility	Types of Material	Detection time (order of magnitude)	Frequency of PIT [a] (per year)	Inspection Mode [b]
Reactors	NU, LEU;	1 year	1	Intermittent
	Pu in irradiated fuel;	1 - 3 months	1	4-6 times/year
	Pu or HEU in fresh fuel	1 - 3 weeks	1 - 4	Intermittent each 2-3 weeks
Critical Facilities	Pu/HEU pure compounds	1 - 3 weeks	1 - 4	continuous or each 2-3 weeks
	Pu/HEU metal;	1 - 10 days	1 - 4	continuous or each week
Separate storage installations	NU, LEU;	1 year	1 - 2	Intermittent 1-2 times/year
	Pu/HEU pure compounds	1 - 3 weeks	2 - 4	Intermittent each 3 weeks
Fuel fabrication and conversion	NU, LEU;	1 year	1 - 2	Intermittent
	HEU, Pu pure compounds;	1 - 3 weeks	2 - 4	Continuous or each 2 weeks
	HEU, Pu metal	7 - 10 days	2 - 4	Continuous or each week
Reprocessing facilities	LEU, NU, DU;	1 year	)	
	Pu in irradiated fuel;	1 - 3 months	)	Continuous for whole plant
	Pu in processing	1 - 3 weeks	)	
6. Isotopic enrichment facilities	LEU HEU	1 - 3 weeks [c]	2 - 4	From intermittent to continuous

[a] Frequencies and mode of PIT are defined in consideration of effective and accurate flow control by the operator, material quantities and conditions for verification at the plant.

[b] Inspection mode for the verification of LEU, NU, DU and Pu in irradiated fuel depends on corresponding conversion times, material distribution at MBAs, and number of the MBAs. Continuous mode of inspections does not necessarily imply the permanent presence of Agency inspectors at the plant, for example the inspection may be on call and be present during operational activities related to nuclear material inventory changes.

[c] For enrichment plants which cannot be converted readily to the production of $\geqslant 20\%$ ^{235}U, a longer conversion time is chosen.

Following each physical inventory verification inspection, a statement is sent to the Member State summarizing the findings. In facilities where short detection time criteria are relevant, a summary report of these additional inspection activities is submitted to the State at quarterly intervals.

Evaluation of the Agency inspection activity is performed in order to make conclusions and statements that Agency safeguards objectives have been met for a specific interval. Once the Agency is satisfied with the effectiveness of safeguards at the facility level, additional tests are undertaken to reach a conclusion with respect to the effective application of safeguards in the State. This evaluation is made with respect to each material balance area to provide assurance to the world community that no diversion of a significant quantity of nuclear material has occurred in the State.

Once each year, safeguards implementation is critically reviewed in a summary report. This Safeguards Implementation Report is submitted to the Agency's Board of Governors for the final judgement of Agency safeguards effectiveness.

5. CONCLUSIONS

The system of technical criteria under development is intended to enhance the technical credibility of IAEA safeguards. This system offers the appeal of interconnecting logic between non-proliferation goals and the application of specific measures.

Over the course of the next year, it is our intent to finalize certain aspects of the system of technical criteria including the following :

1) definitive criteria with respect to significant quantities and detection times, including :

   a) a consideration of previously recommended significant quantity values, with a view towards establishing a final set of values;

   b) development of an appropriate relationship between significant quantities and safeguards verification goal quantities;

   c) development of a working definition of detection time with respect to credible diversion strategies, which takes account of variations in the types and specificity of anomalous indications observed by inspectors, and is useful during the design, operation and review phases of safeguards implementation;

   d) establish detection time values necessary for effective safeguards; and

   e) establish verification time and goal quantity criteria for safeguards implementation for all material types and facility sizes.

2) Nuclear material accountancy verification criteria, including detection sensitivity, probability of detection, false alarm rate, material balance closing frequency, flow verification criteria, and inventory procedures for different materials.

3) criteria for the use and quantification of containment and surveillance in Agency safeguards:

a) as a means to reduce physical inventory re-measurement;

b) in situations where extreme access difficulties are encountered for flow and inventory verification; and

c) in future large scale facilities where achievable material accountancy performance may be insufficient to warrant its acceptance as the safeguards measure of fundamental importance.

REFERENCES

[1] INTERNATIONAL ATOMIC ENERGY AGENCY, INFCIRC/153 (Corrected: "The Structure and Content of Agreements between The Agency and States Required in Connection with the Treaty on the Non-Proliferation of Nuclear Weapons".)

[2] INTERNATIONAL ATOMIC ENERGY AGENCY, INFCIRC/66/Rev.2, "The Agency's Safeguards System".

[3] HOUGH, G., SHEA, T., TOLCHENKOV, D., "Studies of technical criteria for the application of IAEA safeguards", Proc. 19th Annual Meeting of the Institute of Nuclear Materials Management, June 1978.

[4] UNITED NATIONS, UN Doc. A/6858 (Oct. 1968).

[5] See, for example: WILRICH, M., TAYLOR, T., "Nuclear Theft: Risks and Safeguards", Ballinger Publishing Company, Cambridge (1974).

[6] SHEA, T., TOLCHENKOV, D., "The Role of containment and surveillance in IAEA Safeguards", Paper IAEA-SM-231/110, those Proceedings, Vol. I.

DISCUSSION

D.A. HEAD: There seems to be a paradox between "threshold amount" and "significant quantity". If "threshold amount" is the amount of material required in a nuclear explosive, and "significant quantity" is that amount plus an allowance for process losses, it would seem that the "significant quantity" for each nuclear material should be larger than the corresponding "threshold amount". However, the "threshold amount" for plutonium at 95% ^{239}Pu has been stated to be 8 kg Pu, while SAGSI defines the "significant quantity" for all plutonium isotopes to be 8 kg.

T. SHEA: Threshold amounts were indicated to the United Nations by an advisory group in 1968. SAGSI does not consider itself competent to advise the Agency on matters related to nuclear explosives and thus took note of these values, together with somewhat lower values as indicated in recent literature. SAGSI considered the various conversion operations which have parallels in peaceful nuclear activities, the physics of fast neutron assemblies with respect to varying isotopic mixtures, data such as are contained in the United Nations report, and minimum reflected sphere critical mass values, and on that basis recommended specific quantities of safeguards significance.

A.G. HAMLIN: I think your paper conceals certain important difficulties. Once numerical targets are accepted into safeguards, it follows that the operators' accountancy system must be capable of meeting these targets, with or without the help of containment and surveillance, and considerable upgrading may then be required.

If containment and surveillance are upgraded to the levels implied, the operator will become more involved in responding to inquiries about apparent anomalies and will have to accept a higher frequency of false alarms.

While I am all in favour of having numerical standards for safeguards, these secondary effects need to be evaluated carefully or the operator will be dragged more and more into the operation of safeguards rather than being watched by a detached inspectorate.

T. SHEA: One of the basic precepts of IAEA safeguards is minimal intrusion into plant operation. Safeguards approaches are tailored to specific circumstances at each plant, with prime consideration given to the attainment of IAEA objectives. The facility operator is an important partner in establishing an approach which will permit the IAEA to realize its objectives in such a manner that the burden imposed on the facility is acceptable.

H.G. STURMAN: Since allegations that a diversion has taken place would be a most serious matter for the State or commercial organization concerned, it seems essential to ensure that the probability of the overall safeguards system making a wrong allegation is no more than $1 : 10^3$ and preferably even lower.

T. SHEA: The false-alarm values noted in the paper relate to the observation of abnormal circumstances or situations which must be resolved. A decision to notify the appropriate bodies that a diversion had occurred would follow an intensive effort to resolve such abnormalities or to determine whether unresolved abnormalities indicate diversion or some far less serious matter.

W.C. BARTELS: At the opening of this Symposium, Mr. Grümm, the Deputy Director General of the Agency's Department of Safeguards, emphasized the need to improve safeguards acceptability by minimizing the cost and interference factors for plant operators and also by maximizing safeguards credibility throughout the world. In some places nuclear power programmes are a political issue, and a condition for their continuance is the achievement of a convincing level of effectiveness in international safeguards. In the world of sovereign States absolute credibility has its price. If the safeguards system were to conceal the problems arising in safeguards operations, those who felt threatened would never believe that all was well.

I offer you the United States of America as an example. In an effort to maintain the acceptability of nuclear power by demonstrating to the limits of our ability that no nuclear material has been stolen or diverted within the United States, we annually publish the cumulative amount of material unaccounted for at each facility. When these MUF amounts exceed what the IAEA calls a "significant quantity", the Government issues an explanation of what actually happened, and we all live in an atmosphere of openness. You see, we do not believe in setting alarm levels so high that we never – or almost never – report a problem when it seems to have occurred.

R.J. DIETZ: The suggested detection times of 1–3 months for highly radioactive materials – compared with times of 1–3 weeks for decontaminated materials – indicate that chemical reprocessing is the rate-determining consideration. Published experience seems to show that the metallurgical operations common to both contaminated and decontaminated materials are in fact the most time-consuming operations in the preparation of weapons-usable metallic components. This fact suggests that the factor of four difference between these times cannot be supported in the case of a State already having reprocessing capability.

Has any thought been given to raising the lower values, and perhaps lowering the higher values, in order to accommodate these considerations more realistically, and what is the rationale that prompts the apparent discrepancy? This question is extremely relevant to the operational impact and false-alarm concerns voiced by the previous questioners.

T. SHEA: The figures quoted are estimated conversion times rather than detection times. In establishing detection times, consideration of the particular situation is essential. Thus, a reprocessing facility is safeguarded against the introduction of undeclared feed materials as well as against the diversion of declared materials.

The values and ranges indicated represent the recommendations of an advisory group comprised of senior technical specialists.

IAEA-SM-231/33

IMPROVEMENT OF MATERIAL ACCOUNTABILITY VERIFICATION PROCEDURE

K. IKAWA*, M. HIRATA**, H. NISHIMURA
* Division of Power Reactor Project
** Office of Planning
Japan Atomic Energy Research Institute

H. KURIHARA+, S. AOE++
+ Safeguards Division, Nuclear Safety Bureau
++ Policy Division, Atomic Energy Bureau
Science and Technology Agency

S. TAKEDA
Department of Nuclear Engineering,
Tokai University,
Japan

Abstract

IMPROVEMENT OF MATERIAL ACCOUNTABILITY VERIFICATION PROCEDURE.

This paper describes a method of considering the characteristic of the national nuclear fuel cycle quantitatively. This characteristic is presented here by the diversion factor (F_D), which relates to how attractive the nuclear material in the reference fuel cycle is to a diverter. This expression was applied to several nuclear fuel-cycle models. As a result, it has been found that this may be used as a quantitative approach for representing the characteristic of the national nuclear fuel cycle, according to how much it may be later improved.

1. INTRODUCTION

In the report "Safeguards – 1975–1985" [1], Rometsch et al. presented a modified approach to verification based on 'categorization of nuclear material' with the aim of limiting the increased cost of safeguards. Three categories were separately defined for uranium and plutonium, based on the content of fissile nuclides and of radioactive β- and γ-emitters, and the characteristics of the States' nuclear fuel cycles. The report suggested that such categorization might be taken into consideration on setting the values of significant quantities.

Rometsch's approach presents two alternatives – the first to increase the significant quantities for the nuclear material of categories 2 and 3, and the second is to carry out the IAEA safeguards with the mode of Level I assurance for the nuclear material of categories 2 and 3.

In this paper, the authors discuss an approach to justify the inspection effort without using the categorization of nuclear material. To this end the authors studied a method of changing the conventional quantity defined as the "significant quantities" (SQ) into the quantity that takes into account the characteristic of national nuclear fuel cycles. This change is carried out by using a weighting function. It can be obtained as an extension of a formula, which gives a numerical result called 'the diversion factor', denoted by F_D, when applied to a stage of a fuel cycle [2].

This formula is designed to account for various factors such as the type of fissile material, the fissile concentration of that material, the quantity of material available in terms of critical masses, the chemical and physical properties of the material regarding separation from non-fissile material, and how much the material is diluted with non-fissile material. In addition, storage procedures, shipping procedures, shipping and storage time periods, the number of personnel involved in handling the material, and other transport-related data are included in the formula.

Among these factors some are not used under the present international safeguards system, but were thought necessary in order to obtain some counter-measures against the risk of abrupt diversion. Section 2 outlines the background of this study.

Section 3 describes the present problems concerning the verification procedures on material accountancy, and proposals for defining a "significant quantity", a "goal quantity", "detection time" and "detection probability" are made. Section 4 clarifies the definitions of terms used in this paper; and Section 5 shows an application study of our new approach.

2. BACKGROUND

2.1. Requirement on improvement of material accountancy

The technical objective of international safeguards is defined in Article 28 of INFCIRC/153, i.e. it is the timely detection of the significant quantity of nuclear material, using material accountancy as a safeguards measure of fundamental importance, with containment and surveillance as important complementary measure, and it is required that the IAEA obtain the technical conclusion as a statement, in respect of each material balance area, of the amount of MUF over a specific period with its limits of accuracy.

To achieve this objective, the verification procedure of material accountancy has been developed through detailed statistical considerations, through which diversion strategies and inspection strategies as counter-measures are fully studied to make up strong but cost-effective inspection planning.

However, it has recently been felt that these inspection strategies, mainly based on statistical considerations of MUF, may not be sufficient to counter abrupt diversions of nuclear materials of relatively short conversion time and to detect protracted diversions in facilities of relatively large throughputs of special nuclear materials. Consequently, the most important target to improve the present safeguards implementation is to establish effective safeguards measures against these abrupt diversions, and also to provide effective safeguards measures to protect these protracted diversions. To establish such a measure, "timely detection" is the most important aspect to be considered. From this point of view, the near-real-time material accounting concept and the concept of a periodic or semi-continuous physical inventory taking are felt to be desirable. On the basis of these concepts, Cleanout Physical Inventory Taking (CPIT), Draindown Physical Inventory Taking (DPIT), Running Physical Inventory Taking (RPIT) and Running Book Inventory Taking (RBIT) are examined together with Flow Follow-Up (FFU) as a surveillance method.

These measures are recognized as desirable from the viewpoint that the "State System of Accounting for and Control of nuclear material (SSAC)" should ensure that adequate nuclear materials accounting and control are continuously implemented by the operator. It must also be recognized, however, that from the viewpoint of the safeguards objective, as defined in INFCIRC/153, these precise safeguards measures should be applied only for strategic points of strategic facilities.

2.2. Growth of safeguards cost

Improvement of the technical effectiveness of safeguards implementation would require an increase of safeguards cost, should it be carried out without any consideration of cost-effectiveness. Another element to increase the growth of safeguards cost is the increase of the amount of nuclear material in the fuel cycle required to produce nuclear electrical power.

It is well known that the main component of safeguards cost is the inspection manpower required to achieve the technical objective, and the optimization of the other components can only lead to a small cost reduction.

To limit the growth of safeguards costs, a reduction of the inspection effort is the most effective. As an approach, Rometsch et al. [1] presented two alternatives as mentioned in the introduction of this paper. They proposed a substantial increase in the values of the significant quantities of nuclear material of categories 2 and 3, but it was not made clear how this increase could be effected. This paper describes an approach to this problem, after clarifying the definition of significant quantity with respect to a threshold amount, detection capability and a goal quantity (GQ).

2.3. Requirements for non-proliferation

In any nuclear fuel cycle, theoretically speaking, some of the nuclear materials in the cycle can also be used for a nuclear weapon or a nuclear explosive device. The major items considered to protect the nuclear fuel material in the nuclear fuel cycle from the risk of diversion are as follows:

(1) Material accountability
(2) Physical protection
(3) Weapons usability
(4) Non-probability for material to be modified
(5) Non-ability for material to be modified
(6) Interruptability
(7) Sensitive technology protection
(8) Material inaccessibility
(9) National vulnerability
(10) Susceptibility to international control
(11) National industrial level

Among the above-mentioned items are those related to physical protection. To evaluate the physical protection system of a nation under the international safeguards would create difficult problems. This, however, is unavoidable in solving the problem of non-proliferation that confronts us. As one of the methods, we recommend that the function of the physical protection be classified into several parts, and that efforts should be made to look for the part useful for international verification. It is undesirable to claim blindly that physical protection cannot become an object of international verification.

On the basis of the above idea, in this paper the concept of the diversion factor, including elements related to the above items and including physical protection, is adopted as a semi-quantitative approach to express and to evaluate characteristics of a national nuclear fuel cycle. It is self-evident that the diversion factors at present include only a few of the above items in a very simple form. We should make efforts to improve the contents still further.

3. SIGNIFICANT QUANTITY AND VERIFICATION PROCEDURE

3.1. Original meaning of significant quantity (SQ)

The significant quantity is "the quantity to be detected", if missing, and is understood, in the context of the NPT, as a "quantity of nuclear material which is required to manufacture a single nuclear explosive or the quantity needed to produce by appropriate conversion the material required to manufacture such an explosive." From this understanding, the following considerations were made separately on the significant quantity:

$$*SQ = TA + \delta$$

where TA is the threshold amount of manufacture a single nuclear explosive, and δ is conversion losses and process holdups.

$$*SQ \geqslant M_I(\alpha, \beta, \sigma_{MUF})$$

where $M_I(\alpha, \beta, \sigma_{MUF})$ is an a priori defined minimum quantity of material which could be detected by the inspector, if both of the level of significance, α, and the target detection power, $1-\beta$, were set to their fixed values and also the achievable levels of accountancy, σ_{MUF}, were assumed to be known from the design information.

It was expected that the second requirement would be expected to be achievable within a stated time period, i.e. a material balance period. But it has become clear that this condition could not always be satisfied in a facility that has a relatively large throughput or inventory. To solve this problem, the most effective approach is to increase the frequency of physical inventory taking to shorten the definition time for α, β, σ_{MUF} and then M_I. However, for a large-scale facility, a precise physical inventory taking should be carried out monthly or every ten days, if M_I is to be increased to the value less than or equal to SQ. These frequencies must hamper the normal operation of the facility. This is the dilemma related to the definition of significant quantity.

3.2. Significant quantity and goal quantity

From the practical aspects, and to minimize the degree of hindrance on normal operations, it is permitted to take a certain quantity greater than the "significant quantity" as the amount to be detected within a material balance period. Such a quantity is defined as a "goal quantity (GQ)". The following relations between TA, SQ, M_I and GQ are clarified:

(1) $SQ = TA + \delta$ (δ as defined previously)

Within a material balance period,

(2) if $SQ > M_I(\alpha, \beta, \sigma_{MUF})$, $GQ = SQ$,

(3) if $SQ \leqslant M_I(\alpha, \beta, \sigma_{MUF})$, $GQ \doteq M_I(\alpha, \beta, \sigma_{MUF})$.

When the goal quantity is selected as those defined above, the level of significance, α, and the detection probability, $(1-\beta)$, are ensured while, should the inspector detect the diversion with the detection probability $(1-\beta)$ the quantity of diversion might be greater than a "significant quantity" in the case of $M_I > SQ$.

To avoid such a risk occurring, frequent but rough inventory taking should be carried out within a material balance period. Such an inventory taking should be designed for the timely detection of the amount of SQ ($\leq M_I$ defined for material balance period). The frequency of such inventory takings in a material balance period should be chosen from the relations between SQ, M_I and the conversion time of the quantity of SQ. The "conversion time" is understood in this paper as the time necessary to produce the weapon-grade material from the material of the quantity of SQ.

3.3. Detection of significant quantity

In the context of NPT, the concept of significant quantity is meaningful only to a State as a whole and, further, has an important meaning if the quantity could be converted into the weapon-grade material in the State. Therefore, if the State has no means for conversion, the significant quantity to be detected is less meaningful. From this point of view, the authors transform the original significant quantity into the new significant quantity, taking the characteristics of the fuel cycle of the State into consideration. The concept and the method to obtain the new significant quantity is presented later in this report.

4. DEFINITIONS OF TERMS USED IN THIS PAPER

4.1. Threshold amount (TA)

A "threshold amount" of nuclear material is the approximate quantity of special fissionable material in a single nuclear explosive device.

4.2. Significant quantity (SQ)

A "significant quantity" of nuclear material is the approximate minimum quantity of nuclear material necessary for obtaining a "threshold amount". This quantity includes conversion process holdups and losses. Then the following relation between TA and SQ is always valid:

$$SQ \geq TA$$

4.3. Goal quantity (GQ)

When the expected accuracy of a facility material accountancy is expressed by σ_{EA}, the detection capability M_I of the inspector for the facility is approximately obtained by the following equation [3]:

$$M_I = 5\,\sigma_{EA}$$

This is the material quantity detectable (if missing) by the inspector when $\alpha = \beta = 5\%$.

The goal quantity (GQ), or the quantity to be regarded by the inspector as safeguards goal for the facility, is determined by the relation between M_I and SQ as follows:

(i) If $M_I \geqslant SQ$, $GQ = M_I$

(ii) If $M_I < SQ$, $GQ = (C_1 \cdot SQ + C_2 \cdot M_I)/(C_1 + C_2)$

If $M_I < SQ$, it is permissible to change it to GQ = SQ in the context of the Non-Proliferation Treaty (NPT). Despite this, judging from the necessity of effective verification on operator's measurement accuracies, the value satisfying $M_I < GQ < SQ$ is used. Although C_1 and C_2 can be decided arbitrarily to some extent, we used the following values in this study

$$C_1 = 1, \text{ and } C_2 = 3.$$

The relation between these quantities is shown in Fig.1. If the object material is a mixture of plutonium and uranium, we calculated the goal quantities GQ(Pu) and GQ(U) independently, and obtained their weighted averages taking account of the significant quantity, in order to acquire the GQ(Pu + U′) for the object material. The equation used is

$$GQ\,(Pu + U') = GQ(Pu) + \frac{SQ\,(Pu)}{SQ\,(^{235}U)} \times GQ\,(^{235}U)$$

4.4. Diversion factor

In Ref. [1], the categorization of nuclear material was defined as based on the content of fissile nuclides, the content of radioactive β- and γ-emitters and the characteristics of the States' nuclear fuel cycle, the latter being expressed as a combination of the reactor, reprocessing plant and enrichment plant. The property of a certain category of nuclear material is related to the risk of diversion to some extent, that is with weapon usability in terms of fast critical

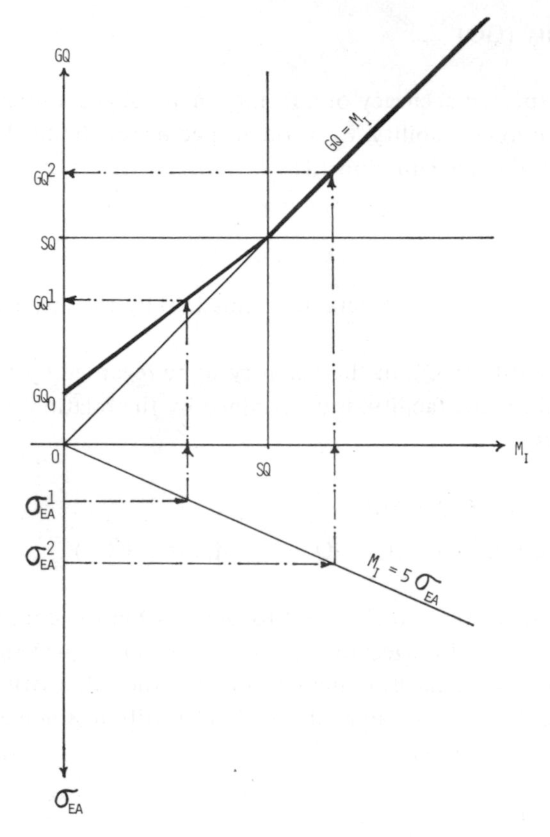

FIG.1. Relations between expected accuracy (σ_{EA}), detection capability (M_I), significant quantity (SQ or WSQ), and goal quantity (GQ). If $M_I \geqslant SQ$, $GQ = M_I$, and if $M_I < SQ$, $GQ = [C_1 \cdot SQ + C_2 \cdot M_I]/(C_1 + C_2)$. $GQ_0 = [C_1/(C_1 + C_2)] \cdot SQ$.

mass, and material accessibility in terms of radioactivity. These relations are, however, stated qualitatively rather than quantitatively.

To express them more quantitatively, the authors adopted the concept of diversion factor which was originally developed by Franklin [2] to assess the risk of the diversion of nuclear material at each stage of a fuel cycle. The diversion factor is a numerical result of a formula where each of the various characteristics of a fuel cycle is expressed as a numerical value as shown in Appendix I. This concept is useful for our present purpose, but some modifications are needed. The modified diversion formula is applied independently for uranium and plutonium at each stage of the nuclear fuel cycle through the following expression:

$$F_D(n) = C_n \cdot W \cdot \{(\Omega \cdot L_i + T \cdot G \cdot L_t)/L_T\}_n$$

n: uranium or plutonium

Each expression in this equation represents the factor associated with a particular type of concern: W is the weapons factor, L_t is the fissile throughput per year, L_i is the fissile inventory, L_T is the total fissile given by $(L_i + L_t)$, Ω the storage factor, T the transport factor, and G the relative vulnerability of transport over storage, or material in process. C_n is a newly introduced factor, a facility correlation factor which is used to represent the importance of the co-existence of facilities for different purposes in a manner similar to the material categorization approach.

This factor is represented as:

$$C_U = C_U (P_1, P_2, P_3) = X^a$$

$$C_{Pu} = C_{Pu} (P_1, P_2) = X^b$$

where

$P_1 = 1$ if a power reactor exists and $= 0$ if not,

$P_2 = 1$ if a reprocessing plant exists and $= 0$ if not,

$P_3 = 1$ if an enrichment plant exists and $= 0$ if not,

and

$$a = P_1 + k_1 P_2 + k_2 P_3$$

$$b = P_1 + k_3 P_2 .$$

X, k_1, k_2 and k_3 are somewhat arbitrary constants which shall be decided by further analyses but, in this paper, $k_1 = k_2 = k_3 = 1$ is selected for simplicity. In addition, the diversion factor is normalized so as to make $F_D(n)$ equal to unity for the worst case in the all fuel cycles under NPT safeguards. Details of the definition and the meanings of the symbols in the diversion formula are presented in Appendix 1, and values calculated for the four fuel cycle models are tabulated in Appendix 2.

4.5. Weighted significant quantity (WSQ)

A "weighted significant quantity" is the amount given by the significant quantity inversely multiplied by the diversion factor, which is estimated for the nuclear fuel cycle in a State under consideration. This diversion factor is to be obtained against all the nuclear fuel cycles covered by IAEA Safeguards, and the worst case among the results is to be normalized in such a way as to become 1. This procedure is described in the following:

Assume that there are L States under the IAEA Safeguards, and the i-th State has M_i nuclear fuel cycles, and the j-th nuclear fuel cycle includes N_j fuel-cycle stages (reactors, reprocessing, etc.) in it. In this case, the diversion factor F_D is shown as follows by using the formula already mentioned (as a value before normalization):

$$F_D^{i,j,k}(n) \quad \begin{cases} n = \text{uranium or plutonium} \\ i = 1, 2, \text{------}, L \\ J = 1, 2, \text{------}, M_i \\ k = 1, 2, \text{------}, N_j \end{cases}$$

The total diversion factor of the nuclear fuel cycle in the i-th State, $TF_D^i(n)$, can be obtained by summing up with respect to all j's and k's and is expressed as follows:

$$TF_D^i(n) = \sum_{j,k} F_D^{i,j,k}(n)$$

The normalization factor N_n is to be selected as follows:

$$N_n = \max\{TF_D^1(n),\ TF_D^2(n) \text{------}, TF_D^L(n)\}$$

Thus, the normalized diversion factor for the i-th State $f_D^i(n)$ is obtained through the following equation:

$$f_D^i(n) = \frac{1}{N_n} \cdot TF_D^i(n)$$

Thus, the value of the "weighted significant quantity, $WSQ^i(n)$", applicable to all the fuel cycles included in the i-th State, can be given by the equation

$$WSQ^i(n) = [f_D^i(n)]^{-1} \cdot SQ(n)$$

This value is applicable to all the stages included in the i-th State's fuel cycle as a whole.

5. APPLICATION TO MODEL FUEL CYCLES

5.1. Fuel-cycle model

To see how the weighted significant quantity can reduce the inspection effort in an effective manner, the authors estimated inspection sample sizes

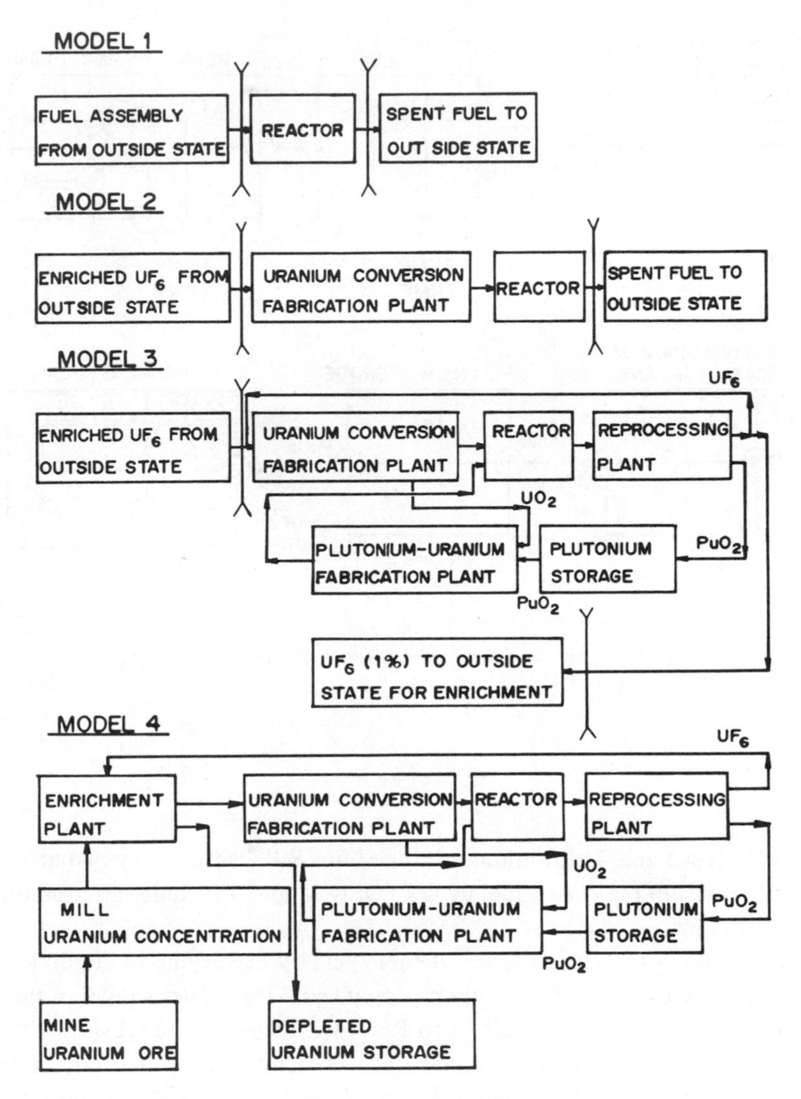

FIG.2. Fuel-cycle models.

for a few specific nuclear fuel cycles. The sample size itself is, of course, only one of indexes but a major one to represent the total inspection effort. Model nuclear fuel cycles were selected so as to be available to show the effect of different facility correlation factors on inspection sample sizes.

Four types of fuel-cycle model were considered as illustrated in Figs 2 and 3. Every fuel cycle includes seven power reactors (LWR) of 1100 MW(e) capacity. In Models 1 and 2, spent fuel from LWRs is stored without reprocessing, and thus no recycling of plutonium is intended in these fuel cycles. On the other

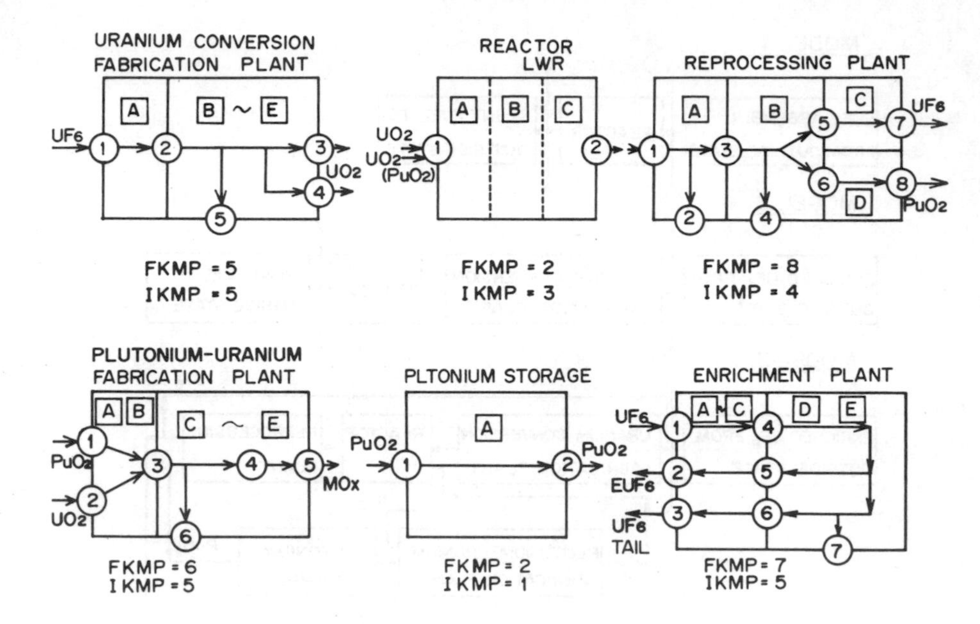

FIG.3. Flow and inventory KMP in facilities used in the models.

hand, in Models 3 and 4 plutonium produced in LWRs is fully recycled after reprocessing within the closed fuel cycles. Only Model 4 includes an uranium enrichment plant.

Table I shows the mass balance of fuel cycle by classifying the figures according to whether or not plutonium is recycled. The values shown in this table have been obtained by referring to Ref. [4]. Tables II and III show the amount of nuclear materials held by each of the facilities and the total amount included in each model. The value of inventory in each facility was appropriately decided, while the value of the flow was automatically decided from the property of the fuel cycle itself.

5.2. Results of calculation

5.2.1. Diversion factors and weighted significant quantities

Diversion factors were calculated for each stage of each nuclear fuel cycle by using the diversion factor formula together with the values and expressions shown in the appendixes.

It is assumed in this study that each of four fuel-cycle models corresponds to a fuel cycle in a State, which has only one fuel cycle within it. Therefore, suffixes in the diversion factor $F_D^{i,j,k}$ are ranged as follows

i	1	2	3	4
j	1	1	1	1
k	1	1, 2	1, 2, 3, 4, 5	1, 2, 3, 4, 5, 6

and values of k correspond to the stage numbers as shown below:

k	Stage No.	
1	5	LWR
2	4	Uranium conversion fabrication plant
3	6	Reprocessing plant
4	2	Transportation
5	3	Plutonium storage
6	7	Plutonium uranium fabrication plant
7	1	Enrichment plant

For each fuel-cycle model three cases for the facility correlation factor (C_U or C_{Pu}) are considered by selecting different values of X in order to see the effect of the correlation factor on the resulting value of the diversion factor. Calculated diversion factors are shown in Table IV. The values for X = 1 are those where there is no facility correlation, which means that the inspection effort necessary for a certain stage (facility) in a fuel cycle is not affected by the existence of any facilities in the fuel cycle.

The diversion factors ($F_D^{i,j,k}$) were totalled with respect to all the stages included in a fuel cycle to obtain the total diversion factor (TF_D) of the fuel cycle, which was normalized after all the total diversion factors were calculated for every fuel-cycle model. Using the reciprocal of the normalized total diversion factor as a weighting function, the weighted significant quantity (WSQ) to be applied for each fuel-cycle model was calculated. Table V shows the resulting values. From this table it is known that some values of weighted significant quantities occasionally exceed the total amount of nuclear material in the relevant fuel cycle. These values are shown in parentheses in the table, and not used in the later part of the calculation. In such a case the total amount of nuclear material in the relevant fuel cycle is temporarily chosen as an alternative value for the fuel cycle instead of the calculated WSQ, but further investigation on the

TABLE I. FUEL-CYCLE MASS BALANCE

	No Pu-thermal Recycling		Pu-thermal Recycling										
	LWR[a] 1100MW(e)	U-Conversion Fabrication	LWR-Pu[a] 1100MW(e)		U-Conversion Fabrication		Reprocessing		Plutonium Storage	Plutonium-Uranium Fuel Fabrication		Enrichment	
Input:	Stage 5	Stage 4	Stage 5		Stage 4		Stage 6		Stage 3	Stage 7		Stage 1	
Type of item	assemblies	cylinders	assemblies		cylinders		assemblies		bird-cage	drum	bird-cage	cylinders	
Items/year	225	177	196	29	155	21	1575		544	21	509	165	647
			UO_2 fuel	MOx fuel									
weight/item	169 kg U	1.5 tU	169 kgU	169kg(Pu+U)	1.5 tU	1.5 tU			4 kg Pu	1.5 tU	4 kg Pu	1.5tU	1.5tU
weight/year	38tU	266tU	33.2tU	4.8t(Pu+U)	232tU	32.0tU	250tU	2200kg	2178kgPu	32.0tU	2036kgPu	248tU	970tU
%enrichment(%Pu)	2.5	2.5	2.5	6	2.5	Depl.(0.3)	1			Depl(0.3)		1	Nat.
Material	UO_2	UF_6	UO_2	PuO_2-UO_2	UF_6	UF_6	UO_2	Pu	PuO_2	UO_2	PuO_2	UF_6	UF_6
Output:												Products	
Type of item	assemblies	assemblies	assemblies		assemblies	drums	cylinder	bird-cage	bird-cage	assemblies		cylinders	
Items/year	225	1575	196	29	1372	21	165	544	509[b]	203(29 x7)		155	657
			UO_2fuel	MOxfuel									
weight/item	162.2kgU	169kgU			169kgU	1.5tU	1.5tU	4kgPu	4kgPu	155.6kgU	9.9kgPu	1.5tU	1.5tU
weight/year	36.5tU	266tU	31.9tU	4.6t(Pu+U)	231.9tU	32.0tU	247.5tU	2178kgPu	2036kgPu	31.6tU	2016kgPu	232tU	986tU
%enrichment(%Pu)	1.0	2.5	1		2.5	Depl(0.3)	1			Depl(0.3)	6	2.5	0.3
Material	UO_2	UO_2	UO_2	PuO_2-UO_2	UO_2	UO_2	UF_6	PuO_2	PuO_2	PuO_2(UO_2)		UF_6	UF_6
Pu(kg)/year	300		300										
discards(kg)/year		100U			100U		2500U	22Pu		370U	20Pu	100	
Annual throughput	24Ekg (Feed)	166Ekg	309Ekg (Feed)		147Ekg		2225Ekg		2178Ekg	2035Ekg		194Ekg (Product)	
Frequency of PIT	1	2	1		2		4		4	4		4	

[a] Seven reactors of this capacity consist of the fuel cycle.

[b] Plutonium in storage will increase by 35 bird cages per year in this model.

TABLE II. FACILITY DATA USED FOR MODEL FUEL CYCLES

Reactor LWR

KMP	Stratum	kgU / kgPu	Items /year
flow			
1	Fresh Fuel Ass.	38000	225
2	Spent Fuel Ass.	36500 / 300	225
inventory			
A	Fresh Fuel Ass.	76000	450
B	Ass. in Core	(149600)	900
C	Spent Fuel Ass.	73000 / 600	450

Reactor LWR-Pu

KMP	Stratum	kgU / kgPu	Items /year
flow			
1	UO_2 Fresh Fuel Ass.	33200	196
	MOx Fresh Fuel Ass.	4512 / 288	29
2	UO_2 Spent Fuel Ass.	31900 / 261	196
	MOx Spent Fuel Ass.	4416 / 538	29
inventory			
A	UO_2 Fresh Fuel Ass.	66400	392
	MOx Fresh Fuel Ass.	9024 / 576	58
B	Ass. in Core	(150232)	900
C	UO_2 Spent Fuel Ass.	63800 / 524	392
	MOx Spent Fuel Ass.	8832 / 1076	58

Uranium Conversion Fabrication

KMP	Stratum	kgU / kgPu	Items /year
flow			
1,2	UF_6 cylinders	266000	177
3	Assemblies	265900	1575
4	-		
5	Waste	100	330
inventory			
A	UF_6 Cylinders	42000	28
B	UO_2 Pellets	25200	3452
C	UO_2 Rods	25141	7287
D	Assemblies	37760	223
E	Scrap	7560	1890

Uranium Conversion Fabrication[a]

KMP	Stratum	kgU / kgPu	Items /year
flow			
1,2	UF_6 Cylinders	232000	155
	DUF_6 Cylinders	32000	21
3	Assemblies	231900	1372
4	UO_2 Drums	32000	21
5	Waste	100	330
inventory			
A	UF_6 Cylinders	42000	28
	DUF_6 Cylinders	5750	4
B	UO_2 Pellets	25200	3542
	DUO_2 Drums	6000	4
C	UO_2 Rods	25141	7287
D	Assemblies	37760	223
E	Scrap	7560	1890

Reprocessing

KMP	Stratum	kgU / kgPu	Items /year
flow			
1	Assemblies	250000 / 2200	1575
2	Waste 1	1250 / 12	330
3	Imput Tank	248750 / 2188	250
4	Waste 2	1250 / 12	330
5,7	UF_6 Cylinders	247500	165
6,8	PuO_2 Bird cages	2176	544
inventory			
A	Assemblies	56250 / 495	354
B	Scrap	12500 / 110	43
C	UF_6 Cylinders	55625	37
D	PuO_2 Bird cages	488	122

Plutonium Storage

KMP	Stratum	kgU / kgPu	Items /year
flow			
1	PuO_2 Bird cages	2178	544
2	PuO_2 Bird cages	2036	509
inventory			
A	PuO_2 Bird cages	4356	1088

Plutonium Uranium Fabrication

KMP	Stratum	kgU / kgPu	Items /year
flow			
1,3	PuO_2 Bird cages	2036	509
2,3	UO_2 Drums	32000	21
4	Output rods	31600 / 2016	9947
5	MOx Assemblies	31600 / 2016	203
6	Waste	370 / 20	300
inventory			
A	PuO_2 Bird cages	508	127
B	UO_2 Drums	6000	4
C	MOx Rods	7901 / 504	2487
D	MOx Assemblies	15567 / 1008	100
E	Waste	300 / 10	300

Enrichment

KMP	Stratum	kgU / kgPu	Items /year
flow			
1	UF_6 Cylinders	$248tU_{1.0}$	165
	"	$970tU_N$	647
2	"	$232tU_{2.5}$	155
	"	$986tU_{0.3}$	657
inventory			
A	UF_6 Cylinder	$120tU_{2.5}$	80
	"	$495tU_{0.3}$	330
	"	$125tU_{1.0}$	83
	"	$486tU_N$	324

[a] For Pu-thermal recycling.

TABLE III. AMOUNTS OF NUCLEAR MATERIALS IN EACH FUEL-CYCLE MODEL (kg)

Fuel Cycle Model	Stage No. & Facility	Flow/year		Inventory		Total Inventory	
		^{235}U	Pu	^{235}U	Pu	^{235}U	Pu
1	5.LWR	7x950	7x300	7x5245	7x1200	36,715	8,400
2	4.U-C-F	6650	-	3442	-		
	5.LWR	7x950	7x300	7x5245	7x1200	40,157	8,400
3	4.U-C-F'	5896	-	3477	-		
	7.MOx-F	96.0	2036	89.3	2030		
	5.LWR-Pu	7x844	7x799	7x4690	7x4350		
	6.Reprocess	2500	2200	1244	1093		
	3.Pu Storage	-	2178	-	4356	37,640	37,929
4	4.U-C-F'	5896	-	3477	-		
	7.MOx-F	96.0	2036	89.3	2030		
	5.LWR-Pu	7x844	7x799	7x4690	7x4350		
	6.Reprocess	2500	2200	1244	1093		
	3.Pu Storage	-	2178	-	4356		
	1.Enrich.	9377	-	9191	-	46,831	37,929

Note: Every fuel cycle produces electricity of 7700 MW(e).

occurrence of such a case is necessary. Incidentally, the calculated WSQs for plutonium are very large for Models 1 and 2, while for Models 3 and 4 each of these is the same as the ordinary significant quantity. This inclination seems to derive mainly from the great difference in the quantity of the weapon factor (W) and the difference in the number of stages included in the fuel cycle.

5.2.2. Goal quantities and resulting inspection sample sizes

Table VI shows values of goal quantities (GQ's) calculated by the obtained 'weighted significant quantities' using the definition of GQ described in Section 4.3. Values for expected accuracies (σ_{EA}) and detection capabilities (M_T), which were used to estimate goal quantities, are also shown in this table.

Inspection sample sizes are calculated by the mixed-attribute variable sampling model proposed by Hough et al. [3]. Some assumptions were made for verification procedures of certain facilities. As for LWR, only item accounting was assumed for the material accountancy verification, and also item accounting was assumed for the inventory taking in the plutonium storage although some measurements by NDA were assumed to be carried out for the verification of

TABLE IV. DIVERSION FACTORS FOR EACH STAGE FOR THREE DIFFERENT VALUES OF X, USED TO REPRESENT THE FACILITY CORRELATION FACTOR

	Model 1 or 2			Model 3 or 4				
Stage No.	5		4	4	5		6	
	LWR S.F.		U-C-F	U-C-F	LWR-Pu S.F.		Reprocessing	
Element	U	Pu	U	U	U	Pu	U	Pu
W	1.16E-5	1.75E-8	2.20E-3	2.20E-3	3.47E-6	1.50E-4	6.76E-4	3.31E-2
T	0.05	0.05	0.04	0.04	0.05	0.05	0.0	0.0
Ω	0.125	0.125	0.125	0.125	0.125	0.125	0.0625	0.0625
G	2	2	2	2	2	2	2	2
L_t	6650	2100	6650	5896	5905	5593	2500	2200
L_i	38628	8400	3442	3477	20126	17234	1244	1093
L_T	45278	10500	10092	9373	26031	22827	3744	3293
if X=1								
C_U or C_{Pu}	1	1	1	1	1	1	1	1
F_D	1.40E-6	2.10E-9	2.10E-4	2.84E-4	4.14E-7	1.78E-7	1.41E-5	6.87E-4
if X=2								
C_U or C_{Pu}	2	2	2	4(8)	4(8)	4	4(8)	4
F_D	2.81E-6	4.19E-9	4.20E-4	1.14E-3	1.66E-6	7.11E-7	5.62E-5	2.75E-3
				(2.27E-3)	(3.31E-6)		(1.12E-4)	
if X=3								
C_U or Pu	3	3	3	9(27)	9(27)	9	9(27)	9
F_D	4.21E-6	6.29E-9	6.29E-4	2.55E-3	3.73E-3	1.60E-6	1.26E-4	6.18E-3
				(7.66E-3)	(1.12E-5)		(3.79E-4)	

	Model 3 or 4					Model 4
Stage No.	2		3	7		1
	Transportation		Pu-Store	MOx - Fab.		Enrich.
Element	U	Pu	Pu	U	Pu	U
W	6.76E-4	3.31E-2	3.31E-2	8.82E-4	3.31E-2	1.69E-3
T	0.03	0.0012	0	0.03	0.03	0.012
Ω	0.0	0.0	6.25E-3	0.125	0.0625	0.025
G	4	4	-	2	2	2
L_t	2475	2176	2178	96	2036	9337
L_i	0	0	4356	89	2030	9191
L_T	2475	2176	6534	185	4066	18568
if X=1						
C_U or C_{Pu}	1	1	1	1	1	1
F_D	8.11E-5	1.59E-4	1.38E-4	8.05E-5	2.03E-3	4.13E-5
If X=2						
C_U or C_{Pu}	4(8)	4	4	4(8)	4	(8)
F_D	3.25E-4	6.36E-4	5.15E-4	3.22E-4	8.11E-3	(3.31E-4)
	(6.49E-4)			(6.44E-4)		
if X=3						
C_U or C_{Pu}	9(27)	9	9	9(27)	9	(27)
F_D	7.30E-4	1.43E-3	1.24E-3	7.25E-4	1.82E-2	(1.12E-3)
	(2.19E-3)			(2.17E-3)		

Note: (1) Values in ( ) are for Model 4.

(2) X is a constant used to represent the facility correlation factor in the diversion factor.

TABLE V. TOTAL DIVERSION AND NORMALIZED DIVERSION FACTORS, AND THE CORRESPONDING WEIGHTED SIGNIFICANT QUANTITIES

These values are shown for three different values of X, which relate to the facility correlation factor

Fuel Cycle Model		1	2	3	4
Included Stages		5	4,5	2,3,4,5,6,7	1,2,3,4,5,6,7
if X=1					
$TF_D(U)$		1.404E-6	2.112E-4	4.598E-4	5.011E-4
$TF_D(Pu)$		2.095E-9	2.095E-9	3.011E-3	3.011E-3
$f_D(U)$		2.802E-3	4.215E-1	9.176E-1	1.00
$f_D(Pu)$		6.959E-7	6.959E-7	1.00	1.00
WSQ	HEU	8.9t	59kg	27kg	25kg
	LEU	26.8t	178kg	82kg	75kg
	Pu	(11500 t)[a]	(11500 t)	8kg	8kg
if X=2					
$TF_D(U)$		2.808E-6	4.224E-4	1.839E-3	4.009E-3
$TF_D(Pu)$		4.190E-9	4.190E-9	1.204E-2	1.204E-2
$f_D(U)$		7.004E-4	1.054E-1	4.587E-1	1.00
$f_D(Pu)$		3.479E-7	3.479E-7	1.00	1.00
WSQ	HEU	35.7t	237kg	55kg	25kg
	LEU	(107 t)	712kg	163kg	75kg
	Pu	(23000 t)	(23000 t)	8kg	8kg
if X=3					
$TF_D(U)$		4.212E-6	6.336E-4	4.138E-3	1.353E-2
$TF_D(Pu)$		6.285E-9	6.285E-9	2.709E-2	2.709E-2
$f_D(U)$		3.113E-4	4.683E-2	3.059E-1	1.000
$f_D(Pu)$		2.320E-7	2.320E-7	1.000	1.000
WSQ	HEU	(80.3t)	534kg	82kg	25kg
	LEU	(241 t)	1602kg	245kg	75kg
	Pu	(34500 t)	(34500 t)	8kg	8kg
Total Material in Fuel Cycle					
(LEU)	^{235}U	36,715kg	40,157kg	37,640kg	46,831kg
	Pu	8,400kg	8,400kg	37,929kg	37,929kg

[a] The value in parentheses exceeds the total amount of nuclear material in the relevant fuel cycle, in which case the total amount of nuclear material in the fuel cycle is finally chosen as the weighted significant quantity for the fuel cycle. The same treatment is made for other values in parentheses.

TF_D: Total unnormalized diversion factor.

f_D: Normalized diversion factor.

WSQ: Weighted significant quantity.

HEU: High-enriched uranium.

LEU: Low-enriched uranium.

TABLE VI. EXPECTED ACCURACIES (σ_{EA}), DETECTION CAPABILITIES (M_I) AND GOAL QUANTITIES (GQ) FOR EACH STAGE IN EACH FUEL-CYCLE MODEL

		No Pu-Thermal Recycling			Pu-Thermal Recycling										
Fuel Cycle		Model 1	Model 2		Model 3					Model 4					
Stage No.		5	4	5	4	7	5	6	3	4	7	5	6	3	1
		LWR	U-C-F	LWR	U-C-F	MOx-F	LWR-Pu	Rep.	Pu-S	U-C-F	MOx-F	LWR-Pu	Rep.	Pu-S	Enrich.
δ_{EA}(%)	^{235}U	0.2	0.3	0.2	0.3	0.5	0.2	0.8	-	0.3	0.5	0.2	0.8	-	0.2
	Pu	-	-	-	-	0.5	0.5	1.0	0.4	-	0.5	0.5	1.0	0.4	-
σ_{EA}(kg)	^{235}U	1.9	10.3	1.9	10.4	0.5	3.4	10.0	-	10.4	0.5	3.4	10.0	-	18.5
	Pu	-	-	-	-	10.2	2.9	10.9	2.2	-	10.2	2.9	10.9	2.2	-
M_I(kg)	^{235}U	9.5	51.6	9.5	52.2	2.2	16.9	49.8	-	52.2	2.2	16.9	49.8	-	92.3
	Pu	-	-	-	-	50.8	14.4	54.7	10.9	-	50.8	14.4	54.7	10.9	-
GQ(kg) [SQM]	^{235}U	25.9	57.5	25.9	57.9	20.4	31.4	56.1	-	57.9	20.4	31.4	56.1	-	92.3
	Pu	-	-	-	-	50.8	14.4	54.7	10.9	-	50.8	14.4	54.7	10.9	-
[MCM]	^{235}U	MC=3	MC=3	MC=3	MC=3	MC=3	MC=3	MC=3	-	57.9	20.4	31.4	56.1	-	92.3
	Pu	MC=3	-	MC=3	-	MC=2	MC=2	MC=2	MC=2	-	MC=2	MC=2	MC=2	MC=2	-
[WSQM] if X=1	^{235}U	6707	83.2	51.6	59.6	22.2	33.2	57.8	-	57.9	20.4	31.4	56.1	-	92.3
	Pu	-	-	-	-	50.8	14.4	54.7	10.9	-	50.8	14.4	54.7	10.9	-
if X=2	^{235}U	9182	217	185	79.9	42.4	53.4	77.8	-	57.9	20.4	31.4	56.1	-	92.3
	Pu	-	-	-	-	50.8	14.4	54.7	10.9	-	50.8	14.4	54.7	10.9	-
if X=3	^{235}U	9182	439	408	100	62.9	73.9	98.6	-	57.9	20.4	31.4	56.1	-	92.3
	Pu	-	-	-	-	50.8	14.4	54.7	10.9	-	50.8	14.4	54.7	10.9	-

Note: SQM: Significant quantity method.
MCM: Material categorization method.
WSQM: Weighted significant quantity method.
X is a constant used to represent the facility correlation factor in the diversion factor.

TABLE VII. COMPARISON BETWEEN METHODS IN TERMS OF SAMPLE SIZE

	No Pu-Thermal Recycling					Pu-Thermal Recycling							
Fuel Cycle Model	Model 2					Model 3					Model 4		
Method	SQM	MCM	WSQM			SQM	MCM	WSQM			SQM	MCM	WSQM
Base of F.C.F.	-	-	X=1	X=2	X=3	-	-	X=1	X=2	X=3	-	-	X=1,2,3
U-Conv.-Fabrication													
flow Na	244	0	191	84	45	219	0	214	178	144	219	219	219
Nv	564	0	93	40	39	296	0	260	102	81	296	296	296
inventory Na	250	0	174	68	34	250	0	242	184	148	250	250	250
Nv	924	0	122	104	104	434	0	354	130	112	434	434	434
total Na	494	0	365	152	79	469	0	456	362	292	469	469	469
Nv	1488	0	215	144	143	730	0	614	232	193	730	730	730
Reprocessing Plant													
flow Na						53	0	53	53	53	53	53	53
Nv						39	0	39	39	39	39	39	39
inventory Na						12	0	12	12	12	12	12	12
Nv						26	0	26	26	26	26	26	26
total Na						65	0	65	65	65	65	65	65
Nv						65	0	65	65	65	65	65	65
Pu-U-Fabrication													
flow Na						58	0	58	58	58	58	9	58
Nv						39	0	39	39	39	39	31	39
inventory Na						184	0	184	184	184	184	28	184
Nv						88	0	88	88	88	88	78	88
total Na						242	0	242	242	242	242	37	242
Nv						127	0	127	127	127	127	109	127
Pu-Storage													
flow Na						166	0	166	166	166	166	166	166
Nv						75	0	75	75	75	75	75	75
inventory Na						1374	0	1374	1374	1374	1374	1374	1374
Nv						616	0	616	616	616	616	616	616
total Na						1540	0	1540	1540	1540	1540	1540	1540
Nv						691	0	691	691	691	691	691	691
Enrichment													
flow Na											114	114	114
Nv											55	55	55
inventory Na											462	462	462
Nv											168	168	168
total Na											576	576	576
Nv											223	223	223
Total													
flow Na	244	0	191	84	45	496	0	491	455	421	610	561	610
Nv	564	0	93	40	39	449	0	413	255	234	504	496	504
inventory Na	250	0	174	68	34	1820	0	1812	1754	1718	2282	2126	2282
Nv	924	0	122	104	104	1164	0	1084	860	843	1332	1340	1332
total Na	494	0	365	152	79	2316	0	2303	2209	2139	2892	2652	2892
Nv	1488	0	215	144	143	1613	0	1497	1115	1076	1836	1818	1836

receipts and shipments of plutonium in bird-cages. As for other facilities, verifications by sampling measurements were assumed for MUF MBA only. The results are shown in Table VII. From this table it can be seen that the weighted significant quantity is useful to reduce inspection sample sizes so as to reflect the importance of the co-existence of sensitive facilities in view of safeguards objectives. It must also be recognized, however, that the effort of the facility correlation factor on the final result of inspection sample size does not appear in a simple manner. The effect seems to vary from case to case. For example, the total attribute sample size for the case X = 1 in Model 2 is 365 and reduces to 79 for X = 3, i.e. the reduction rate is about 78%, while the total variable sample size changes from 215 to 143 in the same situation. In this case the reduction rate is only 33%. The cause of this complexity must be investigated further.

5.3. Conclusion

The use of weighted significant quantity is useful to reduce the inspection effort while maintaining the level I assurance for verification procedure. This method is available not only to reduce the IAEA's safeguards cost but also to reduce the cost of SSAC.

This paper was meant to clarify whether or not the concept of the diversion factor can be used as a method of expressing and evaluating the characteristic of national fuel cycle semi-quantitatively. From this point of view, the results seem to show it to be possible. The reliability of the WSQ, however, depends chiefly on that of the basic data used in calculating the diversion factor. In this respect, the most important future problem for this method is how to improve the reliability of the diversion factor.

ACKNOWLEDGEMENTS

The authors are deeply indebted to Mr. H. Yoshida of the JAERI for his continuous assistance to prepare this paper, and are grateful for the valuable comments from Mr. Y. Kawashima of the Nuclear Material Control Center (NMCC), and to Messrs H. Okashita, H. Natsume and H. Umezawa of the JAERI. The authors wish to express great appreciation to Mr. K. Horino of NMCC for his earnest assistance to make up the fuel-cycle models for this study.

REFERENCES

[1] ROMETSCH, R., et al., "Safeguards – 1975–1985", Safeguarding Nuclear Materials (Proc. Symp. Vienna, 1975) 1, IAEA, Vienna (1976) 3–36.
[2] BATTELLE COLUMBUS LABORATORIES, Final Report on Study of Advanced Fission Power Reactor Development for the United States, 3 BCL-NSF-C-946-2.
[3] HOUGH, C.G., BEETLE, T.M., "Statistical methods for the planning of inspection", Safeguarding Nuclear Materials (Proc. Symp. Vienna, 1975) 1, IAEA, Vienna (1976) 561–80.
[4] GMELIN, W., HOUGH, G., SHMELEV, V., SKJOELDEBRAND, R., "A technical basis for international safeguards", Peaceful Uses Atom. Energy (Proc. 4th Int. Conf. Geneva, 1971) 9, IAEA, Vienna (1972) 487.

APPENDIX I (from Ref. [2])

The weapons factor, W, is given by

$$W = Q_f \cdot F_c \cdot C \cdot S \cdot M$$

Q_f relates to the hazard associated with fissile quantity at a given stage in the fuel cycle and is larger where greater quantities of material are encountered; F_c is the fissile concentration factor and is defined by the fissile isotope enrichment times the abundance of fissile species in the chemical form; S is the separation factor and relates to the difficulty of access by which the material may be reached whether or not it requires separation and, if so, whether this may be performed mechanically or chemically; M is the dilution factor and varies directly as the number of atoms of concern; and C, the contamination factor, relates to the presence or absence of fission-production contamination.

The storage factor, Ω, is given by

$$\Omega = s \cdot A \cdot \theta \cdot P \cdot R$$

s is related to the method of storage and is smaller if the degree of security afforded is greater; A represents the alarm factor and is smaller where the material in storage is alarmed; θ, the storage time factor, relates to the potential safeguards concern associated with long storage. It is greatest when the material is left unchecked for times exceeding one month; P is used to indicate the effect of the personnel having access to the stored material. Where the persons involved have a security clearance or psychological profile, this factor is reduced. The storage factor is multiplied by L_i, the quantity of inventory material.

T represents the transportation factor and is given by

$$T = \tau \cdot t \cdot R \cdot P$$

τ is determined by the method of transport selected and is smaller if the degree of security afforded by the method is greater; t represents the transport time factor and is greater as the transport time increases; R is a factor which relates to the safeguards benefits of radio contact; P indicates the effect of the number of personnel involved in handling the material during a particular transit step. It is assumed that the more people involved in the transfer, the less vulnerable is the material to diversion. The transportation factor, T, is multiplied by a weighting factor, G, which represents the relative vulnerability of transport over storage or material in process. The weighted transportation factor is multiplied by the quantity of material being transported within a year, L_t.

DIVERSION FACTOR FORMULA VALUES

S: Separation factor

(a) No separation required	1
(b) Can be separated mechanically	
(i) Without remote handling	1/2
(ii) With remote handling	1/10
(c) Must be separated chemically	
(i) Without remote handling	1/20–1/10
(ii) With remote handling	1/200–1/100

M: Dilution factor

$$\text{Defined by } M = \frac{\text{number of fissile atoms present in mixture}}{\text{number of fissile and non-fissile atoms present in mixture}}$$

C: Contamination factor

(a) No fission products present	1
(b) Fission products present	1/5

Q_f: Fissile quantity factor

(a) More than two critical masses	1
(b) More than one but less than two critical masses	1/3
(c) Less than one critical mass	1/10

F_c: Fissile concentration factor
Defined by $F_c = E_f \cdot M_f$
where E_f = fissile isotope enrichment
M_f = weight fraction of fissile species in chemical form

τ: Transportation method factor

(a) Non-guarded shipment	1
(b) Guarded shipment	1/5
(c) No shipment required	0

t: Transit time factor

(a) More than one week	1
(b) Three days to one week	4/5
(c) One day to three days	3/5
(d) Six hours to one day	2/5
(e) Less than six hours	1/5

R: Radio contact formula

(a)	No radio contact	1
(b)	Radio contact less frequently than every four hours	1/2
(c)	Radio contact every two to four hours	1/10
(d)	Continuous radio contact	1/20

P: Personnel factor

(a)	Only one person involved in handling step	1
(b)	Two to three persons involved in handling step	1/2
(c)	Four or more persons involved in handling step	1/5

(d) A compensating factor, P = 0.5, may be applied for each person involved who has a security clearance and/or a psychological profile report

s: Storage method factor

(a)	Guarded storage, two or less physical barriers	1
(b)	Guarded storage, more than two physical barriers	1/2
(c)	No storage required	0

A: Alarm factor

(a)	No alarm operating	1
(b)	Alarm operating	1/2

θ: Storage time factor

(a)	More than one month	1
(b)	One week to one month	1/2
(c)	Less than one week	1/10

APPENDIX 2.

VALUES IN THE DIVERSION FACTOR FORMULA AND CALCULATED DIVERSION FACTOR FOR EACH FUEL-CYCLE MODEL

Fuel Cycle	No Pu-Thermal Recycling			Pu-Thermal Recycling											
Model	Model 1 or 2			Model 3 or 4[b]											Model 4
Stage No.	5		4	4	5.		6		2		3	7		1
Facility	LWR S.F.		U-C-F	U-C-F	LWR-Pu S.F.		Reprocessing		Transportation		Pu-Storage	MOx-Fab.		Enrich.
Element	U	Pu	U	U	U	Pu	U	Pu	U	Pu	Pu	U	Pu	U
S	0.01	0.005	0.1	0.1	0.01	0.005	0.1	0.05	0.1	0.05	0.05	0.1	0.05	0.1
M	0.657	0.00321	1.0	1.0	0.657	0.0362	1.0	1.0	1.0	1.0	1.0	1.0	1.0	1.0
C	0.2	0.2	1.0	1.0	0.2	0.2	1.0	1.0	1.0	1.0	1.0	1.0	1.0	1.0
Q_f	1.0	1.0	1.0	1.0	1.0	1.0	1.0	1.0	1.0	1.0	1.0	1.0	1.0	1.0
F_c	0.00881	0.00543	0.022	0.022	0.00264	0.0413	0.00676	0.662	0.00676	0.662	0.662	0.00882	0.662	0.0169
W	1.16E-5	1.75E-8	2.20E-3	2.20E-3	3.47E-6	1.50E-6	6.76E-4	3.31E-2	6.76E-4	3.31E-2	3.31E-2	8.82E-4	3.31E-2	1.69E-3
τ		0.2	0.2	0.2		0.2	0	0	0.2	0.2	0	0.2	0.2	0.2
t		1.0	0.8	0.8		1.0	-	-	0.6	0.6	-	0.6	0.6	0.6
R		0.5	0.5	0.5		0.5	0.5	0.5	0.5	0.05	0.05	0.5	0.5	0.2
P		0.5	0.5	0.5		0.5	0.5	0.5	0.5	0.2	0.5	0.5	0.5	0.5
T		0.050	0.040	0.040		0.050	0	0	0.03	0.0012	0	0.030	0.030	0.012
s		0.5	0.5	0.5		0.5	0.5	0.5	0	0	0.5	0.5	0.5	0.5
A		1.0	1.0	1.0		1.0	0.5	0.5	-	-	0.5	1.0	0.5	0.5
θ		1.0	1.0	1.0		1.0	1.0	1.0	-	-	1.0	1.0	1.0	1.0
Ω		0.125	0.125	0.125		0.125	0.0625	0.0625	0	0	0.00625	0.125	0.0625	0.0250
G	2	2	2	2	2	2	2	2	4	4	-	2	2	2
L_t	6650	2100	6650	5896	5905	5593	2500	2200	2475	2176	2178	96	2036	9337
L_i	38628	8400	3442	3477	20126	17234	1244	1093	0	0	4356	89	2030	9191
L_T	45278	10500	10092	9373	26031	22827	3744	3293	2475	2176	6534	185	4066	18568
if X=1														
C_U or C_{Pu}	1	1	1	1	1	1	1	1	1	1	1	1	1	1
F_D*	1.40E-6	2.10E-9	2.10E-4	2.84E-4	4.14E-7	1.78E-7	1.40E-5	6.87E-4	8.11E-5	1.59E-4	1.38E-4	8.05E-5	2.03E-3	4.13E-5
if X=2														
C_U or C_{Pu}	2	2	2	4(8)	4(8)	4	4(8)	4	4(8)	4	4	4(8)	4	8
F_D	2.81E-6	4.19E-9	4.20E-4	1.14E-3 (2.27E-3)	1.66E-6 (3.31E-6)	7.11E-7	5.62E-5 (1.12E-4)	2.75E-3	3.25E-4 (6.49E-4)	6.36E-4	5.52E-4	3.22E-4 (6.44E-4)	8.11E-3	3.31E-4
if X=3														
C_U or C_{Pu}	3	3	3	9(27)	9(27)	9	9(27)	9	9(27)	9	9	9(27)	9	27
F_D	4.21E-6	6.29E-9	6.29E-4	2.55E-3 (7.66E-3)	3.73E-6 (1.12E-5)	1.60E-6	1.26E-4 (3.79E-4)	6.18E-3	7.30E-4 (2.19E-3)	1.43E-3	1.24E-3	7.25E-4 (2.17E-3)	1.82E-2	1.12E-3

[a] $F_D(n) = C_n W\{(\Omega L_i + T G L_t)/L_T\}_n$.

[b] Values in ( ) are for Model 4.

DISCUSSION

H.W. SCHLEICHER (*Chairman*): Any study which might lead to a reduction of inspection effort without loss of efficiency is of course worth while. I have the feeling, however, that by introducing all the parameters and factors shown in your paper, the impression of an objective result might be created although the numerical values will in any case be somewhat arbitrarily chosen. It seems to me particularly dangerous if goal quantities as high as you indicate are laid down for countries with a small nuclear industry. Can you comment on this problem?

K. IKAWA: I can understand your concern regarding the introduction of so many parameters in the safeguards verification procedure. I, too, consider that it is desirable to use only a few parameters for verification of a national safeguards system. From this point of view, the material categorization method seems to be satisfactory. With this method, however, the implementation of IAEA safeguards by independent verification would vary drastically from one category to another, even when the characteristics of the materials are not very different. For example, uranium of 21% enrichment is to be classed in Category I and verified by the IAEA itself under Level V assurance, while uranium of 19% enrichment is classed in Category II and will not be verified by the IAEA itself. I think a categorization which requires a discrete quantification of the characteristics of materials is not suitable for international safeguards. To solve this problem the introduction of a large number of parameters is unavoidable.

As regards countries with small nuclear industries, you might have misunderstood. In general, a goal quantity depends on the diversion possibility in a fuel cycle, not on the scale of the nuclear industry. For example, even if a country has only one enrichment plant, the goal quantity to be applied to the plant will be 25 kg ^{235}U for HEU and 75 kg ^{235}U for LEU. If a country has only power reactors (LWRs), however, the goal quantities will be higher than the values at present laid down. In any case no practical problem of a dangerous nature will arise.

ANALYSIS OF NATIONAL DIVERSION FROM A LIGHT-WATER REACTOR FUEL CYCLE*

J.E. GLANCY, L. KULL
Science Applications Inc.,
La Jolla, California,
United States of America

Abstract

ANALYSIS OF NATIONAL DIVERSION FROM A LIGHT-WATER REACTOR FUEL CYCLE.

As part of the United States programme of technical assistance to IAEA safeguards, Science Applications Inc. is performing a diversion analysis of a LWR fuel cycle. The analysis identifies and characterizes diversion possibilities that include: (1) diversion of safeguarded nuclear material; (2) misuse of a safeguarded facility to process unsafeguarded material; and (3) modification of a safeguarded facility to process either safeguarded or unsafeguarded material in order to make it more useful for clandestine nuclear weapons fabrication. For each diversion possibility a number of safeguards possibilities are identified and then the concealment possibilities that a State could use to counter these safeguards are identified and characterized. The concealment possibilities include co-operative actions among facilities within the fuel cycle. The analysis concludes with an assessment to determine: (1) safeguards coverage and balance; (2) safeguards performance relative to proposed technical criteria; (3) difficulty of concealing diversion; and (4) technology needs.

1. INTRODUCTION

One of the major concerns about widespread use of nuclear energy is the possiblity that nuclear weapons proliferation will be enhanced by the diversion of nuclear material and technology from facilities in a commercial nuclear fuel cycle. The USA is, therefore, supporting a program of technical assistance to the International Atomic Energy Agency (IAEA) so that more effective safeguards can be developed and applied. As part of this program Science Applications, Inc. (SAI) is performing a diversion analysis of a fuel cycle consisting of power reactors, uranium and plutonium fuel fabrication, isotope enrichment, and irradiated fuel reprocessing facilities.

The results of the diversion analysis will be submitted to the IAEA's systems studies program for use in preparing descriptive models for safeguarding generic types of nuclear facilities. These descriptions will be

* Work performed as part of the United States programme of technical assistance to the IAEA safeguards, under the auspices of the Department of Energy and the International Safeguards Project Office at Brookhaven National Laboratory.

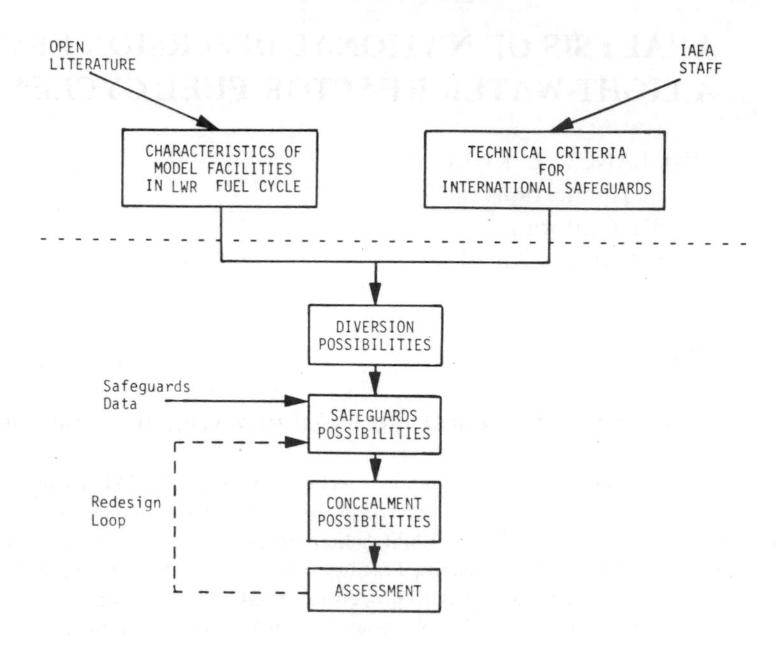

FIG.1. Diversion analysis approach.

useful for: (1) explaining IAEA safeguards concerns and giving member States a preview of international safeguards activities; (2) providing guidance and establishing uniformity in planning of safeguards inspection practices; and (3) providing ideas and direction for the development of new or improved safeguards methods and techniques.

An overview of the diversion analysis approach is shown in Fig. 1. The two necessary inputs for the analysis are the characteristics of the LWR fuel cycle facilities and the technical criteria for international safeguards. The four steps of the analysis are logically sequential. First, the diversion possibilities for each facility are identified and characterized by parameters that are meaningful for a safeguards analysis. Second, from a review of the safeguards possibilities in the literature [1,2,3,4], a reasonable safeguards approach was selected for each facility. Third, a general discussion of concealing diversion from all types of safeguards and the specific concealment possibilities that could be used against each illustrative safeguards approach were identified. Fourth, a subjective assessment was performed both qualitatively and quantitatively.

The following sections of this paper discuss each of the four steps and the two necessary inputs for the analysis that are shown in the six boxes of Fig. 1. In this paper it is only possible to present a brief discussion of the diversion analysis and give a few examples of results for one facility, reprocessing. More details can be found in the technical reports on this work [5,6,7,8].

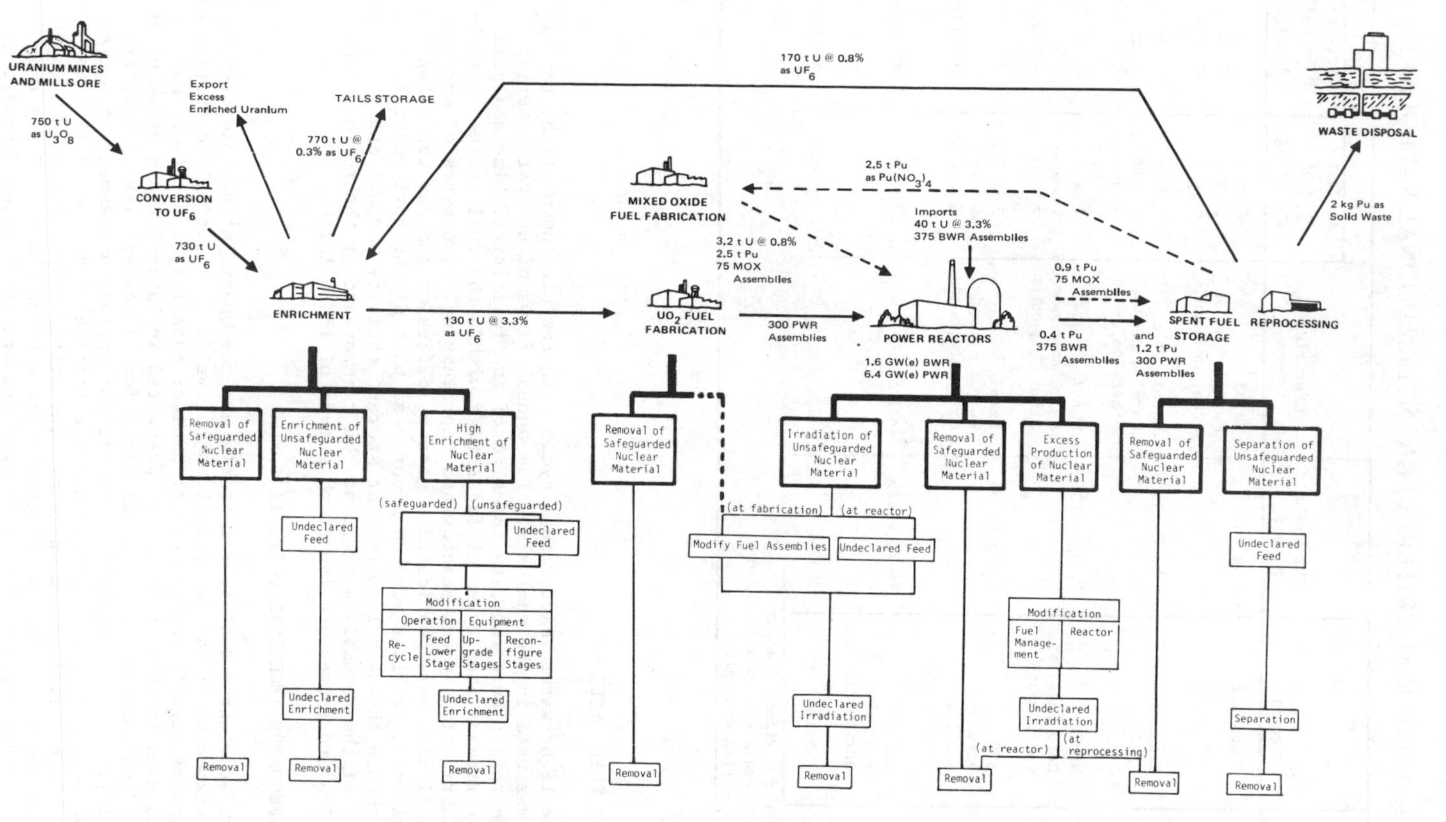

FIG.2. Annual flows and diversion possibilities for the light-water reactor fuel cycle and plutonium recycle.

TABLE I. TECHNICAL CRITERIA FOR INTERNATIONAL SAFEGUARDS

Material Category	Composition	Threshold Amount	Material Form	CRITICAL TIME Activity Less than 10^5 Ci/kg	Activity More than 10^5 Ci/kg
1	Plutonium	8 kg Pu	Metal	2 days	-
			Pure Compounds	2 weeks	-
			Other Compounds	6 weeks	2 months
2	Uranium-233	8 kg U-233	Metal	2 days	-
			Pure Compound	2 weeks	-
			Other Compounds	5 weeks	2 months
3	HEU More than 25% enrichment	25 kg U-235	Metal	2 days	-
			Pure Compounds	2 weeks	-
			Other Compounds	5 weeks	2 months
4	IEU Between 10% and 25% enrichment	50 kg U-235	Metal	2 weeks	-
			Pure Compounds	6 weeks	-
			Other Compounds	2 months	3 months
5	LEU Less than 10% enrichment	75 kg U-235	All	12 months	15 months
6	Thorium	20 t Th	All	12 months	15 months

HEU = Highly enriched uranium
IEU = Intermediate enriched uranium
LEU = Low enriched uranium

2. LWR FUEL CYCLE

The light water reactor fuel cycle with recycle shown at the top of Fig. 2 was used in the analysis. The annual flows of nuclear material shown on this figure indicate the size of the facilities. The analysis was performed on four types of facilities within the fuel cycle: enrichment, fabrication, power reactor and reprocessing. Two types of enrichment facilities were considered: gaseous diffusion and gas centrifuge. The enrichment tailings were considered as part of the enrichment facility and, although the away from reactor storage facility was not analyzed, the analysis of the assembly storage at the reactor and at the reprocessing facilities are representative of this type of installation.

3. SAFEGUARDS TECHNICAL CRITERIA

The safeguards technical criteria are the quantitative interpretation of the statement of the safeguards objective [9]: "...timely detection of diversion of significant amounts of nuclear material." The technical criteria used in this analysis are presented in Table I and consist of the threshold amounts and the critical times for six categories of nuclear material [10]. The threshold amount is the approximate amount of nuclear material needed to make a weapon and the critical time is the time to convert the material to weapon form. The criteria in Table I were under consideration by the IAEA staff at the start of this work. Since then they have undergone a number of changes and are still under discussion.

If, as presently expected, the final criteria do not differ appreciably from those in Table 1, the assessment results may change only slightly and the remainder of the analysis would not be affected.

The technical criteria also usually include a probability of detection, sometimes expressed as a numerical detection probability (e.g., a 95% detection probability). This number has meaning for a safeguards plan that uses verification of nuclear material accounting by independent measurement. However, for approaches that rely on containment and surveillance or on observation of facility measurements rather than independent measurements, a detection probability is difficult to calculate. Because these approaches are common, and are especially effective for detecting abrupt diversion, we used the additional criterion that all diversion paths be under surveillance (unless the path involves penetration of a substantive containment), and that the surveillance devices be highly tamper resistant. We applied another criterion that observation of facility measurements should also include observation of calibrations and the use of surveillance between calibration and measurement to detect tampering.

4. DIVERSION POSSIBILITIES

There are three types of diversion possibilities: (1) diversion of safeguarded nuclear material; (2) misuse of a safeguarded facility to process unsafeguarded nuclear material; and, (3) modification of a safeguarded facility to process safeguarded or unsafeguarded nuclear material so it is more useful in weapons. The reasonableness of the possibilities involving unsafeguarded material is different for NPT states, with all facilities and material safeguarded, than for non-NPT states, with safeguards limited to only some of the nuclear facilities. In non-NPT states unsafeguarded material could be available, but in NPT states the material would have to be diverted from another nuclear facility.

The diversion possibilities for the LWR fuel cycle are shown at the bottom of Fig. 2. Each of the three types of possibilities are broken down into the essential steps required to achieve diversion: (1) introducing undeclared nuclear material into the facility; (2) modifying the process; (3) operating the facility in an undeclared way; and (4) removal of the nuclear material.

Fig. 2 shows there are at least two possibilities that involve collusion among the reactor and either the fabrication or reprocessing facilities. The figure does not show the possibilities that arise for two facilities to cooperate to conceal a diversion. This is discussed in the next section and in Table IV.

The diversion possibilities were described in terms of a set of parameters that indicate the plausibility and attractiveness of each diversion. These parameters also give information that helps in the design of safeguards. The parameters for the diversion of safeguarded nuclear material (removal) are discussed below. Those for the other diversion possibilities are similar. Values of these parameters for some of the materials that could be targets for diversion at the reprocessing facility are listed in Table II.

1. <u>Target Characteristics</u> - There are many locations containing nuclear material and some grouping is necessary to make the analysis manageable. The grouping is determined by: (a) the intrinsic separation of areas within the plant due to process-related containment, and (b) the chemical and physical form of the material.

TABLE II. DIVERSION POSSIBILITIES FOR REPROCESSING FACILITY – REMOVAL OF SAFEGUARDED MATERIAL

TARGET				MATERIAL CHARACTERISTICS				
Location		Form	Type	Threshold Amount	Inventory (1)	Activity	Flow Per Cricial Time (2)	Mass T Obtain Thresho Amount (
SPENT FUEL STORAGE	1	Assembly	U+Pu+FP	8 Kg Pu	1 MT (125)	10^4 Ci/Kg	75	1.5 MT (200)
SEPARATION	4	Hulls	Trace	8 Kg Pu	10-100 g (.001-.01)	0.3 Ci/Kg	0.1-1	40 MT (5000)
SEPARATION	6	High Level Liquid Waste	Trace	8 Kg Pu	--	2×10^3 Ci/ℓ	0.6	500 MT (60,000
SEPARATION	9	Purified Nitric Acid 20 g/l Pu	Pu	8 Kg Pu	25 Kg (3)	300 Ci/ℓ	13	50 Kg (6)
PU STORAGE	12	Concentrated Nitric Acid 250 g/l Pu	Pu	8 Kg Pu	1500 Kg (190)	4×10^3 Ci/ℓ	20	50 Kg (6)
ANALYTICAL SERVICES	15	Samples	U+Pu U Pu	8 Kg Pu or 7500 Kg U	--	2 Ci/smpl 10μCi/smpl 100Ci/smpl	0.1 .01 1.0	4MT(50 300MT(4 125Kg(1
SAMPLING CELL	18	Sampling Line	U+Pu+FP	8 Kg Pu	18 Kg (2.2)	10^4 Ci/ℓ	n/a	4 MT (500
WASTE STORAGE	20	Liquid or Solid Waste	Trace	8 Kg Pu	10-100 Kg (1-12)	2×10^3 Ci/ℓ	1.	50 MT (6000)

TIME CHARACTERISTICS		RESOURCES		IMPACT ON PLANT PROCESS & OPERATION
Critical Time	Acquisition and Removal Time	To Convert	To Remove	
2 M	0.8 hr	RF,CF,WF	Shipping Cask Crane & Vehicle 0-12 manhrs	1-2
2M	3 M	RF,CF,WF	Special Containers Crane & Vehicle 6 man-days	1
2 M	3 M	RF,CF,WF	Special Shielding & Canister, Crane & Vehicle, 100 man-days	1
2 wks	1 hr	CF,WF	Regular 10P Plastic Containers & Vehicle Tap Pu Line 2 man-days	1-2
2 wks	1 hr	CF,WF	Regular 10 p Plastic or SS Canisters Vehicle 2 manhrs	1-2
6 wks 12 M 2 wks	1 yr 100 yr 3 M	CF,WF CF,WF	4-8 manhrs	1
2 M	1 wk	RF,CF,WF	Special Casks & Vehicles, 4 manhrs	1
2M	1 d	RF,CF,WF,	Special Casks Crane & Vehicle 12-15 man-days	1

(1) Threshold amounts units in parentheses.
(2) Flow given in number of threshold amounts per critical time

RF = Reprocessing Facility
CF = Conversion Facility
WR = Weapon Facility
MT = metric ton
M = month

1 – Negligible
2 – Minor
3 – Moderate
4 – Large
5 – Substantial

TABLE III. SAFEGUARDS AND CONCEALMENT POSSIBILITIES FOR PART OF THE REPROCESSING FACILITY

	Target			SAFEGUARDS POSSIBILITIES				CONCEALMENT POSSIBILITIES	ASSESSMENT		
				Accountability							
	Location	Form	Type	MBA	MBA	Inspection Procedure	Containment and Surveillance		Target Appeal Factor F_T	Safeguards Vulnerability Factor, F_S	Target Attractiveness Index, TAI
Spent Fuel	1	Assembly	U+Pu+FP	MBA-1	KMP-1	• Count the number of casks (100%) • Count and identify fuel assemblies (100%) • Verify records with shipper (100%) • Record location in storage pool (100%) • Estimate Pu content from fabricator and reactor data (100%)	• Check seal on shipping cask (if any) • Observe cask opening and unloading (camera) • Re-seal cask • Apply motion monitor to cask handling crane • Observe transfer to storage pool (camera)	• C.1.1 Remove assembly by alternate path, AND • C.1.1 Use alternate method of movement instead of crane, AND • B.1.2 Substitute dummy, AND • B.2.2 Falsify identity of dummy, OR • C.2.1 Remove by shipping cask and avoid camera surveillance, AND	6	0.6 or 0.05	0.4
				MBA-1	KMP-A	• Count and identify assemblies in pool (100%) • Verify records with previous inspection (100%)	• Observe storage with CCTV	• B.1.2 Substitute dummy, AND • B.2.2 falsify identity of dummy			

Separations	2	Assembly	U+Pu+FP	MBA-1	(KMP-1)	• Count number of assemblies to shear (100%)	• Observe assembly transfer (camera)	• C.1.1 Remove assembly by alternate path after transfer to shear, AND • A. Concealment possibilities for MBA-1, OR • C.2.1 Conceal removal from camera, e.g., by loss of lighting, AND • C.1.1 Use alternate method of movement instead of transfer crane, AND • A. Concealment possibilities for MBA-1	5	0.02 or 0.003	0.1
	3	Chopped Assemblies	U+Pu+FP	MBA-1	--	none	• Observe transfer of fuel from shear to dissolver (camera)	• C.1.1 Conceal removal from camera, e.g., by loss of lighting, AND • A. Concealment possibilities for MBA-1	5	0.02	0.1
	4	Hulls	Trace	MBA-1	KMP-3	• Observe weighing of dissolver basket (100%) • Observe NDA of hulls (100%)	• Observe transfer of hulls to waste treatment (camera) • Observe solidification and final packaging (camera)	• C.1.1 Remove hulls after assay but prior to solidification by using alternate path, OR • None (Remove hulls after solidifcation), OR • C.1.1 Remove hulls prior to assay by using alternate path, AND • A. Concealment possibilities for MBA-1	0.5	0.4 or 1 or 0.02	0.5
	5	Dissolver Solution	U+Pu+FP	MBA-1 and MBA-2	KMP-2 and KMP-B	• Observe bulk weight or volume measurement (100%) • Observe homogenization and sampling (100%) • Request and observe sampling (100%) • Correlate input with burnup calculations and use gravimetric method (100%) • Observe calibrations (100%) • Observe analysis of samples in laboratory (<100%)	• Apply seals to valves on lines to and from tanks • Apply seals to secure calibration (where possible) • Observe presence and sense of flow in lines to and from dissolver and accountability tanks (remote indicating instruments) • Apply seals to samples from dissolver	• C.1.1 Remove solution prior to assay by installing alternate pipe, AND • A. Concealment possibilities for MBA-1, OR • A. Concealment possibilities for MBA-2 if removed after assay	4	0.02 or 0.1	0.4

2. Material Characteristics - The important characteristics are: (a) The inventory of threshold amounts; (b) the number of threshold amounts of uranium or plutonium flowing per critical time; (c) the intensity of radioactivity of the material which may require remote handling and special shielding; and, (d) the total mass of material that must be removed to obtain a threshold amount, including the mass of the container required for safe handling.

3. Acquisition and Removal Time - Acquisition time is defined as the time required to accumulate a quantity of material containing a threshold amount. It is zero if the inventory is greater than a threshold amount, otherwise it is an estimate of the time to divert the flow out of the process stream. Removal time is an estimate of the time required to remove the acquired target material from the facility premises, including loading of casks or canisters, use of special equipment and loading on vehicles. These times are important in determining the frequency of surveillance.

4. Resources - The required resources indicate the size of the task the facility is undertaking and the level of activity that could be detected by safeguards. Resources include special shipping containers, equipment needed to move and ship the casks and an estimation of the manhours needed to accomplish the removal. External resources required for further processing of the target material are also identified.

5. Impact on the Plant Process and Operations - The diversion process may result in a disruptive impact on the plant process equilibrium and on other plant activities. These impacts were grouped in five different categories ranging from negligible to substantial impact.

The actual diversion paths are considered in determining the removal time and resources. Alternate paths, such as penetration of substantive containment, are considered in the concealment possibilities.

5. SAFEGUARDS POSSIBILITIES

There are many safeguards possibilities for each facility in the LWR fuel cycle. Every one is comprised of a mix of material balance accounting, item accounting, and containment and surveillance. The approach in this analysis was to select one "reasonable" set of safeguards possibilities for each facility and describe it in detail so that the concealment possibilities could be identified. No attempt was made to design an "optimum" safeguards system.

The safeguards description includes: defining material balance areas (MBA's); defining key measurement points (KMP's); specifying inventory frequency and inspection frequency; listing accounting inspection procedures such as verifying and recording data; differentiating between independent inspector measurement and observation of facility measurement; defining calculations; and describing containment and surveillance practices.

The inspection procedures and containment and surveillance practices for one MBA in a reprocessing facility are listed in Table III. The MBA is one of four at the reprocessing facility and includes spent fuel receiving and storage, dissolution, clarification, and the accountability vessel. This example will be used in the next sections to explain concealment possibilities and assessment.

6. CONCEALMENT POSSIBILITIES

Concealment actions occur in direct response to a safeguard. Table IV presents a list of the general types of concealment possibilities grouped into three categories: (a) Bulk Material Accountability; (b) Item Accountability; and (c) Containment and Surveillance. The safeguards inspection measures are listed in order of increasing effort to implement and the corresponding concealment possibilites are listed in order of increasing difficulty to execute without detection.

The most effective safeguards against concealment are independent measurement by the inspector to verify mass or item accountability, and the application of tamper-resistant surveillance along all diversion paths. In this case, the only way to conceal diversion is to divert an amount less than the measurement uncertainty and to tamper with the surveillance device. If, due to lack of resources or technology a less effective safeguards approach is used, then the concealment possibilities are numerous and less difficult.

The concealment possibilities for the specific example at the reprocessing facility are presented in Table III. Each possibility is identified according to the group and number given on Table IV. In many cases more than one concealment action is required (as indicated by AND in the table) and in other cases there are alternative possibilities (as indicated by an OR in the table).

A frequent entry in Table III is "A. Concealment Possibilities for MBA-1." There are three of these possibilities that will conceal diversion of any of the materials in MBA-1 from verification of material balance accounting. These are:

A.4.1 Substitute biased samples of hulls, dissolver solution, solid waste, or recycle acid, OR

A.4.2 Tamper with calibration of hull NDA, or accountability tank volume or load cell, OR

A.4.3 Defeat C&S applied to secure calibrations by

- C.1.1 Finding alternate route to tamper, or
- C.1.2 Tampering when on-site surveillance is not effective, or
- C.1.3 Falsifying seals used to secure instruments.

Of course, if the diversion is less than the material balance uncertainty then none of these concealment actions is necessary.

7. ASSESSMENT

An assessment of the possibilities of nuclear material diversion takes into consideration the attractiveness of the diversion, the intensity and effectiveness of safeguards, and the difficulty or risk of detection associated with concealing diversion from the safeguards. The assessment provides a statement of the following aspects of diversion:

Coverage and Balance - Coverage means that safeguards will be applied to detect all the diversion possiblities and balance means that the intensity of the safeguards is in proportion to the attractiveness of the diversion possibility.

TABLE IV. CONCEALMENT POSSIBILITIES

INSPECTION MEASURE	CONCEALMENT POSSIBILITY
A. Bulk Material[a] Accountability	A. Conceal Removal of Material* from one Location or Many Within an MBA
A.1 Verify MUF calculation	A.1 Falsify records of amounts of material
A.2 Verify LEMUF calculation	A.2 Falsify records of measurement error (inflate error variances)
A.3 Verify measurement amounts of receipts or shipments by checking shippers or receivers value, respectively	A.3 Shipper or receiver in same state is in collusion and also falsifies records
A.4 Verify measurement amounts and error estimates by observation	
A.4.1 Observe facility measurement	A.4.1 Use biased measurement methods or substitute biased samples
A.4.2 Observe facility measurement calibrations	A.4.2 Tamper with measurements after calibration and replicate measurement
A.4.3 Apply C&S to measurements between observations	A.4.3 See concealment possibilities for C&S
A.5 Verify measurement amounts and error estimates by independent measurement	A.5.1 Substitute nuclear material from another place in the facility (double inventory) or from another facility (safeguarded or unsafeguarded material)
	A.5.2 Falsify an accidental loss or overstate amount lost
	A.5.3 Remove an amount less than the measurement uncertainty of the material balance
B. Item[b] Accountability	B. Conceal Removal of Complete Item** or Partial Amount from an Item
B.1 Count Items	B.1.1 Falsify records of number of items on inventory, received, or shipped
	B.1.2 Substitue a "dummy" item visually identical to the diverted item
B.2 Identify Items	B.2.1 Falsify records of identification markings

[a] This material can be from a single location that is anywhere in the material balance area, or may be from many different places. However, the material cannot have been recorded or associated with a single container or tank, otherwise it would be safeguarded using item accountability techniques.

INSPECTION MEASURE	CONCEALMENT POSSIBILITY
	B.2.2 Substitute a false identity on the "dummy" item
B.3 Verify Seals	B.3.1 Falsify records of seal identity B.3.2 Substitute false seal on the "dummy" item B.3.3 Circumvent seal inspection
B.4 Verify receipt or shipment records of number, identity and seal by checking shipper or receiver records, respectively	B.4 Shipper or receiver in same state are in collusion and also falsify records
B.5 Observe qualitative assay (attributes)	B.5 See A.4
B.6 Independent qualitative assay (attributes)	B.6.1 Remove partial amount B.6.2 Substitute material to falsify measurement (mass for weight, radiation for NDA, sample for chemistry)
B.7 Observe quantitative assay (variables)	B.7 See A.4
B.8 Independent quantitative assay (variables)	B.8.1 Substitute nuclear material from another place within the facility (double inventory) or from another facility (safeguarded or unsafeguarded) B.8.2 Remove a partial amount less than measurement error
C. Containment and Surveillance	C. Conceal Removal from C&S or Circumvent C&S Entirely
C.1 Apply C&S along normal removal path	C.1 Remove by alternate path
C.2 Apply C&S along all paths	C.2 Conceal removal from C&S (e.g. emergency or loss of lights or power)
C.3 Apply C&S to material	C.3 Tamper with C&S device

[b] Includes items on inventory or being transferred as shipments and receipts. Requirement is that an amount of material is recorded and can be associated with a single container or tank.

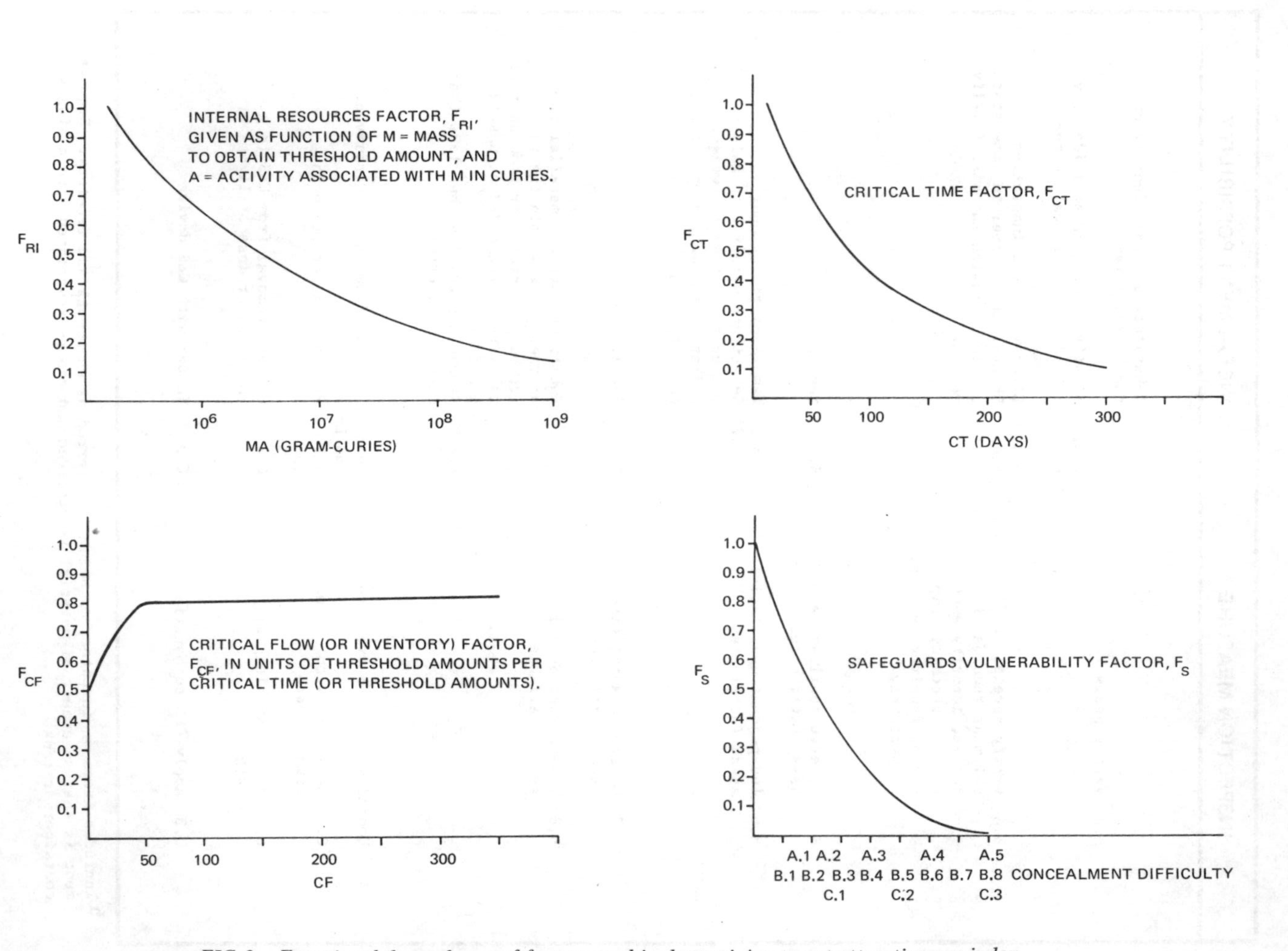

FIG.3. Functional dependence of factors used in determining target attractiveness index.

- Performance - Performance means that the detectable quantity and timeliness match the criteria for detecting threshold amounts within a critical time, that the safeguards measures can be implemented with reasonable effort and minimal intrusion, and that the safeguards can be applied effectively and reliably.

- Concealment Difficulty - Difficult concealment means that a number of concealment actions are required and some of the concealment possibilites rank at the difficult ends of the lists in Table IV.

A qualitative assessment can be made by reviewing data such as that presented in Table III. A quantitative assessment can be made by calculating an overall target attractiveness index (TAI) of diversion as the product of two factors:

$$TAI = F_T \times F_S$$

where,

F_T = Target Appeal Factor
F_S = Safeguards Vulnerability Factor

The target appeal factor is composed of a number of more specific factors.

$$F_T = F_{CF} \times F_{CT} \times F_{RI} \times F_{RX}$$

where,

F_{CF} - The critical flow factor is a function of the flow in threshold amounts per critical time or the inventory in threshold amounts.
F_{CT} - The critical time factor is a function of the critical time for the material type.
F_{RI} - The internal resources factor is a function of the effort required to divert from the facility. This includes the manpower, special equipment and impact on plant operations.
F_{RX} - The external resources factor is a function of the resources required to make a weapon from the material after diversion.

The last factor, external resources, is country specific and is not included here. The form of the other functions is given in Fig. 3.

The target appeal factors, the safeguards vulnerability factors, and the target attractiveness indexes for the reprocessing facility are listed in Table III. The target appeal factors have been normalized to a value of 100 for the most attractive material, plutonium nitrate.

8. SUMMARY

This paper has shown several examples of the specific application of a method of diversion analysis to an LWR fuel cycle. The method itself, however, is generally applicable to any size or type of fuel cycle. The results from this method of diversion analysis can be used to: (1) reveal vulnerabilities in a safeguards system; (2) optimize the design of a safeguards system (through iterative use of the method); (3) document the apparent effectiveness of a safeguards system. As soon as there is more experience in using this diversion analysis it will be possible to report on these kinds of results, that are not available at this preliminary stage of the project.

REFERENCES

[1] NAKIČENOVIĆ, S., "Experience in the application of Agency inspection practices", Safeguarding Nuclear Materials (Proc. Symp. Vienna, 1975) 1, IAEA, Vienna (1976) 379.

[2] ROMETSCH, R., HOUGH, G., "The position of IAEA safeguards relative to nuclear material control accountancy by States", Nuclear Power and its Fuel Cycle (Proc. Conf. Salzburg, 1977) 7, IAEA, Vienna (1977) 441.

[3] INTERNATIONAL ATOMIC ENERGY AGENCY, IAEA Safeguards Technical Manual, IAEA-174, IAEA, Vienna (1975).

[4] INTERNATIONAL ATOMIC ENERGY AGENCY, Review and Analysis of the Status of Safeguards Technology for Uranium Enrichment Facilities, AG-110, IAEA, Vienna, April 1977.

[5] GLANCY, J., GRAHAM, R., HECHT, M., McDANIEL, T., Diversion Analysis of an Enrichment Facility, Draft Rep. ISPO-22 or SAI-78-694-LJ, Science Applications Inc., June 1978.

[6] GLANCY, J., EL-BASSIONI, A., HAGAN, W., BORGONOVI, G., Diversion Analysis of a LWR Reprocessing Facility, Draft Rep. ISPO-23 or SAI-77-956-LJ, Science Applications, Inc., May 1978.

[7] GLANCY, J., HAGAN, W., McDANIEL, T., Diversion Analysis of a Light Water Reactor Facility, Draft Rep. ISPO-23 or SAI-78-704-LJ, Science Applications Inc., June 1978.

[8] BORGONOVI, G., GLANCY, J., McDANIEL, T., Diversion Analysis of a Mixed Oxide Fuel Fabrication Facility, Draft Rep. ISPO-29 or SAI-78-705-LJ, Science Applications Inc., June 1978.

[9] INTERNATIONAL ATOMIC ENERGY AGENCY, The Structure and Content of Agreements Between the Agency (IAEA) and the States Required in Connection with the Treaty on the Non-Proliferation of Nuclear Weapons, INFCIRC/153.

[10] INTERNATIONAL ATOMIC ENERGY AGENCY, Safeguards Criteria for Significant Quantities and Timely Detection, AG-43/8, IAEA, Vienna, 1 August 1977.

DISCUSSION

A.G. HAMLIN: In carrying out this analysis, was an intelligent adversary assumed? In a dynamic system, each move alters the ground state and the possible responses, so that after a few moves, if an intelligent reaction is not evaluated, the situation becomes unreal.

J.E. GLANCY: Yes, the analysis does consider an intelligent adversary. The concealment possibilities step follows the definition of safeguards and the assumption is that the State is fully conversant with the safeguards plan. Thus, the analysis always assumes the State will make the most intelligent choice of concealment.

A.G. HAMLIN: Each end situation constitutes a new starting point, involving new ploys, new strategies and new responses, and is never a final state, so where does one stop?

J.E. GLANCY: We did not include criteria for deciding when to stop the process of safeguards design, concealment countermeasures, safeguards, etc. The analysis provides an approach for the decision maker to determine safeguards effectiveness and inspection effort. His decision to stop will be based on whether the effectiveness is adequate or his resources are exhausted.

D. GUPTA: Have you considered the possibility of allocating different weighting factors to different diversion strategies? This might help the Agency to concentrate its efforts on strategies which are more probable than those which are of academic interest only.

J.E. GLANCY: Yes, we have considered weighting factors for the different diversion strategies. There is a problem in assigning these factors because they are country-specific. I mean the factors cannot be assigned to a generic fuel cycle, but have to be determined by considering a specific country's resources. We have not used these factors and have treated each type of diversion possibility as a separate problem.

Session II (Part 1)

FACILITY DESIGN CRITERIA FOR INTERNATIONAL SAFEGUARDS

Chairman: A. BURTSCHER (Austria)

IAEA-SM-231/7

SAFEGUARDS SYSTEMS PARAMETERS

R. AVENHAUS, J. HEIL
Bundesministerium für Forschung und Technologie,
Bonn, Federal Republic of Germany

Abstract

SAFEGUARDS SYSTEMS PARAMETERS.

In the IAEA model agreement INFCIRC/153 the objectives of international nuclear material safeguards have been specified *qualitatively*. Since that time the statistical procedures for the primary safeguards tools – data verification and material accountability – have been worked out, and a continuous effort has been made to determine *quantitatively* the objectives and the technical effectiveness of the present system which has to take into account the capability of measurement systems as well as the constraints posed by plant operations. In this paper analyses are made of the values of those parameters that characterize the present safeguards system that is applied to a national fuel cycle; those values have to be fixed quantitatively so that all actions of the safeguards authority are specified precisely. Furthermore, those parameter values are shown which can be *calculated* according to some agreed criteria, and those which have to be fixed *subjectively* by taking into account semi-quantitative and qualitative arguments. Finally, the paper outlines the way in which the system should be adjusted stepwise so that safeguards goals and technical possibilities coincide; and what steps should be taken should this procedure not give satisfactory results. The analysis starts by introducing three categories of quantities: The *design parameters* (number of MBAs, inventory frequency, variance of MUF, verification effort and false-alarm probability) describe those quantities whose values have to be specified before the safeguards system can be implemented. The *performance criteria* (probability of detection, expected detection time, goal quantity) measure the effectiveness of a safeguards system; and the *standards* (threshold amount and critical time) characterize the magnitude of the proliferation problem. The means by which the values of the individual design parameters can be determined with the help of the performance criteria; which qualitative arguments can narrow down the arbitrariness of the choice of values of the remaining parameters; and which parameter values have to be fixed more or less arbitrarily, are investigated. As a result of these considerations, which include the optimal allocation of a given inspection effort, the problem of analysing the structure of the safeguards system is reduced to an evaluation of the interplay of only a few parameters, essentially the quality of the measurement system (variance of MUF), verification effort, false-alarm probability, goal quantity and probability of detection.

INTRODUCTION

According to the model agreement of the International Atomic Energy Agency (IAEA) [1] the *objective* of the international nuclear material safeguards system has been formulated as 'the timely detection of diversion of significant quantities of nuclear material from peaceful nuclear activities to the manufacture of nuclear weapons or of other nuclear explosive devices or for purposes unknown, and the deterrence of such diversion by the risk of early detection'. However, it has not been said what 'timely' means, what 'significant' quantities are and so on; in other words, the objective has not been formulated in *quantitative* terms.

In connection with the discussions in 1978 about the non-proliferation resistance of fuel cycles, the importance of having a clear picture of the relation between technical objectives and the effectiveness of IAEA safeguards was again stressed.

There are two possibilities of approaching the technical objectives of IAEA safeguards. The first way, more or less straightforward, is to define the objectives quantitatively without considering the effort which is necessary to reach this objective, or the limitations posed by the uncertainty of measurements etc. This may cause confusion in a technical and political sense, as there is a common understanding that the implementation of safeguards should not lead to an undue burden for the plant operators. In other words, the danger of this way is that, in order to reach fixed *technical* safeguards objectives, one might disturb the *political* idea which should be supported by reasonable safeguards measures.

The second way is to start with the political idea of the Non-Proliferation Treaty (NPT) and try to convince the world community of these ideas, using the internationally accepted structure of the IAEA model agreement [1], and developing the quantification of technical objectives by collecting experience in IAEA safeguards practice. By taking this way, however, one cannot be a hundred per cent sure that no nuclear material is diverted while gaining such experience.

In this paper an attempt is made to analyse the problem of defining technical safeguards objectives, to understand the mutual dependency of the involved technical parameters (verification effort, false alarm and detection probability[1] etc.), and finally to develop a decision scheme for determining the parameter values with the help of internationally accepted criteria.

The analysis is performed for the international nuclear material safeguards system as described in Ref. [1]. This means that the principle safeguards *tool* is material accountability, with containment and surveillance as complementary measures which will not be taken into account explicitly here, and that the *procedure* is such that the plant operator collects all data necessary for the material balance establishment whereas the safeguards authority verifies the data, with the help of independent measurements, and establishes the material balance on the basis of the operator's data if there are no significant differences between the operator's data and the inspector's findings. The statistical analysis of this scheme has been given at various places (see, for example, Refs [3, 4]).

---

[1] The definitions given in this paper follow wherever possible the terminology used in Ref.[2].

DESIGN PARAMETERS

The implementation of the IAEA safeguards system in an national fuel cycle requires the establishment of *material balance areas (MBAs)*, i.e. areas for which material balances are closed at the end of an inventory period. These areas may consist of parts of nuclear facilities, whole facilities or even of whole national fuel cycles. Therefore, the first parameter to be considered is the

(a) Number of material balance areas (MBAs) in a State: The establishment of a material balance (i.e., the comparison of book and physical inventories) in a MBA requires the frequency of those comparisons to be determined; thus, the second parameter to be considered is the

(b) Number of inventory periods per reference time in a MBA: The reference time may be one year; generally, it can be specified only after careful consideration of the aim of safeguards activities. The establishment of a material balance procedure requires the installation of a measurement system. As there is the possibility (within limits) of choosing between different types of measurement systems, which differ by accuracy and consequently, costs, one has to determine the

(c) Quality of the measurement system: This parameter is described concisely by the variance of the *Material Unaccounted For*[2] *(MUF)*, σ^2_{MUF}. In fact, there is not very much choice which means that this value, with a few exceptions, is fixed by the present state of technology.

According to the construction of the safeguards system, the IAEA has to spend some effort (to be measured in man-hours and/or monetary terms) in verying the data provided by the plant operator. Therefore, the IAEA has to fix the

(d) Verification effort in a State[3]: Finally, a prescription has to be developed to determine when an alarm must be raised. As will be seen later, this is equivalent to fixing the value of the

---

[2] It would be better to call this quantity 'book-physical inventory difference', as in most cases the material is accounted for; however, with measurement errors. In fact this name has been used for some time, but the term MUF has already entered so many papers and even agreements and treaties that no chance remains to change this somewhat misleading terminology.

[3] It should be noted that according to Ref. [1] not the verification effort for all nuclear installations in a State together is fixed, but for groups of similar installations.

(e) Overall false-alarm probability: As follows from the discussion so far, a decision must be made about the value of these parameters before the IAEA safeguards can be carried out. Therefore, we will call these parameters *design parameters.*

PERFORMANCE CRITERIA

According to the objective of IAEA safeguards, we take

(f) The probability of detecting a diversion, and furthermore

(g) The expected detection time of a diversion[4] as *performance criteria.* Now, it is clear that the probability of detection and the expected detection time are functions of the amount M of material which will be diverted. Therefore, we have to introduce this quantity, the

(h) Goal quantity[5], explicitly into the list of performance criteria, because we can *either* consider the probability of detection as a function of the goal quantity, *or* the goal quantity for a given probability of detection, as our first performance criterion (the same holds for the expected detection time as a function of the goal quantity).

Frequently, the probability of detection as a function of the goal quantity M (and sometimes also as a function of the false-alarm probability) is called the *effectiveness or efficiency* of the system. This may be justified as long as the detection time is not a major aspect. In the following, for systematic reasons, we will consider separately the three performance criteria given above, keeping in mind that they *cannot be determined independently* once the values of the design parameters have been fixed.

The objective of IAEA safeguards says nothing regarding to what extent a diversion should be *localized* if detected. Nevertheless as an analogy to the expected detection time we give a last performance criterion,

(i) The expected detection location of a diversion.

---

[4] As detection must not take place within the reference time under consideration, it is reasonable to use the expected detection time under the condition that detection will take place within the reference time as performance criterion.

[5] Called 'significant quantity' in Ref. [2].

STANDARDS

So far, we have not discussed the question whether or not a safeguards system, whose design parameter values have been specified, and whose performance criteria values have been calculated, *works satisfactorily*. What does 'satisfactorily' mean? It means that the performance criteria values are consistent with several *standards* which characterize the magnitude of the proliferation problem and which we will formulate as follows.

The exclusive purpose of nuclear material safeguards is 'verifying that such material is not diverted to nuclear weapons or other nuclear explosive devices' (Ref. [1] Para.2). Therefore, the standards for an international nuclear material safeguards system are determined primarily by the strategic value of the fissile material under consideration. We measure this strategic value by

(j) The amount of fissile material needed for the construction of a nuclear weapon (threshold amount)[6] (depending on the physical and chemical forms of the material under consideration); and furthermore, by the

(k) Time (and effort) needed for the production of a nuclear weapon (critical time).

In Table I all design parameters, performance criteria and standards are put together. It is important to realize that Table I represents a list of quantities which determine a safeguards system and which can be expressed by quantitative figures. Table I does not contain *qualitative arguments* (like, for example, compliance with normal plant operations, or quality of the record-report-system of the operator; see, for example Ref. [5]); they will, however, play an important role in determining *values* of parameters given in Table I.

DETERMINATION OF INDIVIDUAL DESIGN PARAMETERS

As outlined in the foregoing section, the values of the design parameters are either given in an existing fuel cycle or have to be fixed before the safeguards system can be implemented in a State. In this section, we demonstrate that no more parameter values have to be fixed (in other words, the subjectivity which cannot be removed is limited to the determination of these values) and furthermore, those arguments that can be given for narrowing down the arbitrariness of the choice of the design parameter values.

---

[6] In some Refs (see e.g. Ref.[6]) a distinction is made between *direct use* and *indirect use* material depending on the effort to extract the fissile material from the available raw material.

TABLE I. LIST OF PARAMETERS

(Performance criterion 9 is only given for completeness)

Design Parameters

(1) Number of Material Balance Areas

(2) Number of inventory periods per reference time

(3) Quality of measurement system (σ^2_{MUF})

(4) Verification effort

(5) False-alarm probability

Performance Criteria

(6) Probability of detection

(7) Expected detection time for a diversion

(8) Goal quantity

(9) Expected detection location for a diversion

Standards

Strategic importance of material, namely

(10) Amount of fissile material needed for the construction of a nuclear weapon (threshold amount)

(11) Time (and effort) needed for the production of a nuclear weapon (critical time)

We start with the *number of material balance areas (MBAs).* It has been shown in Ref. [7] that, from the aspect of the probability of detection, it is optimal not to subdivide a given MBA into several sub-areas. Thus, going to the extreme, it would be optimal to consider the whole fuel cycle of a State as one MBA. However, other criteria or boundary conditions have to be considered, which might lead to different conclusions.

An argument for as few MBAs as possible is as follows. If transfers are made between facilities or parts of facilities at the same location, the subdivision of these into several sub-areas could cause additional measurement and organizational efforts, which could conflict with the principle of 'compliance with normal plant operation'.

An argument for several MBAs in a State is that, in a State with a large fuel cycle, it is impossible to take a complete physical inventory for the whole fuel cycle at the same time. This, however, means that a diversion in one part of the cycle can always be covered by 'borrowing' material from another part. Thus, one concludes that MBAs must be defined in such a way that a complete physical inventory can be taken at one time; otherwise containment and surveillance measures have to be taken that guarantee the impossibility of such a diversion strategy.

With the determination of the number of MBAs in a State we have established a *spatial* frame for that system. We now have to determine the *number of inventory periods* during a reference time, i.e. the *time* frame, for each MBA.

There is an argument similar to that for the number of MBAs if we take the overall probability of detection as the only criterion, then one can show that it is optimal to have only *one* inventory period per reference time in one MBA (see Ref. [8]).

The performance criterion 'expected detection time', on the other hand, does not necessarily provide an argument for numerous inventory periods per reference time. A detection can take place only after a physical inventory taking, i.e. after the material balance has been closed; therefore, one might indeed have numerous inventory periods per reference time. However, because of the measurement errors a diversion cannot be detected with certainty after the material balance has been closed; in fact, because of the limited overall false-alarm probability (or equivalently, the limited inspection effort), the higher the inventory frequency the smaller will be the chance for detection. As a result, the expected detection time, considered as a function of the number of inventory periods, has a minimum, as numerical calculations have shown. This minimum naturally depends on the total amount of material diverted.

In Annex 1 this situation is shown with the help of a single storage inventory verification scheme, where the only error source results from the fact that the limited inspection effort requires a random sampling verification. With the help of a numerical example Fig.3 illustrates that, for a smaller total diversion, the probability of detection decreases rapidly with increasing number of inventory takings; therefore, the expected detection time has its minimum at a smaller number of inventory takings.

Finally, there is the aspect of compliance with normal plant operations. Inventory takings normally cause some production disturbances; therefore, purely for safeguards purposes it will not be easy to perform many more inventories than the plant operators perform for internal plant purposes. The weight of this argument, however, depends on the kind of inventory to be taken; thus, one has to consider the details of the plant under consideration.

The *quality of the measurement system* necessary for the establishment of material balances in a MBA is in most cases given *a priori* and is not subject to major decisions from the safeguards authorities. One might even question the necessity of listing the quality of the measurement system under the design parameters. On the other hand, there might be cases where the system is completely inadequate to meet the 'standard' and where thoughts have to be given to the possibility of improving the measurement system.

If the values of the design parameters are fixed, the safeguards authority has a basis for establishing its data verification system, i.e. a random sampling system which guarantees that the material balance data provided by the plant operator are not falsified. Then the first question to be settled is the total verification effort (spent in inspection man-hours or money, e.g. for chemical analyses). For simplicity we will concentrate only on the *direct* effort to be spent for the verification of data and not take into account base load efforts for administration, travelling etc.

An upper limit of the verification effort is given by the effort to be spent for the verification of *all* operator's data and the collection of information necessary for the material balance establishment; we call this the '100% effort'. As usually in random sampling problems a reasonable security for detecting wrong items is given by sample sizes considerably smaller than 100%, the actual verification effort will be far below the 100% effort. The question is how far below. One might proceed as follows. As the IAEA Safeguards Agreements [1] exclude the possibility of 'lumping' together different kinds of plants (e.g. reprocessing and fabrication), which means that one cannot shift the verification effort from one kind of plant to another in the sense of an optimization, one starts with a fixed total effort below the 100% effort for each group of facilities. In so doing, one takes into account rules for the *maximum routine effort* in single facilities which have been established in the IAEA (see, e.g. Ref. [9]). Thereafter, one allocates the effort for each group of facilities to their MBAs via an optimization procedure, using the overall probability of detection as an optimization criterion. Having determined the optimized probability of detection for each single MBA as a function of the amount of material diverted, one can compare the results with the standards and see whether or not this overall verification effort is satisfactory; otherwise the whole procedure must start with an increased (or decreased) effort from the beginning. The point to be made here is that the verification effort must not be fixed independently for each MBA, but can be determined via an optimization if the effort for several MBAs within a group of similar facilities is fixed.

With the determination of the number of MBAs in a State, the inventory frequencies and the verification effort in the MBAs, the general frame for the implementation of safeguards is fixed. What still has to be done is to settle the details of the safeguards procedures, which is equivalent to determining the values of *false-alarm probabilities.*

Annex 2 illustrates the procedure of fixing values of single false-alarm probabilities for a given overall false-alarm probability at the hand of the (MUF,D)-

scheme for the data verification and material balance establishment procedure in a given MBA[7]. The question of how to fix the value of the overall false-alarm probability (and similarly, the overall goal quantity) can now be solved in a way resembling that of fixing the total verification effort in a MBA via an optimization procedure.

Nevertheless, independent of the problem at what level the optimization should be stopped (one inventory period and one MBA, or one reference time and one MBA, or several MBAs), one must finally determine the value of some overall false-alarm probability. This raises the question of how this value can be found.

To answer this question, two arguments must be taken into account. First, the probability of detection increases with increasing false-alarm probability. Second, as expressed in the name, fixing a value of the false-alarm probability means fixing an average rate of false alarms with all the negative aspects – effort for clarifying the false alarm, plant operation disturbances etc. Therefore, again an *iterative procedure* is necessary for determining appropriate values. One starts with a reasonable value (from the viewpoint of the second argument), goes through the optimization procedure, determines the overall probability of detection, and starts with a new value if the results are not satisfactory.

COMPLIANCE OF PERFORMANCE CRITERIA AND STANDARDS

In the foregoing section we discussed the determination of the values of the design parameters of a safeguards system. The procedure followed was that preliminary values of global parameters (overall false-alarm probability, overall verification effort etc.) had been fixed; furthermore, that values of, for example, single false-alarm probabilities, had been determined by an optimization procedure (taking into account inspection and adversary strategies) and that the values of performance criteria had been calculated, and that this procedure had to be started again from the beginning once the result was not satisfactory. Now we will discuss what 'satisfactory' means.

We start with the notions of *threshold amounts* and *critical time*, which were defined as the amounts of fissile material (depending on its physical and chemical form) and time needed for the production of a nuclear weapon. Rough data for threshold amounts are given in Ref. [11]; second-hand data for threshold amounts and critical times are given, e.g. in Ref. [12]. We have considered these quantities as *standards* for the safeguards problem in the sense that they provide yardsticks for any concrete system to be implemented. The problem arises if

[7] For different data evaluation schemes, e.g., the MUF-D-scheme, see Ref.[10], the procedure would have to be modified appropriately.

one wants to 'operationalize' these yardsticks. This problem can be illustrated by the following questions.

The *first question* is – will the threshold amounts be related to a State, to a group of MBAs (e.g. a so-called *regional fuel-cycle centre),* to a MBA, which may even be only a part of a facility and, furthermore, to one inventory period, or to the reference time? In the sense of non-proliferation, clearly one might argue that it must be related to a State. On the other hand, if a national fuel cycle consists of several plants, located at different places and managed by independent (perhaps even international) companies, then one might also find that the threshold amounts should be related to single facilities (in general equivalent to single MBAs). An argument in the same direction is the fact that, as long as it is only for a facility, not for the whole fuel cycle, the inventory can be taken at one point in time, and the safeguards authority gets from this technical procedure a statement only for that facility. The translation of these technical statements for different facilities into a general statement for the State may require some additional inspection effort the extent of which depends on the specific situation. In cases where the fuel cycle consists only of a few reactors this additional effort will be smaller than in cases where a complete fuel cycle exists.

The *second question* arises with respect to critical times. Published estimates vary between several days and many months. What should the safeguards authority do if the time between two physical inventories in a plant, according to its routine operations, is far larger than the associated critical time? As shown in the foregoing section, an increase of the inventory-taking frequency need not necessarily solve the problem, but may cause severe plant disturbances. Should one be satisfied if there is an indication of a diversion, even though perhaps very late compared with critical time standards? This consideration raises the very difficult question of *second action levels* – what will happen after a diversion has been indicated 'almost with certainty'? Will the plant operation be stopped, or will it continue so that, in the next inventory period, there is a chance of the same thing happening again? How many such cases can be tolerated in the reference time?

Let us assume now that the questions of relating threshold amounts to one or more MBAs, and critical times to inventory periods, are principally settled. Then we can start the procedure of determining the values of design parameters as described in the last section. This means that one starts with preliminary values of the design parameters, and ends with some performance criteria values which now have to be compared with the standards. How can this comparison be performed?

As an example, consider the probability of detection as a function of the goal quantity for a material balance test which has been represented graphically in Fig.1. Let the value of the threshold amount for that inventory period be, e.g. $M_0 = 8$ kg purified Pu. This standard may be met by a goal quantity of 20 kg purified Pu, if a 99% probability of detecting a diversion of 20 kg is considered

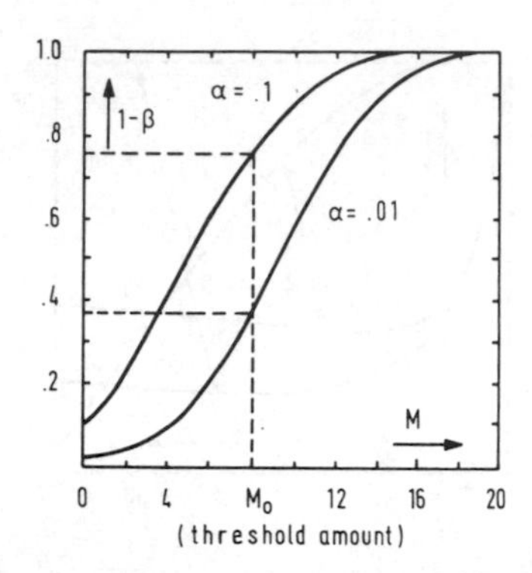

FIG.1. Probability of detection as function of goal quantity M with false-alarm probability α as parameter and σ^2 = 4 according to Eqs (2–4). For explanation of threshold amount see text.

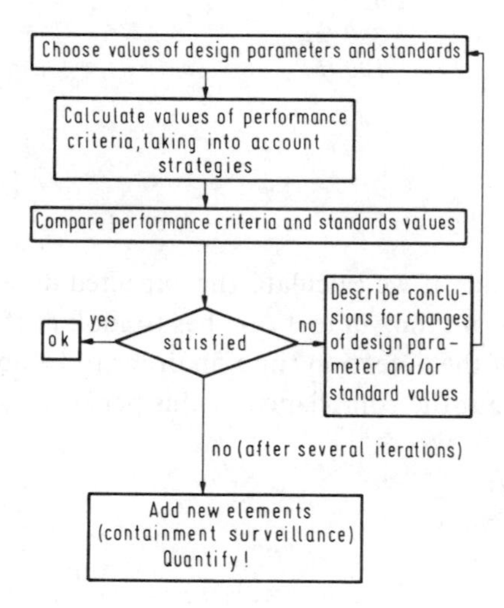

FIG.2. Procedure for fixing safeguards parameter values.

to be equivalent to a 37% probability of detecting a diversion of 8 kg. Here, we have fixed the false-alarm probability $\alpha_1 = 0.01$. Now, do we consider this result as satisfactory? Or can we justify an increase of the false-alarm probability to $\alpha_2 = 0.1$, thus obtaining a detection probability of 76% for a diversion of 8 kg? This example shows that it is not sufficient, but may lead to misunderstandings with respect to the system effectiveness if one gives an isolated figure of, e.g. 8 kg nuclear material as a threshold amount, without mentioning corresponding detection and false-alarm probabilities.

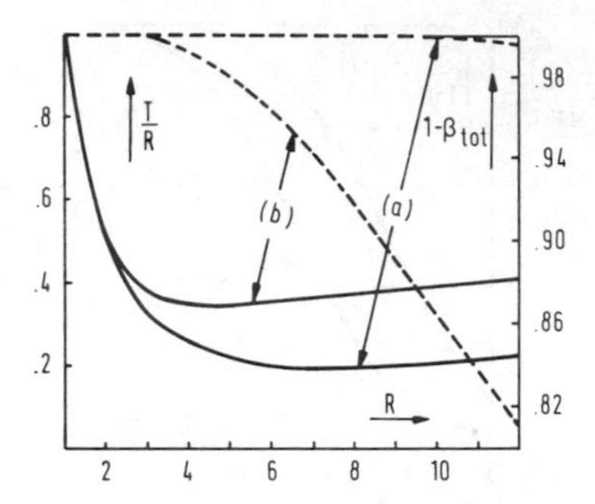

FIG.3. Storage inventory verification.
Expected detection time T/R (expressed in fractions of reference time) and total probability of detection $1 - \beta_{tot}$ (dashed line) as functions of numbers R of inventory taking for total numbers of batches in storage N = 10 000
verified batches n = 2 000
falsified batches r = 330 (a)
100 (b).

This situation is similar if we calculate the expected detection time and compare it with the critical time; in fact one has to look at the distribution, or at least at the variance of the detection time, in order to be able to make a reasonable statement about the compliance of this performance criterion and the associated standard.

CONCLUSION

In Fig.2, the procedure for establishing a safeguards system, which has been discussed, is represented graphically. The Figure shows the iterative nature of the procedure which should not only be considered as a consequence of mathematical difficulties, but even more as an inherent property of problems of that kind. One starts with a preliminary system, observes its effectiveness in meeting some preliminary standards and revises thereafter both the system and the standards until reasonable agreement is reached. As examples from different fields, air and water pollution control systems shall be mentioned.

(It is interesting, to compare Fig.2 with the corresponding figure in Ref. [13] which is much more detailed. That figure was developed before the present IAEA system was established; therefore, one may interpret Fig.2 as the implementation of the ideas given in Ref. [13].)

What can be done if, even after a substantial revision of the standards, no agreement at all can be reached with respect to performance criteria and standards? The 'normal' answer to this question would be the proposal to complement the safeguards system by appropriate containment and surveillance measures equivalent to material accountability. This, however, is not easy for the following reason.

The present IAEA safeguards system is based primarily on the material accountability principle since material accountability is quantifiable, which, in fact, was one of the major reasons for this choice. The quantification of protective mechanisms using containment and surveillance measures – in other words, the specification of the effectiveness of these mechanisms – is difficult and to a certain extent not free from qualitative and subjective considerations (see, e.g., Ref. [14]). Nevertheless, especially in view of specific diversion strategies (like 'abrupt' diversion), one will have to take into account these additional measures.

In this connection, a remark should be made about the objective of safeguards. As outlined in the introduction, it is the 'timely detection of diversion' but, in addition, the 'deterrence of such diversion by the risk of early detection'. But how can this qualitative 'risk of early detection' be quantified? Somehow containment and surveillance measures will enhance this risk.

To prove the usefulness of the supplementing containment and surveillance measures with respect to the problems raised above, we need more serious R and D work, supported by the experience gained in the course of implementing IAEA safeguards according to existing agreements.

Annex 1

STORAGE INVENTORY VERIFICATION ON AN ATTRIBUTE SAMPLING BASIS

We consider a storage containing N batches and assume that the storage operator intends to falsify the data of r of these N batches during the reference time. We assume, furthermore, that the total inspection effort available during the reference time allows the verification of the data of n batches.

Let us assume that R inventories will be performed during the reference time. Then it is intuitively clear (which can be proved exactly) that, from the aspect of the overall probability of detection, the total inspection effort available should be distributed equally on the R inventory takings and, in turn, that the operator should distribute his r falsifications equally on the R time intervals between the inventory takings.

Let β_i, i = 1 ... R, be the probability of not detecting at least one falsification at the i-th inventory taking; it is given by

$$\beta_i = \left(1 - \frac{r}{R \cdot N}\right)^{\frac{n}{R}} =: \beta \tag{1}$$

Then the total probability of detection $1 - \beta_{tot}$ is given by

$$1-\beta_{tot} = 1 - \prod_{i=1}^{R} \beta_i = 1 - \beta^R = 1 - \left(1 - \frac{r}{R \cdot N}\right)^n \tag{2}$$

As can be seen immediately, $1 - \beta_{tot}$ as a function of R takes its maximum value for R = 1.

The expected detection time T (under the condition that detection will take place during the reference time), measured in numbers of inventory periods, is given by the following expression (see Ref. [3]):

$$T = \sum_{i=1}^{R} i \cdot (1-\beta_i) \cdot \prod_{j=1}^{i-1} \beta_j \Bigg/ \left(1 - \prod_{i=1}^{k} \beta_i\right) \tag{3}$$

which leads in our case to the following expression for the expected detection time T/R, measured in fractions of the reference time,

$$\frac{T}{R} = 1 + \frac{1}{R} \cdot \frac{1}{1-\beta} - \frac{1}{1-\beta^R} \tag{4}$$

It can be shown that T/R as a function of R is not necessarily monotone; in Fig.1 a numerical example is given for which T/R takes a minimum.

Annex 2

DETERMINATION OF FALSE-ALARM PROBABILITIES IN THE (MUF, D)-SCHEME

Let us consider the data verification problem in one MBA for one inventory period in a very simplified way. The normal procedure is that the operator has delivered all data necessary for the material balance establishment, and that the safeguards authority now verifies a fraction of these data with the help of independent measurements on a random sampling basis. A numerical example is given in Ref. [3].

As one has to take into account measurement errors, a *statistical test* has to be performed, the purpose of which is to make a statement whether or not differences between the inspector's and the operator's data can be explained by measurement errors. This is usually done with the help of the so-called *D-statistics* – the difference D between the sum of all the inspector's data and the sum of all corresponding operator's data is formed, a *significance threshold* s_1 is chosen, and it is decided that, in fact, the difference can be explained by measurement errors, if $D \leqslant s_1$, whereas the contrary is decided if $D > s_1$. The question arises how to fix the value of the significance threshold s_1.

For this purpose we determine the *false-alarm probability* α_1 (neutrally called the *error first-kind probability*), which is the probability that the statement $D > s_1$ will be made if, in fact, only measurement errors cause the difference. There is a simple one-to-one correspondence between the significance threshold s_1 and the false-alarm probability α_1:

$$1-\alpha_1 = \Phi\left(\frac{s_1}{\sigma_{D/H_0}}\right) \tag{5}$$

where Φ is the Gaussian distribution function, and where σ^2_{D/H_0} is the variance of D 'under the null hypothesis' (i.e. under the assumption that there are only measurement errors); thus the question for the value of s_1 is equivalent to the question for the value of α_1.

Once the value of α_1 is fixed, the data verification scheme is fixed, and one can determine the probability of detecting a diversion of size M_1 (by data falsification, 'alternative hypothesis H_1') approximately as follows:

$$1-\beta_1 = \Phi\left(\frac{M_1-\sigma_{D/H_0}\cdot U_{1-\alpha_1}}{\sigma_{D/H_1}}\right) \tag{6}$$

where U is the inverse of the Gaussian distribution function, and where σ^2_{D/H_1} is the variance of D under the alternative hypothesis. But how can the value of α_1 be fixed?

To answer this question, we go one step further. Let us assume that no significant differences between the inspector's and the operator's data have occurred. Then the *material balance test* is performed with the help of the book-physical-inventory difference MUF such that it is decided that no material is missing if MUF $\leqslant s_2$, where s_2 is the significance threshold, and the contrary, if MUF $> s_2$.[8] As before, the significance threshold s_2 and the false-alarm probability for this test, α_2, are connected by the relation

$$1-\alpha_2 = \Phi\left(\frac{s_2}{\sigma_{MUF}}\right) \tag{7}$$

and the probability of detecting the diversion of the amount M_2 of material is given by the following a relation:

$$1-\beta_2 = \Phi\left(\frac{M_2}{\sigma_{MUF}} - U_{1-\alpha_2}\right) \tag{8}$$

Now we come back to our original question. We can determine the values of α_1 and α_2 if we fix the value of the *overall false-alarm probability* α, defined as the probability of committing a false alarm either with respect to D or to MUF, and which is approximately[9] related to α_1 and α_2 by the equation

$$1-\alpha = (1-\alpha_1)\cdot(1-\alpha_2) \tag{9}$$

by *maximizing* the overall probability of detection $1-\beta$, approximately given by

$$1-\beta = 1-\beta_1\cdot\beta_2 = 1-\Phi\left(\frac{\sigma_{D/H_0}\cdot U_{1-\alpha_1}-M_1}{\sigma_{D/H_1}}\right)\cdot\Phi\left(U_{1-\alpha_2}-\frac{M_2}{\sigma_{MUF}}\right) \tag{10}$$

with respect to α_1 and α_2 under the boundary condition of fixed overall false-alarm probability α.

---

[8] In actual fact, other factors are taken into account such as hidden inventory and unmeasured losses; thus, this scheme is only the first step in the whole evaluation process.

[9] The two random variables D and MUF are stochastically dependent because the operator's data are contained in both these variables. Therefore, the correct formulae for α and β are much more complicated than those given by Eqs (9) and (10). It can be shown, however, that these formulae represent surprisingly good approximations to the correct formulae (see Ref. [3]).

It should be noted that the determination of the values of goal quantities, in our case M_1 and M_2, can be handled quite in the same way. They can be determined for a given total amount $M = M_1 + M_2$ by *minimizing* (as this would be interest of the operator who might want to divert material) the overall probability of detection.

REFERENCES

[1] INTERNATIONAL ATOMIC ENERGY AGENCY, The Structure and Content of Agreements between the Agency and States Required in Connection with the Treaty on the Non-Proliferation of Nuclear Weapons, IAEA, Vienna, INF/CIRC/153 (1971).

[2] INTERNATIONAL ATOMIC ENERGY AGENCY, IAEA Safeguards Technical Manual, Part A, Introduction, Safeguards Objectives Criteria and Requirements. IAEA-174, IAEA, Vienna (1976).

[3] AVENHAUS, R., Material Accountability – Theory, Verification, Applications, Monograph of the Wiley IIASA Series on Applied Systems Analysis, J. Wiley, New York and Chichester (Feb. 1978).

[4] INTERNATIONAL ATOMIC ENERGY AGENCY, IAEA Safeguards Technical Manual, Part F, **1**, Statistical Concepts and Techniques, IAEA-174, IAEA, Vienna (1977).

[5] BENNET, C.A., Factors involved in defining quantities of nuclear materials which are of safeguards significance, unpublished working paper (1976).

[6] See, e.g., the relevant documents of the IAEA Board of Gouvernors.

[7] AVENHAUS, R., FRICK, H., "Vergleich zweier Verfahren der Materialbilanzierung", Operations Research Conf. Aachen, September 1977.

[8] FRICK, H., Game theoretical treatment of material accountability problems, Part II, Int. J. Game Theory **6**, Issue 1, (1977) 41–49.

[9] GMELIN, H., HOUGH, C.G., SHMELEV, V., SKJÖLDEBRAND, R., "A technical basis for international safeguards", Peaceful Uses Atomic Energy (Proc. 4th Int. Conf. Geneva, 1971) **9**, IAEA, Vienna (1972) 488–510.

[10] HOUGH, C.G., BEETLE, T.M., "Statistical methods for the planning of inspections", Safeguarding Nuclear Material (Proc. Symp. Vienna, 1976), **2**, IAEA, Vienna (1976) 561–80.

[11] ROMETSCH, R., LOPEZ-MENCHERO, E., RYZHOV, M.N., HOUGH, C.G., PANITKOV, YU., "Safeguards", ibid., **I** pp 3–14.

[12] HILDENBRAND, G., Kernenergie, Nuklearexporte and Nichtverbreitung von Kernwaffen, Atomwirtsch. Atomtech. **22** 7/8 (1977) 374–80.

[13] HÄFELE, W., Systems analysis in safeguards of nuclear material, Peaceful Uses of Atomic Energy (Proc. 4th Int. Conf. Geneva, 1971) **9**, IAEA, Vienna (1972) 303–22.

[14] NILSON, R., "An evaluation of safeguards inspection techniques – A time for change", AIF Topical Conf. International Commerce and Safeguards for Civil Nuclear Power, New York, March 1977.

DISCUSSION

J.E. LOVETT: Most discussion of containment/surveillance (C/S) techniques relates to their use to preserve the validity of previously verified measurement data. C/S techniques are invaluable for this purpose, significantly reducing the inspection effort which might otherwise be required to re-verify constant data repeatedly. It is my view, however, that the uncertainty in such C/S measures is precisely the uncertainty in the original data and their verification. Do you agree?

R. AVENHAUS: In those cases where C/S measures supplement the material accounting data verification scheme, e.g. if seals are used in order to save verification effort, the uncertainty in C/S measures is indeed that of the original data and their verification. The problem of quantifying C/S measures becomes serious when these measures are used as *alternatives* to material accounting, for example if the uncertainty of the measurements is too large in the sense of my presentation.

MESURES PROPRES A EVITER LES SOUTIRAGES CLANDESTINS DE MATIERES FISSILES (CONCEPT D'USINE TUYAU)

P. AMAURY, J. REGNIER, M. MASSON
Compagnie générale des matières nucléaires (Cogéma),
Le Plessis-Robinson, France

Abstract–Résumé

MEASURES FOR AVOIDING CLANDESTINE DIVERSION OF FISSILE MATERIALS (QUASI-TOTAL CONTAINMENT CONCEPT).

The paper describes a safeguards system designed for plants of the new generation with a view to submitting it for consideration by the national authorities. Since the large masses to be processed make the present system, based on materials accounting, inadequate, it has become necessary to redefine the respective parts played by the three safeguards techniques and within this context containment becomes the main technique. The "usine tuyau" (a quasi-total containment system), in which the fissile material is at all times inaccessible to human beings, is designed for this purpose. Certain cases in which the inaccessibility would seem difficult to ensure are described briefly in order to show that it is possible to have continuity of the containment; for example, the sampling and analytical procedures relating to quantities of plutonium of appreciable weight are remotely controlled. The quasi-total containment system is therefore a means of improving the protection of personnel and preventing diversion of the fissile materials.

MESURES PROPRES A EVITER LES SOUTIRAGES CLANDESTINS DE MATIERES FISSILES (CONCEPT D'USINE TUYAU).

Le mémoire décrit un système de garanties étudié pour les usines de la nouvelle génération, en vue de sa présentation aux autorités nationales. Les masses importantes à traiter rendant insuffisant le système actuel, basé sur la comptabilité-matière, on a été conduit à redéfinir les rôles respectifs des trois techniques de garanties, et le confinement devient dans ce concept la technique principale. L'usine tuyau, dans laquelle la matière fissile est partout inaccessible à l'homme, est conçue dans ce sens. Quelques cas où l'inaccessibilité pourrait sembler difficile à assurer sont brièvement décrits pour montrer qu'il est possible de permettre la continuité du confinement; ainsi les prises d'échantillons et analyses portant sur des quantités pondérales de plutonium sont téléopérées. L'usine tuyau permet donc d'assurer une meilleure protection du personnel et la prévention contre le détournement des matières fissiles.

INTRODUCTION

L'objet de ce mémoire est la préparation, en vue de sa présentation aux autorités nationales, du système de garanties étudié pour les usines de retraitement de la nouvelle génération.

Les masses de matière fissile à traiter dans les grandes usines seront telles que les risques de détournement doivent être étudiés dès la définition de l'usine. La sécurité du personnel et de l'environnement rendent par ailleurs obligatoire un confinement poussé de la matière fissile et plus généralement de la radioactivité.

Les nouvelles usines françaises, dans lesquelles la matière fissile est partout inaccessible à l'homme, sont conçues en fonction de cette double contrainte. Dans cette idée, la prévention est recherchée en premier lieu, le contrôle jouant un rôle complémentaire et d'ailleurs indispensable. C'est le concept d'usine tuyau développé ici, dont les principales conséquences font l'objet d'un examen approfondi.

1. GENERALITES

1.1. Usine de retraitement

La prochaine usine de retraitement, UP3 A, implantée sur le Centre de La Hague, sera la première étudiée et réalisée intégralement selon le concept d'*usine tuyau.* Elle comprendra:
- un stockage des combustibles irradiés,
- la dissolution des combustibles et le traitement chimique de l'uranium et du plutonium,
- la conversion de l'uranium et du plutonium,
- un atelier de vitrification des produits de fission,
- un traitement des effluents liquides,
- des stockages de produits finis.

Le procédé chimique utilisé est le procédé Purex.

Cette usine retraitera les combustibles des réacteurs à eau légère de caractéristiques suivantes:
- irradiation nominale: 33 000 MW·d/t,
- irradiation maximale: 43 000 MW·d/t,
- irradiation spécifique nominale: 33 MW/t,
- irradiation spécifique maximale: 40 MW/t,
- enrichissement initial: 3,5% en ^{235}U ou équivalent,
- durée de désactivation avant retraitement $\geqslant$ 1 an.

Les caractéristiques de l'usine sont les suivantes:
- capacité: 800 t/an,
- produits finis: UO_3 et PuO_2 aux normes Unirep.

Selon les accords commerciaux, l'oxyde de plutonium pourra être livré pur ou mélangé à de l'oxyde d'uranium.

Les méthodes adoptées sont les suivantes:

- dissolution: sur éléments combustibles cisaillés;
- extraction: un cycle de codécontamination uranium et plutonium, trois cycles de purification de l'uranium et deux cycles de purification du plutonium;
- conversion: précipitation à l'ammoniaque pour l'uranium, à l'acide oxalique pour le plutonium;
- effluents: le sous-produit «produits de fission» du premier cycle de purification sera, après concentration, transformé en verres; les autres sous-produits liquides seront, en majeure partie, recyclés après concentration, les effluents de l'usine seront donc réduits en volume et activité, ils seront soit épurés avant rejet soit rejetés directement; les sous-produits solides seront traités pour en éliminer au maximum les matières fissiles, puis seront conditionnés dans le bitume ou le ciment, les déchets obtenus seront stockés dans les stockages nationaux.

1.2. Techniques de garanties

Comme toutes les usines ayant à traiter des matières fissiles, UP3 A sera soumise au contrôle national selon des modalités définies par les organismes de sécurité. Ce contrôle utilise les techniques classiques de garantie:

- la comptabilité-matière, qui est l'application, à des flux de matière, des règles de la comptabilité classique, y compris les inventaires de début et de fin d'opérations,
- le confinement, isolement du produit au moyen de barrières physiques,
- la surveillance, observation humaine ou à l'aide d'instruments, des mouvements de la matière fissile ou de l'intégrité du confinement.

La comptabilité-matière est une technique de contrôle en temps différé de la matière fissile, le confinement est une technique de prévention, et la surveillance une technique de contrôle en temps réel ou différé.

Les usines de retraitement de petite taille se prêtent assez bien à l'application de la technique de la comptabilité-matière: les flux de matière fissile sont relativement faibles par rapport à la quantité de matière fissile considérée comme dangereuse. Les opérations se font par charges limitées. Le suivi de la matière fissile y est assez facile.

Les conditions ont très sensiblement changé pour les usines de grande capacité:

- les flux journaliers de matière fissile représentent plusieurs fois la masse d'un explosif nucléaire,
- la quantité de matière fissile immobilisée dans l'usine est très importante,
- la tendance est à l'abaissement des doses intégrées très en dessous des seuils admissibles.

Par ailleurs, ces usines sont soumises à des impératifs de rentabilité et tout arrêt, quelle qu'en soit la cause, inventaire physique par exemple, représente un manque à produire qu'il faut éviter.

La prévention de conception contre le détournement a donc paru, pour les futures usines, l'objectif à atteindre: la convergence des deux objectifs – éviter le détournement et protéger le personnel – a poussé les concepteurs du projet à rechercher un confinement de la matière fissile de la même efficacité pour toutes les phases du procédé. Cette notion de base correspond à l'image d'usine tuyau.

2. USINE TUYAU

2.1. Principes généraux

2.1.1. Définition

L'*usine tuyau* est un lieu de passage pour la matière fissile tel qu'il fasse évoluer les produits (uranium et plutonium) d'une seule entrée vers un très petit nombre de sorties en garantissant une inaccessibilité quasi totale en tout autre point.

Dans le cas considéré, il y aura une entrée de produit brut (le combustible) et deux sorties de produits finis.

2.1.2. Frontières

Le moyen mis en œuvre dans l'usine tuyau est d'assurer l'inaccessibilité de la matière fissile en utilisant les barrières de confinement nécessaires au procédé et à la radioprotection. L'appareillage de génie chimique constitue en lui-même une première barrière (barrière primaire), et le confinement biologique, une deuxième barrière (barrière secondaire). Il suffit de garantir que l'une, au moins, de ces deux barrières, existe et qu'elle est inaltérable.

Dans le cas de traversée des deux barrières (tuyau de réactifs ou d'instrumentation par exemple, une barrière supplémentaire devra être constituée.

En résumé, on peut dire que les frontières du tuyau sont telles que la matière fissile est toujours à l'intérieur du tuyau et reste inaccessible à l'opérateur qui, lui, est toujours à l'extérieur du tuyau.

2.1.3. Entrée et sortie des produits

L'entrée est constituée par la cellule de cisaillage. En amont de cette cellule, hors usine tuyau, il est considéré que la radioactivité spécifique élevée et la surveillance représentent une protection efficace contre le détournement. Dans l'usine tuyau on s'assurera toutefois que le plutonium ne peut être détourné en retour par cette voie.

La sortie du plutonium s'effectuera vers des stockages dans lesquels la protection contre le détournement aura une qualité de même niveau que dans l'usine tuyau. Il est admis que le stockage est hors usine tuyau. Toutefois, la liaison entre usine et stockage présentera les mêmes garanties d'inaccessibilité que le tuyau, chaque fois que cela sera possible; mais cela ne peut être érigé en principe, du fait de l'éventuel éloignement de ce stockage.

On se ramène alors au cas de la protection contre le détournement pendant le transport.

La fin du procédé uranium est à considérer dans deux cas:
– l'enrichissement est élevé, il faut alors la mettre en œuvre dans l'usine tuyau,
– l'enrichissement est faible et la fin du procédé peut avoir lieu hors usine tuyau mais il faut alors s'assurer que cette sortie n'est pas une voie de détournement pour le plutonium.

Dans les usines de retraitement de la nouvelle génération, où l'enrichissement sera faible, l'usine tuyau s'arrêtera à la sortie du nitrate d'uranyle purifié, mais toutes précautions seront prises dès la conception pour empêcher le plutonium d'être détourné vers la ligne uranium.

2.1.4. Tuyau – lieu de passage

Le vocable *tuyau* a été pris à l'origine dans le double sens de confinement étanche et de ligne sans rupture de charge. On pourrait en effet contribuer à la protection contre le détournement en n'accroissant pas sensiblement les rétentions de matières fissiles dans la grande usine par rapport aux petites par adoption du traitement en ligne.

En fait, l'expérience montre que même si l'on garde comme but de limiter la rétention, on ne peut éviter les stockages tampons. Le concept d'usine tuyau se rapporte donc à un confinement continu. S'y ajoute l'obligation, pour répondre aux exigences du contrôle national, de pouvoir vider l'installation pour inventaire, sauf les stockages, considérés comme durables et conçus comme tels du point de vue du bilan (exemple, stockage de nitrate de plutonium de capacité correspondant au traitement de plusieurs mois).

2.1.5. Conception générale

L'interposition de plusieurs barrières intégrales entre la matière fissile et le personnel est rendue nécessaire pour la protection contre la contamination et l'irradiation.

Elle entraîne la nécessité d'utiliser systématiquement la téléopération. Chaque fois que cela est possible, l'automatisation du procédé est souhaitée aussi pour éviter la dispersion du personnel d'exploitation dans les «coursives».

Celui-ci sera rassemblé dans des salles de contrôle en très petit nombre (si possible une seule, située au barycentre des différentes unités).

Cette conception est rendue possible par les progrès liés à l'expérience et facilite le contrôle des accès dans les zones attenantes aux barrières.

Dans les premières usines de retraitement la première barrière assurait un bon degré d'inaccessibilité dans le cas de la voie aqueuse. Par contre, pour la voie sèche, les produits pulvérulents n'étaient souvent séparés des opérateurs que par la mince barrière des gants de boîtes à gant.

Dans une usine de la nouvelle génération, la radioactivité du plutonium est telle que les boîtes à gant ont complètement disparu au profit de cellules α, β, γ, n et la téléopération y sera ainsi la règle. En même temps, un gros effort est fait pour garantir le premier confinement de la poudre afin d'éviter la contamination de la deuxième barrière. Il en résulte qu'il sera possible de réaliser une inaccessibilité par la première barrière de degré très satisfaisant qui apparente les cellules de voie sèche à celles de la voie humide.

Cet effort est, de toute façon, nécessaire en certains points pour des raisons de procédé. Exemple: éviter la reprise d'humidité par l'oxyde de plutonium entraînant en fait, sur toute la section entre le four et le conteneur de stockage, un confinement primaire quasi étanche.

Il n'est pas utile de rappeler l'intérêt, pour la continuité des opérations, d'éviter les systèmes de transfert nécessitant une maintenance directe comme les pompes. D'autres aspects de la conception seront traités plus loin.

2.2. Cas où l'application directe du principe de la double barrière d'inaccessibilité est mise en défaut

2.2.1. Généralités

On doit examiner un certain nombre de situations connues qui représentent la majorité des cas où le principe de la double barrière d'inaccessibilité est apparemment mis en défaut et donner une idée des mesures qui seront prises pour ramener cette inaccessibilité à un niveau équivalent.

2.2.2. Lignes de réactifs et d'utilités

Il faut empêcher le pompage en retour, dans ces lignes, soit dans la ligne elle-même, soit par introduction d'un tube souple. On dispose pour cela d'un certain nombre de moyens:

1) premier point accessible hors de la deuxième barrière situé à une hauteur empêchant le pompage sous vide,

2) tuyauterie traversant la première barrière non plongeante ou plongeante et perforée ou ventilée,
3) ligne munie de chicanes,
4) ligne ne pouvant être utilisée pour créer un pompage en retour par air-lift.

2.2.3. *Instrumentation*

En voie aqueuse, on retient comme exemple de rupture de la barrière primaire les mesures par bullage. Celles-ci seront réalisées de façon à respecter les règles 1 et 4 du paragraphe précédent. Dans les cas où cela ne sera pas possible (et qui seront exceptionnels) la partie sensible sera placée dans une enceinte protégée par un système de surveillance très élaboré (fig. 1).

La mesure par bullage est probablement l'exemple le plus délicat et peut être considérée comme représentative. Il y a lieu de remarquer que le nécessité de garder dans la barrière secondaire une grande longueur de bulle à bulle est, de toute façon, indispensable pour la radioprotection du personnel.

2.2.4. *Maintenance «active»*

Le cas type nous paraît être représenté par les appareils mécaniques de fin de procédé pour lesquels le principe visé lors de la conception est le suivant:

Dans la plus grande partie des cas les opérations de maintenance se font à l'intérieur de la deuxième barrière par échange standard téléopéré à partir de la salle de contrôle, par des appareils spécifiques intérieurs à la cellule, comme s'il s'agissait d'opérations de production. De plus, avant échange standard l'appareil sera vidé chaque fois que possible. Le matériel échangé transite à l'intérieur du tuyau dans une ou plusieurs cellules permettant de réaliser ou une réparation sommaire ou une élimination de la matière fissile par lavage ou dépoussiérage avant conditionnement précédant la sortie vers l'atelier de décontamination-entretien.

Les points d'entrée de matériel de rechange seront en petit nombre et conçus de façon à ne pas permettre la sortie de matières fissiles en retour.

Les points de sortie du matériel usagé et conditionné seront traités comme les points de sortie des autres déchets solides:

1) Ne sortent que des matériels ou déchets ne contenant plus que des traces de matières fissiles. Ceci revient à installer des unités de prédécontamination sommaire dans l'usine tuyau, ce qui est considéré comme nécessaire pour soulager les ateliers de décontamination centraux qui ne peuvent plus faire face, dans le cadre des méthodes actuelles de gestion des déchets, à l'activité croissante des objets à décontaminer.
2) Les points de sortie sont peu nombreux (ce qui est nécessaire si l'on veut éviter la multiplication des postes de prédécontamination sommaire).

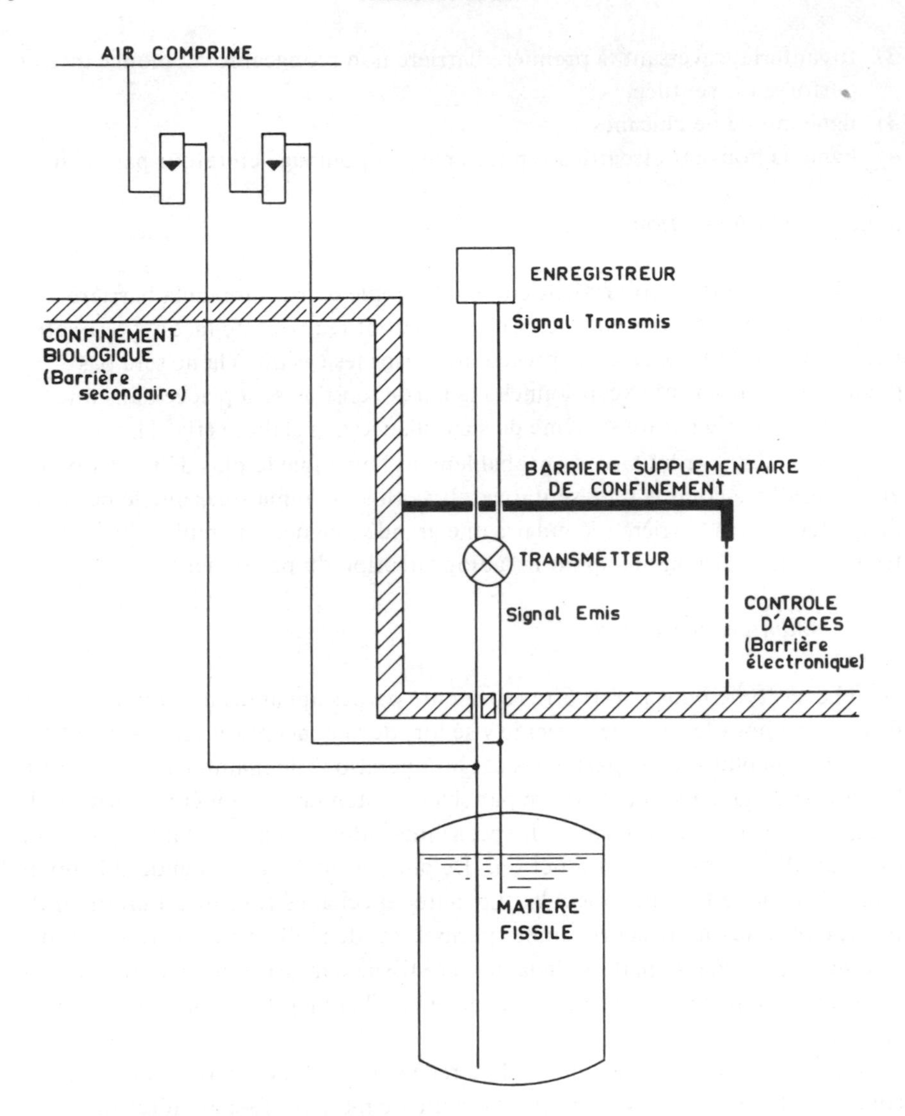

FIG.1. Exemple de montage d'un transmetteur (cas exceptionnel).

3) Les sorties sont précédées d'un tri et d'un contrôle individuel de contamination α, β, γ, n permettant de détecter la présence anormale de plutonium (ce qui conduirait à revenir en amont).
La limitation des points de sortie, en voie sèche, rend la surveillance aisée à organiser du fait que la redondance des mesures peut être envisagée sans grever l'économie du projet.

2.2.5. Echantillonnage – Laboratoire d'analyse

Les prises d'échantillons seront automatisées et, du même coup, téléopérables.

Elles seront situées à l'intérieur de la deuxième barrière et respecteront en général le principe de double barrière et les règles fixées dans les cas particuliers précédents. Les transferts d'échantillons se feront par pneumatique, dans des circuits scellés.

Au laboratoire, les analyses sur des quantités pondérales seront automatisées ou téléopérées selon des principes analogues à ceux de l'usine tuyau. Il en sera de même des préparations de prises d'essai de quantités infinitésimales de plutonium (exemple, filaments pour spectrographe).

Pour respecter ces principes, il est essentiel d'adapter le laboratoire et de rationaliser les analyses, tant dans leurs méthodes que dans leurs moyens, en prenant en compte l'expérience acquise. Les études de faisabilité réalisées en France montrent que sur ces bases, il est possible de considérer la partie du laboratoire gérant des quantités pondérales de plutonium comme appartenant à l'usine tuyau, sans discontinuité.

2.2.6. Sorties d'effluents

L'interposition entre le circuit de traitement principal et les sorties d'effluents, de circuits de recyclage de sous-produits très élaborés crée une barrière quasi infranchissable aux quantités pondérales de plutonium entre ces circuits principaux.

Les sorties seront en très petit nombre et permettront une surveillance poussée.

2.3. Incidents – Opérations exceptionnelles

Bien que tout soit fait pour éviter des incidents ou opérations exceptionnelles entraînant le non-respect des principes et règles évoqués ci-dessus, leur existence ne peut être exclue. Toutefois, leur fréquence sera considérablement plus faible que dans les installations de conception ancienne.

Cet état de fait rend d'autant plus aisée la mise en œuvre de procédures spéciales particulièrement strictes nécessitant un contrôle humain et un appareillage important. Ces procédures devront être rédigées avec un soin tout particulier et être l'objet d'un examen soigné a priori et a posteriori par les responsables de la sécurité.

3. ROLES RESPECTIFS DES TECHNIQUES DE GARANTIES

Les contraintes dues à la quantité et à la qualité des matières traitées, les règles de sécurité de tous genres, ainsi que les impératifs économiques, imposent l'étude rationnelle de l'application des techniques de garanties dès la conception de l'usine. Dans l'usine tuyau, dès la conception, les rôles respectifs des techniques de garanties sont définis:

- le confinement, technique de prévention, est appliqué entre l'entrée du combustible irradié et les sorties des produits,
- la surveillance, technique de contrôle des accès et des mouvements, est utilisée à chaque rupture de confinement et, en complément, chaque fois qu'il paraît nécessaire,
- la comptabilité-matière, technique de contrôle des masses de matière fissile, est utilisée pour la connaissance des entrées et sorties de matière et du stock, de façon aussi précise que le permettent les mesures.

Il n'est plus nécessaire, le confinement étant étendu à toute l'usine et la surveillance jouant un rôle complémentaire, de comptabiliser la matière fissile à toutes les étapes du traitement: la matière est inaccessible, elle ne peut suivre que des voies obligatoires et la suppression des points de mesure intermédiaires devient alors possible.

Le confinement est donc la technique de garanties principale, mais elle reste toujours indissociable des deux autres: dans le concept d'usine tuyau, le rôle dévolu auparavant à la comptabilité-matière est joué par l'ensemble des trois techniques de garantie. Il en résulte une meilleure protection contre les risques de détournement.

CONCLUSIONS

Dans l'usine tuyau, la prévention liée à la nature même des produits traités, et la prévention qui résulte de la nécessité de confiner la matière dans les circuits du procédé, sont recherchées simultanément et apportent une solution efficace au problème de la prévention du détournement de matière fissile.

DISCUSSION

D. GUPTA: Perhaps I did not understand you, but it seems to me that a system such as you describe would be useful mainly for physical protection at the national level. For example, surveillance of the movements of operating personnel into or out of a facility may not be possible under IAEA safeguards. Would you like to comment on this?

J. REGNIER: The safeguards system described is in fact intended for presentation to national authorities. It is based on techniques recognized by the Agency, with the role of containment enhanced. Surveillance was also dealt with because it is complementary to containment and accounting. It is not incompatible with Agency objectives and could even be of great assistance in achieving them.

H.W. SCHLEICHER: With the "usine tuyau" (quasi-total containment system) what possibilities would Euratom and IAEA inspectors have for obtaining input samples or verifying their analysis, bearing in mind that the taking of samples and their transfer and analysis are automated and take place within the containment?

J. REGNIER: A distinction must be made between the possibility of taking samples and the possibility of removing them from the containment. The taking of samples will be determined by agreements between governmental authorities and Euratom or the IAEA. Although sample analysis will be automated and performed within the containment, it will still be possible to verify the results as long as the measuring instruments are periodically calibrated.

There seems to be little purpose in removing samples from the containment, especially in view of the relative instability of the solutions obtained from highly irradiated fuels; the transfer of such samples could expose the workers and population to risks which it is better to avoid.

IAEA-SM-231/10

DESIGN FEATURES RELEVANT TO IMPROVED IAEA SAFEGUARDS

D. GUPTA
Nuclear Research Center Karlsruhe,
Karlsruhe

J. HEIL
Ministry of Research and Technology,
Bonn,
Federal Republic of Germany

Abstract

DESIGN FEATURES RELEVANT TO IMPROVED IAEA SAFEGUARDS.

The IAEA safeguards have become a subject matter of international discussions in connection with the expanding peaceful uses of nuclear energy. Using a decision structure suggested by the Standing Advisory Group on Safeguards Implementation (SAGSI) of the IAEA, this paper discusses the influence of different elements of this structure considered for safeguards implementation. It is assumed that four main elements influence the design and implementation of IAEA safeguards. They are the State's System of Accountancy and Control, the design features of a facility, the safeguards measures available to the IAEA, and the design safeguards effectiveness. It is shown that the most important single factor influencing implementations is the interplay between the design features of a facility and the available safeguards measures. Safeguards effectiveness is influenced strongly by the goals set forth for the IAEA safeguards.

INTRODUCTION

The IAEA safeguards have become a subject matter of intensive international discussions in connection with the expanding peaceful uses of nuclear energy in many countries of the world. The general trend of these discussions has been to define in more definite terms the objective of the IAEA safeguards, establish those elements which influence the successful design and implementation of Agency safeguards and suggest means of improving its effectiveness.

Following a series of discussions at the SAGSI, IAEA (Standing Advisory Group on Safeguards Implementation) it was suggested that a structure containing the various elements relevant to the design and implementation of IAEA safeguards systems be established to facilitate further discussions. This structure is reproduced in Fig. 1. It is seen that after the design goals of the Agency system have been established, four elements may influence directly the design and the implementation.

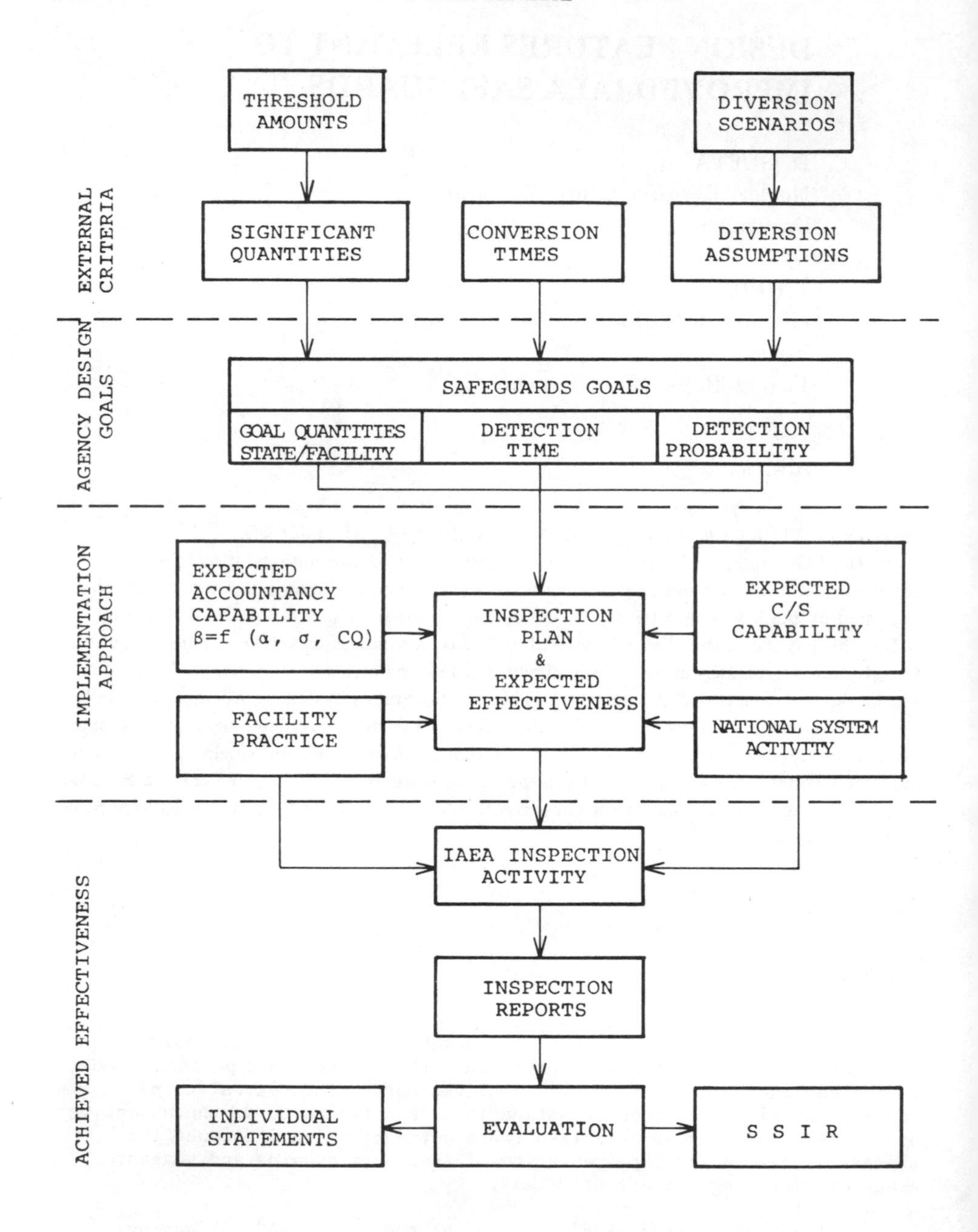

FIG.1. Elements of design and implementation for IAEA safeguards

C/S = containment/surveillance.
SSIR = Spatial Safeguards Implementation Report.

They are the State's system activities, facility design features and practices, available IAEA safeguards measures and the design level of effectiveness which the safeguards system has to attain. These elements are all interrelated and may influence the design and implementation of Agency safeguards in a fairly complicated manner. The present paper discusses briefly the role and influence of these elements in a pragmatic manner.

1. STATE'S SYSTEM OF ACCOUNTANCY AND CONTROL (SSAC)

The SSAC in a State is a necessary prerequisit for the proper implementation of Agency safeguards. The existence of the SSAC is however, not sufficient for the Agency to attain its safeguards goals. The SSAC lays down the legal, administrative and technical framework to enable the Agency to execute its safeguards functions in territories under the control of a State. It also ensures that organizational and functional responsibilities and the required safeguards infrastructure are defined or laid down (and maintained) at the level of a nuclear facility under safeguards in such a way that the Agency can carry out its activities in order to achieve its safeguards goals. It is also the responsibility of a SSAC to provide or make available all the relevant information and data so that the Agency can verify them with a view to ascertain that there has been no diversion of nuclear materials from peaceful uses to nuclear weapons or other nuclear explosive devices.

Some of the technical elements of a SSAC relevant to Agency safeguards have been indicated in |1|. The Agency is engaged in developing general guidelines for a SSAC relevant to international safeguards |2|. The extent and detail which the different elements of a SSAC in a State should have, will depend on the extent and type of nuclear activities in that State. A complete absence of the different elements of a SSAC in a State might force the Agency to make increasing use of subjective judgment for meeting its goal. Partial absence might severely hamper its activities. Thus the main influence of the absence or an incompleteness of SSAC on the design and implementation of the Agency safeguards would be to provide for intensive and additional Agency activities including additional inspections and to make use of subjective judgment factors in achieving its goals.

2. DESIGN FEATURES AND PRACTICES IN A FACILITY

Both the design and the implementation of Agency safeguards are most profoundly influenced by the facility characteristics. They set practical limits to the safeguards performance and determine the level of uncertainty and extent of the Agency knowledge in the time-location-amount-plane in connection with safeguards |3|. The extent of knowledge required by the facility operators for material management practices may not always be sufficient for achieving safeguards goals by the Agency. In any nuclear facility, the layout, the process and operational conditions determining the flow and inventory characteristics and the information system used for nuclear material management, are among the more important facility features influencing Agency safeguards. The possibility of an adaptation of the safeguards measures to these features and vice versa, might be the most important single factor influencing the design and implementation of Agency safeguards. Some design characteristics of enrichment, fabrication (for low enriched uranium) and reprocessing facilities have been presented in Table I. Some of these characteristics have been discussed in the following chapters with the safeguards measures to illustrate this particular point.

TABLE I. SOME DESIGN FEATURES AND PRACTICES IN TYPICAL BULK NUCLEAR FACILITIES RELEVANT TO IAEA SAFEGUARDS

Enrichment (centrifuge facilities)	Fabrication (LEU)	Reproccessing (small-scale)
a) Possibilities of high enrichment inside the cascade area which may not be accessible to the Agency inspectors	a) Some of the process materials may not always be available in a readily measurable form or may not be directly measurable	a) Amounts of nuclear materials in the process area may not always be accessible for direct and instant verification by Agency inspectors
b) Inaccessibility of materials inside the cascade area	b) Materials may not be accessible or available for direct measurement in the product streams (complete fuel elements)	b) Present status: Input measurement accuracy in the range of percentages; product measurements in the range of a tenth of a percentage Waste measurement with low accuracy
c) High accuracy of input and product analysis Low accuracy of waste stream (not tails) analysis	c) High accuracy of input and intermeadiate measurements Low accuracy of inventory materials Low accuracy of waste streams; waste streams may not be readily available in a measurable form	c) All the safeguards relevant information may not be available in readily verifiable form
	d) Generation of information may take place at a different place than that at which it is used	

3. SAFEGUARDS MEASURES AND DESIGN FEATURES OF A FACILITY

It is well known that the Agency can use material accountancy, containment and surveillance measures to fulfill its goals. The extent and relation amongst these measures are laid down in the respective facility attachments for a nuclear facility. One of the most common features particularly for bulk facilities is the fact that in some parts of such a facility, the nuclear materials in question may not be accessible or available in readily measurable form for Agency verification purposes |4|, so that its knowledge on the location and amount for this part of the material might be zero or associated with a very high degree of uncertainty. The facility operator might have additional process information or estimates at his disposal (to which the Agency might not have access) and/or may not require such materials to be available in accessible or measurable form for the purposes of plant operation. The Agency safeguards measures have to be adapted under such conditions to the facility features in such a manner that it can still meet its design goals. This may be possible by making use of a number of "operation indicators" or correlations |4| and combining a number of safeguards measures |5|. Typical example of such a case is the nuclear material content in the piping network of a process area between two successive process tanks in a reprocessing facility. Although the content of the pipeline can not be directly verified by the Agency, it can ensure and verify that this amount can be discharged only into the second process tank. This assurance can be obtained by a combination and correlation of the amounts and flowrates for the two process tanks and the use of sealing and surveillance measures around the piping network in question. Although with such a combination of safeguards measures, the uncertainty in knowledge associated with the amount of material in the pipeline network still remains, the uncertainty in knowledge for the Agency with regard to the <u>use</u> of this particular amount is eliminated.

Another design feature or practice in the bulk facility in general is the high measurement uncertainty of waste streams containing nuclear materials. If the rest of the process treams and inventory amounts are verified with the desired degree of high accuracy, the Agency inspectors can verify the measurement uncertainty of the waste streams and ensure through observation or other surveillance along with sealing measures that the discarded waste streams would no longer be accessible to plant operation without the knowledge of the Agency inspectors. Again with such a combination of measures, the Agency can fulfill its goals in ensuring that nuclear material from this particular stream has not been diverted for nuclear weapon or other nuclear explosive devices, even though the uncertainty in knowledge with regard to the exact amount in these streams may be fairly high and still exists.

In this context it may be worthwhile to analyse the role of containment and surveillance measures as means for meeting the goals of Agency safeguards. It has often been feared that because of the difficulties in quantifying the effects of c/s measures (as is possible in the case of measures associated with material accountancy) their role in IAEA safeguards has been and will remain somewhat vague and will be similar to other subjective elements of such safeguards systems.

Experience at the Agency and elsewhere indicate that this is not the case. The examples given above show clearly that under certain conditions the Agency can ascertain with the help of c/s measures <u>alone</u> that there has been <u>no diversion</u> of nuclear materials from peaceful uses to nuclear weapons or other nuclear explosive devices. It is to be noted that according to Article 7 Ref. |1| this is also the goal of the Agency safeguards (that is to ascertain that no diversion has taken place). The examples given above can be expanded to different reactor systems and critical assemblies and similar statements for such facilities can be made. The difficulty of quantification with the c/s measures arises when the Agency goal is set to establish, along with some other parameters the actual amount which has been assumed to be diverted. In other words c/s measures can, along with the accountancy measures or even alone, provide quantifiable information for achieving the goals of international safeguards provided these goals are defined accordingly, for example according to Article 7, Ref. |1|.

4. SAFEGUARDS EFFECTIVENESS

The effectiveness is an yardstick to assess the performance of the safeguards system. According to Fig. 1, the expected effectiveness of the safeguards system for a given facility is defined and laid down during the planing of the implementation program for the Agency system. This design effectiveness is then compared with the achieved effectiveness (an a-posteriori analysis) after the safeguards system has been implemented and evaluated for that facility.

Definition of the effectiveness of Agency safeguards may be fairly complex and a large number of parameters can influence it. Besides, it may be quite difficult to express it in the form of a single value. It is also understandable that effectiveness depends on the design goal of Agency safeguards as well. The value of effectiveness for example would be different for the case when the goal of Agency safeguards is defined to be the detection in a timely manner of the diversion of a significant quantity of nuclear material, than for the case in which the Agency safeguards goal would be defined to ascertain that no diversion of nuclear material has taken place. An internationally acceptable evaluation procedure has to be established to determine the yardstick for effectiveness.

5. CONCLUDING REMARKS

In Ref. |1| the definiton of the goals of the Agency safeguards contain a number of elements. In Article 7 for example the purpose of the Agency safeguards is to "ascertain .... that there has been no diversion of nuclear material.....". In Article 28 on the other hand in defining the objective of safeguards four elements have been brought into play, namely, the elements of detection, the timeliness, the significant quantity and the deterrence by the risk of early detection. Past experience and projection of this experience to future nuclear activities in peaceful sectors in different countries, tend to show that difficulties tend to arise when too much emphasis is placed on individual elements of the Agency safeguards objectives. This is particularly illustrated by the recent emphasis and discussions on the element of timeliness of detection associated with the significant quantity. Such difficulties can normally be eliminated or reduced by taking into consideration all the elements of the objectives of the Agency safeguards and adapting them in a broadened manner to the real conditions prevailing in a facility in the design and implementation of its safeguards system.

REFERENCES

[1] INTERNATIONAL ATOMIC ENERGY AGENCY, The structure and content of agreements between the Agency and States required in connection with the Treaty on the Non-Proliferation of Nuclear Weapons, INFCIRC/153 (corrected), IAEA (1972).

[2] State's System of Accounting for and Control of Nuclear Material (SSAC), AG-43/12; Working paper on Standing Advisory Group in Safeguards Implementation (SAGSI) Meeting on Guidelines for SSAC, April 1978.

[3] GUPTA, D., "Various aspects of systems analysis", Safeguards Techniques (Proc. Symp. Karlsruhe, 1970) 2, IAEA, Vienna (1970).

[4] GUPTA, D., HEIL, J., "International safeguards in large-scale nuclear facilities", Nuclear Power and its Fuel Cycle (Proc. Conf. Salzburg, 1977) 7, IAEA, Vienna (1977) 465–82.

[5] ROMETSCH, R., HOUGH, G., "The position of IAEA safeguards relative to nuclear material control accountancy by States", ibid, pp.441–54.

DISCUSSION

A.G. HAMLIN: I agree with you that safeguards objectives can be defined in different ways. In certain circumstances an inspectorate could conclude, on the basis of containment and surveillance, that no diversion had occurred from a facility. Unfortunately, in a real world, things may happen in a facility which temporarily cause the output to be larger or smaller than predicted. This will not be detected by containment and surveillance.

The essence of safeguards is that material balance areas (MBAs) should mesh together as a means of detecting abnormalities. When an MBA controlled by accountancy meshes with an MBA controlled by containment and surveillance, the output fluctuations referred to above will appear as discrepancies in the first MBA. Could you comment on this?

D. GUPTA: The proper meshing of adjacent MBAs is certainly a very important factor for the proper functioning of an international safeguards system. Fluctuations in the inputs and outputs of an MUF-controlled MBA can normally be clarified at the time of physical inventory taking. If more frequent clarification is required, a combination of inventory taking and containment/surveillance measures may be necessary. Indeed it may be very important in this connection to consider a combination of different measures after taking into account a reasonable set of diversion possibilities.

IAEA-231/143

КОНСТРУКТИВНЫЕ ОСОБЕННОСТИ И ТЕХНИЧЕСКОЕ ОБЕСПЕЧЕНИЕ АЭС С РЕАКТОРАМИ ТИПА ВВЭР, ПОВЫШАЮЩИЕ ЭФФЕКТИВНОСТЬ ГАРАНТИЙ

Н.С. БАБАЕВ
Институт атомной энергии им. И.В.Курчатова
Государственного комитета по использованию атомной
энергии СССР,
Москва

В.Б. ФОРТАКОВ
Центральный научно-исследовательский институт
информации и технико-экономических исследований
по атомной науке и технике
Государственного комитета по использованию атомной
энергии СССР,
Москва

В.П. КРУГЛОВ, В.К. СЕДОВ
Нововоронежская атомная электростанция
им. 50-летия СССР,
Нововоронеж

Союз Советских Социалистических Республик

Abstract–Аннотация

DESIGN FEATURES AND PROVISION OF TECHNICAL DEVICES FOR NUCLEAR POWER STATIONS WITH WATER-MODERATED AND COOLED POWER REACTORS (VVER) BY WHICH THE EFFECTIVENESS OF SAFEGUARDS CAN BE IMPROVED.

The authors consider the introduction into light-water reactor power plant design of technical devices for improving the effectiveness of the IAEA safeguards system. Using the example of power stations with VVER-440 and VVER-1000 reactors, they examine the main characteristics and design features of these facilities important for safeguards. An appropriate system is determined for the application of safeguards, including accounting measures for fresh fuel (verification of the nuclear material content of assemblies) as well as containment and observation methods, duplicated by accounting measures (verification of burnup and nuclear material content of assemblies) and supplemented by independent control of reactor energy output for irradiated fuel. The authors recommend provision of the following technical devices, which when introduced would facilitate the application of safeguards and improve their effectiveness: sealing of the reactor; an automatic system for surveillance of irradiated fuel, coupled with an irradiated assembly counter; sealing of the shipping containers; a reactor energy output integrator; and rigs for measuring fresh and irradiated fuel assemblies.

КОНСТРУКТИВНЫЕ ОСОБЕННОСТИ И ТЕХНИЧЕСКОЕ ОБЕСПЕЧЕНИЕ АЭС С РЕАКТОРАМИ ТИПА ВВЭР, ПОВЫШАЮЩИЕ ЭФФЕКТИВНОСТЬ ГАРАНТИЙ.

Рассматривается введение в проекты АЭС с легководными реакторами технического обеспечения, позволяющего повысить эффективность гарантий МАГАТЭ. На примере АЭС с реакторами типа ВВЭР-440 и ВВЭР-1000 разбираются основные данные и конструктивные особенности этих установок, существенные для гарантий. Определяется целесообразная схема применения гарантий для свежего топлива, включающая меры учета (проверку содержания ядерного материала в кассетах), а для облученного – меры сохранения и наблюдения, дублируемые мерами учета (проверкой выгорания и содержания ядерного материала в кассетах) и дополняемые независимым контролем энерговыработки реактора. Рекомендуется техническое обеспечение АЭС, внедрение которого облегчает применение гарантий и повышает эффективность: опечатывание реактора, автоматическое наблюдение за облученным топливом в сочетании со счетчиком облученных кассет и опечатыванием транспортных контейнеров, установка интегратора энерговыработки, стендов для измерения свежих и облученных кассет.

1. ВВЕДЕНИЕ

Создание ядерных установок с использованием конструктивных решений и устройств, способствующих применению гарантий МАГАТЭ, является важным и насущным шагом в направлении обеспечения эффективного контроля за ядерными материалами. Уже выдвигалось предложение о рассмотрении в рамках МАГАТЭ международных мер по порядку лицензирования ядерных установок с тем, чтобы способствовать достижению на них нужной эффективности контроля [1]. Среди установок ядерного топливного цикла наиболее многочисленными являются атомные электростанции (АЭС). В настоящее время большую часть эксплуатирующихся и вновь создаваемых АЭС составляют станции с легководными реакторами. Примерами установок такого типа являются АЭС с реакторами типа ВВЭР, сооруженные и сооружаемые при техническом содействии Советского Союза в целом ряде стран, в том числе в ГДР, Финляндии, ЧССР, БНР, ВНР. В настоящее время эти АЭС оборудованы реакторами типа ВВЭР-440 (электрической мощностью 440 МВт), в дальнейшем предполагается также применение реакторов типа ВВЭР-1000, первый из которых установлен на V блоке Нововоронежской АЭС (СССР). В данном докладе рассмотрены конструктивные особенности этих реакторов, существенные для гарантий, предложены технические меры, повышающие эффективность применения гарантий на АЭС данного типа, и приведены результаты испытаний ряда приспособлений, которые могут использоваться для целей гарантий.

2. КОНСТРУКТИВНЫЕ ОСОБЕННОСТИ АЭС С РЕАКТОРАМИ ТИПА ВВЭР-440 И ВВЭР-1000, ОБЛЕГЧАЮЩИЕ ПРИМЕНЕНИЕ ГАРАНТИЙ

Вопросы применения гарантий на АЭС с реакторами типа ВВЭР обсуждались и ранее [2, 3]. Ниже рассматриваются характерные особенности типичной атомной электростанции с двумя реакторами типа ВВЭР-440 на примере III и IV блоков Нововоронежской АЭС и электростанции с реактором типа ВВЭР-1000 на примере V блока этой же АЭС.

ТАБЛИЦА I .

Тип реактора	ВВЭР-440	ВВЭР-1000
Тепловая мощность, МВт	1375	3200
Электрическая мощность, МВт	440	1000
Активная зона:		
высота, м	2,5	3,5
загрузка урана, т	42	66
число кассет	349	151
Рабочая кассета:		
вес урана, кг	120	437
число твэлов	126	317
число полых трубок	I *	14**
диаметр твэла, мм	9,1	9,1
Среднее обогащение свежего топлива, %	3,6	3,3-4,4
Среднее выгорание топлива, МВт-сут/кг урана	28,6	26,5-40,0
Средняя продолжительность работы между перегрузками, час	7000	7000

* Для датчиков внутриреакторного контроля.

** 12–для регулирующих стержней, 1 – для датчика внутриреакторного контроля и 1 – центральная.

Основные данные, существенные с точки зрения гарантий, приведены в табл. I. На рис. 1 представлены схемы движения ядерного материала на этих АЭС: свежие топливные сборки (кассеты) для перегрузки поступают на склад свежего топлива, откуда после кратковременного хранения загружаются группами в контейнер ("чехлы") и перемещаются в бассейн-хранилище реактора, в котором производится перегрузка топлива. Установка кассет в реактор и выгрузка их из реактора производятся с помощью перегрузочной машины, которая также используется для перемещения выдержанных кассет из бассейна-хранилища в транспортный контейнер для отправки с АЭС. Большой вес транспортного контейнера позволяет осуществлять его перемещение по реакторному залу до люка транспортной галереи лишь с помощью мостового крана.

Можно отметить следующие особенности рассматриваемых АЭС, существенные с точки зрения гарантий.

1) Наличие корпуса реактора, исключающего доступ к активной зоне, за исключением периода перегрузки (примерно раз в год), когда в течение нескольких недель при сня-

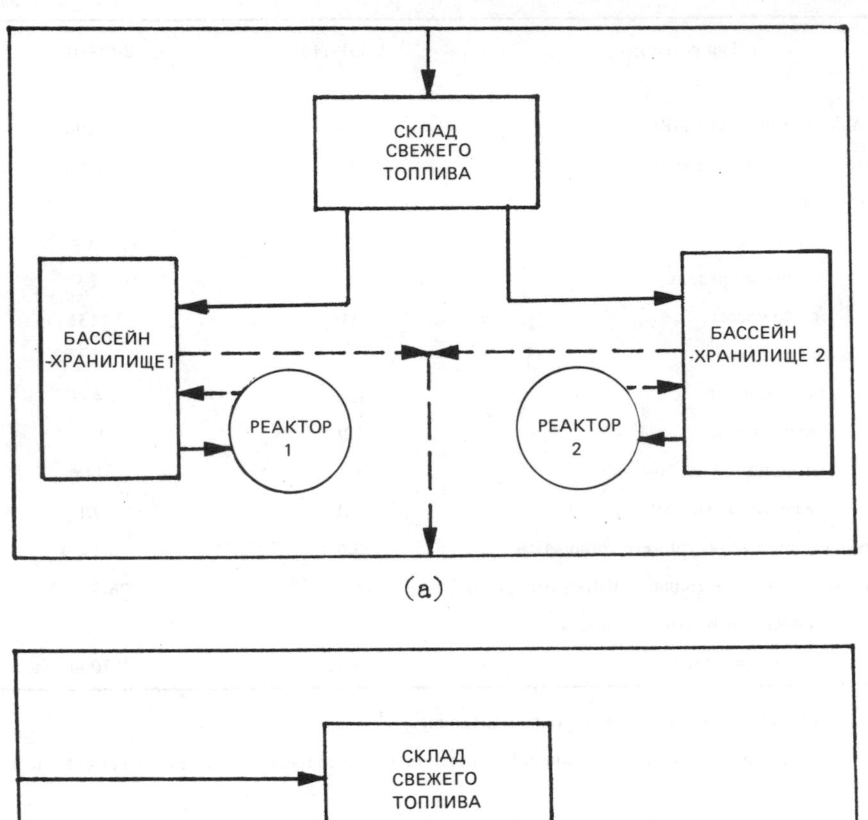

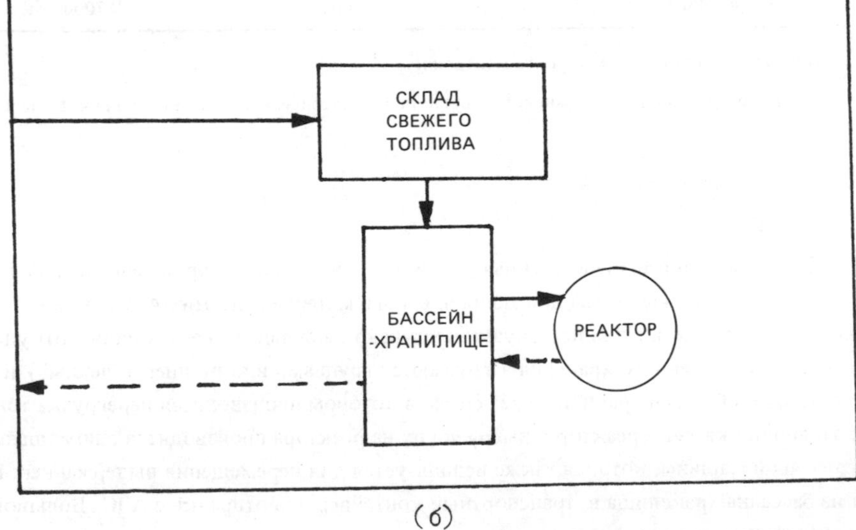

Рис. 1. Схемы движения ядерного материала на АЭС с реакторами типа ВВЭР-440 (а) и ВВЭР-1000 (б).
⟶ свежее топливо, - - → облученное топливо.

той крышке доступ к активной зоне открыт. Наличие корпуса и редкая перегрузка топлива существенно облегчают применение гарантий, так как упрощается контроль за прохождением кассет через АЭС и затрудняется использование реактора для организации незаявленного облучения неучтенных исходных материалов. Доступ к активной зоне при снятой крышке помогает организовать проверку наличия кассет в реакторе во время проведения физической инвентаризации.

2) Весь ядерный материал находится на АЭС в очехлованном виде в крупных кассетах, содержащих, в зависимости от типа реактора, около 120 или 437 кг урана. Каждая кассета имеет идентификационный номер, расположенный сверху на ее головке и различимый даже под слоем воды в бассейне-хранилище и в реакторе, открытом для перегрузки. Это позволяет организовать покассетный учет ядерного материала, облегчить проверку числа кассет во всех местах их возможного пребывания на АЭС, организовать удобный отбор на случайной основе кассет для проведения проверочных измерений, в том числе для всей совокупности кассет в отдельной стране. Следует отметить, однако, что крупные размеры кассет затрудняют проверку содержания в них ядерных материалов, так как из-за эффекта самоэкранировки простейшие пассивные методы измерений позволяют контролировать лишь наружные 1-2 слоя твэлов. Вместе с тем, при условии разработки миниатюрных гамма-спектрометрических детекторов, наличие в кассетах полых каналов может использоваться для измерений твэлов внутренних слоев.

3) Отсутствие необходимости разборки кассет при нормальном использовании топлива на АЭС. Так как замена твэлов (например, с поврежденной оболочкой) в кассетах не производится, существенно упрощается учет ядерных материалов на АЭС и контроль за отсутствием умышленного изъятия отдельных твэлов: хотя операция изъятия и подмены твэлов в кассете технически возможна, оператор установки лишен возможности в случае обнаружения правдоподобно объяснить отсутствие или подмену отдельных твэлов "ошибками" операций по замене твэлов, и вероятность использования такого метода переключения ядерного материала существенно снижается.

4) Сравнительно низкая стратегическая ценность свежего топлива (обогащение ураном-235 не выше 3,6 или 4,4 %, соответственно, для ВВЭР-440 и ВВЭР-1000) и высокая сратегическая ценность облученного топлива, содержащего примерно 8-10 кг плутония на тонну урана (около 1 кг на кассету для ВВЭР-440 и более 4 кг на кассету для ВВЭР-1000), сочетаемая с его высокой радиоактивностью и хрупкостью. Эта особенность позволяет сосредоточить процедуры гарантий в первую очередь на обеспечении эффективного контроля в отношении облученных кассет, причем этот контроль облегчается тем, что их эвакуация требует использования специальных процедур и тяжелых транспортных контейнеров, пути перемещения которых могут быть однозначно определены, что позволяет использовать меры сохранения и наблюдения, основанные на регистрации факта появления в наблюдаемой зоне контейнера, который мог быть использован для транспортировки облученных материалов.

3. СХЕМА ПРИМЕНЕНИЯ ГАРАНТИЙ НА АЭС С РЕАКТОРАМИ ТИПА ВВЭР

Применение гарантий основывается на соблюдении принципа баланса материала, т.е. требует проверки как инвентарных количеств, так и изменений инвентарных количеств в каждой зоне баланса материала (ЗБМ). При определении ЗБМ для целей гарантий на АЭС с реакторами типа ВВЭР следует, в частности, учитывать следующие критерии:

1) обеспечение возможности проведения инвентаризации и сведения баланса материала на одну дату (обязательное условие, вытекающее из самого понятия ЗБМ) и

2) определение по возможности минимального числа ЗБМ (позволяет избежать излишнего увеличения объема отчетной информации из-за перемещений топлива в пределах АЭС).

С учетом этих критериев на рассматриваемых АЭС с реакторами типа ВВЭР могут быть определены одна ЗБМ, и раздельные ключевые точки измерений инвентарного количества для каждого реактора и каждого отдельного хранилища.

Схема применения гарантий, целесообразная для АЭС рассматриваемого типа должна учитывать возможность осуществления переключения ядерных материалов из ЗБМ: во-первых, незаявленного изъятия находящегося под гарантиями реакторного топлива (свежего или облученного) и, во-вторых, облучения в реакторе незарегистрированного исходного материала с последующим удалением его из ЗБМ, аналогично удалению облученного топлива.

В первом случае факт переключения может быть скрыт, например, с помощью завышения количества кассет, отправленных из ЗБМ или имеющихся в наличии в ЗБМ, включая использование имитаторов кассет и пустых "пеналов", где якобы размещены поврежденные кассеты, с целью затруднить обнаружение недостачи ядерного материала. Во втором случае облучение в реакторе значительных количеств незаявленных материалов приводит к понижению достижимой глубины выгорания реакторного топлива, что может быть скрыто, например, путем завышения энерговыработки реактора и отдельных кассет, по сравнению с реально достигнутой. К таким же последствиям привело бы и удаление части ядерного материала из свежих кассет, что придает еще большее значение контролю достигаемого выгорания облученных кассет.

Учитывая изложенные выше факторы, представляется, что следующая схема применения гарантий позволяет эффективно контролировать отсутствие переключений ядерного материала из ЗБМ на АЭС рассматриваемого типа:

1) применение полных мер учета (идентификация и счет числа кассет, выборочная проверка содержания ядерного материала в них) к свежему топливу;
2) применение ограниченных мер учета (идентификация и счет числа кассет) в сочетании с мерами наблюдения и обеспечения сохранности (фото - или теленаблюдением, опечатыванием) в целях проверки отсутствия незаявленных отправлений из ЗБМ облученных материалов;
3) в случае отказа мер обеспечения сохранности и наблюдения (поломка оборудования или трудно интерпретируемые результаты наблюдения) - применение полных мер учета (идентификация и счет числа кассет, выборочная проверка

с помощью неразрушающих методов измерений выгорания) к облученному топливу;

4) проверка соответствия достигаемой глубины выгорания топлива величинам, типичным для данного реактора, и проверка суммарных данных о выгорании кассет с помощью независимого контроля энерговыработки реактора.

Приведенная схема применения гарантий требует проведения измерений кассет, проверки загрузки и опечатывания транспортных контейнеров с облученным топливом, обслуживания приборов наблюдения. Эта схема может быть упрощена, в случае оснащения свежих кассет опечатывающими и идентифицирующими приспособлениями, гарантирующими их целостность и невозможность незаметного удаления отдельных твэлов, что сделает, например, ненужными измерения свежих кассет. Однако меры обеспечения сохранности и наблюдения для целей проверки отсутствия незаявленных отправлений из ЗБМ облученных материалов будут по-прежнему необходимы.

4. ТЕХНИЧЕСКОЕ ОБЕСПЕЧЕНИЕ АЭС, ПОВЫШАЮЩЕЕ ЭФФЕКТИВНОСТЬ ГАРАНТИЙ

Оснащение строящихся АЭС приспособлениями для гарантий может существенно повысить удобство и эффективность контроля. В проектах АЭС может предусматриваться применение следующих мер технического обеспечения гарантий МАГАТЭ:

1) *Приспособление для опечатывания крышки реактора* или подходящих элементов надреакторного оборудования. Опечатывание гарантирует невозможность введения для облучения в активную зону и изъятия из нее топлива и материалов без нарушения печати. Печать располагается в местах, удобных для периодической проверки ее целостности и замены, и защищенных от возможного повреждения при выполнении работ в реакторном зале.

2) *Автоматическое оборудование для фото - или теленаблюдения за облученным топливом*, включая наблюдение за реактором, хранилищем, местами загрузки и путями перемещения транспортного контейнера для облученного топлива. Компановка АЭС должна быть такова, чтобы упростить организацию автоматического наблюдения за облученным топливом *вплоть до его отправки с АЭС.* Применение наблюдения позволяет обнаружить незаявленное изъятие любых облученных материалов и является основным средством обеспечения эффективности гарантий на АЭС данного типа.

3) *Счетчик облученных кассет*, перемещаемых из хранилища для загрузки в транспортный контейнер. Планировка хранилища и места загрузки контейнера должны позволять установку счетчика таким образом, чтобы исключить возможность незарегистрированных перемещений. Счетчик анализирует спектр гамма-излучения кассеты с тем, чтобы надежно отличить ее от возможного имитатора, и устанавливает направление ее движения, аналогично, например, счетчикам, разработанным для реакторов с непрерывной перегрузкой топлива [4] . Сочетание установки счетчика с автоматическим наблюдением за облученным топливом позволяет повысить эффективность и упростить проверку загрузки транспортного контейнера перед отправкой с АЭС.

4) *Приспособление для опечатывания транспортного контейнера,* позволяющее зарегистрировать время опечатывания с тем, чтобы можно было проверить, что контейнер был опечатан *до его удаления из зоны автоматического наблюдения.* Вместе со счетчиком числа кассет оно поможет избежать необходимости проведения инспекции, связанной только с отправкой облученного топлива с АЭС.

5) *Стенд для проведения измерений облученных кассет* в бассейне-хранилище. Компоновка хранилища должна позволять проведение таких измерений на выборочной основе. Стенд позволяет осуществлять использование гамма-спектрометрического прибора с высокой разрешающей способностью для определения выгорания топлива в кассетах с помощью измерения активности осколков деления. Точность определения выгорания должна быть не менее 10 % (относительное стандартное отклонение), при этом неопределенность в определении количества плутония в кассете составит примерно 3%, что представляется достаточным для целей идентификации облученных сборок на АЭС. Основными требованиями к аппаратуре и методике измерений являются возможность независимой проверки их градуировки и отсутствие использования таких параметров, которые нельзя проверить независимо от оператора установки (например, уровни мощности кассеты, характеристики спектра нейтронов и т.д.).

На Нововоронежской АЭС проводились измерения гамма-спектров облученных кассет реактора типа ВВЭР-440 непосредственно под водой в бассейне-хранилище IV блока при помощи специального стенда с использованием германий-литиевого детектора [3]. Усовершенствованная (с коллиматором) схема стенда показана на рис. 2. Стенд представляет собой вертикальную, изолированную от попадания влаги трубу диаметром примерно 150 мм и длиной 6 м со свинцовой защитой и коллиматором длиной около 2 м. Выгорание топлива в кассетах определялось по методике, изложенной в работе [5], с использованием измеренных относительных интенсивностей гамма-линий изотопов цезия-134 и цезия-137. Проведенные исследования показали, что такой стенд может быть удобно использован в бассейне-хранилище легководного реактора и что измерения целых кассет на нем представляют практическую ценность для целей гарантий (относительное стандартное отклонение выгорания кассеты при этом равно 15%), но желательна дальнейшая проверка метода на большом числе кассет. Однако, для таких измерений более удобно использование детекторов, не требующих постоянного охлаждения (например, сверхчистого германия или теллурида кадмия).

6) *Прибор для независимой проверки энерговыработки реактора* за период между перегрузками. Принцип действия прибора, конструкция и способ его установки должны обеспечивать независимую проверку его градуировки и невозможность умышленного искажения его показаний оператором АЭС. Точность измерений энерговыработки реактора должна быть примерно 10 % (относительное стандартное отклонение), что представляется достаточным для целей контроля данных о суммарном выгорании облученных кассет.

Возможность независимой регистрации энерговыработки изучалась на реакторе типа ВВЭР-440 III блока Нововоронежской АЭС при помощи прибора "Казбек" [6], представляющего собой родиевый детектор, соединенный с ртутным капиллярным куло-

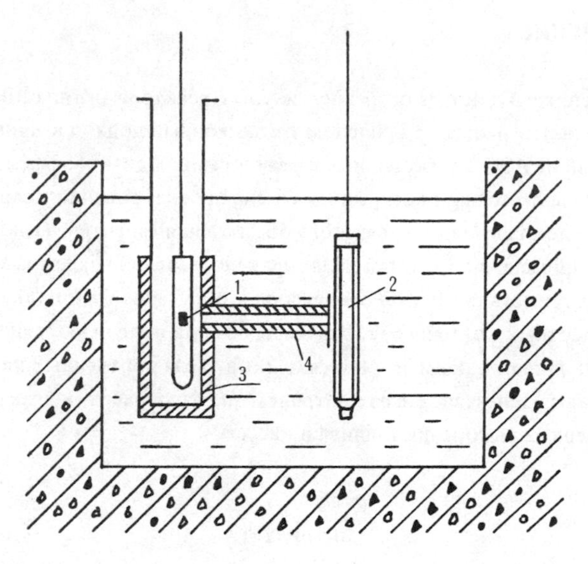

Рис. 2 Схема измерений гамма-спектров облученных кассет в бассейне-хранилище на АЭС с реактором типа ВВЭР: 1 – детектор; 2 – кассета; 3 – свинцовая защита; 4 – коллиматор.

нометром. Детектор устанавливался в экспериментальном ("сухом") канале, проникающем в центральную полую трубку кассеты, находящейся в активной зоне реактора. Компенсирующее устройство, включенное в цепь кулонометра, позволяло исключить влияние гамма-излучения на показания прибора, и последний регистрировал интегральный нейтронный поток в центре кассеты в течение примерно двух месяцев. Ошибка в определении энерговыработки кассеты из-за погрешностей калибровки прибора была оценена экспериментально и составила примерно 8%.

Установка большого количества детекторов для целей гарантий (например, по одному на кассету) позволит увеличить достоверность контроля суммарной энерговыработки реактора, но представляется трудно осуществимой, так как потребует создать систему, в значительной степени дублирующую штатную систему внутриреакторного контроля. Приемлемой может оказаться установка небольшого числа детекторов, например, трех – по одному на каждый из секторов симметрии активной зоны. Однако исследования способов контроля энерговыработки необходимо продолжить.

7) *Стенд для проведения измерений характеристик свежего топлива.* Этот стенд нужен для удобного проведения достаточно эффективных измерений свежих кассет на выборочной основе. Стенд позволяет использовать гамма-спектрометрическую аппаратуру для контроля обогащения урана в условиях низкого фона, а в случае использования активных методов измерений, и для проверки содержания ядерного материала в кассетах. Точность измерения должна обеспечивать обнаружение изъятия из кассеты нескольких твэлов, т.е. быть не менее 5% (относительное стандартное отклонение).

5. ЗАКЛЮЧЕНИЕ

Рассмотренные в докладе особенности АЭС с реакторами типа ВВЭР и схема применения гарантий подтверждают в основном традиционный подход к применению гарантий, осуществляемый на АЭС с легководными реакторами. Однако, эффективность гарантий существенно повышается при внедрении на этапе проектирования и строительства АЭС предлагаемых в докладе мер технического обеспечения гарантий. Основным условием эффективного применения гарантий является использование надежных мер сохранения и наблюдения, осуществляемых по отношению к облученному топливу, включая опечатывание транспортных контейнеров. Эти меры сочетаются с выборочной проверкой свежих кассет, дополняются применением счетчиков числа облученных кассет, подготавливаемых к отправке, и независимым контролем энерговыработки реактора, а также дублируются проверкой выгорания топлива в кассетах.

ЛИТЕРАТУРА

[1] МОРОХОВ, И.Д. и др., "О международных гарантиях нераспространения ядерного оружия", Nuclear Power and its Fuel Cycle (Proc. Int. Conf. Salzburg, 1977) 7, IAEA, Vienna (1977) 345.

[2] СКВОРЦОВ, С.А., МИЛЛЕР, О.А. "О технической возможности применения гарантий на АЭС с реакторами типа ВВЭР", Safeguards Techniques (Proc. Symp. Karlsruhe, 1970) 1, IAEA, Vienna (1970) 339.

[3] СКВОРЦОВ, С.А. и др., "Контроль ядерных материалов на АЭС с реакторами ВВЭР", Safeguarding Nuclear Materials (Proc. Symp. Vienna, 1975) 1, IAEA, Vienna (1976) 597.

[4] WALIGURA, et al., "Safeguarding On-Power-Fuelled Reactors. Instrumentation and Techniques", Nuclear Power and its Fuel Cycle (Proc. Int. Conf. Salzburg, 1977) 7, IAEA, Vienna (1977) 599.

[5] ДЕМИДОВ, А.М., МИЛЛЕР, О.А., "Определение выгорания в твэлах из относительного содержания двух продуктов деления", Труды Симпозиума СЭВ, М., 2, (1968) 340.

[6] КУЛИКОВ, Ю.К. и др., Ат. Энерг. 34 5 (1973) 396.

DISCUSSION

H. MILLER: Could you please describe the neutron source used in your active measurement of fresh fuel bundles?

N.S. BABAEV: Experiments on non-destructive analysis of fresh fuel are planned for the very near future, and the characteristics of the neutron sources to be used have still to be determined. The results reported here were obtained in experiments with irradiated fuel using gamma spectrometry.

THE CAPABILITIES OF PRESENT-DAY SAFEGUARDS TECHNIQUES IN EXISTING NUCLEAR FACILITIES SUCH AS FABRICATION, REPROCESSING AND URANIUM ENRICHMENT

W. BAHM*, R. BERG**, U. BICKING*, W. GOLLY*,
W. RUST*, F. SCHINZER+, R. STRAUSS**,
W. SCHMIDT++, G. TRETTER*, F. VOSS*, D. GUPTA*
* Kernforschungszentrum Karlsruhe GmbH, Karlsruhe
** Gesellschaft zur Wideraufarbeitung von Kernbrennstoffen mbH, Eggenstein-Leopoldshafen
\+ NUKEM GmbH, Hanau
++ URANIT GmbH, Jülich

Federal Republic of Germany

Abstract

THE CAPABILITIES OF PRESENT-DAY SAFEGUARDS TECHNIQUES IN EXISTING NUCLEAR FACILITIES SUCH AS FABRICATION, REPROCESSING AND URANIUM ENRICHMENT.

Three types of nuclear facility (high enriched uranium fabrication plant, small-scale reprocessing plant, enrichment plant with centrifuges) are analysed with regard to the present capabilities of the three basic safeguards measures – material accountancy, containment and surveillance. The possibilities of adapting the existing conditions in these facilities to safeguards requirements are discussed. Overall material balance accuracies for reference campaigns are estimated and compared with the objectives of international safeguards. It is shown that these objectives can be achieved with the existing technologies. Further improvements are considered with a view to increasing the effectiveness of international safeguards and to reducing the burden on plant operation.

INTRODUCTION

The post NPT-IAEA safeguards have started functioning in existing nuclear facilities with currently available measures and techniques. This paper analyses three types of nuclear facility, namely (1) high enriched uranium fabrication, (2) small-scale reprocessing, and (3) uranium enrichment based on centrifuges with a view to assessing the capabilities of the present safeguards techniques used and the possibilities of adapting the existing conditions to safeguards requirements in a facility.

After a short description of the three reference facilities, the basic safeguards measures, relevant to these facilities, are considered. The capabilities of these measures are then analysed with respect to the objectives of international safeguards. It can be seen that, under prevailing conditions in the reference facilities, the objectives of international safeguards can be achieved fairly easily. The paper ends with a discussion of some of the improvements which may be considered for the reference facilities with a view to increasing the effectiveness of IAEA safeguards and to reducing the burden for the facility operation.

1. FUEL FABRICATION FACILITY FOR HIGH ENRICHED URANIUM

The fabrication facility chosen for the present investigation is a Nukem[1] type. In such a facility the main products are uranium/aluminium materials testing reactor (MTR) and uranium/thorium oxide fuels containing high enriched uranium around 90%). The annual throughput for the reference campaign for this facility has been assumed to be around 400 kg enriched uranium. The facility can also manufacture products of low enriched, natural and depleted uranium and has the corresponding storage capacities.

1.1. Process description

Although the Nukem-type reference facility has a wide range of products involving different ^{235}U concentrations, and a number of MBAs may have to be established to take this fact into account, the production line with high enriched uranium (HEU) can be considered to be the most important part from the point of view of IAEA safeguards since it involves large quantities of highly enriched uranium. In this paper the discussions on the capabilities of safeguards measures for a fabrication facility on the Nukem type are concentrated on the HEU part of the facility alone. On the assumption that one safeguards MBA will be established around the HEU part, a simplified flow diagram [1, 2] for this production line with the key measurement points and some other relevant information are presented in Fig.1. Feed for all the enriched uranium production lines is UF_6, which is stored in cylinders in the input storage. In the chemical process area UF_6 is reduced to the intermediate product UF_4 and then converted to uranium metal for the MTR line and U_3O_8 for the THTR pebble production

[1] Nukem GmbH, Hanau.

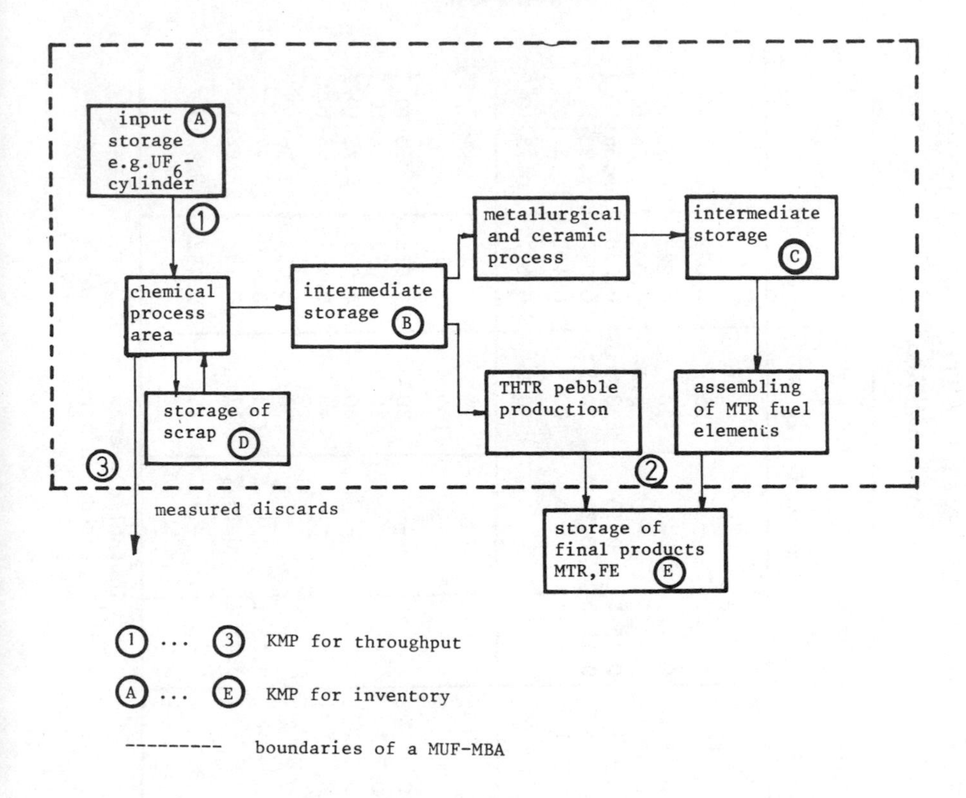

FIG.1. Process steps with Key Measurement Points (KMP) relevant to safeguards for a fabrication facility of the NUKEM type.

line. All the uranium scraps produced in this line are chemically treated and recovered in the chemical process area. The process waste containing uranium leaves the MBA as measured discards.

1.2. Measures relevant to safeguards

The expected measurement errors, both random and systematic, at the key measurement points for throughput and inventory batches are summarized in Table I [1, 3]. On the basis of these measurement errors, the uncertainties of a material balance for a half-yearly reference campaign are established, using the error propagation method as recommended in Part F of the IAEA Safeguards Technical manual [4]. The flow and the inventory amounts for the reference campaign, along with the associated standard deviations (1σ values expressed in kg HEU), are also presented in Table I.

TABLE I. MEASUREMENT UNCERTAINTIES[a] AT KMPs OF A NUKEM-TYPE FACILITY WITH UNCERTAINTIES (1σ IN kg HEU) OF A MATERIAL BALANCE FOR A HALF-YEARLY REFERENCE CAMPAIGN

KMP	weighing		analytical		NDA(γ)[b]		ref.-campaign		
	random	systematic	random	systematic	random	systematic	m[c] (kg HEU)	σ (kg HEU)	σ^2 (kg HEU)^2
throughput									
1 UF_6	0.01 %[d]	-[e]	-	-[f]	-	-	195	0.022	$0.47 \cdot 10^{-3}$
2 UO_2	0.1 %	0.1 %	0.01 %	0.05 %	-	-	20	0.027	$0.70 \cdot 10^{-3}$
metal platelets	0.1 %	0.1 %	0.01 %	0.05 %	-	-	5.1	0.008	$0.06 \cdot 10^{-3}$
ceramic pictures[g]	-	-	-	-	0.15 %	0.15 %	38.5	0.058	$3.36 \cdot 10^{-3}$
U Al pictures[g]	-	-	-	-	0.25 %	0.25 %	30.5	0.077	$5.93 \cdot 10^{-3}$
THTR pebbles	0.1 %	0.1 %	0.1 %	0.1 %	-	-	100	0.102	$10.3 \cdot 10^{-3}$
3 measured discards	0.1 %	0.1 %	≈3 %	≈3 %	-	-	0.9	0.027	$0.73 \cdot 10^{-3}$
inventory									
A UF_6	no measurement		-	-	-	-	100	-	-
B UO_2	0.1 %	0.1 %	0.01 %	0.05 %	-	-	30	0.036	$1.30 \cdot 10^{-3}$
UF_4	0.1 %	0.1 %	0.2 %	0.2 %	-	-	30	0.072	$5.18 \cdot 10^{-3}$
U Al_3	0.1 %	0.1 %	0.2 %	0.1 %	-	-	40	0.065	$4.22 \cdot 10^{-3}$
THTR particles	0.1 %	0.1 %	0.1 %	0.1 %	-	-	40	0.041	$1.68 \cdot 10^{-3}$
C metal	0.1 %	0.1 %	0.01 %	0.05 %			50	0.058	$3.36 \cdot 10^{-3}$
ceramic pictures	-	-	-	-	0.15 %	0.15 %	113	0.170	$28.9 \cdot 10^{-3}$
U Al pictures	-	-	-	-	0.25 %	0.25 %	77	0.193	$37.3 \cdot 10^{-3}$
D scrap	0.1 %	0.1 %	~3 %	~3 %[h]	-	-	15	0.452	$204.3 \cdot 10^{-3}$
E final products	no measurement		-	-	-	-			

a) Ref. |1,3|
b) Relative to standards
c) Ref. |3|
d) Shipper's values
e) Gross-Tare weighing
f) Stoichiometrical composition
g) In fuel elements
h) Dependent on scrap category

σ(MUF) ≈ 0.8 kg HEU = 0.4 % of feed

Most of the items in the input, product and storage areas can be sealed by the IAEA if required. The measurement systems at the different key measurement points (KMPs) are such that the source and batch data can be recorded in a fairly conspicuous manner. The mode of operation permits access to throughput and inventory amounts at any time. The annual throughput permits a continuous presence of the IAEA inspectors at the facility.

1.3. Capabilities of safeguards measures

The material balance for the reference campaign for half a year gives an uncertainty of about 0.8 kg HEU corresponding to about 0.3–0.5% of the input (approx. 200 kg HEU) depending on how the error propagation is calculated. The major part of the uncertainty stems from the scrap amount (approx. 50–60%). Should the amount of high enriched uranium, which would be significant for such a facility, be taken to be 25 kg ^{235}U, the uncertainty in the half-yearly material balance for this facility (approx. 1 kg) may be considered to be small compared with this amount. This would mean that for each material balance period, the diversion of a significant amount can be easily detected in such facilities. Also, the uncertainty in the material balance lies in the range considered by the IAEA as internationally achievable [5]. In view of the unconcealed material flow and inventory management, continuous knowledge on the movement and existence of these materials can also be maintained by IAEA inspectors, who are continuously present in the facility and would place them in a position to cover other diversion strategies (e.g. abrupt diversion of a significant amount of material). In other words, in such a facility the IAEA safeguards can meet the objective of a timely detection of a diversion of significant quantities of materials for a wide range of conceivable diversion strategies.

2. SMALL-SCALE PILOT REPROCESSING FACILITY

The reprocessing facility taken for the present investigation is a pilot plant of the WAK[2] type. It can reprocess irradiated fuel elements with an average burnup of approximately 30 000 MW·d/t. In one case fuel with a burnup of 39 000 MW·d/t has been reprocessed. The nominal throughput is about 35 t of heavy metal per annum corresponding to about 300 kg Pu/a.

2.1. Process steps relevant to safeguards

The WAK uses the classical PUREX process. The relevant process steps are shown schematically in Fig.2. The irradiated fuel elements are brought to the

[2] WAK: Wiederaufarbeitungsanlage Karlsruhe.

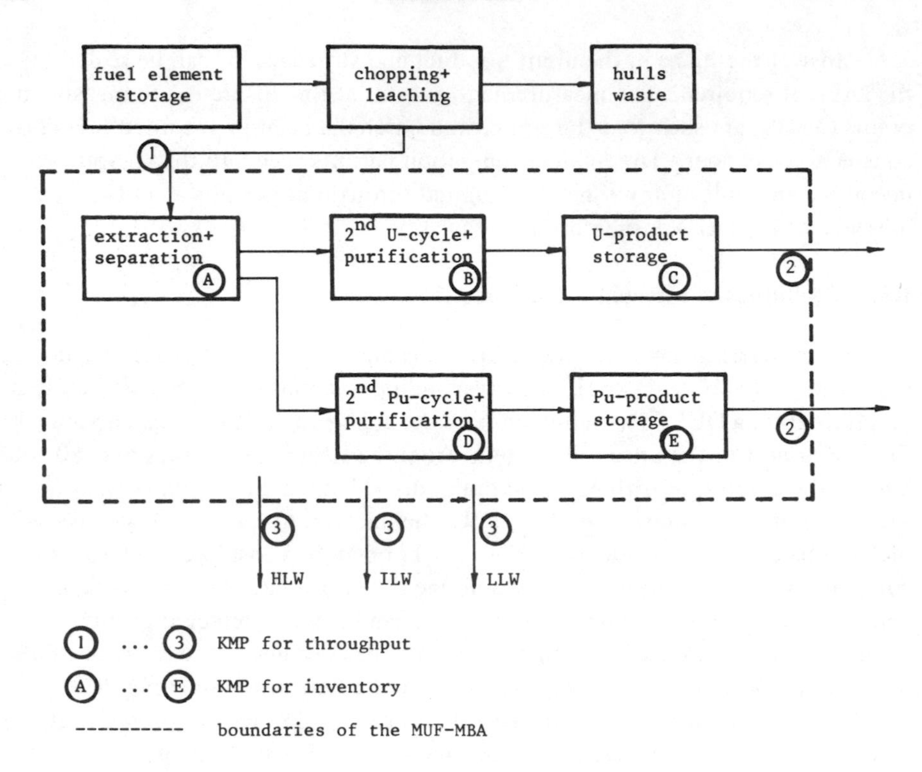

FIG. 2. Process steps with key measurement points for a reprocesssing facility of the WAK type.

facility in shielded casks and are stored temporarily under water in a storage bay prior to processing. The fuel elements are mechanically chopped to about 5-cm-long pieces which are leached with nitric acid. The leached hulls are put into drums which are filled with concrete before final disposal. The leached, highly active acid liquid containing uranium, plutonium and fission products is treated successively with organic solvent in counter-current mixer-settler batteries to remove the fission products and then to separate uranium from plutonium. After a number of final decontamination steps the purified uranium and plutonium are temporarily stored in the product storage areas in the form of uranyl and plutonium nitrate, respectively. Bottled plutonium nitrate products are stored only briefly in the final storage area before shipping. The low and medium active waste solutions containing traces of uranium and plutonium leave the facility after an analysis has been made on the uranium and plutonium amounts. The high-level wastes are stored at the facility and will be specially treated before final disposal.

2.2. Measures relevant to safeguards

In principle a facility of the WAK type could be divided into three MBAs, one with the irradiated fuel storage including the chop leach part of the process, the second mainly involving the process area starting with the accountancy tank and ending before the product storage area, including the laboratories' developing sections etc., and the third including the product storage area with uranium and plutonium only. The main purpose of such a division would be to have the first MBA involving solely the shipper/receiver data, the second only the process data which generate the MUF values, and the third involving an area in which only book values of the stored materials (book inventory) are maintained. The advantage of such a division is that, by separating the MBAs in this manner, the evaluation of data (the three MBA types require different data evaluation procedures) and application of safeguards measures become more simple. For MBAs 1 and 3 the containment, surveillance and digital accountancy measures are mainly applicable. For MBA 2 the material accountancy measures for establishing and verifying material flows, inventories and balances are of primary importance.

A careful analysis of the process flows, layout of equipment and pipelines, and the operational mode in the facility indicates very clearly that all the required sealing, surveillance and other observation measures can be applied adequately at suitable strategic points without any difficulty or modification to the facility. Therefore, any movements unaccounted for or handling of nuclear materials can be detected by the IAEA inspectors by such measures. Actual throughput data for approximately half a year are given in Table II. The measurement uncertainties expected at the KMPs were taken from the literature [6]. Based on these values, the material balance uncertainty was calculated for a reference campaign. For the purpose of physical inventory-taking it was assumed that the process solutions would be brought to various process tanks, the contents of which would be assayed with the accuracies indicated in Table II.

As in the case of a Nukem-type facility, IAEA inspectors can be present continuously at the WAK because of its throughput. All operating data on the flow and inventory of nuclear materials necessary for safeguards can be verified at strategic points by the IAEA inspectors.

2.3. Capability of safeguards measures

Because of the continuous presence of inspectors in the facility, the possibility of applying and verifying various containment and surveillance measures at different strategic points, and the capability of independent verification of the existence of flow and inventory materials, the IAEA inspectors can maintain a continuous knowledge of the existence and amount of these materials.

TABLE II. MEASUREMENT UNCERTAINTIES[a] AT KMPs OF A PILOT REPROCESSING PLANT WITH UNCERTAINTIES (1σ IN kg Pu) OF A MATERIAL BALANCE FOR A HALF-YEARLY REFERENCE CAMPAIGN

KMP	weighing/volume random	weighing/volume systemat.	analytical random	analytical systematic	ref.-campaign m[b)] (kg Pu)	ref.-campaign σ (kg Pu)	ref.-campaign σ^2 $(kg\ Pu)^2$
throughput							
1 feed solution	1 %	1 %	0.7 %	0.7 %	93.2	1.118	1.250
2 Pu product	0.01 %	0.01 %	0.1 %	0.1 %	91.5	0.091	0.008
3 HLW	3 %	3 %	50 %	50 %	0.97	0.489	0.239
ILW	3 %	3 %	50 %	50 %	0.09	0.046	0.002
LLW	3 %	3 %	50 %	50 %	0.003	0.002	0.000
inventory							
A - E	1-3 %	1-3 %	0.1-0.7%	0.1-0.7%	0.64[c)]	0.020	0.000

a) Ref. |6|

b) Jahresbericht 1976 über die Erfahrungen beim Betrieb der WAK (unpublished)

c) Change in the inventory

σ(MUF) = 1.2 kg Pu = 1.3 % of feed

As shown in Table II a half-yearly material balance can be closed with an uncertainty of about 1.3% of the feed corresponding to approximately 1.2 kg Pu. This uncertainty is small compared with the amount of 8 kg Pu considered provisionally by the IAEA as the quantity significant to safeguards. The material balance uncertainty falls within the range of uncertainties assumed by the IAEA to be internationally achievable [5].

According to the normal mode of operation the average residence time and amounts of in-process plutonium are such that any deviation from normal operation which may be considered to be relevant to IAEA safeguards can be established fairly rapidly and easily by the IAEA inspectors with the help of measures mentioned earlier. Because of these inherent characteristics of a small-scale reprocessing plant like WAK, the IAEA safeguards can fulfil its objective, as in the case of the Nukem facility, for a wide range of assumed diversion scenarios.

3. ENRICHMENT FACILITY WITH CENTRIFUGES

Enrichment facilities based on centrifuges are not operated at present on a commercial scale in the Federal Republic of Germany. However, for the

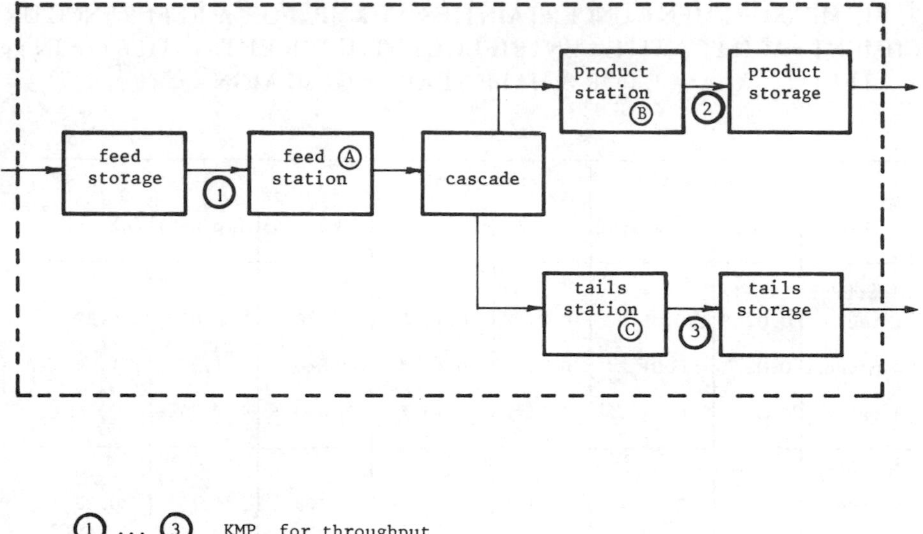

FIG.3. Process steps with key measurement points for an enrichment facility with centrifuges.

present investigations, a centrifuge plant with an annual throughput of about 395 t U-nat and a product rate of about 59 t/a of 3% ^{235}U has been considered in this paper for the sake of completeness. Such a facility has some interest in characteristics which are important from the aspect of IAEA safeguards.

Uranium enrichment is unique among the bulk facilities in a LWR fuel cycle in that the amount of effective kilograms of uranium is increased during the process because of the enrichment. Besides, a facility for the production of low enriched uranium could also be used in principle to produce high enriched uranium which could be relevant to nuclear explosives. For enrichment facilities, in particular with centrifuges, an interesting situation arises because the cascade area for enrichment as well as the information on the centrifuge characteristics, are not accessible to the IAEA or its inspectors.

In view of these factors the IAEA may like to design its safeguards system to cover additional diversion scenarios for a centrifuge facility [7] which would be irrelevant for other bulk facilities. Such scenarios may include (a) high enrichment of the inaccessible cascade material, or material equivalent to the measurement uncertainty of the material balance for safeguarded materials, in a part of the inaccessible centrifuge cascade, and (b) high enrichment of

TABLE III. MEASUREMENT UNCERTAINTIES AT KMPs FOR A REFERENCE URANI ENRICHMENT FACILITY WITH CENTRIFUGES WITH UNCERTAINTIES (1σ IN kg ^{235} OF A MATERIAL BALANCE FOR A HALF-YEARLY CAMPAIGN

KMP	weighing random	weighing systematic	enrichment random	enrichment systematic	ref.-campaign m (kg U-5)	ref.-campaign σ (kg U-5)	ref.-campaign σ^2 (kg U-5)^2
throughput							
1 feed	0.02 %	0.01 %	0.01 %	0.15 %	1370	2.06	4.22
2 product	0.02 %	0.01 %	0.01 %	0.15 %	880	1.32	1.74
3 tails	0.02 %	0.01 %	0.04 %	0.15 %	490	0.74	0.54
inventory							
A feed	0.02 %	0.01 %	0.01 %	0.15 %	<109	<0.16	<0.03
B product	0.02 %	0.01 %	0.01 %	0.15 %	<151	<0.23	<0.05
C tails	0.02 %	0.01 %	0.04 %	0.15 %	< 66	<0.10	<0.01

gas phase inventory in the cascade: $\leq$ 0.4 kg U-5

σ(MUF) = 2.8 kg U-5 = 0.2 % of feed

external material by bringing it inside the cascade and using an inaccessible part of the cascade for this enrichment.

The process characteristics of the 200 t/a SW facility which is analysed below and which is expected to come under IAEA safeguards shortly, indicate clearly that safeguards objectives can be met in such a facility even for the extended diversion scenarios.

3.1. Description of process steps relevant to safeguards

A box diagram indicating the important process steps for the reference centrifuge facility is shown in Fig.3. The whole facility has been assumed to consist of a single MBA and the relevant KMPs for the throughput and the inventory amounts are also shown in Fig.3. Table III summarizes the expected values of measurement uncertainties for the KMPs. Some relevant data on a half-yearly reference campaign with the uncertainties of the material balance are also included in Table III.

Cylinders containing solid UF_6 with natural ^{235}U concentration (0.71%) are received and stored in the feed storage area. Individual cylinders are connected to the feed station when required, heated in autoclaves, and purified of HF and other gaseous impurities before the gaseous UF_6 is introduced into the cascade. The feed station includes autoclaves, buffer containers and UF_6 traps. In the centrifuge cascade the feed is split into two streams, a product stream containing about 3% ^{235}U and a tail stream with about 0.3% ^{235}U concentration. The two streams pass through the respective product or tail stations, where the UF_6 is frozen out and then evaporated in de-sublimers and transferred to storage cylinders. These cylinders, containing solid UF_6 products or tails, are finally stored in the respective storage area.

3.2. Safeguards measures

A centrifuge facility offers an ideal basis for the application of international safeguards. The process gas is purified UF_6 which remains unchanged throughout the process. The feed, product and tail streams are stored in cylinders which can be easily sealed or for which adequate surveillance measures can be applied. The gas phase process inventory is extremely low (about 0.4 kg ^{235}U for the reference facility), whereas the UF_6 inventories at the feed, product and tail stations can be made available at any time to the IAEA inspectors for independent verification. The measurement errors are some of the best obtained under operating conditions in bulk facilities.

3.3. Capabilities of safeguards measures

Table III summarizes the data on uncertainties for a half-yearly material balance. The major contribution comes from the measurement uncertainty (0.15%) of standards for mass spectrometric assay of ^{235}U. The absolute amount corresponding to the 0.2% uncertainty ranges between 2.5–3 kg ^{235}U. For low enriched uranium ($< 20\%$ ^{235}U) the IAEA considers 75 kg ^{235}U to be of safeguards significance with a corresponding detection time of about a year on a provisional basis. The material balance uncertainty is small compared with this amount. If one considers the possibility of high enrichment inside the inaccessible part of the cascade, one would require about 1300 kg uranium with 3% ^{235}U enrichment corresponding to 39 kg ^{235}U to produce 25 kg high enriched uranium with 90% ^{235}U concentration [7]. The uncertainty in the material balance would not be sufficient to be used for this purpose. The process inventories could not be used for this purpose either, since they are verifiable by the IAEA inspectors at any time. The gas inventory is negligibly small. With annual or bi-annual inventories the diversion of 75 kg ^{235}U would therefore be detected with certainty.

The diversion strategy using external material for higher enrichment inside the cascade has often been considered by the IAEA. Besides the fact that such a possibility can be covered fairly easily by simple containment and surveillance measures and other perimeter control systems, such strategies must be associated with very low probabilities of occurrence since a State would have a number of other more attractive and less risky strategies by which to acquire significant quantities of nuclear material.

In view of the above analysis the reference enrichment facility with centrifuges can fulfil the requirements of international safeguards and the objectives of such safeguards can be met fairly easily.

4. POSSIBILITIES OF IMPROVEMENT

Although in all the three reference facilities discussed in this paper international safeguards objectives can be met with present-day technologies available to the plant operators and safeguards organizations, the plant operators and the Government of the Federal Republic of Germany are interested in continuously improving the safeguards capabilities of international safeguards. A number of areas are discussed below in which such an improvement may lead to more effective international safeguards.

4.1. Uncertainty in material balance

The major contribution to material balance uncertainty comes from the systematic error components in different measurement systems in the three facilities. In Nukem it stems from the systematic error component of the analysis for scrap. It is proposed that the conversion process in this facility be changed from an organic to an inorganic based system which permits the discard amounts to be reduced and brings the scrap into a more easily recoverable form. In the WAK the major uncertainty contribution comes from the systematic error components of the volumetric and the plutonium concentration measurement systems (1% and 0.7%, respectively). These two combined correspond to 1.2%, i.e. more than 90% of the total uncertainty of the material balance. An improvement in the overall measurement system at this single KMP could cause a considerable improvement in the overall uncertainty of the material balance in such a facility. For the centrifuge enrichment plant the major uncertainty, as mentioned earlier, arises from the uncertainty in the measurement standards. No further improvement at this point is required.

4.2. Transparency of the information system

The most important contribution towards improvement in the effectiveness of safeguards implementation may be considered to be the availability of a clear, easily verifiable information system. The actual conditions under which it has to be operated differ considerably from facility to facility. For example, in

Nukem the introduction of the first step of such a system to establish and verify the physical inventory led to a significant reduction in the associated effort and errors.

4.3. Plant layout and access to nuclear materials

Since international safeguards have started operating in existing nuclear facilities, a major part of the initial difficulties was associated with the requirement of back-fitting measures. For Nukem a new facility has been designed with particular attention to, among others, the requirement of international safeguards [2]. Important points are (a) streamlining the layout of the plant to provide conspicuous evidence of process flows and inventories; (b) possibilities of sealing complete storage areas to provide simple inventory-taking and verification possibilities; (c) improved conversion process (see section 4.1) etc.

The existing plant layout was not a particular obstacle to safeguards implementation for the WAK, particularly because the actual amounts of plutonium present at any particular time in the process area were of the same order of magnitude as the amount considered significant to safeguards by the IAEA. Besides, the purified plutonium stream remains accessible all the time for verification by IAEA inspectors.

The enrichment facility discussed earlier could be difficult for ensuring proper perimeter control. A provision was therefore made to adjust the inspection frequency in such a manner as to ensure a continuity of knowledge with regard to the flow and inventory of UF_6 in the facility [7]. A newly designed facility would not create any such difficulty.

5. CONCLUDING REMARKS

Three operating facilities (fabrication for high enriched uranium, reprocessing and enrichment with centrifuges) involving sensitive technologies were analysed with regard to the capabilities of present-day techniques relevant to the three safeguards measures of material accountancy, containment and surveillance for international safeguards. Initial experiences indicate that all the three facilities, although basically different from each other, provide adequate bases for the use and application of international safeguards technologies and that in all these cases the existing technologies are sufficient to meet the objective of the timely detection of the diversion of significant quantities of nuclear material. Further improvements are possible and even desirable, mainly to ease the application of safeguards measures or to reduce the undue burden on plant operation.

The most important single factor which makes the capabilities of the present-day safeguards techniques appear adequate or not are the diversion strategies which the IAEA may consider to be covered by these technologies. The capabilities of such technologies may have to be extended if extremely unlikely diversion strategies are used by the IAEA and are to be detected under actual operating conditions in such facilities. It may be necessary for a broad international consensus to be obtained before a set of diversion strategies is taken over as an operating hypothesis for the implementation of IAEA safeguards.

REFERENCES

[1] SCHINZER, F., GUPTA, D., KRAEMER, R., Expected measurement accuracies and limits of error in a fabrication plant for high enriched uranium, KFK 1695 (1973).

[2] BICKING, U., SCHINZER, F., Auslegung und Planung einer Überwachungs- und Sicherheitstechnisch Optimierten Fabrikationsanlage, Abschlußbericht der Firma Nukem, NUKEM-389 (1978).

[3] "JAHRESBERICHT 1975" Projekt Spaltstoffflusskontrolle, KFK 2295 (1976).

[4] INTERNATIONAL ATOMIC ENERGY AGENCY, IAEA Safeguards Technical Manual, Part F "Statistical concepts and techniques" **1**, IAEA-174 (1977).

[5] ROMETSCH, R., LOPEZ-MENCHERO, E., RYZHOV, M.N., HOUGH, C.G., PANITKOV, Yu., "Safeguards – 1975–1986", Safeguarding Nuclear Materials (Proc. Symp. Vienna, 1975) **1**, IAEA, Vienna (1976).

[6] HAGEN, A., NENTWICH, D., KRAEMER, R., GUPTA, D., HAEFELE, W., Development of Safeguards Procedures for a reprocessing plant similar to the WAK type, KFK 1102 (1970).

[7] INTERNATIONAL ATOMIC ENERGY AGENCY, IAEA Working Paper, "Safeguarding uranium enrichment facilities", Review and Analysis of the Status of Safeguards Technology for Uranium Enrichment Facilities, AG-110 (1977).

DISCUSSION

A.M. BIEBER: In your pilot reprocessing plant how do you ensure that no significant quantities of fissionable material remain in the leached hulls? In particular, how can Agency inspectors verify these measurements?

F. VOSS: Perhaps Mr. Gupta would like to answer the questions.

D. GUPTA: A leached hull monitor has been foreseen for determining the content of fissile materials. The calibration data for such a monitor can be verified.

R.J. DIETZ: Your conclusions regarding ability to meet the proposed safeguards criteria are based on studies of facilities with rather limited throughput. Would you care to comment on the effect of increased scale of operation on the ability to meet these criteria? Commercial-sized facilities would be at least one, and perhaps two, orders of magnitude larger in throughput.

D. GUPTA: The main purpose of the present paper was to establish a simple fact of life, namely that Agency safeguards goals can be achieved in existing facilities. It is realized that additional efforts might be necessary with increasing throughputs. Naturally, detailed investigations will be required for the Agency to achieve its goals in large-scale facilities. I should like to refer to the excellent work being carried out in the United States of America and other countries which indicates that these goals can also be achieved in larger facilities by combining different safeguards measures.

W.C. BARTELS: With reference to the enrichment plant, I should like to know how your analysis projects the variation of accountability with increasing plant size. For example, with increasing production of separative work units per year, at about what level of production would MUF values approach quantities of significance in regard to Agency safeguards?

D. GUPTA: In a centrifuge facility of the type considered here, the uncertainty in the material balance is almost a linear function of the throughput. This means that a material balance uncertainty equivalent to the significant quantity of 75 kg ^{235}U would be reached in a facility with a throughput of about 10 000 t of natural uranium per year. In facilities with such large throughputs the material is likely to be handled in a number of parallel enrichment cascades so that such high ranges of uncertainties will not be present in a single cascade.

Session II (Part 2)

ELECTRONIC PROCESSING OF SAFEGUARDS INFORMATION

Chairman: A. BURTSCHER (Austria)

EXPERIENCE AT THE INTERNATIONAL ATOMIC ENERGY AGENCY IN PROCESSING SAFEGUARDS INFORMATION

Y. FERRIS, W. GMELIN, J. NARDI, V. SHMELEV
Department of Safeguards,
International Atomic Energy Agency, Vienna

Abstract

EXPERIENCE AT THE INTERNATIONAL ATOMIC ENERGY AGENCY IN PROCESSING SAFEGUARDS INFORMATION.

The International Atomic Energy Agency (IAEA) has the unique responsibility and authority for obtaining safeguards information under the terms of its agreements. Safeguards information in this paper refers to accountancy data, design information and quantitative inspection data. When discussing a computerized information system for processing safeguards data one must distinguish between a facility, a State, and an international information system. All data fed into the international system and all output from the system are treated with the utmost confidence. The computerized data are encrypted and the printed reports controlled. As the amount of data increased and the user requirements became more complex, the Department of Safeguards realized in 1976 that the current operating system eventually would not be able to meet the demands. A team was established to determine the long-term needs of IAEA Safeguards and to implement a system to satisfy those needs.

1. INTRODUCTION

1.1. The International Atomic Energy Agency (IAEA or Agency) with its headquarters in Vienna, Austria, has the unique responsibility and authority for obtaining information on all nuclear and other materials subject to safeguards under the terms of the general types of agreements. Four basic documents pertinent to the Agency in general and to safeguards information in particular are the Statute [1], the Non–Proliferation Treaty [2], INFCIRC/153 corrected [3] and INFCIRC/66/Rev.2 [4].

1.2. Safeguards information and the treatment thereof is a broad subject. In this paper it is confined to accountancy data as defined in INFCIRC/153 and the specific agreements based on INFCIRC/66/Rev.2, to design information, and to data resulting from the verification activities of the inspectorate.

1.3. The processing and analysis of safeguards information can be explored on a facility level, a State level and an international level. Each information system has different safeguards aims and different methods of achieving them. This paper confines its scope to the international safeguards information system and discusses the other types only for the purpose of contrast with the international system.

1.4. Actual output reports and statistics arising from operating the system are presented. Because the Agency is contractually bound to retain the confidence of all safeguards data submitted to it, all data presented in this paper are coded. Traceability to a particular Member State is not possible. Within the Agency's Department of Safeguards, the data from Member States are treated with the utmost care. From the encrypted data base to the printed output, information is released only to those who have a "need to know".

1.5. Even with a currently operating information system, the Department of Safeguards realized in 1976 that the amount of data was growing so fast that the system eventually would not be able to handle the volume. A team was immediately established to study the problem and to design and implement a safeguards information system which would satisfy the long-term requirements of Agency Safeguards. Its progress to date is discussed below.

2. BACKGROUND INFORMATION

2.1. Under the terms of the various Safeguards Agreements concluded with Member States, the International Atomic Energy Agency receives safeguards information for all nuclear and other materials subject to safeguards. The IAEA's authority to request and obtain such information can be traced back to its Statute [1] Article III.A.5. The Non-Proliferation Treaty (NPT) [2] in Article III.1 reinforces this authority and the model agreement (INFCIRC/153) [3] between the IAEA and Member States, which are a party to the NPT, provides details and definitions relative to the implementation and application of safeguards. For Member States which are not a party to the NPT, there are approximately 40 individual agreements for applying safeguards to nuclear and other material. The Agency's Safeguards system for States reporting under non-NPT agreements is described in INFCIRC/66/Rev.2 [4]. When discussing safeguards information, therefore, the distinction must be made between data submitted in compliance with NPT requirements and all other data.

2.2. The Agency is, *inter alia*, responsible not to disclose safeguards information which may contain proprietary data on inventories and flows of nuclear materials. This obligation is of particular importance for the development and operation of a computerized system. Particular types of data security measures, which have been implemented at the Agency include (1) administrative measures as contained in the safeguards manual, (2) computer hardware measures, i.e. a restricted access remote entry station for exclusive use by safeguards personnel, and (3) software measures such as issuing protection passwords and cyphering computer files.

2.3. Safeguards information can mean many different things. Inspection plans, operating records, accounting reports, design information, man-days of inspection effort etc., all provide information relevant to safeguards. For the purpose of this paper, the expression "safeguards information" will include accountancy data as defined either in INFCIRC/153 or the individual safeguards agreements, design information, and data resulting from the verification activities of the inspectorate.

2.4. Not only the distinction between NPT and non-NPT-type reporting should be kept clear, but also the distinction between a Facility Information System, a State Information System and an International Information System should be emphasized. A Facility Information System is for the purpose of serving the facility. Such a system can be used to control the inventory, manage the nuclear material, analyse Material Unaccounted For (MUF), provide calculations for calibration of measurement systems, provide cost/benefit analysis, balance shipments and receipts, etc. A Facility Information System often controls material only because control results from other specific management decisions considering the content of the information system. Facilities cannot afford the luxury of having a nuclear material control system for the sole purpose of answering the needs of a State regulatory agency. The facility system must pay for itself by providing valuable production, accounting and management information to the facility.

2.5. A State Information System is also concerned with inventory control, measurement systems, material management and MUF, but for the State as a total entity. The State system should follow all shipments until they are received at their proper destination and should be able to identify or locate all nuclear material under its jurisdiction within a very short time ($\leqslant$ 1 day). A State system should analyse shipper/receiver differences and indicate when resolution at the State level is required. The State system should co-ordinate the safeguards information of all its facilities. It is often a regulatory system ensuring that supply and demand of precious nuclear resources are kept in balance among its several facilities. It can be a schedule, determining regional output requirements within the State. Further, the State system has the responsibility to provide proper safeguards information from all its facilities to the international system in a timely and efficient manner, adhering to previously agreed upon parameters as stated in subsidiary arrangements and facility attachments.

2.6. The international system is concerned with inventory changes, material balance and MUF for the entire State and shipments and receipts between States. Activities within a State are of concern only to the extent that they affect the safeguards position of that State regarding inventory and unresolved domestic shipments. An international organization is not a regulatory body nor does it concern itself with efficient or economical use of nuclear resources. It establishes, rather, the requirements for the receipt of timely, complete and consistent Safeguards information in order to be able to report abnormalities, failures and/or diversions of nuclear material to the appropriate authorities as provided for in existing agreements. Materials management, health and safety, production rates, cost analysis and all other sources of information not directly related to safeguards are not included in the Agency's information system.

2.7. The safeguards aim of each of these three general types of system is also different. The facility level is concerned primarily with diversion within the facility by a single individual or by several individuals working in collusion. The State level of safeguards normally is concerned with preventing diversion of nuclear material by a single-purposed group. As its purpose an international safeguards information system must verify that safeguarded material is not diverted to any non-peaceful purpose. INFCIRC/153 asserts strongly that accountancy is a safeguards measure of fundamental importance. The international system, therefore, concerns itself primarily with inventory, changes to inventory, composition of inventory, shipments in transit between and within States, MUF analyses, compatibility of State's data with design information, inspector's verification results etc., to determine whether or not nuclear material has been diverted for use in non-peaceful activities.

3. INFORMATION FLOW

3.1. A system, regardless of whether it is a facility, State or international system, must have data fed to it in order to exist. The international system either receives its data from the Member State or generates them by its safeguards activities. Those data originating in the Member States consist broadly of Accounting Reports and Design Information. Each of these is described differently in INFCIRC/153 and INFCIRC/66/Rev.2.

In INFCIRC/153 the three main accounting reports are

3.1.1. **Inventory Change Reports** showing changes in the inventory of nuclear material

3.1.2. **Material Balance Reports** showing the material balance based on a physical inventory of nuclear material actually present in the material balance area

3.1.3. **Physical Inventory Listing** showing material identification and other batch information for each batch of nuclear material physically present in the facility at a given time.

In INFCIRC/66/Rev.2 reporting is less structured than in INFCIRC/153. Accounting reports must simply show the receipts, transfer out and use of all safeguarded nuclear material.

3.2. Design information is required by the Agency under both INFCIRC/153 and INFCIRC/66/Rev.2. The latter requires a design review with the State submitting sufficient information for the Agency to carry out its reponsibility regarding safeguards. The NPT model agreement, in contrast, specifies that design information in respect of each facility shall include its general character, purpose, nominal capacity, geographical location, the form, location and flow of nuclear material, the general layout of important items of equipment which use, produce or process nuclear material, a description of features of the facility relating to material accountancy, containment and surveillance and many more such detailed requirements. Design information is submitted at the time when the facility first comes under safeguards and is updated as required by operational changes.

3.3. The Agency, too, provides information to the Safeguards System in the form of inspection working papers and analytical results of samples taken at the facilities. The working papers contain both qualitative and quantitative information. The latter is what is currently being processed and consists of comparisons between planned versus actual activities, results of audits, and various types of evaluation.

3.4. Safeguards data, which are generated by the State system of accounting and control of nuclear material, are sent through regular channels for classified mail, passed through the various mail registries and deposited eventually in the Section for Data Processing Operations (DPO) within the Department of Safeguards. Safeguards information, which originates because of inspection activities, passes directly from the inspectorate to DPO. Figure 1 shows the flow of information from the input to the output stage.

3.5. Computer listings are not an end unto themselves but are used for evaluation. There are essentially two levels of evaluation within the Agency [5]. The first is that performed by DPO for the inspectorate and by the inspectors themselves. This evaluation encompasses but is not limited to determining types of formal error and their means of correction, preparing listings for an inspector to take with him to the facility, preparing formal reports for the Member States

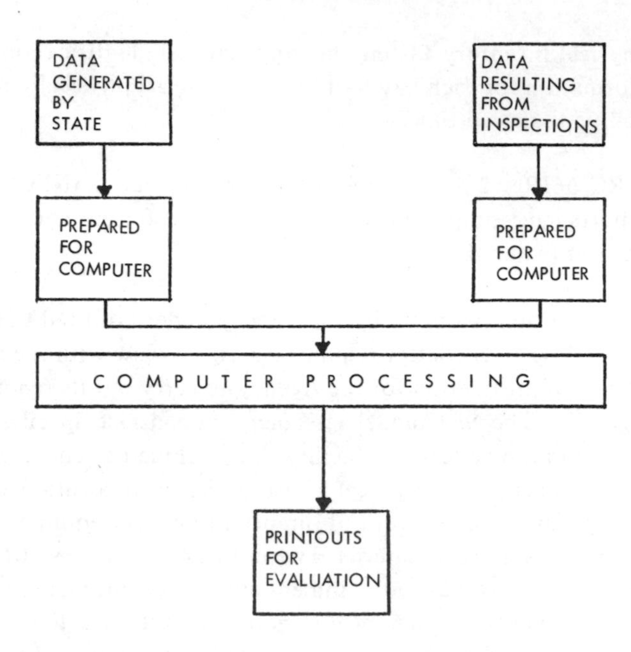

FIG.1. Flow of safeguards information.

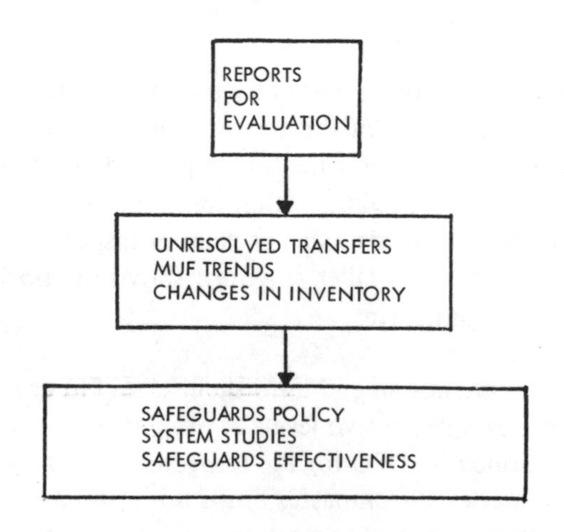

FIG.2. Data evaluation processes.

STATISTICS NON NPT

Type of Information	Number of Reports received by International Atomic Energy Agency			
	1975	1976	1977	1978
ADVANCED NOTIFICATIONS	1	4	12	3
JOINT NOTIFICATIONS	190	125	274	330
ACCOUNTING REPORTS	362	435	1890 *)	1415 **)
LETTERS	31	35	38	37

*) Outstanding - 3 Reports 3 Facilities from 3 Countries as of 1977-12-31

**) Outstanding - 6 Reports 4 Facilities from 3 Countries as of 1978-06-31

FIG.3. Amount of non-NPT information generated by Member States.

which show official inventories, and providing other Sections within the Department of Safeguards with special listings for an individual, one-time type of evaluation.

3.6. The second level of evaluation is done for management purposes and involves determining safeguards policy and its effectiveness, providing input for the safeguards implementation report and producing general safeguards statements. Figure 2 shows the flow of information through the evaluation phases.

4. PRACTICAL EXPERIENCE IN HANDLING SAFEGUARDS DATA

4.1. As stated previously safeguards information received by the Agency is different for non-NPT-type reporting States and for NPT reporting States. For non-NPT-type agreements, the Agency receives from the State system:

4.1.1. Advanced norifications of exports and imports of nuclear material

4.1.2. Joint notification of international shipments of nuclear material between States party to an agreement

4.1.3. Semi-annual or annual, as stated in the agreement, account summary reports of the nuclear material under each agreement showing the book inventory and the summary of the transactions which took place.

4.2. The NPT-type reporting consists of providing the Agency with the reports described in paragraph 3.

STATISTICS NPT

CATEGORY:	NUMBER OF ITEMS IN CATEGORY 1972 through 1977-08-31	1978-08-31
REPORTING		
Countries	31	36
MBA's	326	765
ACCOUNTING		
Reports	9.500	15.258
Records	190.593	364.560
INSPECTION		
Reports	157	389
Records	692	2.072
AVERAGE DELAY OF ACCOUNTING REPORTS IN DAYS	65 *)	60 *)
AVERAGE LENGHT OF PROCESSING CYCLE	16.81 **)	8.8 **)

*) The range of these delay times vary between 17 days and 182 days for 1977 and 15/100 days for 1978.

**) The ranges vary between some 30 days in 1975 and 1 day in 1978 for tape input.

FIG.4. Statistics concerning NPT data base and report processing.

5. STATISTICS

5.1. Accountancy data have been received under non-NPT agreements since 1963. The number of records in the data base reached 24 448 by September 1978 from 18 countries. Figure 3 shows the number and type of non-NPT accountancy reports provided to the Member States from 1975 through 1978.

5.2. Adequate safeguards imply that data are received and processed in a timely manner. This unfortunately is not always the case for non-NPT data.

5.3. The NPT system has been operating since 1974. Originally it provided five types of output report. Today it provides 30 pre-defined reports plus two highly flexible reports which can be designed according to the specific need of the inspector at the moment. Figure 4 shows the growth in the data base resulting from the increase in the number of reporting countries in the MBAs and in the number of accounting reports and records from 1975 to date.

5.4. Until 1977 the bulk of the accounting reports was provided on hard copy which in turn had to be punched on computer cards, an expensive, cumbersone, error-prone process. Since 1977, however, the main part of the information is provided on magnetic tapes. Cost-saving for conversion resulted on the one hand, but on the other, some new problems resulted because no direct human intervention is possible prior to computer processing. Nevertheless, the Safeguards Department encourages Member States to use magnetic tapes as information carrier.

5.5. Figures 5 and 6 show two typical printouts obtained from the computerized system. The first displays a material balance evaluation report and the second the semi-annual statement to the States.

5.6. Inclusion of inspection working papers in the IAEA Safeguards System data base is a rather recent undertaking. The processing of inspection data, therefore, has been implemented to a limited extent only. Figure 4 lists the type and number of inspection data processed to date. The main use of the output has been to compare data contained in the inspection reports with those in the accounting reports. Figure 7 shows an example of such a comparison.

6. SYSTEM DEVELOPMENT ACTIVITIES

6.1. The anticipated growth in volume of safeguards data, the expansion of the scope of data processing to cover a wider variety of types and sources of data, the growing needs for more complex analysis, and the evolution of international awareness of the importance of safeguards has necessitated a continuing review of the long-range needs of the IAEA for processing safeguards information. Thus, while current needs are being met, there is a continuous development effort to ensure the longer-term needs.

6.2. A study was initiated to identify the crucial areas that a safeguards information system would have to contend within the late 1970s and early 1980s. These included:

6.2.1. The volume of accounting data expected on the basis of anticipated growth in all parts of the nuclear fuel cycle

6.2.2. The volume of data which should be expected as a result of IAEA inspection activities

6.2.3. Identification of individual data elements and their characteristics

6.2.4. An estimation of the computer hardware capacity and software characteristics required to process and evaluate the resulting information

PSI2 USER REPORT TYPE 94 S A F E G U A R D S C O N F I D E N T I A L PPPPP SSSSS AAA
DATE 780921 PAGE 2 INFORMATION MAY BE CONTAINED IN THIS REPRINT P P S A A
PPPPP SSSSS --- AAAAA
P S A A
P SSSSS A A

REQUESTED BY: ==> BY WHOM? <==
REQUEST NO: ____
REQUEST DATE: 780921

MATERIAL BALANCE EVALUATION REPORT
MBA: PS-A
REQUESTED FOR DATE: 771025
AVAILABLE INFORMATION: REPORTS- 1-- 6 DATES-770512--780218
-TOLERANCE FOR MATERIAL BALANCE: +-01 STAND.UNITS.
-FOR D,N,P AND T ELEMENT CODE THE BALANCE OF FISSILE ISOTOPES IS NOT CHECKED.
-THE BALANCE OF PURE U-233 IS NOT CHECKED.
-THE BALANCE OF ITEMS IS NOT CHECKED.
-ERROR PROPAGATION FOR MUF MBAS ONLY.
-TYPE OF MBA: BOOK.
-CORRECTIONS ARE INCLUDED. PRINCIPLE APPLIED: VRCP
-DATE CONTROLLED MODE SPECIFIED

A. SUMMARY

ELEMENT CATEGORY	MATERIAL BALANCE FROM	TO	NO.	EVALUATION AND STATUS
URANIUM DEPLETED	770513	771025	2	OPEN DUE MBR VS.PIL/ICR
URANIUM NATURAL	770513	771025	2	OPEN DUE MBR VS.PIL/ICR
URANIUM ENRICHED/UNIFIED	770513	771025	2	OPEN DUE MBR VS.PIL/ICR
PLUTONIUM	770513	771025	2	**** CLOSED ****
THORIUM	770513	771025	2	**** CLOSED ****

B. DETAILS

MAT.BALANCE ENTRY TYPE	ACCOUNTS SHOWS REPORTED IN	ITEMS	URANIUM DEPLETED (KG)	URANIUM NATURAL (KG)	URANIUM ENR/UNIF (G)	URANIUM E/U-FISSIL (G)	URANIUM U-233,ISC=K (G)	PLUTONIUM TOTAL (G)	THORIUM TOTAL (KG)
PB	TOTAL MBR	0	10659	40754	567518	47327	0	197	22
	TOTAL IPIL/MBR	930	10661	641	466438	35430	0	196	21
	DIFF. MBR-IPIL/MBR	-930	-2	40113	101080	11897	0	1	1
SF	TOTAL MBR	0	0	40111	60704	1785	0	0	0
	TOTAL ICR	8298	0	40111	60704	1785	0	0	0
	DIFF. MBR-ICR	-8298	0	0	0	0	0	0	0
SD	TOTAL MBR	0	10659	643	466519	35437	0	197	22
	TOTAL ICR	941	10659	643	466519	35437	0	197	22
	DIFF. MBR-ICR	-941	0	0	0	0	0	0	0
BA	TOTAL MBR	0	0	0	40295	10105	0	0	0
	TOTAL ICR+PB	-8309	2	-40113	-60785	-1792	0	-1	-1
	DIFF. MBR-ICR+PB	8309	-2	40113	101080	11897	0	1	1
PE	TOTAL MBR	0	0	0	40295	10105	0	0	0
	TOTAL PIL	79	0	0	40295.139	10105.220	0	0	0
	DIFF. MBR-PIL	-79	0	0	-0.139	-0.220	0	0	0
MF	TOTAL MBR	0	0	0	0	0	0	0	0
	TOTAL ICR-PIL	-8388	2	-40113	-101080.139	-11897.220	0	-1	-1

FIG.5. Example only of material balance evaluation report.

PSI2 USER REPORT TYPE 181 SAFEGUARDS CONFIDENTIAL
DATE 780921 PAGE 2 INFORMATION MAY BE CONTAINED IN THIS REPRINT

REQUESTED BY: ==> BY WHOM? <==
SEQUENCE NO: ____
REQUEST DATE: 19780921

DATE OF PRODUCTION: 1978.09.21

INVENTORY FOR OF NUCLEAR MATERIAL
SUBJECT TO SAFEGUARDS AS OF 1978.09.21

A. CONSOLIDATED BOOK INVENTORY AS OF 1978.09.21

		ACCOUNTANCY DATA					LIMITS OF ONE SIGMA ERROR	
MBA	FACILITY(IES)	ELEMENT	WEIGHT OF ELEMENT	UNIT KG/G	WEIGHT OF FISSILE ISOTOPE(S) IN G (URANIUM ONLY)	ISOTOPE CODE	WEIGHT OF ELEMENT	WEIGHT OF FISSILE ISOTOPE(S)
PS-A	PSA-	D	2	KG				
		N	-40113	KG				
		E	48327	G	16539	G,J		
		E	0	G	0	K		
		P	-1	G				
		T	-1	KG				
PS-C	PSC-	D	10655	KG				
		N	640	KG				
		E	465755	G	35354	G,J		
		E	0	G	0	K		
		P	198	G				
		T	95	KG				

FIG.6. Example only of inventory report.

PSI2 USER REPORT TYPE 121 S A F E G U A R D S C O N F I D E N T I A L
DATE 780921 PAGE 3 INFORMATION MAY BE CONTAINED IN THIS REPRINT

UPDATING OF BOOK INVENTORY AND COMPARISON WITH THE INSPECTION REPORT

REASON FOR PRODUCTION: R E Q U E S T

COUNTRY :
FACILITY CODE(S):
MBA CODE :
UPDATED BOOK INVENTORY AS OF 19
LATEST REPORT FROM MBA INCLUDED: NO. 3 ,DATE(TO)78.0 .31

REQUESTED BY: ==> BY WHOM? <==
SEQUENCE NO : 0
REQUEST DATE: 1978.09.21

(FAC) (YEAR) (NO)
INSPECTION REPORT:____/ 19__ / ____
DATE OF INSPECTION:__/__/__
PERIOD COVERED BY UPDATING:FROM____/__/__ TO ____/__/__

MATERIAL	ADJUSTED BOOK INVENTORY CALCULATED ON THE BASIS OF ACCOUNTING REPORTS					OPERATOR'S DATA FROM PARA.4.2.A OF INSPECTION REPORT					DIFFERENCES	
	ELEMENT			FISS.ISOTOPE		ELEMENT			FISS.ISOTOPE		ELEMENT	FISS.ISOT.
CATEGORY	CODE	WEIGHT	G/KG	WEIGHT (GRAMS)	CODE	CODE	WEIGHT	G/KG	WEIGHT (GRAMS)	CODE	WEIGHT	WEIGHT
ENRICHED UR.	E		G		J,G	E		G		J,G		
	E		G		K	E		G		K		
PLUTONIUM	P		G			P		G				
NATURAL UR.	N	0	KG			N		KG				
DEPLETED UR.	D	0	KG			D		KG				
THORIUM	T	0	KG			T		KG				

N O T E S :

1.ADJUSTED BOOK INVENTORY VALUES TAKE INTO ACCOUNT:
-SHIPPER/RECEIVER DIFFERENCES AND
-ANY INVENTORY CHANGES WHICH TOOK PLACE ON THE DATE TO WHICH THE BOOK INVENTORY IS UPDATED.

2.IF THE DATE TO WHICH THE BOOK INVENTORY IS UPDATED CORRESPONDS TO THE ENDING DATE OF MATERIAL BALANCE PERIOD,THE ADJUSTED BOOK INVENTORY VALUES CALCULATED ON THE BASIS OF ACCOUNTING REPORTS TAKE INTO ACCOUNT THE ROUNDING ADJUSTMENTS TO INVENTORY CHANGES

FIG.7. All identification and quantities have been removed to preserve confidentiality.

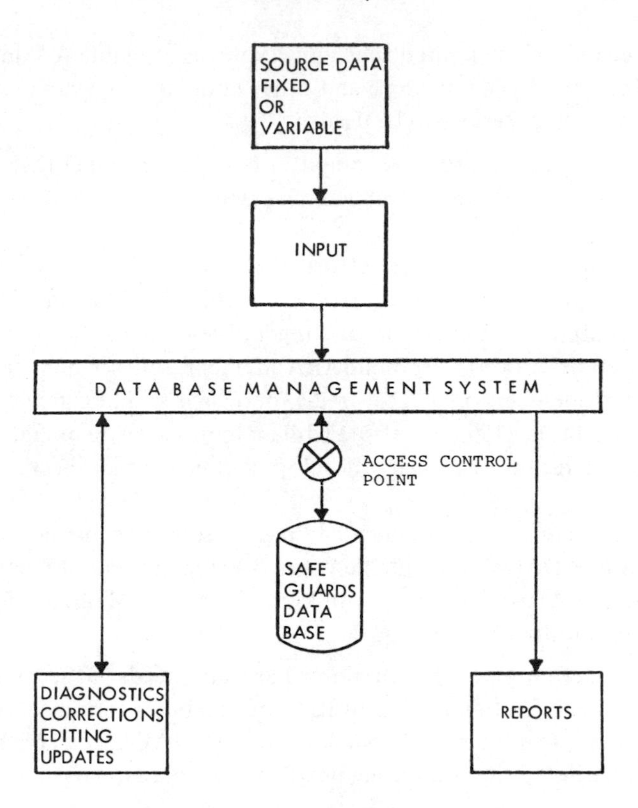

FIG.8. Safeguards information system.

6.2.5. An identification of the information needs of inspectors, of safeguards staff responsible for evaluation of effectiveness of safeguards, and of safeguards management

6.2.6. A consideration of the expected continuing changes in IAEA staff which would be involved in the use, maintenance, and extensions of the information system in the future.

6.3. From an analysis of these characteristics a number of conclusions were drawn which have shaped the direction of the development effort to meet future needs.

6.3.1. The system must be easily maintainable. Staffing policy of the IAEA results in frequent changes in the professional staff. Since long-term staff continuity is unlikely, the system must be such that a new staff member can understand the structure and functions of the system in a relatively short time and can contribute accordingly.

6.3.2. Because of the limited time and resources available for the development effort, reliable commercial and other available software meeting essential system criteria is to be used.

6.3.3. The expansion of IAEA computer hardware to meet the current and future needs of safeguards is and will be essential.

6.4. Responsibility for development efforts rests with the Section of Data Processing Development, one of three sections within the Division of Safeguards Information Treatment. The task of building a development staff has been augmented by contributions of several IAEA Member States which have supplied professional staff members on a "cost-free-expert" basis. They are contributing much of the current development effort while a more permanent staff is acquired. To date results of the early development effort have been:

6.4.1. The evaluation of commercial "data base management systems", selections of ADABAS (Adaptable DAta BAse management System), acquisition of ADABAS as a "gift-in-kind" through a Member State, and its installation on the IAEA computer;

6.4.2. The definition of the computer hardware needs for support of safeguards, specification and acquisition of computer terminals as a "gift-in-kind", and the recent installation by the IAEA of a larger computer based primarily on the computing needs of safeguards;

6.4.3. A conceptual design of the safeguards information system required to accommodate the characteristics previously mentioned (See Fig.8);

6.4.4. The establishment and implementation of the necessary procedures to ensure the security of data;

6.4.5. The initial design and completion of portions of the necessary software to load, validate and maintain a safeguards data base;

6.4.6. The creation and loading of a data base to include data reported under the NPT through 1977, all current data in-house for non-NPT reporting and data reported by a Member State under a special exercise to test newly developed formats and procedures.

6.5. The first phase of reports generation is currently in progress and it is anticipated that parts of the system will be in operation by the end of 1978.

6.6. Initial development efforts have been concentrated in particular in the area of accountancy and design data. Future program development will be extended to include inspector's data, laboratory analysis, and evaluations, equipment, and forecast data.

REFERENCES

[1] INTERNATIONAL ATOMIC ENERGY AGENCY, Statute of the International Atomic Energy Agency as amended up to 1 June 1973, IAEA, Vienna.

[2] INTERNATIONAL ATOMIC ENERGY AGENCY, Treaty on the Non-Proliferation of Nuclear Weapons INFCIRC/140, 22 April 1970, IAEA, Vienna.

[3] INTERNATIONAL ATOMIC ENERGY AGENCY, The Structure and Content of Agreements between the Agency and States required in connection with the Treaty on the Non-Proliferation of Nuclear Weapons, INFCIRC/153 (corrected) June 1972, IAEA, Vienna.

[4] INTERNATIONAL ATOMIC ENERGY AGENCY, The Agency's Safeguards System (1965, as provisionally extended in 1966 and 1968), INFCIRC/66, Rev.2, IAEA, Vienna.

[5] FARRIS, G., GMELIN, W., SHMELEV, V., "The IAEA Safeguards evaluation system", Am. Nucl. Soc. Winter Meeting, San Francisco (1977).

DISCUSSION

D. GUPTA: Do you have any capability for evaluating the information collected in your information system?

Y. FERRIS: Yes, the Agency has such a capability. The responsibility for the *data* evaluation lies with the Data Processing Operations Section and the responsibility for *statistical* evaluation lies with the Data Evaluation Services Section. In addition, a section has been created which evaluates safeguards in general.

D. GUPTA: Can your system measure up to the new objectives proposed for Agency safeguards, for example the fast detection of a sudden diversion. People are talking about diversion times of the order of weeks, whereas the accountancy reports for your information system have an average delay time of about 60 days.

Y. FERRIS: The Agency's means for timely detection of diversion include both inspection activities at facilities and accountancy reports, as required by INFCIRC/66 and INFCIRC/153. The inspection activities include the examination of facility records, the measurement of interim inventories and the use of optical surveillance. Inspection frequency is based on the strategic value of the material under safeguards.

Accountancy supplements the inspection effort by providing a means of evaluating and verifying the reports submitted by States to the Agency. The accountancy reports are useful documents which can later be used for evidence in case of apparent attempts at diversion or deception. The Agency can also use the documented accounting statements to perform further analysis of a less timely but more sensitive nature.

Reporting under INFCIRC/153 requires reports to be dispatched to the Agency as soon as possible, but at least within 30 days after the end of the month in which the reportable event took place. For the special case where a State must first report to a multinational, centralized reporting system, the period of time is increased to 60 days. No State or system, however, is legally bound to take this long in reporting and some States reported within an average of 17 days in 1977. The delay time for reporting to the Agency is constantly decreasing, and the Member States are continuously encouraged by the Agency to report more quickly.

UNIFORM DATA DESCRIPTION FOR A GENERALIZED REAL-TIME NUCLEAR MATERIALS CONTROL SYSTEM

V. JARSCH, S. ONNEN, F.-J. POLSTER, J. WOIT
Kernforschungszentrum Karlsruhe GmbH,
Karlsruhe, Federal Republic of Germany

Abstract

UNIFORM DATA DESCRIPTION FOR A GENERALIZED REAL-TIME NUCLEAR MATERIALS CONTROL SYSTEM.

One of the basic problems in designing nuclear materials control systems is to handle and store the information necessary for plant operation and safeguards purposes. The method presented in this paper describes all kinds of nuclear material having different characteristics with a uniform and efficient data scheme. It is used to build up the data base of the nuclear materials control system for the Karlsruhe Nuclear Research Centre.

1. INTRODUCTION

The R+D programme on safeguards at the Karlsruhe Nuclear Research Centre includes the development of a data-processing system for accounting and the control of the nuclear materials stored and handled in the various laboratories and reactors of the Centre. The great variety of processes and materials, which is typical for a research centre, demands a general concept, which should be applicable not only to a distinct facility, but to all the different institutes and facilities of the research centre, and consequently to other nuclear plants subject to safeguards control.

The most important precondition for a generalized system's concept is to find a way to describe the very inhomogeneous batches of nuclear materials. This paper shows an approach to a uniform and structured data description for nuclear materials.

2. HOW TO ACHIEVE A UNIFORM DATA REPRESENTATION

The abstraction from the objects existing in reality (fuel elements, pellets, liquid waste, etc.) leads to the uniform scheme of the 'logical objects', which contain the information about all characteristic properties of any 'real object' as values of attributes. The set of attributes associated with a logical object contains a complete description of a real-world object.

More precisely, a logical object may be regarded as a 'snapshot', which has been taken from a real-world object at a certain moment, because it describes the object's existence at the very moment or during a time interval where its existence remains unchanged.

Hence, one of the logical object's attributes is the time when 'the snapshot has been taken', Any modification in the existence of the real-world object – be it by transport or by physical changes – is documented by a new snapshot, a new logical object, which differs from the old one in the value of at least one attribute.

This results in a certain number of logical objects generated during the life-cycle of a real object. Containing the dates and type of any alteration they consitute a complete history of the real object, which serves for reporting, inventory, and material follow-up purposes.

3. DATA DESCRIPTION

The logical object is described by a set of attributes. There are two kinds of attribute: first, those that describe the existence of an object in the real world and the operation associated with the object, and second, those that specify the object independently of its physical existence.

With these two kinds of attribute we are able to map the different nuclear materials existing in all types of nuclear facility into the logical object representation.

The first kind of attribute contains information about the identification and physical existence of the nuclear material, e.g. name, location, container, weight, physical form and geometrical dimensions and, in addition, the date, operator, and type of the operation which led to the object's current existence, and the objects immediately preceding and succeeding this object.

The second kind of attribute, called material specification, has no direct regard to the physical existence of a nuclear material. A material specification describes the composition, the relative weights of elements and isotopes, the basis of measurement, and the particular safeguards obligation of a nuclear material.

The fact that the material specification is rather extensive on the one side, and that, on the other side, an object changes its material specification not very frequently and that a large number of real objects may have identical material specifications, leads to the next abstraction step. The attribute 'material specificiation' is separated from the logical object description for reasons of efficiency, and more than one object may share the same material specification.

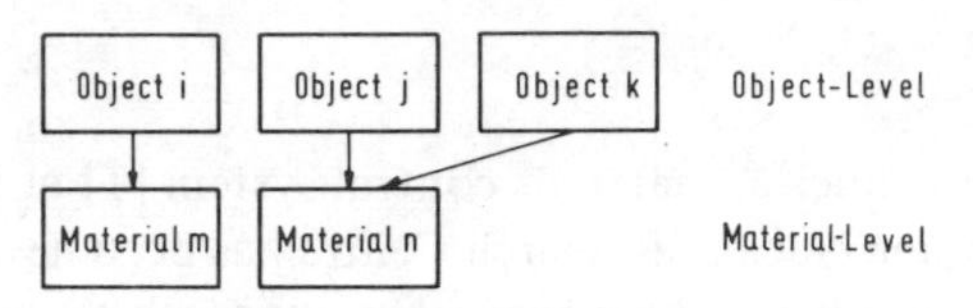

FIG.1. Relations object-material.

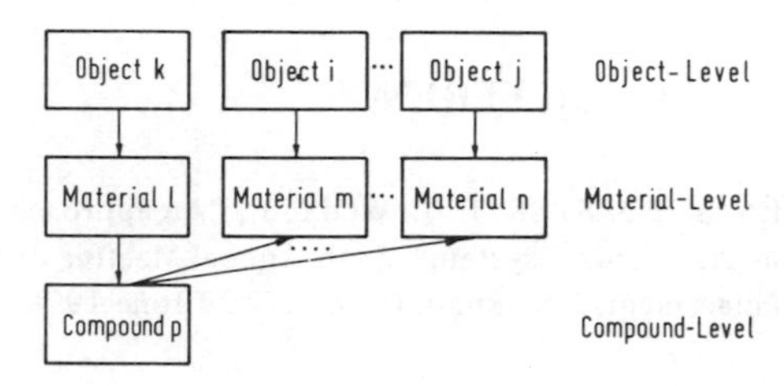

FIG.2. Relations object-material compound.

Each object references a material specification. Object and material specification together represent a certain quantity of nuclear material. The relations between objects and material specifications are shown in Fig.1.

The association of a material with different objects demands that the weights of the elements and isotopes of the material are related to a standard, called a material unit. A material unit may be a fixed quantity, volume or length of nuclear material, or a distinct piece like a plate or a block.

An object is associated with a material specification by giving the name of the material specification and the number of material units the object contains. The weights of elements and isotopes of an object may be calculated by multiplying the weights of the material's elements and isotopes by the number of material units in the object.

If an object consists of more than one distinct material, its material specification contains the sums of the weights of the elements and isotopes of the according materials and a reference to a structure called a compound. A compound contains the names and numbers of material units of the materials the object consists of (Fig.2).

Example: In SNEAK, a fast zero-power plutonium reactor facility in Karlsruhe, a fuel element consists of plates of different plate types. Each plate type references a single material specification and the fuel element references a material specification which refers to a compound associated with the material specifications of the plate types.

4. CONCLUSIONS

The concept of the nuclear materials control system [1] at present being implemented at Karlsruhe Nuclear Research Centre has been developed on the basis of the method presented in this paper. The approach shows that it is possible to describe all safeguards and operator-relevant information of different facilities with a unique and efficient scheme.

REFERENCE

[1] JARSCH, V., ONNEN, S., POLSTER, F.-J., WOIT, J., "An approach to a generalized real-time nuclear materials control system", 19th Annual Meeting of the Institute of Nuclear Materials Management, Cincinnati, Ohio, 27–29 June 1978.

DISCUSSION

J.E. LOVETT: To what extent is your system compatible with a continuous process in which logical objects disappear into the process and totally unrelated logical objects appear at the product end of the process? How would the system work in this case?

V. JARSCH: Any distinct state of a real object can be described by a new logical object. When a process – continuous or discrete – changes the material specification of a real object, the plant operator's booking dialogue contains the definition of the new material specification. The data base is then updated by the application programme, thus representing the new state of the process.

F. SCHINZER: Do you think your system will work at fabrication plants where continuous recycling of small amounts of nuclear material takes place? In some cases you will have to follow the material for years and years.

V. JARSCH: Yes, it will. The problem can be reduced to the standard problem of continuously increasing quantities of data. This is solved using different levels of data storage: long-term mass storage on tapes and short-term storage on discs. By means of special procedures data are transferred regularly from disc to tape, or they are entirely "forgotten" when they are no longer needed.

A. BURTSCHER *(Chairman)*: What are the staffing requirements for a fully implemented system?

V. JARSCH: We shall have one engineer to maintain the equipment, look after the tape archives, and undertake possible future extensions of the system. Data acquisition is performed by the plant operators themselves, who have terminals as near as possible to their work places. Their work load will not increase, because the computer system will replace the manual booking system which they are using at the moment.

AN ACCOUNTANCY SYSTEM FOR NUCLEAR MATERIALS CONTROL IN RESEARCH CENTRES

R. BUTTLER, H. BÜKER, J. VALLEE*
Kernforschungsanlage Jülich GmbH,
Jülich, Federal Republic of Germany

Abstract

AN ACCOUNTANCY SYSTEM FOR NUCLEAR MATERIALS CONTROL IN RESEARCH CENTRES.

The Nuclear Accountancy and Control System (NACS) was developed at KFA Jülich in accordance with the requirements of the Non–Proliferation Treaty. The main features are (1) recording of nuclear material in inventory items. These are combined to form batches wherever suitable; (2) extrapolation of accounting data as a replacement for detailed measurement of inventory items data. Recording and control of nuclear material are carried out on two levels with access to a common data bank. The lower level deals with nuclear materials handling plus internal management while on the upper level there is a central control point which is responsible for nuclear safeguarding within the entire research centre. By keeping the organizational and technical infrastructure it was possible to develop a system which is both economical and operator-oriented. In this system the emphasis of nuclear safeguarding is placed on the acquisition of the nuclear material inventory. As much consideration has been given to the interests of the various operational levels and organizational units as to internal and national regulations. Since it is part of the safeguarding and control system, access to the NACS must be restricted to a limited number of users only. Furthermore, it must include facilities for manual control in the form of records. Authorization for access must correspond with the various tasks of different user groups. All necessary data are acquired decentrally in the organizational units and entered via a terminal. It is available to the user groups on both levels through a central data bank. To meet all requirements, the NACS has been designed as an integrated, computer-assisted information system for the automated processing of extensive and multi-level nuclear materials data. As part of the preventive measures entailed with nuclear safeguarding, the accountancy system enables the operator of a nuclear plant to furnish proof of non-diversion of nuclear material.

1. INTRODUCTION

In the countries of the European Community (EC) nuclear material is subject to the safeguards of the European Atomic Energy Community (EURATOM). In accordance with the Non-Proliferation Treaty those EC countries without nuclear weapons are also subject to supervision by the IAEA - the International Atomic Energy Agency. Once the Verification Agreement between the non-nuclear weapon countries of the EC, EURATOM and the IAEA comes into force (1), operators of nuclear plants will have to install a suitable system for controlling the flow of fissionable material.

---

* UBA-Unternehmensberatung, D-5100 Aachen.

At the request of the Federal Ministry for Research and Technology the Nuclear Safeguards Project Group at the Jülich Nuclear Research Centre (KFA-Jülich) has developed a system for safeguarding nuclear material within research centres (2). This Nuclear Safeguards System (NUSS) is based on the computer-assisted Nuclear Accountancy and Control System (NACS). It has been developed in close cooperation with the internal units affected and with the authorities responsible for safeguarding. Presently, operation of the NACS is being tested in the Hot Cells, a typical section of the KFA.

2. PREREQUISITES AND DEVELOPMENT OBJECTIVES

In 1975 the Nuclear Safeguard Project Group started the first studies of the present situation in the field of nuclear safeguarding (3) and began to develop ideas to meet the new international treaty requirements (4). The NUSS and its most important sub-system, the NACS, were developed from a particular, manually run accounting system (5). This met the requirements of the old EURATOM Regulations 7 and 8 as well as the requirements of national Licencing and Supervisory Authorities. The new system has to satisfy not only the new imperative of international safeguards but also the plant operator's demands for an economic, user-oriented system. In other words, the operator must be able to meet all requirements laid down by the authorities and must have the capability to make in-plant evaluations. In addition, the system has to meet the following secondary conditions:

- uniformity for all operating sections
- the organizational and technical infrastructure of a large research centre must be retained
- information must be generated for the authorities and for the various operating levels and organizational units within the centre
- provision for appropriate data protection and security measures.

One of the characteristics of a large research centre is the broad and constantly changing scope of work. Different research and development tasks are assigned to various divisions which to a large extent operate autonomously. The various KFA divisions are: research institutes, projects, programming groups, common facilities, central departments and other departments. They operate the nuclear facilities, such as reactors, critical systems, laboratories and stores plus an experiment for reprocessing of HTR fuel elements which is currently under construction. Handling of nuclear material is done in the licence areas of these divisions which are physically separate. However, the Board being the licence holder, it has the overall responsibility for all such licence areas. Furthermore, it is typical for a nuclear research centre that the nuclear material in stores and laboratories is distributed in numerous small quantities and spread over numerous and constantly changing locations. Consequently, the emphasis in nuclear safeguarding is not on following the flow of fissionable material but on covering the inventory of nuclear material.

Not only containment and surveillance but also material balancing is of particular interest. The following prerequisites must be met in order to obtain a material balance which can be verified in accordance with international nuclear safeguards (6):

- sectionalization of a nuclear plant into material balance areas (MBA);
- establishment of key measurement points (KMP);
- parcelling of the nuclear material into batches
- inventory taking.

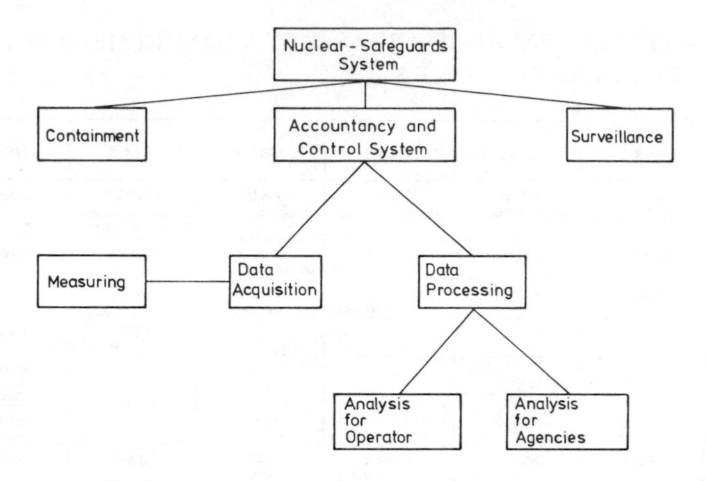

FIG.1. Nuclear safeguards system.

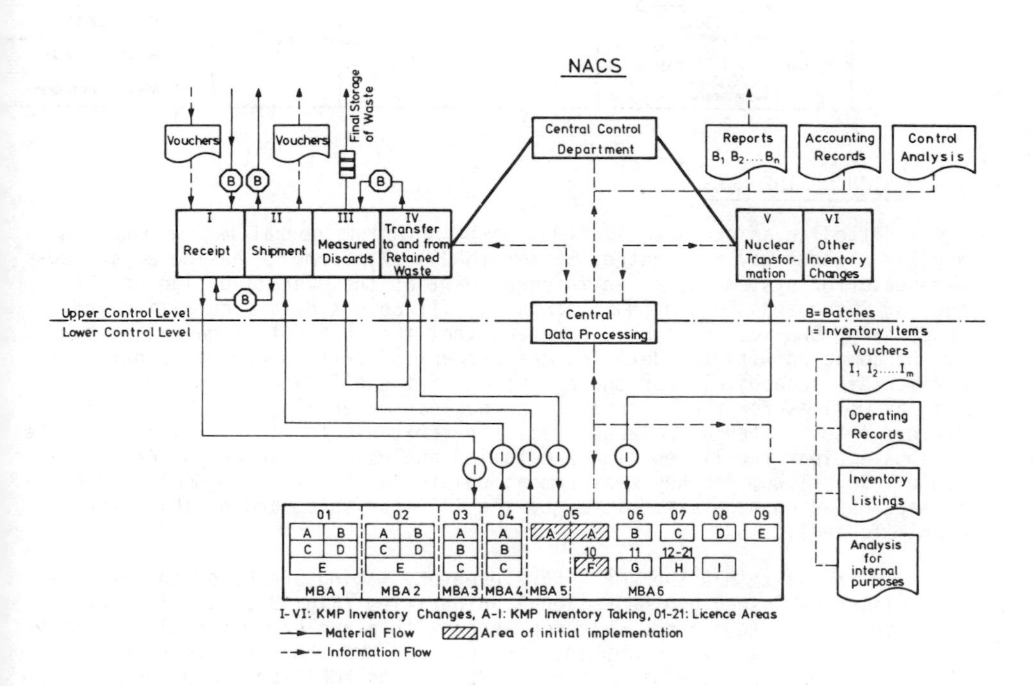

FIG.2. Nuclear accountancy system.

TABLE I. MATERIAL BALANCE AREAS, KEY MEASUREMENT POINTS AND LICENCE AREAS IN THE KFA

Material Balance Area (MBA)	MBA 1	MBA 2	MBA 3	MBA 4	MBA 5	MBA 6
Key Measurement Point (KMP)	LA 01 Research Reactor	LA 02 Research Reactor	LA 03 Critical Assembly	LA 04 Critical Assembly	LA 05.1 Storage Pool and Sorter for Irradiated Fuel	LA 05.2 and 06-21 Research Laboratories
KMP-CH Inventory changes	KMP 1	KMP 1	KMP 1 Receipt of nuclear material	KMP 1	KMP 1	KMP 1
	KMP 2	KMP 2	KMP 2 Shipment of nuclear material	KMP 2	KMP 2	KMP 2
					KMP 3 Measured discards	KMP 3
						KMP 4 Transfer to and from retained Waste
	KMP 5 Nuclear transformation	KMP 5				KMP 5 Nuclear transformation
	KMP *	KMP *	KMP * Other inventory changes	KMP *	KMP *	KMP *
KMP-T Inventory taking	KMP A Fresh fuel storage and storage block inactive part	KMP A	KMP A Fresh fuel storage	KMP A	KMP A LA 05.1	KMP A LA 05.2
						KMP B LA 06
	KMP B Reactor core	KMP B	KMP B Assembly core	KMP B		KMP C LA 07
	KMP C Pool site fuel storage active part	KMP C Storage block active part	KMP C Other locations of nuclear material in MBA	KMP C		KMP D LA 08
						KMP E LA 09
	KMP D Storage pool for irradiated fuel	KMP D				KMP F LA 10
						KMP G LA 11
	KMP E Other locations of nuclear material in MBA	KMP E				KMP H LA 12-21
						KMP I Irradiation

LA — Licence Area

3. DESIGN OF THE NACS

Fitted in an organizational framework which is operator-oriented, the Nuclear Accountancy and Control System (NACS) (7) forms - beside a particular measuring system (8) - an integral part of the overall design of the Nuclear Safeguards System (NUSS) (Fig. 1). It covers data acquisition and processing. One must remember, however, that the NUSS data acquisition cannot be compared with the detailed measurement of data. A central control department is responsible for the operation of the NACS and the measuring system, as well as for the control of the nuclear material (Fig. 2). The KFA is divided into 6 material balance areas. The criteria for this division are the various nuclear facilities plus a detailed analysis of nuclear material inventory and flows. The key measurement points for the first 5 MBA's are established based on models submitted by the IAEA for safeguarding this type of plant (Table I).

In the sixth MBA the KMP's for inventory taking are based on the handling licences issued by the national authorities (Fig. 3). Each licence area corresponds to one KMP. Licence areas with a nuclear material inventory of less than 0.5 kg_{eff} are combined to form one KMP. For inventory changes the KMP's should be consistent for all MBA's. The KMP's are to be established according to plant configuration (Table I). Particular attention is paid to the internal conditions for irradiation of nuclear materials and to the treatment of waste.

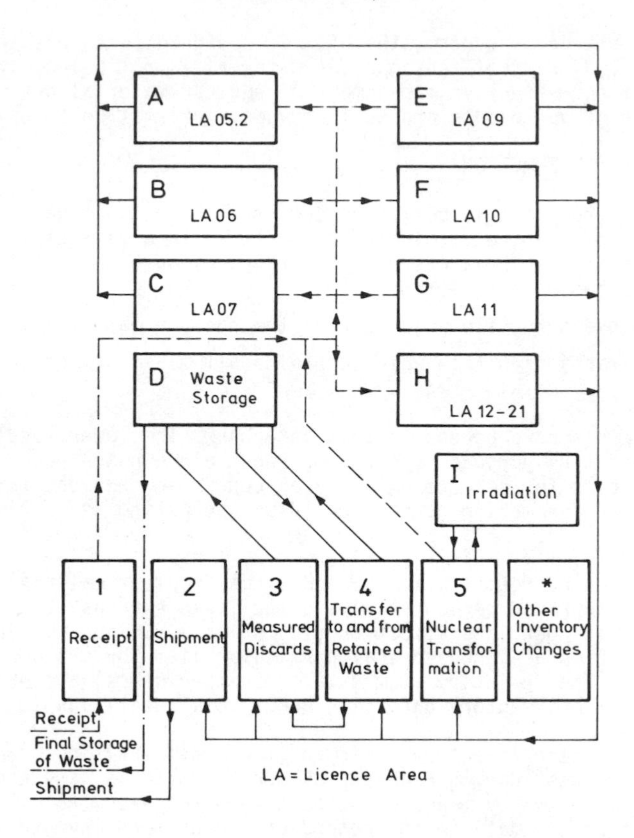

FIG.3. Nuclear material flow between KMPs in MBA 6.

Recording and monitoring of nuclear material is carried out on two levels having access to a common data bank. Internal management of nuclear material in licence areas is handled on the lower level. All lots of nuclear material which can be differentiated and defined as to usefulness on this level are called inventory items. The central control department operates on the upper level; this department combines the inventory items into batches and is responsible for generating accounting records and reports to be sent to national and international authorities (Fig. 2).

The method used for collecting data on nuclear material is based on proven "operational procedures" commonly employed for handling and management of nuclear material. These procedures are, for example, receipts and shipments, division, changes in material and/or inventory item data due to processes or administrative transactions. Because of the large number of inventory items and the limited capacity of existing measurement facilities, updating of existing accounting data must be used. In this fashion it is possible to verify nuclear material items at any time. In contrast, the measurement system can only be used in special cases and even then only for random sampling. In order to meet all requirements, the NACS is designed as an integrated computer-assisted information system for automated processing of extensive and multi-level nuclear material data. As an integral part of the nuclear safeguards system, the NACS is designed to enable a nuclear plant operator to furnish proof that nuclear material is not being diverted. The prerequisites for effective management of nuclear material are met by the possibilities of up-to-date and well arranged information at any time.

4. STRUCTURE AND FUNCTIONS

All required data is collected decentrally in the organizational units and then entered via the terminal. It is stored in a data bank on the central computer where it may be accessed by user groups on both levels. This data bank contains two data files:

- the inventory file which stores the nuclear material inventory
- the transaction file which contains all movements of and changes in the material.

Both levels have access to this data bank. The lower level users have access for updating purposes and for making their own analyses. Upper level users have access for performing analyses concerning nuclear material control as well as for making corrections plus generating accounting records and reports.

Recording and registration of individual nuclear material items is carried out by authorized personnel within the licence areas. Each inventory item is given its own identification (ID) which is unique within the research centre. This ID is assigned to that particular item for the duration of its working life. The inventory item data (characteristics) are entered in a record within the inventory data file under this identification.

Changes in the inventory items and their data arising from "operational procedures" during handling of nuclear material are registered in the accounting system by means of accounting procedures (AP). Each accounting procedure alters the data in the record of one or more inventory items. Likewise, for each accounting procedure a record is generated in the transaction file. This new record lists all alterations which have been carried out. These records are characterized by parameters required for the registration of operational procedures.

By means of the two data sets, both present and previous states of all inventory items may be reproduced. A previous state can be reproduced by means of the accounting transactions filed. The history of an item can be retraced using a sequence of such transformations. Since all data is stored completely and processed quickly, up-to-date information concerning the fissionable fuel inventory or information required for tracking the flow of fissionable fuel may be recalled at any time.

Since for in-plant uniformity other radioactive materials should be included in the accountancy system the following framework of estimated quantities was used as a basis:

- approx. 3o,ooo inventory items in the entire centre, half of which are in the Hot Cells
- 33 - 99 characteristics per inventory item, with a mean of about 7o
- average length of a characteristic: 4 characters
- approx. 3 alterations per inventory item p.a., i.e. 9o,ooo alterations per year
- an average working life per inventory item of 5 years.

5. HARDWARE AND PROGRAMMING LANGUAGE

The organization of the data bank has been developed by the Central Institute of Applied Mathematics of KFA. This institute is also responsible for the programming of the information system. The system runs on the hardware already existing at the KFA, i.e. IBM/37o-168 computer operating under TSS, IBM 2741 console typewriters, DEC-writers, tape- and disk-units.

APL is used as the programming language since it is easy to use and can be easily programmed; furthermore, it is well suited for interactive applications. Additional APL applications are planned in the KFA making possible a later integration into a general data bank system of the KFA.

Outside software was not considered since most of these systems require different hardware. Also, considerable effort must be expended to adapt them to the special requirements of nuclear safeguarding.

6. DATA FILES

Inventory File:

Here, the entire nuclear material inventory is stored as individual inventory items. For each inventory item there is a record fixed length with its own ID as the most important search parameter. Other characteristic fields in this record describe location, type and quantity, state and administrative assignment for each inventory item. A separate retrieval programme realizes direct access to each of these characteristics. The size of the record is 315 bytes and the file is stored on a disk.

Transaction File:

Each accounting procedure is saved in the transaction file. This is for the following reasons:

- to store the history of an inventory item
- to be able to update the inventory file in case of a system crash
- to run analyses of the inventory changes.

The transaction file is structured as follows, in order to achieve a standard record length despite the variety of accounting procedures (each such transaction is characterized by a varying number of specifications). First, a header is generated for each accounting procedure. Apart from data essential to each transaction (accounting procedure-ID, date, inventory change code, date of operational procedure and name of user), it also contains data specific to the transaction (e.g. receiver and shipper data when changing locations). The header has the same size as the record, and it is followed by the appropriate inventory item records already updated. These inventory items are the ones involved in the transaction. It is true that one header and a varying number of inventory item records must be stored for each accounting procedure, depending on the type of transaction, but header and inventory item record are of the same length. By using a certain delimiter between the header and the inventory item record the new "transaction record" can be identified. In addition, by storing the current inventory item record in the transaction file as well, it is possible to reconstruct the inventory file quickly, should this become necessary. The transaction file will probably be stored for 3 months on disk and then on tape. However, the final decision can only be made after completion of the test operation.

KERNFORSCHUNGSANLAGE JUELICH GMBH
NUCLEAR MATERIAL ACCOUNTANCY

INVENTORY ITEM CARD IC

NAME OF USER: BUTTLER DATE: 08.03.78 09.59 H

ID: AE502300 HANDLING NAME: AVR-STCHPR/50/23/0/0/

MBA	KMP-T	INSTITUTION	LICENCE AREA	BUILDING	ROOM	POSITION
6	A	GHZ	H1	01.02	505	

NUCLEAR MATERIALS CATEGORY	MASS (G)	ERROR (PC)	MEASUREMENT- ORIGIN	METHOD	DATE
D-U-DEP					
N-U-NAT					
L-U-LOW					
H-U-HIGH	67.50	10.0	F	MB	141275
P-PU-TOT					
T-TH-232	966.00	10.0	F	MB	141275

NUCLIDE	MASS/ACTIVITY	DIMENS	ERROR (PC)	MEASUREMENT- METHOD	DATE
U-233	10.150	G	10.0	NA	191277
U-235	50.300	G	10.0	NA	191277
SP-PROD	6125.000	CI	50.0	CC	100178

NUMBER OF ITEMS	MAT-DESIGNATION PH	CH	TYPE OF CONTAINER	STATE	QUALITY	DATE OF IRRADIATION
100	S1	NN	0	I	0	191277

MAT-DESCRIPTION	BATCH-REF	OBLIGATION	CONTRACT	USE	OPERATING RECORD
E00I	6AHTS1CH	A	27	CH	H1B1

TOTAL WEIGHT (G)	ENRICHMENT	M/X	ADDITIONAL DESIGNATION	REF	SUB-ITEM
20415.00	89.56 PC	6	THTR-2	B1	

FIG.4. Inventory item card.

The files are supplemented by tables which are used for analysis and search as well as for re-coding.

7. PROGRAMME FUNCTIONS AND EVALUATIONS

On-line updating of the inventory via terminal, by means of accounting procedures (see Chapter 11), occurs daily. If needed, this will be done several times a day or immediately upon receipt of the information. This represents the major programme function. In addition there are various evaluation programmes which are usually run under batch-mode. The upper level requires information on inventory, condition, changes and movements of nuclear material in order to meet reporting and internal control commitments. As per the amended EURATOM Regulations, the following evaluations are required: reports on inventory changes and material balance; inventory listings according to batches and KMP's; accounting records plus special reports. With the exception of the special reports, the system automatically generates the evaluations. This also applies to information on yield, production, purchases, deliveries and inventory. This information has to be submitted to national authorities in accordance with national legislation. If needed, e.g. in case of inspection, calls for particular inventory items, for the mass of a particular material per process location or for sorting programmes for individual characteristics of the inventory item and transaction record can be processed immediately.

The system meets the evaluation requirements of both upper and lower levels. Requirements of the latter are often very application-oriented and thus quite specific to the particular enquiries. The most important evaluations for Hot Cells are: inventory item cards (Fig. 4); accounting records; individual characteristics of particular inventory item groups; inventory lists according to handling location or experiment; criticality tests; batch follow-up and shipping vouchers.

The system is designed such that both levels can, if needed, undertake further direct enquiries using additional APL instructions. With the exception of data acquisition and special operating records the system generates all necessary records.

8. CONTROLS AND DATA PROTECTION

The NACS is equipped with a large number of control functions. Prior to entry data is checked for completeness and format by the terminal operator, drawing on his experience and specialized knowledge. During the interactive accounting session the computer checks the input for plausibility and logical correlation.

Nuclear material accounting data must be protected. To prevent unauthorized access the NACS is operated as a closed system with no direct access possible via APL variables. The data bank is accessed by an initialization procedure which blocks all programme control and interrupt facilities on the user side. It also completes the programme run in case of error. The user identifies himself by an individual password. Access authoritiy is granted according to function and position of the user.

The workspace concept serves to protect data against accidental destruction. Initially, all operations are performed on a copy of the data bank. Only after verification by the user are they entered in the data bank proper. Parallel to this, all operations are recorded in the transaction file and are thus reproducible. Magnetic tape copies of the data bank are produced at regular intervals. In case of a system crash or if the data bank is destroyed, the current status of the data bank can be reproduced using the last copy and the transaction file.

9. INVENTORY ITEMS

Inventory items represent individual, defined parts of the material to be safeguarded. Make-up and quantity of these items are defined by a single sequence of characteristics. An inventory item is the smallest handling unit and the only accounting item. It can be broken down into several lots with the same condition or purpose.

Each existing inventory item is given a definite identification (ID). The ID acts as file reference and identification symbol for all inventory items. It fulfills the following conditions:

- the ID is definite within the entire system
- the ID does not change
- the ID consists of a sequence of 8 characters.

Designation of inventory items occurs only when new inventory items are generated or when batches are accessed externally at the KMP "Receipt". The first accounting procedure assigns each batch one or more ID's in addition to its external batch number. The ID is assigned decentralized, i.e. each licence area uses its own assignment programme. However, the central department assigns certain letter combinations to the first two positions of all ID's in a particular area. This ensures non-ambiguity throughout the system. In addition, a computer check is made as to whether this ID has already been assigned.

An experimenter may wish to call an inventory item by using its handling name. This consists of a character sequence of varying length which yields some information about the type and structure of the inventory item in question. These character sequences are generated in different areas according to various systems. The selection of such a system is left to the user. Unlike the ID the handling name can be changed according to operational requirements.

In order that the handling name can also be used for calling and recording an inventory item, the ID must be correlated with the handling name.

A multiplicity of data (inventory item characteristics) describes the actual condition of an inventory item. This description additionally includes accounting and administrative relationships. All data are recorded for every inventory item (see Fig. 4). An updated version is stored under its ID within an inventory item record in the inventory file.

The data are divided into blocks. Block 7 contains data specific to an area - with the exception of the last characteristic. These can be adapted to the requirements of each sub-area. Code tables required for some characteristics are available to the users.

10. DATA ACQUISITION

Before putting the system on-line the data of all inventory items is entered on special records. For this, the existing book inventory is used as a basis; this is checked by visual controls. The recorded data is punched so as to correspond with the later records, and then a special input programme stores it in the inventory file. When all inventory items have been entered this programme will be deleted and once again access to the system will only be possible via agreed upon accounting procedures. These will map the actual inventory changes into the system.

Various arrangements are available in the KFA for detailed measurement of data associated with approx. 15,000 nuclear material items. Even if these relatively complex measuring facilities are fully utilized - a personnel requirement of 10 man years p.a. - for the purpose of nuclear material safe-

TABLE II. RELATION OF OPERATIONAL PROCEDURES AND ACCOUNTING PROCEDURES

<table>
<tr><th rowspan="2">Operational Procedures</th><th rowspan="2">Accounting Procedures</th><th colspan="2">Obligatory Reporting</th></tr>
<tr><th>EURATOM</th><th>National Authority</th></tr>
<tr><td>Import</td><td rowspan="2">Receive</td><td colspan="2" rowspan="2">yes</td></tr>
<tr><td>Receipt</td></tr>
<tr><td>Receipt inside KFA</td><td rowspan="2">Internal Receipt</td><td>yes</td><td rowspan="2">yes</td></tr>
<tr><td>Receipt inside MBA</td><td>no</td></tr>
<tr><td>Receipt internal</td><td rowspan="2">Internal Transfer</td><td colspan="2" rowspan="2">no</td></tr>
<tr><td>Delivery internal</td></tr>
<tr><td>Delivery inside KFA</td><td rowspan="3">Internal Shipment</td><td>yes</td><td rowspan="3">yes</td></tr>
<tr><td>Delivery inside MBA</td><td>no</td></tr>
<tr><td>Waste delivery</td><td>yes</td></tr>
<tr><td>Export</td><td rowspan="2">Shipment</td><td colspan="2" rowspan="2">yes</td></tr>
<tr><td>Shipment</td></tr>
<tr><td>Separation</td><td rowspan="3">Division</td><td colspan="2" rowspan="3">no</td></tr>
<tr><td>Withdrawal</td></tr>
<tr><td>Other losses</td></tr>
<tr><td>Compilation</td><td rowspan="4">Concentration</td><td rowspan="2">partial</td><td rowspan="4">no</td></tr>
<tr><td>Collection</td></tr>
<tr><td>Other losses</td><td rowspan="2">no</td></tr>
<tr><td>Withdrawal</td></tr>
<tr><td>Nuclear transformation</td><td rowspan="7">Update</td><td colspan="2">yes</td></tr>
<tr><td>Structural change</td><td colspan="2">no</td></tr>
<tr><td>New measurement</td><td colspan="2">partial</td></tr>
<tr><td>Accidental loss</td><td colspan="2">yes</td></tr>
<tr><td>Correction</td><td colspan="2">partial</td></tr>
<tr><td>Adjustment</td><td colspan="2" rowspan="2">no</td></tr>
<tr><td>Other modification</td></tr>
</table>

guarding, only a small part of the inventory items could be measured for isotope vectors and mass of nuclear material. The measuring effort necessary for acquiring and continuously updating the inventory item data would by far exceed the value of the nuclear material.

The operator of a nuclear plant requires an economical method for determining the inventory item data. This can largely be achieved by updating the accounting data. For this purpose, all changes in certain inventory item characteristics must be determined and entered in the computer.

TABLE III. CHARACTERISTICS OF ACCOUNTING PROCEDURES

Characteristics		Inventory Item Card	Accounting Procedures																		Reports to Euratom		
			EZ	IZ	IA	IT	EA	TL				ZL					UD							
								Division				Concentration					Update							
			Entry	Internal Receipt	Internal Shipment	Internal Transfer	Shipment	Simplifiede Division		Detailed Division		Remaining Quantity	Transferred Quantity	Data of Destination Item Known	Simplifiede Concentration	Detailed Concentration	Structural Change	Nuclear Transformation	New Measurement	Correction	Procedures Cancelation	Index of Characteristic in the ICR	Index of Characteristic in the PIL	Index of Characteristic in the MBR
								old	new	old	new			new	new	new								
1.1 I D		×	□	○/□	○/□	○/□	○/□	○/□	□	○/□	[ ]	○/□		○/□	□	□	○/□	○/□	○/□	○/□	○/□	8	6	
1.2 Handling Name		×	⊗	□/●	□/●	□/●	□/●	□/●	⊗	□/●	⊗	□/●		□/●	⊗	⊗	□/●	□/●	□/●	□/●	□/●			
2.1 MBA Code	Shipper		○'	○'	○'		○														⊗'	2/7		
	Receiver		○	○'	○'		○'														⊗'	2/7		
	User	×							▽		▽				▽	▽						2	2	4
2.2 KMP-T	Shipper			○	○		○																	
	Receiver		○	○	○																⊗			
	User	×							▽		▽				▽	▽							4	
2.3 Institution	Shipper		○	○	○		○														⊗*	1		
	Receiver		○	○	○		○														⊗*	1		
	User	×							▽		▽				▽	▽						1	1	1
2.4 License Area	Shipper			○	○		○																	
	Receiver		○'	○'	○'																⊗'			
	User	×							▽		▽				▽	▽								
2.5 Building	Shipper				○	○	○																	
	Receiver		○	○		○															⊗			
	User	×							▽		▽				▽	▽								
2.6 Room	Shipper				○	○	○																	
	Receiver		○	○		○															⊗			
	User	×							▽		▽				▽	▽								
2.7 Position	Shipper				⊗[b]	⊗[b]	⊗[b]																	
	Receiver		⊗[b]	⊗[b]		⊗[b]															⊗			
	User	×							▽		▽				▽	▽								
3.1 Nucl. Materials Category		×	○						▽	□	⊗[b]	□	⊗[b]	□'	□	□		⊗*	○'		⊗*	11	9	3/7
3.2 Mass		×	○					□	□	□	⊗	□	⊗	□	□	□		⊗*	○'		⊗*	12/13	10/11	8/9
3.3 Error		×	○						▽	□	⊗	⊗		⊗	□	○		⊗	○	⊗	⊗			

3.4 Measurement Origin	x	○	□					▽	⊗	⊗			⊗	⊗	○		⊗	○	⊗	⊗			
3.5 Measurement Method	x	⊗b						▽	⊗	⊗	⊗		⊗	⊗	○		⊗	○	⊗	⊗			
3.6 Measurement Date	x	○						▽	⊗	⊗	⊗		⊗	⊗	○		⊗	○	⊗	⊗			
4.1 Nuclide	x	⊗b						▽	□	⊗b	□	⊗b	□	□	□		⊗+	○'		⊗+	14	12	10
4.2 Mass/Activity	x	⊗b					□	□	□	⊗	□	⊗	□	□	□		⊗+	○'		⊗+	15	13	11
4.3 Dimension	x	⊗b						▽	□	⊗		⊗	□	□	□		⊗	○		⊗	16	14	12
4.4 Error	x	⊗b						▽	□	⊗	⊗		⊗	□	⊗b		⊗	○	⊗	⊗			
4.5 Measurement Method	x	⊗b						▽	⊗	⊗	⊗		⊗	⊗	⊗b		⊗	○	⊗	⊗			
4.6 Measurement Date	x	⊗b						▽	⊗	⊗		⊗	⊗	⊗	⊗b		⊗	○	⊗	⊗			
5.1 Number of Items	x	○					○	○	○	○	⊗		⊗	⊗	○	⊗	⊗		⊗	⊗	9	7	
5.2 Material Designation Ph	x	○					⊗	⊗	⊗	⊗	⊗		⊗	▽	○	⊗	⊗		⊗	⊗			
5.3 Material Designation Ch	x	○						▽	⊗	⊗	⊗		⊗	▽	○	⊗	⊗		⊗	⊗			
5.4 Type of Container	x	○	⊗	⊗	⊗	⊗	⊗	⊗	⊗	⊗	⊗		⊗	▽	○	⊗	⊗		⊗	⊗			
5.5 State of Material	x	○						▽		▽			⊗	▽	○	⊗	⊗'		⊗	⊗			
5.6 Quality	x	○						▽	⊗	⊗			⊗	▽	○	⊗⁻	⊗⁻		⊗	⊗			
5.7 Date of Irradiation	x	⊗b						▽		▽			⊗	⊗	○	⊗	○			⊗			
6.1 Material Description	x	□	□	□	□	□	□	□	□	□	□		□	□	□	□	□		□	□	10	8	
6.2 Batch Reference	x	□	□				□	□	□	□	□		□	□	□	□	□		□	□		6	
6.3 Obligation	x	○						▽		▽				▽	▽				⊗'	⊗'	17	15	
6.4 Contract	x	○						▽		▽			□	▽	□				⊗	⊗			
6.5 Use	x	○						▽		▽				▽	▽				⊗'	⊗'	18	16	
6.6 Operating Record	x	⊗b					⊗	⊗b	⊗	⊗'	⊗		⊗	▽	⊗b	⊗	○	○	⊗	⊗	24/25	18	14
7.1 Total Weight	x	⊗b					□	□	□	○	□	⊗b	□	□	□	⊗	⊗	⊗	⊗	⊗			
7.2 Enrichment	x	□					□	□	□	□	□		□'	□'	□'		□	□					
7.3 M/X-Category	x	○	⊗				⊗	⊗	⊗	⊗	⊗		⊗	▽	○	⊗	⊗	⊗	⊗	⊗			
7.4 Additional Designation	x	⊗b	⊗	⊗	⊗		⊗	⊗b	⊗	⊗b	⊗		⊗	▽	⊗b	⊗	⊗	⊗	⊗	⊗			
7.5 Reference	x	⊗b	⊗	⊗			⊗	⊗b	⊗	⊗b	⊗		○	▽	⊗b	⊗	⊗	⊗	⊗	⊗			
7.6 Sub-Item	x							▽											⊗				
8.1 Type of Account Proced		○	○	○	○	○	○		○				○			○	○	○	○	○	6		
8.2 Name of User	x	○	○	○	○	○	○		○				○			○	○	○	○	○			
8.3 Date of Account. Proced		□	□	□	□	□	□		□				□			□	□	□	□	□	3		
8.4 Original Date		○	○	○	○	○	○		○				○			○	○	○	○	□	22		
8.5 Inventory Change Type		□	□	□		□							□				□	□	□	□	6		
8.6 KMP-CH		□	□	□		□											□	□			4		
9.1 External Batch		⊗b																			8		
9.2 Extern. Mat. Description		⊗b																					
9.3 Concise Notes		⊗b																					
9.4 KFA Order No.		○				○																	
9.5 Reception No./Shipment No.		○				○																	
9.6 Receiver's Consent						○																	
9.7 Relation							○						⊗b										
9.8 Type of Update																○	○	○	○	○	6		
9.9 Reason for Correction																			○	○	6		
9.10 Batch Name				□	□	□															8		

○ input data, have to be provided by operator

⊗ input data, may be provided if necessary

□ data, have to be provided from computer storage

▽ data, copied by program

x data, have to be audited with actual status

"Operational procedures" have been defined on the basis of nuclear materials handling. An operational procedure is one which is carried out when the material covered by the mandatory safeguards is handled. Alternatively, it can be a procedure involving the inventory item, e.g. change of location, change in condition, chemical process etc. It also covers procedures which bring about a change in inventory item characteristics although no handling of this material is involved. These latter procedures are more of an administrative nature. For example, they cause a change in obligations, a change of the handling name or reference particulars.

Changes in the nuclear material of the inventory items are calculated from existing accounting data using a ratio. This ratio is determined by estimation or "measurement", i.e. measuring the length of a rod or weighing powder. These primary data and other data relevant to the operational procedure are recorded on data acquisition slips. These are used as the basis for data input. Detailed measurement of nuclear material data is only resorted to if the above data are not at hand or do not appear sufficiently reliable. However, such measurement can be undertaken if measuring equipment involving no great effort is available at the handling point. In addition, for control purposes provision is made for random sampling of nuclear material inventory items, with each sample being subjected to a detailed measuring analysis.

Each operational procedure initializes accounting procedures which record and process them in the accountancy system. The operational procedures and their assignment to the accounting procedures are detailed in Table II.

11. ACCOUNTING PROCEDURES (AP)

The definition of an accounting procedure is as follows: every accounting procedure is the accounting sequel of an operational procedure. It comprises the following steps:

- generating a data acquisition voucher
- entering the data based on the data acquisition voucher and logging the data input
- posting within the computer
- output of an accounting record.

Table II shows the accounting procedures laid down for the system. Each accounting procedure is given a definite AP identification.

With respect to the general course of an AP it should be stated that accounting is necessary as a result of operational procedures. These accounting consequences of the operational procedures may consist of several single steps:

- issuing of new ID's and setting up new inventory item records.
- deleting of ID's and their corresponding records from the inventory file and entering the ID's in the 'Previous ID' table.
- changing the form of individual characteristics of inventory items in the inventory item record.

These steps can occur in the accounting procedure either individually or in combination.

For each AP a data acquisition voucher is generated which records all information required for posting. These details are obtained from the operational procedure itself, from the papers relating to the inventory items in question, and from analysis certificates, experiment descriptions and from other documents. The data acquisition vouchers serve as a basis for the input via terminal. After initializing the system the user must identify himself by means of a password. He then calls the desired accounting procedure and interactively enters the necessary data. During this interactive input session the computer prompts with the characteristics which must be specified. Where possible, data is checked for plausibility and completeness during entry; in case of input error the computer prompts the user to re-enter the data. The user can also check his input himself from the terminal log. After completion of the input the computer processes the accounting procedure and then outputs the accounting record.

Table III lists the AP characteristics specific to the transactions. This is the characteristic matrix for the accounting procedures. For each transaction it tells which specifications are required for processing of the posting, what data the user may specify and what he must detail, and which forms of characteristics the computer will generate based on set algorithms. When the accounting procedures and their characteristics were established special importance was accorded to the fact that the user would have to enter only those characteristics which the computer could not compute itself or which are required for control purposes. For example, the posting of the simplest type of concentration necessitates specification of only 4 characteristics plus the ID of the inventory item concerned.

12. INVENTORY TAKING AND BALANCING

Inventory taking of the physical stock of nuclear material is carried out sequentially at the KMP's of each MBA. Each inventory item may be verified by checking it against the inventory item data given in the inventory lists generated by the accountancy system. These inventory lists show all inventory items of a certain batch and/or all inventory items at a particular handling location of the KMP in question, broken down into batches. For control purposes, inventory items or inventory item samples can be taken at random by an inspector or by the central control department and examined using the existing measuring system. A special computer programme selects the inventory item to be tested based on a statistical and thus objective criterion. KMP's whose inspection has been completed can then return to normal operation.

Within the context of nuclear material balancing the material unaccounted for (MUF) is determined by comparing the physical inventory with the final book inventory .

When using nuclear material, particular in research, processing losses can occur. These losses are not necessarily the result of an accident; for the most part they take the form of waste products, e.g. contamination of a hot cell. For this reason cumulative loss and waste accounts can be established in the accountancy system. The sum of the processing losses and correction differences which accumulates in these accounts during one inventory period corresponds to the sum of the explainable differences between the final book inventory and the physical inventory. Ideally these differences are equal to the MUF assuming no diversion and no unnoticed loss has occurred.

Due to the updating of accounting data and the fact that corrections are carried out immediately after discrepancies have been found, even between taking inventory, the difference between physical inventory and the final book inventory is very small. In practice it is the sum of discrepancies found during random inspection of inventory items. The total MUF may be broken down into two components:

- nuclear material in the cumulative accounts
- discrepancies found during inventory taking.

From the closing balance the operator can furnish proof of non-diversion of nuclear material and he can also obtain information about the operability of his accountancy system.

REFERENCES

[1] Deutscher Bundestag, Verification Agreement, Bill No.7/995 (1973).

[2] BUTTLER, R., et al., "Safeguards system for large research centres with computerized accountancy system", Reactor Meeting, Hanover, April 4–7, 1978.

[3] KOTTE, U., et al., Procedures for the accounting and control of nuclear materials in large research centres as related to the needs of international safeguards, IAEA Research Contract No.1574/RB (1978).

[4] BÜKER, H., et al., "Conceptual design of a system for nuclear material control in a research centre according to the IAEA safeguards requirements", Safeguarding Nuclear Materials (Proc. Symp. Vienna, 1975) **1**, IAEA, Vienna (1976) 137.

[5] BUTTLER, R., KOTTE, U., Description of Manual Nuclear Materials Accounting in a Subsection of the Jülich Nuclear Research Centre within the Scope of Existing Nuclear Materials Control, SFK-500376 (1976).

[6] INTERNATIONAL ATOMIC ENERGY AGENCY, INFCIRC/153, IAEA, Vienna (1971).

[7] BUTTLER, R., et al., An Accountancy and Control System for Research Centres, KFA-Jülich, September 1978.

[8] STEIN, G., et al., Non-destructive Measurement Systems for Nuclear Materials Control in the KFA-Jülich, JÜL-1473, December 1977.

DISCUSSION

V. JARSCH: You said that your system is being operated on a time-sharing basis on your "public" central computer. Surely confidentiality of the highly sensitive safeguards data cannot be guaranteed if the physical storage devices and the computer system are shared with other users not controlled by your system.

H. BÜKER: The system runs as a closed system. Any user has to pass through different kinds of access controls specified by user selection, operation selection and item locations. It is guaranteed that only authorized users have access to the data bank, and every access and transformation of data are recorded in a separate file.

W. FRENZEL: Have you checked whether your accountancy system is compatible with that of the Agency? For example, does an inventory item card contain all the information an Agency inspector needs for doing his job?

H. BÜKER: The accountancy system is compatible with the Euratom system and, so far, also with the Agency system. It is capable of producing all the reports, records and listings needed by Euratom, and by the Euratom and Agency inspectors. An inventory item card contains all the information an Agency inspector needs for his task. Euratom and Agency codes are on all reports, records and listings.

Session III (Part 1)

SAFEGUARDS TECHNOLOGY FOR URANIUM ENRICHMENT FACILITIES

Chairman: D. GUPTA (Federal Republic of Germany)

A URANIUM ENRICHMENT FACILITY SAFEGUARDS TECHNOLOGY BASED ON THE SEPARATION NOZZLE PROCESS

W. BAHM, J. WEPPNER, H.-J. DIDIER*
Kernforschungszentrum Karlsruhe GmbH,
Karlsruhe,
Federal Republic of Germany

Abstract

A URANIUM ENRICHMENT FACILITY SAFEGUARDS TECHNOLOGY BASED ON THE SEPARATION NOZZLE PROCESS.

Within the framework of Federal German-Brazilian co-operation for the peaceful uses of nuclear energy the Federal Republic of Germany is going to export to Brazil a plant for the enrichment (approx. 3%) of ^{235}U. Under the Trilateral Agreement between Brazil, the Federal Republic of Germany and the IAEA this plant, which will operate on the basis of the separation nozzle process, will be safeguarded under INFCIRC/66/Rev.2. This paper describes a first safeguards concept based on a reference plant with an installed separative capacity of 200 t SWU/a, which takes into account the present guidelines of international safeguards. For nuclear materials balancing purposes the plant has been subdivided into 17 key measuring points to assess the nuclear material flow and the nuclear material inventory. Preliminary studies have indicated that the balancing accuracy required for safeguards purposes cannot be achieved by only using the foreseen in-plant measuring systems, since considerable quantities of enriched uranium cannot be covered in this way. This fraction will merely be estimated by the operator and thus cannot be verified by the inspection authorities. The plant components, whose inventories could not be verified in the first estimate of the balancing accuracy referred to above by means of the in-plant measuring systems, also include the low-temperature separators of the cascade shoulder and the product. Assessing and verifying the inventories of these key measuring points is particularly important because of the enrichment (some 3% ^{235}U for the product) and the relatively large inventory and, hence, the considerable contribution to the balancing inaccuracy. This report describes the methods elaborated for inventory-taking and verification at this key measuring point. An estimate of the balancing inaccuracy on the basis of the measuring uncertainties to be expected in the light of the present status of technology indicated values between 0.2 and 0.3% relative to the feed flow with semi-annual inventory-taking. However, this is based on the condition that the experiments planned to determine the inventories of cryogenic separators confirm the measuring uncertainties underlying the calculation.

* INTERATOM GmbH, Bensberg, Federal Republic of Germany.

1. INTRODUCTION

Within the framework of Federal German-Brazilian cooperation for the peaceful uses of nuclear energy the Federal Republic of Germany will export to Brazil a plant for the enrichment (approx. 3 %) of U-235.
This plant will work by the separation nozzle process developed at the Karlsruhe Nuclear Research Center and tested on a semi-technical scale together with industry /1/. Under the Trilateral Agreement between Brazil, the Federal Republic of Germany and the IAEA the plant will be safeguarded on the basis of INFCIRC/66/Rev. 2.

This paper describes a first safeguards concept of such a plant on the basis of a reference plant with an installed separative capacity of about 200 t SWU/a, which concept is in accordance with the present guidelines of international nuclear safeguards. Early cooperation with the architect-engineers of the plant made it possible to design the materials management system in such a way as to take into account both the requirements of plant management and those of nuclear material safeguards.

2. DESIGN CHARACTERISTICS, PROCESS AND PLANT DESCRIPTION

The reference plant is based on the following design characteristics:

Separative capacity:	200 t SWU/a	
Feed:	500 t U/a;	0.72 % U-235
Product:	60 t U/a;	3 % U-235
Tails:	440 t U/a;	0.37 % U-235

The natural uranium delivered as solid UF_6 in containers is flanged onto the feed line in autoclaves and then heated.
The amount of feed required is taken from the vapor phase generated above the liquefied UF_6 and then supplied to the cascade.

In the separation stages of the cascade the process gas (a mixture of 4.2 mole % UF_6 and 95.8 mole % H_2) is subdivided into two streams with different U-235 and H_2 concentrations. The principle by which the different separation stages are interconnected so as to constitute a cascade allows a U-235 concentration gradient to be built up in the cascade, with the maximum U-235 concentration reached at the top of the cascade (product removal), the minimum U-235 concentration at the bottom of the cascade (tails removal).
At the top of the cascade UF_6 is separated from the hydrogen carrier gas. This is done by means of a special preseparation stage for crude separation (gas dynamic separation in separation nozzles) and by down-stream cryogenic separation (fine separation) /2/.
In the UF_6 discharge stage, the pure UF_6 gas coming from the cryogenic separation stages is liquefied by UF_6 compressors and fed to liquid buffers. From one of the liquid buffers UF_6 is fed back to the head end of the cascade to establish the theoretical concentration needed at that point. The net surplus UF_6 is kept and flows into the product containers.
The tails removal is organized accordingly (Fig. 1 + 2).

In a simplified site plan of the enrichment plant of approx. 260 x 480 m^2 the dimension and the physical arrangement of the major plant areas can be seen (Fig. 3).

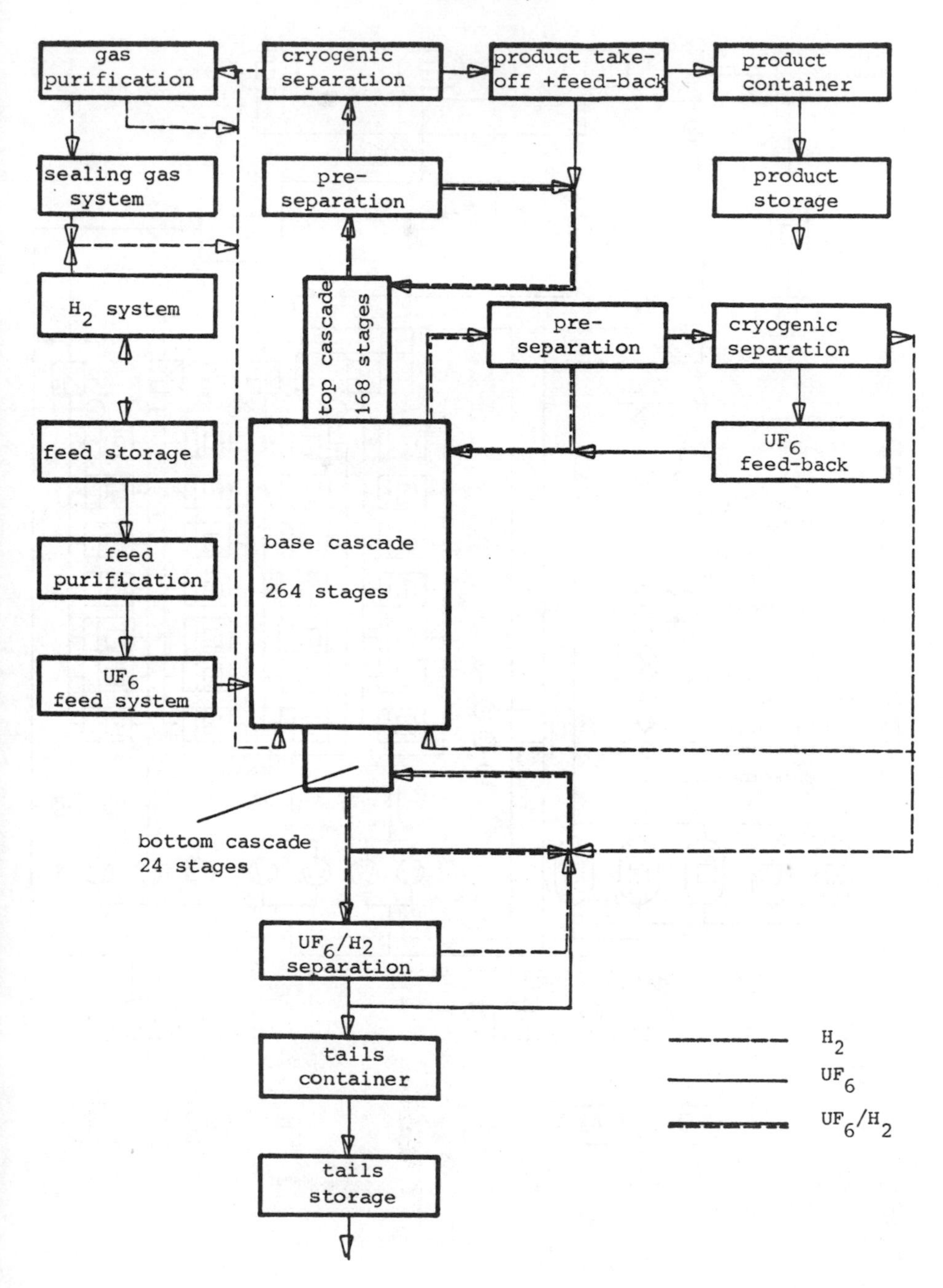

FIG.1. UF_6 flow sheet.

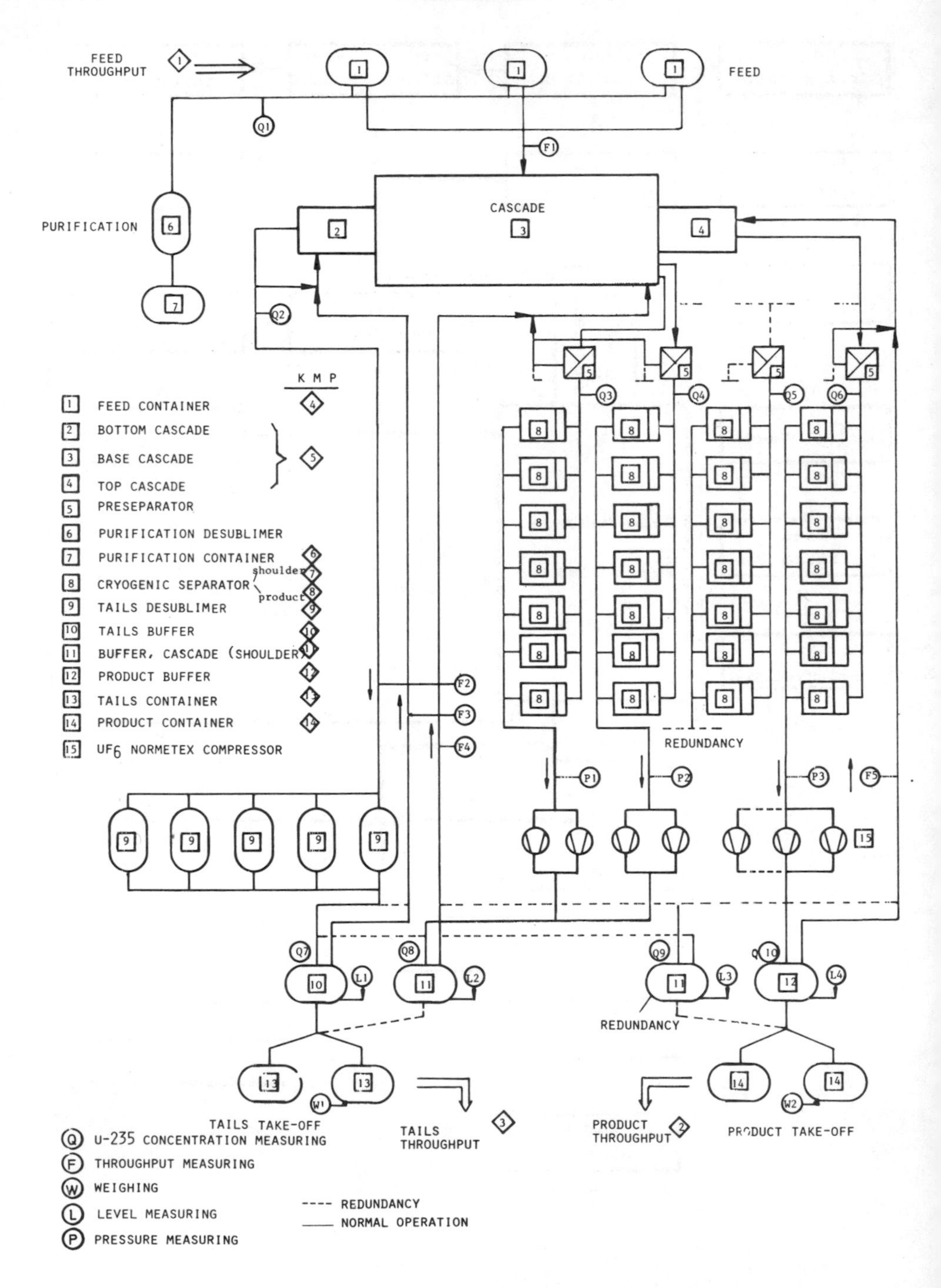

FIG.2. UF_6 flow sheet. Demonstration facility based on separation nozzle process.

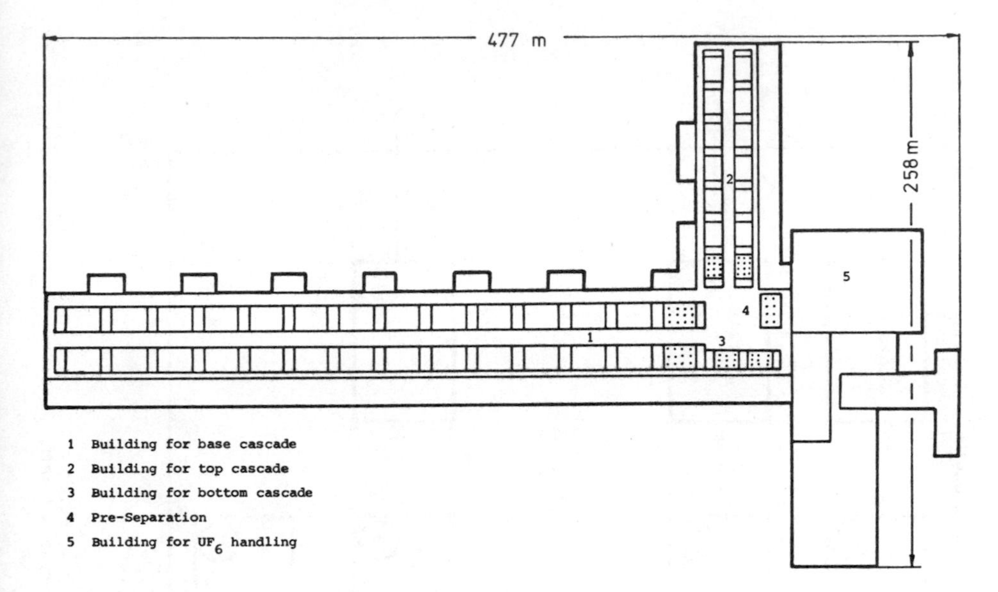

FIG.3. Site plan.

3. SAFEGUARDS INFRASTRUCTURE

To allow suitable international safeguards measures to be executed, a proper infrastructure must be developed and made available. The main components of such safeguards infrastructure are the Design Information and a report and record system. The safeguards concept outlined in this report includes both the development of a model safeguards infrastructure, the performance of experiments on implementing and ensuring this concept, and tests of elements of the infrastructure developed /3/.

Preparing the Design Information the relevant data were compiled for the reference facility introduced above, namely

- design data
- flow and inventory data
- methods and accuracies of measurement of in-plant measuring systems.

Furthermore, the data relevant to records and reports (source data, material balance data, loss and MUF data etc.) which are the basis of nuclear material accountancy were compiled. Besides, possibilities of using surveillance and containment measures had been analysed especially with respect to the timely detection of a diversion.

For assessment of the nuclear material inventory the plant has been subdivided into so-called inventory key measurement points (IKMP) under the following aspects:

- The nuclear material must be present in a form which allows it to be measured for inventory determination.
- Maximum accuracy of information for the individual areas.

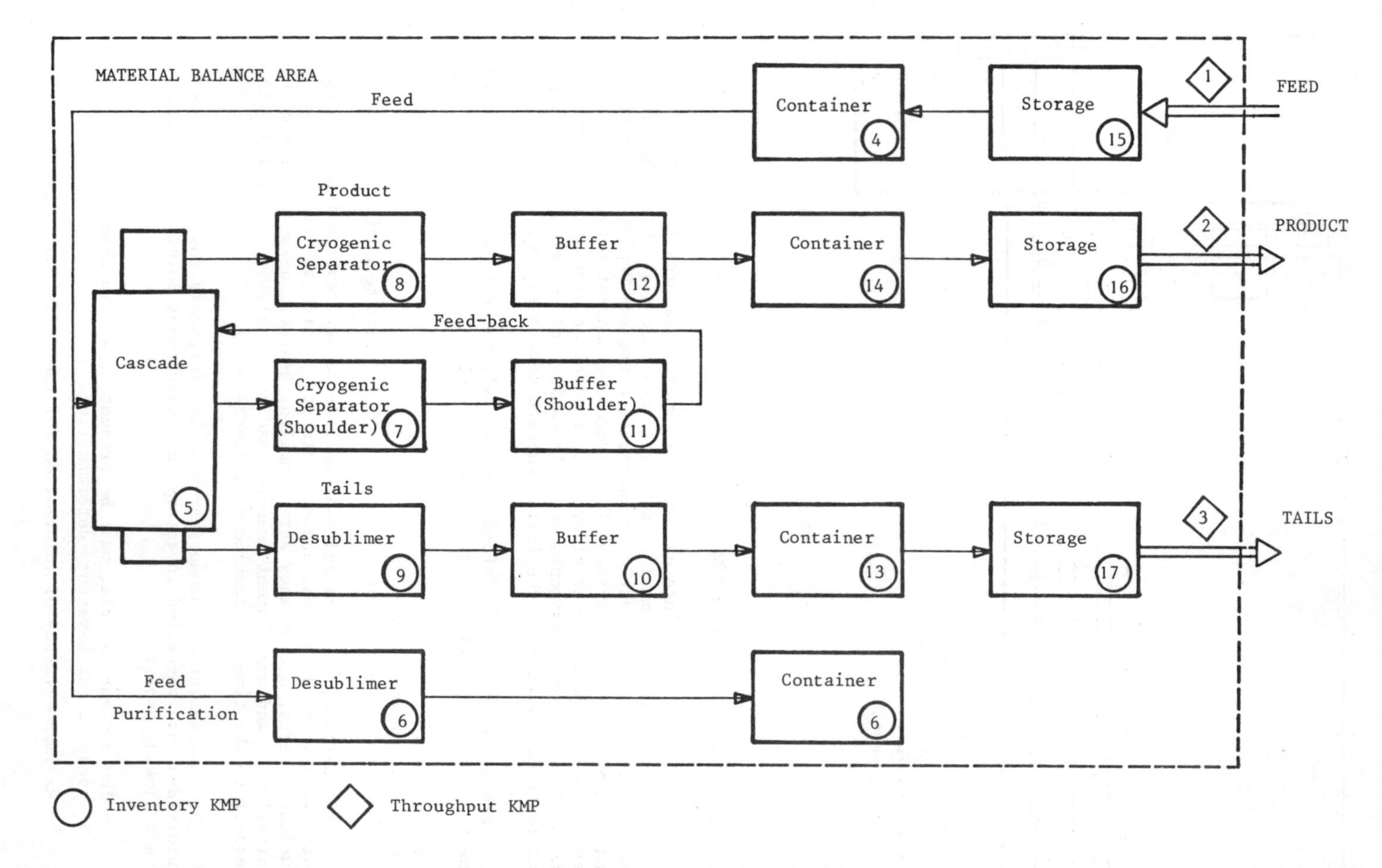

FIG.4. UF_6 flow sheet and key measurement points.

- Combination of subsystems so that sizable quantities of nuclear material are covered.
- Standardization of recording methods in order to cover as many sub-areas as possible by the same technique.
- Adaptation of key measurement points to planned in-plant instrumentation which, at the same time, can be used for inventory taking.
- Minimization of the expenditure in terms of time and costs for inventory taking with respect to intrusion in normal plant operation.
- Verification capability of the inventory by the inspection authority.

The flow of nuclear material (feed, product, tails) is recorded at 3 throughput key measurement points (TKMP) (Fig. 4). In accordance with this layout the reference plant can be subdivided into 17 key measurement points listed in Fig. 4.

On the basis of accounting records a material balance can be established for a specific balancing period and a special material balance area (comparison of book inventory (BI) with the physical inventory (PI)) which, given the necessary precise and complete assessment of inventory and flow, allows information to be extracted which will tell whether a significant amount was diverted within the given balancing period.

On the basis of present practice and for reasons of economy one material balance area has been provided for the reference facility.

For this plant a model balance has been established on the basis of a reference campaign in which only the in-plant measuring systems were used to assess the PI and the flow of nuclear material /4/.

It was found that considerable quantities of enriched uranium cannot be covered in this way. This fraction is merely estimated by the operator and thus cannot be verified by the inspection authority.

Since it is not possible to reach the balancing accuracy required by the inspection authority in this way, further studies dealt particularly with the possibility of measuring and verifying the inventory at the key measurement points wherever this was not possible by only using the in-plant measurement system.

4. DETERMINATION AND VERIFICATION OF THE CRYOGENIC SEPARATORS INVENTORY

The plant sections whose inventories could not be measured and, hence, not be verified by the in-plant measuring systems in the first estimate of the balancing accuracy, as mentioned above, also include the cryogenic separators of the cascade shoulder and the product. Determining and verifying the inventory of this key measurement point is particularly important because of the enrichment (some 3 % of U-235 in the product) and the relatively large inventory and, hence, the considerable contribution to the balancing inaccuracy. The methods elaborated for inventory taking and verification at these key measurement points will be represented below.

The cryogenic separators serve the purpose of separating, with a high separation factor, the UF_6/H_2 gas mixture continuously produced in cascade operation /5/.

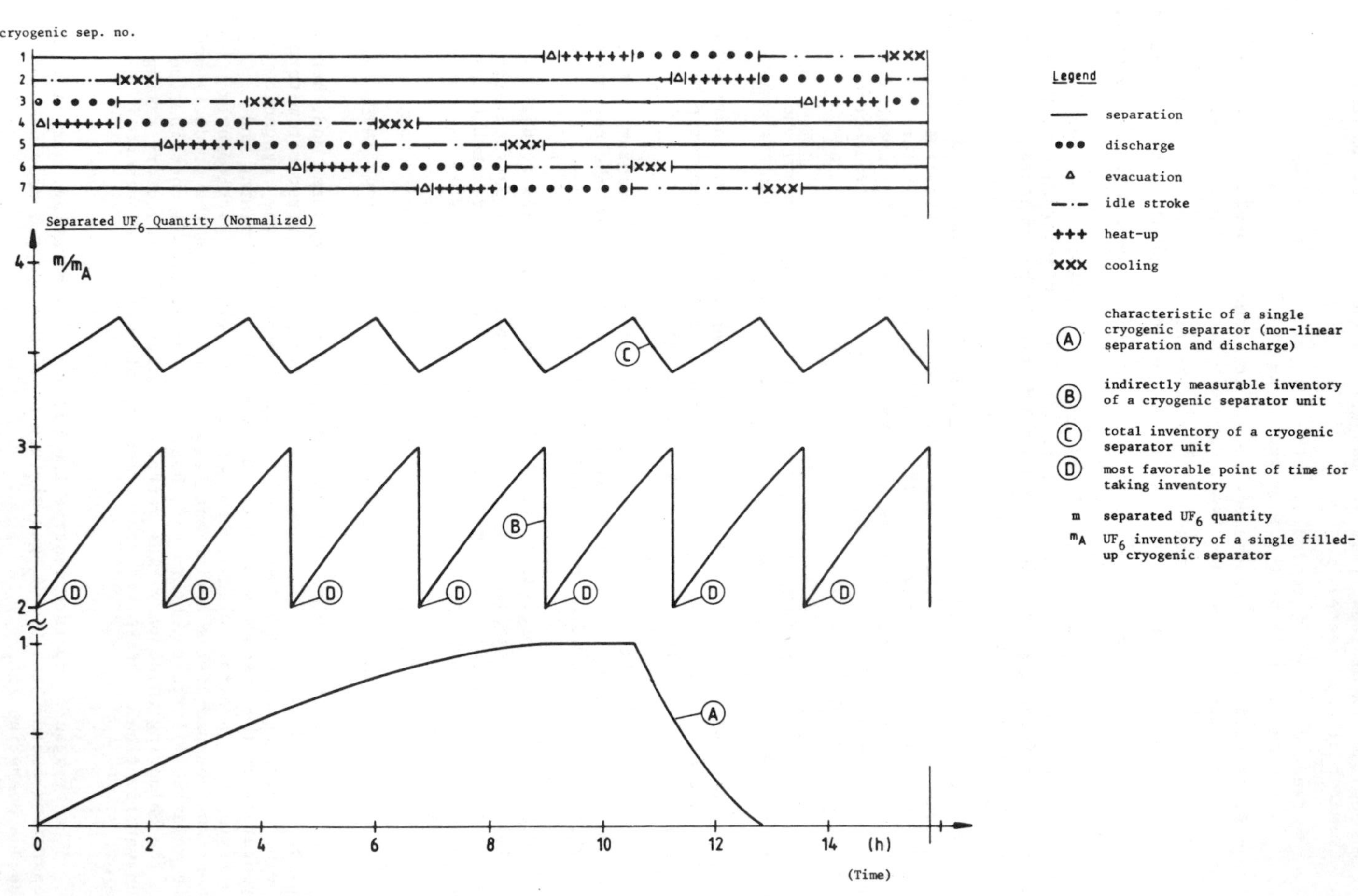

FIG.5. Fixed-cycle control schedule of a cryogenic separator unit.

In the plant facility a total of four units of cryogenic separators consisting of seven cryogenic separators each have been designed, two units being in constant operation in the cascade shoulder, one unit for the product.
The fourth unit serves for redundancy purposes.
For inventory taking the seven cryogenic separators of a unit coupled in one working cycle are taken together.
The time required for the different working cycles (cooling, separation, heating and discharging) and the control scheme of a unit of cryogenic separators can be seen from Fig. 5 .
Four cryogenic separators out of the seven combined in a unit are continously operated in the separation phase, one cryogenic separator in the discharging phase.

The inventories of the seven cryogenic separators of one unit cannot be determined by the same method, i.e. we must distinguish between separating (separation: separation of UF_6 as a solid from the UF_6/H_2 process gas) and discharging cryogenic separators (discharge: discharge of the UF_6 out of the cryogenic separators).

Filled and Discharging Cryogenic Separators

A full cryogenic separator is discharged within one cycle (2.25 h) after evacuation and heating. UF_6 is present as a gas fed to the buffers by means of two parallel UF_6 compressors.
The inventories of these cryogenic separators will be determined through the pump flow which, according to the current technology, is a linear function of the prepressure by way of the pumping behavior of the compressors used.
Measuring and summing up of the prepressure signal throughout the entire discharge time of a cryogenic separator will provide a signal for the filling quantity. If the inventory of a filled cryogenic separator is known, the same method, and difference formation, also allow the inventory of partly filled cryogenic separators to be determined.

Separating Cryogenic Separators

The simplest possibility of determining the inventory of separating cryogenic separators would be the discharge of UF_6 by way of the UF_6 compressors, which would allow a direct measurement of the inventory to be made as described above. However, since this would require plant operation to be interrupted, this possibility of inventory taking cannot be applied.

Direct measurement of the quantities flowing to the cryogenic separators during plant operation is not possible either, because only very low pressure losses can be admitted for process operation.

Hence, a direct measurement of the inventories of separating cryogenic separators is not possible according to the present state of the art without interfering with plant operation.

Examining the possibility of indirect assessment of the inventories of these cryogenic separators by means of the separation characteristic resulted in the following findings:

The flow resistance of a cryogenic separator changes during the separation phase as a function of the degree of loading of the four

TABLE I. MEASURING METHODS AND MEASURING POINTS; RELATIVE STANDARD DEVIATIONS (RSD) FOR SYSTEMATIC AND RANDOM ERRORS

Plant component	Measuring method	δMr	Measuring point (quantity) /M/	δNr	Measuring point (isotop. concentr.) /N/	δNs
feed container	weighing	0,0002	scales	0,0012	Q 1	0,0015
feed container	throughput	0,03	F 1	0,0012	Q 1	0,0015
cascade	estimation	0,10	-	0,005	-	0,0015
purification desublimer/container	weighing	0,0004	scales	0,0012	Q 1	0,0015
cryogenic separators (separation phase)	characteristic of separation	0,05	-	0,0008	Q 3 - Q 6	0,0015
cryogenic separator (discharge phase)	characteristic of the Normetex compressors	~0,02 - 0,03	P1, P2, P3	0,0008	Q 3 - Q 6	0,0015
tails desublimers (separation phase)	throughput	0,05	F 2	0,0036	Q 2	0,0015
tails desublimers (discharge phase)	level +	0,03-0,05	L 1	0,0036	Q 7	0,0015
	throughput	0,05	F 3			
tails buffer	level	0,03-0,05	L 1	0,0036	Q 7	0,0015
buffer cascade shoulder	level	0,03-0,05	L 2	0,0008	Q 8	0,0015
product buffer	level	0,03-0,05	L 4	0,0008	Q 10	0,0015
tails container	weighing	0,0004	scales	0,0036	Q 7	0,0015
product container	weighing	0,0004	scales	0,0008	Q 10	0,0015

δMr = RSD for the random error concerning the mass.
δNr = RSD for the random error concerning the isotope concentration.
δNs = RSD for the systematic error concerning the isotope concentration.

separating cryogenic separators in a group. The separation characteristic (inventory of a cryogenic separator plotted over the time) thus is not linear. Corresponding control of the flow resistance of the different cryogenic separators for purposes of linearization of the separation characteristic is not possible on the plant side.

Such linear behavior would offer the advantage that, if the inventory of a cryogenic separator is known (which is the case as a result of applying the method mentioned above), the whole separation characteristic is known and, hence, the inventories of separating cryogenic separators can be taken at any time, provided that the cycling behavior is known.

In order to get a first tentative idea of the actual separation characteristic, the separation characteristic and the separation rates characteristic (produced by simple differentiation of the separation characteristic for the time) the cryogenic separator operated in an inter-connected mode, i.e. within a cryogenic separator group, was determined for various flow resistance characteristics reflecting approximately the true behavior with the flow resistance of a cryogenic separator in actual plant operation.

The knowledge which was generated about the qualitative development of the separation characteristic is not sufficient, however, for precise determination of the inventory of separating cryogenic separators.

For this reason it is planned to record the true separation characteristic in experimental plant operation of the First Cascade.

For verification of the inventory of the cryogenic separators, both verification of the characteristics of the UF_6 compressors and verification of the separation characteristic by the inspection authorities is required.

5. EXPERIMENTAL PROGRAM

The following four test facilities are available for the experiments necessary to back up and implement the safeguards concept:

Cryogenic separator section, liquefier-evaporator unit, S33 loop, and the First Cascade.

Experiments are mainly carried out for the following purposes:

- Assessment of UF_6 deposits in the cryogenic separators during first filling and in design operation (cryogenic separator section).
- Recording the characteristic of the UF_6 compressors, the level characteristic of the liquid buffers (scaled down) and the UF_6 deposits in the liquid buffers (liquefier-evaporator unit).
- Determining the gas inventory and the deposits in the pipework and the other components (S 33 loop).
- Recording the separation characteristic of a cryogenic separator in compound operation and the level characteristic of the liquid buffers (First Cascade).

6. ESTIMATING THE UNCERTAINTY

After working out possibilities of inventory taking and verification at the key measurement points where this is not possible by the in-plant measurement systems, another model balance was established on the basis of a reference campaign.

Under the assumption that the experiments yet to be carried out will confirm the underlying uncertainties in measurement (Table I /6/), it is possible to establish a balance /3/ for the reference plant with an uncertainty on the order of approx. 4-5 kg of contained U-235 (for semi-annual inventory taking).

This corresponds to a relative uncertainty on the order of 0.2-0.3 % relative to the feed.

ACKNOWLEDGEMENT

The authors would like to thank Dr. Schubert, IKVT, for useful discussions and assistance in the collection of basic information for the paper.

REFERENCES

/1/ BECKER, E.W.; BIER, W.; EHRFELD, W.; SCHUBERT, K.; SCHÜTTE, R.; SEIDEL, D., "Uranium Enrichment by the Separation Nozzle Process", Naturwissenschaften 63, 407-411 (1976)

/2/ BECKER, E.W.; BIER, W.; SCHUBERT, K.; SCHÜTTE, R.; SEIDEL, D.; SIEBER, U., "Technological Aspects of the Separation Nozzle Process", AIChE SYMPOSIUM SERIES, Volume 73, Number 169

/3/ BAHM, W.; DIDIER, H.J.; KLOS, A.; SCHWEGMANN, P.; WEPPNER, J., "Entwicklung eines Überwachungskonzeptes für Trenndüsenanlagen", Abschlußbericht der Phase 1 (1978)

/4/ BAHM, W.; DIDIER, H.J.; GUPTA, D.; SCHWEGMANN, P., "Entwicklung eines Überwachungskonzeptes für Trenndüsenanlagen", Abschlußbericht der Projektdefinitionsphase (1976)

/5/ SCHMID, J.; SCHÜTTE, R., "UF_6-Abscheidungsanlagen in Trenndüsen-kaskaden", KfK 2428

/6/ BEYRICH, W.; DÜRR, W.; GROSSGUT, W., "UF_6-Interlaboratoriumstest", KfK 2340 (1977)

/7/ KRASKA, E.; OTTO, R.; WENK, E.; AVENHAUS, R.; GUPTA, D., "Assessment of Material Unaccounted For (MUF) and Inspection Efforts in a Centrifuge Plant", KfK 1696 (1972)

DISCUSSION

K. NIENHUYS: What is the contribution of the cryogenic separators to the relative uncertainty of 0.2 – 0.3% mentioned in your paper?

J. WEPPNER: About a third.

K. NIENHUYS: You assume that the underlying uncertainties used to calculate this 0.2 – 0.3% uncertainty will be confirmed by experiment. What are the uncertainties in these uncertainties?

J. WEPPNER: The separation characteristic referred to in my presentation is assumed to have an uncertainty of about 5%.

W. FRENZEL: Is it correct that the safeguards are based on a material balance verification concept? And did you consider the possibility of applying containment/surveillance (C/S)?

J. WEPPNER: The concept provides the basis for establishing a material balance. As you know, the facility is completely accessible for Agency inspections so that, if required, C/S measures can always be applied.

J.E. LOVETT: In most facilities the need to shut the facility down and clean out process equipment is the "rate-determining" consideration as regards physical inventory (PI) frequency. Since the inventory system described does not involve such shutdown, what are the rate-determining considerations?

J. WEPPNER: Perhaps Mr. Gupta would like to reply.

D. GUPTA (*Chairman*): The main rate-determining factor for PI-taking would, in my opinion, be the estimation and verification of the UF_6 content of the cryogenic system. A very frequent verification of this content may be a severe hindrance to the operation of the facility.

W. FRENZEL: I should like to extend Mr. Lovett's comment by asking whether any physical inventory taking is necessary.

D. GUPTA (*Chairman*): This is an interesting question. I suppose the answer has to come from the Agency, since it is responsible for fixing the frequency of inventory-takings for safeguards purposes.

IAEA-SM-231/43

APPLICATION DES TECHNIQUES DE GARANTIES A L'USINE DE DIFFUSION GAZEUSE D'EURODIF

J.H. COATES
Commissariat à l'énergie atomique,
Paris

J.R. GOENS
Eurodif,
Bagneux,
France

Abstract–Résumé

APPLICATION OF SAFEGUARDS TECHNIQUES TO THE EURODIF GAS DIFFUSION PLANT.

So far safeguards techniques have not been applied to gas diffusion plants, hence their use at the Eurodif plant cannot be based on previous experience. Nevertheless, the characteristic features of diffusion plants are such that safeguards procedures specifically suited for this technique can be proposed. The first of these features is the fact that appreciably altering the enrichment level of the plant product is not possible without making easily detectable changes either in the plant structure itself or in the movement of incoming and outgoing materials. Furthermore, because of the size of gas diffusion plants large stocks of uranium are present in them. Although inventory differences may be small in relative terms, they are large in absolute terms and exceed the quantities of low-enriched uranium considered significant from the standpoint of safeguards. Lastly, the impossibility for economic reasons of taking a physical inventory of the plant after it has been emptied prevents a comparison of the physical inventory with the book inventory. It would therefore seem that the safeguarding of a gas diffusion plant should be focused on the movement of nuclear material between the plant and the outside world. The verification of inputs and outputs can be considered satisfactory from the safeguards standpoint as long as it is possible to make sure of the containment of the plant and of the surveillance for the purpose of preventing clandestine alterations of structure. The description of the Eurodif plant and the movement of materials planned there at present indicate that the application of such a safeguards technique to the plant should be acceptable to the competent authorities. For this purpose a monitoring area has been set aside in which the inspectors will be able to keep track of all movements between the outside world and the enrichment plant; the fact that these movements are intermittent and hence easier to quantify will facilitate the task.

APPLICATION DES TECHNIQUES DE GARANTIES A L'USINE DE DIFFUSION GAZEUSE D'EURODIF.

Les techniques de garanties n'ont pas été appliquées jusqu'ici aux usines de diffusion gazeuse et leur mise en oeuvre à l'usine d'Eurodif ne peut donc pas se baser sur une expérience antérieure. Les caractéristiques propres aux usines de diffusion permettent néanmoins de proposer des modes de contrôle spécifiquement adaptés à cette technique. La première de ces caractéristiques est qu'il n'est pas possible de modifier dans une mesure importante le niveau d'enrichissement du produit de l'usine sans procéder à des modifications aisément décelables, soit de la structure même de l'usine, soit des mouvements de matières qui y entrent et en sortent. D'autre part, la taille des usines de diffusion gazeuse conduit à des inventaires importants en uranium. Les différences d'inventaire, tout en étant faibles en valeurs relatives, sont importantes en valeurs absolues et dépassent les quantités d'uranium faiblement enrichi considérées comme significatives du point de vue du contrôle des garanties. Enfin, l'impossibilité, pour des raisons économiques, de faire un inventaire physique de l'usine après vidange interdit une comparaison de cet inventaire physique avec un inventaire comptable. Il apparaît donc que le contrôle d'une usine de diffusion gazeuse devrait se concentrer sur les mouvements des matières nucléaires entre l'usine et le monde extérieur. Ces contrôles des entrées et des sorties peuvent être considérés comme satisfaisants du point de vue de l'application des garanties pour autant que des assurances aient pu être acquises sur le confinement de l'usine et sur sa surveillance destinée à prévenir les modifications clandestines de structure. La description de l'usine d'Eurodif et des mouvements des matières qui y sont actuellement prévus montrent que l'application d'une telle méthode de contrôle à cette usine devrait être acceptable pour les autorités compétentes dans le domaine. Une zone de contrôle y a en effet été réservée d'où les inspecteurs pourront suivre tous les mouvements entre le monde extérieur et l'usine d'enrichissement, et cela d'autant plus aisément que ces mouvements ont un caractère discontinu, ce qui facilite leur quantification.

Il n'y a pas jusqu'ici d'exemple d'application des techniques de garanties aux usines de diffusion gazeuse. L'usine d'EURODIF sera la première à être affectée uniquement à des besoins civils et à être soumise à des contrôles non seulement nationaux mais également internationaux.

Les Etats-Unis ont indiqué au cours des travaux de l'INFCE qu'ils n'envisageaient pas actuellement de porter tout ou partie de leurs usines de diffusion gazeuse sur la liste des installations qui seraient soumises à l'inspection de l'A.I.E.A. Quant aux installations d'enrichissement soviétiques, il n'a jamais été déclaré officiellement qu'elles appliquent la technologie de la diffusion gazeuse et leur contrôle n'a jamais été envisagé jusqu'ici.

On se trouve donc dans le cas d'EURODIF devant une innovation dont les modalités ne sont pas à l'heure actuelle définies dans le détail puisque les accords particuliers de contrôle de cette usine ne sont pas encore établis, que ce soit avec l'EURATOM ou avec l'A.I.E.A.

Il ne sera donc donné ici que des indications préliminaires sur la façon dont il apparaît possible de contrôler une usine d'enrichissement par diffusion gazeuse, compte tenu de ses caractéristiques techniques propres. L'application des mesures ainsi esquissées sera ensuite rapportée au cas concret de l'usine d'EURODIF.

Une première caractéristique des usines de diffusion gazeuse est qu'il n'est pas possible de modifier dans une mesure importante le niveau d'enrichissement du produit de l'usine. Celui-ci peut varier en fonction du débit d'extraction et l'une des réponses américaines au questionnaire INFCE sur l'enrichissement fait état de ce qu'une cascade dessinée pour produire de l'uranium enrichi à 3 ou 4 % pourrait produire de l'uranium enrichi à plus de 50 % si le débit d'extraction était quasi nul. Il faut immédiatement remarquer que cette modification du taux d'enrichissement se heurterait à une difficulté physique insurmontable dans le cas d'une usine optimisée pour les bas enrichissements : les tailles d'équipements des étages et d'alimentation et d'extraction de la cascade y sont telles que l'augmentation de teneur des produits enrichis conduirait à des problèmes de criticité. En outre, du point de vue contrôle, un fonctionnement à reflux quasi total serait décelé par une modification des mouvements de matières qui constituent précisément l'un des paramètres qui, à nos yeux, est parmi les plus importants à suivre par les contrôleurs.

Une autre technique considérée également pour atteindre les hauts enrichissements serait le recyclage d'uranium de bas enrichissement et le fonctionnement discontinu. On se rapproche du fonctionnement en reflux total avec les mêmes problèmes de criticité et la même possibilité de détection par les variations de mouvements de matières.

Il est, d'autre part, impossible de modifier aisément la structure des usines de diffusion gazeuse pour atteindre les hauts enrichissements : tous les étages sont en effet installés en série et l'allongement de la cascade ne peut donc être réalisé par réarrangement de la disposition des étages, la seule possibilité consistant à rajouter des étages, ce qui est très aisément détectable.

En conclusion, on peut dire que l'on voit mal comment une usine de diffusion gazeuse conçue pour produire de l'uranium de bas enrichissement puisse être détournée de son type de production pour être clandestinement consacrée à la production d'uranium fortement enrichi.

Une deuxième caractéristique essentielle des usines de diffusion gazeuse du point de vue contrôle est leur taille. L'optimum économique conduit à des usines de très grande capacité et donc de très grand inventaire. L'expérience américaine indique que la limite d'erreurs sur l'inventaire de l'uranium-235 est de l'ordre de 0,2 % du débit intégré de la cascade. Ce débit intégré n'est pas publié mais il a été signalé que les différences d'inventaire des cascades d'enrichissement américaines ont été de l'ordre de 0,1 % de ce débit, avec une valeur moyenne annuelle à Oak Ridge de 119 kg d'uranium-235 et à Portsmouth de 94 kg d'uranium-235. Ces différences d'inventaire si elles sont faibles en valeurs relatives sont importantes en valeurs absolues et dépassent les quantités d'uranium faiblement enrichi qui sont considérées comme significatives du point de vue contrôle des garanties.

Enfin, le procédé par diffusion gazeuse fonctionne en continu ; le temps de remplissage et de mise en équilibre de la cascade est relativement long, de l'ordre de mois et il est donc exclu, pour des raisons économiques, de faire un inventaire physique après vidange et, partant, une comparaison de cet inventaire physique avec un inventaire comptable.

Les mouvements des matières dans l'usine donnent lieu à de très nombreuses opérations par an et leur intérêt d'un point de vue contrôle est limité.

On est donc ainsi conduit à proposer de concentrer les activités de contrôle sur les entrées et les sorties des matières nucléaires, c'est-à-dire sur les mouvements entre le monde extérieur et l'usine. Pour qu'une telle concentration soit satisfaisante du point de vue de l'application des garanties, il est indispensable que deux assurances puissent être acquises :

(i) que le confinement de l'usine vis-à-vis du monde extérieur du point de vue mouvement des matières est parfaitement réalisé, à l'exception du transit par une zone de contrôle,

(ii) qu'une surveillance a pu être établie permettant de vérifier que dans la zone des usines aucune installation clandestine d'enrichissement n'a été installée ni aucune modification de la cascade réalisée.

L'usine d'EURODIF a déjà été décrite par ailleurs [1, 2] et l'on ne reprendra ici que les caractéristiques qui paraissent importantes d'un point de vue contrôle des garanties.

La capacité annuelle de l'usine du Tricastin lorsqu'elle sera en pleine production sera de $10,8 \; 10^6$ UTS/an. Exprimées en débits d'introduction et de soutirage, un exemple de caractéristiques du régime nominal est le suivant :

Débits	Teneur en U^{235} (%)	(tUF_6/a)
. alimentation	0,711	26 000
. production	3,25	2 700
	2,60	900
	1,90	750
	1,50	650
. rejet	0,25	21 000

Les caractéristiques réelles de fonctionnement pourront s'écarter, dans une certaine mesure, des valeurs citées ci-dessus.

L'usine ne fournit qu'un service d'enrichissement : les clients apportent leur propre quantité d'UF_6 naturel et reprennent l'UF_6 enrichi ; l'UF_6 appauvri peut être repris ou bien laissé sur le site selon les clauses particulières du contrat passé avec chaque client d'EURODIF.

La figure 1 donne l'implantation générale de l'usine qui est située sur un site de 250 Ha et la figure 2 le détail de la zone d'entrée-sortie.

La signification des principaux bâtiments participant à la production d'uranium enrichi est donnée ci-après.

Les quatre bâtiments repérés 1 regroupent les étages de diffusion et les liaisons interétages. Ils constituent la cascade de diffusion

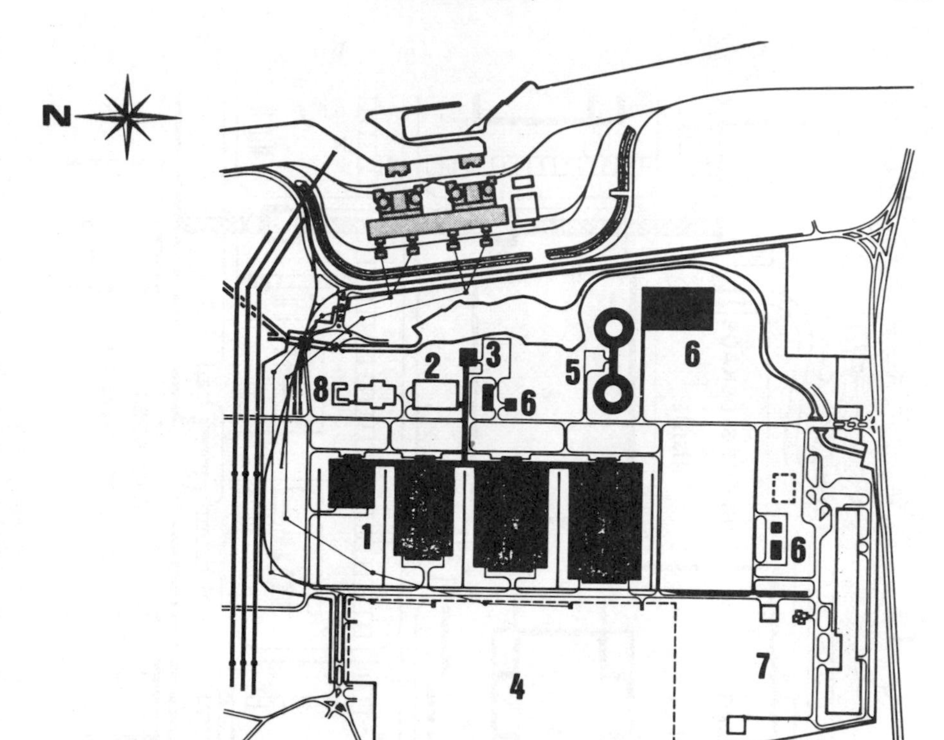

1. CASCADE
2. ANNEXE DE PROCEDE
3. POSTE DE COMMANDE
4. POSTE ELECTRIQUE
5. REFRIGERANTS ATMOSPHERIQUES
6. AUXILIAIRES
7. ATELIERS
8. ZONE REC

FIG.1. Usine du Tricastin: plan de masse.

proprement dite. Celle-ci est constituée de 1 400 étages de trois tailles dont une répartition typique est la suivante :

- petite taille	: 220	section d'enrichissement
- taille moyenne	: 280	
- grosse taille	: 300	
- grosse taille	: 420	section d'appauvrissement
- taille moyenne	: 120	
- petite taille	: 60	

Les débit et capacité nominaux des étages sont :

	débit (kg UF_6/s)	capacité (UTS/an)
- petite taille	28,7	2 270
- taille moyenne	81,3	5 640
- grosse taille	183,8	13 050

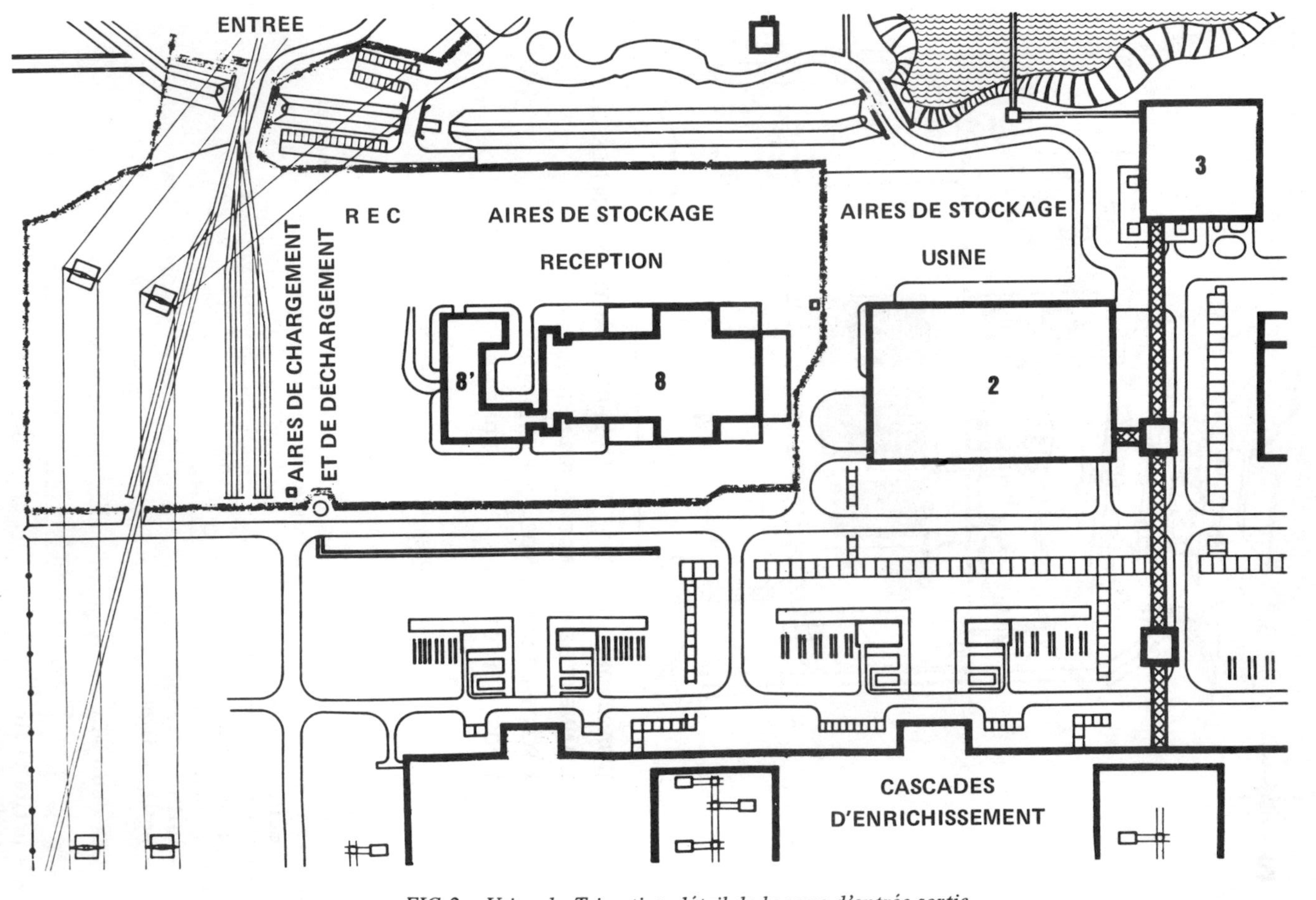

FIG.2. Usine du Tricastin: détail de la zone d'entrée-sortie.

Les étages sont disposés par groupes de vingt reliés entre eux par des conduites débouchant dans des collecteurs d'usine et interusines. Ces collecteurs aboutissent dans un bâtiment (2) dénommé "Annexe U".

. Cette "Annexe U" regroupe tous les moyens permettant d'assurer toutes les manipulations d'UF_6 qui précèdent, suivent ou vont de pair avec le processus d'enrichissement et comporte les unités fonctionnelles suivantes :

- L'unité d'alimentation qui assure l'introduction en continu dans le circuit de diffusion de l'UF_6 à enrichir. Cette introduction est réalisée par vaporisation directe à partir de conteneurs d'approvisionnement placés dans des étuves individuelles dont l'air est chauffé électriquement.
- L'unité de soutirage de l'UF_6 enrichi qui assure, à partir de différents points de la cascade dont la position dépend de la teneur désirée, l'extraction de l'UF_6 enrichi. Quatre teneurs différentes peuvent être simultanément extraites en connectant quatre collecteurs à la cascade. Cette extraction est basée sur l'utilisation des propriétés du diagramme d'équilibre de phase des différents corps présents dans le gaz de procédé. On utilise des cristallisoirs, la coulée dans les conteneurs étant finalement faite en phase liquide.
- L'unité de soutirage de l'UF_6 appauvri qui assure, à partir de deux collecteurs reliés au pied de la cascade, le prélèvement de l'UF_6 appauvri en vue de son conditionnement pour stockage.
- L'unité de purge qui assure l'élimination en tête de cascade des différentes impuretés contenues dans le flux de procédé.
- L'unité d'extraction-remplissage qui permet de réaliser tous les transferts d'UF_6 ou d'autres agents fluorants nécessités par les opérations de mise en service et d'entretien des différents circuits constitutifs de l'usine de diffusion gazeuse.
- L'unité de destruction des résidus fluorés et de lavage des évents qui assure le traitement des résidus fluorés liquéfiés et des résidus gazeux de l'ensemble des unités de l'Annexe U.

C'est dans l'"Annexe U" que sont également localisés les analyseurs en ligne permettant les contrôles analytiques du procédé.

. Le bâtiment 3 est le poste de commandement où sont centralisés toutes les informations et les moyens nécessaires à la conduite de la cascade et de l'"Annexe U".

. Le bâtiment 8 et l'aire qui l'entoure constituent la zone "Réception, Expédition, Contrôle" (REC). Cette zone, d'environ 33 000 m2, abrite tous les dispositifs permettant le contrôle de qualité et de quantité de l'UF_6 entrant et sortant de l'usine d'une part et d'autre part le conditionnement de l'UF_6 dans les emballages "clients". Tous les échanges avec l'extérieur passent par cette zone, véritable station de réception de la matière de base et, de ce fait, accessible aux clients d'EURODIF et aux contrôleurs internationaux.

La zone REC permet également l'accueil des véhicules de transport des conteneurs d'UF_6. Elle est équipée de moyens de déchargement et de chargement pour les conteneurs et les coques de protection. Elle comporte, en outre, des aires de vérification de la conformité des emballages.

Les parcs de stockage situés dans la zone REC sont réservés au parcage de tous les conteneurs UF_6 en transit sur le site. Ils permettent de régulariser les mouvements de matière. Un parc pour le stockage de l'appauvri non repris par le client y est également prévu.

. Les autres bâtiments figurant sur le plan de masse concernent essentiellement des unités auxiliaires ne participant pas directement à la production : centrales calorifique , frigorifique, air comprimé, azote.

Le site est délimité par une clôture de protection d'au moins 2 mètres de haut. Une route périphérique épousant approximativement les formes de la clôture permet d'en assurer la surveillance au cours de rondes. L'entrée principale du site est située au nord-est et donne accès à la zone REC et à la zone "usines".

Les différents mouvements de matière entre les principaux bâtiments cités plus haut peuvent se résumer comme suit :

- Produit d'alimentation F

  . Approvisionnement

  Le produit est approvisionné à l'aide de conteneurs de 12 tonnes (type 48 Y). Ceux-ci sont réceptionnés dans la zone REC. On procède au déchargement, à la vérification externe, à la pesée et à un échantillonnage. On prévoit de réceptionner en moyenne 40 à 50 conteneurs par semaine. Chaque conteneur est ensuite mis sur un parc de stockage en attente.

  . Introduction dans la cascade

  Les conteneurs de produit F sont repris et amenés dans l'unité d'Alimentation du bâtiment 2 dans laquelle s'effectue le transfert de l'UF_6 des conteneurs vers le circuit de la cascade de diffusion. Le régime d'introduction est évidemment identique, en moyenne, à celui d'approvisionnement, soit 40 à 50 conteneurs/semaine.

- Produit enrichi

  . Soutirage

  Le produit enrichi prélevé à partir de la cascade de diffusion est mis dans des conteneurs "usine" de 13,5 tonnes, type 48 Z. L'opération s'effectue dans le bâtiment 2. Ensuite, les conteneurs sont acheminés vers un parc de stockage en attente de livraison du produit ; on prévoit de manipuler 8 à 10 conteneurs par semaine.

  . Livraison

  Les conteneurs "usine" remplis du produit enrichi sont amenés dans la zone REC pour que leur contenu soit échantillonné, pesé et transféré dans des conteneurs "clients" de 2,25 tonnes, type 30 B. Ces derniers sont ensuite vérifiés, munis de leur coque de protection et chargés sur un véhicule de transport. Le nombre de conteneurs à manipuler ainsi est estimé à 50/semaine.

- Produit appauvri W

  . Soutirage

  Le produit appauvri prélevé dans la cascade de diffusion est mis dans des conteneurs "clients" ou "usines" type 48 Y. L'opération est réalisée dans le bâtiment 2. Les conteneurs sont ensuite amenés sur parc de stockage. Le nombre de conteneurs manipulés est de 35 à 40 par semaine.

  . Livraison

  Les conteneurs "clients" remplis de produit appauvri sont amenés dans la zone REC pour être pesés et échantillonnés. Ils sont ensuite vérifiés et chargés sur un véhicule de transport. On prévoit de manipuler de 10 à 15 conteneurs "clients" par semaine.

  . Stockage définitif

  Les conteneurs "usines " remplis de produit appauvri sont amenés dans la zone REC pour être pesés et échantillonnés (par lot). Ils sont ensuite acheminés vers le lieu du stockage définitif. Le nombre de conteneurs manipulés est de 25 à 30 par semaine.

  . Reconversion UF_6 en un composé uranifère plus stable

  Cette reconversion est actuellement à l'étude. Dans l'hypothèse où elle serait réalisée, les conteneurs "usines" seraient acheminés, après pesée et échantillonnage en zone REC, vers le lieu de reconversion.

- Effluents liquides

  . Les effluents liquides susceptibles de contenir des traces d'uranium pourront être traités dans une installation existante située en dehors des limites de l'usine du Tricastin. L'étude de ce projet est en cours. Dans cette éventualité, on prévoit que l'enlèvement des effluents au niveau des unités productrices d'effluents (atelier de traitement de surface, unités de traitement des évents) et leur livraison, s'effectueront à l'aide de citernes mobiles dont le contenu sera pesé et échantillonné dans la zone REC.

- Echantillons d'UF_6

  . L'échantillonnage de l'UF_6 en transit sur le site est réalisé dans la zone REC. Les quantités prelevées d'UF_6 sont mises soit dans des bouteilles de 1,7 kg, soit dans des tubes pour échantillons gazeux contenant moins de 10 g et sont ensuite acheminées au laboratoire et/ou envoyées aux clients pour analyse ou encore conservées dans un local de la zone REC comme échantillons témoins. Les prélèvements non utilisés sont ensuite réintroduits dans les différents circuits de procédé.

Les analyses des différents produits effectuées sur place au laboratoire d'EURODIF (8') ont pour rôle essentiel d'assurer le contrôle chimique et isotopique de l'hexafluorure transitant par le site. Elles seront limitées à celles requises par les contrats, les contrôles nationaux et internationaux et la gestion des matières. Il est prévu d'analyser un échantillon en phase liquide par 5 conteneurs d'alimentation, par 6 conteneurs de produit et 5 conteneurs de rejets. En phase gazeuse, 4 conteneurs sur 5 tant d'alimentation que de rejets seront échantillonnés. Il est actuellement estimé qu'un peu plus de 1 500 échantillons liquides et près de 4 000 échantillons gazeux seront analysés chaque année.
Les équipements d'analyse utilisés permettent une précision de l'ordre du pour cent, à l'exception des mesures de poids réalisées à 0,01 % et des contrôles isotopiques et spectrométriques dont la précision sera supérieure à 0,1 %.

Les pertes qu'il est prévu de rencontrer lors de l'exploitation auront essentiellement deux origines :

- les fuites d'UF_6 qui se produiront fatalement malgré les normes très strictes imposées à la construction de la cascade

et

- la fixation de l'UF_6 adsorbé sur les surfaces en contact par la décomposition de celui-ci en présence de certains matériaux.

L'ordre de grandeur de ces pertes, qui devrait être vérifié lors de la mise en exploitation, est actuellement estimé à quelques parts pour 10 000 du débit d'alimentation de l'usine et serait donc identique aux erreurs de pesée. D'autres pertes peuvent être entraînées sous forme de contamination des pièces de l'usine qui devraient être déposées. Ces pertes pourront être mesurées lors des décontaminations, qui seront effectuées soit dans des ateliers initialement créés pour le premier montage et partiellement transformés en ateliers de démontage "rouge", soit au centre voisin de Pierrelatte. Les modalités d'échantillonnages et de mesures associées à ces décontaminations ne sont toutefois pas définies à ce jour.

Enfin, la teneur en uranium des rejets de l'unité de destruction des résidus fluorés et des lavages des évents de l'"Annexe U" ne devrait pas dépasser quelques centaines de grammes par an. Les solutions qui les contiennent seront soumises à analyse avant d'être expédiées vers une station de traitement chimique extérieure au site du Tricastin.

Les quantités qu'il est actuellement prévu de stocker, indépendamment des matières en exploitation, sont de 1 000 tonnes d'UF_6 enrichi entre 1 et 5 %, 6 500 tonnes d'UF_6 à teneur naturelle et 40 000 tonnes d'UF_6 partiellement appauvri.

L'inventaire de la zone "usines" est actuellement estimé à 5 000 tonnes d'uranium environ dont la moitié dans la cascade, l'autre moitié représentant un en-cours qui pourrait d'ailleurs être plus élevé pendant la phase de démarrage au cours de laquelle le fait de disposer d'une cascade tronquée conduit à stocker dans cette zone "usines" des rejets partiellement appauvris.

La gestion de l'uranium sera assurée à l'aide d'un ordinateur travaillant en temps réel. Le système mis en place vise à la constitution d'un inventaire permanent pour la gestion des conteneurs et des opérations déterminant leur contenu. Ce système permettra la restitution des informations concernant les mouvements des conteneurs, les pesées et les analyses des matières contenues et les diverses opérations de transfert.

En notant ainsi toutes les informations relatives au cycle parcouru par les conteneurs, on pourra suivre les matières de base tout au long de leur trajet sur le site.

La disposition de l'usine telle que décrite ci-dessus et son mode d'exploitation rendent possible un contrôle basé sur le confinement et la surveillance de la zone "usines" couplés au suivi des entrées et sorties des matières nucléaires de la zone REC.

Du point de vue confinement, la cascade offre une première barrière : l'accès aux matières qu'elle contient ne peut se faire que par l'"Annexe U", à l'exception des prises d'échantillons accessibles aux éléments mobiles d'analyses (EMA). Ces prises d'échantillons sont réparties tout au long de la cascade, à raison d'une prise par groupe de 20 étages. Elles ne sont pas utilisées de façon continue mais seulement au cours des périodes de mise en exploitation ou de modification importante de régime, ou pour contrôler mensuellement le profil isotopique de la cascade.

Une deuxième barrière entre les matières et le monde extérieur est constituée par les bâtiments de l'"Annexe U" et des cascades. L'accès du personnel à ces dernières ne peut se faire que par une unique galerie débouchant à l'étage de la salle conduite du poste de commande. Enfin, le site est entouré d'une clôture se prolongeant par un voile bétonné enterré et pourvu des moyens modernes de surveillance.

Les éléments paraissent donc réunis pour établir à la satisfaction des inspecteurs que les matières nucléaires sont bien confinées dans la zone des usines. Des modalités devront être établies pour le contrôle d'opérations exceptionnelles, comme l'évacuation, le cas échéant, vers des installations extérieures, de pièces susceptibles de transporter des quantités non négligeables d'uranium sous forme de contamination de surface. Ceci entre toutefois dans le cadre des dispositions particulières de contrôle qui restent à établir.

Le dénombrement aisé des étages de diffusion, le fait que leur arrangement en série correspond à la configuration de cascade conduisant à l'enrichissement maximum, l'importance des travaux à engager pour modifier cette situation devraient permettre aux contrôleurs de vérifier par des visites épisodiques des usines que celle-ci ne peuvent produire que de l'uranium faiblement enrichi.

Le caractère discontinu des transferts de matières nucléaires entre la zone REC et celle des usines devrait largement faciliter le travail des inspecteurs. Aucune limitation d'accès n'est prévue pour eux dans la zone REC. Avec l'accord de l'exploitant, ils y trouveront tous les moyens de mesure, d'analyse et de comptabilité nécessaires à l'exercice de leur mission et pourront y ajouter leurs moyens propres dans les locaux qui leur seront réservés.

Il y a donc lieu de croire que les dispositions particulières de contrôle qui seront établies pour l'usine d'EURODIF concilieront les soucis de non-prolifération de la Communauté internationale et le désir de l'exploitant de ne pas être entravé dans ses activités de production et de voir protégés certains aspects confidentiels de celles-ci.

REFERENCES

[1] BESSE, G., «Le programme EURODIF: état actuel du projet», Nuclear Energy Maturity, Proceedings of the Paris Conference, Invited Sessions, Prog. Nucl. Energy, Series 1976, Pergamon Press, Oxford and New York (1976) 202.

[2] ERGALANT, J., LEBRUN, C., LEDUC, Ch., PERRAULT, M., Bull. Inf. Sci. Tech. (Paris) n° 206 (1975) 111.

DISCUSSION

D. GUPTA *(Chairman):* You suggested in your paper that physical inventory-taking by measurement of the amounts of UF_6 inside the cascade of the diffusion plant should be completely replaced by C/S measures. This seems to be an interesting idea, since there may be very large amounts of UF_6 and their fluctuations may be very high. However, it will not then be possible to carry out any MUF analysis nor will it be possible to draw any conclusions with regard to diversion on the basis of a material balance. How are you thinking of equating this suggestion with the basic structure of international safeguards?

J.R. GOENS: We consider that MUF analysis of the toll (REC) area is adequate for safeguarding purposes and we shall propose that such a concept be used in the facility attachment for the Tricastin gaseous diffusion plant.

W.C. BARTELS: I believe you said the feed rate is to be 26 000 tonnes per year and you correctly quoted the United States experience that adsorption and chemisorption of uranium will be a few tenths of one per cent of throughput. These figures indicate a build-up in the plant at the rate of hundreds of kilograms of ^{235}U per year. This is a serious consideration and in the United States of America we are making a careful study as to how international safeguards might be applied to such a plant. We will be prepared to discuss our results in the near future.

J.E. GLANCY: I understood from your presentation that criticality problems constitute one of the factors preventing the production of highly enriched uranium (e.g. 50%). Is it true that criticality is also a problem for enrichments of 20% ^{235}U?

J.R. GOENS: All safety analyses have been conducted on the hypothesis that no enrichment higher than 5% would be allowed.

W. FRENZEL: Would inspectors have access to the sampling stations in order to ensure qualitatively that the plant product is not uranium of higher enrichment?

J.R. GOENS: As indicated in my paper, sampling is performed in the toll (REC) area. The inspectors have full access to this area and the analysis required by them can be performed there.

SAFEGUARDING A GAS CENTRIFUGE PILOT PLANT IN JAPAN

T. MINATO
Power Reactor and Nuclear Fuel
Development Corporation,
Tokyo, Japan

Abstract

SAFEGUARDING A GAS CENTRIFUGE PILOT PLANT IN JAPAN.

The first part of the Japanese gas centrifuge pilot plant for uranium enrichment at Ningyo-toge will be in full operation in August 1979. The nominal plant capacity is 50 t SWU/a, and has been designed to allow the effective application of safeguards, although a "non-access area" will be established to protect "commercially sensitive technology". The plant will be divided into two MBAs – a process MBA and a storage MBA, and will have 12 KMPs. It has been suggested that "running physical inventory-taking" of the process area be introduced at the end of each accounting period, which might enable continuous operation of the plant. This will not affect the accuracy of total plant inventory, because the inventory of the "running" area will be quite small. An improvement in measuring accuracy and a decrease of MUF are essential to make safeguards more effective. For this purpose the simulation analysis of nuclear material flow in the plant was conducted, from which the required accuracies and the measuring intervals at each measuring point were established. An on-line mass-spectrometer and a portable enrichment analyser for the UF_6 cylinder have been developed. Measurements of uranium deposits on metal surfaces of equipment are carried out to estimate the concealed inventory more accurately.

1. PILOT PLANT

The construction work of the Japanese gas centrifuge pilot plant for uranium enrichment was begun at Ningyo-toge about 800 km west of Tokyo in 1977 (Fig. 1). The first part of the plant is scheduled to start operation in August 1979, and additional centrifuges will be installed in succession. This plant is based on the development work of uranium enrichment technology conducted by the PNC at Tokai Works and other organizations over the past 20 years. The future construction of a demonstration plant will be based on the experience gained from this pilot-plant operation.

The pilot-plant site of about 40 000 m^2 is located in the area of the PNC Ningyo-toge Works. The main facilities under construction include the process building, which contains gas centrifuge cascades and UF_6 feed and withdrawing systems, administration offices, UF_6 storage, a waste-water treatment facility etc. (Fig. 2).

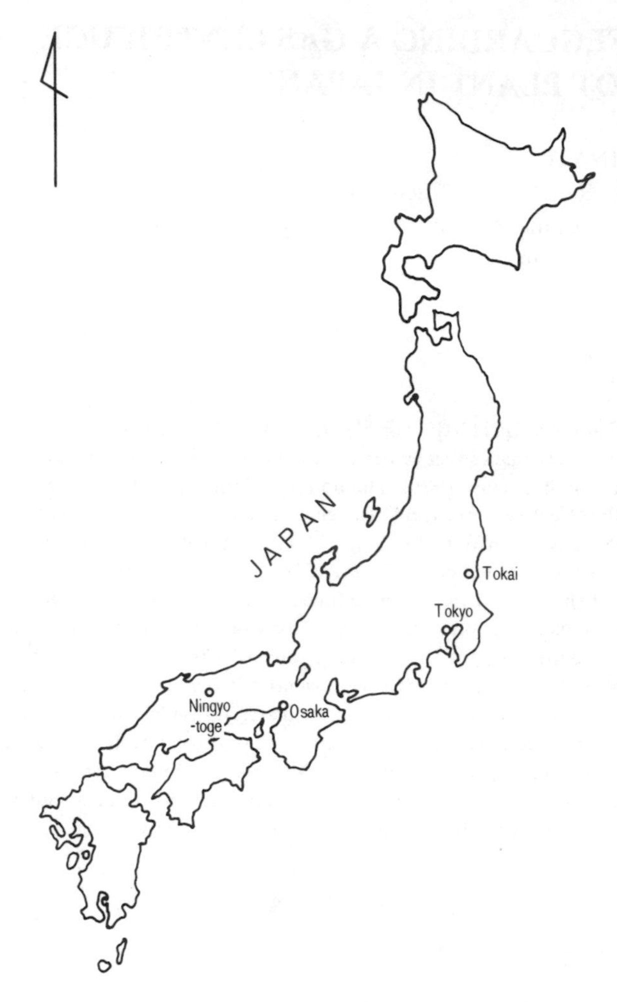

FIG.1. Location of PNC Ningyo-toge works.

The plant has the nominal capacity of 50 t SWU/a and is designed to produce slightly enriched uranium of about 3% ^{235}U to be used as LWR fuel. Japan has paid keen attention to the international Non-Proliferation Treaty and, as a plant design basis, we have made the protection of the commercially sensitive centrifuge technology sufficiently compatible with international safeguards measures.

2. SAFEGUARDS SYSTEM

This section is based on the draft proposal to the Japanese Government and is subject to change.

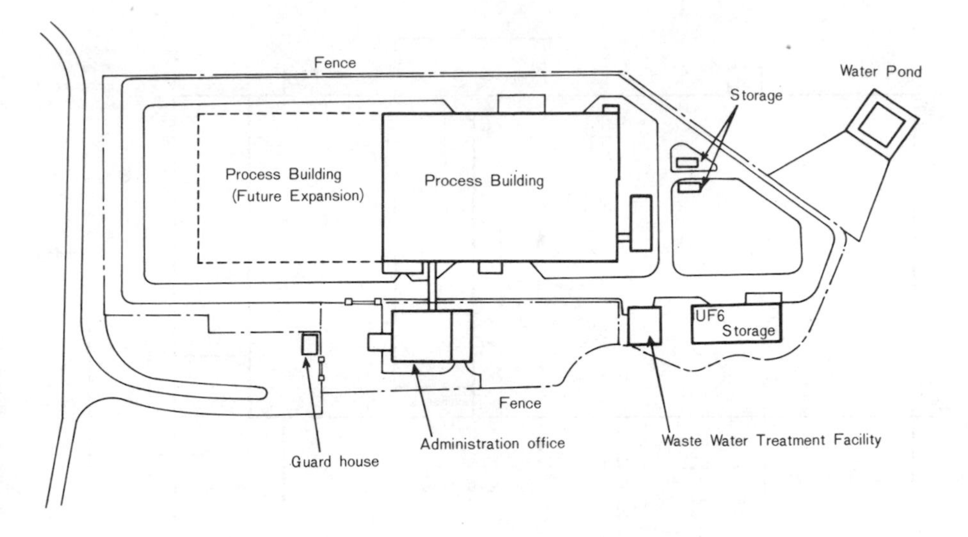

FIG.2. Uranium enrichment pilot plant.

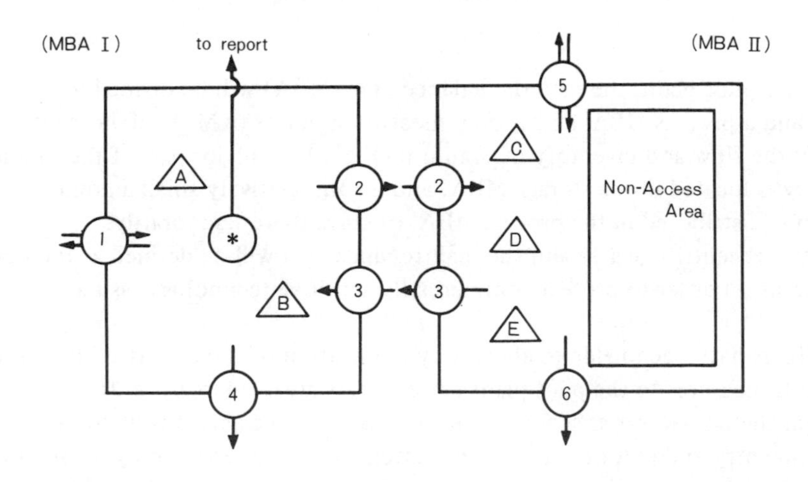

Flow KMPs
1) Receipt & return
*) SRD determination
2) Transfer to Process Area
3) Transfer to Storage Area
4) Shipment
5) Receipt and shipment of small quantitiy and Standard sample
6) Measured discards and retained wastes

Inventory KMPs
A) Storage (Shipper's data)
B) Storage (Operator's measurement basis)
C) In-process inventory
D) Analytical Laboratory inventory
E) Others

FIG.3. MBAs and KMPs of uranium enrichment pilot plant.

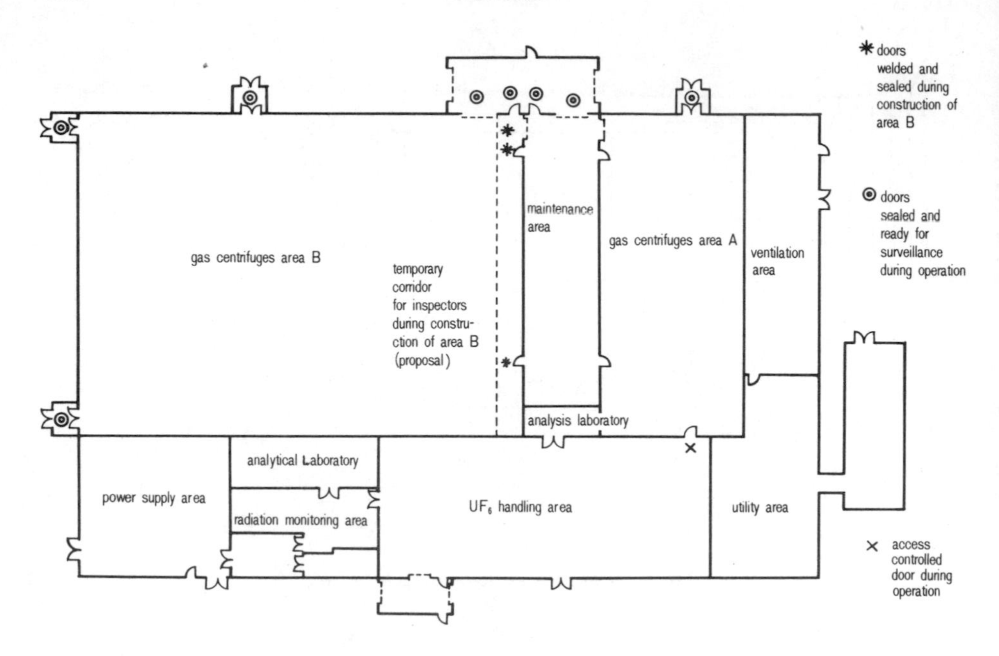

FIG.4. Layout of process building.

In the pilot plant, the material balance area (MBA) will be divided into a storage and a process MBA, and 12 key-measuring points (KMP) will be established to verify the flow and inventory of uranium (Fig. 3). A major part of the uranium inventory is located in the storage MBA, and a comparatively small amount of uranium is distributed in the process MBA, or stays there temporarily.

The gas centrifuge area and the maintenance area will be defined as the non-access areas, in order to protect commercially sensitive technology, as mentioned earlier.

Efforts have been made to allow easy application of the effective international safeguards measures to the pilot plant's non-access area – namely, entrance to and exit from the non-access area will be limited via only one access control gate, where the entry and exit of persons and materials will be recorded by a computer-aided weight detector. Further, a surveillance video camera, which has been developed by PNC, will be installed for monitoring from the inspector's office. Other exits (emergency exits) of the non-access area will have double doors to facilitate sealing for containment, and surveillance cameras will be installed.

After operation commences in August 1979, a temporary corridor between non-access areas A and B (the latter under construction) will be provided to give the inspectors access to the entire zone around the non-access area during operation (Fig. 4). The uranium inventory within the non-access area of the process MBA is quite small.

Weighing scales, flow meters, mass-spectrometers and chemical analysis will be used to determine various uranium values. The uranium measurement at KMPs will be mainly made by off-line weighing and mass-spectrometric analysis, which have high measuring precision. The uranium flow-rate, weight and ^{235}U enrichment will be measured by on-line meters to determine the uranium distribution on time in the process.

Running physical inventory-taking (RPIT)

In the centrifuge enrichment area, the plant must be operated continuously over a long period; therefore, as a PIT approach, it is conceivable to apply RPIT at the end of each accounting period.

Actual RPIT procedures will be as follows:

(1) The amount of contained uranium in the spare feed UF_6 cylinder and the cold traps for product and tail are measured, and the cylinder and cold traps are made ready for use;

(2) All the feed UF_6 cylinders and cold traps for product and tail in operation are simultaneously switched to the spare ones;

(3) The amount of uranium contained in the cylinders and cold traps, which are separated from the continuous operation, are measured;

(4) The amount of uranium contained in cylinders and other containers or vessels, which are not used in the continuous operation, is also measured;

(5) The amount of uranium within the cascades and auxiliary absorption traps in operation (sampling and weighing uranium during operation are impossible) are estimated in accordance with a specified calculation procedure;

(6) The "running-in-process inventory" is the total of items (1)–(5).

In the centrifuge plant, the UF_6 inventory within the cascades is quite small (usually about 10 kg per operation unit in this plant) as compared with those in other UF_6 containing areas. Therefore the accuracy of RPIT is supposed to be almost the same as that when the "clean-out physical inventory-taking" is conducted.

3. PHYSICAL PROTECTION

In the pilot-plant design, the physical protection functions are stronger than considered necessary by the original requirements for Category III material in use and storage in INFCIRC/225. Ningyo-toge is located near the westernmost end of mountains running lengthwise through the Japanese mainland, and has a greater snowfall in winter compared with the sea-coast areas near Tokyo. In view of the snowfalls, the outer fence surrounding the plant has been made rigid and provided with a detection system against trespassers.

Entrance to the pilot plant area is only permitted to those who have obtained advance permission, and check gates, etc. are provided at each facility

TABLE I. PRECISION OF ^{235}U ISOTOPE RATIO MEASUREMENTS

Reference[a] (^{235}U mol%)	Reference value ($^{235}U/^{238}U$, × 10^{-2})	Measured value ($^{235}U/^{238}U$, × 10^{-2})	Precision (%)	95% confidence limits (× 10^{-2})
2.656	2.728	2.829	0.30	2.823 ~ 2.835
1.804	1.837	1.892	0.22	1.889 ~ 1.895
1.483	1.505	1.546	0.26	1.543 ~ 1.548
1.287	1.303	1.340	0.30	1.337 ~ 1.343

[a] All samples were imported through the USAEC.

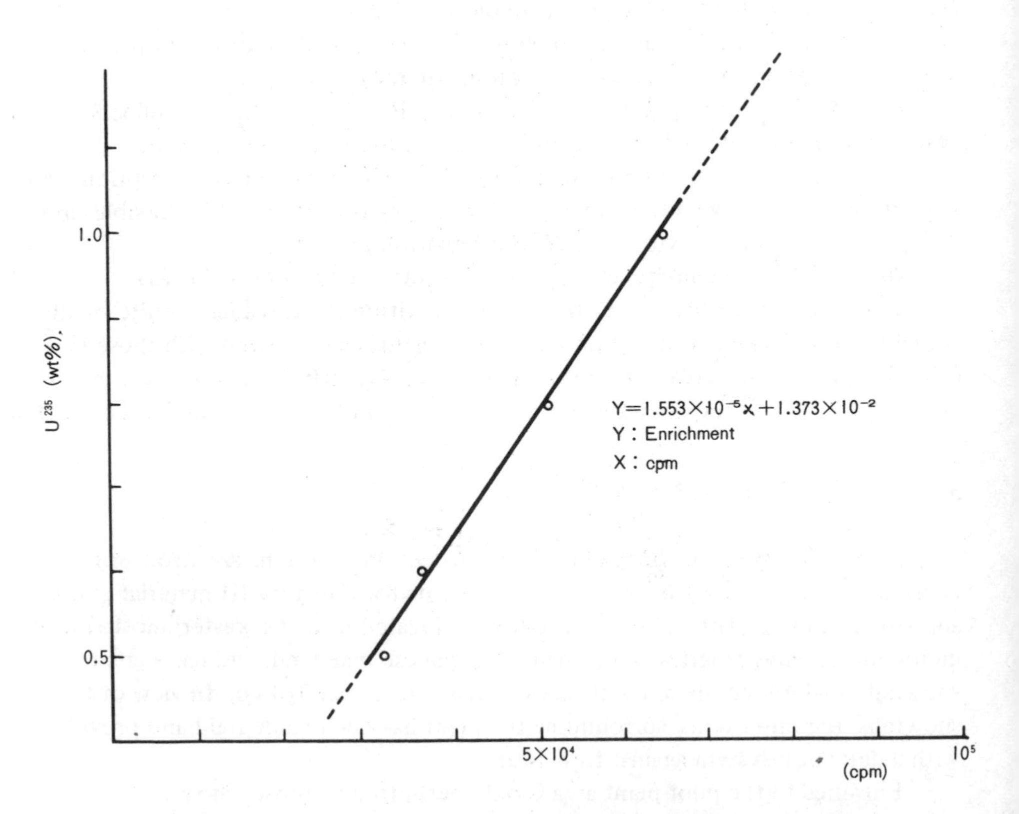

FIG.5. Empirical equation for uranium enrichment of UF_6 in the cylinder.

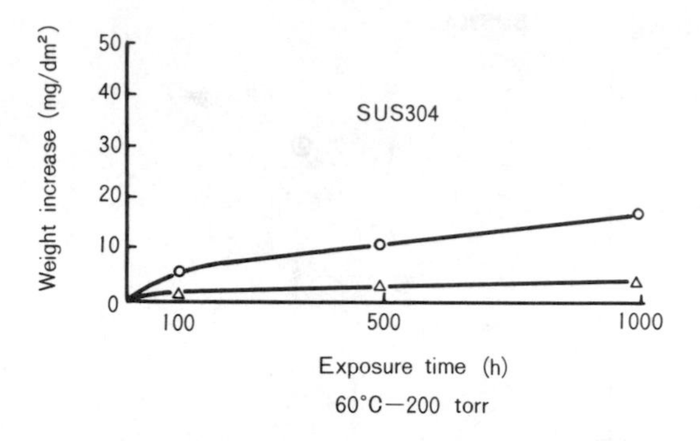

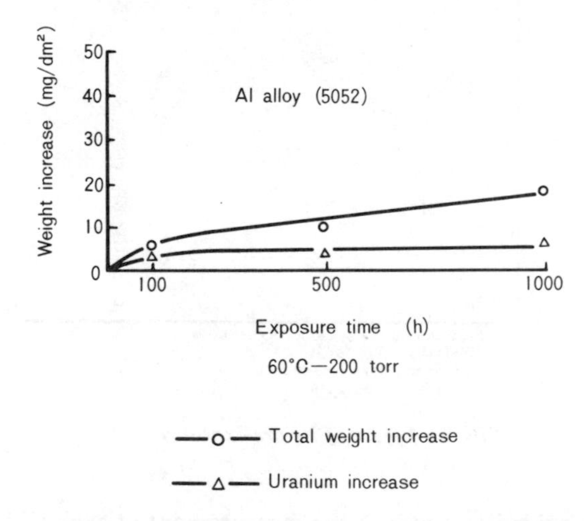

FIG.6. Weight increase on the metal surface.

to check qualified persons by an identification card. These protection systems are under centralized control and monitoring, and designed so that any event is recorded.

4. SAFEGUARDS TECHNOLOGY

Prior to the plant operation, various research and development programmes have been under way to render the safeguards measures of the pilot plant more effective.

A simulation analysis of nuclear material flow in the plant was conducted; operation loss due to changes in the plant operation conditions, hidden inventory and MUF were estimated; and the measuring precision and intervals at each measuring point were established to optimize the total measuring accuracy. As a

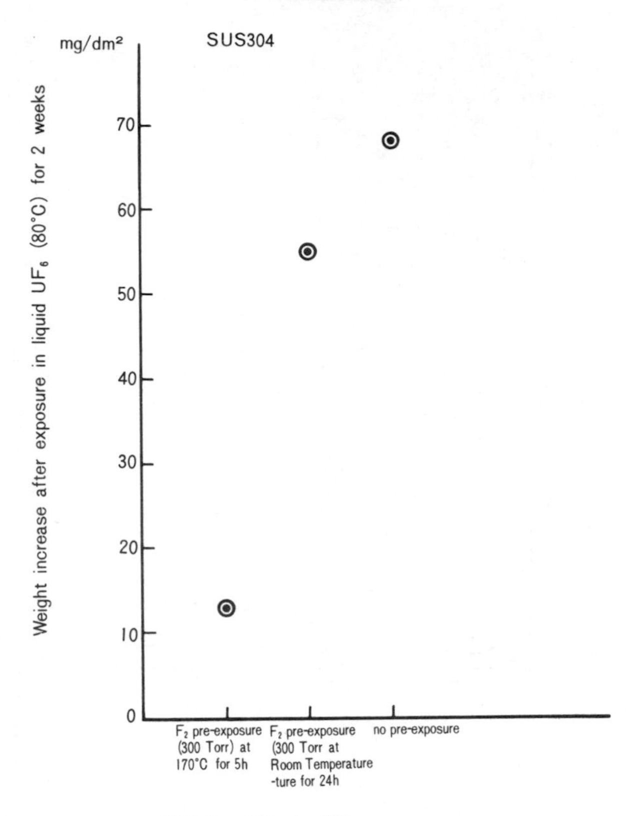

FIG.7. Effect of F_2 pre-exposure.

result, it has been decided to conduct the measurement at each KMP by off-line measurements; however, the material flow is designed so as to be controlled continuously and comparatively accurately through on-line measurement at the other auxiliary measuring points.

For auxiliary measuring means, an on-line mass-spectrometer and a non-destructive enrichment measurement method, based on gamma-ray measuring, have been developed.

The on-line mass-spectrometer is a quadrupolar-type mass-spectrometer, which continuously analyses the uranium isotopes ^{235}U and ^{238}U with a guaranteed precision of ± 0.30%. Table I shows an example.

The non-destructive enrichment measurement method measures the gamma ray of 185 keV and background, using two single-channel analysers, and determines the enrichment with the aid of a computer. Figure 5 shows an example of the measured curves obtained by this method, and it can be seen that the relative precision is as good as 5%. Differences of the per cent order of magnitude in enrichment are detectable by this method, so that the UF_6 enrichment in the UF_6 cylinder can be rapidly confirmed.

On the other hand, studies on the measurement of uranium deposit on metal surfaces, and pretreatment effects, are being made to decrease the amount of MUF and to estimate more accurately any hidden uranium which had stuck to the inner walls of process pipings. Uranium deposits on metal surface so far revealed by these studies are shown in Fig.6. In the case of stainless-steel SUS 304, the uranium deposit is saturated in about 500 hours and amounts to about 3.7 mg U/dm^2. This value does not differ much in the case of aluminium or other metal alloys. If the value 5 mg U/dm^2 is used, the uranium deposit is 500 g/1000 m^2 of the inner area of the process piping. As the amount of the deposit greatly depends on the metal surface condition, some positive measures have been taken to obtain a type of surface condition that would reduce the uranium deposit. One of these measures is to flow F_2 gas in advance on the surface in contact with UF_6 gas. The result is shown in Fig. 7. In this experiment, the sample was exposed to liquid UF_6 under an accelerated condition as compared with UF_6 gas. The weight increase was reduced by 80% when treated with F_2 at a high temperature (170°C), and by about 20% with F_2 treatment at room temperature. The uranium deposit showed a similar result.

DISCUSSION

K.J. QUEALY: Are we to understand that access to the plant is such that IAEA inspectors are not able to satisfy themselves, at first hand, as to the volume and enrichment of material held up in the cascade or in process?

T. MINATO: IAEA inspectors do not have access to the cascades, the uranium inventory of which is in any case quite small. However, IAEA inspectors have access to the process outside the cascades, and the volume and enrichment of material entering and leaving the cascades will be verified.

J.E. LOVETT: How frequently do you expect to use the running physical inventory-taking (RPIT) technique?

T. MINATO: We are planning to use it once a year.

W. FRENZEL: I have two questions. Firstly, as a plant operator, what would be your estimate of the MUF determined by such an annual running plant inventory? Secondly, in the event of a MUF figure of more than 0.4% of the annual throughput, would you clean out the cascades?

T. MINATO: We believe the MUF will be small. The measures needed if MUF is comparatively large will be determined by negotiation between the IAEA and the Japanese Government.

W. FRENZEL: I should just like to add, as a purely private opinion, that I think the IAEA ought to take into account the operator's point of view when designing its safeguards for such plants.

USE OF MINOR URANIUM ISOTOPE MEASUREMENTS AS AN AID IN SAFEGUARDING A URANIUM ENRICHMENT CASCADE

S.A. LEVIN, S. BLUMKIN, E. VON HALLE
Oak Ridge Gaseous Diffusion Plant,
Union Carbide Corporation,
Oak Ridge, Tennessee,
United States of America

Abstract

USE OF MINOR URANIUM ISOTOPE MEASUREMENTS AS AN AID IN SAFEGUARDING A URANIUM ENRICHMENT CASCADE.

Surveillance and containment, which are indispensable supporting measures for material accountability, do not provide those charged with safeguarding an installation with the assurance beyond the shadow of a doubt that all the input and output uranium will in fact be measured. Those who are concerned with developing non-intrusive techniques for safeguarding uranium enrichment plants under the Nuclear Non-Proliferation Treaty have perceived the possibility that data on the minor uranium isotope concentrations in an enrichment cascade withdrawal and feed streams may provide a means either to corroborate or to contradict the material accountability results. A basic theoretical study has been conducted to determine whether complete isotopic measurements on enrichment cascade streams may be useful for safeguards purposes. The results of the calculations made to determine the behaviour of the minor uranium isotopes (^{234}U and ^{236}U) in separation cascades, and the results of three plant tests made to substantiate the validity of the calculations, are reviewed briefly. Based on the fact that the ^{234}U and ^{236}U concentrations relative to that of ^{235}U in cascade withdrawal streams reflect the cascade flow-sheet, the authors conclude that the use of the minor isotope concentration measurements (MIST) in cascade withdrawal streams is a potentially valuable adjunct to material accounting for safeguarding a ^{235}U enrichment cascade. A characteristic of MIST, which qualifies it particularly for safeguards application under the NPT, is the fact that its use is entirely non-intrusive with regard to process technology and proprietary information. The usefulness of MIST and how it may be applied are discussed briefly.

INTRODUCTION

Material accountability is the basic and most direct means for safeguarding nuclear facilities against clandestine diversions of nuclear materials. Surveillance and containment, which are indispensable supporting measures for material accountability, do not provide those charged with safeguarding an installation with the assurance beyond the shadow of a doubt that all of the input and output uranium will in fact be measured. For this reason, some method to complement material accountability is desirable.

Those who are concerned with developing non-intrusive techniques for safeguarding uranium enrichment plants under the Nuclear Non-Proliferation Treaty have perceived the possibility that data on the ^{234}U concentrations, and that of ^{236}U when present, in addition to that of the ^{235}U in the cascade withdrawal and feed streams may provide a means to either corroborate the material accountability results or indicate that the integrity of the declared cascade operation is suspect. A basic theoretical study has been conducted to determine whether complete isotopic measurements on enrichment cascade streams may be useful for safeguards purposes. Numerous steady-state multicomponent concentration gradient and productivity calculations have been made to develop the fundamental data for this study. The results of the calculations have been reported in a series of five reports.[1] In this paper, a brief review and appraisal is made of the potential value of utilizing the measurement of ^{234}U and ^{236}U concentrations as well as that of ^{235}U in enrichment cascade feed and withdrawal streams as a safeguard technique.

SUMMARY AND CONCLUSIONS

The results of the calculations made to determine the behavior of the minor uranium isotopes in separation cascades and the results of the three plant tests made to substantiate the validity of the calculations are reviewed briefly. Based on the fact that the ^{234}U and ^{236}U concentrations relative to that of ^{235}U in cascade withdrawal streams depend on the cascade flowsheet, it is concluded that the use of minor isotope concentration measurements (MIST) is a potentially valuable adjunct to material accounting for safeguarding a ^{235}U enrichment cascade. Another characteristic of MIST which qualifies it for safeguards applications under the NPT is the fact that its application would be entirely non-intrusive on process technology and proprietary information.

REVIEW OF THE BEHAVIOR OF THE MINOR URANIUM ISOTOPES IN ^{235}U ENRICHMENT CASCADES

Minor Isotope Concentration Dependence

The study has shown that the steady-state concentrations of the uranium minor isotopes relative to that of the ^{235}U in the withdrawal streams of a separation cascade are determined by the following parameters:

A. The ^{235}U top product enrichment. The ^{234}U concentration relative to that of ^{235}U in both the product and tails streams increases with increasing ^{235}U concentration in the product stream. This is illustrated in Figure 1 which is a plot of the ^{235}U to ^{234}U concentration ratios in the product and tails streams of a squared-off cascade, designed to yield product at 4% ^{235}U and tails at 0.25% ^{235}U from natural uranium, when the cascade is operated to withdraw product with ^{235}U concentrations greater than the design product concentration.

The ^{236}U concentration relative to that of ^{235}U decreases in the product and tails streams with increasing ^{235}U product concentration. Figure 2 is shown to illustrate this. The ratios of the ^{235}U to ^{236}U concentrations in the product and tails streams represent the results of calculations for various ^{235}U product

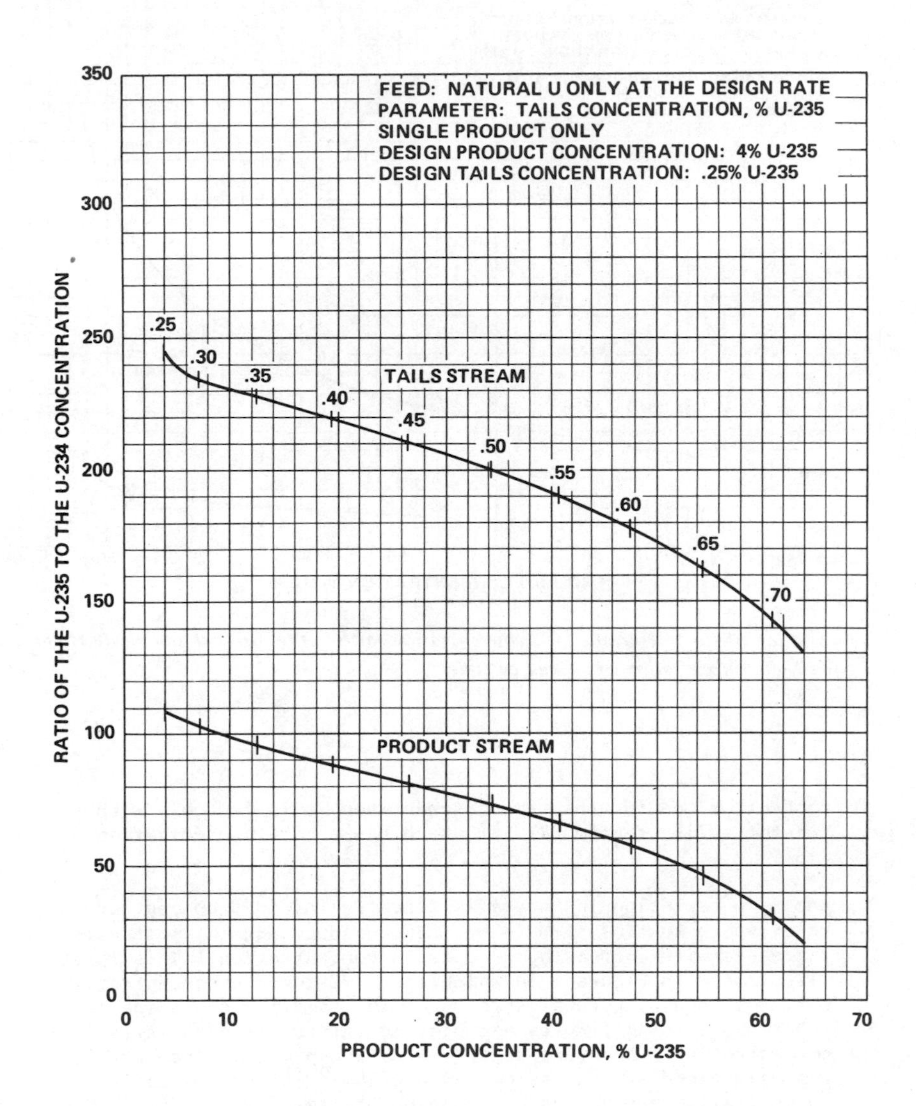

FIG.1. Variation of the ^{235}U to the ^{234}U concentration ratio in the product and waste streams with the product concentration of a fixed cascade.

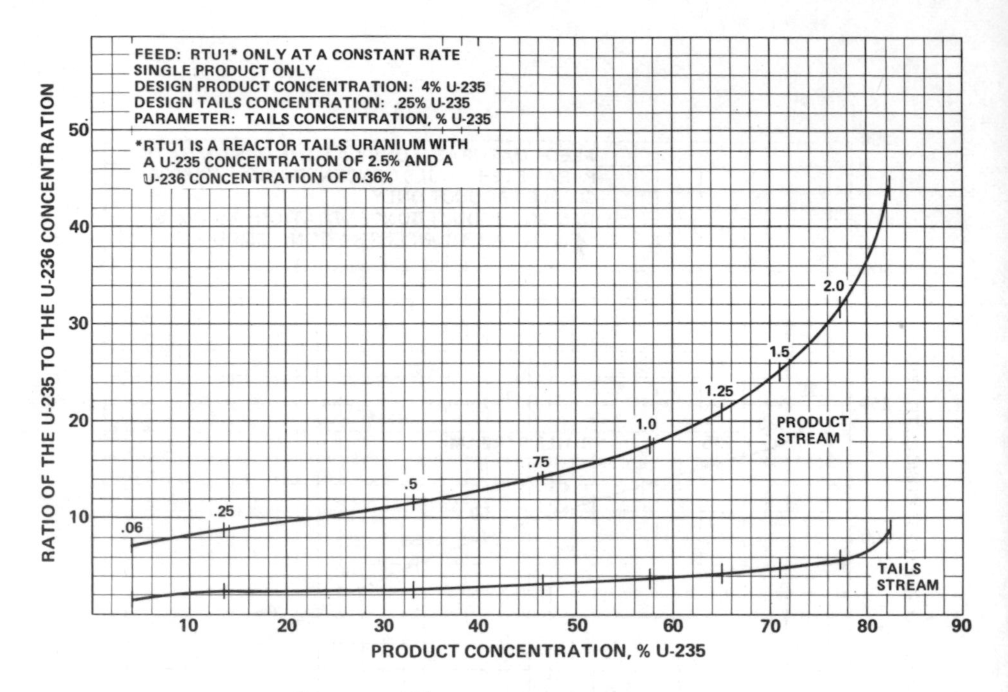

FIG.2. Variation of the ^{235}U to the ^{236}U concentration ratio in the product and waste streams with the product concentration of a fixed cascade.

concentrations for the same cascade when it is fed only with a reactor tails uranium (RTU1)[1] which has a ^{235}U concentration of 2.5%.

B. The ^{235}U depletion in the tails stream. The ^{234}U concentration relative to that of ^{235}U in both the product and tails streams increases with increasing ^{235}U tails concentration. This is illustrated in Figure 3 in which the ^{235}U to ^{234}U concentration ratios in the withdrawal streams of the same fixed cascade of the two preceding figures are plotted versus the ^{235}U tails concentration. For this set of data, the product concentration was maintained at its design value of 4% ^{235}U and the tails concentration was varied by changing the feed rate. The ^{236}U concentration relative to that of ^{235}U increases in the product stream but decreases in the tails stream with increasing ^{235}U tails concentration. Figure 4 demonstrates this behavior. The feed to the cascade is comprised of reactor tails U (RTU1) and natural U in constant proportions while the total rate of feed is varied to obtain the various ^{235}U tails concentrations.

[1] RTU1 is one of two species of reactor tails uranium used in this study to calculate the effects of non-normal uranium feeds.

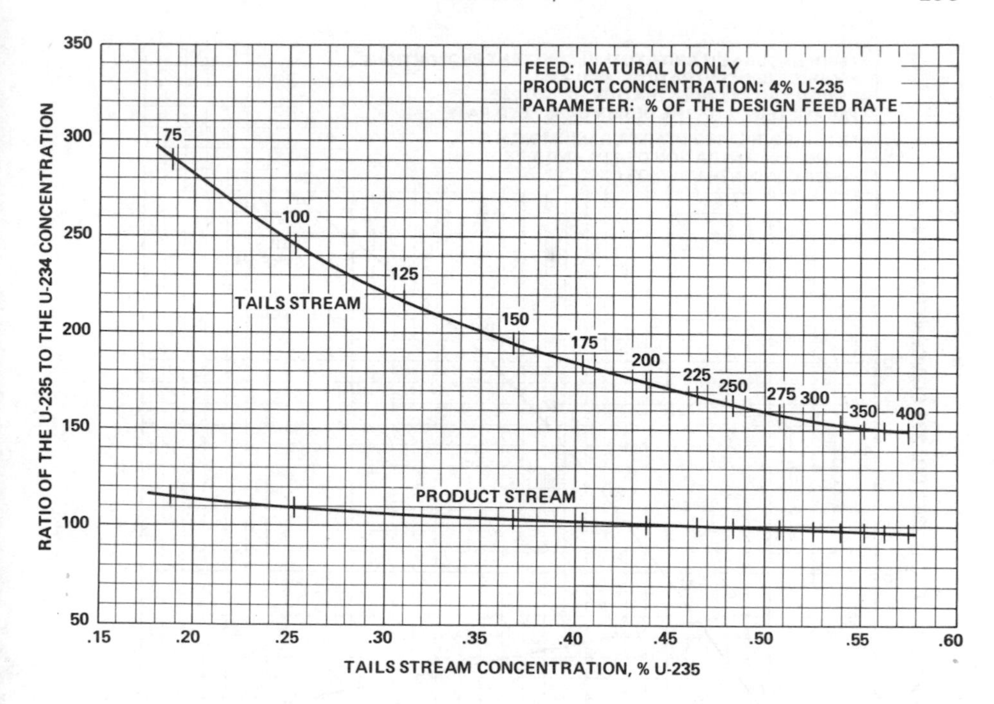

FIG.3. Variation of the ^{235}U to the ^{234}U concentration ratios in the product and tails streams with the ^{235}U tails concentration of a fixed cascade.

C. <u>The use of uranium other than natural as a feed to the cascade.</u> Obviously ^{236}U will appear in cascade withdrawal streams only if uranium containing the isotope, such as uranium recovered from spent reactor fuel, is or has been fed to the cascade. When non-normal uranium is fed, the ^{234}U and ^{236}U concentrations relative to that of ^{235}U in cascade withdrawal streams are affected by both the ratio of the concentration of the ^{235}U to that of the minor isotope and the ^{235}U concentration in the non-normal uranium, and by the fraction of the total feed it comprises.

The ^{234}U concentration relative to that of ^{235}U may be depressed or enhanced by the use of the non-normal uranium depending upon its isotopic composition, that is, upon the ^{235}U to ^{234}U concentration ratio and the ^{235}U concentration. Figure 5 demonstrates the effect of feeding one or the other of two species of reactor tails uranium in various proportions with natural uranium on the ^{235}U to ^{234}U concentration ratios in the withdrawal streams of a fixed cascade. The reactor tails uranium, designated as RTU1, has a considerably higher ^{235}U concentration and ^{235}U to ^{234}U concentration ratio than does that designated as RTU2. Figure 6 is the counterpart to Figure 5 illustrating the behavior of the ^{235}U to ^{236}U concentration ratios in the product and tails streams. The ^{235}U to ^{236}U concentration ratio in RTU1 is also considerably greater than it is in RTU2.

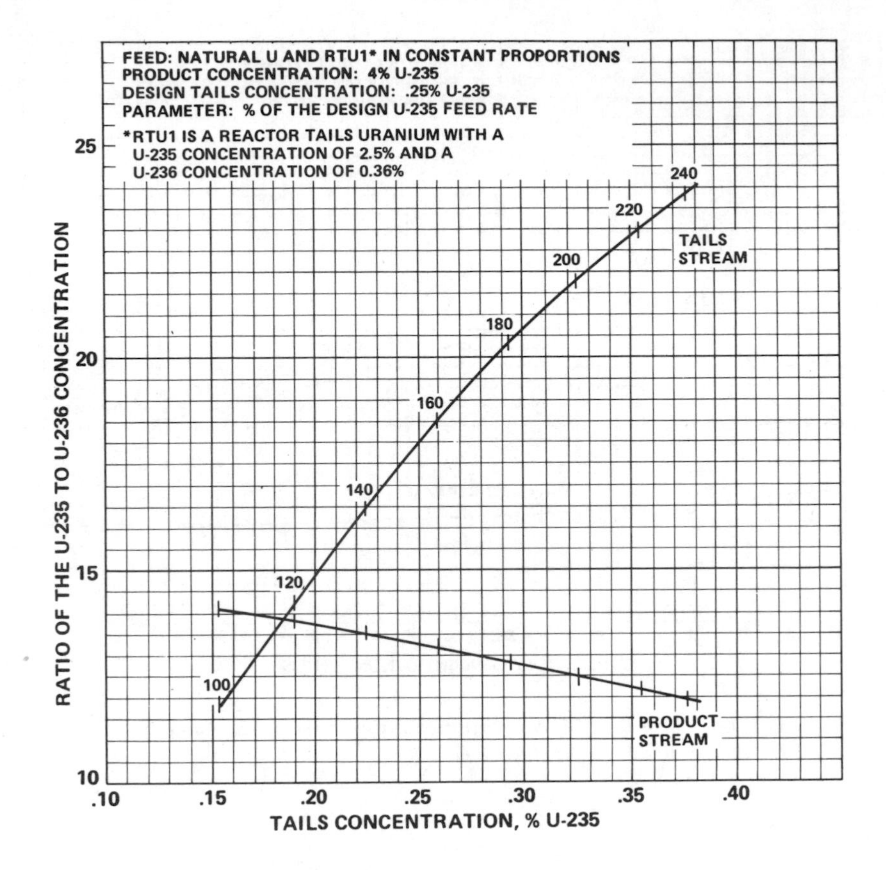

FIG.4. Variation of the ^{235}U to the ^{236}U concentration ratios in the product and tails streams with the ^{235}U tails concentration of a fixed cascade.

D. The existence of an additional product stream. When the withdrawal of a second product stream is begun, the minor isotope concentrations in the first product stream will be affected even if the ^{235}U concentration in it remains unchanged. The minor isotope concentrations relative to that of ^{235}U in the tails stream are also affected. The magnitude of this effect depends upon the degree of enrichment of the additional product and the fraction this product represents of the total cascade production. Figure 7 is presented to illustrate how the ^{235}U to ^{234}U concentration ratios in the first product stream and the tails stream are affected by the existence of an additional product stream.

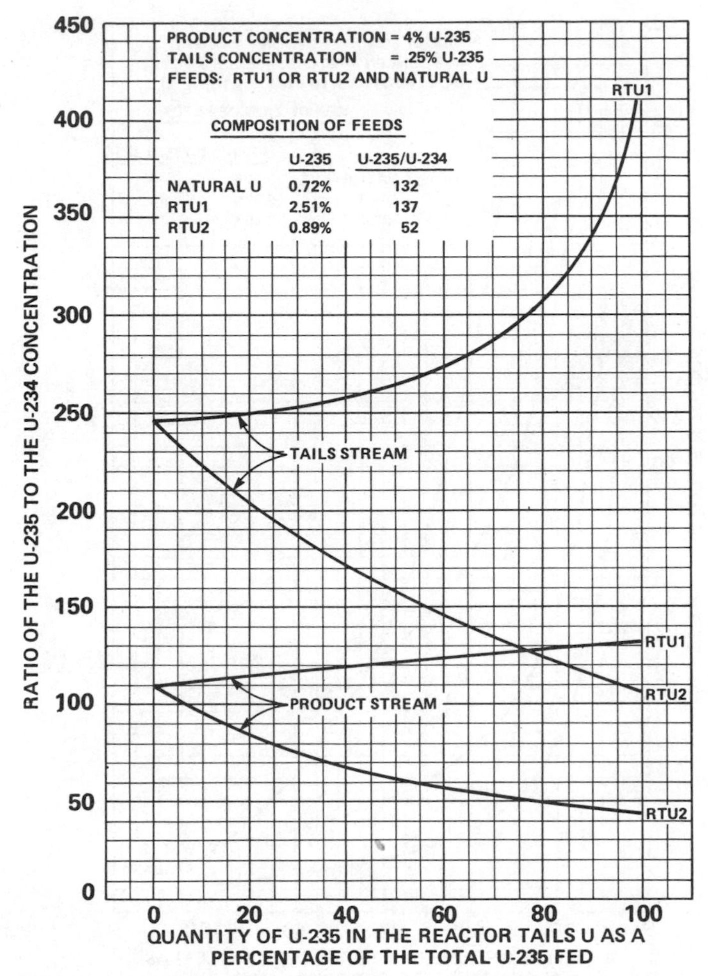

FIG.5. Comparison of the effects of two species of reactor tails uranium on the ^{234}U concentrations in the withdrawal streams of a fixed cascade.

E. The ratio of feed streams and/or the ratio of product streams in the case of more than one feed and/or more than one product stream. When the cascade mode of operation normally consists of more than three external streams, that is, more than one feed stream and/or more than one product stream, there is an infinite set of combinations of external stream rates for which the ^{235}U concentration in the cascade withdrawal streams can be maintained constant. However, the minor isotope concentrations in the withdrawal streams will be affected by the changes in the stream rates. Figures 5 and 6, which have already been presented for item C, above, illustrate this. While the ^{235}U concentrations in the cascade product and tails stream are constant at their design values, and the yield of product is increasing as the proportion of reactor tails in the feed is increasing, the minor isotope concentrations are changing.

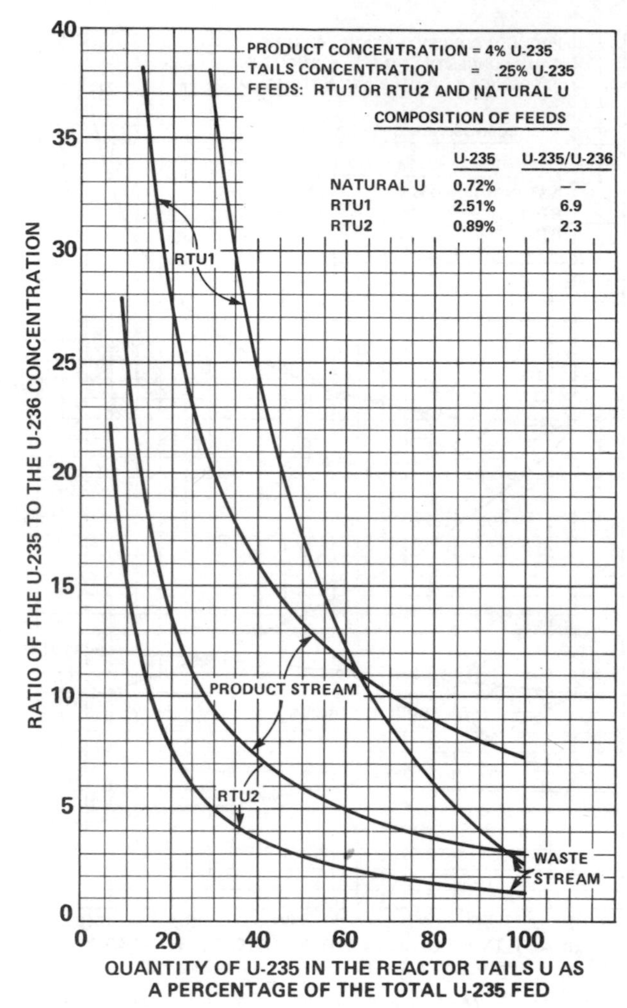

FIG.6. Comparison of the effects of two species of reactor tails uranium on the ^{236}U concentrations in the withdrawal streams of a fixed cascade.

F. The cascade design only when it deviates considerably from ideality. In general, the steady-state ^{235}U concentrations relative to that of ^{235}U in the withdrawal streams of a squared-off cascade differ measurably from those of the corresponding ideal cascade only when the separative efficiency of the squared-off cascade is less than 90%. The ^{236}U concentrations are somewhat more sensitive, becoming measurably different when the separative efficiency of the cascade is less than 96%.

G. Process gas losses. Calculations show that the minor isotope concentrations relative to that of ^{235}U in cascade withdrawal streams are affected by process gas losses that are uniformly distributed along the cascade. This effect becomes barely measurable only when the loss rate is greater than about 1.5% of the cascade feed rate.

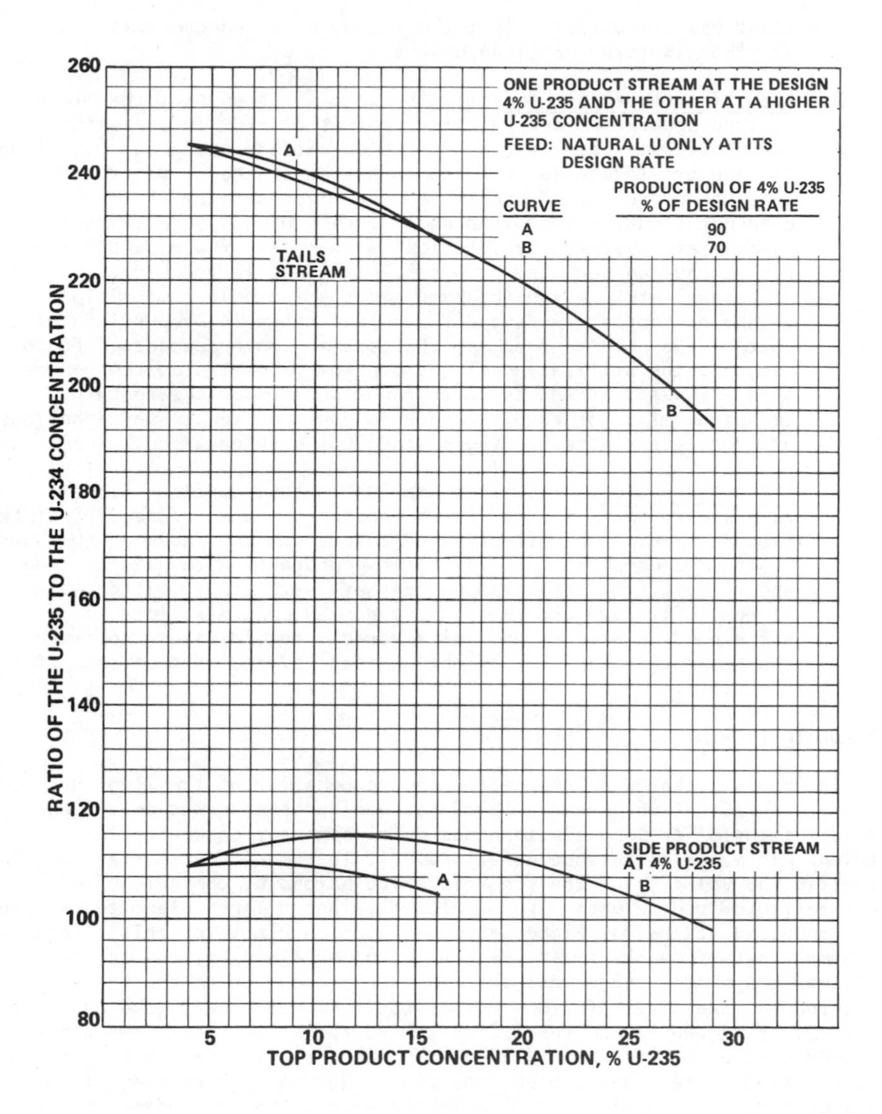

FIG. 7. The ^{235}U to ^{234}U concentration ratios in the design product stream and the tails stream for two-product operation.

Minor Isotope Concentration Independence

The study has demonstrated that the steady-state concentrations of the uranium minor isotopes are independent of:

A. The stage separation factor. It has been shown that the minor isotope concentrations relative to that of ^{235}U in the withdrawal streams of a gaseous diffusion cascade and a gas centrifuge cascade are independent of the magnitude of the $^{235}U/^{238}U$ separation factor. Furthermore, the minor uranium isotope concentrations in the withdrawal streams of a gas centrifuge cascade are essentially identical with those of a gaseous diffusion cascade when the ^{235}U concentrations in the corresponding withdrawal streams are the same and the isotopic composition of any non-normal feeds and their feed rates in proportion to that of the natural feed are the same for both cascades. Consequently, the utilization of the uranium minor isotope concentrations in cascade feed and withdrawal streams as a safeguard technique will not require any disclosure of sensitive technology or proprietary data for the cascade being safeguarded.

B. The cascade size. The minor uranium isotope concentrations relative to that of ^{235}U in cascade streams are independent of the cascade size. Two well-designed cascades using feeds of the same isotopic composition and in the same proportion to natural feed and yielding the same ^{235}U concentrations in their withdrawal streams will have the same ^{234}U and ^{236}U concentrations in corresponding product and tails streams regardless of how much the separative capacity of one cascade is greater than that of the other.

CASCADE TEST DATA

The basic background information on the behavior of the minor uranium isotopes in separation cascades which has been presented was developed from solutions of sets of simultaneous multicomponent concentration gradient and material balance equations. It is therefore important to establish the validity of the reported calculations by comparisons of measured isotopic gradients with predicted values. Three plant tests have been conducted at the Oak Ridge Gaseous Diffusion Plant for this purpose. The sample data from the first test which was obtained by standard routine analyses on a single stage mass spectrometer proved to be too badly scattered to make a satisfactory comparison. However, very good agreement was obtained in the second test[2] and third tests[2] between the measured and predicted values for the ^{235}U to ^{234}U concentration ratios when the determinations were made by high precision scans on a three-stage mass spectrometer. The results for ^{234}U in the second test are presented in Figure 8. The observed ^{235}U to ^{236}U concentration ratios in the second test on the other hand showed a large deviation from the predicted values. This was due to the ^{236}U concentrations not being at steady-state when the samples were drawn from the cascade. Unfortunately, the ^{236}U concentrations in the several cylinders containing the partially enriched uranium fed to the cascade during the three days just prior to the time of sampling had varied by a factor of three. At the time of the cascade sampling at the start of the third test, the primary objective of which was to make an indirect measurement[3] of the cascade inventory by means

[2] Results on cascade concentration data for the third plant test have not been published.

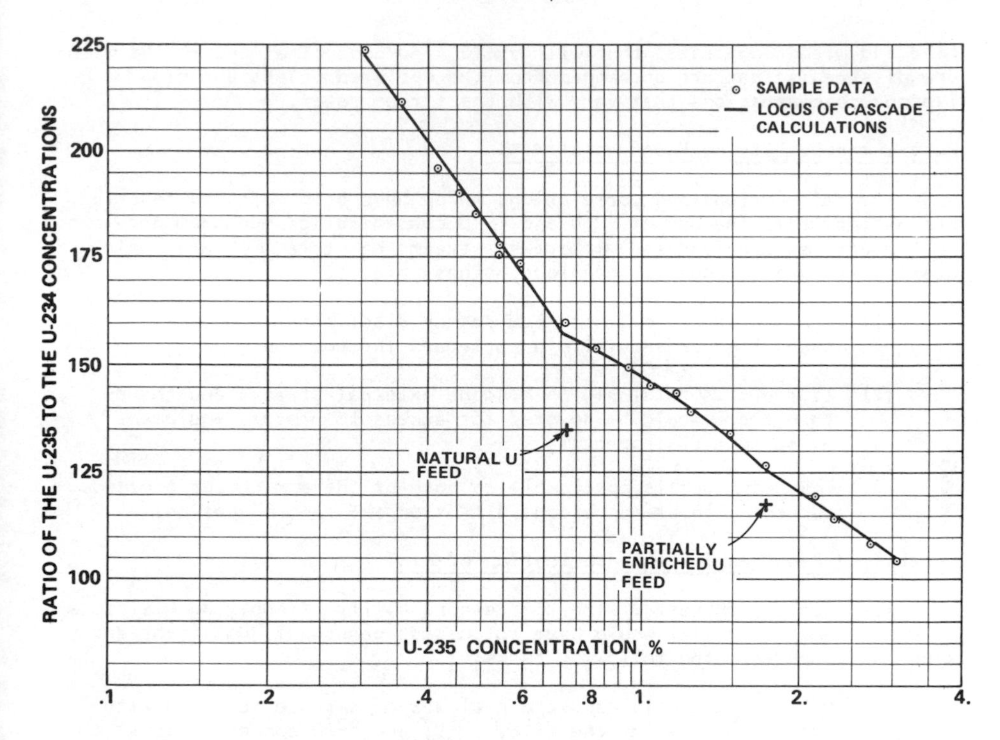

FIG.8. Comparison of measured and calculated ^{234}U and ^{235}U cascade concentrations.

of transiently enhanced ^{236}U concentrations, the ^{236}U concentrations along the cascade were everywhere < 10 ppm. Due to the imprecision of the analyses at such low concentrations, the ^{236}U data did not closely match the calculated cascade curve, but was scattered on both sides of it, thus grossly verifying it.

APPRAISAL AND APPLICABILITY OF MIST

The subject of utilizing measurements of the uranium minor isotopes as a safeguard technique has come to be known by its acronym: MIST. The data which have been obtained in this study definitely indicate that MIST has the potential of being a useful adjunct to, but not a substitute for, material accountability for safeguarding a ^{235}U enrichment facility. Because the ^{234}U and ^{236}U concentrations relative to that of ^{235}U in the cascade withdrawal streams reflect the cascade flowsheet[3] in a specific way, MIST can be quite valuable in confirming or contradicting the declared flowsheet and can serve as a means of determining the actual one when it fails to confirm the declared one. In order to utilize MIST, the flow rate of at least one stream (product, feed or tails) must be known. The most useful application of MIST would be for a multi-feed and/or multi-product case wherein there exists unlimited combinations of stream

[3] In its usage in this study, the word *flowsheet* refers to the flow rate and the ^{235}U concentration of each of the cascade feed and withdrawal streams.

feed and withdrawal rates that will yield ^{235}U concentrations in the withdrawal streams that are unchanged from the declared values but minor isotope concentrations that vary with the stream rates.

Particular Qualifying Features of MIST

The calculations and correlations which have been made and reported demonstrate that the use of ^{234}U and ^{236}U concentration measurements in enrichment cascade feed and withdrawal streams has three features which qualify it as a safeguards technique. These are:

(1) The procedure does not require any disclosure of process technology or sensitive proprietary information.

(2) It requires data only on cascade external streams and therefore there would be no need for access to process equipment by safeguards inspectors.

(3) The technique is applicable to both of the enrichment processes currently in use: gaseous diffusion and centrifugation.

Capability Required for the Application of MIST

An organization vested with the responsibility of applying MIST to a ^{235}U enrichment facility would have to possess personnel, equipment and know-how to perform the following tasks:

(1) Make sufficiently accurate analyses on cascade feed and withdrawal samples for the ^{234}U, ^{235}U, and ^{236}U concentrations therein.

(2) Design an optimum or nearly optimum squared-off cascade to do the separation job corresponding to the design flowsheet declared for the cascade under safeguards supervision.

(3) Carry out four-component concentration gradient and productivity calculations for various flowsheets for any specified cascade.

Application Costs

No attention has been given in this study to the cost of utilizing MIST. Matters such as the number of samples which will have to be analyzed, the cost of high precision isotopic analyses, the number of inspectors, requirements for back-up personnel and the cost of computing will determine the application cost. The number of inspectors and the numbers of samples required should not be in excess of what is necessary to carry out material accounting procedures. The incremental cost of MIST will largely be due to analytical costs and the requirements of back-up personnel to set up the appropriate calculations and interpret the results.

REFERENCES

[1] Blumkin S., and Von Halle, E., The Behavior of the Uranium Isotopes in Separation Cascades, Union Carbide Corporation, Nuclear Division, Oak Ridge Gaseous Diffusion Plant, Oak Ridge, Tennessee, K-1839, Part 1, November 21, 1972; K-1839, Part 2, August 29, 1973; K-1839, Part 3, March 22, 1974; K-1839, Part 4, December 18, 1974; and K-1839, Part 5, January 19, 1976.

[2] Blumkin, S., and Von Halle, E., A Preliminary Report on the Minor Isotopic Measurements in a Gaseous Diffusion Cascade: MIST Test II, Union Carbide Corporation, Nuclear Division, Oak Ridge Gaseous Diffusion Plant, Oak Ridge, Tennessee, K/OA-2388, November 12, 1973.

[3] Blumkin, S., and Von Halle, E., A Method for Estimating the Inventory of an Isotope Separation Cascade by the Use of Minor Isotope Transient Concentration Data, Union Carbide Corporation, Nuclear Division, Oak Ridge Gaseous Diffusion Plant, Oak Ridge, Tennessee, K-1892, January 13, 1978.

DISCUSSION

W. FRENZEL: In your paper you state that the MIST technique can be applied to both diffusion and centrifugation processes. Can it be used for control of a plant using the nozzle process?

S.A. LEVIN: Yes, MIST can be applied to any cascade of stages.

J.E. LOVETT: Have any sensitivity studies been performed to estimate the quantity of separative work which might be diverted within the uncertainty of the isotope ratio measurements?

S.A. LEVIN: Not as yet.

D. GUPTA *(Chairman)*: I should like to utter a few words of caution in connection with the use of MIST for safeguards purposes in uranium enrichment facilities. The basic assumption for its application is that the *same* cascade arrangement would be used for clandestine production of highly enriched uranium as for normal low-enriched uranium production. This may not be the case for a centrifuge plant.

Secondly, the accuracy of the ^{234}U or ^{236}U measurements must be very high compared to the normal fluctuations of these isotopes on account of process variations. Such high accuracies may not be possible under normal operating conditions.

S.A. LEVIN: With regard to your first comment I would say that if a centrifuge plant is composed of many independent cascades, then the possibility exists for separate production of HEU.

With regard to your second comment, I believe the results of the second and third cascade tests indicate that the accuracy of the ^{234}U measurements was sufficient, as shown by Figure 8. In the case of ^{236}U, if feed containing this component is supplied in a steady fashion, the same result as that for ^{234}U should be obtained.

A.G. HAMLIN: You state in your paper that one experiment was perturbed by the use of feed different from that which had been expected. Could an operator, who wished to nullify inspection by this method, confuse the results by apparently legitimate feed mixing to such an extent that the technique becomes totally useless?

S.A. LEVIN: In normal operation of an enrichment cascade, feeding would be steady, and as long as isotopic measurements of the feeds are made, MIST will still give valid results. If the operator continually changed the feed composition, thereby making operation more difficult, without any apparent reason, he would be suspect and an explanation would be required.

Session III (Part 2)

SAFEGUARDS TECHNOLOGY FOR FUEL FABRICATION FACILITIES

Chairman: D. GUPTA (Federal Republic of Germany)

AN INTEGRATED UNITED STATES/IAEA SAFEGUARDS EXERCISE AT A LOW-ENRICHED URANIUM FUEL FABRICATION FACILITY

A.M. BIEBER, Jr.
Brookhaven National Laboratory,
Long Island, New York

C.M. VAUGHAN
General Electric Company,
Wilmington, North Carolina,
United States of America

J. HORNSBY, G. HOUGH, E. KOTTE,
V. SUKHORUCHKIN
Department of Safeguards,
International Atomic Energy Agency, Vienna

Abstract

AN INTEGRATED UNITED STATES/IAEA SAFEGUARDS EXERCISE AT A LOW-ENRICHED URANIUM FUEL FABRICATION FACILITY.

The exercise was broken down into three basic activities: (1) preparation of basic IAEA documents describing the facility and the IAEA safeguards to be applied; (2) a reporting exercise in which for about six months the facility submitted all IAEA reports which would have been submitted had the facility actually been under IAEA safeguards; and (3) preparation and testing of IAEA inspection and verification procedures for the facility. The IAEA documents prepared included an example Design Information Questionnaire (DIQ) and Facility Attachment. The reporting exercise flow of data was from the facility to the US Nuclear Materials Management and Safeguards System (NMMSS) and then from NMMSS to the IAEA using magnetic tape suitable for direct input to the IAEA safeguards computer system. Facility physical inventory taking and verification by the United States Nuclear Regulatory Commission (USNRC) was observed by inspectors and inspection procedures were tested. Important facility records that would simplify implementation of safeguards were defined.

1. INTRODUCTION

In order to provide the United States of America and the IAEA with additional practical experience in the application of IAEA safeguards to a typical large LWR uranium fabrication facility, an integrated safeguards exercise has been conducted as part of the U.S. program for technical assistance to the IAEA.[1] An existing commercial fuel fabrication facility was used as the example for this exercise.

TABLE I. MATRIX OF KMP VERSUS LOCATION

KMP \ Material Location	Receiving Pad	Process Support Area	Warehouse A	Warehouse B	Rod Loading Area and Furnace Room	Special Fuel Shop	Bundle Assembly Room	Bundle Storage Area	Shipping Area	Laboratories and Misc.
A. Materials at Shipper's Values	X									
B. Misc. Samples, Displays, etc. Not in A or C-J	X	X	X	X	X	X	X	X	X	X
C. UF_6 Cylinders	X	X							X	
D. UO_2 Powder and Green Pellets	X		X	X		X			X	X
E. UO_2 Pellets and Unscanned Fuel Rods	X				X	X			X	X
F. Scanned Fuel Rods	X					X	X		X	X
G. Fuel Assemblies	X						X	X	X	
H. U_3O_8 Powders	X	X							X	
I. Solid or Sludge Scrap and Waste	X	X				X			X	
J. Recoverable Liquid	X	X							X	

NOTE: X in a square indicates that the type of material may be present at the location.
A blank in a square indicates that the type of material is never found at that location.

The exercise provided valuable experience to all participants and demonstrated that an investment in suitable interface computer programs by the operator, state and Agency will simplify implementation of safeguards, improve effectiveness and reduce operator interference. In addition, the use of a "real time" book inventory system at the facility may result in savings of the available inspection manpower.

Details of various facets of the exercise may be found in the appropriate documents listed in the References. In addition, one or more Agency reports will be issued that deal with the reporting exercise, inspection approach and facility records that facilitate safeguards. These documents should be of benefit to member states wishing to establish safeguards systems at such plants, and also to the Agency in order to enable inspectors to achieve maximum safeguards effectiveness with the minimum of interference to the plant operator. It is important to note that documents such as the Design Information Questionnaire and Facility Attachment which were used in the exercise were intended as examples and do not necessarily imply endorsement by either the U.S. government or the IAEA.

2. DESIGN INFORMATION AND FACILITY ATTACHMENT

The first stage of the exercise was preparation of the basic IAEA documents describing the facility and the IAEA safeguards to be applied. The aim of this stage was to check the Agency's standardized forms of the Design Information Questionnaire (DIQ) and Facility Attachment (FA) for LEU fuel fabrication plants as applied to a large modern commercial facility, and to provide examples in order to facilitate preparation of such documents for similar installations in the future.

The DIQ(2) describes the facility, its mode of operation, and its accounting, sampling and measurement techniques and procedures in detail. Sampling and measurement uncertainties for each flow and inventory material are given separately for element, isotope, and bulk determinations and are subdivided into random and systematic parts. Based on these data the calculations of MUF and LEMUF for a typical material balance period have been done.(3) The level of detail of the information in the DIQ permitted successful preparation of the forms used by the IAEA for input of design information to the IAEA computerized safeguards information system.

Among the unique features of the Facility Attachment(4) is the definition of inventory key measurement points in terms of material types rather than locations and the inclusion of the facility's material type codes in the data reported to the IAEA. Table I summarizes the final inventory KMP definitions arrived at. The selection of the inventory KMPs by material types made stratification of the inventory for verification easier, having in mind that the objective of stratification is to place all items on inventory into categories, where all items have the same chemical and physical form and have a similar range of measurement errors. Figure I also identifies the possible locations in which any given material type may be found. Locations are further subdivided into about 200 stations. Facility computer printouts summarizing the amounts of material by location and material type are designed to facilitate sample plan calculation, item counting and rapid location of individual items for remeasurement.

Facility material type codes as well as IAEA material description codes were included in the information reported on PILs and ICRs (as part of the batch name). This was done because at any given KMP, there may be several forms of material consisting of basically the same material type but measured with different errors. These materials have different facility codes but the same IAEA code. Unless these materials can be separated in reporting, it will be impossible for the IAEA to calculate correctly an independent estimate of LEMUF. Reporting by facility codes and IAEA codes partly solves this problem, by use of measurement uncertainties keyed to the more discriminating facility material description codes.

3. REPORTING EXERCISE

3.1 Background and Objectives

The basic objective of the reporting exercise segment of the overall integrated exercise was to test methods for transmission of safeguards data from a U.S. facility to the IAEA via the U.S. Nuclear Materials Management and Safeguards System (NMMSS), located in Oak Ridge, Tennessee. In particular, the reporting exercise allowed testing of a new flexible format(5) for providing data to the IAEA on computer-readable magnetic tape suitable for direct input to the IAEA Safeguards Information System (ISIS). The reports submitted were the three basic types specified in INFCIRC/153: the Physical Inventory List (PIL); the Inventory Change Report (ICR); and the Material Balance Report (MBR). The ICR is roughly equivalent to the U.S. form NRC/DOE-741, and the MBR, to the U.S. NRC/DOE-742; no U.S. equivalent of the PIL now exists.

3.2 Computer Programs for Conversion of Data from Facility to NMMSS Formats

Two sets of computer programs have been prepared: one set for conversion of facility data to NMMSS formats and codes; and one set for preparation of IAEA-format tapes from NMMSS data. Three different problems had to be solved by the facility-NMMSS conversion software. First, the basic data on nuclear materials at the facility had to be converted from item-by-item data to batch data. Second, facility codes and reporting units had to be converted to U.S. NMMSS codes and units, and the data reformatted according to NMMSS standards. And third, data such as batch names, KMP identifications, and measurement basis codes which are required by the IAEA but not by the U.S. had to be developed for each batch.

Of these three problem areas, the grouping of data on items into batch data proved to be the most difficult. The batch definition given in INFCRIC/153, while precise, is not helpful in deriving an operational definition of batch for a facility where accounting is on an item rather than a batch basis and there is really no "natural" batch for most materials. Likewise, the descriptions of typical batches given in Code 4 of the example Facility Attachment proved difficult to use. Eventually, items were grouped together as a batch if they satisfied all of the following conditions: 1) all items in the same MBA and KMP; 2) all items have the same facility material type code; 3) all items have the same NMMSS material type code; 4) all items have the same measurement basis code; 5) all items have the same measurement error code (Table II). This proved not to be completely correct in all cases but it was sufficiently good to be practicable for the purposes of the exercise.

TABLE II. MEASUREMENT ERROR CODE

Error %	ME Code	Error %	ME Code
>0-0.1	A	>10.0-15.0	J
>0.1-0.3	B	>15.0-20.0	K
>0.3-0.5	C	>20.0-30.0	L
>0.5-1.0	D	>30.0-40.0	M
>1.0-2.0	E	>40.0-60.0	N
>2.0-3.0	F	>60.0-80.0	O
>3.0-4.0	G	>80.0-100.0	P
>4.0-5.0	H	>100.0	Q
>5.0-10.0	I	Not available	Z

$$\text{Error \%} = \frac{\text{Limit of Error (Isotope)}}{\text{Total Isotope Wt.}}\ 100 \text{(for transactions only)}$$

The error % is determined using limit of error and isotope wt. information from the form 741 input data. If no limit of error is available, the ME code is assumed to be Z (zero).

NOTE: The MEC for transactions, since it is based on the total (random and systematic) error for a line, will be different from the MEC for the same material in inventory, which is based on random error only.

Conversion of facility codes to NMMSS codes and formats proved relatively easy to do automatically. The extra data required by the IAEA but not by current U.S. reporting in most cases also proved fairly easy to develop. The requirement for unique batch names on ICRs was difficult to meet, and because U.S. regulations do not require detailed data on measurement basis to be reported, the measurement basis codes were fairly difficult to derive correctly and it proved to be impossible to derive them correctly in some cases.

Two computer program systems were written for conversion of facility data to NMMSS formats. One takes as input facility reconciled item inventory data and produces a tape in NMMSS format listing batches of material on inventory with all data necessary for NMMSS to construct a PIL. This is referred to as the inventory conversion system. The other takes as input punched cards containing basic transaction (shipment, receipt, discard) data required by U.S. regulations, and produces a printed transaction document, a transaction history file, and a file which may be written to magnetic tape for transmission of the data to NMMSS. This is referred to as the transaction conversion system.

Briefly, the inventory conversion system works as follows. The facility maintains an internal current information computer accounting system which stores important data on all discrete items containing nuclear material in the facility. At the time of a physical inventory, the data in this system are reconciled with the results of the inventory, and a magnetic tape is prepared which lists all items in the inventory.

This reconciled inventory tape, which is the primary input to the inventory conversion system, contains a list of all discrete items on inventory with a location code, a two-character material description code, item identification number, gross, tare and net weight, weight of element and isotope and other data listed for each item. Data on batches of material not in the form of discrete items (principally holdup in process equipment) are input to the system on cards. Such material typically makes up less than 1% of the total inventory. The conversion system basically reformats the input data, derives (from tables input on cards) a NMMSS inventory composition code, a measurement error code, a KMP code, and an MBA code for each item, derives a measurement basis code, aggregates items into batches, and writes a magnetic tape containing the batch data in NMMSS format.

A novel feature of the inventory conversion system is the use of the batch name to convey useful information. On the inventory list sent to NMMSS, the batch name is composed of eight characters. For items other than fuel assemblies, the first two characters are the facility material type code, the next is the KMP, then the IAEA element code (D, N, or E), a code indicating the approximate range of the measurement error for a single itme in the batch (see Table I), the IAEA measurement basis code (M, N, T, or L), and last, a two digit sequence number which can be incremented to distinguish between two otherwise identical batches. For fuel assemblies, the first two characters of the batch name are the facility material type code, and the remaining six are the assembly identification number. The final output of the inventory conversion system is a magnetic tape containing a list of all batches of material on inventory.

United States transaction reporting regulations require only transfers between facilities to be reported, while the IAEA requires reporting of transfers between those MBAs defined for IAEA reporting. For the purposes of the exercise, the IAEA divided the facility into two MBAs: U-WS, the shipper/receiver MBA, containing all material in the facility held at shippers' values; and U-WM, the process MBA, containing all material which has been measured by the facility. The act of measurement of material (or acceptance of shipper' values) constitutes the transfer from U-WS to U-WM. Since the form 741 contains both shippers' data and receivers' data, and the date (as the date of signature by the receivers' responsible officer) of the measurement by the facility, NMMSS can construct ICRs for transfers between the two MBAs at the facility from the data contained on 741s without requiring additional reports to satisfy IAEA reporting requirements.

The primary inputs to the transaction conversion system are cards containing virtually all the information required on 741 forms. The batch name (containing the same data in the same format as for inventory reporting), inventory change type, flow KMP number, and measurement basis are generated internally by the computer program, and shipper and/or receiver addresses are looked up in an external (tape or disk) file by the program. Subtotal and/or total weights for the transaction are computed. The program writes a transaction history file which contains essentially the data input plus some dates. Modified 741 forms containing the extra information necessary for construction of ICRs are printed. Data on each transaction are written out on magnetic tape for transmission to NMMSS.

3.3 NMMSS Software System and Data Base

Data base operations(6) for the exercise were performed in parallel with the on-going operations of the NMMSS. The flow of information in the conventional data systems was not disrupted at any stage of the exercise. The expanded forms of the data representing the exercise were processed and interacted with the conventional system to verify that there was no loss of information content for U.S. national requirements under existing regulations.

Four integrated software subsystems were developed to process the data: the reference data subsystem; the PIL subsystem; the ICR subsystem; and the MBR subsystem.

3.3.1 Reference Data Subsystem. The reference data subsystem consists of software which creates and maintains the materials descriptors file and generates formatted listing of the file. The file is used to edit the PIL and ICR data and to convert the material descriptors from the facility/NMMSS codes to the IAEA codes.

3.3.2 PIL Subsystem.(7) The PIL subsystem consists of an edit/update program and an NMMSS-IAEA data converter and tape generator. The edit/update program verifies that all the data elements meet the requirements (content and form) of both the U.S. and the IAEA, and updates the file containing the edited data. The program also generates a data list and an error list.

The NMMSS-IAEA data converter and tape generator accesses the inventory file and produces a PIL on magnetic tape which conforms to the specifications of the IAEA Document SISD/9. It also generates two lists of the data showing the IAEA codes (one in readable format and one in the form of the data as written on the tape).

3.3.3 ICR Subsystem.(8) The ICR subsystem consists of eight software units which perform the following functions:

1) Edit the data to verify that it satisfies IAEA requirements. Produce data and error listings.

2) Edit the data to verify that it satisfies U.S. requirements. Produce data and error listings.

3) Merge the data edited above.

4) Update the transaction file, produce error listing.

5) Convert the data file structure from the queued sequential access method to the basic direct/indexed sequential access method.

6) List the data in a readable format similar to that of the data input forms.

7) Convert the data from the NMMSS structure and format to the IAEA structure and format. Identify all corrections. Produce data listings showing IAEA codes (readable format, tape images).

8) Supply backward referencing information (where available) for corrections. Generate an ICR tape as specified in SISD/9.

3.3.4 MBR subsystem.[9] The MBR subsystem consists of two software units which perform the following functions:

1) Access the inventory data file and for given MBAs and dates generate beginning and ending inventories for each nuclear material at each MBA.

2) Using the beginning and ending inventories generated above, for each MBA and material type, summarize all ICRs representing events which took place during the material balance period, compute shipper-receiver differences, compute adjusted book ending inventories, and compute MUF. Generate and MBR tape as specified in SISD/9. Produce data listings showing IAEA codes (readable format, tape images).

TABLE III. REPORTING EXERCISE REPORTS FOR COUNTRY: USA MBA: U-WM, YEAR 1977

Report Type	Report No.	From Date	To Date	No. of Entries	Status	Report Date
PIL	000001	770314		1209		770517
ICR	000002	770315	770329	104	N	770822
ICR	000003	770322	770607	640	N	770902
ICR	000004	770601	770809	712	N	771024
ICR	000005	770601	770805	115	C	771026
ICR	000006	770420	770807	11	N	771219
PIL	000007	770812		1761		771219
MBR	000008	770314	770812	28	N	771219

3.4 IAEA Computer Programs

The data provided during the reporting exercise were used to test various parts of the new programs comprising the IAEA Safeguards Information System.[10] Integration of the exercise into the development process, use of the data from a complex facility for testing and debugging, and the necessary interactions with those supplying the information provided extremely valuable support for the development effort. Table III shows the reports received from NMMSS on magnetic tape for the process MBA. The number of entries indicates how many lines were contained in the report in question. The status is coded by "C" in the case of corrections.

The information processing system is composed of three major parts: input processor; postload processor; and output processor. The input processor section of the information system provides for all functions of transforming data from the various forms and formats in which it reaches the computerized part of the system into a standard form for loading into the data base. Data are expected to arrive, in the future, on a variety of media including hard copy, punched copy, punched cards, paper tape and magentic tape. A wide variety of structures and formats of data must also be accommodated. The input processor is designed in a modular fashion so that new structures and forms of data can be accommodated with minimum effort. All functions (modules) of the input processor have already been programmed. For the purpose of the exercise, the functions "registration", "preload and load", and "archive" were applied.

The post-load processor section of the information system is composed of many modules for performing routine operations on the data after it is loaded into the data base. One of the principles embodied in the system is that essentially all data are stored in the data base as reported -even if they contain syntactical or logical errors. Error analysis, correction processing, and maintenance of a historical trace of errors and corrections are major functions of this part of the system. It also includes such operations as dimensional units conversion and conversion of numbers from character representation into binary numbers for use internally by the computer.

The functions of the post-load processor are again represented by various modules, of which the accounting data processor is almost completely implemented and the design data processor is partly completed. Therefore the functions "record retrieval", "header operations", "diagnostic checking", "data conversion", "record activation and deactivation", and "correction chaining" were applied to the exercise accountancy data. Some more work has to be done for the function "correction chaining" and for the functions of the design data processor.

The output processor includes the modules which produce summaries or detailed reports for use by the safeguards staff. Some modules consist of operations for selecting, sorting, and printing summaries of data elements from the data base as specified by users of the data. Other modules consist of operations on specified sets of data to derive results for use in evaluation of various aspects of safeguards functions.

The major conclusions drawn from the exercise, in terms of the mechanics of reporting, are that the new flexible format approach to reporting and use of a data base management system as a major part of the information system are viable and practical. More detailed analyses using both the accountancy and design data will be carried out in the future as the necessary programs are developed.

4. INSPECTIONS

4.1 Physical Inventory Taking

Staff of the Agency's Department of Safeguards observed a physical inventory taking by facility personnel, and the independent verification by Nuclear Regulatory Commission (NRC) inspectors. Before each semiannual inventory taking the operator converts the majority of material into reliably measurable and identifiable forms (e.g., minimum amount at green pellet stage) in order to obtain the most accurate possible figure for his inventory. All material on the inventory is measured during the pre-inventory period and part is stored under seal. Each location and each item is tagged and easily identifiable. The computer input is made by plant operators by means of 75 terminals. An inventory list can be established for each location within a few hours.

During the actual inventory taking the plant is closed down. In order to take the inventory the operator employs about 25 teams to cover all the groups of inventory items. The NRC inspection team consisted of the team leader and six inspectors. Much effort was expended during the reconciliation period to finalize the inventory list and to evaluate the data collected during the inspection before the material balance could be confirmed.

The careful advance preparations by the operator greatly facilitated the inventory taking which was executed very well. The existance of an effective national system of safeguards and the cooperation of the operator greatly assisted the NRC inspectors. The ready availability of an itemized list, by location and type (see Table I), enabled stratification of the inventory and permitted timely application of sample plans. The operators measurement basis is well defined and all items on the inventory were measured before the end of the physical inventory taking. Of special value was the use of the operator's installed NDA equipment to measure difficult materials such as fuel rods and waste boxes.

4.2 Inspection Effort

IAEA and U.S. NRC inspection activities are essentially the same. However, the goals are different. U.S. NRC has to check whether the operator complies with federal regulations to assure that the facility has the capability to meet safeguards requirements specified in the license to operate and that performance is adequate. The Agency has to provide for timely detection of the diversion of significant quantities of nuclear material and the deterrence of such diversion by the risk of early detection. This results in different inspection frequencies for flow and inventory verification.

For full effectiveness at a plant such as that chosen for the exercise, which has a throughput of approximately 600 t/year of LEU and semiannual inventory takings, it is probable that continual inspection by the IAEA would be necessary. However, this may prove to be impractical and implementation could, for example, consist of 12-24 equally spaced inspections per year. The semiannual physical inventory taking would perhaps require the presence of 5-6 inspectors at the plant for a period of 10 days. The remaining inspections would perhaps require the presence of 2 inspectors for 3 days to verify the flow of material within and into and out of the plant by means of independent samples and measurements, to update the facility book inventory and to audit the various facility records. This makes a total of 160 - 252 man-days per year.

4.3 Important Facility Records

A special effort was made to define facility records that would facilitate inspections. There are three basic categories of records at the facility which are relevant for IAEA inspection: 1) source data for the physical inventory; 2) source data for inventory change reports, and 3) source data for LEMUF calculations. Records for the first two categories are similar and are summarized as follows:

1) Dictionary of Material Description Codes. A cross index of facility, NMMSS and IAEA material description codes which is maintained in NMMSS and made available to the inspector 1-2 weeks before the inspection.

2) Prelist of Items. Contains important identification, source and batch data for every item organized by location and material type. The list is divided up among each inspection team to facilitate item counting and random selection for physical checks.

3) Prelist Summary. Provides total element and isotope weight and number of items for each material type and location (Table I). Available 1-2 weeks before inspection to facilitate stratification and preparation of sample plans.

4) Source Documents. Supporting documents such as shippers documents, lab analysis reports, bulk weight or volume records, fuel rod scanner reports, and basis for average factors used for element and isotope concentration. Used to check consistency of records on a statistical sample basis.

5) Final Item List. Same as 1) above after verification is complete and all corrections are made. Usually available 2-3 weeks after inspection and is organized into batches for each element code (D, N, E).

6) Final List Summary. Same as 2) above after verification is complete. Organized by batch totals which becomes the basis for error propagation and reconciliation of facility records with PIL or ICR.

In the case of inventory change transactions some of these records such as the prelist and final list summaries are not available from the facility transaction conversion system but are available from NMMSS. Consideration is being given to providing some of these lists on magnetic tape for direct input to the IAEA computer. This would greatly reduce the records audit burden on the inspector in cases where 30,000 inventory items are not unusual.

Two additional facility records are needed for inventory change transactions:

1) Shipper-Receiver Difference Record. Provides a list of the important source and batch data for each shipper-receiver transaction and may include limit of error calculations. This record is available from NMMSS.

2) Liquid Waste Discharge Documentation. A summary of important laboratory results and total uranium discharges for a given period. (Solid waste discards are measured using NDA equipment and can be easily verified by the inspector using item count and random sampling methods).

4.4 LEMUF Calculations

If the LEMUF exceeds 0.5% (two standard deviations) of additions to or removals from the process (whichever is the larger quantity) a lack of control is indicated under U.S. NRC regulations. Both random and systematic error are determined for bulk, analytical and NDA methods. Only random error is determined for sampling methods. These errors are expressed as relative standard deviations (RSD).

Systematic RSDs are calculated from data obtained from measurements of standard weights or materials. Standards must be traceable to the National Bureau of Standards and their assigned value must cover the range of the process operation.

Determination of the random RSDs for each measurement system requires a minimum of fifteen replications, or 100% replication if the population is less than fifteen, for each material balance period. Replicate measurements are made on material generated, added to, and removed from the process. These measurements are made at random throughout the period in a manner such that all operating conditions are covered. Materials subjected to sampling and analysis have multiple samples per container to estimate sampling error, and duplicate analyses per sample to estimate analytical error.

The measurement control or replicate values for distinct material types or measurement systems and the operator's values, when available, are analyzed by a hierarchical analysis of variance computer program for gross weights, tare weights, U-factors, E-factors, and scan values for U-238 and for U-235. The mean and variance are calculated for all of the above variables; in addition to the mean and variance of the difference between the replicate value and operator's value are calculated when these data are available. Weight data are processed by another computer program. Output of this program is a listing of the mean net weight, and the random and systematic RSD of net weight for each scale.

A different computer program is used to compute operator sampling or within-container variance for the materials routinely sampled by the shop operators for U-factor and E-factor. This program also processes the relevant random and systematic components for each U-factor and E-factor. These components as well as the corresponding error components for weight or volume are stored in a file to be read by the computer program which calculates LEs for shipment, receipts and in-process ending inventory.

The variance of MUF was determined manually at the time of the exercise, taking into account correlation between beginning and ending material balance components (by deleting the common items), the number of samples measured (sampling matrix) and systematic error correlations for items measured by the same technique. Since then, computer programs have been developed for these calculations.

These records of the error estimates and the error propagation for each material balance are available for examination by the inspector and methods of summarizing the important data are under consideration.

5. SUMMARY OF PROBLEMS

One of the major purposes of the exercise was to uncover any problems that may exist in U.S. reporting to the IAEA and in the implementation of inspections. It was not intended to find solutions to all problems within the scope and time frame of the exercise. However, the listing of these problems may be one of the most important results of the exercise. Some of the more significant problems are listed below without explanation:

1. U.S. reporting conventions require reporting of quantities to the nearest whole unit (gram or kg) and quantities which round to less than one whole unit are replaced with an asterisk, which means that NMMSS reports may differ from facility records. Also rounding adjustments are included in the overall MUF instead of as separate entires. The Agency prefers reporting of unrounded data and separate entries for each rounding adjustment that can be related to its origin.

2. The NMMSS assumes that all normal uranium is 0.711%. Thus the U-235 totals reported to the IAEA may differ from those in facility records. This problem could be avoided by using the unified uranium concept.

3. One of the most difficult problems encountered in the exercise was that of formulating an operational definition of batch. The definition of batch presented in INFCIRC/153 is difficult to apply in practice. Facility records are based on items and the concept of batch is artificially created in NMMSS to satisfy Agency requirements. Requirements for unique batch names also present problems. This is done by NMMSS but if the same batch were to reappear on two successive reports, it would be incorrectly assigned different batch names in reports to the IAEA.

4. Fuel assemblies shipped contain rods of several different enrichments. NRC instructions call for listing the number of rods in a shipment separately for each enrichment. The Agency requires total U and U-235 for each assembly. The number of lines of data reported for a typical shipment of 32 assemblies increases from 10-15 lines to several hundred if both requirements are met. However, the Agency would require only 32 lines, one for each assembly.

5. Under U.S. regulations cumulative shipper/receiver difference accounts are maintained by NMMSS rather than by facilities. The Agency requires such records to be maintained at the facility. NMMSS can create these records and provide them for use by inspectors.

6. The IAEA requires that a correction apply to the original line entry (batch). U.S. procedures require a transaction to adjust the next book inventory. Thus, the Agency and NMMSS book inventories may not agree.

7. There was no reliable method of assigning the correct measurement basis code (M, N, T, L) and measurement error code, especially for items that are measured by the shipper or at a referee laboratory. This also results in incorrect batching.

8. Waste containers with measured element or isotope weight of zero are classed as depleted uranium which may result in the incorrect total numbers of items in each enrichment category. The unified uranium concept would solve this problem.

9. Measurement error codes for inventory are derived from the random error only. For transactions they are derived from

random and systematic error. Thus the same material type may be assigned different measurement error codes.

10. It was difficult to obtain a clear understanding of measurement error calculations, even though the methodology applied by the operator is very advanced. Better summaries are needed to enable the Agency to examine the records quickly and extract the minimum data for calculation of LEMUF on the IAEA computer.

11. In general, the records examination is a very time consuming activity that requires many man-hours before and after an (inventory) inspection. Better computer methods are needed to reduce the burden of verifying operator sub-totals for material types and locations.

12. Although the inventory is relatively easy to verify due to the use of NDA equipment by the operator for difficult to measure materials, the Agency will encounter difficulty to verify the activity of 25 inventory teams with 5-6 inspectors.

ACKNOWLEDGEMENTS

We gratefully acknowledge P.N. Denison, A. Caldwell, and J.M. Smith for their assistance in preparation of this paper and their excellent support throughout the exercise.

REFERENCES

(1) "Second Program Plan," International Safeguards Project Office. Brookhaven National Laboratory (June 1, 1978).

(2) "Design Information Questionnaire for a Model LWR Fuel Fabrication Plant," Technical Support Organization, IAEA-STR-76, Brookhaven National Laboratory, revised October 1977.

(3) IAEA Safeguards Technical Manual, Part F "Statistical Concepts and Techniques," Vol. I, Chapter 6, IAEA 174, Vienna (1977).

(4) "Facility Attachment for a Model LWR Fuel Fabrication Plant," IAEA-STR-75, Vienna (May 1977).

(5) "Preparation and Recording of Accounting Reports on Magnetic Tape with Use of Labelled Data Elements," Data Processing Development Section, IAEA, SISD/9, Vienna (1977-06-03).

(6) FOREMAN, E.H. et al., "Nuclear Materials Information System Transaction Data Base," UCCND-CSD-INF-63, Computer Science Department, Union Carbide Corporation Nuclear Div., Oak Ridge (September 1975).

(7) MOORE, B.L., CALDWELL, A., "NMMSS-IAEA Inventory Data Interface," NMMSS Reference Manual H4-1, Computer Science Department, Union Carbide Corporation Nuclear Div., Oak Ridge (August 1977), Draft.

(8) STONE, E.M., CALDWELL, A., "NMMSS-IAEA Transaction Data Interface," NMMSS Reference Manual H4-2, Computer Science Department, Union Carbide Corporation Nuclear Div., Oak Ridge (August 1977), Draft.

(9) STONE, E.M., "NMMSS-IAEA Inventory Change Report Software," NMMSS Reference Manual H4-3, Computer Science Dept., Union Carbide Corp. Nuclear Division, Oak Ridge (September 1977), Draft.

(10) FARRIS, G., et al., "The IAEA Safeguards Information System," Paper presented at the ANS Winter Meeting (1977).

DISCUSSION

W.C. BARTELS: I wish to commend you on your outstanding contribution to this important effort. The United States of America undertook this work mainly to help the Agency, but also derived great benefit itself therefrom. During and after this exercise we were negotiating our Subsidiary Arrangement for the application of Agency safeguards in selected United States facilities. The results were useful in arriving at Code 10, the material description code, and also in other respects. This will be evident when the United States make the Subsidiary Arrangement widely available in the near future.

E.A. VAN DER STRICHT: I want to stress that the availability of computer listings giving item data during taking and verification of the physical inventory in bulk material plants is of extreme importance for efficient operation. Electronic equipment used in the field prior to and during physical inventory taking and verification has proved to be one of the most important factors in determining the quality and utility of the resulting verification data. My experience is similar to that reported in this paper, and I fully support the conclusions.

A.M. BIEBER: Thank you for your support. I would like to emphasize, however, that while complete item lists are necessary for effective IAEA inspection, lists which give totals for material categories (i.e. inventory key measurement points, defined as they were during the exercise) and for locations within the facility are also necessary for practical stratification and preparation of sampling plans.

K.J. QUEALY: You stated that each individual item is separately called up and specified, some 40 000 of them, in a printout four inches thick. Can these be reduced to fewer categories to facilitate verification?

A.M. BIEBER: In order to select items to sample for verification of inventory, IAEA inspectors must have available to them a complete list of all items on inventory, broken down and listed item by item. As I mentioned in the paper, the inventory KMPs selected for the exercise are basic material types, so they form the basis of stratification for preparation of a sampling plan. Computer programs can be written (and were, for the exercise) grouping items into batches and/or giving totals for different material types (e.g. IKMPs) and/or locations.

A.G. HAMLIN: I am rather surprised at the size of the inventory (30 000–40 000 items), particularly in view of the statement in the paper that "all items on the inventory were measured before the end of the physical inventory taking". Do you mean that each of these 30 000–40 000 items was actually measured? If not, could you say what was actually done?

A.M. BIEBER: United States regulations for commercial facilities require complete measurement of all material on inventory unless it has been measured previously and the integrity of the measurement assured by use of tamper-indicating seals. In the facility used for the exercise, much of the inventory was pre-measured and sealed, so that only seal verification was required at the time of the inventory. However, a substantial portion of the 30 000–40 000 items was indeed measured at the time of the inventory.

D. GUPTA *(Chairman)*: Did you carry out this experiment purely to establish the physical inventory, or did you also establish a balance, and if the latter, what is its uncertainty?

A.M. BIEBER: Both initial and final physical inventories were taken, and a material balance was drawn up as part of the exercise. The material balance reported to the IAEA agreed quite well with the facility operator's calculated material balance, taking into account the known problems discussed in my paper. United States facilities are required to calculate and document a limit of error on material unaccounted for (LEMUF), and all data necessary for the IAEA to calculate an independent estimate of LEMUF were supplied to the IAEA. I do not know off-hand the exact value of the LEMUF for the period covered by the exercise.

W. FRENZEL: During the physical inventory taking, were all the items counted by the inspectors or was counting performed only at selected locations?

A.M. BIEBER: The NRC verification is somewhat different from that of the IAEA. The NRC's objective is only to ensure that the facility operator is complying with NRC regulations. Thus, the NRC need not verify the entire inventory, but only satisfy itself that the operator's inventory procedures are in compliance with regulations. In contrast, the IAEA must independently verify the entire inventory, making the assumption that the operator and the State may be in collusion to divert material. Having said this, the direct answer to your question is that the NRC inspectors checked item counts and drew samples for measurement checks in selected locations only.

DEVELOPMENT AND APPLICATION OF A SAFEGUARDS SYSTEM IN A FABRICATION PLANT FOR HIGHLY ENRICHED URANIUM

M. CUYPERS
Joint Research Center,
Euratom, Ispra, Italy

F. SCHINZER
Nukem GmbH, Hanau,
Federal Republic of Germany

E. VAN DER STRICHT
Directorate of Euratom Safeguards,
Luxembourg

Abstract

DEVELOPMENT AND APPLICATION OF A SAFEGUARDS SYSTEM IN A FABRICATION PLANT FOR HIGHLY ENRICHED URANIUM.

This paper gives a general view of the safeguards activities performed at the Nukem Fabrication plant (Hanau, Federal Republic of Germany) during the last seven years. The main safeguards-relevant features of the plant are given and discussed. The importance is stressed of a good working relationship between the three principal partners, viz. the operator, the safeguards authority and the latter's technical support service. The definition, implementation and improvement of safeguards equipment and activities are outlined. The paper describes the internal organization established by the operator to fulfil his responsibilities, the safeguards philosophy, the Non-Destructive Assay equipment permanently installed by Euratom Safeguards, the results obtained, and the evaluation of the material balances. Conclusions are drawn (and specific comments made throughout the paper) from the experience gained over this period of seven years.

1. INTRODUCTION

Since Euratom Safeguards were first implemented in the Community in 1959, the Nukem plant at Hanau, Federal Republic of Germany has been of major interest. Considerable safeguards effort has been made there and, although Nukem has, like most fabrication plants, many characteristics unique to itself, the experience gained there has been of decisive importance in clarifying ideas and testing theories about safeguards implementation in bulk-material-handling facilities and in identifying in which directions verification techniques and equipment should be developed. The situation which evolved has been one in which those responsible for this development had the possibility of direct contact with the reality of plant operation and had access to all necessary information.

The situation has permitted – and in the authors' view this is essential – the achievement of a development programme which is effective, in the sense that the inspectors really make use of the technical means produced, such as instruments, seals, procedures and programmes for evaluation etc.

It is difficult to characterize Nukem in its safeguards aspects by merely one sentence saying that the plant was laid out to produce fuel elements for material testing reactors and for pebble-bed high-temperature reactors.

Although this is the main production activity in terms of flow of highly enriched uranium the safeguards dimensions of Nukem cannot be understood, without mentioning some other important features. These are briefly:

(a) The starting product for the main fabrication lines is highly enriched uranium hexafluoride;
(b) There is a large scrap recovery unit in which scrap recovery campaigns are also run for clients;
(c) Various types of chemical, metallurgical and mechanical treatment of uranium are possible and to a large extent the technical means and skills are available;
(d) Large stocks of uranium of all enrichments are at hand in a variety of forms and dimensions. Part of it is stored for clients;
(e) The stock of uranium is split into several thousands of accountancy units spread over a large number of locations;
(f) Part of the uranium stock is mixed with thorium, which is the major component
(g) The production units are not linked so that the production activity in the plant considered as a whole never comes to a complete standstill.

To set up good nuclear material management, not to say an appropriate safeguards scheme in these circumstances, is not easy. Furthermore, the rapid evolution in nuclear regulations in relation to growing public concern about the industrial use of nuclear materials means an additional pressure and requirements for those who had to set up this management and to attend to its implementation. The continual flow of information between the partners enabled them to adapt and improve ongoing projects according to the changing requirements.

2. SHORT DESCRIPTION OF THE FABRICATION LINES

For highly enriched uranium the production area can be subdivided into the following four main fuel preparation and processing steps:

The chemical processing areas
The MTR lines (alloy and cermet lines)
The HTR line (kernel, particle and pebble lines)
The uranium oxide line (pellet and rod line), which will be operated completely in the future, i.e. with pellet pressing, rod loading and fuel-element assembling.

For depleted uranium: Production of shielding parts by melting and machining depleted uranium metal.

For natural and low enriched uranium: Occasionally orders are received to re-treat, process and store small amounts of these two categories of uranium. Large amounts of low enriched uranium hexafluoride are stored.

Figure 1 gives the flow chart of the nuclear material. From this chart it is apparent that the nuclear material is brought back to the store almost after each processing step. This is one of the fundamental characteristics of the nuclear accountancy system of the Nukem plant.

3. NUCLEAR MATERIAL (NM) MANAGEMENT AND ACCOUNTANCY

For security reasons the actions taken to ensure physical protection of the nuclear material are not dealt with in this paper.

As in many companies the general policy is to allow for a strict separation of responsibility between production activities and control and safety functions. This means in practice that both the control/safety departments on the one side and the production departments on the other are directly subordinate to the management.

At various levels in the organization the responsibility of the individuals are stated in a letter of appointment, which must be signed for agreement by both the management and the appointed person.

The main safeguards-relevant responsibilities are:

The supervisor for NM in storage and for accountancy: He is in charge of the continuous recording and monthly reporting of all incoming and outgoing NM. He has to ensure that all batches in the store are correctly labelled with external or internal shipper's data. He has to record the movement of NM batches from production account to production account each time the NM batches do not go back immediately to storage. Finally, he has to check the information provided on the tag for the material batches which are brought back to the store.

Responsibility for NM in the process: The person who has to handle NM is responsible for this material after registration of the movement in the general

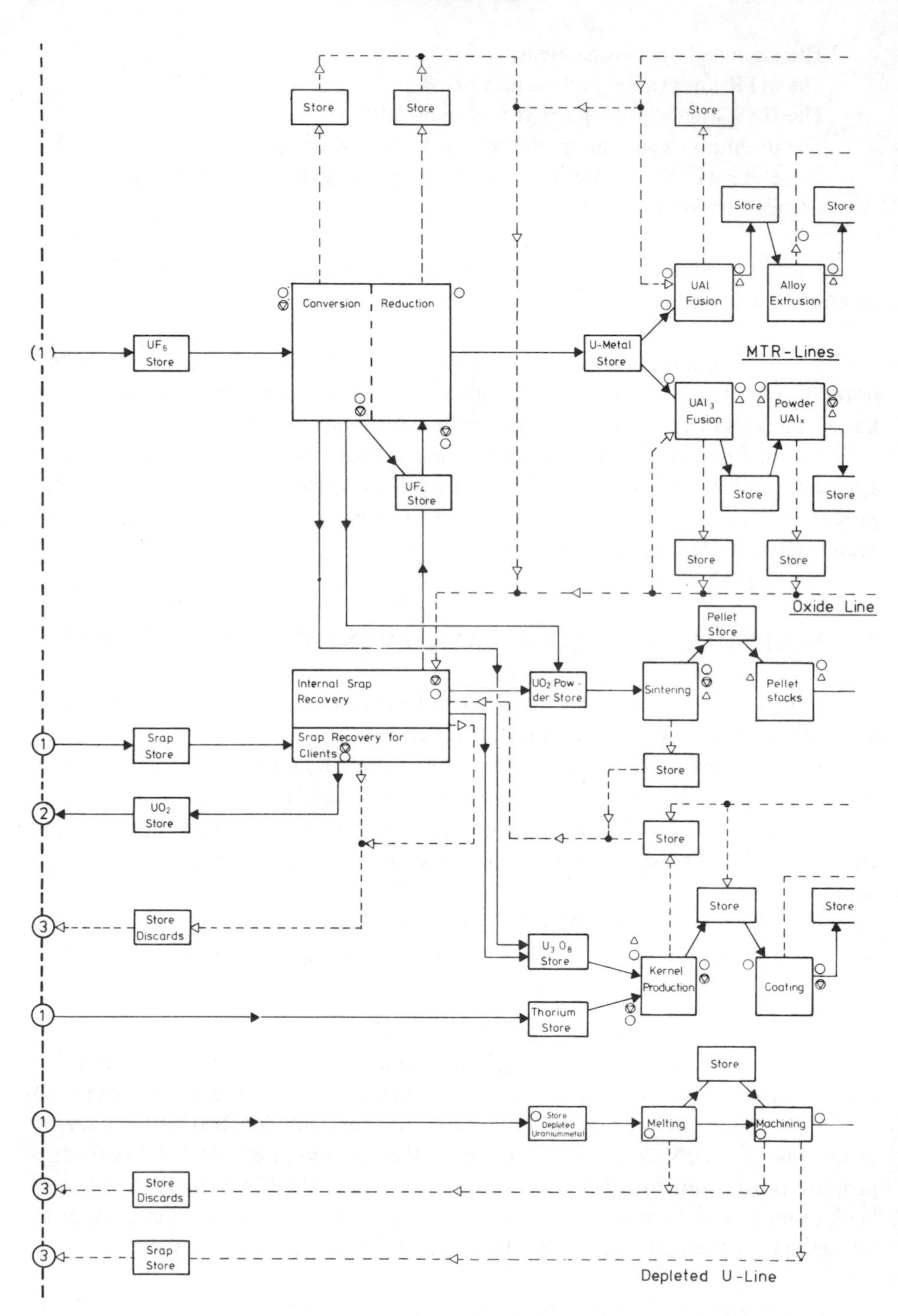

FIG.1. Nuclear material flow chart and nuclear material control points.

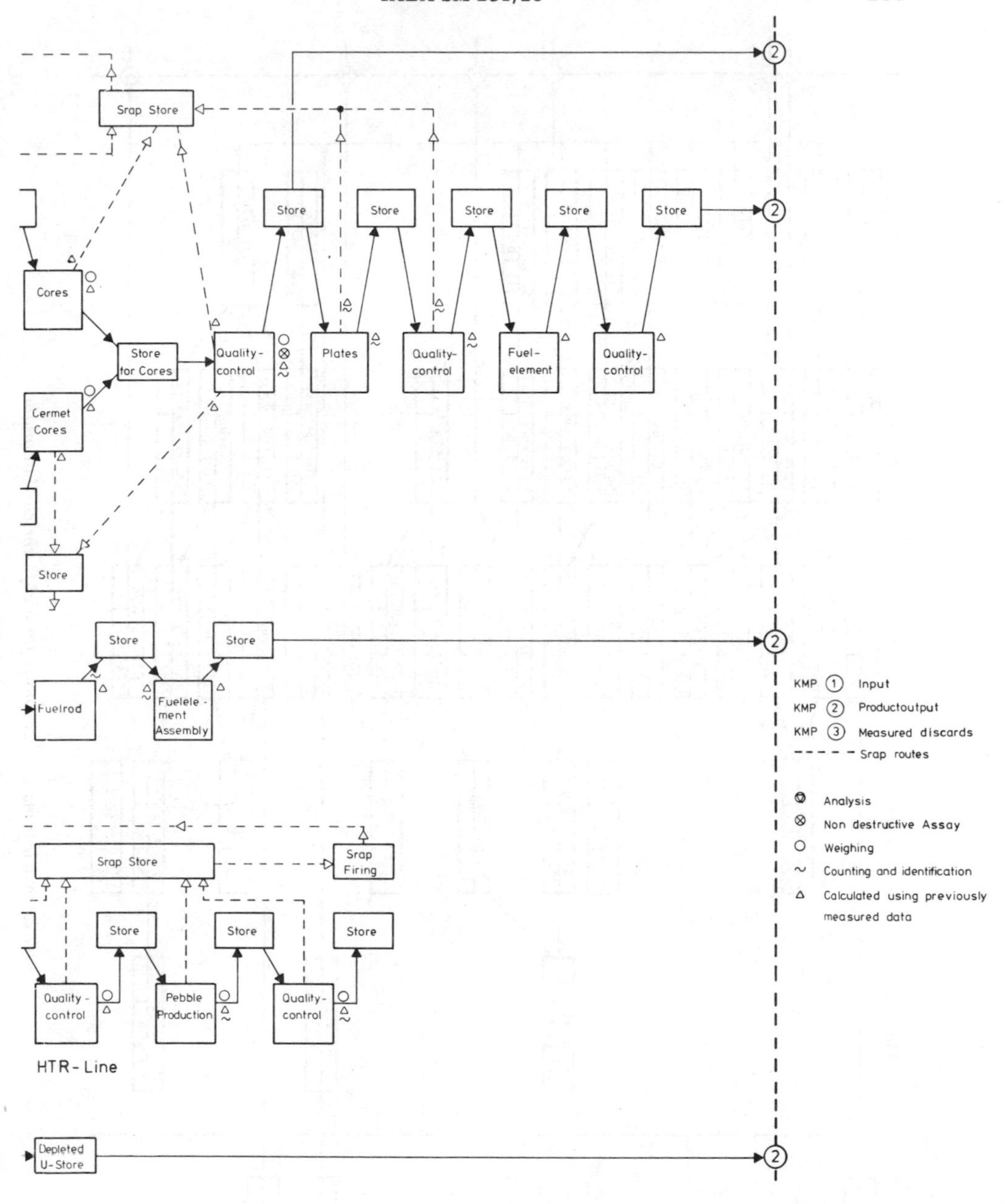
Srap Store
Store
Store
Store
Store
Store
Cores
Store for Cores
Quality-control
Plates
Quality-control
Fuel-element
Quality-control
Cermet Cores
Store
Store
Store
Fuelrod
Fuelele-ment Assembly
Srap Store
Srap Firing
Store
Store
Store
Quality-control
Pebble Production
Quality-control
HTR-Line
Depleted U-Store
KMP 1 Input
KMP 2 Productoutput
KMP 3 Measured discards
Srap routes
Analysis
Non destructive Assay
Weighing
Counting and identification
Calculated using previously measured data

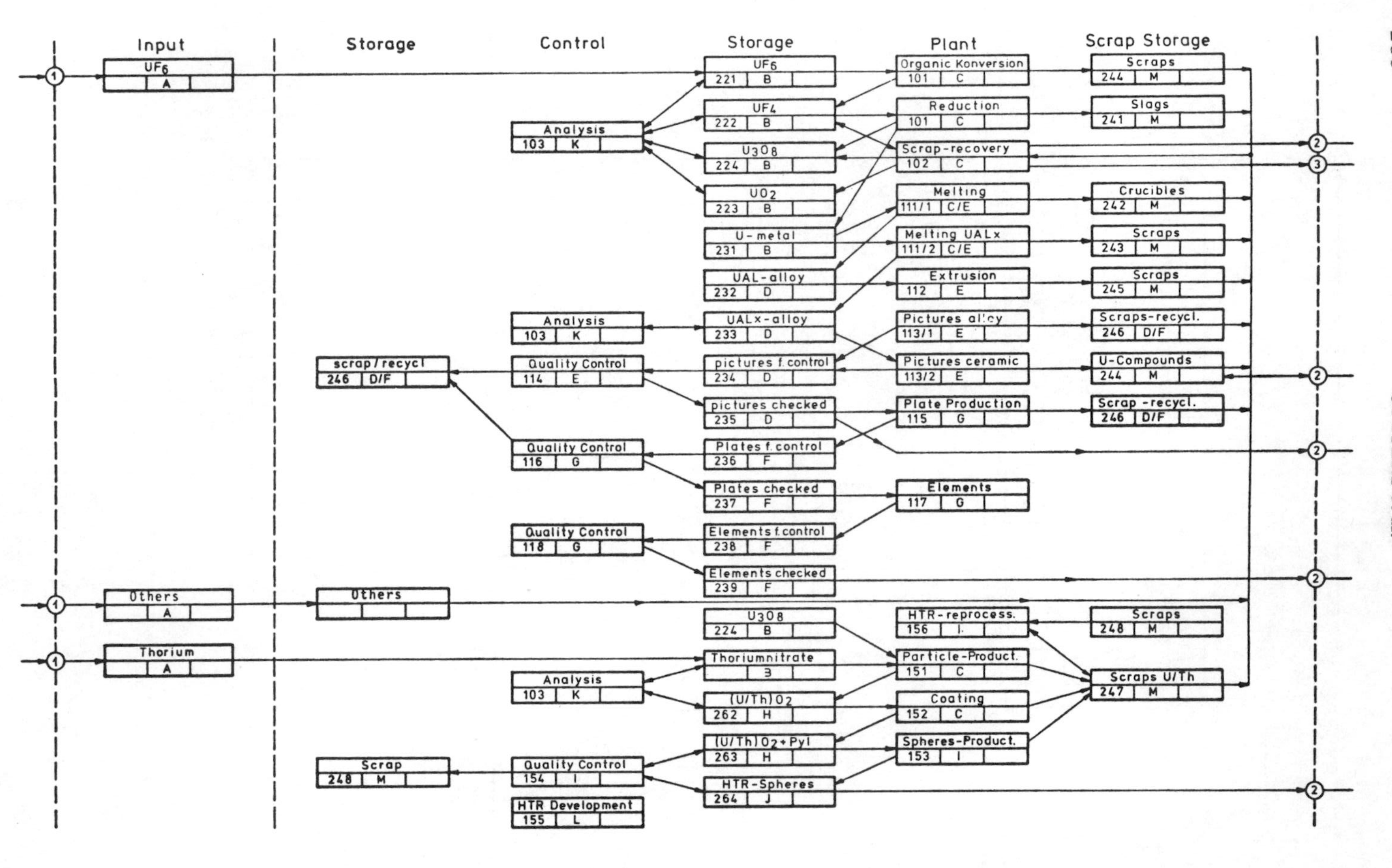

FIG.2. Nuclear material accounts for highly enriched uranium.

ledger. He is relieved of this responsibility after he has handed over the material to the storage supervisor, or to the next production account, in which case he must inform the supervisor of the movement.

Quality control by non-destructive measurement: The controller is responsible for the accuracy of the measurements and compliance with the specifications. Calibration of scales is also part of his responsibility.

Destructive determination of uranium content and enrichment by chemical analysis and mass-spectrometry: The analyst is responsible, under the supervision of the head of the analytical laboratory, for the accuracy of the determinations performed.

Sampling: Utmost care is taken that no (or negligible) sampling errors are made. Accordingly sampling is performed by an expert with a good analytical background who follows written procedures.

Certification of end products: To guarantee the exclusion of erroneous data changing all source data obtained in quality control and/or analytical laboratory are checked once more by the certificate department before compilation of the certificate.

Material accounting: Nuclear material accountancy is divided into two parts. One part is used for the monthly reports to Euratom, and consists of a card register in which the input and output of the plant is recorded on a monthly time basis. The second is a records-system which follows all movements between the storage and the plant and, if necessary, within the plant on a daily basis. For this purpose the plant is divided into several responsibility areas and for each area an account is kept at the central storage book-keeping section (Fig. 2). A continuous supervision of all nuclear materials is possible here because generally the uranium is put into the store between two processing or control stages. In addition, the persons responsible for the plant areas report their uranium stocks once a day to permit cross-checking. Each batch is identified by an accompanying card, showing, inter alia, the amount of uranium and ^{235}U. The accounting system has run since 1975 by means of an electronic data banking system which makes it possible to draw at any time physical inventory listings for all material on storage.

4. THE SAFEGUARDS APPROACH

From what has been dealt with up to now the complexity of the safeguards object is clear. Up to 1000 kg of highly enriched uranium hexafluoride is converted annually according to the orders received. In the scrap recovery unit up to 100 kg of highly enriched uranium is processed in 10 to 12 campaigns annually for internal scrap recovery and, depending on the orders received, from 50 to 100 kg of highly enriched uranium for clients. The inventory of nuclear material is about 15.4 t ^{235}U of which about 10% is of high strategic value. This inventory is split up into roughly 3000 inventory lines. Ninety per cent of the ^{235}U stock remains stored and untouched during a material balance period.

4.1. Since 1971 all the highly enriched uranium in this installation has been subject to a reinforced inspection scheme, whereby the inspector has to strike balances which are, to the greatest extent possible, independent. Therefore, it was decided that inspectors be present in the plant on an almost permanent basis with the duty of verifying by sample taking and analyses, and by non-destructive assay, all receipts and issues of highly enriched material. Furthermore, during their stay at the plant the inspectors had to carry out unannounced inventory verification in small process areas. Total physical inventory verifications were performed twice a year for the MBA with material enriched to 5% and more and once a year for all other nuclear material categories.

The start of the application of the reinforced inspection scheme coincided with the time in which the application of NPT was first envisaged so that, in order to cope with the two sets of problems, it was decided to start a contractual collaboration programme, the partners of which were Nukem, the Joint Research Center of the Commission with its Ispra establishment, and the Safeguards Directorate of the Commission in Luxembourg. In this way the instrumental developments made at Ispra on behalf of the Safeguards Directorate could be tested and perfected in the field. All the projects were discussed jointly from the planning stage onwards.

The contract furthermore guaranteed that any information additional to that then legally required, but which was needed for the system study, be given by Nukem.

The main study objects which eventually became safeguards tools in routine use are:

(i) Neutron interrogation and delayed neutron counting (sigma) for the verification of the HTR fuel pebbles, including a tamper-resistant sampling device [1];

(ii) Photoneutron source interrogation device (PHONID or Sb-Be device) for the verification of inventory items [6];

(iii) The development of tamper-proof ultrasonic seals both for general purpose and for MTR fuel elements [2] including identification equipment;

(iv) A sampling device for sampling UF_6.

Concerning the system studies the major effort was directed at solving practical problems such as the application of stratification to inventory items, which has been reported [3]. These studies also established the requirements for the inventory listings in order to reduce the duration of the inventory verification, which is a major safeguards burden to Nukem.

These activities have led to a better understanding of the way safeguards should be implemented at Nukem. As a result further important joint developments have been undertaken and completed:

(i) The production with KfK of a manual describing the control of the nuclear material flow in the Nukem plant [4];

(ii) A study on the safeguards aspects in the design of a new Nukem plant with KfK [4];

(iii) Development and programming of electronic accountancy data processing with KfK [4];

(iv) The electronic treatment of accountancy data by the operator and, in the field, by Euratom Safeguards during physical inventory takings;

(v) The implementation of programmes to acquire common reference materials to be used for the non-destructive assay of nuclear material by both the inspector and the operator;

(vi) The electronic treatment of the data of a material balance, including the variance of MUF calculation;

(vii) The improvement of NDA inspection data reliability by use of microprocessors imposing the sequence of events in the measurement procedure and the use of computer compatible outputs.

4.2. The MBA subdivision

The first step in an attempt to work out a comprehensive safeguards scheme for the plant took account of the fact that, access to all records and to the plant being granted to the inspector at any time, the plant could be considered basically as one material balance zone. A modification to this view was made to account for the fact that, at the time this concept was set up, the enriched uranium was but one material category. This entailed the blending of low enriched uranium with highly enriched uranium not giving rise to an inventory change report. Therefore, in the first safeguards concept the plant was divided into two MBAs according to the enrichment of the uranium, the barrier being set at 5% enrichment. Since the system was designed to identify those batches which appear twice in the material balance it was not felt necessary to separate the input and output material stores from the process area by creating MBAs around these parts of the plant.

The usefulness of a shipper-receiver MBA was considered but it was eventually decided not to include it in the scheme.

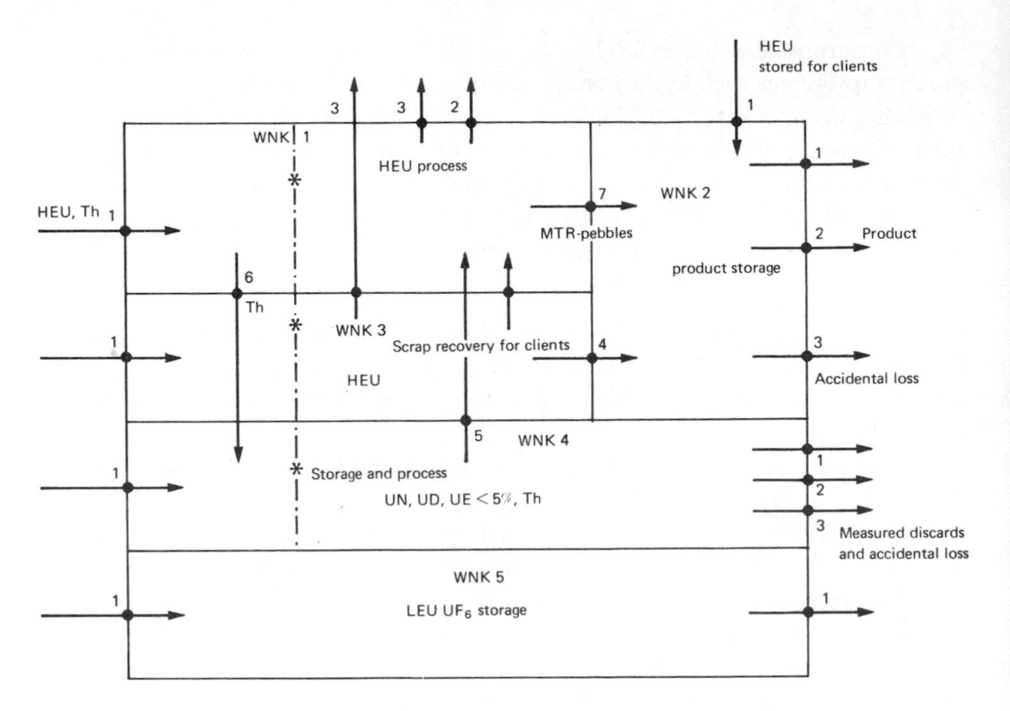

FIG.3. Material balance area scheme of Nukem. ·—·——·—: KMP* transfer from input store to process; S.R. difference.*

Recent developments related to the changing goals to be achieved by the safeguards authorities have led to a review of the MBA scheme, the plant now being divided into five MBAs (see Fig. 3). The most important changes are the introduction of a MBA for scrap recovery for clients and the book MBAs for enriched material stored for clients and for the final product store.

4.3. The inspection activities

The inspection effort can be quantified only when the information provided about the plant activity includes, per typical input or output, the amount of material per material balance period and per shipment, the number of shipments, the typical item and the accuracy of the data. Once the goals of the inspectorate are quantified it is an easy matter to apply the well-known formulas for attribute and variable sampling. It is, however, our experience that each typical flow product deserves some special attention.

At Nukem the situation is as follows:

A. Typical HEU feed material

(1) UF_6	Check delivery notes and seals Take a sample out of each bottle (16 kg HEU per bottle) Check gross weight – seal bottle
(2) U metal	Check delivery notes and seals Observe weighing Take sample in each container Seal container
(3) Scrap cores and plates and other scrap	Check delivery notes and seals Count and identify cores and plates Weigh Seal container Sample homogeneous batches at the recovery

In addition, for material (2) and (3) immediate verifications can be performed with the Sb-Be interrogation device and the ^{235}U amount determined. On feed material statistical sampling is rarely justified and offers little advantage.

B. Typical product material

(1) MTR production	The number of elements of the same type produced during one material balance period is small so that all MTR elements are measured with the γ-scanner, all for consistency checks, some for which a standard is available on an absolute basis. MTR elements for reactors in the European community are provided with a rivet tamper-resistant identification seal. All element containers are sealed for shipment.
(2) RHF cores	Common standards for the cores are available. The cores are verified with the γ-scanner in fixed geometry. Although a large number of cores are produced the measurement time is so short (10 s) that all cores are measured.

(3) U metal UO_2

Population numbers being small no statistical sampling is performed;

- Compound weight is verified
- Samples are taken from each homogeneous batch
- Items are measured (100% basis) with the Sb-Be interrogation device
- Seals are put on all containers.

(4) HTR pebbles

In this case statistical sampling under tamper-resistant conditions has been applied to the total production [1].

C. Items on inventory (highly enriched uranium)

In addition to the typical materials listed above the full intermediate "good" product spectrum is at hand; typical items are UA1 billets, boxes with cores, sandwiches, plates, cans with $UA1_x$ buttons, UAl_x powder, UF_4 powder, containers with kernels, uncoated particles, coated particles, UO_2 powder, U_3O_8 powder.

These materials are mostly in such containers that their ^{235}U content determination in an attribute in variable mode is possible with the Sb-Be activation device.

Furthermore, as can be seen from the flow chart, scraps and waste are generated in small but relevant amounts, part of it being of the directly recyclable type. Except for a few items containing minor amounts of ^{235}U in large and ill-defined matrices the same measurement device (Sb-Be) is used.

At this point an important question arises about the sampling effort to be made on such an inventory. Obviously the most reasonable way seems to be to divide the whole set of batches to be verified into a number of subsets or strata of like objects. This stratification is performed on the basis of the information available under the heading "material description" in the PIL or in the items lists. The existing codes have proved to be unsatisfactory and a better definition is certainly recommended. Since the practical application of this concept left a large number of items uncovered, or resulted in unwanted mixing of different items in a class, other ways have been explored. In a nutshell, the total field is divided into a number of weight classes and a number of item type classes, the latter only for those types of material which unmistakably constitute distinct classes of material. In the case of Nukem the latter are the fuel pebbles, the UF_6 cylinders, the uranium metal and the plates for the MTR fuel elements. Weight classes are chosen so that the distribution of the elements in the classes is such that to speak of standard deviations of the elements still makes some sense.

The idea underlying this subdivision is that, in principle, the field is made up by measured or measurable units. It is clear then that one of the governing criteria is to be found in the ability a measuring device has to confirm that a certain, defined attribute is present or not. In the ideal situation one instrument should have this ability for *all* the elements in the field and they would thus belong to a single class. In the case of Nukem this instrument is the Sb-Be device. The Nukem inventory is accordingly divided into ten classes of objects.

It is also possible to define an arbitrary uranium or ^{235}U amount unit as the elements of the field of arbitrary units. In the case of Nukem the resulting total sampling effort is close to the effort which is found when the assumptions in manual F of the IAEA for attribute sampling, i.e. the *total* goal quantity is diverted from *each* stratum, are accepted. Which technique to use is mainly dependent on the effort required to make the stratification. This effort is large when no electronic data processing is available. The most efficient way to tackle the verification problem is that where due account is taken of the specific characteristics each installation possesses.

This means that the approaches, which are defined for example in manual F, are to be adequately adapted so that the actions of the inspectorate are to the greatest possible extent rapid and not burdensome – the location of material; than storage, management and handling procedures used by the operator; and physical protection measures, are all factors which affect operations during verification. An important aspect will undoubtedly be the means and possibilities which are available to the inspector so that his knowledge of the amount of nuclear material in the items on inventory is maintained for as long as possible. Suitable temporary seals have to be more and more applied in order to keep the inspection effort within acceptable limits.

5. MUF AND VARIANCE OF MUF EVALUATION

The scheme to evaluate the balance of material has been computerized at the Ispra establishment of the JRC on the basis of the model provided by Jaech in TID 26298.

In this model the variances attributed to all the individual uncorrelated error sources are calculated. There are five error sources considered:

Bulk
Uranium factor
Sampling error for the uranium factor
Isotopic composition determination
Sampling error for the isotopic composition.

Identical lines in the balance record are cancelled before the calculation of the variance of the algebraic sum is performed.

The allocation of errors is done through the concept of the error path which has to be attributed to each line in the material balance record. A new error path is created each time the value for the random and/or systematic error is different from the values in a previous path in at least one error source. The values for the random and systematic errors for all scales and methods used and the values for the sampling errors are all put in the data base, together with the particular combination of error sources for each identified error path. The contribution to the total variance for each error source is calculated and these contributions are then summed giving the final result.

Up to now the evaluation program has been tested on the data of two material balance periods starting with April 1977. The results for the material category uranium enriched to 20% and more are given here. The two data sets result from declarations made by the operator in accordance with Euratom regulation No. 3227/76.

The amount of uranium per physical inventory exceeds 2000 kg. The sum of the inventory changes in each period is about 130 kg U. The values for the "book MUF" are less than 100 g U, after subtraction of about 2900 g U which have been declared as new measurements (NM), and for which a verification was made and an explanation given.

The square root of the variance of MUF was found to be about 550 g U, the major contributors to the total variance being the weight random variance for one scale, the uranium alloy random variance for the uranium factor and the systematic error contribution for the non-destructive measurement of the uranium-aluminium cermet production.

All correlated batches (totalling about 1500 kg U and about 350 entry lines out of about 3000) are rejected for the variance of MUF evaluation.

These tests showed the validity of the programs and it is planned to use them also in routine operation for other bulk facilities. A publication is foreseen in the near future.

6. INSTALLED NON-DESTRUCTIVE ASSAY EQUIPMENT

6.1. Sigma

The production of the fuel pebbles is sampled with tamper-proof devices mounted on the shipping containers and the samples obtained are measured with a system which has been described elsewhere [1]. Let us remember that the pebbles are first irradiated with a ^{252}Cf neutron source and the delayed neutrons measured with ^{3}He detectors. More than 400 000 pebbles have been verified in this way. Calibration curves have been established and reference pebbles characterized. Owing to the high number of pebbles verified the random error on the total amount of ^{235}U produced is negligible. The overriding contributor to the uncertainty is the systematic error on the chemical determination of the uranium contents of the pebbles of the calibration curve; this error is estimated at 0.15%.

6.2. NaI (Tl) for MTR elements and plates. This measurement and the measurement procedures have been described [5].

The main problem here is to build adequate reference elements for the different types of element produced.

Up to now the Safeguards Directorate's policy has been to purchase specially manufactured elements. This policy is too costly to cover the full range of element types produced particularly since the ^{235}U amount changes frequently, most of the other characteristics remaining equal. Therefore, another arrangement has been made with Nukem, which foresees that, when the safeguards authority wishes to have a reference element for a production series, it can choose a number of plates from the first group which are all measured with the scanner. A calibration curve is made relating response of the scanner to ^{235}U amount found after analysis and the remaining plates are assembled into the reference element.

This element is kept by the inspector and returned to the operator when the production is stopped. Independently of the policy followed (purchase or temporary hire) the procedure foreseen is the one mentioned above and results in an uncertainty of 0.2% on the ^{235}U content of the element.

6.3. PHONID Sb-Be[1]

The scope of this instrument is to provide for a rapid verification of the ^{235}U content for attribute verification of inventory items without taking them from their containers. Fourteen different calibration curves have been established taking account of per cent uranium, enrichment, nature of matrix and uranium weight ranges.

The mean difference from the calibration curves observed never exceeds 5% and generally lies around 2.5%.

This means that the instrument is capable of variable verifications in attribute mode. As far as the sampling plans are set up for the discovery of total emptying of items, the verification effort more than covers all diversion strategies which would consist in taking away more than, say, 15% from the items in inventory.

---

[1] A Euratom Report is in preparation.

7. FINAL REMARKS

The authors hope that this very condensed review of the safeguards situation in this one particular plant, Nukem, has shown that a practical and effective safeguarding of nuclear materials in this plant is possible. Of course, owing to the complexity of the plant, quite heavy technical investment has been necessary. It has also been shown that the fundamental choices made in order to implement safeguards in this plant were not profoundly modified by changes in the targets of international safeguards. These changes required no discontinuity in the ideas and actions concerning Euratom safeguards. The experience gained at Nukem is now being used in order to increase the efficiency and effectiveness of safeguarding other bulk facilities.

Finally, the authors wish to stress that what has been achieved has not been achieved by them alone. They hope that it has been made clear that these results are the fruit of the help and collaboration of a large number of people with very different backgrounds and fields of action. We wish to express our sincere thanks to them all.

REFERENCES

[1] CUYPERS, M., et al., in Safeguarding Nuclear Materials (Proc. Symp. Vienna, 1975) **2**, IAEA, Vienna (1976) 521–31.

[2] CRUTZEN, M.S.J., et al., ibid., pp. 305–38.

[3] ROTA, A., ibid., **1**, p. 443–60.

[4] Annual progress reports of Kernforschungszentrum Karlsruhe, Projekt Spaltstofflusskontrolle, e.g. Rep. KFK 2465 (1976).

[5] MIRANDA, U., et al., in Safeguards Techniques (Proc. Symp. Karlsruhe, 1970) **1**, IAEA, Vienna (1970) 105.

MODEL FOR THE APPLICATION OF IAEA SAFEGUARDS AT MIXED-OXIDE FUEL FABRICATION FACILITIES

W. ALSTON, W. BAHM, H. FRITTUM,
T. SHEA, D. TOLCHENKOV
Department of Safeguards,
International Atomic Energy Agency, Vienna

Abstract

MODEL FOR THE APPLICATION OF IAEA SAFEGUARDS AT MIXED-OXIDE FUEL FABRICATION FACILITIES.

Current Agency criteria and practices are presented for safeguards at mixed-oxide fuel fabrication facilities. The paper includes a description of typical process activities and the types of materials normally encountered. Credible diversion possibilities and related concealment activities are discussed and Agency criteria for such facilities are reviewed. Requirements and the approach being pursued to counter protracted and abrupt diversion strategies are presented, with a discussion of specific verification requirements necessary to satisfy the short detection time criteria.

I. Introduction

This paper is intended to serve two purposes : (1) to introduce a series of documents under preparation within the Agency detailing standard practices for the application of IAEA safeguards to different types of nuclear facilities; and (2) to describe the state of affairs for one very important type of facility, mixed oxide fabrication.

The generic models under preparation are intended to provide a reference for designing, applying and evaluating safeguards at specific facilities. The outline for the model documents serves as the basis for this abbreviated presentation.

In determining the safeguards approach for a facility, consideration is first given to the State level requirements necessary to counter multiple source [series and parallel] diversion and cross-concealment possibilities. These matters are described in a separate paper [1]. Consideration of these possibilities in a frame of reference specific to the State with a mixed oxide fabrication plant leads to performance requirements for the Agency's safeguards as applied at that facility.

II Facility Description

Mixed oxide fuel fabrication plants manufacture PuO_2 - UO_2 fuels principally for breeder reactor development programmes, for recycle in thermal power reactors and for use in advanced thermal reactors. Figure 1

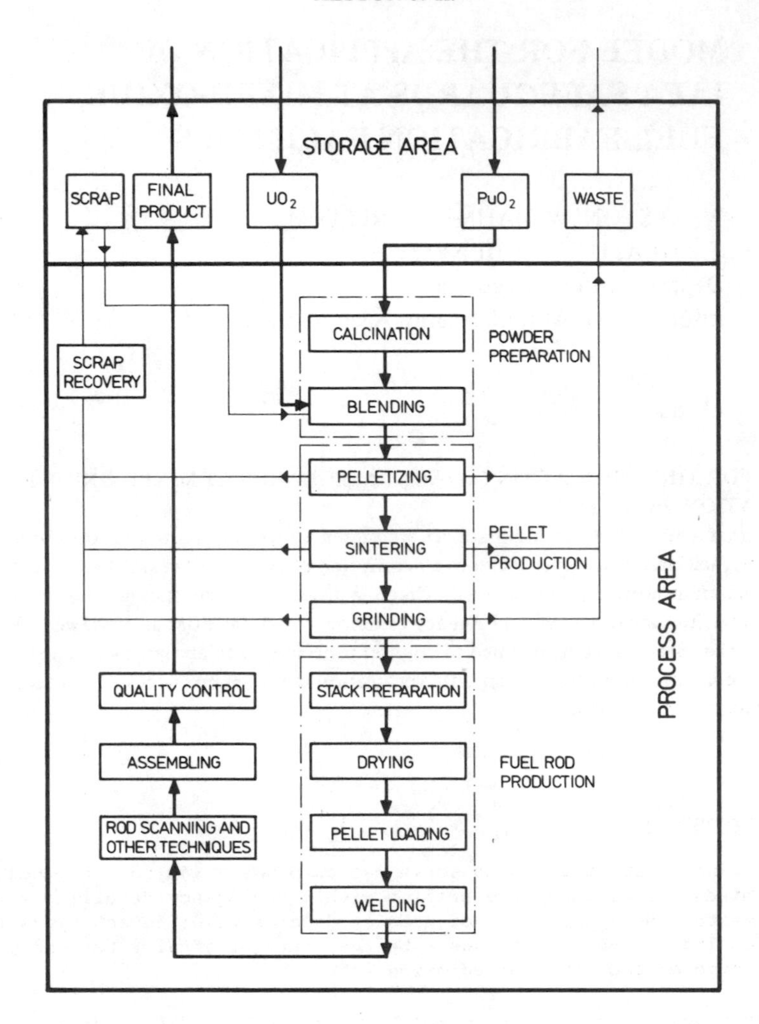

FIG.1. Manufacturing steps in a model MOX fabrication facility.

illustrates the typical manufacturing steps. In addition to these activities, the Agency is asked to apply safeguards to associated Pu $(NO_3)_4$ - PuO_2 conversion and plutonium chemical scrap recovery operations. At MOX fabrication plants then, the feed may be either PuO_2 or $Pu(NO_3)_4$, plus the UO_2 blend material. The principal products are completed fuel assemblies, but the facility may also ship fuel rods and on occasion, pellets and waste. Table 1 illustrates typical inventory and material flow representative of current facilities for a hypothetical plant. This plant serves as the model or reference case for subsequent considerations.

In this paper it is assumed that natural uranium is the only blend material encountered and thus emphasis is very strongly placed on the plutonium. When this is not the case, the provisions are modified accordingly.

III Diversion Analysis

As noted in the facility description, two types of nuclear material are to be found at this reference facility : plutonium and natural uranium. Plutonium is categorized as a "direct-use" material, while natural uranium is described as an "indirect-use" material [1]. Diversion strategies considered credible related to this facility are :

1. abrupt diversion of plutonium for prompt conversion into nuclear explosives;
2. protracted diversion of plutonium at a rate chosen by the divertor to minimize the likelihood of premature detection; and
3. protracted diversion of natural uranium for feed material in either an enrichment or a transmutation process to produce direct-use nuclear materials.

In conjunction with these diversion strategies, it is assumed that concealment activities would be undertaken to avoid or prolong detection. For example, relevant to the abrupt diversion strategy, it is assumed that :

a) record entries may be dropped or intentionally altered;
b) substitute materials may be introduced;
c) materials may be moved within the facility for double verification;
d) container seals may be by-passed; and
e) material may be mis-identified as being in transit

Relevant to the protracted diversion strategy, for example, it is assumed that :

a) material may be diverted into MUF - that is, no attempt may be made at concealment, but detection will be limited by the accuracy associated with closing and verifying the material balance;
b) input accountability may be biased low, intentionally;
c) product or scrap or waste measurements may be biased high, intentionally;
d) records may be altered, after inventory verification, to obscure the true value of MUF, thereby rendering this possible diversion indicator inoperative;
e) materials may be moved within the facility for double verification; or
f) receipts may be processed or shipments dispatched without a verification opportunity.

IV Safeguards Criteria

Safeguards at MOX fabrication plants are intended to counter both abrupt and protracted diversion strategies. To counter these strategies it is essential that the Agency be in a position to verify information on the content and location of nuclear material, as declared by the facility operator. It is imperative that such information is current and made available as necessary for verification.

Countering the abrupt diversion strategy requires the Agency to detect diversion in a time frame which corresponds to the estimated conversion time for that material (i.e. conversion from safeguarded material forms to components of a metal core nuclear explosive). For the materials identified in

TABLE 1. MATERIAL FLOW AND INVENTORY IN A MODEL MOX FUEL FABRICATION PLANT

	Material Form	UO_2 (kg) FBR	UO_2 (kg) LWR	UO_2 (kg) ATR	PuO_2 (kg)	MOX (kg) FBR	MOX (kg) LWR	MOX (kg) ATR	PuO_2 content in MOX (kg)
1. Throughput/yr.									
1.1. Input	powder	1.100	12.000	49.500	500	-	-	-	-
1.2. Output	assemblies	-	-	-	-	1.575	12.300	49.200	492
	waste	-	-	-	-	25.0	200	800	8
2. Average Inventory									
2.1. Storage assumed half year's input and production are on store.	) powder	580	6.000	24.750	250	-	-	-	-
	) assemblies	-	-	-		830	6.250	25.000	250
	) waste	-	-	-	-	14	100	400	4
assumed 10 % of throughput, recycling once per month.	scrap	-	-	-	-	14	100	400	4
2.2. Process Inventory									
Calcination	powder	-	-	-	15	-	-	-	-
Blending	powder	-	-	-	-	50	335	1.500	15
Pelletizing	green pellets	-	-	-	-	14	100	400	4
Sintering	sintered "	-	-	-	-	10	75	300	3
Grinding	grinded "	-	-	-	-	14	100	400	4
Stack Preparation	pellet stack	-	-	-	-	7	50	200	2
Drying	dried pellet stacks	-	-	-	-	10	75	300	3
Pellet loading	rods	-	-	-	-	10	75	300	3
Welding	welded rods	-	-	-	-	3	25	100	1
Rod scanning	rods	-	-	-	-	3	25	100	1
Assembling	rods/assembly	-	-	-	-	60	500	2.000	20

Table 1, these estimates range from 1 - 3 weeks [1]. Accordingly, at mixed-oxide fabrication plants, the Agency attempts to form a conclusion at two-week intervals that an abrupt diversion has not occurred. The procedures being used to derive this conclusion are described later in this paper.

Nuclear materials accountancy, including records audits and verification of material balance closings, serves as the primary mechanism to counter protracted diversion. By auditing records and reconciling reports submitted to the Agency with those records, and by verifying receipts and shipments, the Agency is able to establish the amount of material which should be available for verification on inventory. By verifying the operator's material balance, the Agency is then able to provide quantitative assurance with respect to the maximum amount of material which could have been diverted without detection.

It is necessary to establish verification goal quantities in conjunction with nuclear materials accountancy. This circumstance results from the fact that irreducible measurement errors constitute a base line beyond which additional measurement efforts do not yield additional detection sensitivity. The accountancy verification goal for each type of nuclear material is selected in reference to the significant quantity for that material type, international measurement standards and other considerations including cost-effectiveness, minimal intrusion, and inspector safety. The relationship between significant quantities, State-level detection sensitivity and MBA (material balance area) level verification goals remains a subject of continuing discussion. Until resolved, verification goals should be a maximum of one significant quantity of material, per year, per facility. The significant quantities for the model facility are 8 kg plutonium, and 75 kg U-235 in the natural uranium.

Nuclear material accountancy centers on closing a mass balance over a period of time. The familar closure equation

$$MUF = BI + R - S - EI \qquad (1)$$

is the basis for determining whether materials are adequately controlled. In this equation, measured values of the beginning inventory (BI) plus receipts (R), less shipments (S) and the measured ending inventory constitute the operator's declaration of his ability to account for materials over the interval spanned by two successive inventories. MUF, material-unaccounted-for, would be zero if all materials were measured and if all measurements were free of errors. In reality, some material always remains in and about process equipment, some may be lost to the environment, and measurements of bulk process materials are never free of errors. Diverted material would contribute to this MUF, assuming that records audits make it impossible to mis-state records relating to the amount of material to be verified.

In essence, accountancy tests are made on an MBA basis to assure, with 95% confidence and 5% false alarm probability, that the amount of MUF attributable to material not presented for verification [including in-process hold-up or losses] is first consistent with statistical expectations, and second, is less than the selected verification goal quantity (GQ). A separate verification goal quantity is established for plutonium and for the natural uranium blend material in each MBA. The material balance data resulting from a physical inventory is reported to the Agency within 30 days of each inventory taking. Assembling this data, including the formally transmitted physical inventory listing and the operator's declared MUF value, plus Agency verification data obtained at the facility and from Agency approved analytical laboratories, the questions are posed :

1. Is the operator's reported MUF value consistent with limits of error estimated on the basis of international measurement standards? (For mixed oxide fabrication, the uncertainty in operator material balance should be less than 0.5% of the larger of throughput or or inventory [1]).

2. Is the difference D in Agency and operator measurements, when applied to all material balance components, significant?

3. Is the detection sensitivity adequate? That is, is

$$t_{1-\alpha,1-\beta}, \ \sigma_{MUF-\hat{D}} \leq GQ \qquad (2)$$

$$\text{or } 3.29\ \sigma_{MUF-\hat{D}} \leq GQ$$

for 95% probability of detection with 5% false alarm rate. And finally

4. Is the bias-adjusted MUF value within statistical tolerance? That is, is

$$|\ MUF-\hat{D}\ | \leq 3.29\ \sigma_{MUF-\hat{D}} \qquad (3)$$

When these questions are resolved satisfactorily, the Agency is in a position to conclude that no diversion of a significant quantity of material could have occurred during the material balance period.

Note that verification activities must be planned in order to meet condition 2, including sample sizes and verification accuracy requirements.

This accountability test is the primary safeguards test. It must be supported by satisfactory abrupt diversion detection tests, by satisfactory records audits, and by effective containment and surveillance [2] in order for safeguards to be acceptable at the facility level. This basic conclusion is necessary before safeguards can be evaluated at the State level, as described in Reference 1.

V Facility Organization for IAEA Safeguards

The basic model for material balance areas at mixed oxide fabrication facilities permits a clean demarcation between storage and process operations, that is, between item control and bulk processing areas. A separate MBA is established for receipt, storage and re-shipment of feed materials, and storage of intermediate product and process scrap. A second MBA encompasses the process area for powder-to-rod manufacturing, recovery of scrap generated within the facility and the analytical laboratory. Fuel element assembly, rod and fuel element storage and shipment are separated in a third MBA.

Note that when the facility has a nitrate-to-oxide conversion operation that operation is considered for safeguards purposes as an additional, separate MBA. Also, when the facility includes more than one process line, when the facility operator recovers scrap generated at other facilities, or when a separate R & D activity is maintained, such operations are also normally partitioned as additional MBAs for safeguards purposes.

Strategic points which are key measurement points for flow and inventory verification represent locations where material will be made available for Agency verification. In addition to these, strategic points which provide the access necessary for the Agency to maintain the short detection time capability necessary to counter the abrupt diversion strategy are established to reflect a particular approach at each facility. Consultations involving the facility operator are essential during the period when these matters are decided, to assure that the approach selected and the strategic points established provide the Agency with the ability to meet its objectives, while intruding into plant operations to the minimum, providing proprietary protection, and of course, not creating safety risks.

VI Facility Records and Reports

The underlying basis for the application of Agency safeguards is the verification of information declared by the operator and the State. As set forth in the Agency/State Agreement and clarified in specific facility attachments, the facility operator is required to maintain a system of records and the State is required to submit different reports which together serve as the basis for verification inspection activities. Requirements identifying the types of records and reports, their contents, and the time for recording and reporting information are set forth in the facility attachments.

Operator records of data recorded during measurement or calibration, or needed to derive empirical relationships which identify nuclear material and provide batch data, are considered as source data from which the information contained in reports to the Agency is extracted and summarized. Inspection activities include records audits to ensure that the basis for verification is complete and accurate, and consistent with data provided from other sources.

VII Short Detection Time Activities

During, or at the end of each two week interval, Agency inspectors go through the following verification activities :

1. verify and seal all feed material received at the facility since the last visit, by item counting, weighing and sampling, and/or NDA;
2. verify all material in the vault by item counting, identifying containers, checking the integrity of previously applied Agency seals, and by re-measuring a sample of items selected at random;
3. verify the contents of fuel rods manufactured since the preceding check interval using high accuracy non-destructive assay methods; and
4. verify the contents of fuel assemblies completed since the preceding check to assure that contents correspond to charge logs identifying all fuel rods included in each assembly, by item identification and non-destructive assay.

Two approaches are being developed to achieve the required short detection time capability for the material in process (MBA2). This will be accomplished with little or no disruption of normal process activities. Most mixed oxide facilities have an on-line computer used to maintain information of the location of material within the plant; this information is essential for criticality control especially, and provides an excellent basis for Agency verification.

One approach is to verify ongoing process operations related to nuclear material accountancy and control. This requires access to relevant operator production, criticality and quality control records. Independent measurements will be made as the material becomes available and according to a random verification strategy. The strategic points are chosen to provide access to material and data necessary to conclude that material has not been diverted.

In the other approach, the in-process material is made available for Agency verification at designated locations within MBA 2 at bi-weekly intervals by the operator. This approach makes it possible to identify particular locations, e.g. specific glove boxes within the facility, as strategic points for this purpose, and to provide for inspector access one or two days each two weeks.

According to this second approach, the in-process inventory will be verified using attribute methods to check the operator's flow control system.

For our model plant, assume that during each two week interval 20 kg of plutonium is introduced into the process together with the amount of natural uranium required to produce the mixture ratio for the fuel being manufactured. Assume that the process line is charged with 50 kg Pu and that 20 kg of plutonium product [plus scrap] is produced during that period. The inprocess inventory for the model facility presented for verification consists of the forms and nominal quantities of nuclear material indicated in Table 2. (These values are related on the average inventory supposing a stationary condition. It does not mean, for example in the case of FBR, that 3 assemblies are containing 7 rods.)

Each material form is considered as a separate verification strata. The sample size for each material form is computed using the relationship :

$$n = N \left[1 - \beta^{1/r}\right] \qquad (4)$$

where n is the sample of N items selected for verification, β is the probability of non-detection [set at 0.05], and r is the number of items required for 8 kg Pu [1]. The sample sizes, methods of verification [3] and

[1] Note that when a diversion of 8 kg Pu is distributed over the strata, the combined probability of detection $(1-\beta)$ will still be 95%, provided the diversion is through gross defects [i.e. removal of whole items] or the detection methods are sufficiently sensitive to establish that items selected for verification which have had part of their contents diverted, are in fact defective. The combined probability of attribute non-detection is then :

$$\beta = (\beta_{design})^{\text{amount diverted/Goal amount}}$$

$$\text{for all strata,} \quad \beta_{Total} = \prod_i \beta_{\text{strata } i}$$

$$= \prod_i (\beta_{design})_i^{\text{amount diverted-strata i/goal for strata i}}$$

If all β design values are equal (0.05) and the goal for each strata is the same (8 kg Pu), then the total β is

$$\beta_{Total} = 0.05^{\left(\sum_i \text{amount diverted-strata i}\right)/8\text{kg Pu}}$$

Thus, if there is a diversion of 8 kg Pu from MBA2, regardless of how the diversion is apportioned, $\beta_{Total} = 0.05$

TABLE 2. IN-PROCESS INVENTORY FORMS AND NOMINAL QUANTITIES OF NUCLEAR MATERIAL

Material Form	No. of Items			Pu/Item [kg]			Total Pu
	FBR	LWR	ATR	FBR	LWR	ATR	[kg]
PuO_2	3-6	3-6	3-6	5-2,5	5-2,5	5-2,5	15
MOX Powder	5	38	150	3	0,4	0,1	15
Pellet-Trays	27	200	800	0,6	0,08	0,02	16
Rods	7	50	200	0,6	0,08	0,02	4
Assemblies	3	1	10	6	20	2	20

approximate verification times for the model facility are illustrated in Table 3.

The fundamental hypothesis for this short detection time approach is that at the end of each physical inventory verification inspection, the operator's material control system is assumed to be accurate. This assumption may be changed when the accountancy tests are complete, some two-to-three months later. Assuming the operator's flow control system to be an accurate extrapolation from the last physical inventory, Agency inspectors test each of the material form strata illustrated in the tables for defects. Defects discovered must be resolved to the satisfaction of the inspectors, on the spot. For the tests to be meaningful, the total amount of material not presented for verification at these bi-weekly verifications must be maintained at a low value.

It is intended that by performing these activities on a two-week schedule, the Agency will be in a position to conclude with 95% confidence that no more than 8 kg of plutonium could have been diverted undetected during the two week period since the previous check.

VIII Flow Verification

Precise measurements of feed and product are essential to meeting the protracted diversion detection criteria. For this purpose, verification of all plutonium receipts is essential, prior to processing the material. When the PuO_2 is shipped from another safeguarded facility, accurate measurements will be made by the shipper and verified by an Agency inspector, and the container will be sealed prior to shipment. At the mixed oxide fabrication facility, under this arrangement, the number of items received will be compared to shipper documents [and reports to the Agency], the seals will be checked, and attribute checks will be made to ensure that the containers are intact. A shipper/receiver difference will be determined by the operator, and his value will be used as the input quantity when the difference is not statistically significant. When the material does not originate from another safeguarded facility, verification procedures must include 100% non-destructive measurements and weighing and sampling for off-site chemical analysis.

TABLE 3. SAMPLE SIZES, VERIFICATION METHODS AND ESTIMATED VERIFICATION TIME TO VERIFY THE IN-PROCESS INVENTORY OF THE MODEL FACILITY

Material Form	Sample Size			Verification Method	Estimated Verification time (h) [a]		
	FBR	LWR	ATR		FBR	LWR	ATR
PuO_2 powder	3–5	3–5	3–5	In-Box verification by weight check and gamma ray analysis; out-of-box by HLNCC, gamma ray, calorimetry [b]	2–3	2–3	2–3
MOX powder	3–4	5–7	6–8	" " "	2–3	3–4	4–5
Pellets	6–8	6–8	6–9	" " "	2–3	2–3	2–3
Rods	1–2	1–3	1–3	HLNCC, Gamma Ray Analysis, Calorimetry	1–2	1–2	1–2
Assemblies	3	1	5	HLNCC, Gamma Ray Analysis [c]			

[a] It is assumed that the equipment is calibrated and ready for use.
[b] Plus occasional samples for off-site analysis.
[c] No quantitative methods yet available.

Verification of the product must be by non-destructive assay. This will be done by measuring the plutonium content of rods with methods described in another paper [3]. Representative calibration standards are a most critical need here.

IX Physical Inventory

For mixed oxide fabrication facilities, the Agency requires a minimum of four physical inventories per year as the normal case. In recognition of effective and accurate operator flow control, the number of physical inventories may be reduced to two physical inventories per year. The Agency will verify more inventories if the operator performs more, for example, in response to domestic safeguards requirements.

Each physical inventory is conducted for all MBAs at the facility. Feed material is stopped and material in-process is processed to accurately measurable forms. All in-process material is collected in containers for assay, and process equipment, duct work, glove box surfaces and other areas where nuclear material collects are surveyed and deposits collected for assay to the extent practicable. The nuclear material content of all containers is to be determined through appropriate assay procedures including bulk determination (weight or volume) plus sampling for destructive analysis and through NDA.

X Conclusions

The data obtained from the operator, including the physical inventory listing and his MUF determination, together with the data obtained through verifying receipts, shipments and inventory, are used in the accountancy tests described above. Adequate safeguards are achieved when the design goals are consistent with the Agency's safeguards objectives, when the short detection time tests are satisfactory, and when accountancy performance meets the design goals.

The approach described in this paper is being developed to implement technical criteria recently approved for use by the Director General as the basis for applying Agency safeguards. Negotiations are nearing completion on facility attachments at three mixed oxide facilities, implementing these criteria. No experience has been gained to date in their application - - particularly the short detection time criteria. We are optimistic that the approach is fundamentally sound, and that implementation could be greatly facilitated, for example, by improved seals to permit prompt re-verification of the static stores, and improved NDA measurements, especially product materials.

In this paper we have limited our presentation to those aspects of safeguards relating only to the Facility. In evaluating safeguards in a State, this constitutes the first and most critical step.

REFERENCES

[1] HOUGH, G., SHEA, T., TOLCHENKOV, D., "Technical criteria for the application of IAEA safeguards", IAEA-SM-231/112, these Proceedings, Vol.I.

[2] SHEA, T., TOLCHENKOV, D., "Role of containment and surveillance measures in IAEA safeguards", IAEA-SM-231/110, these Proceedings, Vol.I.

[3] DE CAROLIS, M., "Non-destructive assay of large quantities of plutonium", IAEA-SM-231/107, these Proceedings, Vol.II.

DISCUSSION

G. PHILLIPS: Do you count pellets by number rather than weight, and would you attempt to count mixed-oxide pellets prepared by the gel precipitation route, which are only 100 μm in size?

T. SHEA: No. The verification of pellets at short detection time intervals will be achieved in a variety of ways: by counting the number of pellet boats and comparing that number with operator records; by checking the gross weight of a sample of these boats using mass balances checked with Agency calibration standards; by checking the net weight of a sub-sample of this sample by transferring the pellets into a pre-weighed empty boat, then measuring the filled boat; and by checking the gamma-ray and neutron emissions from this sub-sample of boats, all within the glove gox. In addition, individual pellets may be requested for out-of-box measurements, including off-site analysis. Finally, when containers of pellets are removed from the glove boxes for whatever purpose, they may be measured using, for example, HLNCC instruments or calorimeters, for corroboration.

Session III (Part 3) and Session IV

SAFEGUARDS FOR NUCLEAR POWER REACTORS

Chairman (Session III): D. GUPTA (Federal Republic of Germany)
Chairman (Session IV): C. CASTILLO-CRUZ (Mexico)

Rapporteur summary: *NDA measurements on irradiated fuel assemblies*

Papers IAEA-SM-231/19, 29, 47, 117, 129, 132, 135 and 136 were presented by R. MARTINC as Rapporteur

Rapporteur summary: *Safeguards for fast critical facilities*

Papers IAEA-SM-231/139, 140 and 141 were presented by V.M. GRYAZEV as Rapporteur

ДЕСТРУКТИВНЫЕ И НЕДЕСТРУКТИВНЫЕ МЕТОДЫ АНАЛИЗА ДЕЛЯЩИХСЯ ВЕЩЕСТВ ДЛЯ ЦЕЛЕЙ ВНУТРИГОСУДАРСТВЕННЫХ ГАРАНТИЙ В ГДР

К. ВИЛЛУН, В. ГРУНЕР, Х.-У. ЗИБЕРТ, Д. ХОФФМАН
Государственное управление по атомной безопасности
и защите от излучения ГДР,
Берлин-Карлсхорст,
Германская Демократическая Республика

Abstract–Аннотация

DESTRUCTIVE AND NON-DESTRUCTIVE METHODS OF MEASURING THE QUANTITY AND ISOTOPIC COMPOSITION OF FISSILE MATERIALS FOR PURPOSES OF NATIONAL SAFEGUARDS IN THE GERMAN DEMOCRATIC REPUBLIC.

The authors give a brief description of the destructive and non-destructive methods of measuring the quantity and isotopic composition of fissile materials used in the nuclear materials accounting and control system of the German Democratic Republic. They cite examples of the use of gamma-spectrometry, X-ray fluorescence analysis, neutron activation, radiochemical techniques, mass-spectrometry and alpha-spectrometry.

ДЕСТРУКТИВНЫЕ И НЕДЕСТРУКТИВНЫЕ МЕТОДЫ АНАЛИЗА ДЕЛЯЩИХСЯ ВЕЩЕСТВ ДЛЯ ЦЕЛЕЙ ВНУТРИГОСУДАРСТВЕННЫХ ГАРАНТИЙ В ГДР.

Дается краткое описание деструктивных и недеструктивных методов анализа делящихся веществ, применяемых в системе учета и контроля ядерного материала в ГДР. Приводятся примеры применения γ-спектрометрии, рентгенофлуоресцентного анализа, активных нейтронных методов, радиохимических методов, масс-спектрометрии и α-спектрометрии.

1. СИСТЕМА УЧЕТА И КОНТРОЛЯ ЯДЕРНОГО МАТЕРИАЛА В ГДР

Система учета и контроля ядерного материала в ГДР является частью обширной системы, включающей защиту от излучения, ядерную безопасность и физическую защиту ядерного топлива, предназначенной для предотвращения тех опасностей для человека, установок и окружающей среды, которые могли бы возникнуть при использовании атомной энергии.

В связи с Соглашением о контроле между ГДР и МАГАТЭ, внутригосударственная система учета и контроля ядерного материала [1], которую возглавляет Государственное управление по атомной безопасности и защите от излучения, основывается на структуре зон баланса материалов (ЗБМ), включая:

– законодательство;
– учет материала;
– отчет в МАГАТЭ и
– внутригосударственные ревизии.

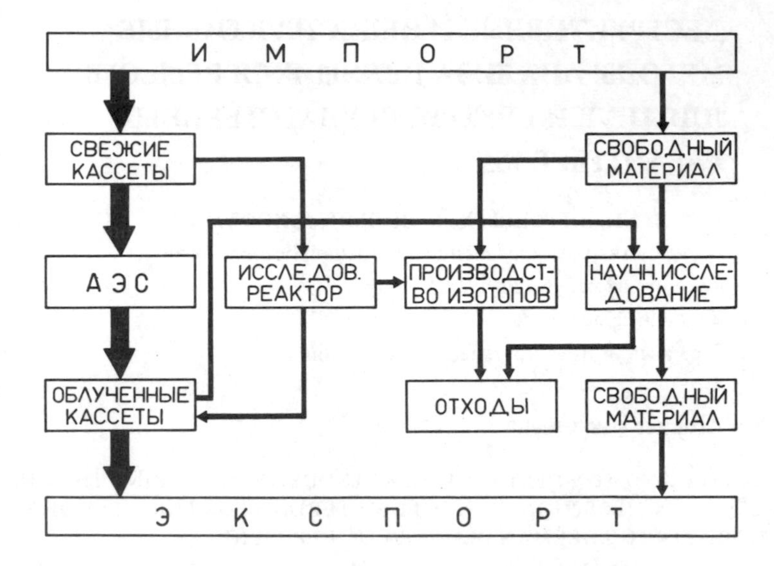

Рис. 1. Общая схема потока ядерного материала в ГДР.

При этом обеспечивается ответственность за тщательное обращение с ядерным материалом и его учет в установках, в которых он хранится или используется, соответственно национальным директивам и международным требованиям.

Помимо учета наличности ядерного материала и его изменений, система учета и контроля ядерного материала включает мероприятия по измерительной технике, направленные:

– на создание измерительного комплекса для определения количеств ядерного материала, имеющегося, поступающего, образовавшегося и израсходованного в ЗБМ, а также передаваемого или другим путем выходящего из ЗБМ;

– на оценку точности применяемых методов учета;

– на оценку имеющегося материала и его потерь, не охваченных или не охватываемых путем измерительных методов;

– на определение, проверку и оценку разницы в количестве материала в данных поставщика и получателя.

2. ТРЕБОВАНИЯ К ИЗМЕРИТЕЛЬНЫМ МЕТОДАМ КОНТРОЛЯ ЯДЕРНОГО ТОПЛИВА

Требования к измерительным методам контроля ядерного материала вытекают из наличия и развития отдельных звеньев ядерного топливного цикла. На рис. 1 показана типичная для ГДР схема потока делящихся веществ. ГДР не располагает установками

ТАБЛИЦА I. ИЗМЕРИТЕЛЬНЫЕ ОБЪЕКТЫ ДЛЯ ИНСПЕКЦИИ ЯДЕРНОГО МАТЕРИАЛА В ПОТОКЕ ДЕЛЯЩИХСЯ ВЕЩЕСТВ В ГДР

АЭС	Производство изотопов	Научные исследования	Особые применения
необлученные кассеты с топливом, облученные кассеты с топливом	необлученный уран, облученный уран, остаточный уран с продуктами деления	необлученный уран, облученный уран, отходы, содержащие U и Pu, отходы, содержащие U, Pu и продукты деления	нейтронные источники, торий, используемый для неядерных целей

для обогащения урана, производства твэлов или переработки облученного топлива. Около 50 % всего имеющегося в ГДР ядерного материала представляет собой кассеты с ядерным топливом для АЭС. Часть ядерных материалов используется в исследовательском реакторе для производства изотопов. Для этих целей используются ядерные материалы в свободном виде с различным обогащением. Применение ядерного материала в свободном виде для научных целей имеет тенденцию к возрастанию. Это применение включает в себя работы, при которых возможны изменения состава и состояния ядерного материала. В связи с этим, требования к представляемым методам экспериментального контроля ядерного материала повышаются.

На основе упомянутой схемы потока делящихся веществ в табл. I приводятся возможные объекты контроля ядерного материала в АЭС, при производстве изотопов, в научных исследованиях и при особых применениях.

3. ЭКСПЕРИМЕНТАЛЬНЫЕ МЕТОДЫ АНАЛИЗА ДЕЛЯЩИХСЯ ВЕЩЕСТВ В СИСТЕМЕ УЧЕТА И КОНТРОЛЯ ЯДЕРНОГО МАТЕРИАЛА

С целью государственного контроля отчетно-учетной документации потребителей в системе учета и контроля ядерных материалов в ГДР применяются разработанные или находящиеся в стадии разработки экспериментальные методы (рис. 2), выбираемые в зависимости от вида проб, для которых должны быть получены данные о характеристиках ядерного материала.

Исходя из того, что при определении характеристик делящихся веществ в пробах, отбираемых Инспекцией по ядерному материалу в ГДР, форма, состав и уровень их активности изменяются в больших пределах, нужные результаты анализа могут быть получены в большинстве случаев только в комбинации различных методов измерения. При этом деструктивные методы применяются, когда:

– исследуемый материал находится в свободном виде;
– требуется высокая точность анализа;
– возможно репрезентативное взятие проб.

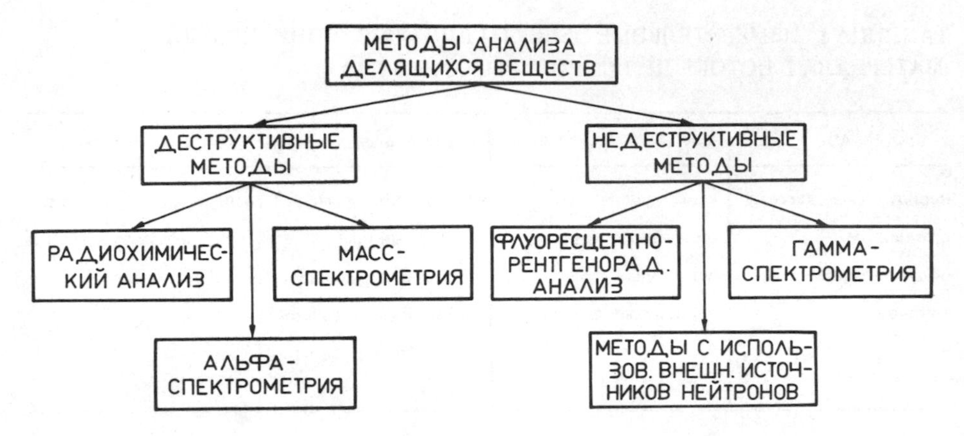

Рис. 2. Экспериментальные методы в системе учета и контроля ядерного материала в ГДР.

Недеструктивные методы определения характеристик делящихся веществ предпочтительно проводятся:

– на готовых изделиях (кассеты и твэлы);

– на облученных пробах;

– на неоднородных пробах (контейнеры, содержащие отходы).

Применяются активные и пассивные методы измерения. В то время, как при пассивных методах измеряется естественное радиоактивное излучение, что связано с ограничением получаемой информации, при активных методах, при которых с помощью внешнего источника излучения индуцируется ядерное деление в веществе проб, может быть получена более полная и целенаправленная информация о содержании делящихся веществ в пробах.

Вследствие относительно малых затрат на оборудование из всех недеструктивных методов анализа одним из широко принятых методов определения характеристик делящихся веществ является рентгено- и гамма-спектрометрия.

Основные ограничения в применении пассивного метода происходят:

– от поглощения излучения внутри пробы и зависимости интенсивного излучения от геометрии проб большого объема;

– от наложения излучения присутствующих продуктов деления.

Активные методы, при которых используется внешний источник нейтронов, имеют, несмотря на требуемые повышенные затраты, ряд преимуществ:

– хорошую проникающую способность быстрых нейтронов;

– нечувствительность относительно интенсивного гамма-излучения пробы;

– способность различить и, в некоторых случаях, дискриминировать отдельные делящиеся нуклиды.

В связи с этим, активные методы особенно пригодны для недеструктивного определения характеристик делящихся веществ в твердых пробах высокой плотности, в крупногабаритных и облученных пробах, а также в пробах с существенным содержанием других веществ.

4. НЕДЕСТРУКТИВНЫЕ МЕТОДЫ ОПРЕДЕЛЕНИЯ ХАРАКТЕРИСТИК ДЕЛЯЩИХСЯ ВЕЩЕСТВ

4.1. Определение характеристик делящихся веществ при помощи гамма-спектрометрии

В простом устройстве, состоящем из коллимированного NaJ (Tl) -детектора и двухканального анализатора, недеструктивным методом определяется степень обогащения твердых проб урана на основе измерения интенсивности собственного излучения ^{235}U с энергией 185,7 кэВ. Степень обогащения определяется при помощи калибровочного уравнения, в случаях, когда толщина пробы больше радиуса действия излучения вышеуказанной энергии. Относительная погрешность измерения для таблеток и порошка урана с 2%-ым обогащением составляет $\sim$ 0,5 за время измерения 30 мин.

Спектрометры на основе Ge (Li) -детекторов применяются без использования стандартных образцов урана при знании абсолютной фотоэффективности детектора, измеренной в определенной геометрии при помощи радиоактивных препаратов известной активности, для определения содержания урана, а также для определения степени обогащения. Для проб с низкой степенью обогащения относительная погрешность составляет 2% за два часа измерения. Без применения каких-либо калибровочных образцов степень обогащения проб низкообогащенного топлива в некоторых случаях определяется недеструктивным образом с относительной погрешностью 3-4% путем учета разницы участков кривых, соответствующих относительной фотоэффективности источников ^{235}U и ^{238}U с одинаковой геометрией.

С целью определения содержания ^{235}U в 200-литровых бочках с отходами, создается устройство, позволяющее проводить сканирование вращающейся пробы при помощи коллимированного Ge (Li) -детектора. Поправки на поглощение излучения в пробе проводятся измерением коэффициента поглощения (^{169}Yb). Имеется возможность регистрации ≥ 5 г ^{235}U за время измерения 30 мин в 200-литровых бочках с отходами, плотность которых составляет $< 0{,}6$ г/см^3.

4.2. Определение выгорания топлива при помощи гамма-спектрометрии

Имеющиеся в ГДР устройства и накопленный опыт позволяют определить степень выгорания облученных кассет топлива АЭС при помощи гамма-спектрометрического измерения концентрации продуктов деления. Эти исследования, начатые из реакторофизических соображений, позволяют получить результаты, применимые непосредственно для целей контроля ядерного материала. Проводились недеструктивные гамма-спектрометрические измерения степени выгорания на кассетах и отдельных твэлах в специальном из-

ТАБЛИЦА II. ТОЧНОСТЬ ГАММА-СПЕКТРОМЕТРИЧЕСКИХ МЕТОДОВ ОПРЕДЕЛЕНИЯ ВЫГОРАНИЯ В КАССЕТАХ ИЛИ ТВЭЛАХ

Объект измерения	Метод определения выгорания	Точность
кассета	недеструктивные гамма-спектрометрические измерения	± 6%
твэл	то же – в комбинации с деструктивным методом	± 4%
кассета	абсолютные гамма-спектрометрические измерения	± 8%

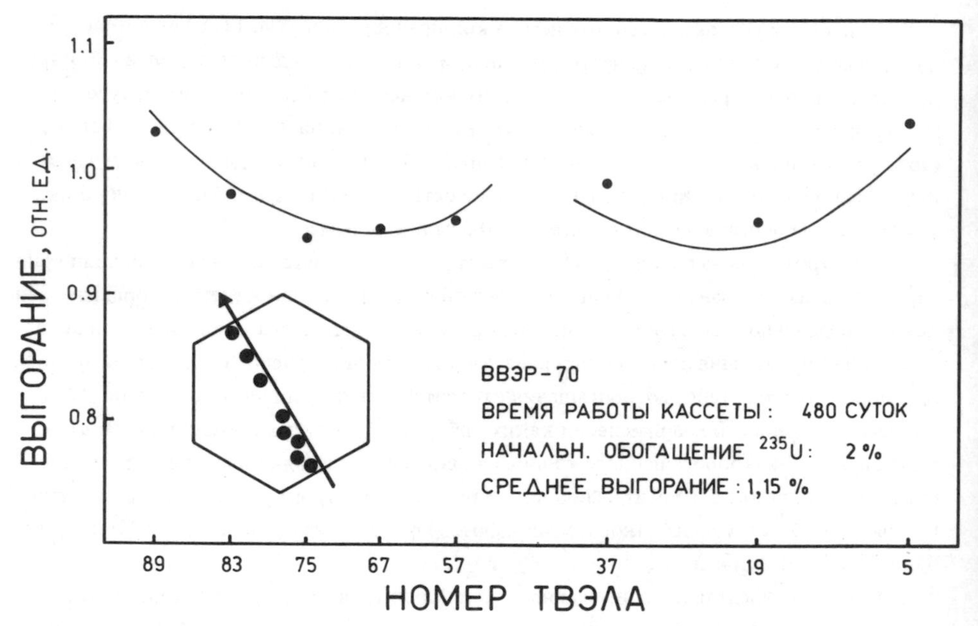

Рис. 3. Распределение измеренных и интерполированных значений полного выгорания по сечению кассеты.

мерительном контейнере при помощи коллимированного Ge(Li)-спектрометра [2]. Было показано, что даже на таких сложных объектах измерения возможно абсолютное определение выгорания путем гамма-спектрометрии. Точность определения выгорания гамма-спектрометрическим методом повышалась при дополнительном использовании деструктивных методов (табл. II).

Для определения радиального распределения выгорания в кассетах исследовались некоторые репрезентативные твэлы на их относительное выгорание и при помощи двухмерной интерполяции определялось выгорание в кассетах. Экспериментально определяемые значения выгорания по сечению кассеты совпадают с интерполированной кривой в пределах ±3,5% (рис. 3) и подтверждают правильность схемы интерполяции, основанной:

– на учете круговой симметрии отдельных типов кассет и

– на учете горизонтального макрораспределения выгорания в реакторе.

При помощи экспериментально определенного распределения степени выгорания по сечению кассеты возможна поправка абсолютных гамма-спектрометрических измерений степени выгорания на кассетах с учетом распределения активности.

Для того, чтобы избежать необходимости абсолютной калибровки спектрометра гамма-излучения, при определении степени выгорания применяется техника изотопной корреляции. Прежде всего исследуются отношения концентраций продуктов деления $^{134}Cs\,/\,^{137}Cs$ и $^{154}Eu\,/\,^{137}Cs$ в зависимости от выгорания [3,4].

4.3. Определение характеристик делящихся веществ методом рентгенофлуоресцентного анализа

С целью прямого недеструктивного определения содержания Pu в облученных реакторных кассетах проводились исследования рентгенофлуоресцентного излучения U и Pu, инициированного гамма-излучением продуктов деления [5]. Предпосылками для определения содержания Pu, в большой степени не зависящего от предшествовавшего облучения, являются:

– содержание U в топливе должно быть известно, как минимум, с той же точностью, которая желательна для определения содержания Pu, – требование, легко выполнимое при низкообогащенном топливе;

– существование однозначного соответствия между содержанием Pu в измеряемом поверхностном слое топлива и его содержанием по сечению кассеты (возможность учета этого соответствия поправочным фактором);

– проведение измерений по оси кассеты с целью учета неравномерности распределения содержания Pu по длине кассеты.

Рис. 4. Зависимость времени, затрачиваемого на измерения, необходимого для определения содержания 1 % Pu в кассетах с относительной погрешностью 1%, от времени выдержки.

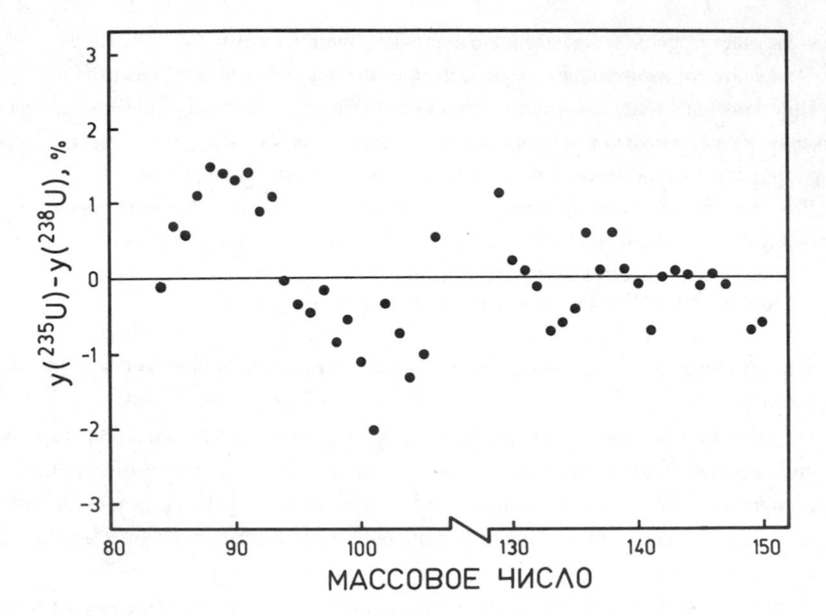

Рис. 5. Распределение по массам разностей выходов продуктов деления ^{235}U и ^{238}U для нейтронов с энергией 14,8 МэВ.

На кассетах исследовательского реактора проводилось определение содержания Pu:

– при длительном времени выдержки (250 дней) – энергоселективным анализом К-линий U и Pu с помощью коллимированного Ge-рентгеновского детектора;

– при коротком времени выдержки (75 дней) – комбинированным дифракционно-энергоселективным анализом К-линий U и Pu на основе системы, состоящей из дифракционного кристалла и Ge-рентгеновского детектора.

В то время, как энергоселективный анализ является расширением гамма-спектрометрического метода на диапазон энергии до 100 кэВ, дифракцией на подходящих кристаллических решетках проводится выделение излучения с энергией 90-130 кэВ на Ge-детектор и, тем самым, снижение нагрузки детектора высокоэнергетическими и рассеянными гамма-квантами. Рис. 4 показывает оцененное время, затрачиваемое на измерения, необходимое для определения содержания 1% Pu с относительной погрешностью ± 1% в зависимости от времени выдержки и дает представление о возможных областях применения обоих методов.

4.4. Определение характеристик делящихся веществ при помощи нейтронных методов с использованием внешних источников излучения

Устройство с использованием источника Sb-Be

Испускаемые источником Sb-Be фотонейтроны индуцируют деления нуклидов без порога деления (^{233}U, ^{235}U, ^{239}Pu), образующиеся нейтроны деления регистрируются

BF_3-счетчиком. Метод предназначен для определения остаточного содержания ^{235}U в контейнерах с отходами объемом до 4 л, содержащих продукты деления активностью 30 Ки. Предварительные измерения показывают возможность регистрации 0,4 г ^{235}U в течение 20 мин с относительной погрешностью ~ 10%.

Излучение гамма-квантов короткоживущими продуктами деления

Разница распределения по массам выхода продуктов деления для ^{235}U и ^{238}U, облученных нейтронами с энергией 14,8 МэВ, настолько мала, что разделение обоих нуклидов по излучению гамма-квантов, испускаемых продуктами деления после кратковременной выдержки (<30 мин), затруднительно (рис.5) [6]. С целью разделения нуклидов проба облучается попеременно нейтронами с энергией 14,8 МэВ и замедленными нейтронами. При замедлении, около 60% быстрых нейтронов тормозятся до энергии ниже порога деления ^{238}U. Высокоэнергетическое излучение гамма-квантов (>1,5 МэВ), испускаемых продуктами деления, регистрируется NaJ (Tl)-детектором после 100 с облучения и 300 с выдержки. В настоящее время за 12 час полного цикла анализа определяются 0,02 г ^{235}U и 1 г ^{238}U с относительной погрешностью 3 %.

Измерение эмиссии запаздывающих нейтронов

Применяется метод регистрации запаздывающих нейтронов для определения делящихся веществ, причем:

– метод выхода основан на зависимости длины цепочки распадов осколков деления от абсолютного выхода запаздывающих нейтронов;

– метод временного анализа использует различие во временных характеристиках и в интенсивности излучения запаздывающих нейтронов для различных делящихся нуклидов.

При помощи ^{3}He-счетчиков, находящихся в замедлителе из парафина, измеряется выход или распределение во времени запаздывающих нейтронов за время перерыва между импульсами нейтронов (14,8 МэВ) после 20 мс выдержки.

При методе выхода разделение ^{235}U и ^{238}U зависит от выбора ширины импульса, которая обуславливает испускание различных групп запаздывающих нейтронов. При ширине 5,8 мс и частоте 10 Гц 1 г ^{235}U определяется в 1,8 раз эффективнее, чем 1г ^{238}U; эффект усиливается при облучении замедленными нейтронами. Измерения на таблетках урана показали возможность регистрации 1 г U с относительной погрешностью 2% за 5 мин измерения [7]. Определение обогащения и содержания U проводится сравнением со стандартными образцами.

Так как метод выхода не обеспечивает удовлетворительного разделения нуклидов, имеющих похожую зависимость сечения от энергии нейтронов (^{238}U и ^{232}Th; ^{235}U и ^{239}Pu), испытывался метод временного анализа запаздывающих нейтронов. В обработку временных спектров входят, помимо экспериментальных параметров (мощность источника нейтронов, эффективность регистрации нейтронов), ядерные данные определяемых делящихся нуклидов (сечение деления, абсолютный выход задержанных нейтронов на акт деления). Неопределенность используемых ядерных данных является причиной больших ошибок при определении содержания делящихся веществ. При использовании различных оцененных наборов данных возможны, например, для сечения деления ^{235}U нейтронами 14,8 МэВ, ошибки величиной 13% (рис. 6). При помощи стандартных образцов найдено наилучшее согласие с экспериментальными результатами для набора данных, приведенного в работе [8]. Относительная погрешность при определении содержания делящихся веществ в необлученных урановых таблетках и твэлах составляет 1% за 15 мин измерения. С целью освобождения метода от применения стандартных образцов для проб массой в несколько грамм требуется повышение точности ядерных данных.

5. ДЕСТРУКТИВНЫЕ МЕТОДЫ ОПРЕДЕЛЕНИЯ ХАРАКТЕРИСТИК ДЕЛЯЩИХСЯ ВЕЩЕСТВ

5.1. Подготовка проб

Растворение твердых, чаще всего окисных проб, проводится в концентрированных минеральных кислотах (HNO_3, HNO_3 – NF). Ядерный материал с большим содержанием продуктов деления растворяется в "горячих" камерах и очищается при помощи метода ионообмена или экстракционной хроматографии. Для выделения продуктов деления из растворов урана служит модифицированный метод ионообмена [9] на DOWEX 1Х8. Накоплен опыт по разделению смеси U – Pu в системах ТБФ/силикагель и ТБФ/ПТФЭ [10], причем, при использовании синергических эффектов при выборе стационарной фазы возможно улучшение эффективности разделения. Подготовка к анализу, а также сам анализ проб, содержащих Pu, проводится в закрытых боксах.

5.2. Радиохимические методы

Определение масс U, Th и Pu проводится испытанными методами анализа, причем выбор отдельных методов зависит от требуемой точности анализа. Для ориентировочных измерений, а также при определении масс на неоднородных пробах достаточно применение методов с относительной погрешностью > 1%.

Полярография

При помощи применения полярографии с пилообразным прямоугольным напряжением в 1 M растворе HNO_3, определяется U в диапазоне массы от 100 мкг до 3 мг с относительной погрешностью 1%. В том же диапазоне массы полярографическое определение Th основано на методе косвенного определения Pu [11] замещением Cd^{2+} в комплексе Cd – ЭДТА плутонием с последующим полярографическим определением Cd.

Спектрофотометрия

При массах < 100 мкг применяются различные спектрофотометрические методы с относительной погрешностью 1-5 %. Определение содержания урана проводится мето-

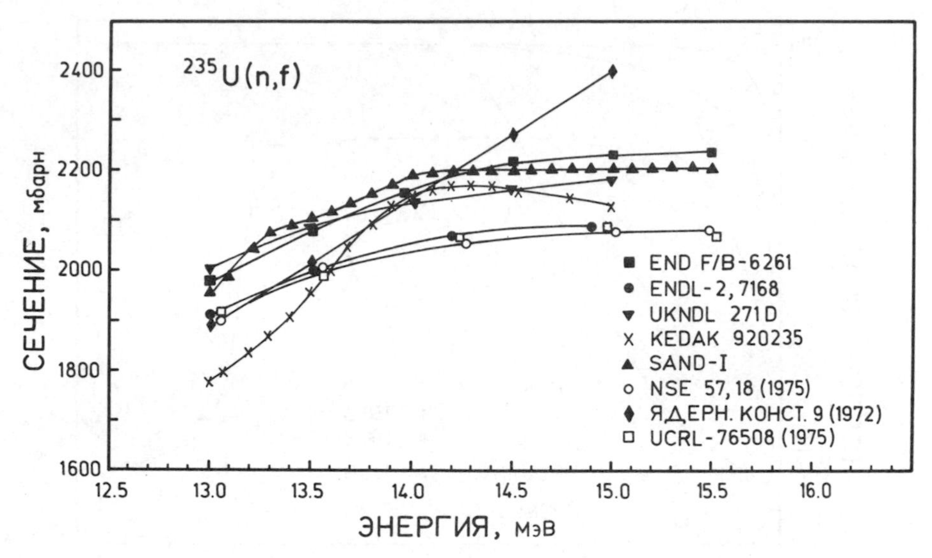

Рис. 6. Оцененные наборы данных для сечения деления ^{235}U для нейтронов с энергией 13,0-15,5 МэВ.

дом [12], похожим на метод определения содержания Pu в азотнокислом растворе при помощи арсенацо-III. Для определения содержания Th в качестве металлохромного индикатора применяется торин. При определении U, Th и Pu с повышенной точностью (< 1%), помимо того, что при отборе проб требуются довольно большие количества анализируемых веществ, сами эти вещества должны быть более чистыми. Это особенно характерно для методов гравиметрии и титрования, когда помехи при кулонометрическом анализе U и Pu легко устраняются предварительным восстановлением.

Потенциостатическая кулонометрия

Уран определяется восстановлением U(VI) до U(IV) на ртутном электроде при потенциале восстановления, равном 335 мВ НКЭ в диапазоне масс от 1 до 10 мг с относительной погрешностью 1-0,4%. Предварительное восстановление проводится при +85 мВ НКЭ. Электролитом для анализа служит 1 M раствор H_2SO_4. Плутоний восстанавливается на платиновом электроде при +265 мВ НКЭ и окисляется при +735 мВ НКЭ.

Титрование

Потенциометрическое титрование при использовании Ce(IV) проводится в диапазоне масс 4-20 мг для U и Pu с относительной погрешностью 0,2% и 0,3-1,3%, соответт-Например, для определения 4 мг Pu относительная погрешность составляет ~ 0,5% [13].

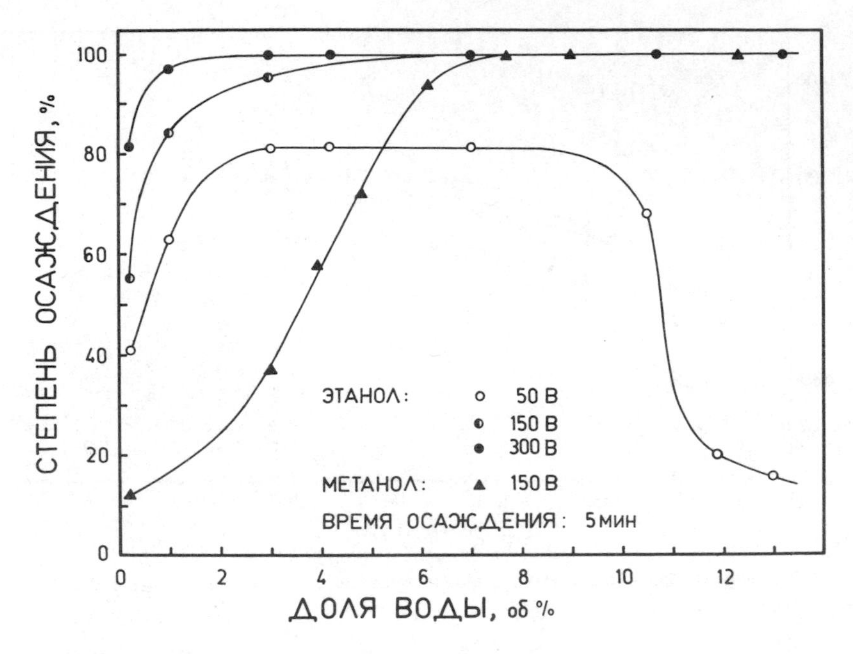

Рис. 7. Степень осаждения U в метаноле и этаноле после 5-минутного осаждения в зависимости от процентного содержания воды в растворе.

5.3. Масс-спектрометрия

Определение изотопного состава U проводится на масс-спектрометрах CH-6 и TH-5 (Varian MAT) и имеет относительную погрешность в пределах от 0,1 до 1%, в зависимости от степени обогащения для проб, содержащих несколько мкг U. Этот метод был применен для определения изотопного состава в облученных твэлах [15]. Из величины содержания ^{235}U и ^{236}U и из линейной зависимости между делением и захватом нейтронов ^{235}U определялось начальное содержание ^{235}U в твэлах, содержащих непереработанное топливо, путем экстраполяции на содержание ^{236}U, равным нулю.

5.4. Альфа-спектрометрия

Помимо масс-спектрометрии, α-спектрометрия позволяет абсолютное определение содержания α-активных делящихся нуклидов с высокой чувствительностью ($^{239}Pu > 10$ пг, $^{238}U > 10$ мкг). Электролитически-осажденные пробы измеряются в вакууме кремниевыми поверхностно-барьерными детекторами. При помощи системы диафрагм угол между пробой и детектором устанавливается с высокой точностью, так, что точность определения массы ($\gtrsim 1\%$) ограничивается в основном статистикой. Исходя из значения качества проб для α-спектрометрии, исследовалась пригодность различных электролитов для осаждения урана, причем применение органических электролитов выгодно отличается в отношении времени осаждения при одинаковом качестве проб в сравнении с неорганическими электролитами. В метаноле и этаноле уже за 5 мин была достигнута 100 %-ая степень осаждения при напряжении 150 и 300 В, соответственно для каждого электролита. Важным фактором для такого короткого времени осаждения является установление оптимального содержания воды в органическом растворе (рис. 7).

6. ЗАКЛЮЧЕНИЕ

С учетом наличия и развития отдельных звеньев ядерного топливного цикла и применения ядерных материалов в ГДР созданы измерительные устройства, позволяющие осуществить экспериментальное измерение количества и изотопного состава делящихся веществ в пробах, отбираемых Инспекцией по ядерному материалу в ЗБМ. При этом ориентировались на применение международных испытанных методов. В то время, как определение характеристик ядерного материала в облученных и необлученных пробах (порошок, растворы, таблетки, кассеты) уже проводится, методы для определения делящихся веществ в объемных, неоднородных пробах находятся в стадии разработки.

ЛИТЕРАТУРА

[1] RÖHNSCH, W., GEGUSCH, M., Safeguarding Nuclear Materials, V.I (Proc. Symp. Vienna, 1975) IAEA, Vienna (1976) 71.

[2] ГРАБЕР, Х. , ХЕРМАНН, А., БАРАНЯК, Л., ХОФМАНН, Г., ХЮБЕНЕР, С., МЭНЕР, Х.-Х., НЕБЕЛЬ, З., ГЮНТЕР, Х., Доклад на IV Симпозиуме СЭВ по "Исследованиям в области переработки облученного топлива", Карловы Вары (ЧССР), КВ 77/А 9 (1977).

[3] GRABER, H., KEDDAR, A., Доклад IAEA-SM-231/129 на данной конференции.

[4] HERMANN, A., MEHNER, H.-H., Доклад IAEA-SM-231/20 на данной конференции.

[5] БОТЕ, Х.-К., КЕРНЕР, Г., ТЮММЕЛЬ, Х.-В., Доклад на IV Симпозиуме СЭВ по "Исследованиям в области переработки облученного топлива", Карловы Вары (ЧССР), КВ 77/А 22 (1977).

[6] MEEK, M.E., RIDER, B.F., NEDO-12154 (1972).

[7] BLUMENTRITT, G., BREITENFELD, R., RICHTER, K., SCHMITT, W., WIETZKE, W., ZfK RPP-1/73 (1973).

[8] EAST, L.V., AUGUSTON, R.H., MINLOVE, H.O., MASTERS, C.S., LA-4605-MS (1971).

[9] STOEPPLER, M., Jül-633-CA (1969).

[10] HERMANN, A., BARANIAK, L., HÜBENER, S., NEBEL, D., NIESE, U., TREBELJAHR, S., Isotopenpraxis 12 7/8 (1976) 277.

[11] ELLIOT, F., FOREMAN, I.K., "Advances in polarography" ed. by I.S. LONGMUIR, Pergamon-Press, V. II (1960) 538.

[12] КЛЫГИН, А.Е., СМИРНОВА, И.Д., КОЛЫАДА, Н.С., Жур. Неорг. Хим. 15 (1970) 3304.

[13] BRANDT, W., Z. Kristallchem. (1969) 193.

[14] NEBEL, D., HERMANN, A., TREBELJAHR, S., J. Radioanal. Chem. 28 (1975) 133.

[15] STEPHAN, H., BELL, G., Z. Kristallchem. (1976) 312.

DISCUSSION

M.R. IYER: How does the waste content in the 200-litre drums measured using non-destructive gamma spectrometric methods compare with results obtained by destructive analysis?

H-U. SIEBERT: Destructive methods are not used for determining the amount of fissionable material in large samples, which are usually of non-uniform structure, because of the difficulty of obtaining representative samples. With the aid of prepared samples of known composition it is possible to verify the efficiency of the non-destructive gamma-spectrometric method of determining the content and characteristics of fissionable materials in such cases.

APPLICABILITY OF NON-DESTRUCTIVE ASSAY TECHNIQUES TO SPENT FUELS FROM BOILING-WATER REACTORS*

S. TSUJIMURA
Japan Atomic Energy Research Institute,
Tokai-mura, Ibaraki-ken, Japan

Abstract

APPLICABILITY OF NON-DESTRUCTIVE ASSAY TECHNIQUES TO SPENT FUELS FROM BOILING-WATER REACTORS.

Non-destructive gamma-ray spectrometry was carried out on the spent fuel assemblies of a whole core of JPDR-I which is a boiling-water reactor. The results were analysed, taking into consideration the spatial variation of the neutron spectrum and the history of reactor operation. The relationship between the $^{134}Cs/^{137}Cs$ ratio and Pu/U was explained reasonably as well as that between ^{137}Cs activity and burnup. The build-up of plutonium in the assemblies can be estimated if the NDA data are calibrated in combination with the results of the destructive assay. The applicability of the NDA techniques to fuel identification in the field is also discussed.

1. INTRODUCTION

A sufficiently accurate and rapid method for determining the amount of plutonium accumulated in spent fuel is indispensable for strengthening the safeguarding of nuclear materials at reprocessing plants and reactor sites. Burnup and accumulation of transplutonium elements are also interesting from the viewpoint of fuel management. To estimate these parameters, spent fuels must be measured non-destructively in a storage pool. Furthermore, an appropriate identification method for individual spent-fuel assemblies should be established for safeguards implementation.

Techniques usable for this purpose are limited [1]. So far gamma-ray spectrometry is the most known and reliable, so its effectiveness must be fully studied.

Gamma-ray spectra of spent-fuel assemblies of the Japan Power Demonstration Reactor (JPDR)-I, a BWR, were measured prior to shipment to the Power Reactor and Nuclear Fuel Development Corp. (PNC) Tokai Reprocessing Plant. The results were analysed in comparing these with the calculated and destructive data. The applicability of the NDA to the fuel identification in the field is also discussed.

---

* This work was performed under a joint research programme between the JAERI and the PNC, and part of the work under IAEA Research Agreement No.2040/CF.

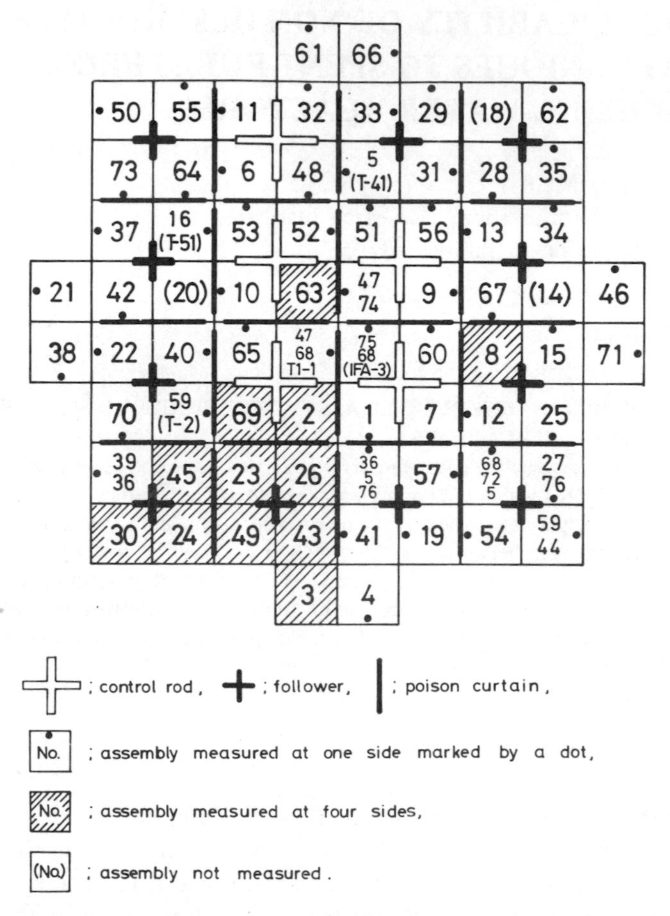

FIG.1. Plan of JPDR-I core and assembly identification number.

2. BACKGROUND

In a previous study, gamma-ray spectrometry had been carried out on eight separate fuel rods from one selected assembly (A20) of the JPDR-I core [2]. Chemical analysis was performed on the 24 representative points measured by the NDA for heavy nuclides and fission products.

The $^{134}Cs/^{137}Cs$ or $^{154}Eu/^{137}Cs$ ratio corresponding to a given burnup varied with the position in the assembly. This has been understood in connection with the difference in irradiation history and with spatial variations of the neutron spectrum. This interpretation is also supported by the fact that the Pu/U ratio does not always correspond to the burnup, particularly in the BWR. It has been concluded that the non-destructively measurable ratios of fission products can be used as a better indicator of Pu/U ratio than burnup. There still remain some problems in applying the gamma-ray spectrometry to the evaluation of plutonium buildup in spent fuels.

3. JPDR-I CORE AND OPERATION HISTORY

The reactor is a natural-circulation BWR of 45 MW thermal output, loaded with 2.6 wt% enriched UO_2 fuels. The core consists of 72 fuel assemblies, 16 cruciform control rods and 24 burnable poison curtains. The plan of the core and the identification number of the assembly are shown in Fig.1. An assembly consisting of 36 fuel rods is arrayed in a 6 × 6 square lattice in a channel box. Each rod is formed of two segments – upper and lower – with the same active length of 721 mm. The fuel pellet is 12.5 mm in diameter and clad in a Zircaloy-2 sheath 0.76 mm thick.

The core was operated from October 1963 to August 1969 but had a long shutdown from June 1968 to June 1969. Most assemblies were loaded in the initial loading position during irradiation. The horizontal burnup distribution was almost symmetrical to the central axis of the core. Consequently, the assemblies could be classified into several groups by the burnup levels and the irradiated positions. From operation data, the burnup of individual assemblies ranged from 110 to 5640 MW · d/t and the cooling time from 8 to 13 years.

4. GAMMA-RAY SPECTROMETRY AT A SPENT FUEL STORAGE POOL

Gamma-ray spectra were measured on 71 assemblies with a gamma-scanning apparatus which was installed temporarily in a JPDR storage pool. The apparatus consisted of a reclining fuel bed, a collimator set, a two-dimensional scanning mechanism, and a gamma-ray spectrometer. The arrangement is depicted in Fig.2.

An assembly was held horizontally on the bed. Gamma-rays were collimated by a vertical beam guide with 20 mm inner diameter and 5.5 m long and measured with a co-axial type of Ge(Li) detector (88 cm^3). Spectra data obtained were recorded on cassette magnetic tapes and analysed with a code BOB-73 [3]. The overall energy resolution was 2.1 keV in FWHM for the 662-keV gamma-ray. The gamma-rays ranging from 200 to 2200 keV were measured. The count-rate was less than 13 000 counts/s and the spectrum distortion was negligible.

The measurement positions were selected from the mesh-points in the burnup calculation as shown in Fig.3. Of the 71 assemblies, eleven forming one octant of the core were measured on ten positions at every side, while the rest were measured in four positions (2, 3, 4 and 5 in Fig.3) at one side only.

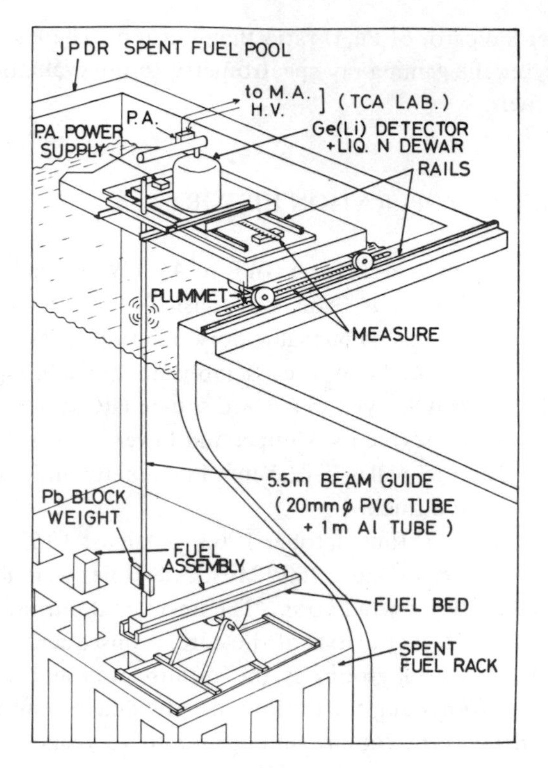

FIG.2. Arrangement of gamma-scanning apparatus at the JPDR pool.

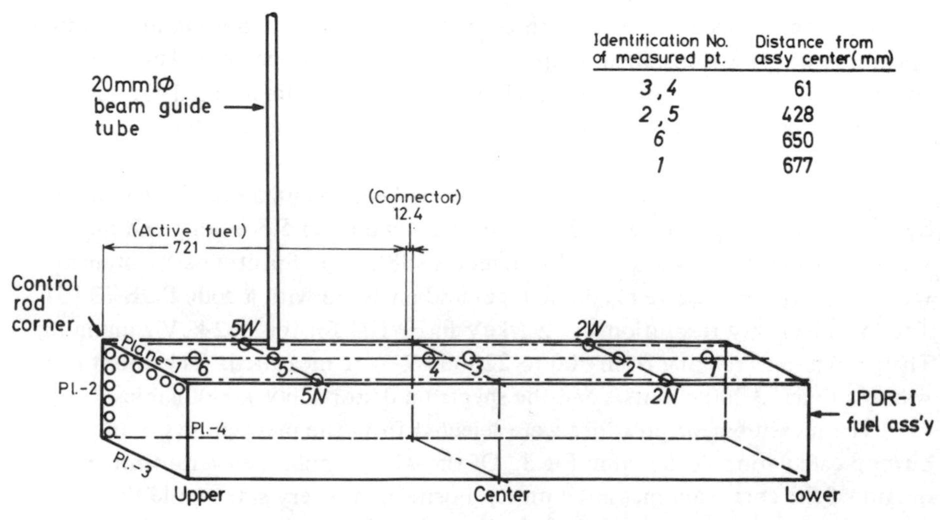

FIG.3. Measurement positions for gamma spectrometry.

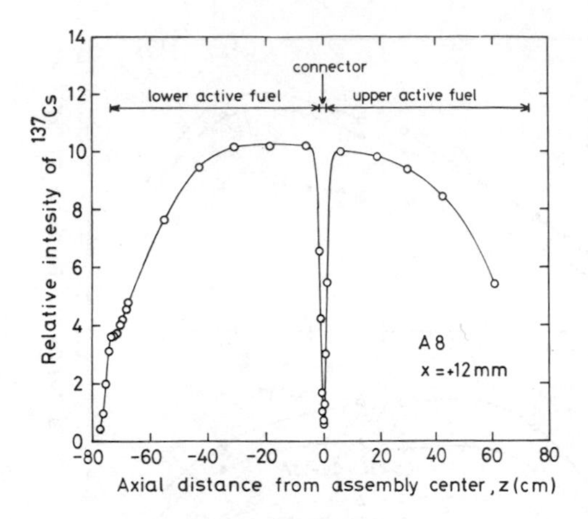

FIG.4. Axial distribution of gamma activity of the fuel assembly.

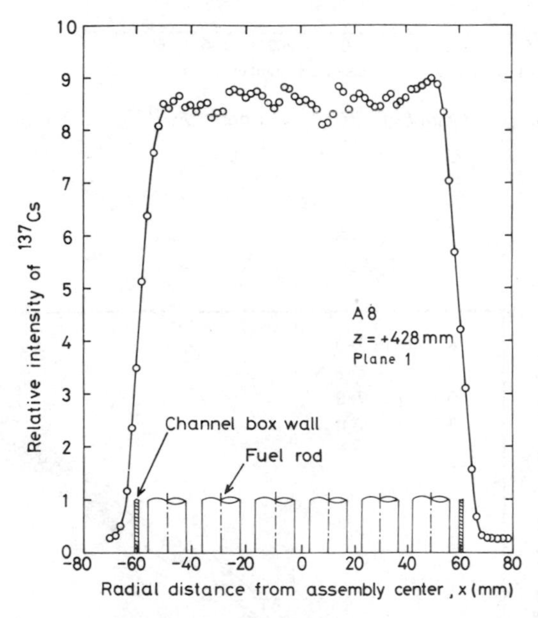

FIG.5. Horizontal distribution of gamma activity of the fuel assembly.

A fuel assembly of the BWR is usually covered by a channel box which screens the fuel rods from view. Both axial and horizontal distributions of gamma activities were measured by scanning as shown in Figs 4 and 5, respectively. The axial centre could be determined by a sharp dip due to a fuel vacancy at the rod connector and the horizontal positions were determined as referring to the rising points in the distribution.

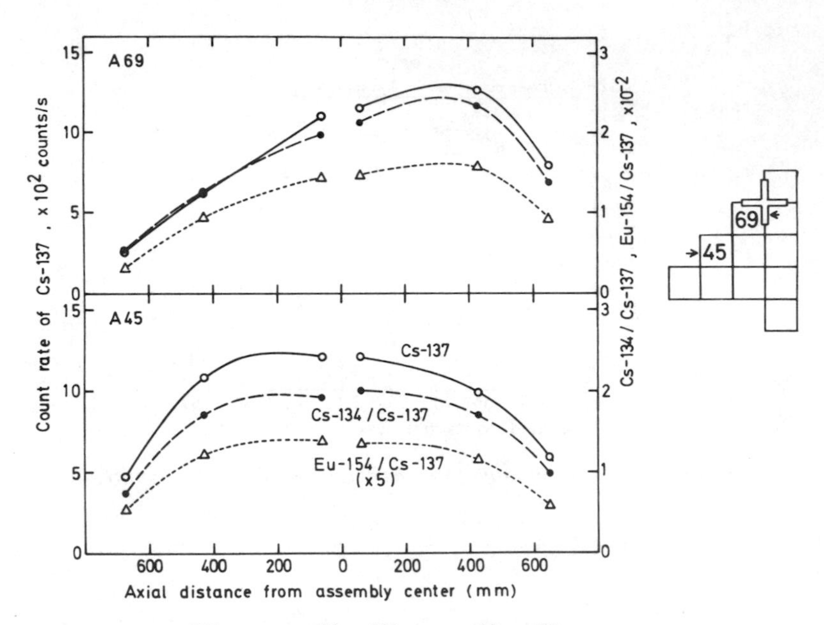

FIG.6. Distribution of ^{137}Cs and of $^{134}Cs/^{137}Cs$ and $^{154}Eu/^{137}Cs$ ratios upon different loading positions.

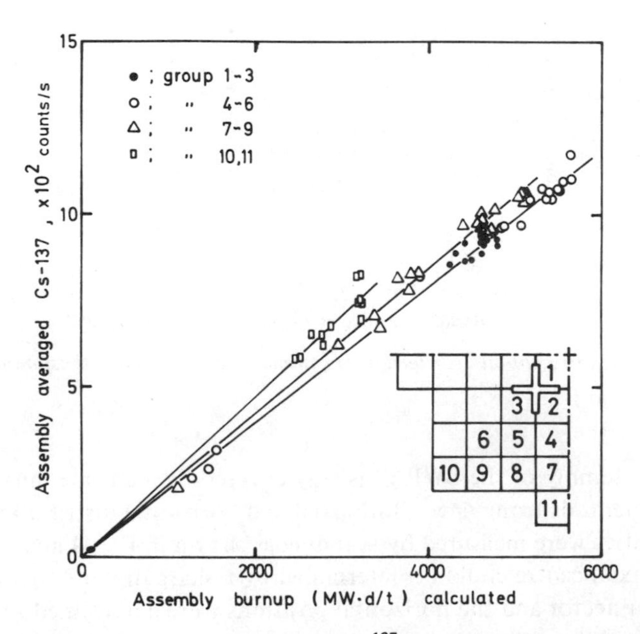

FIG.7. Correlation between assembly-averaged ^{137}Cs activity and calculated burnup.

5. RESULTS OF MEASUREMENT ON THE JPDR FUEL

Photopeaks of ^{137}Cs (662 keV), ^{134}Cs (796 keV) and ^{154}Eu (1275 keV) were identified with sufficient statistics by 400-s counting. But those of ^{106}Ru-Rh and ^{144}Ce-Pr were too small for use. The migration of caesium was neglected because no abnormality in the axial distributions of ^{134}Cs and ^{137}Cs was observed in comparison with that of ^{154}Eu.

The shape of the axial distribution of the fission products strongly depended on the place where the assembly had been loaded during irradiation. Figure 6 shows the distributions of ^{137}Cs and of ^{134}Cs/^{137}Cs and ^{154}Eu/^{137}Cs ratios. Assembly A69 had been adjacent to a partially inserted control rod and the distribution of both burnup and Pu build-up were skewed to the upper part of the assembly. On the other hand, the adjacent control rod of assembly A45 had been fully withdrawn and the skewnesses in those distributions were small. This fact must be taken into account when the assembly-averaged burnup and Pu build-up are evaluated at limited points. Instead, it makes it possible for an assembly to be distinguished from others irradiated under different conditions.

As shown in Fig.7, there is a reasonable correlation between the assembly-averaged activity of ^{137}Cs and the burnup calculated but there is a small systematic deviation between different groups. The deviation may be due either to errors in burnup calculation or to those when averaging NDA results.

Dependence of the relations between the ^{134}Cs/^{137}Cs ratio and the ^{137}Cs activity on the spatial variation of neutron spectrum, which was caused by the steam void formation and water gap in the BWR core, was also recognized in the present measurement of the assembly as shown in Figs 8 and 9. A good correlation can be established by taking into consideration the spatial dependence of the neutron spectrum.

6. EVALUATION OF Pu BUILD-UP IN SPENT FUEL

Figure 10 shows the correlation between the assembly-averaged ^{134}Cs/^{137}Cs ratio and the calculated Pu/U ratio. There was again a systematic deviation between different groups. The larger deviations were found with groups 1, 2 and 3, which had been adjacent to a partially inserted control rod. Comparing the result with the correlation shown in Fig.7, the calculated value of the Pu build-up was possibly underestimated in such assemblies.

In Fig.11, the ^{134}Cs/^{137}Cs ratio obtained by the NDA is correlated with the Pu/U ratio by destructive analysis of the input solution to reprocessing. Each batch of reprocessing consisted of two or four assemblies which had a similar irradiation history. The value of the ^{134}Cs/^{137}Cs ratio was also averaged so as to correspond to the reprocessing batch. The correlation observed in Fig.11

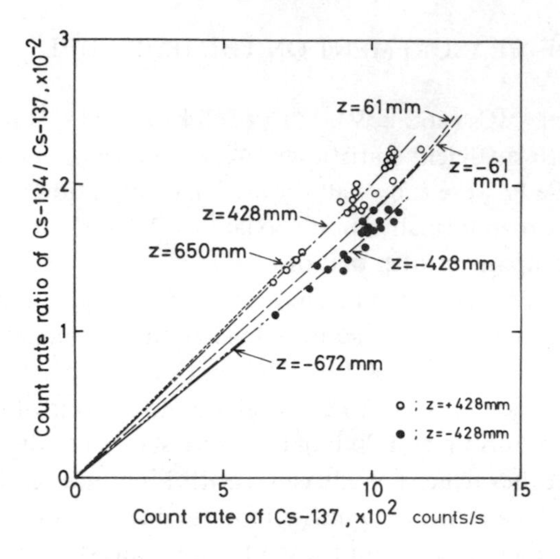

FIG.8. Correlation between $^{134}Cs/^{137}Cs$ and ^{137}Cs of individual positions.

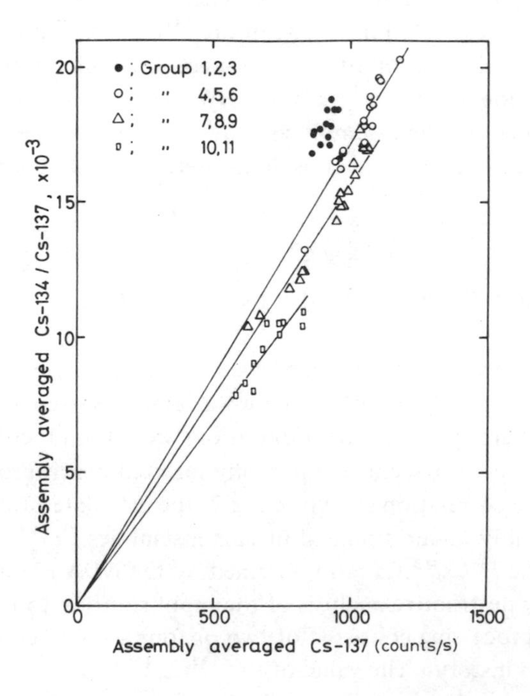

FIG.9. Correlation between assembly-averaged $^{134}Cs/^{137}Cs$ and ^{137}Cs.

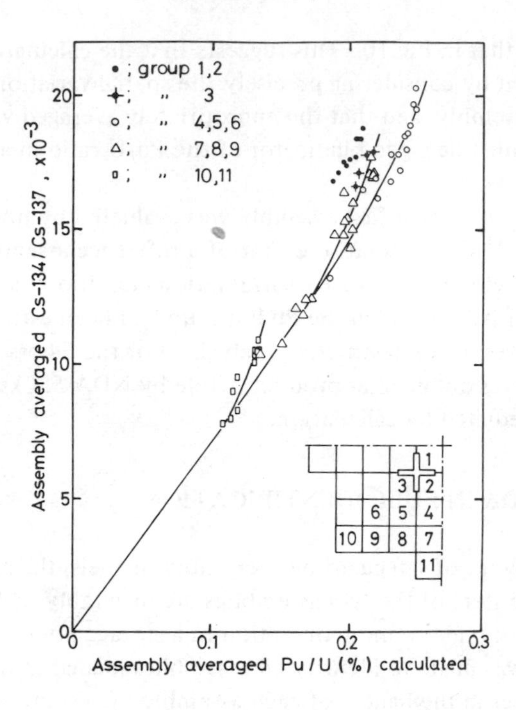

FIG.10. Correlation between assembly-averaged $^{134}Cs/^{137}Cs$ and Pu/U calculated.

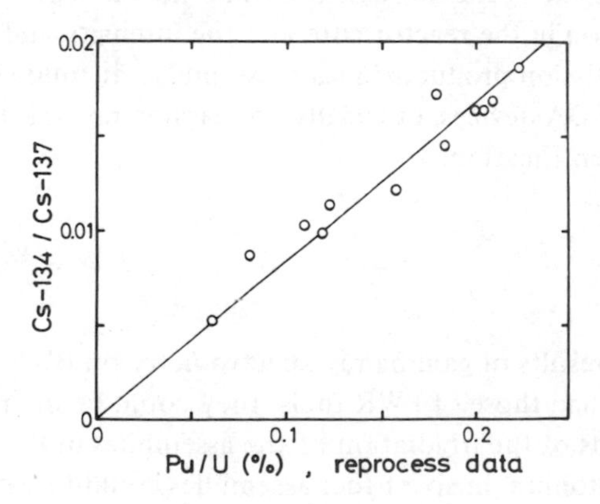

FIG.11. Correlation between $^{134}Cs/^{137}Cs$ determined by NDA and Pu/U measured in reprocessing on individual batches.

seems better than that in Fig.10. This suggests that the calculation method should be improved by considering precisely the spatial variation of the neutron spectrum in the assembly, and that the appropriately averaged value of the $^{134}Cs/^{137}Cs$ ratio must be a good indicator of the Pu/U ratio in an assembly even for a BWR.

The Pu build-up in each fuel assembly was evaluated by normalizing the assembly-averaged $^{134}Cs/^{137}Cs$ ratio to that of a reference assembly. Assembly A8 was chosen as the reference because its irradiation condition had been almost identical to that of A20 in which the Pu build-up had been estimated in combination with the results of destructive analysis. For the 54 assemblies reprocessed, 5.4 kg Pu was obtained as product, while by NDA 5.7 kg was obtained, and 5.6 kg was predicted by calculation.

7. ROLE OF NDA IN FUEL IDENTIFICATION

In the framework of safeguarding spent nuclear fuels, the accounting and identifying of each item of the fuel assemblies are inevitably included. The difficulty is to work under remote operation in a storage pool.

An attempt was made to identify all assemblies loaded in the JPDR-I core by carving a number in the handle of each assembly. On some assemblies, which amounted to several per cent of the whole, the numerals carved were very hard to read because of fur forming on the metal surface. Assemblies with a similar irradiation history may not be distinguished from each other by NDA. Unidentified assemblies, however, were at most several per cent of the whole and could be cleared by referring to the NDA results.

It was thus found that the NDA technique is quite effective for completing the identification of the spent-fuel assemblies on the basis of the consistency of the loaded position in the reactor core and the intensity and distribution of the gamma-rays of fission products in each assembly. It follows from this that the installation of NDA devices, especially for gamma-ray spectrometry, is effective for ensuring fuel identification.

8. CONCLUSION

Although the results of gamma-ray spectrometry on BWR spent fuels are more complicated than those of PWR fuels, they could be interpreted in connection with the records of the irradiation of the assemblies in the reactor core. The build-up of plutonium in spent-fuel assemblies could be estimated, provided that the results of the gamma-ray measurements were carefully analysed and calibrated by direct analysis (DA) of some typical ones. A good combination of NDA, DA and burnup calculation was necessary to ensure the accounting and identification of spent fuels. If a certain number of assemblies for calibration standards are accumulated and the burnup calculation can be made with better accuracy, one may derive substantial results of plutonium build-up in spent fuels from the NDA and related information prior to the reprocessing of the fuels.

ACKNOWLEDGEMENTS

The author is grateful for the helpful contributions from Mr. G. Fukuda and his associate staffs of the PNC, and from Mr. T. Hidaka and his associate staffs of the JPDR-JAERI.

The author is also indebted to Messrs H. Natsume, S. Matsuura, H. Okashita and H. Umezawa for their assistance in preparing this paper.

REFERENCES

[1] HSUE, S.T., CRANE, T.W., TALBERT, L.W., Jr., LEE, J.C., Non-destructive Assay Methods for Irradiated Nuclear Fuels, LA-6923 (Jan.1978).

[2] NATSUME, H., MATSUURA, S., OKASHITA, H., UMEZAWA, H., EZURE, H., Collection of gamma spectra data of irradiated light-water-moderated reactor spent fuel and study of the applicability of the method for fuel identification, IAEA, Research Contract 1119/R1/RB (Oct.1975); and
MATSUURA, S., TSURUTA, H., SUZAKI, T., OKASHITA, H., UMEZAWA, H., NATSUME, H., Non-destructive gamma-ray spectrometry on spent fuels of a boiling water reactor, J. Nucl. Sci. Technol. 12 (1975) 24; and
NATSUME, H., et al., Gamma-ray spectrometry and chemical analysis data of JPDR-I spent fuel, J. Nucl. Sci. Technol. **14** (1977) 745.

[3] BABA, H., SEKINE, T., BABA, S., OKASHITA, H., A Method of the Gamma-ray Spectrum Analysis: FORTRAN IV Programs "BOB 73" for Ge(Li) Detectors and "NAISAP" for NAI(Tl) Detectors, JAERI 1227 (1973).

DISCUSSION

A. RAMALHO: I should first like to make a comment on your paper. Since thc paper includes results from both non-destructive and destructive analysis, it represents a unique situation for the evaluation of NDA measurements. In view of this, I think it would have been extremely valuable if you could have included tables of results to supplement the figures. I also have two questions: First, what is the order of magnitude of the differences between measurements made on the four sides of the assembly? Second, how well can the differences in Fig.9 be explained in terms of the radial variations in the neutron spectrum?

S. TSUJIMURA: The order of magnitude of the differences is 10%. The differences in Fig.9 can be explained by considering the neutron spectrum for the groups shown in Fig.7. The spectrum for group 1–3, located in the centre of the reactor and adjacent to the control rod, is harder than that for group 10–11 in the outer part of the reactor. The hardness of the spectrum for groups 4–6 and 7–9 is intermediate. A hard spectrum favours the capture reaction, increasing the $^{134}Cs/^{137}Cs$ ratio.

J. BOUCHARD: On the basis of our experience with PWRs we think that measurement on all four faces of a fuel assembly is essential in order to obtain a representative average value for the assembly as a whole. This is because of the radial heterogeneity and the small depth of detection of the gamma measurements. Actually, for energies of around 1 MeV, the measured radiation comes exclusively from the first two or three rows of pins seen by the detector.

IAEA-SM-231/47

CONTROLE DES COMBUSTIBLES IRRADIES DES REACTEURS A EAU PAR GAMMAMETRIE SUR LES SITES DES CENTRALES

R. BERAHA*, J. BOUCHARD**, G. FREJAVILLE**, V. ZEČEVIĆ*
* Framatome, Paris-La-Défense
** CEA, Centre d'études nucléaires de Cadarache,
Saint-Paul-lez-Durance,
France

Abstract–Résumé

MONITORING OF IRRADIATED WATER-REACTOR FUELS BY GAMMA SPECTROMETRY AT THE POWER PLANT.

Direct, non-destructive measurement on irradiated fuel of the gamma activity associated with the radioactive decay of fission products provides important data on the state of the fuel and its irradiation history, i.e. power distribution, physical state of the pins and assemblies, and burnup. For the past six years regular measurements have been made on the assemblies from the PWR reactor at the Ardennes nuclear power plant by means of prototype equipment (constructed and used by the French Atomic Energy Commission (CEA)) which is immersed in the fuel cooling pond. Concurrently, Framatome has designed and constructed devices for examining irradiated water-reactor fuel which are based, apart from gamma spectrometry, on such inspection methods as visual observation, measurement of dimensions and analysis of possible deformation. The two organizations have pooled their efforts and are continuing the development work in order to gain from experience already acquired and to supplement it with laboratory tests, thereby improving the measurement technique and broadening its range of application. The present development of the technique is geared first to systematically monitoring the operation of the reactors and, second, to identifying irradiated assemblies by measurement of the burnup and cooling time.

CONTRÔLE DES COMBUSTIBLES IRRADIES DES REACTEURS A EAU PAR GAMMAMETRIE SUR LES SITES DES CENTRALES.

Les mesures directes et non destructives sur les combustibles irradiés des activités gamma liées à la décroissance radioactive des produits de fission apportent des informations importantes sur l'état du combustible et son histoire d'irradiation: distributions de puissance, état physique des crayons ou assemblages, taux de combustion. Depuis six ans des mesures systématiques sur les assemblages du réacteur à eau sous pression de la Centrale nucléaire des Ardennes ont été effectuées à l'aide d'un appareillage prototype réalisé et utilisé par le Commissariat français à l'énergie atomique, et immergé dans la piscine de désactivation des combustibles. Parallèlement, Framatome a étudié et réalisé des dispositifs d'examen des combustibles irradiés pour les réacteurs à eau. Ces dispositifs comportent, outre la gammamétrie, d'autres moyens d'examen tels que : observations visuelles, mesures dimensionnelles, étude de déformations éventuelles, etc. Les deux organismes ont joint leurs efforts pour poursuivre des travaux de développement afin de profiter des expériences précédentes et de les compléter par des essais en laboratoire pour améliorer la technique de mesure et élargir son champ d'application. Les

développements actuels de cette technique sont orientés vers les applications pour un suivi systématique du fonctionnement des réacteurs d'une part et pour l'identification d'assemblages irradiés à partir de la mesure des taux de combustion et du temps de refroidissement d'autre part.

1. INTRODUCTION

Le contrôle des combustibles irradiés en piscine de désactivation ou de stockage fait partie intégrante des actions associées à la notion de garantie des matières nucléaires. La gammamétrie de ces combustibles, sur site, constitue l'un des moyens.

Cette technique non destructive, dont l'application est pleine de promesses, est déjà utilisée à d'autres fins et a dépassé l'étape des premières mises au point [1].

Avec un appareillage adéquat et un système de traitement des données adapté aux besoins, cette technique peut, en effet, apporter deux catégories de renseignements majeurs :

– des informations sur l'état physique du combustible et les événements nucléaires correspondant à l'irradiation qu'il a subie; ces informations permettent de s'assurer de l'intégrité du combustible et d'effectuer un suivi du fonctionnement;
– une garantie sur la composition des assemblages à partir du contrôle des taux de combustion; l'application concerne alors la gestion des assemblages irradiés dans les piscines de stockage, soit des réacteurs eux-mêmes lorsque ce stockage est de longue durée, soit des unités de retraitement.

Nous développons ici dans le cas des réacteurs à eau légère:

– les procédés utilisés pour les premières expériences françaises à la Centrale nucléaire des Ardennes (CNA) avec le bilan de plusieurs campagnes;
– les développements immédiats et futurs qui en découlent sur les appareillages et les traitements de données pour application sur assemblages entiers et sur crayons amovibles;
– les applications envisagées dans le cadre des contrôles de garanties.

2. EXPERIENCE ACQUISE A LA CENTRALE NUCLEAIRE DES ARDENNES

2.1. Principes

Dans le cadre des expériences de qualification des codes de calcul neutronique pour les réacteurs à eau, en particulier du système Apollo-Neptune [2], un important programme expérimental a été exécuté à partir de 1972 sur des combustibles irradiés de la Centrale nucléaire des Ardennes. Des analyses de composition isotopique, des mesures physiques par spectrométrie alpha ou gamma, des mesures d'émission neutronique et la détermination d'effets en réactivité par oscillations ont été effectuées sur des crayons prélevés dans trois assemblages déchargés respectivement à la fin des trois premiers cycles de fonctionnement du réacteur. Il est apparu clairement alors qu'il serait intéressant de pouvoir compléter cet ensemble de résultats par des distributions relatives de quelques produits de fission caractéristiques de la puissance ou du taux de combustion, distributions que l'on ne pouvait envisager d'obtenir que par gammamétrie.

Une expérience préliminaire était réalisée dès juillet 1972 lors de l'arrêt pour rechargement à la fin du deuxième cycle. Les moyens mis en œuvre pour cette expérience étaient très simples – détecteur NaI, défilement vertical de l'assemblage devant le détecteur à l'aide du pont de manutention – et le but était essentiellement de vérifier la validité de la méthode et son intérêt.

C'est à partir de cette expérience préliminaire qu'ont été effectués les principaux choix qui devaient conduire à la réalisation de l'appareil actuellement en service:

– appareil immergé dans la piscine de stockage des éléments irradiés,
– mécanisme autonome pour la scrutation axiale et périphérique,
– utilisation d'un détecteur Ge-Li ou Ge intrinsèque, afin de permettre une discrimination correcte des différents pics,
– collimation «large» permettant d'intégrer plusieurs crayons sur chaque face d'assemblage.

Cet appareil, conçu et réalisé en 1973, a été mis en place avant l'arrêt pour rechargement d'août 1974 (fin du quatrième cycle) et a été utilisé jusqu'à présent pour quatre campagnes aux arrêts de fin du quatrième, du cinquième, du sixième et du septième cycle, ainsi que pour des mesures sur des assemblages plus refroidis.

2.2. Conception et réalisation

Afin de limiter le développement de l'appareil et de réduire au minimum la manipulation des assemblages combustibles, la conception mécanique retenue comprend un panier qui reçoit l'élément à mesurer et qui est fixé sur une simple plate-forme tournante et un ensemble de détection qui se déplace verticalement sur toute la hauteur de l'assemblage et qui peut également être déplacé horizontalement suivant l'axe perpendiculaire à l'axe de visée.

L'installation (fig.1 et 2) comprend:

– un châssis en acier inoxydable soudé d'une hauteur totale de 4,70 m avec une plaque de base de 1,0 × 2,0 m^2 qui supporte le panier porte-assemblage et la colonne de guidage du château-compteur; quatre stabilisateurs escamotables qui augmentent la surface d'appui (2,2 × 1,5 m^2) permettent d'assurer une bonne stabilité pour un encombrement minimal;
– le panier qui reçoit l'assemblage et qui est fixé sur une plate-forme rotative pouvant occuper huit positions régulièrement espacées de 45°; les dimensions

FIG.1. Vue d'ensemble de l'appareil avant la descente en piscine.

intérieures de ce panier sont de 206 X 206 mm; en plus du mouvement de rotation commandé par une perche depuis la passerelle supérieure, la plate-forme possède un réglage angulaire fin de ±45' pour faciliter les alignements entre la machine de manutention et l'appareil de gammamétrie;
le château-compteur en plomb, gainé extérieurement en acier inoxydable, dont le diamètre extérieur maximal est de 405 mm et la hauteur de 800 mm; ce château est fixé sur un tablier support qui se déplace verticalement le long d'une colonne guide, poutre en I en acier inoxydable maintenue de façon rigide sur toute sa hauteur grâce à une vis fixée le long de la colonne; le mouvement de la vis est assuré par un moteur situé sur la plate-forme supérieure;

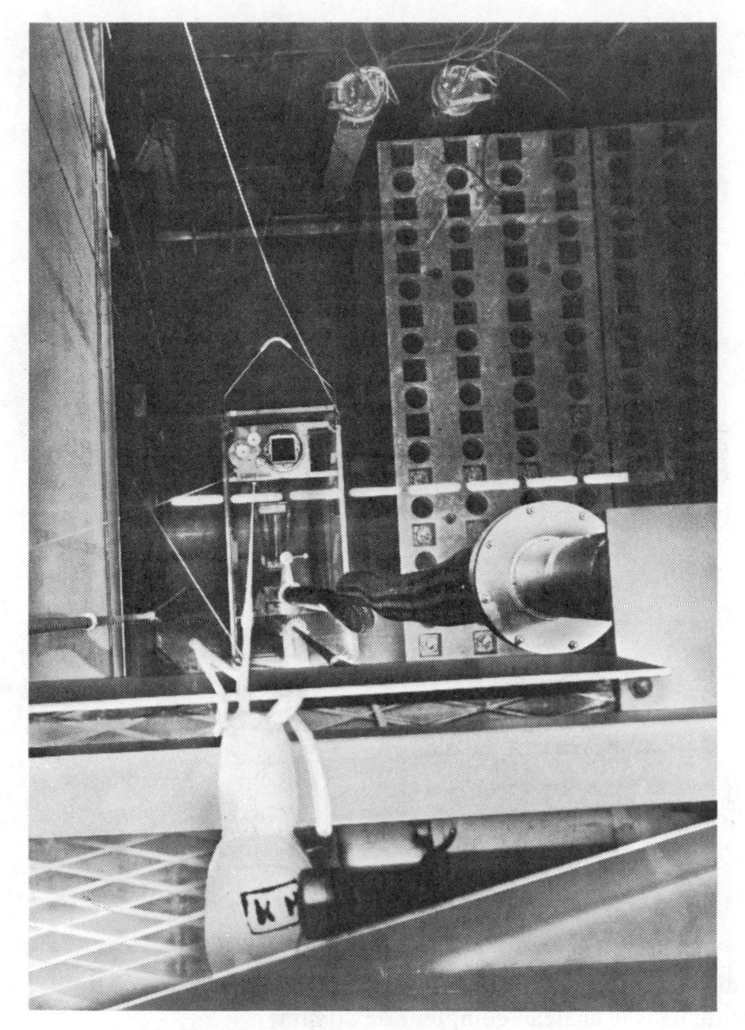

FIG.2. Vue d'ensemble du dispositif en piscine.

– l'ensemble de détection, introduit dans le château-compteur grâce à un tube étanche souple en acier inoxydable qui remonte jusqu'à la passerelle supérieure;
– la passerelle supérieure indépendante du châssis de l'appareil, fixée sur le bord de la piscine; elle reçoit les différents équipements hors de d'eau.

Une coupe du château-compteur montrant la disposition du blindage et le principe de la collimation est donnée sur la figure 3.

En raison de la disposition particulière dans un angle de la piscine, la protection arrière de plomb est limitée. Le canal vertical du château qui reçoit le détecteur est donc excentré et a un diamètre de 110 mm.

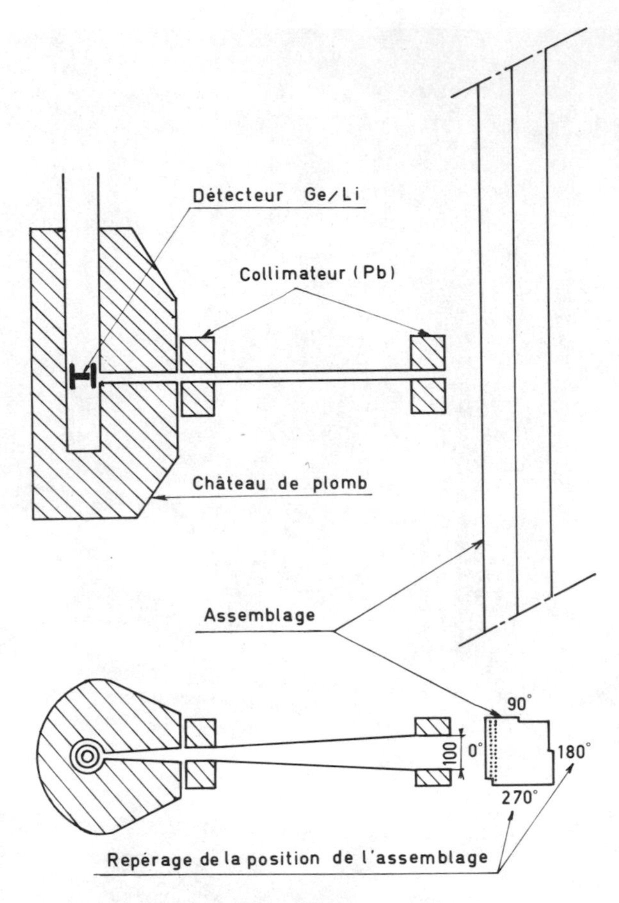

FIG.3. Schéma de principe de l'installation.

Le dispositif comprend:

- une fente dans le château-compteur de 20 mm;
- deux blocs de collimation espacés de 540 mm chacun, ayant une épaisseur de 100 mm de plomb; l'épaisseur de ces fentes est réglable de 0 à 20 mm par un seul dispositif mécanique qui agit simultanément sur les deux fentes;
- un tube d'air qui sépare les deux blocs de collimation;
- un écran supplémentaire de 70 mm de plomb d'épaisseur au maximum peut être intercalé entre l'assemblage et le dispositif de collimation.

Ce dispositif de visée a été conçu pour obtenir des valeurs «intégrées» radialement (largeur de fente 100 mm) avec une finesse de scrutation axiale à définir en fonction du but recherché et du taux de comptage acceptable pour le détecteur.

Comme nous l'avons dit précédemment, le choix d'un détecteur de haute résolution, type Ge-Li, s'imposait pour obtenir une bonne discrimination des différents pics d'émetteurs gamma liés aux produits de fission.

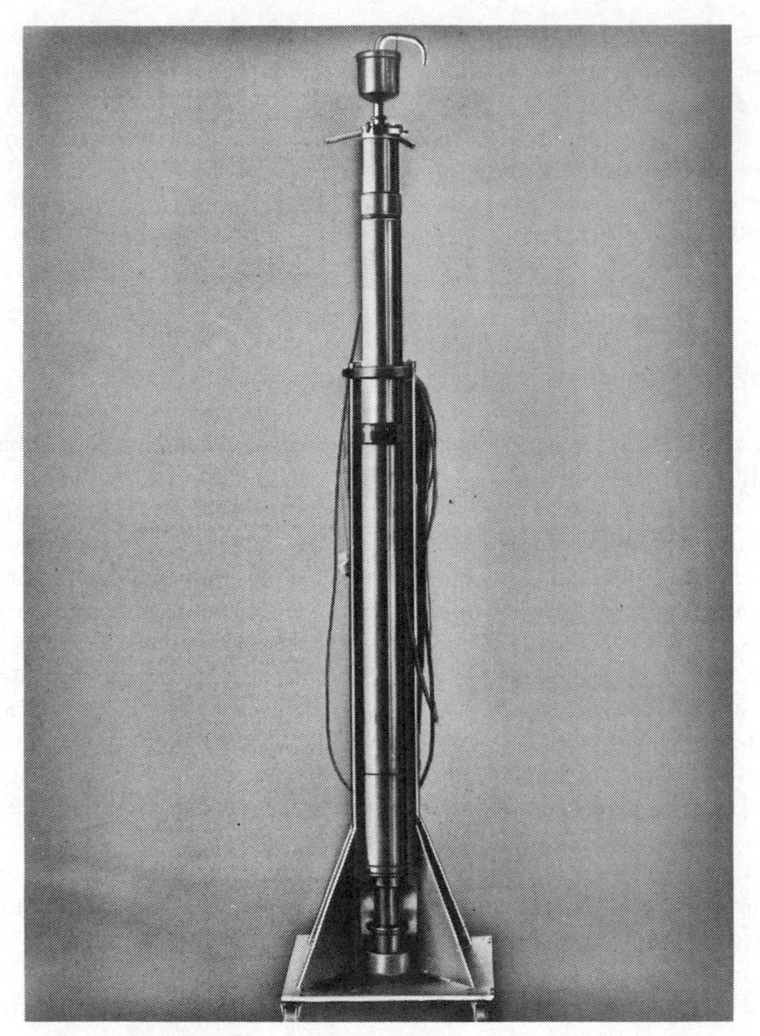

FIG.4. Ensemble de détection avec Ge-Li.

La taille du détecteur a évolué au cours des différentes campagnes de mesure, ce qui nous a fourni quelques éléments pour confirmer les études d'optimisation qui demeurent difficiles en raison du nombre de paramètres mis en jeu: énergie des pics que l'on souhaite mesurer, activité du combustible, ouverture de la collimation, activité de l'environnement, épaisseur du blindage, taux de comptage maximal admissible, etc.

Le détecteur dans son cryostat (fig.4), avec une réserve d'azote liquide lui donnant une autonomie de 60 heures, est disposé dans un tube étanche flexible qui relie le canal vertical du château-compteur à la plate-forme supérieure. La souplesse de ce tube est suffisante pour lui permettre de suivre le mouvement du château-compteur dans les scrutations axiales c'est-à-dire sur plus de trois mètres de déplacement vertical.

Le préamplificateur est placé dans l'ensemble détecteur-cryostat à l'intérieur du tube étanche et est relié à l'électronique située au bord de la piscine. Un sélecteur multicanaux avec perforateur de bandes et imprimante permet de recueillir les spectres pour dépouillement.

2.3. Expérience d'utilisation

Au cours des différentes campagnes de mesure une centaine d'assemblages combustibles ont été examinés.

Aucun incident notable n'a été relevé au cours de ces campagnes ni sur la manutention des combustibles, ni dans le fonctionnement de l'appareillage.

Avec l'appareil tel qu'il est actuellement le temps moyen d'une mesure est d'une heure. Pour cette durée on obtient par exemple pour un assemblage ayant un temps de refroidissement de 30 jours, avec une collimation de 1,5 mm et sans écran, un spectre duquel on peut extraire l'activité du ^{140}La avec une précision de ± 2%, celle du ^{95}Zr avec ± 1,5% et celle du ^{144}Pr avec ± 10%. La précision de cette dernière valeur pourrait être améliorée aux dépens des autres moyennant l'adjonction d'un écran et d'une collimation plus ouverte.

De l'expérience actuelle avec cet appareil nous pouvons d'ores et déjà tirer quelques conclusions:

– La réalisation pratique d'un appareillage précis de gammamétrie sur assemblages équipés de détecteurs de haute résolution est possible et son utilisation ne soulève pas de problèmes insurmontables.

– L'optimisation de l'ensemble de détection avec ses protections et le dispositif de visée nécessitent une attention toute particulière et dépendent dans une large mesure de l'objectif poursuivi.

– L'utilisation d'un calculateur en ligne pour le dépouillement des spectres est un problème aujourd'hui résolu; toutefois il nous paraît indispensable de conserver en parallèle un système d'inscription de spectres sur un support (papier, bande perforée, bande magnétique) qui puisse être repris ultérieurement pour un dépouillement plus élaboré, système qui tient lieu également de sauvegarde en cas de défaillance du calculateur.

– Quelques jours après l'arrêt du réacteur nous pouvons obtenir une carte de distributions de puissance en fin de cycle à partir du ^{140}La et des valeurs plus intégrées sur l'ensemble du cycle avec les ^{95}Zr et ^{144}Pr. Au bout de quelques mois la mesure des isotopes du césium permet de comparer les différents taux de combustion.

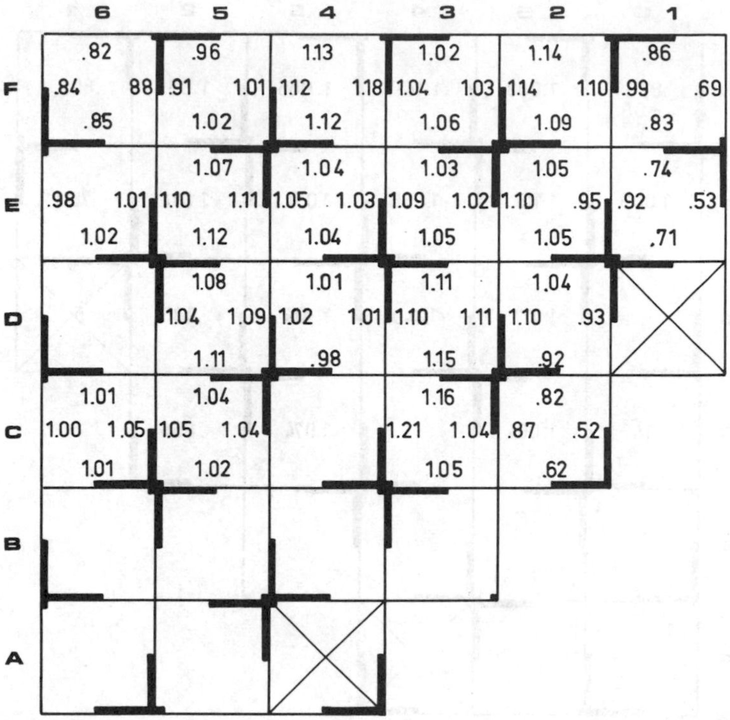

	6	5	4	3	2	1
F	.82 .84 88 .85	.96 .91 1.01 1.02	1.13 1.12 1.18 1.12	1.02 1.04 1.03 1.06	1.14 1.14 1.10 1.09	.86 .99 .69 .83
E	.98 1.01 1.02	1.07 1.10 1.11 1.12	1.04 1.05 1.03 1.04	1.03 1.09 1.02 1.05	1.05 1.10 .95 1.05	.74 .92 .53 .71
D		1.08 1.04 1.09 1.11	1.01 1.02 1.01 .98	1.11 1.10 1.11 1.15	1.04 1.10 .93 .92	X
C	1.01 1.00 1.05 1.01	1.04 1.05 1.04 1.02		1.16 1.21 1.04 1.05	.82 .87 .52 .62	
B						
A			X			

FIG.5. Activités du ^{140}*La mesurées sur les faces des assemblages à la fin du quatrième cycle.*

2.4. Dépouillement des mesures

Les spectres γ obtenus dépendent du temps de refroidissement de l'assemblage combustible lors des mesures.

Pour des temps de refroidissement courts (dizaine de jours) on peut dépouiller sans problème les pics correspondant au couple ^{95}Zr-^{95}Nb, au ^{140}La et au ^{144}Ce. Les pics de ^{134}Cs et ^{137}Cs ne deviennent mesurables qu'après quelques mois de refroidissement.

Le problème majeur du dépouillement des résultats réside dans les conditions de transmission du rayonnement gamma à travers l'assemblage. Pour résoudre ce problème nous avons dû mettre au point un code qui permet d'effectuer les calculs de transmission à partir d'une représentation correcte de l'assemblage. Le code Trajet [3] détermine les distances parcourues dans les différents matériaux rencontrés entre chaque point source et le détecteur puis effectue un calcul d'atténuation en ligne droite pour des photons d'énergie correspondant à la source considérée. A partir de la distribution des sources qui lui est fournie en entrée le code détermine ainsi un taux de comptage du détecteur.

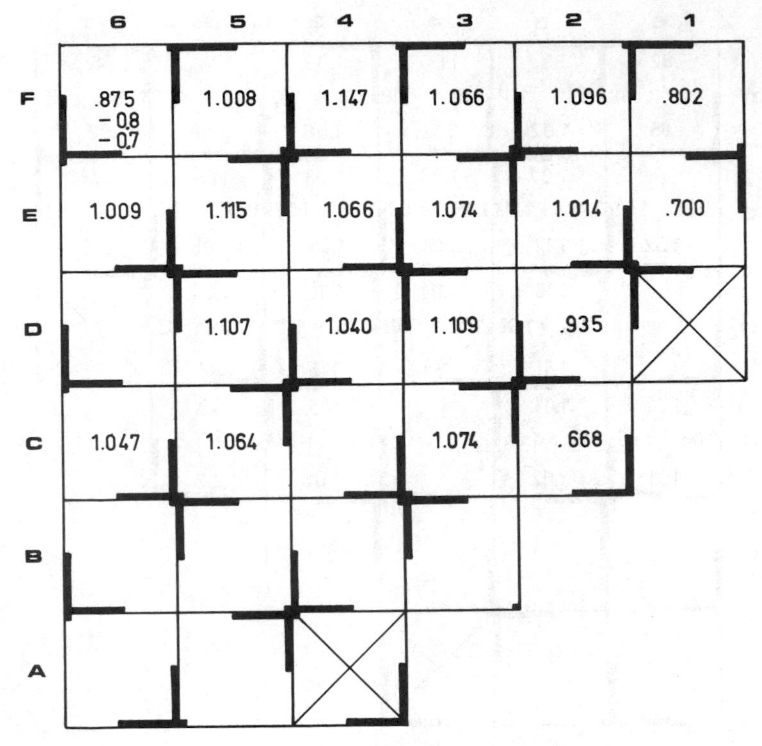

FIG. 6. Puissance par assemblage obtenue par gammamétrie à partir du ^{140}La à la fin du quatrième cycle.

Le code Trajet nous a permis de chiffrer correctement les facteurs de transmission et donc l'importance des divers crayons vis-à-vis du détecteur quelle que soit la géométrie du comptage. De plus il nous permet de passer des résultats expérimentaux à des activités moyennes sur l'assemblage ou pour une zone de l'assemblage. Pour cette opération il est nécessaire de lui fournir un certain nombre d'informations qui peuvent se résumer ainsi:

– rendement de fission moyen du produit de fission considéré pour le combustible particulier,
– distribution microscopique de puissance dans l'assemblage,
– approximation de la distribution de puissance macroscopique dans le cœur.

Il est clair que ces informations peuvent être données de façon plus ou moins détaillée et que la précision finale en dépend partiellement.

Les incertitudes dues au dépouillement sont discutées en détail dans [4]; aussi nous ne ferons ici qu'un bref rappel des conclusions de cette étude. La détermination des puissances moyennes par assemblage à partir de l'utilisation du

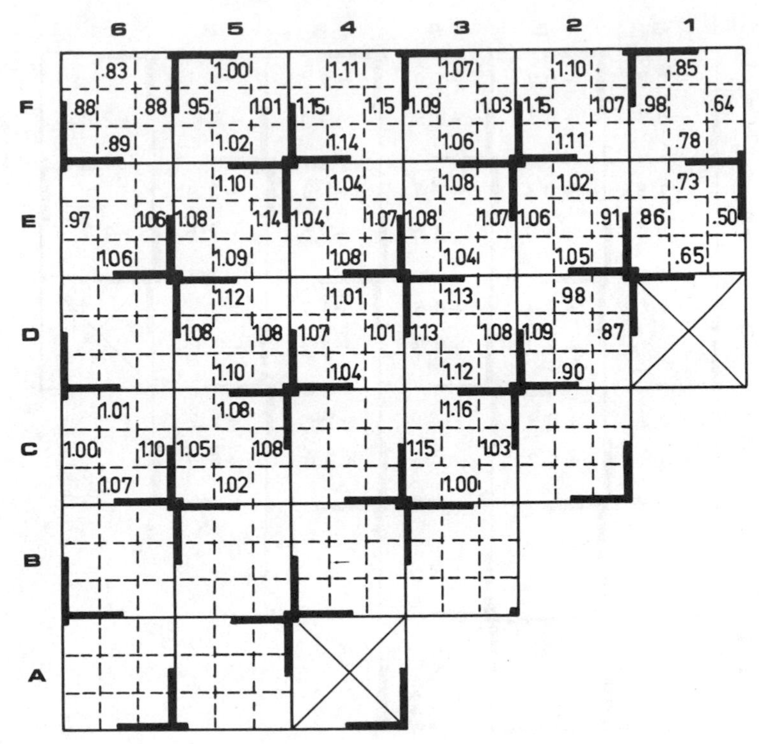

FIG. 7. Puissance par zone obtenue par gammamétrie à partir du ^{140}*La à la fin du quatrième cycle.*

lanthane-140 est effectuée avec une précision (20 σ) inférieure à 2%, l'utilisation des pics Zr-Nb conduirait à des précisions du même ordre de grandeur alors qu'elle serait nettement moins bonne avec le cérium-144.

2.5. Exemples de résultats

A titre d'exemple nous présentons des résultats obtenus à la fin du quatrième cycle. Sur la figure 5 apparaissent les activités du ^{140}La mesurées sur les quatre faces d'une vingtaine d'assemblages. Ces activités ont été corrigées pour être directement comparables (reproductibilité et temps de décroissance).

Ces résultats ont été exploités avec le code Trajet pour en déduire deux types d'informations: la puissance moyenne (en fin de cycle) par assemblage et la puissance moyenne des quatre zones de chaque assemblage. Les résultats correspondants apparaissent sur les figures 6 et 7. Pour la première, puissance par assemblage, l'incertitude affectant chaque valeur est de 1,5% comme nous l'avons vu précédemment. Dans la carte de puissance par zone elle est un peu plus

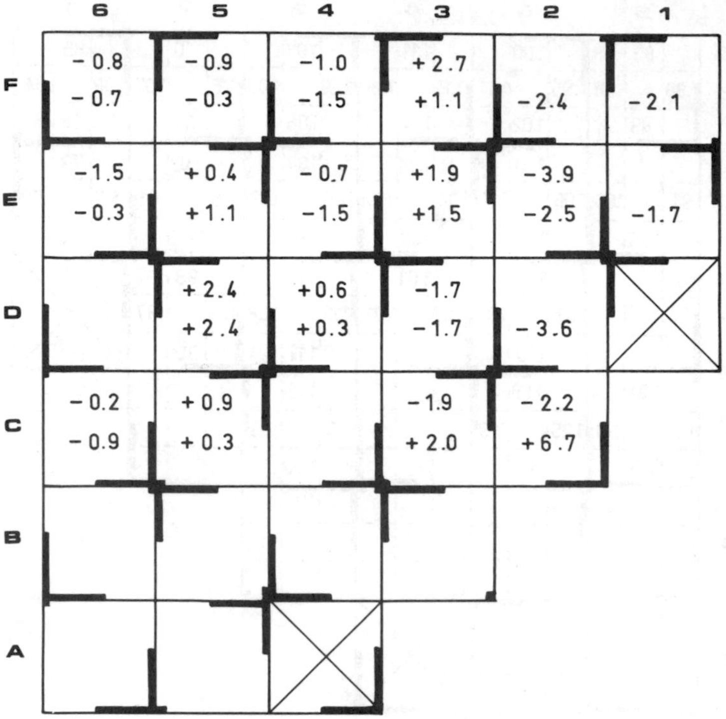

	6	5	4	3	2	1
F	− 0.8 − 0.7	− 0.9 − 0.3	−1.0 −1.5	+ 2.7 +1.1	 −2.4	 −2.1
E	−1.5 −0.3	+ 0.4 + 1.1	− 0.7 −1.5	+1.9 +1.5	−3.9 −2.5	 −1.7
D		+ 2.4 + 2.4	+0.6 +0.3	−1.7 −1.7	 − 3.6	☒
C	− 0.2 − 0.9	+ 0.9 + 0.3		−1.9 + 2.0	− 2.2 + 6.7	
B						
A			☒			

FIG.8. Comparaison entre les résultats obtenus par gammamétrie d'une part, et les résultats obtenus par Aéroball et les valeurs calculées d'autre part (écarts en %).

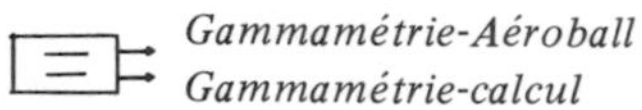
Gammamétrie-Aéroball
Gammamétrie-calcul

élevée, environ 2,5% (2 σ), la différence tenant essentiellement à la statistique des comptages.

Les valeurs obtenues à partir des activités du ^{140}La correspondent à des puissances en fin de cycle intégrées sur environ une période du baryum-140, c'est-à-dire 12 jours. Nous les avons comparées d'une part à une distribution de puissance fournie par l'instrumentation interne avant l'arrêt du réacteur, et d'autre part aux résultats d'un calcul complet effectué avec le système Neptune [2], les distributions de puissance dans ce cas provenant du module de diffusion aux éléments finis à trois dimensions Trident. Ces comparaisons apparaissent sur la figure 8. On constate une bonne cohérence de l'ensemble compte tenu des marges d'incertitudes attendues pour chaque cas.

Cette cohérence se retrouve d'une manière générale pour les différentes campagnes, y compris celles qui concernaient en particulier des assemblages à combustible plutonifère. Quatre assemblages à îlots et deux assemblages «tout Pu» ont été contrôlés de cette manière.

3. APPAREILLAGES EN DEVELOPPEMENT

3.1. Dispositif d'examen de crayons amovibles

Pour améliorer les connaissances du comportement du combustible sous irradiation, Framatome a introduit dans le cœur de certaines centrales de puissance à réacteur à eau sous pression (Tihange 1, Fessenheim 1 et 2, Bugey 2 et 3) un ou plusieurs assemblages dits *à crayons amovibles.* Ces assemblages sont réalisés différemment des assemblages standard: ils permettent un accès direct aux crayons à travers l'embout supérieur. On peut ainsi pendant la durée d'un arrêt de la centrale, grâce à un appareillage adéquat et autonome, enlever les crayons et leur faire subir des examens non destructifs. Après examen, ces crayons peuvent être, soit remis dans l'assemblage pour poursuivre leur irradiation, soit envoyés vers les laboratoires chauds pour des examens destructifs (dans ce dernier cas, ils sont remplacés dans l'assemblage par des crayons neufs d'enrichissement adaptés ou par des crayons constitués de barres en acier inoxydable ou en Zircaloy).

L'installation d'examen des crayons amovibles, placée dans la piscine de désactivation (fig.9), forme un ensemble constitué principalement des éléments suivants:

- un bâti support (fig.10), composé de trois châssis (chacun d'une longueur de 4,5 m environ) verrouillés les uns sur les autres, comprenant deux alvéoles de stockage, les structures de supportage des différents postes d'examens, et le système de déplacement du crayon propre au poste de gammamétrie;
- un outil d'extraction et d'insertion des crayons (fig.11) avec son système de déplacement (rails et chariots), équipé de nombreuses sécurités automatiques qui interdisent toute action manuelle intempestive;
- les armoires de commande contrôle.

Les postes d'examen (démontables sous eau) permettent les opérations suivantes:

- profilométrie (mesure en continu du diamètre des crayons sur deux génératrices avec une précision de ±0,02 mm pouvant absorber une flèche du crayon de 2 mm sur 700 mm de longueur; mesure de longueur du crayon avec une précision de ±0,1 mm);
- prélèvement de dépôt (enlèvement des dépôts non adhérents: les particules détachées sont aspirées par une pompe à travers une cartouche filtrante de faible porosité, les analyses chimiques s'effectuant ultérieurement dans les laboratoires spécialisés);
- observation visuelle par caméra de télévision de l'extérieur des gaines du crayon s'effectuant avant et après le prélèvement de dépôt (un jeu de miroirs permet l'examen complet sur 360° sans rotation du crayon);
- gammamétrie à l'aide d'une installation en partie immergée et en partie hors eau, décrite plus en détail ci-dessous.

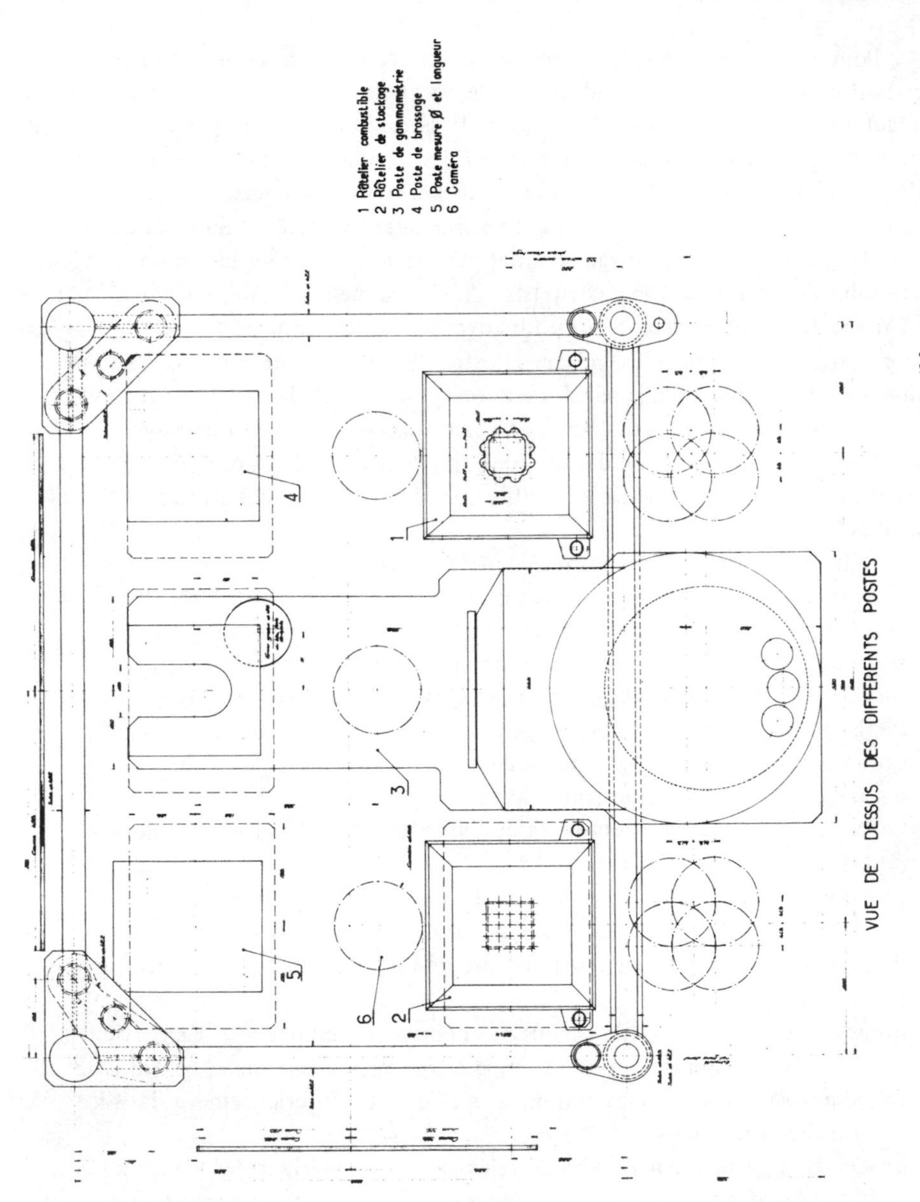

FIG.9. Schéma de l'ir ulation d'examen des crayons amovibles.

FIG.10. Vue d'ensemble de l'installation.

L'ensemble de détection est fixé sous 7 mètres d'eau et le crayon est déplacé verticalement devant la fente du collimateur. Cet ensemble de détection comprend:

- une enceinte étanche en acier inoxydable (pressurisée en azote) contenant le détecteur, son cryostat et le préamplificateur;
- un ensemble de blindage réalisé en Denal avec une épaisseur minimale équivalente à 25 cm de plomb;
- un ensemble de collimation comportant un bloc à fente variable, un atténuateur mobile, un collimateur et un bloc à fente fixe: le bloc à fente variable placé contre le crayon permet de régler la collimation à une épaisseur comprise entre 0 et 4 mm par pas de 0,1 mm; la commande de cette collimation est assurée depuis le plancher de service; l'atténuateur, taillé en échelons, est disposé derrière le bloc à fente variable et permet d'interposer l'équivalent de 0 à 60 mm de plomb dans le faisceau; le collimateur est un tube étanche de 40 cm de long environ placé entre l'atténuateur et le bloc à fente fixe; enfin le bloc à fente fixe, dont l'ouverture est fixée à 4 mm, est placé contre le détecteur et délimite le faisceau de photons qui est vu par ce dernier.

Le détecteur (Ge-Li ou Ge intrinsèque) est placé dans son cryostat à l'intérieur de l'enceinte étanche et relié à un réservoir extérieur de 200 litres qui assure une alimentation automatique en azote liquide (autonomie de 10 jours environ).

FIG.11. Dispositif de gammamétrie vu de dessus.

D'autre part, le détecteur est connecté au calculateur PDP 11 et à un analyseur 4000 canaux.

Le déplacement vertical du crayon, assuré par un dispositif autonome et indépendant de l'outil, peut être, soit continu à vitesse constante fixée entre 0 et 2 m/min, soit par pas élémentaires de 0,02 mm. Les commandes peuvent être manuelles ou exécutées par le calculateur.

Le retour à une position quelconque du crayon est réalisé avec une précision de ±0,05 mm.

Le mouvement de rotation du crayon est possible dans toute position figée verticalement et sur 360°, l'angle de rotation étant connu à ± 1°.

Les examens effectués avec un assemblage à crayons amovibles à la fin du premier cycle de la centrale Tihange 1 (début 1977) ont été limités à l'extraction des crayons, à leur examen visuel, à l'insertion d'une partie des crayons extraits et à l'insertion des crayons neufs (remplacement des crayons destinés à être soumis aux examens destructifs). Lors de l'arrêt du deuxième cycle (début 1978), les opérations ont été complétées par les examens définis ci-dessus.

3.2. Dispositif d'examens multiples des assemblages

Un des problèmes, et non le moindre, que l'on rencontre lors des examens non destructifs du combustible irradié est la place disponible dans la centrale, particulièrement dans la piscine de désactivation.

Il est parfois impossible d'effectuer ces examens sans recourir à des manipulations complexes, ce qui rend les examens très longs et incompatibles avec la durée de l'arrêt de la centrale pour le déchargement-rechargement.

Le dispositif d'examens multiples, actuellement construit par Framatome, répond à un double souci: d'une part, mettre en œuvre dans les piscines de désactivation un éventail aussi large que possible d'appareillages de mesure et de contrôle non destructif et, d'autre part, intégrer ces équipements dans des zones d'encombrement aussi réduit que possible par suite de l'exiguïté des piscines.

Afin de réaliser ces objectifs, il était séduisant de regrouper en un seul dispositif tous les systèmes d'examen et contrôle nécessitant la mise en œuvre de divers déplacements: on a donc réalisé une structure de base commune, appelée bâti commun, constituée d'un châssis support, d'un chariot à déplacements tridirectionnels recevant tour à tour les différents postes d'examen et enfin d'un support d'assemblage permettant son maintien et sa rotation.

3.2.1. Conception d'ensemble

Le châssis est une structure très rigide, fixée par l'intermédiaire d'un socle au fond de la piscine d'une part, et par l'intermédiaire d'adaptateurs d'ancrages placés sur la paroi de la piscine d'autre part.

Il est équipé en outre de rails verticaux parallèles servant au guidage du chariot, ces rails pouvant supporter sans déformation une charge de 3000 daN environ (poids du poste de gammamétrie et du chariot).

Le déplacement du chariot selon l'axe vertical (course supérieure à 10 m) s'effectue soit à une vitesse continue (0 à 2 m/min), soit pas à pas (pas élémentaire de 0,02 mm). Les déplacements dans le sens parallèle à la face d'assemblage (course de 800 mm) s'effectuent à vitesse constante fixe entre 0 et 2 m/min.

Un système de mesure qui combine des codeurs industriels spéciaux avec un système précis et rigide de pignon et crémaillère permet de déterminer la position du chariot dans les trois directions. La précision est de ±0,05 mm dans le cas des déplacements horizontaux et de ±0,1 mm lors des déplacements selon l'axe vertical. L'orthogonalité des trois directions est de ±1°.

Le support d'assemblage, destiné à maintenir l'assemblage en position verticale et à assurer sa rotation lors des examens, est positionné sur le même socle, placé au fond de la piscine qui sert à la fixation du châssis.

Le dispositif d'examens multiples du combustible irradié permet d'effectuer dans sa version actuelle les opérations suivantes:

- observation par caméra des crayons périphériques de l'assemblage: grâce aux mouvements du chariot, la plage d'observation peut varier de deux crayons (observation détaillée) à toute la largeur de l'assemblage (observation globale);
- visualisation de la base et de la face supérieure de l'assemblage (par caméra et à l'aide de deux miroirs escamotables placés de part et d'autre de l'assemblage);
- mesures de déformations sur les crayons périphériques (longueur, espace entre crayons et embouts, espace entre deux crayons, flèche) et sur les assemblages (torsion, flèche, position relative des grilles);
- observation directe détaillée des quatre faces verticales de l'assemblage par périscope.
- gammamétrie.

D'autres types d'examens compléteront ultérieurement les possibilités actuelles.

3.2.2. Gammamétrie

Cette installation permet les analyses par spectrométrie gamma d'un crayon d'angle, d'une rangée de crayons ou d'une face d'assemblage.

La partie immergée est semblable à celle qui est utilisée pour les examens des crayons amovibles, les différences les plus notables étant:

- le déplacement du poste de gammamétrie et non celui du combustible; on se sert des déplacements du chariot selon l'axe vertical et dans le sens parallèle à la face de l'assemblage, ce qui permet de viser, sur toute la hauteur de l'assemblage, n'importe quelle rangée de crayons ainsi que les crayons d'angle;
- l'existence de deux ensembles de collimation, l'une avec la largeur des fentes de visée nettement plus importante pour les mesures de face de l'assemblage, l'autre avec la largeur des fentes de visée nettement plus petite pour les crayons d'angle;
- l'existence d'un équipement particulier caméra-projecteurs fixé sur l'enceinte étanche (qui suit donc les mouvements bidirectionnels du chariot), dont le rôle est double: identification de l'assemblage par la lecture du numéro gravé à sa partie haute d'une part, vérification du bon positionnement de la collimation par rapport à la zone combustible à examiner.

La partie électronique est constituée de l'ensemble de spectrométrie nucléaire Cosinus γ qui permet l'acquisition de données, leur traitement et la visualisation des résultats sous le contrôle d'un calculateur PDP 11 relié à un ensemble de modules Camac et Nim.

Les conditions d'environnement sur le plancher de service, donc à proximité de la piscine de désactivation, étant très sévères (par exemple, humidité de 100%), le calculateur et l'analyseur 4000 canaux sont placés dans le local d'électronique de la centrale qui se trouve à une distance de 300 m environ. Les équipements de commande-contrôle se trouvent donc implantés à deux endroits distincts. Ceci, joint à des conditions particulières d'application de gammamétrie, a nécessité l'adjonction à l'équipement standard Cosinus γ:

- d'un module prolongateur des lignes, fonctionnant en émetteur-récepteur pour des distances supérieures ou égales à 300 m;
- d'un *floppy* disque qui permet une acquisition et un traitement plus rapides;
- d'une télétype imprimante rapide permettant le dialogue avec le calculateur et l'impression des résultats du traitement des données;
- d'une bande magnétique permettant le traitement ultérieur dans un centre de calcul;
- d'un ictomètre permettant une visualisation du signal au niveau du plancher de service;
- d'un deuxième poste de commande situé sur le plancher de service.

Les déplacements du chariot, donc de la partie immergée de la gammamétrie, peuvent être effectués, soit par action manuelle depuis les baies de contrôle-commande situées sur le plancher de service, soit par l'intermédiaire du calculateur (les déplacements ne peuvent être commandés que d'un seul lieu à la fois).

Un programme d'exploitation gère les différents périphériques et les acquisitions, effectue les traitements, prend en charge les déplacements.

La structure de ce logiciel est modulaire et permet de choisir et d'organiser sur simple appel et pour chaque programme de mesure les différentes séquences de prise et de traitement des informations.

4. APPLICATION POUR LE CONTROLE DES GARANTIES

Parmi les méthodes de contrôle non destructif des combustibles irradiés aux fins de garanties sur les matières fissiles, la gammamétrie est certainement la plus attrayante en raison de la richesse des informations qu'elle procure. Son utilisation pour le contrôle des distributions de puissance en fin de cycle et de l'état du combustible qui a été développée dans les paragraphes précédents est déjà susceptible d'apporter des informations intéressantes sur le plan des garanties en permettant de justifier d'une certaine intégrité du combustible ainsi que des conditions de fonctionnement du réacteur dans lequel il est irradié.

Cependant, l'application la plus importante envisagée actuellement sous cet angle concerne le contrôle des taux de combustion. Le CEA a entrepris en 1976 une étude d'avant-projet pour une installation qui permettrait d'effectuer ce contrôle à l'entrée d'une piscine de stockage de combustibles usés ou d'une unité de retraitement.

Technologiquement les solutions retenues diffèrent peu de celles qui ont été présentées dans les applications précédentes; l'appareil reste caractérisé par l'utilisation en piscine de détecteurs Ge-Li ou Ge intrinsèque correctement blindés et recevant le rayonnement en provenance de l'assemblage disposé dans un panier vertical à travers un système de collimations réglables.

C'est donc essentiellement en ce qui concerne les principes de ce contrôle et les conditions pratiques d'application qu'il est intéressant d'examiner les conclusions de cette étude d'avant-projet.

4.1. Principe du contrôle

La mesure du taux de combustion est effectuée à partir de l'activité de produits de fission dont la période radioactive est longue vis-à-vis de la durée d'irradiation. Deux possibilités ont été retenues pour le contrôle d'assemblages de réacteurs à eau irradiés en moyenne pendant trois ans:

- la mesure du rapport césium: activité relative de ^{134}Cs et ^{137}Cs,
- la mesure du ^{137}Cs en absolu.

Dans les deux cas, il est nécessaire de connaître:

- le temps de refroidissement, surtout pour le rapport césium; celui-ci est mesuré à partir d'autres produits de fission comparés au ^{137}Cs (^{95}Zr, ^{144}Ce),
- un minimum d'informations sur la nature et les caractéristiques du combustible ainsi que sur l'histoire d'irradiation.

Ce dernier point, qu'il serait très long de developper ici, est une des conclusions les plus importantes. En effet elle conduit à remplacer une notion de contrôle absolu et isolé du taux de combustion, qui ne peut être valablement défendue pour n'importe quel combustible ou toute histoire d'irradiation, par une notion de contrôle d'informations accompagnant le combustible, qui ne nécessite pas beaucoup plus de mesures, au moins deux ou trois paramètres cependant pour éviter les coïncidences fortuites, mais qui est plus sûre et plus riche dans la mesure où elle permet non seulement de confirmer les conséquences de l'irradiation mais également les principales caractéristiques du combustible soumis à ce contrôle.

4.2. Application pratique

La solution proposée à la suite de l'étude précitée consiste à relever le spectre gamma moyen ou par zones de l'assemblage (cette dernière notion étant importante pour les problèmes de criticité, sources d'autres applications de ce contrôle), à le dépouiller sur calculateur en ligne pour extraire les raies des principaux émetteurs et à procéder ensuite à une analyse des résultats comprenant:

- des tests de cohérence entre deux raies d'un même émetteur,
- la prise en compte des informations portées sur la fiche suiveuse du combustible (caractéristiques, enrichissement, poids, dates d'irradiation, etc.),

– l'interprétation des activités de produits de fission en termes de temps de refroidissement et de taux de combustion, compte tenu des informations précédentes,
– la vérification de concordance de ces deux paramètres dans des fourchettes indiquées avec les valeurs annoncées suivie d'un accord ou rejet suivant le résultat.

A cette procédure s'ajoutent des tests internes de bon fonctionnement de l'appareillage, détecteur, électronique et calculateur, afin d'assurer la fiabilité du système.

4.3. Limitations

Une première limitation est liée au temps de refroidissement, qui ne peut être inférieur à quelques mois compte tenu de l'utilisation du césium. Pour des temps inférieurs il serait nécessaire de baser le contrôle sur le ^{144}Ce, mais il en résulterait une contrainte sur la durée d'irradiation qui limiterait beaucoup l'intérêt pratique.

Par ailleurs, le contrôle portant sur les paramètres d'irradiation, les résultats sont peu sensibles aux données initiales du combustible, masse et enrichissement. Ici la notion de fiche suiveuse globalement validée à partir de deux paramètres est certainement discutable bien qu'elle constitue déjà un progrès par rapport au contrôle absolu. Il est envisagé actuellement d'étudier la possibilité d'adjoindre un contrôle de masse par pesée de l'assemblage ainsi que d'examiner le bénéfice à attendre de l'utilisation simultanée d'une méthode d'interrogation neutronique pour confirmer les teneurs en matière fissile.

5. CONCLUSION

Les possibilités offertes par ces mesures non destructives de distribution de puissance et de taux de combustion et par les autres contrôles que permet la gammamétrie sont importantes et peuvent tout à la fois faciliter le travail des exploitants de réacteurs ou d'usines de retraitement et permettre l'application des garanties sur les matières nucléaires contenues dans les combustibles irradiés.

L'expérience française, acquise principalement par des réalisations sur sites, confirme que l'on peut effectuer des mesures précises sur les assemblages de réacteurs à eau. Les applications orientées vers le suivi du fonctionnement et des examens visant à s'assurer de l'état physique du combustible sont plus avancées que celles qui concernent les contrôles de garanties. Néanmoins les développements technologiques actuels bénéficient à ces deux catégories d'utilisation et les études théoriques sur les principes comme sur l'interprétation des résultats montrent que rien ne s'oppose à une mise en œuvre concrète dans le domaine des garanties.

REFERENCES

[1] ZEČEVIĆ, V., Contribution to the Consultants' Meeting on Non-Destructive Techniques on Water Reactor Fuel Characteristics, IAEA, Vienna, 23–25 Nov. 1977 (to be published).

[2] KAVENOKY, A., *NEPTUNE,* un système modulaire pour le calcul des réacteurs à eau légère, Bull. Inf. Sci. Tech. (Paris) n° 212 (1976).

[3] BERTUOL, M., Spectrométrie gamma dans les réacteurs à eau légère, Thèse, Université de Paris-Orsay (1977).

[4] BOUCHARD, J., FREJAVILLE, G., ROBIN, M., Mesures de distribution de puissance et de taux de combustion par spectrométrie gamma sur les combustibles irradiés de réacteurs à eau, Note CEA-N-1982 (1976).

DISCUSSION

K. NIENHUYS: I understand that there is a minimum cooling time before the $^{134}Cs/^{137}Cs$ ratio can be used to calculate Pu build-up or burnup.

J. BOUCHARD: We have indeed shown from our experience that a minimum cooling time of several months is necessary before burnup measurements on the basis of the $^{134}Cs/^{137}Cs$ ratio can be carried out.

K. NIENHUYS: In view of the lack of reprocessing capacity I wonder whether it is still possible to make these calculations if the fuel elements are ten years old or so. If it is not possible, does that mean that one has to rely solely on containment and surveillance as the means of safeguarding the fuel elements until they reach the dissolver tank in the reprocessing plant?

J. BOUCHARD: We have no experience with long cooling times of around ten years. However, if measurements based on the caesium ratio were too imprecise, we could resort to direct measurement of the ^{137}Cs. This is more complicated and less practical, since it is an absolute measurement, but in principle there is nothing against this method and, if necessary, we then have several hundred years available.

REACTOR-RA FUEL BURNUP MEASUREMENT BY SPATIAL POWER DISTRIBUTION AND THE GAMMA-SPECTROMETRY METHOD

R. MARTINC, V. BULOVIĆ
Boris Kidrič Institute of Nuclear Sciences,
Belgrade, Yugoslavia

Abstract

REACTOR-RA FUEL BURNUP MEASUREMENT BY SPATIAL POWER DISTRIBUTION AND THE GAMMA-SPECTROMETRY METHOD.

The reactor RA is a heavy-water channel-type research reactor. Regarding the accountability of nuclear material, the specific features of reactor RA are: The segmental form of fuel elements; numerous fuel elements in the reactor core and numerous elements to be reloaded annually; and the flexibility and complexity of the in-core fuel management. Consequently, there are many fuel elements in the spent fuel storage with an incoherent and rather complex irradiation history. To account for the nuclear material in the irradiated fuel elements, under the given circumstances, some "fast" procedure for the fuel burnup evaluation should be used. In this respect, the fuel work integration procedure, based on the power distribution evaluation, has been investigated. The radial power distribution is measurable by the channel coolant flow-rate and the inlet/outlet temperature difference. The axial distribution is calculated by the two-group homogeneous diffusion approach. The gamma spectrometry method is used to verify the "power distribution procedure". The intercomparison is made on 17 fuel elements from two fuel channels with different in-core history. The differences between obtained results were within the experimental error associated with both methods. This indicates that the "power distribution method" is a useful tool for routine application on the irradiated fuel of the RA reactor. However, to correlate it successfully further measurements by gamma spectrometry should be done.

INTRODUCTION

The correct fuel burnup evaluation is important from the aspects of safety, economy and nuclear material accountability. However, while the economy and safety predictions deal rather with averaged values, the actual realized fuel burnup in the individual accounting elements (fuel assemblies, channels, slugs, etc.) is of interest from the aspect of nuclear material (NM) accountability. The significant effects of irregularities in the reactor core (e.g. irregularities in the control-rod lattice, sample irradiation holes and loops etc.), and the transients in the in-core fuel management on the individual fuel-element burnup in particular, might occur in the case of research reactors.

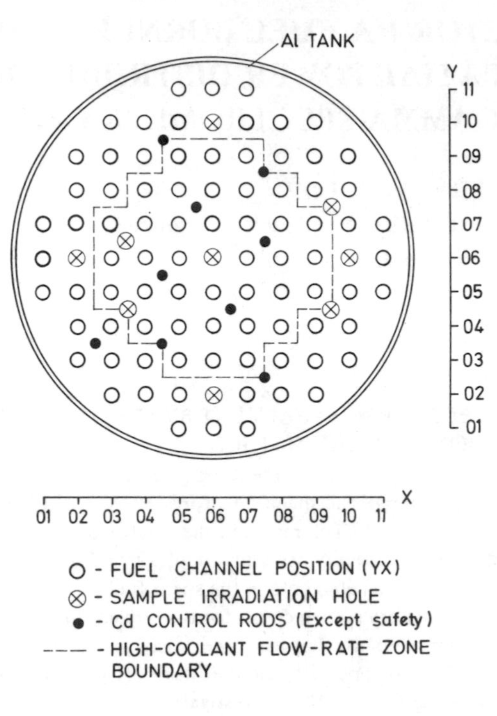

FIG.1. Reactor RA core layout.

The RA reactor is a general-purpose research reactor of the Boris Kidrič Institute of Nuclear Sciences at Vinča. It has been in operation since 1959, working at the nominal power of 6.5 MW. It is a heavy-water-moderated and cooled reactor system, fuelled by tubular low enrichment (2% ^{235}U) metallic uranium fuel in the form of slugs [1]. Since December 1976 a highly enriched (80% ^{235}U) uranium oxide dispersion fuel has gradually been introduced into the reactor core, in a mixed fuel (2% and 80% enriched) lattice mode [2]. The fuel elements of both types are geometrically identical. After 1980 only 80% enriched fuel will be used.

The full core configuration consists of 924 fuel-element positions, i.e. 84 fuel channels (Fig.1) and 11 fuel-element positions per channel. Such core configuration made the fuel management improvements possible, resulting in a considerable fuel consumption reduction. However, the number of fuel elements to be discharged per year is still very large, owing to the segment form of the fuel. Consequently, to account for the nuclear material in the spent fuel elements, some "fast" procedure for burnup estimation should be applied.

TYPE OF FUEL ELEMENT IN-CORE HISTORY	FRESH FUEL „INFLOW" ZONE
1. SINGLE POSITION HISTORY	

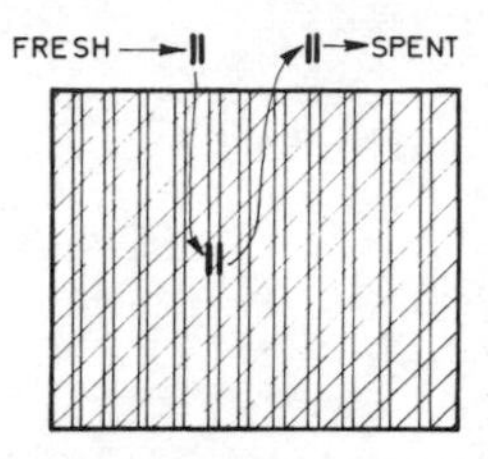

ALL FUEL CHANNELS IN THE CORE (SELECTION CRITERION - MAXIMUM REALISED CHANNEL BURN-UP)
FRESH FUEL : 2% ENR.
SPENT FUEL : 2% (INITIAL)

2. TWO-POSITION HISTORY

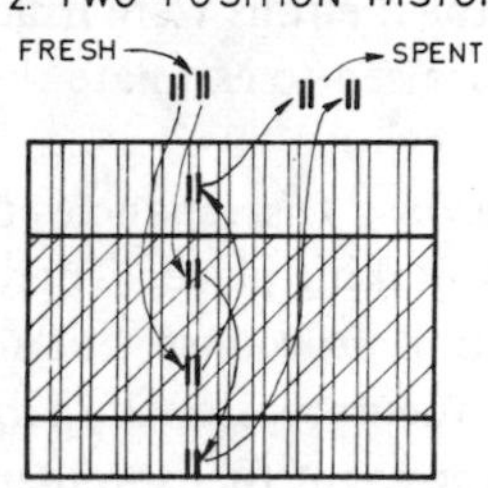

CENTRAL ELEMENT POSITIONS (5 OR 6) IN ALL FUEL CHANNELS IN THE CORE
FRESH FUEL : 2% ENR.
SPENT FUEL : 2% (INITIAL)

3. MULTI-POSITION HISTORY (THE BEST FUEL ECONOMY)

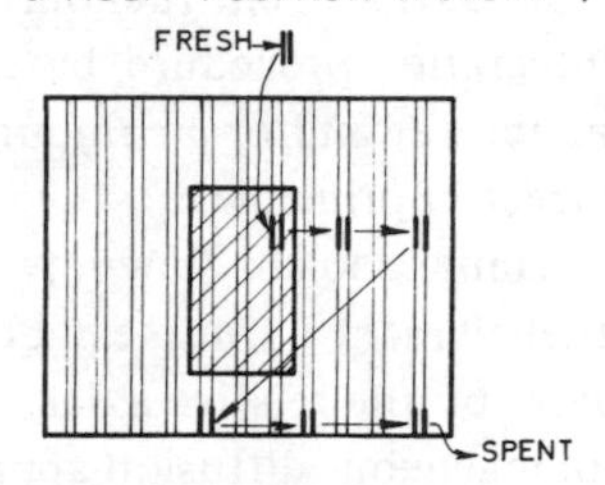

CENTRAL ELEMENT POSITIONS IN THE REACTOR CORE
FRESH FUEL : 2% ENR.
SPENT FUEL : 2% (INITIAL)

4. MULTI-POSITION HISTORY (TRANSIENT IN INTRODUCTION OF 80% ENRICHED FUEL)

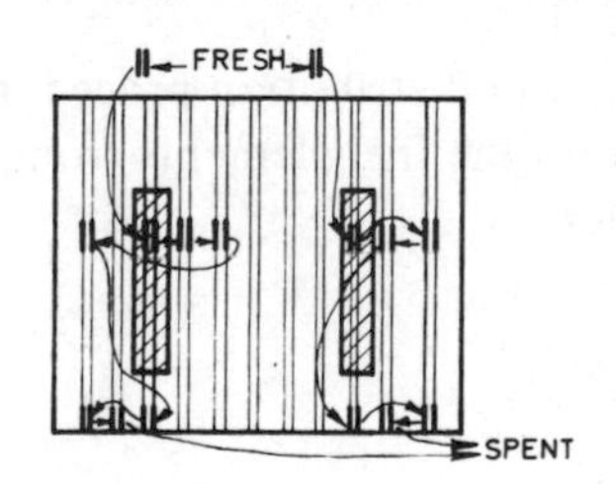

CENTRAL ELEMENT POSITIONS
INTERMEDIATE CHANNEL POSITIONS
FRESH FUEL : 80% ENR.
SPENT FUEL : 2% (INITIAL)

FIG.2. Reactor RA in-core fuel management.

Until 1975 the reactor operating records were supplied with the fuel-channel burnup data, obtained by the procedure based on the following formula [3]:

$$A_c = \frac{P_r(MW)}{N_c} \, t \, \frac{\phi_c}{\overline{\phi}_c} \tag{1}$$

where $P_r(MW)$ – nominal reactor power
t – time of fuel irradiation
N_c – number of fuel channels in the core
$A_c(MW \cdot d/channel)$ – work of the fuel channel

The relative thermal neutron flux distribution $\phi_c/\overline{\phi}_c$ was measured by the foil-activation technique. In the cold unpoisoned reactor the foils were irradiated in the sample irradiation holes. A similar procedure was used to estimate the fuel-element (slug) burnup.

The obtained results were incorrect for several reasons. Discrepancies of up to ± 40% in the channel-averaged burnup values were taking place. Therefore, it was necessary not only to replace the burnup estimation procedure, but also to reconstruct the content of the operating records for the exceptionally large number of elements in the spent fuel location, before the spent fuel shipment would take place.

Therefore, another semi-empirical procedure for routine spent fuel burnup determination was investigated – the fuel work integration procedure, based on the power distribution evaluation. It is not a manpower and/or a computer time-consuming procedure, but it is much more correct in principle.

The fuel-channel power is measurable by the channel coolant flow-rate and the inlet/outlet temperature difference. The fuel-element burnup can be evaluated from the measured fuel-channel work values by the relative axial power distribution, calculated by the two-group homogeneous diffusion approach.

The verification and correlation of the "power distribution method" are to be made mainly by the gamma-spectrometry method.

This paper discusses the verification of the "power distribution procedure" by gamma-spectrometry measurement, carried out on the fuel elements from two fuel channels with different irradiation histories.

1. RA REACTOR IN-CORE FUEL MANAGEMENT

The complexity of the fuel irradiation history is one of the specific characteristics of reactor RA. Three basic types of fuel in-core history, which have been applied since 1959, are:

1.1. Single-position history

There is no fuel-element transfer in the reactor core. In principle, the entire reactor core is a fresh fuel "inflow" zone (see Fig.2). All fuel elements from the selected fuel channel are reloaded. This type of refuelling procedure was used exclusively until 1967.

1.2. Two-position history

The fuel elements are transferred axially, but not in a radial direction. The fresh fuel elements are introduced into the central element positions in the fuel channel. The spent fuel elements are discharged from the peripheral positions (Fig.2). For this purpose a special refuelling machine (for axial fuel-element reshuffling) was designed [4] and manufactured by the reactor RA Department of the Boris Kidrič Institute.

This type of refuelling was dominantly used in the 1967–1975 period.

1.3. Multi-position history

The three-dimensional fuel flow is realized by the combination of axial and radial fuel-element transfer. The individual fuel element occupies 4–10 positions during its lifetime in the reactor core.

In the 1975–1976 period the fresh fuel "inflow" zone was restricted to the most central fuel-element positions (Fig.2), while the spent fuel elements were discharged from the most peripheral positions in the core [2].

In December 1976 the fresh fuel "inflow" zone was transferred to radially intermediate positions (Fig.2). The in-core fuel flow was planned to meet safety requirements of the transient in the introduction of highly enriched fuel into the RA reactor core by the mixed-fuel lattice mode [2, 5].

Another version of the multi-position history is planned for the equilibrium burnup regime with highly enriched fuel (after 1980).

2. THE POWER DISTRIBUTION METHOD FOR FUEL BURNUP MEASUREMENT

All fuel channels are instrumented by pairs of thermometres to measure the coolant inlet/outlet temperature differences. The total coolant flow-rate in the primary circuit is measured by a membrane-type flow-meter. A typical figure is 250 m^3/h. Regarding the coolant flow-rate the fuel channels are divided into central (36) and peripheral (48), with the coolant flow-rate ratio of 1.66 (Fig.1).

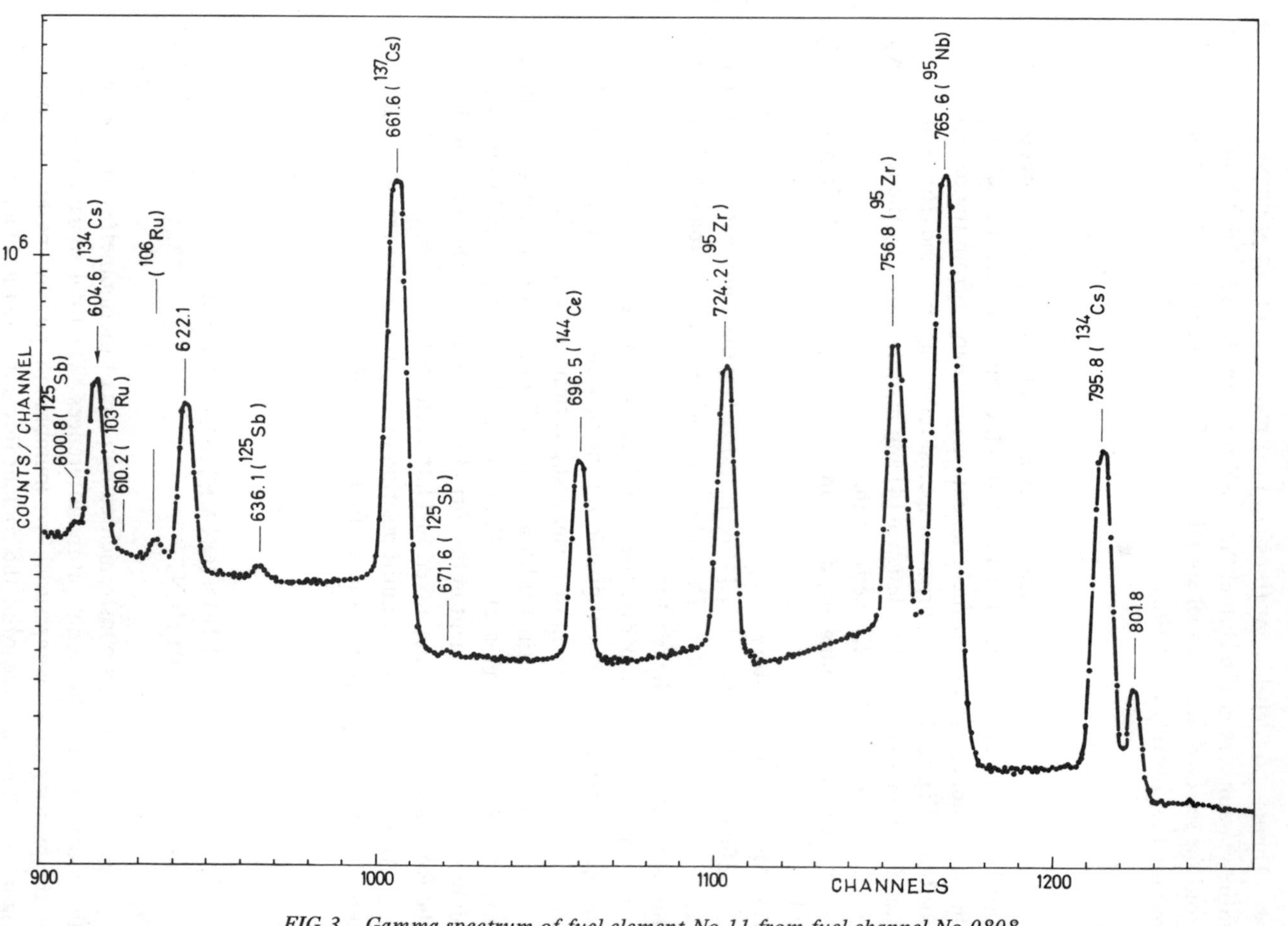

FIG.3. Gamma spectrum of fuel element No.11 from fuel channel No.0808.

The fuel-channel power is derived from the measured channel inlet/outlet temperature difference and the coolant flow-rate, with an experimental error estimated at ± 8%.

The error in the total coolant flow-rate (through the channels) and the error in the estimated part of the fission energy produced in the reactor core have a rather systematic character (over long reactor operation periods). These parameters could be determined by a correlation procedure, based on gamma-spectrometry measurements.

On knowing the monthly campaign averaged channel power values, the integrated channel work is then easily derived. The measured individual fuel-channel work values involve the effects of irregularities and heterogeneities in the reactor core, which could hardly be calculated satisfactorily.

To evaluate the fuel-element burnup, the measured fuel-channel work values should be corrected by the relative axial power distribution. It is calculated by a version of the standard two-group homogeneous diffusion approach. The effect of the control rods is taken into account by the super-cell concept. The control zone has no circular contour, but this is not of great importance, since the calculated radial (channel) power distribution and the reactivity effect are practically of no interest in this case.

The axial power distribution is calculated by the one-dimensional code HORA [5]. The axial distribution depends to some extent on the burnup distribution in the surrounding media. Therefore, the over-all calculation is made in the first step of the calculation, with a relatively rough axial and radial active zone region division. The set of the obtained radial buckling values is then used for one-dimensional fuel-channel calculation, with more detailed axial active zone division. This simple and fast calculational procedure is repeated until the difference in the input and output axial burnup distribution vanishes. The calculation is initiated with the measured channel work values distribution.

The cell parameters, as function of burnup, are taken from Ref.[6].

It should be noted that, in the case of the RA reactor, there is a relatively strong influence of the surrounding non-multiplying media (multiple radial and axial reflectors, the reactor tank and the pressure chamber, the upper Al shielding blocks, etc.). The effects of these material zones, particularly on the power distribution shape in the peripheral fuel layers, are larger than the effects of some basic approximations inherent to the calculational procedure described. Therefore, the material zone division includes the mentioned non-multiplying media.

3. GAMMA-SPECTROMETRY METHOD

The correlation of the fuel burnup with the contents of some gamma radio-active fission products (^{137}Cs, ^{134}Cs and ^{106}Ru) in the fuel is done by a computer

TABLE I. IRRADIATION HISTORY AND THE AVERAGE BURNUP OF DISCHARGED FUEL ELEMENTS

Fuel position		In the reactor		Type of the fuel in-core history	Burnup (MW·d/kg)	
Channel	Element	From	to		Power distr.	Gamma spectr.
0808	All	Feb.69	Feb.71	Single-position	6.80	6.58
0305	Central Periph.	Sep.68 Sep.69	Aug.69 Jun. 70	Two-position	6.66	6.82
0709	Central	Oct.73	Feb.75	Multi-position		
1005	Central	Feb.75	May 75			Not
0709	Periph.	May 75	Dec.76		11.5	measured
0501	Periph.	Dec.76	Sep.77			

program [7], accounting for the qualitative complexity of the burnup process in low enriched uranium fuel, as well as for the in-core fuel management and the reactor operation history. The ^{106}Ru, ^{137}Cs and ^{134}Cs contents in the fuel were determined by the fuel gamma-radiation measurement by a semiconductor Ge(Li) spectrometer.

The quality of the gamma-spectrometry measurements is illustrated by the spectrum shown in Fig.3.

The Ru and Cs isotope contents were determined in two steps:

First, the fission-product activity quotients were determined by the procedure [8] according to which all necessary calibrations and corrections were derived from the same spectrum as the activity quotients.

Second, the ^{137}Cs content was determined by a variant of the standard procedure [9], comparing the magnitudes of its photopeak in the spectrum of the fuel and the spectrum of the standard.

4. RESULTS

The average burnup results for the fuel elements discharged from three selected fuel channels, with different irradiation history, are presented in Table I.

Regarding the fuel in-core lifetime and the necessary cooling time the fuel with typical multi-position history is not yet available for experimental investigation. However, the selected fuel (Table I) is a good enough example to illustrate this type of fuel management.

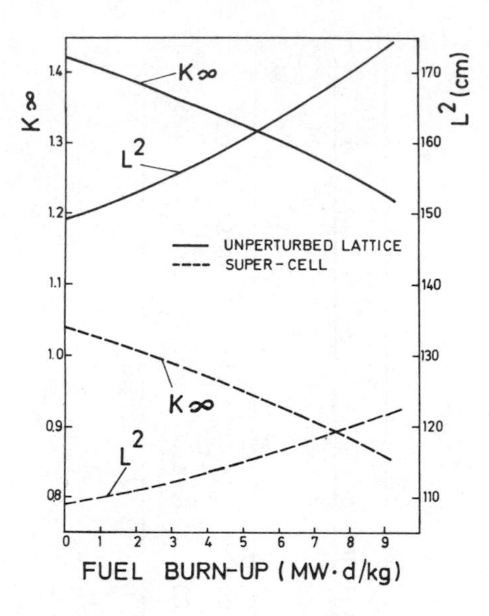

FIG.4. Unperturbed and perburbed lattice parameters.

The groups of six fresh fuel elements were introduced into the 0305 and 0709 channels and their histories were followed until discharging (Table I). The following standard refuelling scheme for axial fuel-element shuffling is applied:

	Fuel-element position					
Before shuffling	6	7	8	9	4	5
After shuffling	1	2	3	4	10	11

Some fuel elements from the channels listed in Table I belong to the "control zone". The perturbed (super-cell) and unperturbed lattice parameters, as a function of burnup, are presented in Fig.4.

The results of the axial power distribution calculation are presented in Table II.

The calculation is made with the same control-rod insertion of 22.5 cm for both the 0808 and the 0305 fuel channels. However, the actual average effective insertions were 25 ± 3 cm and 18 ± 3 cm, respectively. The error associated is not an error in the control-rod positioning, but in the effective

TABLE II. CALCULATED RELATIVE AXIAL POWER DISTRIBUTION
(Single-position and two-position in-core history)

Channel	Fuel-element positions in the channel										
	1[a] (top)	2[a]	3	4	5	6	7	8	9	10	11 (bottom)
0808	0.35	0.50	0.70	0.86	0.95	1.00	0.99	0.95	0.88	0.78	0.69
0305	0.24	0.36	0.52	0.66	0.88	0.96	1.00	0.98	0.92	0.76	0.65

[a] Control-rod insertion depth is equal to the length of two fuel elements.

control rod length, which was "shortened" owing to the significant degree of Cd control-rod burnup (the actual logarithmic boundary distribution among the burned control rods is simplified by the assumption of an "effective control-rod shortening").

The obtained results of the discharged fuel-element burnup are presented in Table III.

The results obtained by the procedure described by Eq.(1) are incorrect regarding either the channel-averaged value or the shape of the axial burnup distribution (Table III – method 3).

The first results obtained by the power distribution procedure [10] were incorrect, particularly for the uppermost fuel elements from channel 0808 ("single-position" history) because of neglecting the control rods' influence (Table III – method 2(b)). However, the results for the fuel elements from channel 0305 ("two-position" history) were fairly good, mainly owing to the cancelling effect of the upper element power overestimation, and the power underestimation of lower elements (Table III – method 2(b)).

The results obtained by the power distribution method (including the effect of control rods) are presented in Table III – method 2(a).

The differences in results, obtained by gamma-spectrometry and power distribution methods, are within ± 3% for channel-averaged values, i.e. within ± 7% for individual fuel elements.

5. CONCLUSIONS

The results obtained by the power distribution procedure and gamma-spectrometry measurements are in very good agreement for both types of fuel in-core history.

The power distribution procedure is particularly useful in the case of axial fuel-element transfer ("two-position" and "multi-position" history) because in that case it is not very sensitive to the errors in the axial power distribution.

In the future only the use of the multi-position in-core fuel management is planned. Therefore, the power distribution procedure will be routinely used for the spent fuel element burnup measurement.

However, the experimental evidence available is not sufficient to draw definite and general conclusions on the power distribution method utility. In this respect, more measurements by gamma-spectrometry (and some other techniques) should be done on the properly selected fuel; also further work on the gamma-spectrometry method adjustment to multi-position irradiation history should be done. The obtained results would also make it possible to correlate the power distribution procedure, particularly regarding the total coolant flow-rate and the part of the fission energy balance realized in the reactor core, which are otherwise roughly estimated.

TABLE III. SPENT FUEL-ELEMENT BURNUP

Channel	Method	Fuel-element burnup (MW·d/kg and differences (%) from gamma-spectr. results											Channel-averaged
		Fuel-element position											
		1 (top)	2	3	4	5	6	7	8	9	10	11 (bottom)	
0808	1	3.08	4.21	5.66	6.92	7.62	8.18	8.49	7.95	7.62	6.82	5.86	6.59
	2(a)	3.02	4.33	6.02	7.30	8.13	8.53	8.49	8.18	7.57	6.69	5.90	6.80
		−2	3	6	6	7	4	0	3	−1	−2	1	3
	2(b)	3.86	5.19	6.34	7.26	7.90	8.20	8.24	7.98	7.43	6.58	5.80	6.80
		25	23	12	5	4	2	−3	0	−2	−5	−1	3
	3	3.10	4.56	6.27	8.00	9.14	9.65	9.81	9.65	8.87	8.03	7.19	7.66
		1	8	11	16	20	18	16	21	16	18	23	16
0305	1	6.06	7.09	7.10	7.13						6.62	6.90	6.82
	2(a)	5.78	6.56	7.02	7.26						6.35	6.99	6.66
		−5	−7	−1	2						−4	1	−2
	2(b)	6.11	6.68	7.02	7.23						6.28	6.63	6.66
		1	−6	−1	1						−5	−4	−2

1. Gamma-spectrometry method; 2. Power distribution method; ((a) – control rods' effect included, (b) – without control rods);
3. Evaluated by Eq.(1).

REFERENCES

[1] INTERNATIONAL ATOMIC ENERGY AGENCY, Directory of Nuclear Reactors **1**, IAEA, Vienna (1966) 217.

[2] MARTINC, R., "Mixed fuel lattice as an optimal solution for the transient regime of the highly enriched uranium fuel introduction into the reactor RA core", Proc. Conf. Utilisation of Nuclear Reactors in Yugoslavia, May 1978, Belgrade, to be published (in Serbo-Croatian).

[3] KRSTIĆ, D., BULOVIĆ, V., MILOŠEVIĆ, D., MAKSIMOVIĆ, Z., J. Eng. Phys. **13** (1975) 19.

[4] MILOŠEVIĆ, D., Design of the axial fuel inversion refuelling machine, Int. Rep. (unpublished).

[5] MARTINC, R., MILOŠEVIĆ, M., BULOVIĆ, V., CUPAC, S., "Research reactor RA safety analysis", Proc. Conf. Utilisation of Nuclear Reactors in Yugoslavia, May 1978, Belgrade, to be published (in Serbo-Croatian).

[6] STRUGAR, P., STANČIĆ, M., Rep. IBK-876 (1969). In Serbo-Croatian.

[7] BULOVIĆ, V., Rep. IBK-1020 (1971). In Serbo-Croatian.

[8] BULOVIĆ, V., Rep. IBK-1196 (1973). In Serbo-Croatian.

[9] BULOVIĆ, V., KRSTIĆ, D., MAKSIMOVIĆ, Z., MILOŠEVIĆ, D., BORELI, F., "Non-destructive reactor RA fuel burn-up determination", Proc. Third SEV Symp. Investigation in the Field of Irradiated Fuel Reprocessing, 1974, Marianske Lazni) **3** (1974) 28. In Russian.

[10] MARTINC, R., BULOVIĆ, V., ŠOTIĆ, O., EKARV, R., PETRUNIN, D., DELEGARD, C., STEVOVIĆ, J., JAĆIMOVIĆ, Lj., MAKSIMOVIĆ, Z., "Non-destructive testing of fresh and irradiated fuel of research reactor RA", Proc. Conf. Utilisation of Nuclear Reactors in Yugoslavia, May 1978, Belgrade, to be published (in Serbo-Croatian).

GAMMA-SPECTROMETRIC DETERMINATION OF BURNUP AND COOLING TIME OF IRRADIATED ECH-1 FUEL ASSEMBLIES

H. GRABER, A. KEDDAR*, G. HOFMANN, S. NAGEL
Central Institute for Nuclear Research,
Rossendorf, Dresden,
German Democratic Republic

Abstract

GAMMA-SPECTROMETRIC DETERMINATION OF BURNUP AND COOLING TIME OF IRRADIATED ECH-1 FUEL ASSEMBLIES.

Under an IAEA Technical Contract gamma-spectrometric measurements on irradiated ECH-1 fuel assemblies from the Rossendorf research reactor (WWR-SM type) were carried out. Attention was especially focused on determination of burnup from concentrations of ^{137}Cs and ^{144}Ce; determination of cooling time from concentration ratios for $^{144}Ce/^{137}Cs$ or $^{95}Zr/^{95}Nb$; and investigation of correlations between burnup and fission-product concentration ratios. This report describes the measurement system and the techniques employed, together with the results of the evaluation. For purposes of safeguards it is recommended that relative gamma-spectrometric measurements be carried out and burnup be determined from the established correlations with $^{134}Cs/^{137}Cs$ or $^{154}Eu/^{137}Cs$ concentration ratios, and cooling time from the $^{144}Ce/^{137}Cs$ ratio.

INTRODUCTION

Exact knowledge of the nuclear transformations taking place in the nuclear fuel during reactor operation is an essential prerequisite to the optimization of the reactor campaigns and the balancing of the fuel. For experimental investigation of burnt-up fuel non-destructive and destructive methods are being developed in the Central Institute for Nuclear Research Rossendorf [1].

For measurements on a larger number of objects the gamma-ray spectrometry of fission products has the undisputed advantage of permitting detailed investigation on fuel assemblies [2] and rods [3, 4] with comparatively little expenditure by using commercial measuring devices. Therefore, this method is nowadays the most widely applied procedure for the non-destructive investigation of burnt-up nuclear fuel.

* International Atomic Energy Agency, Vienna.

Under a Technical Contract [5] between the International Atomic Energy Agency and the Central Institute for Nuclear Research Rossendorf, gamma-spectrometric measurements were made on fuel assemblies from the Rossendorf research reactor. The aim of the investigations was to develop non-destructive methods for determining burnup and cooling time of irradiated fuel assemblies on the basis of high-resolution gamma spectrometry.

In preparation for the measurements, the programme system for evaluating gamma-spectrometric burnup measurements [6–8] was revised. At the same time reconstruction was carried out of the measurement equipment used for studying spent fuel at the Rossendorf Institute.

This report gives a general picture of the work performed. A short description of the measuring object and equipment is followed by experimental details and results.

1. ECH-1 FUEL ASSEMBLIES

The Rossendorf research reactor is operating with ECH-1 fuel assemblies. Figure 1 shows an assembly in section. The assembly consists of three concentrically arranged tubular fuel elements containing fuel in the form of a UO_2-Al dispersion. Each assembly contains 108 g U, the initial enrichment is 36%. The length of the fuel is 600 mm, but the axial fuel distribution is not uniform. Measurements made on 26 unirradiated assemblies [9] produce the typical fuel distribution curves shown in Fig.2. The mean non-uniformity factor (maximum/mean fuel thickness ratio) derived thereby is 1.11. Furthermore, Fig.2 contains a medium distribution of the axial neutron flux density [10]. The axial non-uniformity factor of the neutron flux density is 1.32, the normalizing factor is equal to $0.51 \times 10^{13}\ cm^{-2}\ s^{-1}\ MW^{-1}$. The positions of the nine measuring points along the assembly axis are also given in Fig.2.

2. MEASUREMENT SYSTEM

The measurement system set up in the reactor hall to study the irradiated ECH-1 assemblies consists of the following parts, as shown in Fig.3:

Measuring vessel with a scanning device;

Collimator system between assembly and detector;

Ge(Li) detector and electronic devices for recording gamma intensity profiles and gamma spectra.

The assemblies are transported from the cooling pool to the measuring vessel in a special transfer flask, which acts as an upper shield during the measurements.

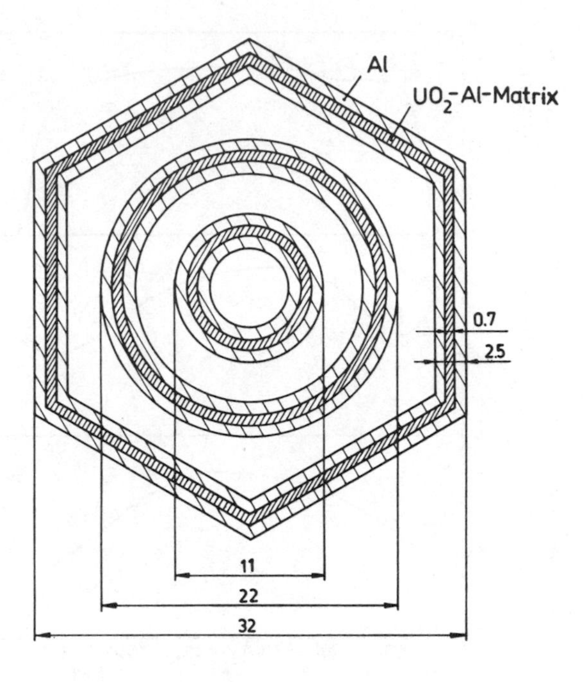

FIG.1. Single ECH-1 assembly.

The assembly can be moved inside the measuring vessel in two directions, the movements being either separate or combined:

Vertical movement at a rate of 23 mm · min^{-1};
Rotation at a rate of 70–127 rpm.

The vertical position of the assembly has an uncertainty of ±1 mm. The azimuthal position of the assembly (angular position) can be read off on a graduated scale with an uncertainty of ±7.5°. The collimator system consists of the following units:

Two collimators with constant slot height of 5 mm to provide the radiation geometry;
One collimator with an adjustable slot height for regulating the radiation intensity.

The shape and width of the slots were chosen to produce a conic radiation beam such that the detector sees the full cross-section of the assembly. The collimator system is adjusted by means of a theodolite and the height of the slot is read from a dial gauge.

The detector used was a coaxial Ge(Li) detector produced inthe Institute for Nuclear Research Řež (Czechoslovakia) with a sensitive volume of 18 cm^3

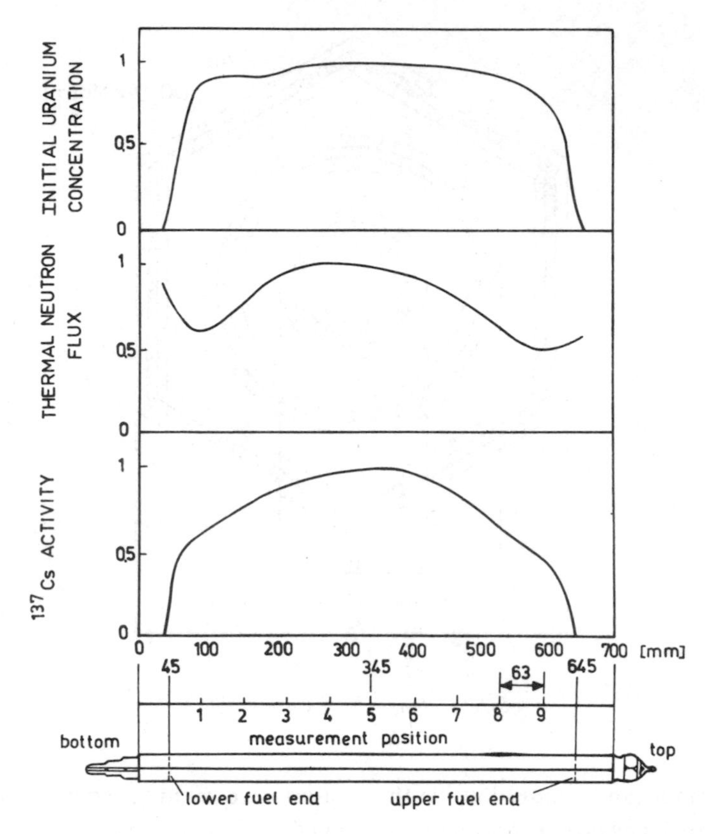

FIG.2. Normalized axial distributions of initial uranium concentration, thermal neutron flux density and ^{137}Cs activity.

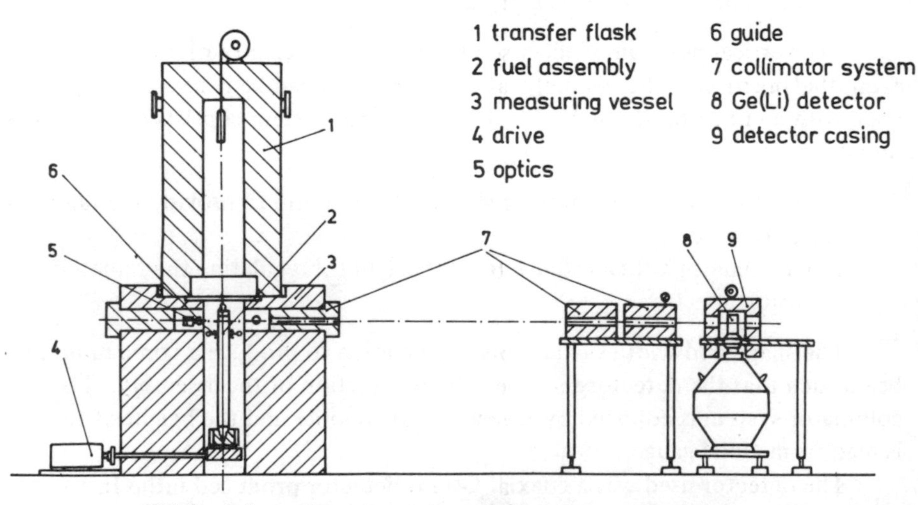

FIG.3. Schematic representation of the measurement system.

and a resolution of 2.1 keV, as measured for the 1332-keV line of ^{60}Co. The lead casing, which shields the detector against the radiation in the reactor hall, kept the background down to about 5 counts/s, whereas the full-range gamma intensity was of the order of 10^3 counts/s.

3. MEASUREMENT PROCEDURES

For the measurements 15 assemblies were selected. The energy output and cooling time of these assemblies are listed in Table I. The following three kinds of measurement were performed:

(1) Recording of gamma spectra at nine equidistant measuring points along the assembly (see Fig.2);
(2) Recording of a cumulative spectrum by measurements at nine equidistant points and recording of a separate gamma spectrum in the activity maximum (measuring point 5);
(3) Recording of a gamma spectrum in the activity maximum of the assembly.

The first procedure is to determine the axial distribution of the fission product concentrations and burnup.

The second procedure makes it possible to establish the maximum/mean burnup ratio, while the third procedure serves for a quick determination of the maximum burnup.

The gamma spectra were recorded over the 450–1550 keV energy range. The measurements were made in air, with the assembly being rotated and with a slot height of 1 mm for the collimator on the detector side. To reduce the gamma intensity a 6-mm lead absorber was used. For three assemblies with cooling times of less than one year the absorber thickness was increased to 20 mm.

For each of the two measurement conditions (6 and 20 mm of lead) we carried out a calibration of the measurement system, using a certified cylindrical ^{137}Cs source (activity 353 mCi ± 2.7%).

For separate spectra the measurement time was usually 400 s, and for cumulative spectra, 200 s at each measurement point. A few long-term measurements (up to 18000 s) were made in order to establish the relative efficiency.

4. EVALUATION OF MEASUREMENTS

For the evaluation a program system [6] was used that included the following sub-programs:

PEAK, to determine photopeak areas P_{ij};
ABSINT, to calculate the relative efficiency ϵ_i^{rel};
ABBRAND, to determine the fission product concentration n_j and burnup β.

TABLE I. OPERATIONAL DATA OF ASSEMBLIES INVESTIGATED AND BURNUP VALUES DETERMINED

		Operational data			Experimental results		
Ass. No.	Energy output (MW·d)[a]	t_{ir}[b] (a)	t_{res}[c] (a)	t_c[d] (a)	$\bar{\beta}(^{137}Cs)$ (%fima)	$\bar{\beta}_{cum}(^{137}Cs)$ (%fima)	$\bar{\beta}(^{144}Ce)$ (%fima)
24	25.9	2.4	6.1	2.4	20.3		31.6
39	21.5	2.3	5.8	3.9	21.5		26.8
40	17.6	2.0	5.1	4.6	17.3		16.3
44	26.2	2.9	7.3	2.4	22.5		34.4
45	21.5	2.4	5.9	3.3	22.8		17.1
81	20.3	2.0	4.9	2.3	20.9		31.9
94	13.0	0.9	2.2	2.0	15.6	15.5	19.4
113	15.8	1.0	2.4	1.5	18.5	19.0	19.8
114	15.4	1.0	2.4	1.5	18.2		20.8
115	15.4	1.0	2.4	1.5	18.0	18.6	22.0
116	12.5	0.8	2.0	2.0	14.6		16.2
118	14.7	1.0	2.4	1.5	18.1		22.1
121	16.3	0.8	1.9	0.6	18.6		18.0
125	16.3	0.8	1.9	0.6	20.4	20.5	21.1
126	16.3	0.8	1.9	0.6	18.0	19.1	18.2

[a] Energy output of assembly, estimated from reactor energy output.
[b] Irradiation time (power $>$0.1 MW).
[c] Residence time for assemblies in reactor.
[d] Cooling time as written in operational log (referred to 4 July, 1977).

The PEAK program [11] calculates peak areas by totalling the channel contents. Background behaviour is specified both linearly and stepwise. For good single peaks, the error in the peak area is between 1 and 2%.

The ABSINT code [6, 8] is used to calculate the relative efficiency of the whole measurement system directly from the spectra of the objects measured. For this purpose the lines from the isotopes ^{106}Rh, ^{134}Cs, ^{144}Pr and ^{154}Eu are used. This relative efficiency contains both relative self-absorption in the object measured and relative absorption in external absorbers. It is described by an energy-dependent function $\epsilon(E)$ that is represented on a logarithmic scale by two parabolas. The error in the relative efficiency is only slightly energy-dependent and amounts to about ±2% in the energy range under consideration.

The sub-program ABBRAND [6] calculates absolute fission product concentrations n_{ij} per unit of length of the assembly section, as follows:

$$n_{ij} = \frac{N_{ij}}{b} = \frac{P_{ij}}{\gamma_{ij} \cdot \varepsilon_i^{rel} \cdot t_m \cdot \lambda_j \cdot \exp(-\lambda_j \cdot t_c)} \cdot \frac{1}{S_e \cdot \varepsilon_e \cdot b} \quad (1)$$

where

N_{ij} is the amount of fission product j calculated from gamma transition i (at the end of irradiation),
b is the effective height of the volume under consideration,
γ_{ij} is the quantum yield of the i-th gamma transition of fission product j,
t_m is the measuring time,
t_c is the cooling time,
λ_j is the decay constant of fission product j,
S_e is the self-absorption factor for calibration energy e, and
ε_i^{rel} is the relative efficiency of the measuring system for energy E_i of gamma quantum i (normalization $\varepsilon_e^{rel} = 1$).

The product $(\epsilon_e \cdot b)$ is determined by a calibration, which is made under exactly the same conditions as the assembly measurements. Assuming that the length of the source is greater than the height b of the assembly section considered when seen through the collimator, we find

$$\varepsilon_e \cdot b = \frac{P_e^{cal}}{\gamma_e \cdot t_m^{cal} \cdot a} \quad (2)$$

where the index cal means calibration, the index e means the calibration energy e, and a is the specific activity of the calibration source. An explicit knowledge of the height b of the assembly section considered is not necessary [12].

The self-absorption factor S_e was calculated [7]. The error for the given energy range is ±2%.

From all the photopeaks recorded for one nuclide we derived a mean error-weighted value for the corresponding nuclide concentration upon termination of irradiation.

From the concentration of fission product atoms present, n_j, we are able – correcting for the loss of fission product atoms during irradiation through neutron capture and radioactive decay – to calculate the concentration of all atoms of one kind formed, n_j/C_j. To calculate the C_j factors we need to have, in addition to the decay constants, effective cross-sections and fission yields for the nuclides under consideration, the neutron flux density versus time history of irradiation [2, 12]. Assuming proportionality between flux and energy output, these pairs of values can be obtained from the operational data by averaging over periods for which the irradiation history has remained sufficiently constant.

The burnup β is obtained from the concentration n_j of primary fission products as

$$\beta(\text{fima}) = \frac{n_j}{n_o\,(\overline{YC})_j} \quad \text{with} \quad (\overline{YC})_j = \frac{\beta^5}{\beta}(YC)_j^5 + \frac{\beta^9}{\beta}(YC)_j^9 \tag{3}$$

where

n_o is the initial concentration of heavy nuclei per unit of length,

$Y_j^{5,9}$ is the yield of fission product j from fission of ^{235}U or ^{239}Pu, and

$\beta^{5,9}$ is the burn-up of ^{235}U or ^{239}Pu.

The $\beta^{5,9}/\beta$ ratios used as weight factors for calculating $(\overline{YC})_j$ are calculated from the $^{106}Ru/^{137}Cs$ concentration ratio [6]. Given the high initial enrichment of the ECH-1 assemblies, however, the proportion of ^{239}Pu in the total number of fissions is small.

The cooling time can be determined on the basis of the ratio of two primary fission products with sufficiently different half-lives (for example, $^{144}Ce/^{137}Cs$), or else with the aid of the ratio between two genetically interdependent nuclides (for example, $^{95}Zr/^{95}Nb$).

The ratio of the two primary fission products $^{144}Ce/^{137}Cs$ gives us the cooling time as

$$t_c = -\frac{1}{\lambda_4 - \lambda_7} \cdot \ln\left[\frac{N_4(t_c)\cdot Y_7^5\cdot(\lambda_4' - \lambda_5)(e^{-\lambda_5\cdot t_{ir}} - e^{-\lambda_7'\cdot t_{ir}})}{N_7(t_c)\cdot Y_4^5\cdot(\lambda_7' - \lambda_5)(e^{-\lambda_5\cdot t_{ir}} - e^{-\lambda_4'\cdot t_{ir}})}\right] \tag{4.1}$$

where

t_{ir} is the irradiation time,

$\lambda_i' = \lambda_i + \sigma_i^a \cdot \phi$ is the loss parameter of fission product i, and

$\lambda_5 = \sigma_5^a \cdot \phi$ is the loss parameter of ^{235}U, with σ_5^a as the corresponding cross-section.

Since a reactor normally operates at intervals it is appropriate to correct for the irradiation history by using the C_j factors:

$$t_c = -\frac{1}{\lambda_4 - \lambda_7} \cdot \ln \frac{N_4(t_c) \cdot Y_7^5 \cdot C_7^5}{N_7(t_c) \cdot Y_4^5 \cdot C_4^5} \qquad (4.2)$$

Thus, either the course of the irradiation history (Eq.(4.2)) or at least the length of the irradiation time t_{ir} and the mean neutron flux density ϕ (Eq.(4.1)) is required. In both cases, only ^{235}U fission is considered; ^{239}Pu fission may be taken into account by introducing factors $(\overline{YC})_{7,4}$ in Eq.(4.2). If we use the two genetically interdependent fission products "1" and "2", we get the cooling time

$$t_c = -\frac{1}{\lambda_1 - \lambda_2} \ln \left[\left(\frac{N_2(t_c)}{N_1(t_c)} - \frac{\lambda_1}{\lambda_1 - \lambda_2} \right) \Big/ \left(\frac{N_2(t_{ir})}{N_1(t_{ir})} - \frac{\lambda_1}{\lambda_1 - \lambda_2} \right) \right] \qquad (5)$$

with

$$\frac{N_2(t_{ir})}{N_1(t_{ir})} \approx \frac{\lambda_1}{\lambda_2 - \lambda_5} \quad \text{for} \quad \lambda_1 \cdot t_{ir}, \lambda_2 \cdot t_{ir} \gg 1 \qquad (6)$$

Thus, no knowledge of irradiation history is required.

5. MEASUREMENT DATA

Figure 4 shows two gamma spectra from assemblies with different cooling times. Lines for the nuclides ^{95}Zr, ^{95}Nb, ^{106}Rh, ^{134}Cs, ^{137}Cs, ^{144}Pr and ^{154}Eu were observed and the corresponding photopeaks analysed. For cooling times below one year the ^{134}Cs peaks at 563 and 569 keV are located so unfavourably against the background that correct peak analysis is impossible. The 724-keV line can be attributed for certain to ^{95}Zr only in the case of cooling times less than a year because, after longer cooling times, the contribution made by the 723-keV transition of ^{154}Eu to this line increases notably.

5.1. Efficiencies

The relative efficiencies obtained are shown in Fig.5. The calibration gives, in accordance with Eq.(2), the two normalizing factors:

For a 6-mm lead absorber

$\epsilon \cdot b(662\ \text{keV}) = 3.82 \times 10^{-8}$ mm ±4.2%

For a 20-mm lead absorber

$\epsilon \cdot b(662\ \text{keV}) = 7.35 \times 10^{-9}$ mm ±4.4%

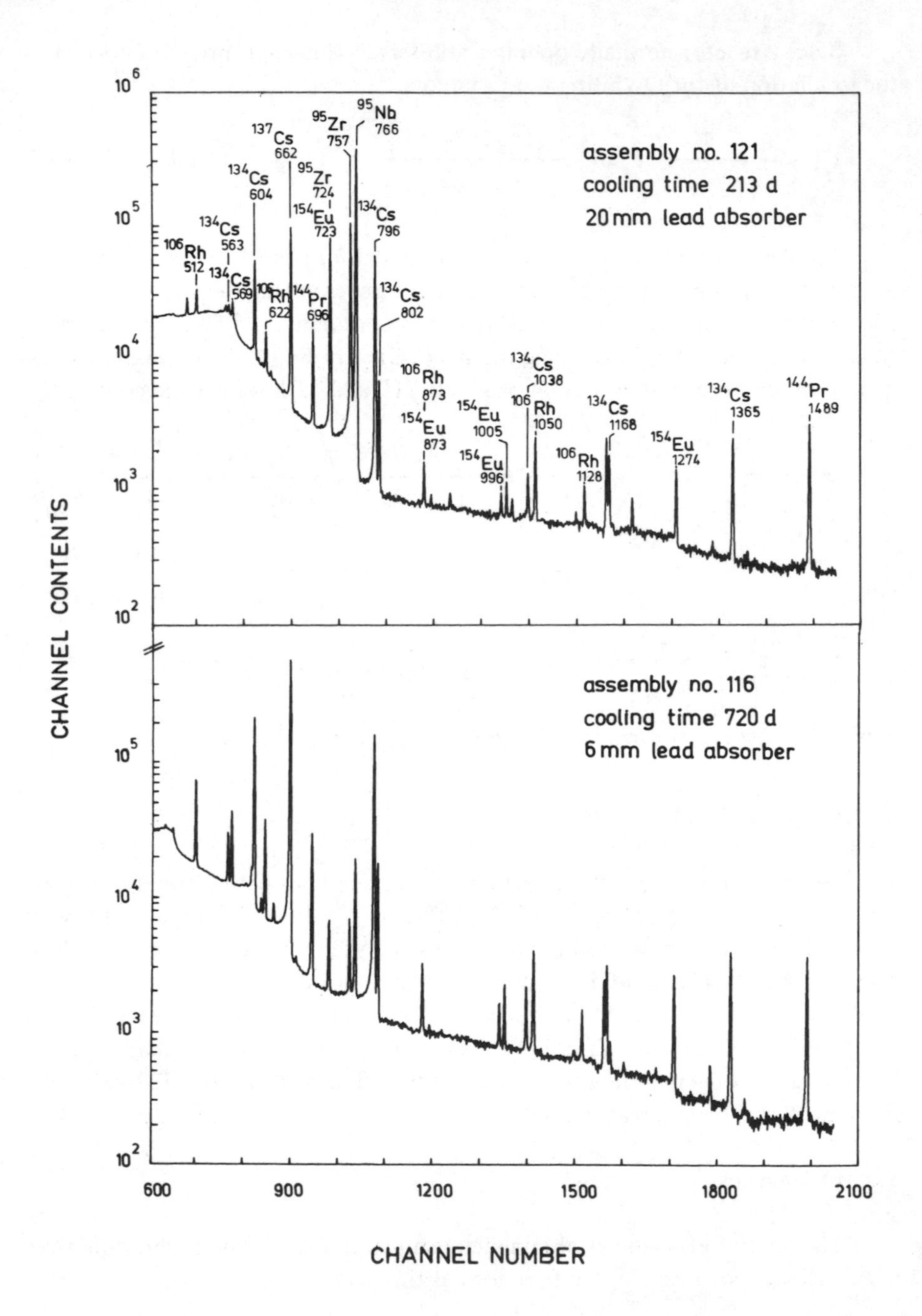

FIG.'4. Gamma-ray spectra for two assemblies with different cooling times.

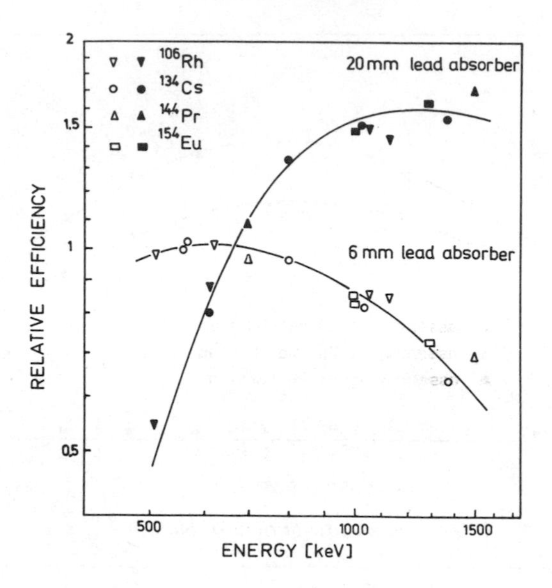

FIG.5. Relative efficiency of the measurement system.

5.2. Burnup and fission product concentrations

Figure 6 shows two measured absolute axial burnup distributions. The contribution of ^{239}Pu to total burnup ranges from 1.1 to 3.3%; the mean value is 2.5%. The error of absolute burnup values is ±12%, but the initial concentration of uranium alone contributes an error of ±10%.

If converting the β_{max} values obtained at measurement point 5 (maximum burnup) to mean values $\overline{\beta}$ for the whole assembly, we have to remember that, in the case of non-uniform initial uranium concentration, the measured ^{137}Cs content is not proportional to burnup. Therefore, in accordance with Eq.(3), we should assign to each ^{137}Cs content a uranium concentration corresponding to the particular measurement point. This leads to the following relationship:

$$\overline{\beta} = \beta^{(5)} \cdot \frac{U_z}{K_z} \qquad (7)$$

where

U_z is the non-uniformity factor for the initial uranium concentration, and

K_z is the non-uniformity factor for the ^{137}Cs distribution.

For the $\overline{\beta}$ shown in Table I the numerical values U_z= 1.11 and K_z = 1.32 (average from seven measured ^{137}Cs distributions) are used.

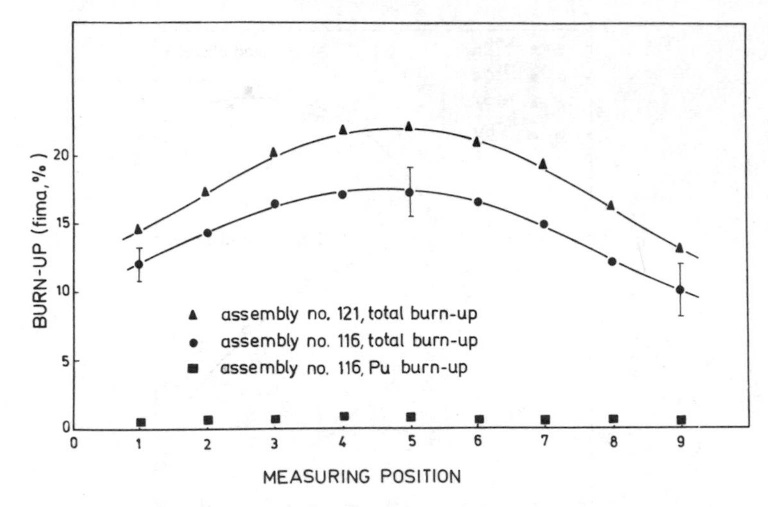

FIG.6. Axial distributions of burnup.

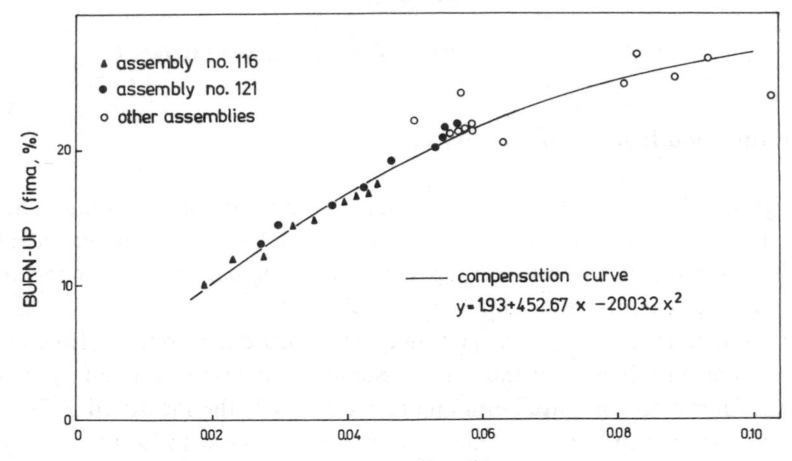

FIG.7. Concentration ratio $^{134}Cs/^{137}Cs$ (corrected).

Further, Table I contains the burnup values $\bar{\beta}_{cum}$ for the whole assembly, derived from cumulative measurements and calculated as follows:

$$\bar{\beta}_{cum} = \frac{\sum_{i=1}^{9} n_{137}^{(i)}}{\frac{1}{L} \int_0^L n_o(x)\, dx \cdot (\overline{YC})_{137}^{(5)}} \tag{8}$$

where

i is the cell number, and

L is the fuel length.

Instead of one representative value $(\overline{YC})_{137}^{ass}$ for the whole assembly the value $(\overline{YC})_{137}^{(5)}$, pertaining to measurement point 5, was used, for the sake of simplicity. The resultant error in the case of the Rossendorf reactor is about 10^{-2}%.

The deviations of the cumulative measurement data (second procedure) from those obtained by spectral recording at maximal axial distribution (third procedure) amount to as much as -3.5%, the mean deviation for five cumulative measurements is -0.7%. The cumulative measurement gives average burnup values with relatively little trouble; it is not necessary to know the axial distributions of the initial uranium concentration and the ^{137}Cs distribution.

For operational periods of up to about two years the primary fission product ^{144}Ce ought to be suitable for burnup determination. Table I also includes burnup values derived from ^{144}Ce, using a corresponding non-uniformity factor for the ^{144}Ce distribution of $K_Z = 1.26$ (irradiation history corrected). ^{144}Ce is definitely suitable as a burnup monitor for irradiation periods up to around two years.

5.3. Correlations between burnup and fission product concentration ratios

Correlations between burnup and fission product ratios are of particular interest in that they enable us, given the availability of appropriate "calibration curves", to determine burnup solely by measuring a ratio. Bi-unique functional relationships cannot, however, be obtained until after correction for the effect of the irradiation history [12].

Figure 7 shows the correlation between burnup and the corrected ^{134}Cs/^{137}Cs concentration ratio. The compensation curve shown in the figure was obtained through a parabolic approach, the validity of which is supported by theoretical considerations. The degree of certainty is B = 0.999 and the relative residual spread 4.1%.

Figure 8 shows the corresponding correlation between burnup and ^{154}Eu/^{137}Cs. Also in this case a parabolic approach is best for the compensation curve. In this instance the degree of certainty is B = 0.997 and the relative residual spread 6.9%.

5.4. Cooling times

From the ratio of the primary fission products ^{144}Ce/^{137}Cs cooling times were calculated – given knowledge of the irradiation history – from Eq.(4.2) with a mean deviation of ±5% from the cooling time written in the operational log. Maximum deviation is -20% for assemblies with operating times of five years or or more.

Without knowledge of the irradiation history (Eq.(4.1)) the deviations from actual cooling time are considerably greater and may in the most unfavourable cases

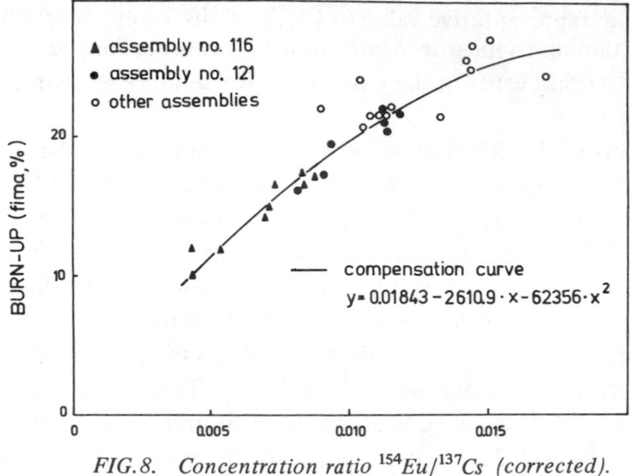

FIG.8. Concentration ratio $^{154}Eu/^{137}Cs$ (corrected).

cases attain as much as -50%. The reason for this is that, given the irradiation conditions applying to the Rossendorf research reactor, the condensation of many years of irradiation into a single interval is too rough an estimation. Table II shows the corresponding values.

Because of the half-lives of 65 and 35 days only cooling times of barely up to a year from the $^{95}Zr/^{95}Nb$ ratio are definable. For the three assemblies under consideration – Nos 121, 125, and 126 – we obtained values that were 15–50% above the figures from the operating log. The reason for these large deviations is mainly the uncertainty involved in analysis of the photopeaks at 757 keV (^{95}Zr) and 766 keV (^{95}Nb) by means of our PEAK program. In this case, a more sophisticated evaluation code is preferable.

6. CONCLUSIONS

The gamma-spectrometric studies on ECH-1 fuel assemblies from the Rossendorf research reactor suggest the following conclusions:

(a) After correction for irradiation history, burnup can be obtained from the ^{137}Cs concentration with an uncertainty of about ±12%. The largest contribution to this error is in the initial uranium concentration (±10%);

(b) For operating times of up to about two years the ^{144}Ce concentration can also be used to determine burnup;

(c) The mean burnup of the assemblies was determined by transformation from the axial maximum burnup on the one hand, and, on the other, by recording a cumulative spectrum along the assembly axis. Results obtained by these two procedures tally well, showing a mean deviation of less than 1%;

TABLE II. DETERMINATION OF COOLING TIME
(Referred to date of measurement)

Ass. No.	$^{144}Ce/^{137}Cs$					$^{95}Zr/^{95}Nb$	
	t_c [a] (d)	t_c [b] (d)	t_c [c] (d)	b/a	c/a	t_c [d] (d)	d/a
24	886	702	455	0.79	0.51		
39	1427	1345	1012	0.94	0.71		
40	1679	1719	1445	1.02	0.86		
44	889	714	321	0.80	0.36		
45	1208	1346	1089	1.11	0.90		
81	835	668	474	0.80	0.57		
94	717	646	609	0.90	0.85		
113	562	555	523	0.99	0.93		
114	561	527	494	0.94	0.88		
115	720	699	638	0.97	0.89		
118	561	499	466	0.89	0.83		
121	213	229	237	1.08	1.11	237	1.16
125	206	213	221	1.03	1.07	313	1.53
126	206	223	231	1.08	1.12	266	1.30

[a] From operational log.
[b] Determined by Eq.(4.2) (with correct irradiation history).
[c] Determined by Eq.(4.1) (irradiation history condensed to a single time interval with mean flux value).
[d] Determined by Eq.(5).

(d) In the burnup range from 10 to 30% (fima) a parabolic dependence of burnup on the corrected $^{134}Cs/^{137}Cs$ concentration ratio was found. The burnup can be determined by relative gamma-spectrometric measurements, on the basis of this correlation, with an error of around ±5%. Similar results were obtained with respect to the correlation between burnup and the $^{154}Eu/^{137}Cs$ ratio;

(e) The $^{144}Ce/^{137}Cs$ concentration ratio, following correction for the irradiation history, appears to be a suitable monitor for determining cooling times between 0.5 and 5 years with a mean error of ±5%. Description of the irradiation history by approximation in terms of only one interval of time and a mean neutron flux makes the uncertainty in determining the cooling time greater (it is a function of the validity of the approximation and the cooling time itself);

(f) For purposes of safeguarding nuclear material it is advisable to make relative gamma-spectrometric measurements and to determine burnup from the established correlation with the $^{134}Cs/^{137}Cs$ concentration ratio and the cooling time from the $^{144}Ce/^{137}Cs$ ratio. Corrections for irradiation history must be introduced, however.

ACKNOWLEDGEMENTS

The authors wish to acknowledge Messrs R. Berndt and H.-C. Mehner from the Rossendorf Institute for their collaboration in the performance of the experiments. They also would like to express their thanks to the Academy of Sciences of the GDR and to the International Atomic Energy Agency for supporting this work.

REFERENCES

[1] GRABER, H., HERMANN, A., HÜTTIG, W., GÜNTHER, H., SCHIFF, B., Kernenergie **20** (1977) 98.

[2] GRABER, H., HOFMANN, G., NAGEL, S., GÜNTHER, H., Kernenergie **17** (1974) 73.

[3] HOFMANN, G., GRABER, H., MEHNER, H.-C., NAGEL, S., GÜNTHER, H., SCHIFF, B., Kernenergie **20** (1977) 128.

[4] MEHNER, H.-C., GRABER, H., HOFMANN, G., NAGEL, S., GÜNTHER, H., Kernenergie **20** (1977) 242.

[5] INTERNATIONAL ATOMIC ENERGY AGENCY, IAEA Technical Contract No.1965/TC (1977).

[6] HOFMANN, G., MEHNER, H.-C., Programme for evaluation of gamma spectrometric burn-up measurements (in German), Working Rep. ZfK-RPR-2/74 (July 1974).

[7] BERNDT, R., HOFMANN, G., MEHNER, H.-C., Calculation of gamma self-absorption factors for ECH-1 fuel assemblies (in German), Working Rep. ZfK-RPM-4/77 (Feb. 1977).

[8] MEHNER, H.-C., Kernenergie **19** (1976) 3.

[9] BERNDT, R., Determination of ^{235}U distribution in ECH-1 fuel assemblies (in German), Working Rep. ZfK-RPM (May 1974).

[10] SCHNEIDER, B., Central Institute for Nuclear Research Rossendorf, private communication.

[11] BERNDT, R., PEAK – A gamma-spectrum evaluation programme for small-size computers (in German), Working Rep. ZfK-RPM-2/77 (Feb. 1977).

[12] HOFMANN, G., NAGEL, S., Gamma Spectrometric Studies on Spent Assemblies from a Power Reactor, Dissertation, Academy of Sciences of the German Democratic Republic (1977). In German.

SAFEGUARDS VERIFICATION OF SPENT MATERIALS TESTING REACTOR (MTR) FUEL USING GAMMA-RAY SPECTROMETRY

G.L. HANNA
Australian Safeguards Office,
Coogee, New South Wales,
Australia

Abstract

SAFEGUARDS VERIFICATION OF SPENT MATERIALS TESTING REACTOR (MTR) FUEL USING GAMMA-RAY SPECTROMETRY.

Attempts to exploit gamma-ray interrogation as a method of safeguards verification of spent nuclear fuel have aimed at obtaining information on fission product concentrations that can be related to burnup and cooling times of individual fuel assemblies. In this study, measurements of gamma count-rate ratios of selected fission product nuclides were made at the centre plane of a number of fuel assemblies irradiated in the HIFAR materials testing reactor and were compared with similar published measurements obtained previously by IAEA safeguards inspectors. Results were consistent with those of the IAEA but it is concluded that the technique used was capable of verifying burnup to only ± 20–50% and cooling time to only ± 100 days of the nominal values over the ranges investigated (25–50% burnup and 450–800 days' cooling). Calculations were made to investigate the potential for improving the precision of the method. In particular, two factors emerged. First, it was shown that none of the fission product ratios available from those nuclides which can be measured most readily are sufficiently independent of irradiation history and burnup to be used as an unambiguous and independent monitor for cooling time. Second, it was shown that for a given burnup the value of a particular fission product count-rate ratio depends significantly on neutron flux density and hence burnup rate. This is important in materials testing reactors because considerable flux gradients occur along the fuel element and severe perturbations can occur adjacent to in-pile experiments. Thus, to achieve higher precision in gamma spectrometry verification, detailed information on irradiation history of individual assemblies must be available to inspectors and information such as neutron flux density might also need to be subject to verification.

1. INTRODUCTION

Verification of the fissile content of spent fuel assemblies from materials testing reactors (MTRs) is an essential step in providing data upon which a quantitative evaluation can be made of materials control in these facilities. For the timely and economic gathering of data, it is also essential that they be obtained by non-destructive methods. Although both gamma-ray and neutron techniques are potentially useful in this respect, gamma-ray methods are attractive to safeguarding authorities because they are potentially capable of high accuracy and present day equipment is reasonably if not readily portable.

Gamma-ray methods, however, are not capable of directly measuring the uranium and plutonium contents because of interference from intense gamma radiation from fission products. Despite this, the method has been investigated as a means of burnup measurement and hence indirect measurement of residual and produced fissile material content.

A serious problem in applying gamma-ray interrogation to water-moderated power reactors is the high level of self attenuation in individual fuel rods, and the so-called 'black-out zone' in the assembly centre where gamma activity cannot be measured. Materials testing reactor fuel assemblies suffer from this problem to a much lesser extent and so the application of this technique is more favourable in the MTR case.

Dragnev, Diaz-Duque and Pontes [1] of the IAEA developed an empirical method of gamma-ray interrogation of spent fuel from the HIFAR materials testing reactor. HIFAR fuel assemblies in current and recent use are of a concentric fuel tube design (Mark 4/5) and a radial plate design (Mark 3) respectively, both of which have a hollow tubular centre to permit insertion of hardware for neutron irradiation experiments. The empirical method of Dragnev, et al. was aimed at permitting measurement of burnup and cooling time of both types of assembly and the integrity of the Mark 4/5 type. (The active sections of the latter are cropped from non-active sections before storage in such a manner that fuel tubes become detached from one another and must be placed in a simple tubular container to prevent their separation. Thus, the measurement is intended to indicate whether or not all four fuel tubes are present in a container.) In principle, the method would be applicable to the box-type MTR fuel geometry although the empirical relationship would need to be established for that case.

This paper reports on a re-examination of the method proposed by Dragnev, et al. for verifying burnup and cooling time. The objective was to assess the precision of the technique and to this end both experimental measurements, using Mark 4/5 assemblies, and a theoretical assessment were made. The work is discussed in that order.

2. THE EMPIRICAL METHOD

Based on earlier work done at the BR-2 reactor in Belgium (Beets, Dragnev & Hecq, [2]) and NPD in Canada (Dragnev & Burgess [3]), Dragnev, et al. [1], have claimed success in using certain fission product activity ratios to monitor the operator's declared data on the burnup and cooling time of Mark 3 elements and the integrity of Mark 4/5 fuel elements irradiated in the HIFAR reactor. The parameters used were:

(i) the ratio of count rate of ^{144}Ce (696 keV) to the count rate of ^{137}Cs (662 keV), as an indicator of cooling time (denoted here as R^{696}_{662});

(ii) the ratio corrected to zero cooling time of the count rate of ^{134}Cs (796 keV) to the count rate of ^{137}Cs (662 keV), as an indicator of burnup (denoted here as R^{796}_{662}); and

(iii) the correlation between ^{137}Cs count rate and R^{796}_{662} as a monitor of assembly integrity for Mark 4/5 elements.

In summary, Dragnev, et al. reported that their measurements resulted in estimates of cooling time and burnup which agreed with operator's data within the statistical accuracy of the gamma-ray measurements. They derived an empirical relationship between burnup (i.e. the operator's declared burnup) and R^{796}_{662} to enable estimation of burnup in other fuel elements for which R^{796}_{662} is measured. For the elements measured, an average discrepancy of 4.9 per cent (expressed as a percentage of the operator's figures) was claimed between the operator's figure and burnup estimated from R^{796}_{662}.

For the present work, the experimental arrangement described by Dragnev, et al. was used. Fuel assemblies were placed on a rack at the bottom of the cooling pond and viewed by a small (0.3 cm^3) pure germanium detector mounted at the top of a vertical air-filled collimator 32 mm diameter and 5.4 m long. Assemblies were viewed through their centre planes and no longitudinal scans were made. The multi-channel analyser was adjusted to cover the energy range 0-1.33 MeV and calibrated with ^{137}Cs and ^{60}Co sources.

Elements available for measurement fell into two categories, those that had cooled for 15-17 months and those that had cooled for 2-3 months. Even with the longer cooled elements, the spectrometer deadtime was very excessive (50-70 per cent) unless a filter (4-5 mm lead plus 3 mm cadmium) was placed before the detector; this reduced the deadtime to less than 20 per cent. Elements cooled for 2-3 months were far too active to count, even with the filter in place, and consequently were not measured.

It is recognised that filters may not be the best way to reduce the intensity of radiation falling on the detector. Collimators are preferable but were not available at the time. The use of filters, however, was consistent with the practice followed by Dragnev, et al. and in IAEA inspections.

Measurements were made on seven Mark 4/5 elements (150 g ^{235}U before burnup) and four Mark 4/14 (115 g ^{235}U) elements. Operator's data on burnup and cooling time are summarised in Table I.

All spectra corresponded well with those observed by Dragnev, et al. or reported elsewhere in the literature. A notable feature is the extremely high Compton scatter at energies below about 600 keV; the high spectrometer deadtime arises almost entirely from this low energy radiation and is of no interest whatsoever in the analysis. Reduction of this low energy contribution is highly desirable in this type of work.

2.1 Burnup measurement

The ratio of the ^{134}Cs count rate to ^{137}Cs count rate was evaluated as a burnup index, using both the 604 keV (R^{604}_{662}) and 796 keV (R^{796}_{662}) ^{134}Cs radiations. Results are presented in Figures 1 and 2. Although Dragnev, et al. [1] correlated their count ratios with burnup, expressed as grams ^{235}U, correlation with burnup expressed as per cent ^{235}U is used here to facilitate comparison of results, particularly as it permits elements of different fissile content to be compared. For comparison, therefore, the data of Dragnev, et al. were adapted to this form and are included in Figure 2.

TABLE I IRRADIATION HISTORY OF FUEL ELEMENTS MEASURED BY GAMMA SPECTROMETRY

Serial No.	Weight ^{235}U (g)	Burnup % U	Cooling Time (days)
414/12	115	41.8	476
414/13	115	51.1	476
414/16	115	44.5	476
414/18	115	37.3	450
45A/211	150	48.2	511
45A/216	150	46.6	511
45A/223	150	36.1	511
45A/227	150	38.2	511
45A/228	150	43.7	450
45A/229	150	41.5	476
45A/223	150	39.8	454

The mathematical relationship between burnup and the count rate is relatively complex and, as was done by Dragnev, et al., the results were evaluated assuming that a linear relationship holds over the burnup range of interest. From the point of view of applying the method in the field, it is most convenient to regard burnup, B (the figure to be verified), as the independent variable, and the count rate ratio, R, as the dependent variable, so that the simple linear relationship will be of the form

$$R = mB + n.$$

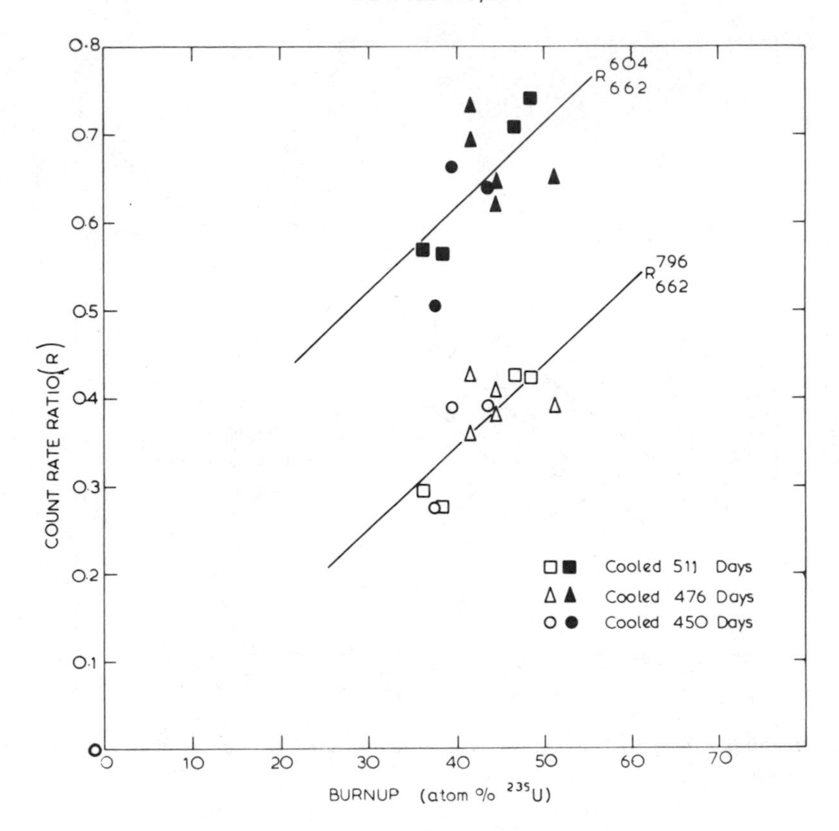

FIG.1. Count-rate ratios R^{604}_{662} and R^{796}_{662} plotted against burnup.

From the experimental results, the following expressions are derived by the linear least squares method:

This work

$$R^{604}_{662} = 0.010\ B + 0.230 \qquad (1)$$

$$R^{796}_{662} = 0.009\ B - 0.030 \qquad (2)$$

Dragnev, et al.

$$R^{796}_{662} = 0.010\ B - 0.040 \qquad (3)$$

(Dragnev, et al., in fact, derived expressions of the form B = m'R + n'.)

Consideration of branching ratios suggests that the slopes of equations (1) and (2) should be the same and the intercepts different. Comparison of confidence intervals for the 'true slopes' and 'true intercepts' shows this to be true, in fact, at the 95 per cent confidence levels, although the ratio between the observed intercepts is greater than the value of 1.11 expected from the tabulated branching ratios for 604 keV and 796 keV radiation from ^{134}Cs. This could be due, at least in part, to the use of the lead-cadmium filter.

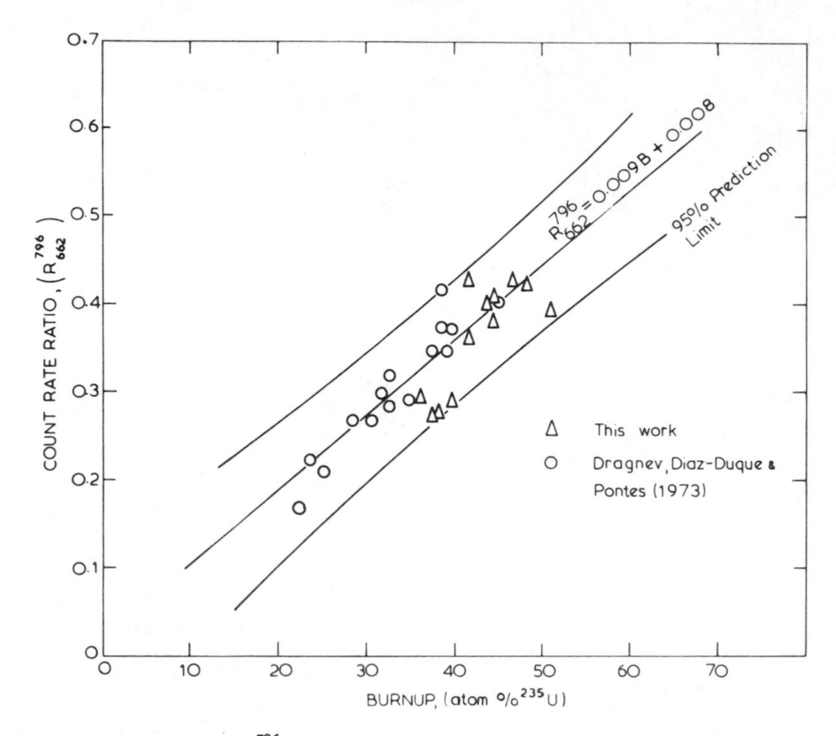

FIG.2. Measurements of R^{796}_{662} combined with those measured by Dragnev et al. [1] *and plotted against burnup.*

Likewise, there are no statistically significant differences at the 95 per cent confidence level (and higher) between the slopes and intercepts of equations (2) and (3). Although fuel assembly geometry differed from that used by Dragnev, et al. (Mark 3 elements, radial fuel plates), the quality of measurement is such that it does not show possible differences in count rate ratios arising from geometry. Thus, accepting the statistical equivalence of the present data and that of Dragnev, et al., the two sets of results may be combined to give the following regression equation:

$$R^{796}_{662} = 0.009\ B + 0.008 \qquad (4)$$

with a correlation coefficient of 0.89. The combined data are plotted in Figure 2, together with the line corresponding to equation (4) and the prediction limits giving the range in which future measurements of R for a given burnup B can be expected, with 95 per cent confidence, to fall. Prediction intervals are broad and within the range of data plotted in Figure 2, burnup estimated from a measured value of R^{796}_{662} could be in error by 20-50 per cent.

2.2 Cooling time measurement

Data on the cooling time parameters R^{696}_{662} (^{144}Ce) and R^{724}_{662} (^{95}Zr) are given in Figures 3 and 4.

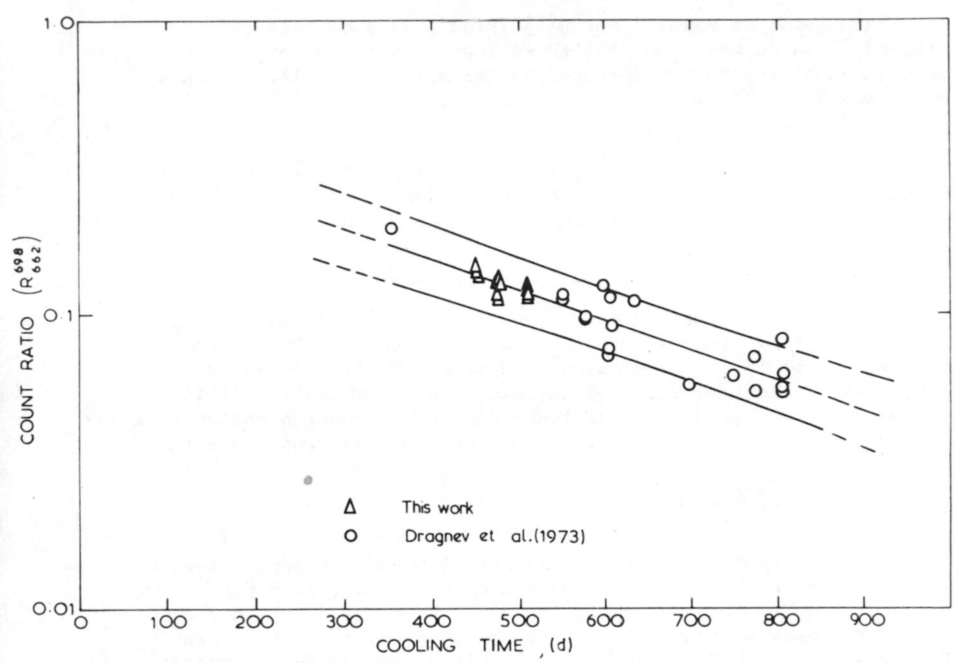

FIG.3. Count-rate ratios R^{696}_{662} combined with results of Dragnev et al. [1] and plotted against cooling time.

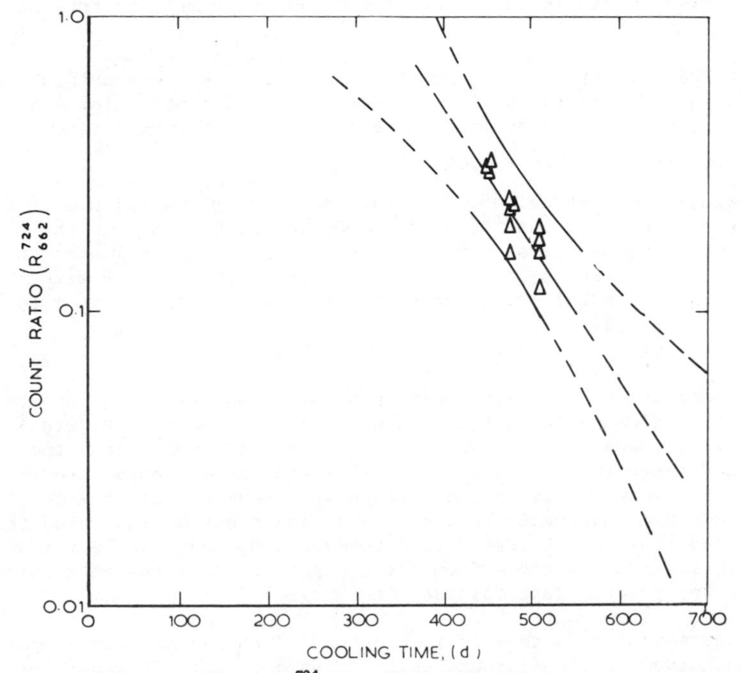

FIG.4. Count-rate ratios R^{724}_{662} plotted against cooling time after irradiation.

Inspection of results for R_{662}^{696} (Figure 3) shows that the cooling time of elements measured in this work were, with one exception, shorter than those of assemblies measured by Dragnev, et al [1]. Regression equations of the form

$$\log R_{662}^{696} = aT + b$$

(where T is cooling time in days), derived from the results of Dragnev, et al., and those of this work are statistically equivalent on the basis of 95 per cent confidence intervals. The combined results yield the regression equation

$$\log R_{662}^{696} = -0.001\ T - 0.418 \qquad (5)$$

with a correlation coefficient of -0.93. The 95 per cent confidence interval for the slope of equation (5) is -0.001 ± 0.0002 which embraces the true slope of -0.00102 calculated from tabulated decay constants. Thus, taking the true value of the slope and substituting it in the normal equations of the linear least squares regression analysis yields the following expression relating cooling time and count rate ratio R_{662}^{696}:

$$\log R_{662}^{696} = -0.00102\ T - 0.405 \qquad (6)$$

The combined experimental data are plotted in Figure 3 together with the line corresponding to equation (6). Calculated prediction limits, assuming that variances are equal to those determined from linear regression analysis of the combined data, are also shown in Figure 3 as an indication of the possible range for future measurements. It is to be noted that over the range pertaining to the combined experimental data, and with the present measurement precision, cooling times can be verified to within only about ± 100 days from measurements of R_{662}^{696}.

Since Dragnev, et al., did not measure ^{95}Zr peaks, comparison and/or pooling of results on R_{662}^{724} measurements is not possible. The regression line for the present results is given by the expression

$$\log R_{662}^{724} = 1.518 - 0.0047\ T \qquad (7)$$

with a correlation coefficient of -0.84. The true slope calculated from decay constants of ^{95}Zr and ^{137}Cs is -0.00459, which is within the 95 per cent confidence interval for the observed value. Substituting the true slope in the least squares normal equations, as was done with the R_{662}^{696} data, gives the following expression relating cooling time T to count rate ratio R_{662}^{724}:

$$\log R_{662}^{724} = 1.531 - 0.0046\ T \qquad (8)$$

The experimental data, the line corresponding to equation (8), and the 95 per cent confidence level prediction limits are shown in Figure 4. Unfortunately, the number of experimental points is small (and, the cooling time range in particular, is small) and, as expected, prediction intervals broaden markedly in the extrapolated regions either side of the experimental data. Nonetheless, within and close to the data, the cooling times could be indicated to ± 50 days compared to ± 100 days for R_{662}^{696}. This advantage derives from the steeper slope of the regression line (i.e. the greater rate of decay of ^{95}Zr).

Another advantage with R_{662}^{724} is that the ^{95}Zr peak exhibits higher peak-to-background ratios (around unity) than does the ^{144}Ce peak (around 0.16) and can be measured with slightly higher precision. On the other hand, the ratio R^{696}_{662} offers promise of verifying a wider range of cooling times because of the longer half-life of ^{144}Ce. Furthermore, the theoretical calculations described below indicate that R^{696}_{662} is somewhat less dependent on irradiation history than is R^{724}_{662}. Further measurements involving as wide a range of cooling times and burnup as possible would be necessary to establish which ratio would be better.

2.3 Measurement precision

Since prediction intervals based on both the present work and that of Dragnev, et al. [1] are very broad, precision of measurement should be improved if the method is to have adequate quantitative significance for burnup and cooling time verification. This could probably be done by using longer counting times, by taking steps to reduce Compton scatter and X-ray production in the experimental hardware, and by standardising the rotational position of the fuel elements. (Assemblies are cropped before measurement and the initially concentric fuel tubes are displaced and secured to one side of the storage container resulting in asymmetric geometry.) The improvement resulting from such precautions may not be large, however, and it is necessary to understand from the theoretical point of view whether the method is inherently capable of improved precision.

3. THEORETICAL EVALUATION

3.1 General considerations

Broadly, the gamma-emitting nuclides most suitable for measuring irradiated fissile material are fission products with high fission yields and moderate to long half-lives. Several workers have used individual fission products as an indirect measure of burnup, but such measurements require careful calibration against standards, and careful attention to geometry, sample size and shape. The use of ratios of fission product nuclides, both nuclides being measured simultaneously with the one detector and spectrometer, avoids the need for making absolute measurements. Consequently, it is a desirable approach for safeguards verification work, provided that there is adequate correlation with the parameters to be verified. Several workers other than Dragnev, et al. (Rasmussen, Sovka & Mayman [4], Hick & Lammer [5], Oden & Christensen [6], and Heath [7]), have considered the use of nuclide ratios for measuring irradiation parameters, but only the use of ratios is examined here.

For burnup verification, a useful nuclide ratio (R_B) must vary with increasing burnup, and the sensitivity of the verification capability will depend on the rate of change of R_B with burnup. Ideally, the following conditions should be met.

(a) The half-life of at least one of the nuclides should be long in comparison to the irradiation time, so that the ratio does not saturate, and so that it is relatively independent of irradiation time, including shutdowns in an interrupted irradiation.

(b) Fission yields should be high and neutron absorption cross sections low, so that high net yields will result and facilitate measurement.

TABLE II FISSION PRODUCT NUCLIDES, DECAY CONSTANTS, FISSION YIELDS AND NEUTRON ABSORPTION CROSS-SECTIONS

Nuclide Ratio	Decay Constants		Fission Yields		Neutron Absorption Cross-Sections		$\sigma \cdot \phi$ ($\phi=10^{14}$ neutrons cm^{-2} s^{-1})	
A_1/A_2	λ_1	λ_2	Y_1	Y_2	σ_1 $x10^{-24}$ cm^2	σ_2 $x10^{-24}$ cm^2	$\sigma_1\ \phi$	$\sigma_2\ \phi$
$^{95}Zr/^{137}Cs$	$1.2x10^{-7}$	$7.3x10^{-10}$	6.5	6.3	1*	0.11	$1x10^{-10}$	$1.1x10^{-11}$
$^{106}Ru/137Cs$	$2.2x10^{-8}$	$7.3x10^{-10}$	0.39	6.3	0.146	0.11	$1.46x10^{-11}$	$1.1x10^{-11}$
$^{134}Cs/^{137}Cs$	$1x10^{-8}$	$7.3x10^{-10}$	0	6.3	134	0.11	$1.34x10^{-8}$	$1.1x10^{-11}$
$^{144}Ce/^{137}Cs$	$2.8x10^{-8}$	$7.3x10^{-10}$	5.4	6.3	1*	0.11	$1x10^{-10}$	$1.1x10^{-11}$
$^{95}Zr/^{106}Ru$	$1.2x10^{-7}$	$2.2x10^{-8}$	6.5	0.39	1*	0.146	$1x10^{-10}$	$1.46x10^{-11}$
$^{134}Cs/^{106}Ru$	$1x10^{-8}$	$2.2x10^{-8}$	0	0.39	134	0.146	$1.34x10^{-8}$	$1.46x10^{-11}$
$^{144}Ce/^{106}Ru$	$2.8x10^{-8}$	$2.2x10^{-8}$	5.4	0.39	1*	0.146	$1x10^{-10}$	$1.46x10^{-11}$
$^{134}Cs/^{95}Zr$	$1x10^{-8}$	$1.2x10^{-7}$	0	6.5	134	1	$1.34x10^{-8}$	$1x10^{-10}$
$^{134}Cs/^{144}Ce$	$1x10^{-8}$	$2.8x10^{-8}$	0	5.4	134	1*	$1.34x10^{-8}$	$1x10^{-10}$
$^{95}Zr/^{144}Ce$	$1.2x10^{-7}$	$2.8x10^{-8}$	6.5	5.4	1*	1*	$1x10^{-10}$	$1x10^{-10}$

* Assumed value. Tabulated values of $\sigma\phi$ are generally much less than λ. Hence calculations will only be significantly affected if true value is much greater, say 100 times, than the assumed value.

(c) Half-lives should be long enough to permit measurement to be made over the cooling times during which safeguards verification might be required (for HIFAR fuels, up to say 600 days).

(d) The net accumulation rate of each nuclide must be different (to yield a variation of R_B with burnup), but this will almost inevitably result from differences in fission yield, half-life and neutron absorption cross section.

For verification of cooling time, the requirements are different:

(e) The ratio (R_C) should be independent of burnup and irradiation time (including shutdown times), hence the two nuclide concentrations should saturate early in the irradiation. (The remote possibility that the two could increase during irradiation at rates which would preserve a constant ratio is not discussed here.)

(f) The nuclides must have different half-lives to give a variation of R_C with cooling time, but half-lives should be reasonably long (say not less than 60 days) to permit verification of long cooling (storage) time (i.e. 10 half-lives is 600 days or longer).

(g) The nuclides should have high neutron absorption cross sections to cause saturation early in the irradiation (see (e) above) despite the need (f) above) for long half-lives.

Fission product nuclides which have been used by workers in this field are as follows (fission yields, half-lives and decay gamma energies, in that order, are also shown):

^{141}Ce	6%	32.5 d	145 keV
^{140}Ba	6.35%	12.8 d	537 keV
^{140}Ba-^{140}La*	6.35%	12.8 d	1600, 490 & 815 keV
^{106}Ru-^{106}Rh*	0.24%	1.01 y	622 keV
^{137}Cs	6.15%	30.2 y	662 keV
^{144}Ce-^{144}Pr*	4.5%	284 d	696 & 1489 keV
^{95}Zr	6.1%	64 d	724 keV
^{95}Zr-^{95}Nb*	6.1%	64 d	766 keV
^{154}Eu	0.164%	16 y	1004 & 1274 keV.

(*Secular equilibrium)

To these may be added ^{134}Cs (2.06 y, 604 and 796 keV) which is not a fission product, but is produced in measurable yields from neutron capture by the stable fission product ^{133}Cs (5.78 per cent fission yield).

Consideration here is confined to those nuclides which could actually be measured with the experimental technique described above, namely those with half-lives of 65 days and longer and having gamma energies above 600 keV. These are $^{106}Ru/Rh$, ^{134}Cs, ^{137}Cs, $^{144}Ce/Pr$ and ^{95}Zr. Ratios of these nuclides are tabulated together with decay constants, fission yields and neutron absorption cross sections in Table II. Also

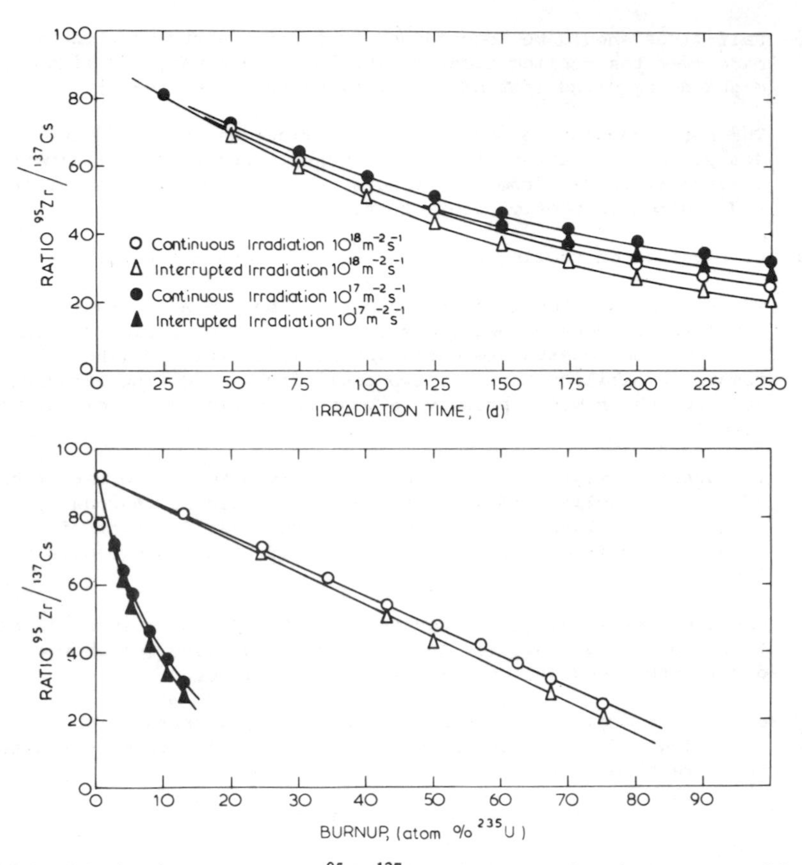

FIG.5. Calculated count-rate ratio $^{95}Zr/^{137}Cs$ as functions of irradiation time and burnup.

tabulated is the product of absorption cross section and a neutron flux of 10^{14} cm^{-2} s^{-1} which is an index of the rate of burnup by neutron absorption. Comparison of the latter term with the decay constant indicates whether the loss of nuclide during irradiation occurs mainly by radioactive decay or by burnup. (The comparison shows that in most cases decay is the dominant factor.)

The data in Table II suggest (bearing in mind the requirements (a) to (g) listed above, and for typical irradiation times of 100 to 175 days) that the most favourable ratio for verifying cooling time would be $^{144}Ce/^{106}Ru$ followed in order by $^{95}Zr/^{106}Ru$, $^{95}Zr/^{137}Cs$ and $^{144}Ce/^{137}Cs$. The most suitable burnup verification ratios are not apparent from a simple examination of Table II. In either case, cooling time or burnup ratios using ^{95}Zr will not be suitable where very long cooling times are involved because of the decay of ^{95}Zr.

To permit a more thorough evaluation based on nuclides that are readily measurable, a computer program was written to show the variation of ratio with burnup, irradiation time, cooling time and neutron flux.

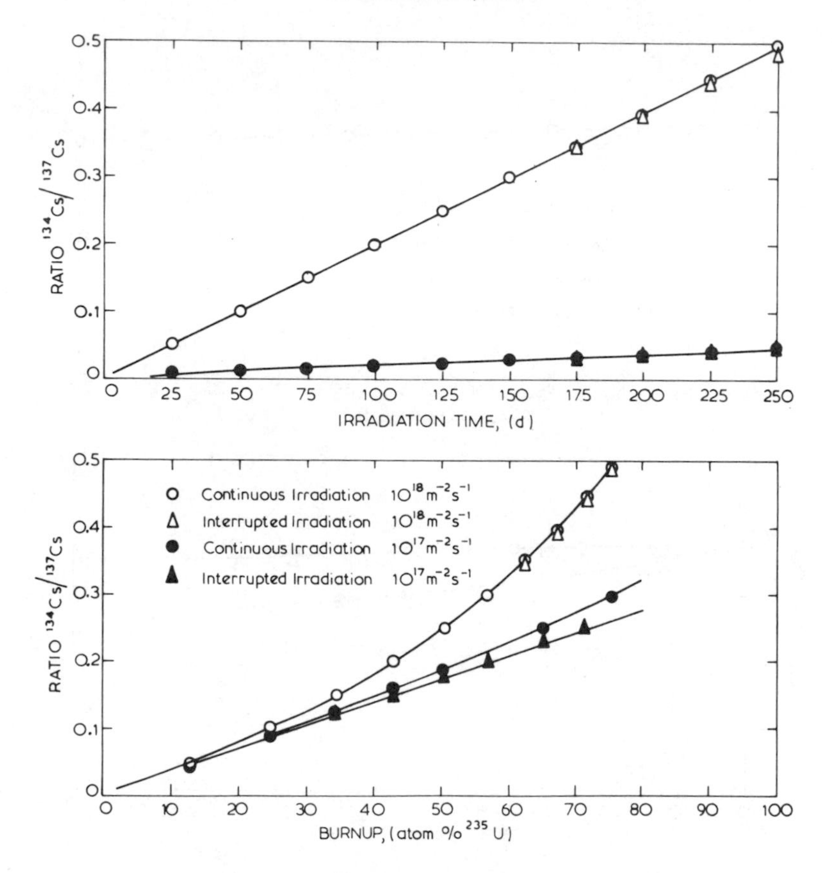

FIG. 6. Calculated count-rate ratio $^{134}Cs/^{137}Cs$ as functions of irradiation time and burnup.

It was possible to examine the more basic relationships between nuclide ratios and irradiation parameters, and to compare the experimental measurements on fuel assemblies with values calculated from irradiation history data supplied by the reactor operator.

The mathematical approach was based on the fundamental equation for fission product formation,

$$\frac{dN_f}{dt} = Nu\ \sigma_f\ \lambda - N_f\ \lambda_f - N_f\ \sigma_f\ \phi$$

This was developed to calculate a count rate ratio for two cases: firstly when the two nuclides contributing to the ratio were radioactive fission products and secondly when one nuclide was the neutron activation product of a stable fission product (i.e. ^{134}Cs).

3.2 General relationships

The predicted variation of three ratios with irradiation time and burnup are shown in Figures 5 to 7. Figure 5 for the ratio $^{95}Zr/^{137}Cs$,

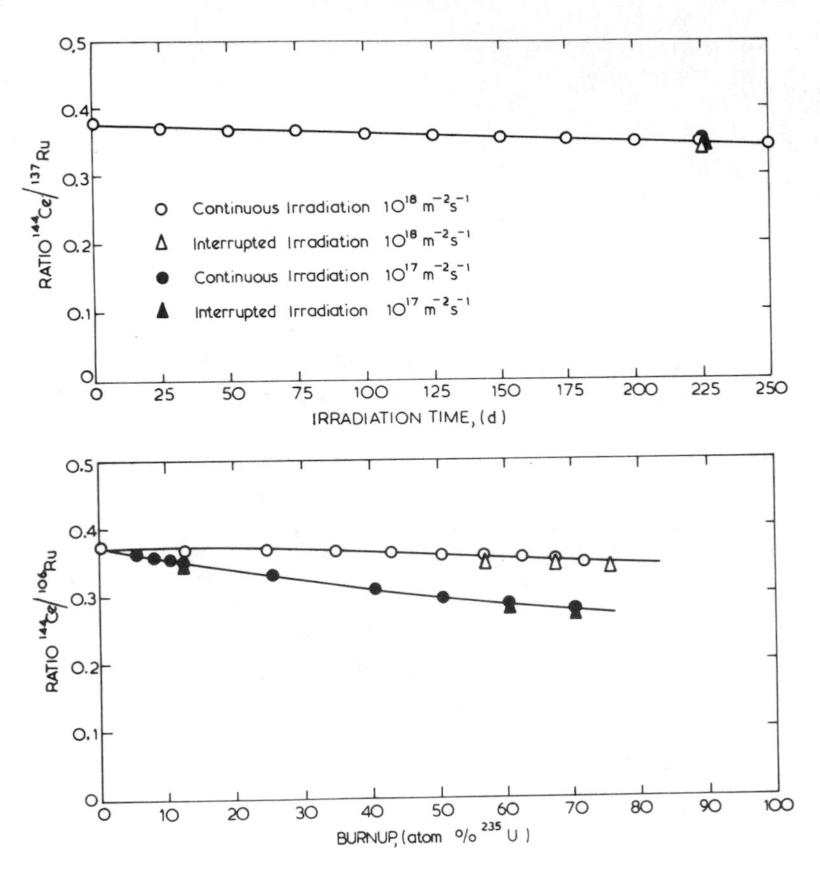

FIG. 7. Calculated count-rate ratio $^{144}Ce/^{106}Ru$ as functions of irradiation time and burnup.

is also illustrative of the behaviour of the ratios $^{95}Zr/^{137}Cs$, $^{106}Ru/^{137}Cs$, $^{144}Ce/^{137}Cs$ and $^{95}Zr/^{106}Ru$, and the behaviour of the ratio $^{134}Cs/^{95}Zr$ was similar to that shown in Figure 6 for the ratio $^{134}Cs/^{137}Cs$. These figures also show the effect of neutron flux by allowing comparison of curves for fluxes of 10^{13} cm^{-2} s^{-1} and 10^{14} cm^{-2} s^{-1}, the latter approximating closely to that obtaining in practice at core centreplane in HIFAR. In general, the effect of a ten-fold increase in flux is a markedly higher count rate ratio for a given burnup. The ratio which appears least affected by flux density is the $^{134}Cs/^{137}Cs$ ratio (i.e. the one chosen by Dragnev, et al. [1] for empirical development); but even here, Figure 6 suggests, for example, that a measured ratio of 0.25 could mean a burnup of either 50 per cent or 70 per cent depending on whether the flux was 10^{13} or 10^{14} cm^{-2} s^{-1}. Paoletti Gualandi & Peroni [8] and Foggi, Frenquellucci & Perdisa [9] have also observed this ratio to be dependent on power level (hence neutron flux) in light water reactor (LWR) fuel pins.

Except for $^{134}Cs/^{137}Cs$, all ratios studied would more nearly serve as indicators of irradiation time than burnup but, given that ratios could be measured very accurately, the data suggest that within the range of neutron flux studied, estimates of irradiation time could be in error by 20 per cent. It has been suggested by Hick & Lammer [5] that the $^{134}Cs/^{137}Cs$ ratio could be used to measure, albeit with low accuracy, the integrated neutron flux. The present calculations suggest that neither this ratio nor any of the others considered would serve for this purpose.

The only ratio showing promise as a cooling time monitor is the $^{144}Ce/^{106}Ru$ ratio. Figure 7 indicates that this ratio decreased only very slightly with burnup and irradiation time at an irradiation flux of 10^{14} cm^{-2} s^{-1}, although the rate of change with burnup (but not irradiation time) is considerably greater at a flux of 10^{14} cm^{-2} s^{-1}. Unfortunately the ratio changes only very slowly with time after reactor shutdown and measurement of the ratio to a precision (one standard deviation) of two per cent (the best level achieved in the experimental results described) would verify cooling time to only ± 100 days. This is comparable to the capability of the present empirical method and, if measurement precision could be improved, say to one per cent standard deviation, the ratio $^{144}Ce/^{137}Cs$ might prove useful.

3.3 Comparison of calculated and measured ratios

The computer program was used also to calculate fission product count rate ratios expected from a knowledge of the irradiation history (i.e. operator's data) of the HIFAR fuel assemblies measured and reported above. Calculations were done for the central region of the assembly corresponding to the peak neutron flux and the region measured by gamma spectrometry. Calculated ratios were adjusted approximately for differential attenuation effects of the lead-cadmium filter, and for the variation of detector efficiency with gamma energy, by determining correction factors from data compiled by Hanna, Walker & Beach [10] for a Ge-Li detector of similar volume to the one used in this work.

Reasonable to good agreement between calculated and observed values was obtained for ratios utilising ^{137}Cs as one nuclide but, in other cases, agreement was less satisfactory (see Table III). Greater attention to the detail of irradiation times and neutron fluxes, together with more precise measurement of count rates might result in improved agreement. The scatter in the present experimental results precludes the estimation of systematic differences between measured and calculated figures which might arise from inaccurate data for branching ratios, or fissions yields, etc.

It should be stressed that although the best available flux data were used for the calculations, it is not known whether these data adequately reflected flux perturbations resulting, for example, from irradiation experiments. Such perturbations can be quite severe and, if present, could give rise to large discrepancies between calculated and observed data.

4. CONCLUSIONS

The most important point emerging from this study is that ratios of count rates from the most readily measured fission product nuclides in

TABLE III COMPARISON OF CALCULATED AND OBSERVED COUNT RATE RATIOS

FUEL ASSEMBLY NO.	PEAK BURNUP % ^{235}U	COUNT RATE RATIOS											
		^{134}Cs*/^{137}Cs		$^{144}Ce/^{137}Cs$		$^{95}Zr/^{137}Cs$		$^{144}Ce/^{106}Ru$		$^{95}Zr/^{106}Ru$		$^{95}Zr/^{144}Ce$	
		obs	calc	obs	calc	obs	calc	obs	calc	obs	calc	obs	calc
45A-211	56.6	0.423	0.221	0.395	0.324	26.85	33.0	2.605	.299	176.8	30.7	67.9	101.1
216	54.1	0.428	0.206	0.403	0.360	35.99	35.0	1.675	.304	149.5	32.0	89.2	105.2
223	41.95	0.294	0.145	0.436	0.350	39.43	45.5	1.840	.304	166.7	39.3	90.5	128.2
227	44.9	0.276	0.163	0.380	0.345	43.40	42.9	1.695	.304	193.5	37.7	114.5	122.5
228	50.4	0.40	0.185	0.411	0.334	35.55	38.0	1.760	.302	151.9	34.3	86.5	113.4
229	47.8	0.428	0.172	0.398	0.345	30.55	40.3	1.840	.304	141.2	35.9	76.7	117.4
233	45.8	0.410	0.163	0.410	0.345	42.35	42.1	1.605	.304	165.8	37.0	103.5	117.4
414- 12	50.7	0.398	0.1875	0.435	0.339	40.75	37.7	1.310	.304	122.8	34.3	93.5	111.8
13	62.1	0.391	0.259	0.350	0.308	25.54	28.3	1.855	.298	130.2	27.3	70.0	91.2
16	51.35	0.409 0.380	0.191	0.360 0.395	0.334	34.35 34.85	37.2	1.290 1.650	.304	145.6	33.7	95.8 88.5	110.5
18	44.1	0.273	0.156	0.406	0.345	35.80	43.5	1.567	.304	138.3	38.1	88.5	124.1

* 796 keV line for ^{134}Cs.

HIFAR spent fuel are markedly dependent on the neutron flux density during irradiation. Hence, it must be concluded that:

(i) safeguards verification of burnup based on measurement of fission product ratios has little meaning without verification of neutron flux;

(ii) any empirical relationship between fission product ratios and burnup must be determined for a particular neutron flux and applied only where flux is known to conform to this value; and

(iii) much of the scatter in experimental results is probably due to differences in neutron flux in the different core positions in which fuel elements were irradiated.

It appears also that none of the nuclides considered permit a choice of a ratio which is suitable for cooling time verification because, with one exception, they are very dependent on burnup and/or neutron flux. The ratio $^{144}Ce/^{106}Ru$ appears to be relatively independent of burnup at a flux of 10^{14} cm^{-2} s^{-1}, but this is not true at a flux of 10^{13} cm^{-2} s^{-1}. Furhermore, its rate of change with time after irradiation is low and estimates of cooling time derived from it (for irradiations at fluxes giving independence from burnup) would be approximate (e.g. ± 100 days or worse) at the best counting precisions so far achieved with the available measurement technique.

On the other hand, the agreement achieved in this preliminary work between calculated and measured values is encouraging and suggests that a more detailed study, aiming at more precise measurement of ratios, more precise knowledge of irradiation parameters and perhaps a more detailed mathematical model, might show the way to a method which will be more valuable in safeguards verification. It seems highly likely, however, that information of quantitative value to the safeguards inspector will only be obtained if he has knowledge of much of the detail of irradiation history of individual fuel assemblies and this might complicate safeguarding operations even further.

REFERENCES

[1] DRAGNEV, T., DIAZ-DUQUE, R., PONTES, B., Safeguards Gamma Measurements on Spent MTR Fuel, IAEA/STR-41 (1973).

[2] BEETS, C., DRAGNEV, T., HECO, R., Trans. Am. Nucl. Soc., 15 2 (1972) 673.

[3] DRAGNEV, T., BURGESS, K., Gamma Measurements on Spent Fuel at NPD, Canada, IAEA/STR-39 (1973).

[4] RASMUSSEN, N.C., SOVKA, J.A., MAYMAN, S.A. "The non-destructive measurement of burnup by gamma ray spectroscopy". IAEA Symposium on Nuclear Materials Management, Vienna (1965).

[5] HICK, H., LAMMER, M. "Interpretation of gamma-spectrometric measurements on burnt fuel elements". IAEA Symposium on Safeguards Techniques, Karlsruhe (1970).

[6] ODEN, D.R., CHRISTENSEN, D.E. Application of Gamma-ray Spectrometry as a Supplementary MIST Technique. BNWL-SA-4059 (1971).

[7] HEATH, R.L. The Potential of High Resolution Gamma Ray Spectrometry for Assay of Irradiated Reactor Fuel. WASH-1076 (1967).

[8] PAOLETTI GUALANDI, M., PERONI, P., "Determination of burnup and plutonium content by gamma-spectrometry measurements of radioactive fission products". IAEA Symposium on Safeguarding Nuclear Materials, Vienna (1975).

[9] FOGGI, C., FRENQUELLUCCI, F., PERDISA, G. "Isotope correlations based on fission product nuclides in LWR irradiated fuels". IAEA Symposium on Safeguarding Nuclear Materials, Vienna (1975).

[10] HANNA, G.L., WALKER, D.G., BEACH, P.M. Full Energy Peak Efficiencies for Three Gamma Ray Detectors. AAEC/TM384 (1967).

К ВОПРОСУ ОБ ОПРЕДЕЛЕНИИ ВЫГОРАНИЯ ТОПЛИВА ИЗ ОТНОШЕНИЯ АКТИВНОСТЕЙ $^{134}Cs/^{137}Cs$

Б.А.БИБИЧЕВ, В.П.МАЙОРОВ, Ю.М.ПРОТАСЕНКО,
П.И.ФЕДОТОВ
Радиевый институт им.В.Г.Хлопина,
Ленинград

М.А.САНЧУГАШЕВ
Нововоронежская атомная электростанция
им.50-летия СССР,
Нововоронеж,
Союз Советских Социалистических Республик

Abstract–Аннотация

THE PROBLEM OF DETERMINING FUEL BURNUP FROM THE $^{134}Cs/^{137}Cs$ ACTIVITY RATIO.

The authors cite data on measurement of the burnup and the $^{134}Cs/^{137}Cs$ activity ratio for VVER-365 and VVER-440 reactor assemblies with different initial enrichment (3.0, 3.3 and 3.6%) and with different mean burnup of the fuel. They describe a method of calculating the ^{137}Cs and ^{134}Cs coefficients, making allowance for the loss of ^{137}Cs and ^{134}Cs over the period preceding irradiation of the assemblies under consideration. When calculating these coefficients allowance was made for variation in the thermal neutron flux density and hardness of the neutron spectrum as the burnup in the fuel assemblies proceeded. A correlation analysis is made for each assembly and for the experimental data as a whole. It is demonstrated that there is a linear dependence between the burnup and $^{134}Cs/^{137}Cs$ activity ratio throughout the burnup range studied (0.7–4.1 at.%) that, within the limits of measurement error, is not a function of the initial fuel enrichment.

К ВОПРОСУ ОБ ОПРЕДЕЛЕНИИ ВЫГОРАНИЯ ТОПЛИВА ИЗ ОТНОШЕНИЯ АКТИВНОСТЕЙ $^{134}Cs/^{137}Cs$.

Приведены результаты измерения выгорания и отношения активностей $^{134}Cs/^{137}Cs$ в кассетах реакторов ВВЭР-365 и ВВЭР-440 с различным начальным обогащением (3,0; 3,3 и 3,6%) и с различным средним выгоранием топлива. Описан способ вычисления коэффициентов C_{137} и C_{134}, учитывающих убыль ^{137}Cs и ^{134}Cs, произошедшую в течение предшествовавшего облучения исследуемых кассет. При вычислении этих коэффициентов учитывалось изменение плотности потока тепловых нейтронов и жесткости спектра нейтронов по мере выгорания топлива в кассетах. Проведен анализ корреляции для каждой из кассет и для всей совокупности экспериментальных данных. Показано, что имеет место линейная зависимость выгорания от отношения активностей $^{134}Cs/^{137}Cs$ во всем исследуемом диапазоне выгорания от 0,7 до 4,1 ат %, которая в пределах погрешности измерений не зависит от начального обогащения топлива.

1. ВВЕДЕНИЕ

Определение выгорания топлива из отношения активностей $^{134}Cs/^{137}Cs$ имеет ряд важных преимуществ [1] по сравнению с обычным методом, основанным на измерении концентраций осколочных изотопов [2]. Основное преимущество этого метода состоит в том, что для измерения отношения активностей $^{134}Cs/^{137}Cs$ не нужно определять абсолютную эффективность измерительной установки. Это особенно важно в том случае, когда измеряемым объектом является твэл или кассета энергетического реактора. Однако для практического использования этого метода необходимо установать существование однозначной зависимости выгорания от отношения активностей $^{134}Cs/^{137}Cs$, по крайней мере, для топлива реакторов данного типа.

При исследовании выгорания топлива в кассетах реакторов типа PWR и BWR была найдена зависимость этой корреляции от предшествовавшего облучения [3, 4], от обогащения топлива [4, 5] и от спектра нейтронов [3]. Влияние предшествовавшего облучения можно устранить введением поправки на убыль ^{134}Cs и ^{137}Cs за время облучения данной кассеты [4]. Влияние же обогащения топлива и спектра нейтронов учесть значительно сложнее и это ограничивает возможность применения отношения активностей $^{134}Cs/^{137}Cs$ для измерения выгорания топлива энергетических реакторов.

2. МЕТОДИКА И ТЕХНИКА ИЗМЕРЕНИЯ

Для исследования корреляции выгорания и отношения активностей $^{134}Cs/^{137}Cs$ в топливе реакторов типа ВВЭР были измерены концентрации ^{137}Cs и ^{134}Cs в пяти кассетах реакторов ВВЭР-365 и ВВЭР-440 с различным обогащением топлива и с различным средним выгоранием. Номера кассет, тип реактора, обогащение топлива по ^{235}U, число циклов облучения с постоянной мощностью реактора и расчетное среднее выгорание топлива в кассетах приведены в табл. I.

В каждой кассете исследовалось 26 твэлов, расположенных по двум взаимно перпендикулярным диаметрам кассеты. Картограмма расположения исследуемых твэлов в одной из кассет приведена на рис. 1. В десяти точках по высоте этих твэлов измерялись гамма-спектры осколочных изотопов. Измерения гамма-спектров проводились с щелевым свинцовым коллиматором и Ge(Li)-детектором [6]. В качестве эталонного твэла использовался один из исследуемых твэлов каждой кассеты. После окончания измерений кассеты из эталонного твэла в месте измерений вырезался образец, растворялся и в аликвоте раствора измерялись концентрации урана, ^{137}Cs и ^{134}Cs. Погрешность измерения содержания ^{137}Cs и ^{134}Cs на единицу веса урана в растворах составила около 2% и погрешность измерения концентраций этих продуктов деления в данной точке по высоте твэлов – около 7%.

ТАБЛИЦА I. ОСНОВНЫЕ ПАСПОРТНЫЕ ДАННЫЕ О ПРЕДШЕСТВОВАВШЕМ ОБЛУЧЕНИИ ИССЛЕДУЕМЫХ КАССЕТ

Номер кассеты	Тип реактора	Обогащение топлива по ^{235}U (вес %)	Число циклов облучения с постоянной мощностью	Расчетное среднее выгорание топлива (ат %)	Время облучения (календарные сутки)
РП-3 № 223	ВВЭР-365	3	7	2,88	1032
ДР-3 № 80	ВВЭР-365	3	1	1,23	342
ОИ-3М-5 № 12	ВВЭР-365	3	3	3,41	1004
РП-3,3 № 71А	ВВЭР-440	3,3	5	1,43	416
РП-3,6 № 213	ВВЭР-440	3,6	3	3,11	1213

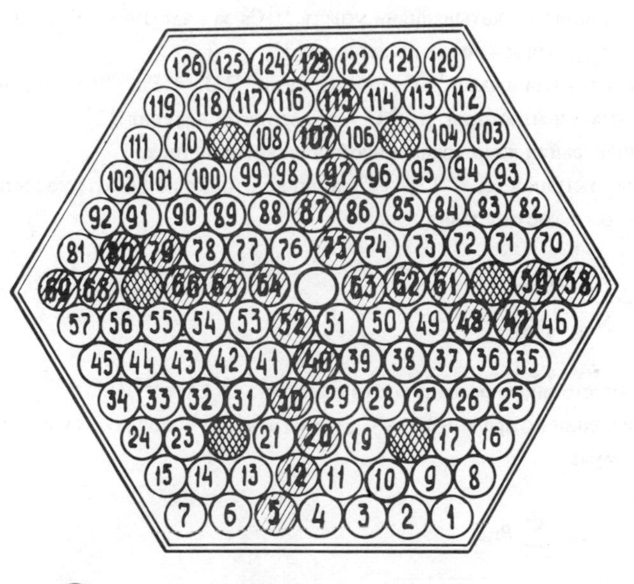

ПОГЛОЩАЮЩИЕ ЭЛЕМЕНТЫ

ИССЛЕДУЕМЫЕ ТВЭЛЫ

Рис.1. Картограмма расположения исследуемых твэлов в кассете РП-3,3 № 71А.

3. ОБРАБОТКА РЕЗУЛЬТАТОВ ИЗМЕРЕНИЙ

Измеренные в десяти точках по высоте 26 твэлов значения концентраций ^{137}Cs и ^{134}Cs – N_{137} и N_{134} использовались далее для вычисления средних значений этих величин по сечению кассеты – $\overline{N}_{137}$ и $\overline{N}_{134}$. При вычислении $\overline{N}_{137}$ предполагалось, что имеется радиальная зависимость $\overline{N}_{137}$ и $\overline{N}_{134}$ от расстояния между осью кассеты и осью твэла, а азимутальной зависимостью этих величин пренебрегалось. Погрешность, связанная с таким способом вычисления $\overline{N}_{137}$ и $\overline{N}_{134}$, проверялась экспериментально. Для этого в кассете РП-3,6 № 213 в одной точке по высоте были измерены значения N_{137} и N_{134} во всех 126 твэлах. Оказалось, что $\overline{N}_{137}$ и $\overline{N}_{134}$, вычисленные по результатам измерения этих величин в 26 и 126 твэлах, отличаются менее чем на 1%, в то время как суммарная погрешность определения $\overline{N}_{137}$ и $\overline{N}_{134}$ в данной точке по сечению кассеты составила около 6%.

Выгорание топлива в данной точке по высоте кассеты W вычислялось из следующего соотношения:

$$W = \frac{\overline{N}_{137} \cdot C_{137}}{\overline{Y}_{137} \cdot N_U^0} \cdot 100 \ [\text{ат}\ \%] \tag{1}$$

где C_{137} – коэффициент, учитывающий убыль ^{137}Cs за кампанию облучения данной кассеты в данной точке по высоте [2],

$\overline{Y}_{137}$ – эффективный выход ^{137}Cs при делении ^{235}U, ^{239}Pu, ^{241}Pu, облучаемых тепловыми нейтронами, и ^{238}U – нейтронами деления [2],

N_U^0 – концентрация ядер урана в необлученном топлива.

Отношение активностей $^{134}Cs/^{137}Cs$ в данной точке по высоте кассеты вычислялось из следующего соотношения:

$$\eta = \frac{\lambda_{134} \cdot \overline{N}_{134} \cdot C_{134}}{\lambda_{137} \cdot \overline{N}_{137} \cdot C_{137}} \tag{2}$$

где λ_{134}, λ_{137} – постоянные распада соответственно ^{134}Cs и ^{137}Cs.

Для вычисления коэффициентов C_{137} и C_{134} часто используется следующая приближенная формула:

$$C_i = \frac{\lambda_i \cdot \sum_{n=1}^{N} \overline{R}_n t_n}{\sum_{n=1}^{N} \overline{R}_n e^{-\lambda_i \Theta_n} \cdot (1 - e^{-\lambda_i t_n})} \tag{3}$$

где $\overline{R}_n$ – средняя относительная тепловая мощность кассеты за n-й цикл облучения,

λ_i – постоянная распада i-го продукта деления,

t_n – длительность n-го цикла облучения,

Θ_n – интервал времени от конца n-го цикла облучения до начала измерения.

ТАБЛИЦА II. РЕЗУЛЬТАТЫ ВЫЧИСЛЕНИЯ КОЭФФИЦИЕНТОВ C_{137} и C_{134} (НА ДАТУ ОСТАНОВКИ РЕАКТОРА) ДЛЯ КАССЕТЫ РП-3 № 223 С ИСПОЛЬЗОВАНИЕМ ПРОГРАММЫ, ИЗЛОЖЕННОЙ В РАБОТЕ [7], И ФОРМУЛЫ (3)

Положение по высоте	C_{137}	C_{134}
1	1,037	1,425
2	1,037	1,465
3	1,037	1,473
4	1,037	1,476
5	1,037	1,474
6	1,037	1,473
7	1,037	1,468
8	1,037	1,465
9	1,037	1,448
10	1,037	1,410
Расчеты по формуле (3)	1,032	1,560

При выводе этой формулы предполагается, что скорость образования данного продукта деления пропорциональна тепловой мощности кассеты, и пренебрегается убылью данного изотопа в ядерных реакциях.

Следует отметить, что ^{134}Cs не является прямым продуктом деления, и кроме того скорость образования ^{134}Cs должна зависеть не только от мощности реактора, но и от спектра нейтронов, поскольку ^{133}Cs имеет большой резонансный интеграл захвата нейтронов. Поэтому для более корректного вычисления коэффициентов C_{137} и C_{134}, а также для вычисления $\bar{\bar{Y}}_{137}$ использовался полуэмпирический метод расчета изотопного состава топлива [7]. Для упрощения вычислений реальная кампания облучения каждой кассеты разбивалась на несколько циклов облучения с приблизительно постоянной тепловой мощностью, которая известна в каждом цикле из теплофизических измерений.

Начальные значения плотности потока тепловых нейтронов Φ_0 и жесткости спектра нейтронов α_0 в данной точке по высоте кассеты находились с помощью процедуры подгонки измеренных значений N_{137} и $N_{134}/(N_{137})^2$ и вычисленных значений этих величин. Найденные таким образом значения Φ_0 и α_0 использовались далее для вычисления C_{137}, C_{134} и выгорания (W).

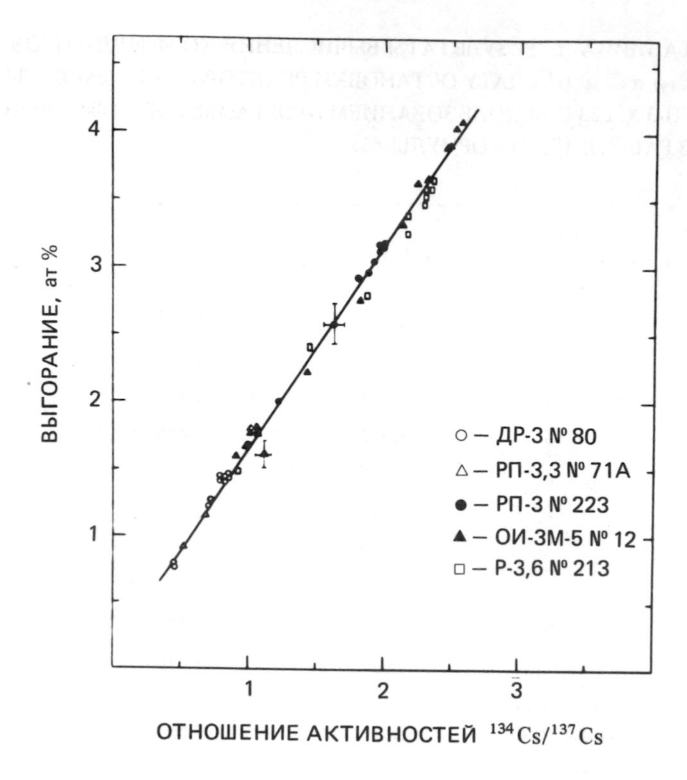

Рис. 2. Зависимость выгорания топлива от отношения активностей $^{134}Cs/^{137}Cs$.

В качестве примера в табл. II приведены значения C_{137} и C_{134} (на дату остановки реактора), вычисленные в десяти точках по высоте кассеты РП-3 № 223 с помощью программы, изложенной в работе [7]. В последней строке таблицы приведены также значения C_{137} и C_{134}, вычисленные по формуле (3). Из таблицы видно, что значения C_{137} не изменяются по высоте кассеты и в пределах 0,5 % совпадают со значением C_{137}, вычисленным по формуле (3). В то же время значения C_{134} изменяются по высоте и отличаются на 6-9 % от значения C_{134}, вычисленного по формуле (3). Для кассет с меньшим средним выгоранием топлива это различие в коэффициентах C_{134} меньше.

Следует также отметить, что описанный выше способ вычисления коэффициеттов C_{137} и C_{134} с помощью программы, изложенной в работе [7], требует измерения концентрации ^{137}Cs. Это лишает метод определения W из отношения активностей $^{134}Cs/^{137}Cs$ его основного преимущества по сравнению с обычным методом, основанным на измерении концентраций ^{137}Cs и ^{106}Ru. Однако эти коэффициенты можно вычислить также с достаточной точностью и по измеренному отношению N_{134}/N_{137}. Для этого, например, можно использовать в качестве α_0 значение этой величины, вычислен-

ТАБЛИЦА III. РЕЗУЛЬТАТЫ АНАЛИЗА КОРРЕЛЯЦИИ ВЫГОРАНИЯ И ОТНОШЕНИЯ АКТИВНОСТЕЙ $^{134}Cs/^{137}Cs$

Номер кассеты	Число экспериментальных точек	Значения параметра "a"	Значения параметра "b"	Коэффициент корреляции
РП-3 № 223	10	1,58 (2)	0,02 (3)	0,999 (74)
ДР-3 № 80	10	1,75 (3)	– 0,02 (2)	0,999 (72)
ОИ-3М-5 № 12	10	1,71 (3)	– 0,29 (6)	0,999 (57)
РП-3,3 № 71А	10	1,65 (4)	0,842 (33)	0,998 (76)
РП-3,6 № 213	10	1,47 (4)	0,15 (7)	0,997 (59)
РП-3 № 223, ДР-3 № 80, ОИ-3М-5 № 12	30	1,54 (2)	0,10 (2)	0,997 (23)
Все кассеты	50	1,52 (2)	0,12 (2)	0,997 (19)

ное с помощью программы РОР, которая используется в настоящее время для расчета нейтронно-физических характеристик реакторов типа ВВЭР-365 и ВВЭР-440 [8]. В этом случае программу, изложенную в работе [7], следует использовать в режиме подгонки по одному параметру.

4. АНАЛИЗ КОРРЕЛЯЦИИ

На рис.2 приведены измеренные значения выгорания (W) для соответствующих значений отношения активностей $^{134}Cs/^{137}Cs$ (η) для всех исследуемых кассет. Видно, что все экспериментальные точки в пределах погрешности измерений ложатся на общую прямую, приблизительно проходящую через начало координат. Результаты анализа [9] корреляционных прямых $W = a\eta + b$ приведены в табл. III. В таблице приведены значения параметров "a", "b" и коэффициентов корреляции для каждой из кассет, для кассет с 3 % обогащением топлива и для всей совокупности экспериментальных точек. Из табл. III видно, что различие параметров "a" и "b" для различных кассет несколько превышает погрешности определения этих величин. Это различие может быть связано с погрешностью калибровки измерительной установки по эффективности, с различным обогащением топлива в кассетах, с различием спектра нейтронов и динамики его изменения для каждой из кассет, а также с отсутствием точных сведений о предшествовавшем облучении.

Из табл. III следует также, что значения параметров "a", "b" и коэффициента корреляции для трех кассет с одинаковым обогащением топлива и для всех исследуемых кассет совпадают в пределах погрешности определения этих величин. В связи с этим можно отметить, что сильной зависимости корреляции W и η от обогащения топлива, которая была найдена для кассет реактора типа ВВЭР-70 с обогащением 0,72; 1,5 и 2,0% [4], не наблюдается.

Таким образом, результаты измерения выгорания топлива и отношения активностей $^{134}Cs/^{137}Cs$ в кассетах реакторов типа ВВЭР-440 показали, что имеется линейная зависимость выгорания и отношения активности $^{134}Cs/^{137}Cs$ во всем исследуемом диапазоне выгорания от 0,7 до 4,1 ат%. Корреляция в пределах погрешности измерений не зависит от начального обогащения топлива и может быть использована для измерения выгорания топлива в целых кассетах этих реакторов.

ЛИТЕРАТУРА

[1] EDER, O.J., LAMMER, M., "Influence of uncertainties in fission – product nuclear data on the interpretation of gamma-spectrometric measurements on burnt fuel elements", Nuclear Data in Science and Technology (Proc. Symp. Paris, 1973), V. I, IAEA, Vienna (1974) 233.

[2] GRABER, H., HOFMANN, G., NAGEL, S., Kernenergie **17** (1974) 73.

[3] MATSUURA, S., TSURUTA, H., J. Nucl. Sci. Technol. **12** 1 (1975) 24.

[4] GRABER, H., MEHNER, H.C., Z. Kristallchem., Okt. (1977).

[5] FOGGI, C., FRENGUELLUCCI, G., PERDISA, G., "Isotope correlations based on fission-product nuclides in LWR irradiated fuels", Safeguarding Nuclear Materials (Proc. Symp. Vienna, 1975), 2, IAEA, Vienna (1976) 425.

[6] БИБИЧЕВ, Б.А., ГОЛУБЕВ, Л.И., КОВАЛЕНКО, С.С., МАЙОРОВ, В.П., СУНЧУГАШЕВ, М.А., ФЕДОТОВ, П.И., Доклады совещания специалистов стран-членов СЭВ по вопросам транспортирования отработавших твэлов и неразрушающим методам определения в них содержания делящихся материалов, М., (1976) 12.

[7] БИБИЧЕВ, Б.А., МАЙОРОВ, В.П., РАЗУВАЕВА, М.А., ФЕДОТОВ, П.И., ЛОВЦЮС, А.В., СТЕПАНОВ, А.В., "Определение выгорания и изотопного состава урана и плутония методом гамма-спектрометрии продуктов деления", Доклад SM-231/135 на настоящем Симпозиуме.

[8] СИДОРЕНКО, В.Д., БЕЛЯЕВА, Е.Д., Препринт ИАЭ-1171, М., (1966).

[9] FOGGI, C., ZIJP, W.L., "Data treatment for the isotopic correlation technique", Safeguarding Nuclear Materials (Proc. Symp. Vienna, 1975), 2, IAEA, Vienna (1976) 405.

ОБ ИСПОЛЬЗОВАНИИ СОБСТВЕННОГО И ИНДУЦИРОВАННОГО НЕЙТРОННОГО ИЗЛУЧЕНИЯ ДЛЯ НЕРАЗРУШАЮЩЕГО АНАЛИЗА ОТРАБОТАВШЕГО ТОПЛИВА ВВЭР

А.А.ВОРОНКОВ, Б.Я.ГАЛКИН,
Н.М.КАЗАРИНОВ, П.И.ФЕДОТОВ
Радиевый институт им. В.Г.Хлопина,
Ленинград

Л.И.ГОЛУБЕВ
Нововоронежская атомная электростанция
им. 50-летия СССР,
Нововоронеж,
Союз Советских Социалистических Республик

Abstract–Аннотация

THE USE OF INHERENT AND INDUCED NEUTRON RADIATION FOR NON-DESTRUCTIVE ANALYSIS OF SPENT FUEL FROM VVER POWER REACTORS.

The authors measured the neutron radiation of spent fuel resulting from the (α, n) reaction and the spontaneous fission of transuranian isotopes; in addition, they measured the photoneutron radiation generated by gamma-quanta of ^{106}Ru and ^{144}Ce on D_2O and Be targets surrounding the fuel. It was shown that the neutron radiation from fuel with a burnup of more than 20 000 MW · d/t of uranium after a cooling time of over two years is determined essentially by ^{244}Cm. A correlation is found between the neutron yield and the burnup. Fuel elements from a VVER reactor assembly were subjected to neutron scanning and the results were compared with data obtained by gamma scanning. The shape of the curve describing the photoneutron intensity distribution along the fuel element is in good agreement with the curve for distribution of ^{106}Ru and ^{144}Ce over the element.

ОБ ИСПОЛЬЗОВАНИИ СОБСТВЕННОГО И ИНДУЦИРОВАННОГО НЕЙТРОННОГО ИЗЛУЧЕНИЯ ДЛЯ НЕРАЗРУШАЮЩЕГО АНАЛИЗА ОТРАБОТАВШЕГО ТОПЛИВА ВВЭР.

Измерено нейтронное излучение отработавшего топлива, возникающее в результате (α, n)-реакции спонтанного деления трансурановых изотопов, а также фотонейтронное излучение, генерируемое гамма-квантами ^{106}Ru и ^{144}Ce на мишенях из D_2O и Be, расположенных вокруг топлива. Показано, что нейтронное излучение для топлива с выгоранием больше 20000 МВт · сут/тU и выдержкой более двух лет в основном обусловлено наличием ^{244}Cm. Обнаружена корреляция между выходом нейтронов и глубиной выгорания. Проведено нейтронное сканирование твэлов кассеты ВВЭР. Результаты нейтронного сканирования сравниваются с данными гамма-сканирования. Форма кривой, характеризующей распределение интенсивности фотонейтронов по длине твэла, хорошо согласуется с кривой распределения ^{106}Ru и ^{144}Ce по твэлу.

1. ВВЕДЕНИЕ

Определение выгорания и накопления изотопов трансурановых элементов в кассетах энергетических реакторов – одна из важнейших проблем ядерной энергетики, от решения которой в значительной степени зависит задача оптимизации топливного цикла и понимание процессов, происходящих в активной зоне реактора во время кампании облучения топлива. Определение этих характеристик отработавшего топлива играет весьма существенную роль в осуществлении системы гарантий.

Наиболее точные измерения выгорания и накопления трансурановых элементов получают при использовании радиохимических методов в сочетании с масс-спектрометрией. Однако, использование этой методики для решения указанных выше задач с практической точки зрения весьма проблематично ввиду большой ее трудоемкости. Большое число анализов, которое должно быть проведено для решения поставленной задачи, требует разработки и применения достаточно простых экспрессных методов анализа без разрушения твэлов и кассет.

Практическое осуществление количественного контроля делящихся материалов также может быть осуществлено лишь при наличии экспрессных методик анализа. Одной из наиболее важных задач, связанных с учетом всего инвентарного количества делящихся веществ, участвующих в топливном цикле, является применение аналитического контроля непосредственно в отработавших кассетах АЭС, конечно, при условии достижения необходимой точности измерений. По-видимому, только методы неразрушающего контроля или их сочетание с расчетными методами могут обеспечить необходимую производительность паспортизации облученных кассет.

2. НЕРАЗРУШАЮЩИЕ МЕТОДЫ

При разработке неразрушающих методов анализа отработавшего топлива энергетических реакторов (ВВЭР и легководных) необходимо учесть следующие особенности твэлов и кассет, как объектов такого анализа:

1) сложный изотопный состав топлива, что ограничивает применение многих методов;

2) интенсивное гамма- и нейтронное излучение обусловливает необходимость автоматизации процесса измерения и создает весьма специфические условия для работы детектирующей аппаратуры;

3) большая длина, а в случае кассет и большой диаметр измеряемых объектов, вследствие чего твэлы и кассеты имеют неодинаковое распределение делящихся изотопов по длине и диаметру. Это приводит к необходимости использовать аппаратуру сканирующего типа;

4) высокая плотность ядерного материала позволяет использовать для анализа лишь излучение с большой проникающей способностью;

5) постоянство геометрии измеряемых твэлов и кассет для АЭС данного типа может существенно уменьшить при массовом контроле влияние методических погрешностей, связанных с геометрией объекта;

6) большие количества делящихся материалов в измеряемых объектах позволяют получить необходимый статистический материал за относительно малые времена измерения.

Наибольшее распространение среди методов неразрушающего анализа отработавших твэлов и кассет в настоящее время получил гамма-спектрометрический метод определения выгорания и содержания делящихся изотопов. К настоящему времени опубликовано несколько десятков работ, посвященных как обсуждению результатов, полученных с помощью этого метода, так и исследованию специфических особенностей, характерных для него. Наиболее существенными факторами, влияющими на результаты измерений, получаемые гамма-спектрометрическим методом, являются поглощение гамма-квантов и миграция осколочных элементов. Учет поглощения гамма-квантов осложняется тем фактом, что распределение активности осколочных элементов внутри твэла или кассеты в общем-то неизвестно, хотя для легководных реакторов это утверждение, по-видимому, звучит не столь категорично. Влияние эффекта поглощения может быть продемонстрировано данными о поглощении гамма-квантов в кассетах легководных реакторов. Лишь около 25% гамма-квантов, испускаемых ^{137}Cs с энергией 661 кэВ, выходит из кассеты. Очевидно, что информация, получаемая о выгорании внутренних твэлов в кассете, весьма ограниченна.

Измерение отношения активностей, например отношения активности ^{134}Cs к активности ^{137}Cs ($^{134}Cs/^{137}Cs$), вместо измерения концентраций этих изотопов упрощает проблему, связанную с поглощением, но это отношение кроме выгорания зависит еще и от соотношения потоков эпитепловых и тепловых нейтронов. Это связано с тем, что сечение захвата тепловых нейтронов и резонансный интеграл для ^{133}Cs сильно различаются и равны 30 и 450 барнам, соответственно. Образование же ^{137}Cs, в основном, связано с потоком тепловых нейтронов.

Использование для определения выгорания вместо отношения $^{134}Cs/^{137}Cs$, отношения $^{154}Eu/^{137}Cs$ является более предпочтительным с точки зрения меньшей зависимости от предшествовавшего облучения и от соотношения потоков эпитепловых и тепловых нейтронов (сечение захвата тепловых нейтронов и резонансный интеграл для ^{153}Eu равны 450 и 1500 барнам, соответственно). Однако, в связи с тем, что выход ^{153}Eu на акт деления значительно меньше выхода ^{133}Cs, определение активности ^{153}Eu требует длительных измерений и может быть выполнено только после большого времени выдержки облученного топлива.

Однако, следует отметить еще раз, что несмотря на упомянутые сложности, метод гамма-сканирования является наиболее разработанным методом неразрушающего анализа отработавшего топлива.

В последнее время почти одновременно были выполнены работы, в которых исследовалось собственное нейтронное излучение отработавших твэлов и кассет [1-3]. В этих работах изучалась зависимость полного нейтронного потока от величины выго-

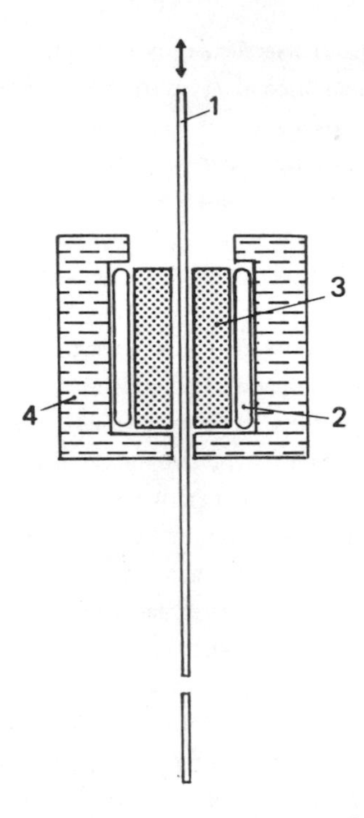

Рис.1. Схема установки для нейтронного сканирования твэлов. 1 – твэл; 2 – счетчик тепловых нейтронов; 3 – сменные мишени; 4 – замедлители нейтронов.

рания и от времени выдержки твэлов. Анализ этой зависимости показывает, что пассивный нейтронный метод анализа отработавших кассет может оказаться весьма перспективным методом определения выгорания и содержания делящихся изотопов в кассетах легководных реакторов.

Привлекательность методов, основанных на регистрации собственного нейтронного излучения топлива (нейтроны спонтанного деления и нейтроны, образующиеся в (α, n)-реакции на ^{18}O) и индуцированного нейтронного излучения (фотонейтроны, образующиеся в (γ, n)-реакции на Be и D_2O), обусловлена следующими причинами:

1) простота и надежность аппаратуры, используемой для регистрации нейтронов;

2) высокая проницаемость нейтронов, что существенно упрощает проблему, связанную с учетом эффектов поглощения при прохождении излучения через вещество кассеты;

3) высокая скорость измерений, обусловленная большой интенсивностью нейтронных потоков, испускаемых облученным топливом, и большим телесным углом $(> 2\pi)$, в котором детектируются нейтроны;

4) нейтронные измерения могут быть выполнены сразу после извлечения кассеты из реактора, в то время, как гамма-спектрометрические измерения выгорания могут быть проведены после определенного времени выдержки кассеты;

5) нейтроны, возникающие при спонтанном делении и в результате (α, n)-реакции связаны с долгоживущими изотопами урана и трансурановых элементов. Поэтому на результаты нейтронных измерений практически не влияет предшествовавшее облучение кассет в реакторе;

6) в случае использования (γ, n)-реакций на Be и D_2O фотонейтроны образуются при помощи γ-квантов (испускаемых в основном ^{144}Ce и ^{106}Ru) с энергией 1,8-2,3 МэВ, поглощение которых в веществе кассеты в 5-6 раз меньше, чем поглощение γ-квантов с энергией 500-800 кэВ, которые используются обычно в традиционных гамма-спектрометрических методах анализа.

Отмеченные преимущества нейтронных измерений стимулировали проведение нами работы по изучению характеристик нейтронного излучения отработавшего топлива ВВЭР с целью выяснения возможности использования нейтронной эмиссии для определения выгорания и накопления трансурановых элементов в топливе.

Уже первые экспериментально полученные величины потоков нейтронов при измерениях на кусках твэлов ВВЭР оказалось много выше того, что можно было ожидать в результате спонтанного деления и (α, n)-реакции от накопившегося в топливе плутония. Наблюдаемые нейтронные потоки можно объяснить лишь накоплением в отработавшем топливе заметных количеств изотопов кюрия.

Представляет также интерес измерение зависимости интенсивности потока фотонейтронов, возникающих в (γ, n)-реакции на мишенях из Be и D_2O, от величины выгорания отработавшего топлива.

3. МЕТОД НЕЙТРОННОГО СКАНИРОВАНИЯ

Суть метода состоит в следующем: если участок твэла окружить веществом-мишенью (рис.1), для которого энергия порога (γ, n) реакции равна $E^n_{\gamma, n}$, то γ-кванты продуктов деления, содержащихся в топливе, с энергией $E\gamma > E^n_{\gamma, n}$ будут генерировать в мишени фотонейтроны, которые регистрируются счетчиками фотонейтронов. В том случае, когда в качестве мишени используется замедлитель нейтронов, будет регистрироваться собственное нейтронное излучение топлива.

Собственное нейтронное излучение отработавшего топлива обусловлено, как отмечалось выше, спонтанным делением тяжелых ядер и нейтронами, образующимися в результате (α, n)-реакции на ^{18}O при α-распаде этих ядер. Нейтроны спонтанного деления можно отделить от нейтронов (α, n)-реакции, пользуясь тем обстоятельством, что при спонтанном делении в одном акте деления испускается несколько нейтронов. Используя счетчики нейтронов в режиме совпадений, можно определить среднее число нейтронов на акт спонтанного деления, что можно применить для изотопного анализа делящегося вещества. Данные по выходам нейтронов спонтанного деления и нейтронов, образующихся в результате (α, n)-реакции для различных изотопов тяжелых ядер приведены в табл. I. Из табл. I видно, что весьма интенсивными излучателями нейтронов спонтанного деления являются изотопы ^{240}Pu, ^{242}Pu, ^{242}Cm, ^{244}Cm. Эти же

ТАБЛИЦА I. НЕЙТРОННЫЕ ВЫХОДЫ ДЕЛЯЩИХСЯ ИЗОТОПОВ, НАКАПЛИВАЮЩИХСЯ В ТОПЛИВЕ

Изотоп	$T_{1/2}$ (с.д.) (лет)	$\overline{\nu}$	Число с.д. (1/г.с.)	Число нейтронов (1/г.с.)	$T_{1/2}$ (α) (лет)	Число α (1/г.с.)	Число нейтронов (окислы) (1/г.с.)
^{235}U	$3,5 \cdot 10^{17}$	1,7	$1,6 \cdot 10^{-4}$	$2,9 \cdot 10^{-4}$	$7,03 \cdot 10^{8}$	$8,00 \cdot 10^{4}$	
^{236}U	$2,9 \cdot 10^{16}$	1,9	$2,0 \cdot 10^{-3}$	$4,6 \cdot 10^{-3}$	$2,34 \cdot 10^{7}$	$2,40 \cdot 10^{6}$	
^{238}U	$8,6 \cdot 10^{15}$	1,99	$6,5 \cdot 10^{-3}$	$1,1 \cdot 10^{-2}$	$4,47 \cdot 10^{9}$	$1,25 \cdot 10^{4}$	
^{239}Pu	$5,5 \cdot 10^{15}$	2,3	$1,0 \cdot 10^{-2}$	$2,3 \cdot 10^{-2}$	24065	$2,33 \cdot 10^{9}$	38
^{240}Pu	$1,34 \cdot 10^{11}$	2,14	419	946	540	$8,43 \cdot 10^{9}$	140
^{242}Pu	$7,2 \cdot 10^{10}$	2,12	780	1600	$3,7 \cdot 10^{5}$	$1,45 \cdot 10^{8}$	2
^{241}Am	$1,15 \cdot 10^{14}$	2,4	0,50	1,15	433	$1,27 \cdot 10^{11}$	3700
^{242}Cm	$6,6 \cdot 10^{6}$	2,50	$8,2 \cdot 10^{6}$	$2,1 \cdot 10^{7}$	0,447	$1,22 \cdot 10^{14}$	$1,1 \cdot 10^{7}$
^{244}Cm	$1,34 \cdot 10^{7}$	2,69	$4,0 \cdot 10^{6}$	$1,1 \cdot 10^{7}$	18,11	$3,0 \cdot 10^{12}$	$9 \cdot 10^{4}$

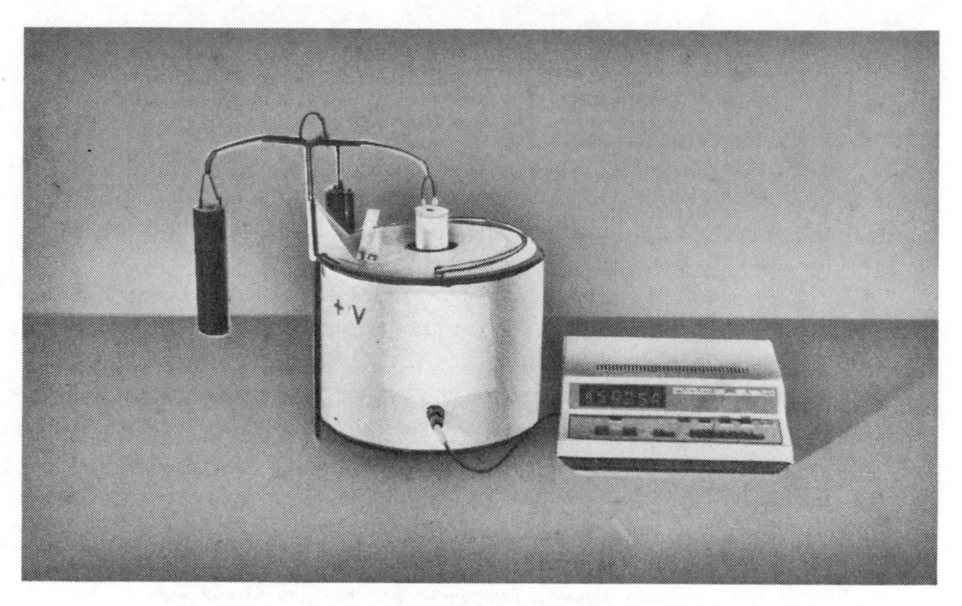

Рис. 2. Общий вид установки УНИТ-2.

изотопы вместе с ^{239}Pu и ^{241}Am являются наиболее активными источниками нейтронов, возникающих в (α, n)-реакции. В отработавшем топливе накапливаются продукты деления, которые являются интенсивными излучателями γ-квантов. Ряд изотопов излучает γ-кванты с энергиями выше порога (γ, n)-реакции на ядрах ^9Be и ^2H. Однако, для топлива с выдержкой более полугода практически только $^{106}Ru - {}^{106}Rh$ вызывает генерацию фотонейтронов на дейтерии, а $^{106}Ru - {}^{106}Rh$ и $^{144}Ce - {}^{144}Pr$ — на бериллии.

4. АППАРАТУРА

В зависимости от характера исследований нами было разработано и изготовлено несколько вариантов аппаратуры для измерения интенсивности нейтронного потока, излучаемого образцами отработавшего топлива. В установке УНИТ-1 для регистрации нейтронов использовались промышленные гелиевые счетчики нейтронов типа СНМ-12. Детектор нейтронов выполнен в виде герметичной конструкции, которая обеспечивает возможность его дезактивации без нарушения работоспособности. Детектор включает в себя кожух с замедлителем нейтронов, сборку счетчиков СНМ-12, свинцовую защиту, сменные фляги-мишени, содержащие H_2O, D_2O и Be, а также предусилители.

Регистрирующая аппаратура включает в себя два счетных канала: контрольный

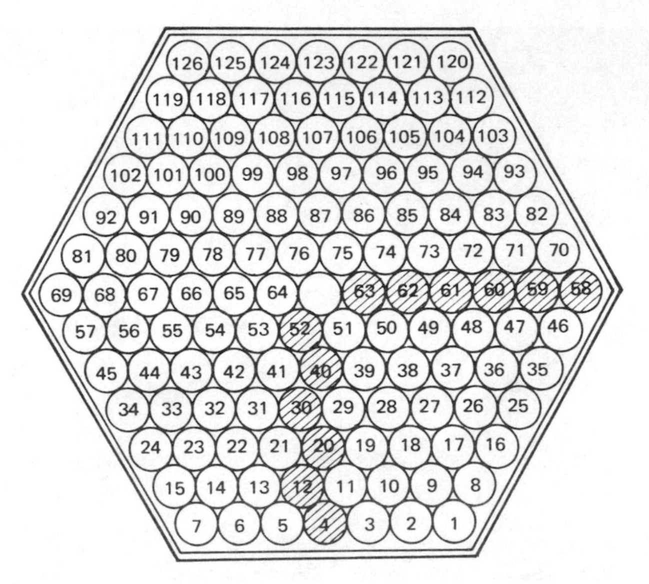

Рис. 3. Расположение измеренных твэлов в кассете. (⊘ — измеренные твэлы).

и рабочий. В рабочий канал включена схема с временным окном, фиксирующая число кратных совпаданий в интервале 100 мкс. Возможности установки УНИТ-1 ограничены использованием счетчиков СНМ-12, которые чувствительны к γ-излучению, в результате чего приходится вводить свинцовую защиту, ограничивающую поток γ-квантов.

Основное отличие установки УНИТ-2 от УНИТ-1 заключается в том, что в ней вместо счетчиков СНМ-12 использовались промышленные камеры деления типа КНТ-31. Такая замена уменьшает эффективность установки примерно в 2,5 раза по сравнению с УНИТ-1, но снимает проблему защиты детекторов нейтронов от γ-квантов. Дополнительное требование при создании УНИТ-2 состояло в обеспечении возможности недеструктивного нейтронного сканирования твэлов.

Детектор нейтронов состоит из кожуха, изготовленного из нержавеющей стали, внутри которого вокруг центральной сквозной трубки расположены камеры деления КНТ-31. Внутри кожуха расположены и предусилители. В качестве замедлителя нейтронов используется парафин. Общий вид установки УНИТ-2 показан на рис. 2.

5. РЕЗУЛЬТАТЫ ИЗМЕРЕНИЙ

Измерение выхода нейтронов и фотонейтронов (табл. I) производилось в "горячей" камере с помощью описанных выше установок.

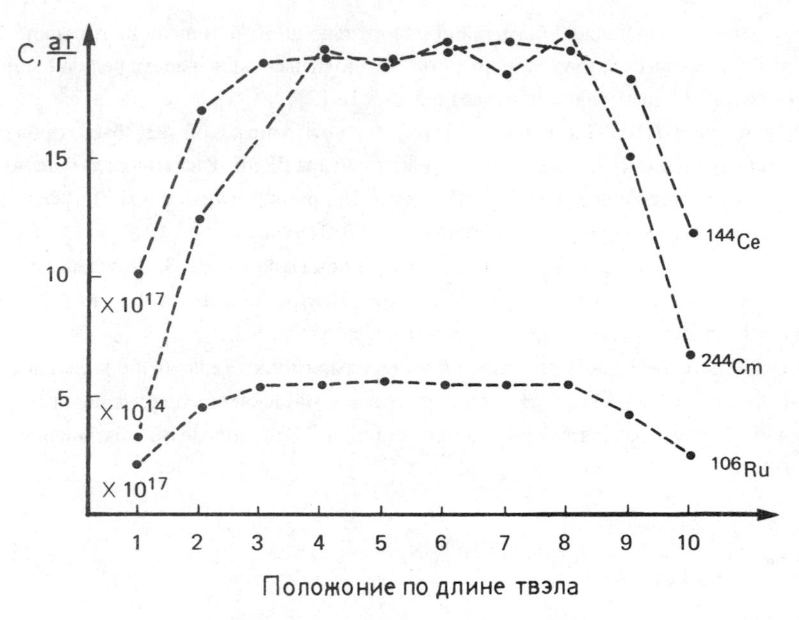

Рис. 4. Распределение концентраций ^{244}Cm, ^{106}Ru и ^{144}Ce по длине твэла.

Исследуемые образцы представляли собой куски твэлов длиной 12 см, вырезанные из твэлов кассеты ВВЭР. Начальное обогащение по ^{235}U – 3%.

Сначала измерялась интенсивность собственного нейтронного излучения топлива. Затем в измерительную установку помещались мишени из Be и D_2O и измерялся выход фотонейтронов.

Оценка величины нейтронного потока, обусловленного изотопами плутония, накопившимися в топливе, приводит к значению $\sim 3 \cdot 10^3$ н/с·кг. Эксперимент дает величину $\sim 10^5$ н/с·кг. Следовательно, в топливе должны присутствовать более интенсивные излучатели нейтронов, чем изотопы плутония. Такими изотопами могут быть изотопы ^{242}Cm ($T_{1/2} = 164$ дня) и ^{244}Cm ($T_{1/2} = 18$ лет). Повторные измерения, проведенные через два месяца, не показали уменьшения собственного потока нейтронов. Следовательно, более 90% всех излучаемых нейтронов должен испускать ^{244}Cm. В случае такого нейтронного потока количество накопившегося в топливе ^{244}Cm должно составить ~ 20 мг/кг топлива для топлива с расчетным выгоранием 28 000 МВт · сут/тU. Измеренное среднее число совпадений также подтверждает, что источником нейтронов в топливе, в основном, являются изотопы кюрия.

Была измерена зависимость интенсивности собственного нейтронного потока от выгорания топлива. При изменении глубины выгорания в образцах примерно в 2 раза (от 12 000 до 28 000 МВт· сут/тU) удельный выход нейтронов меняется приблизительно в 60 раз. По упрощенной формуле расчета содержания трансурановых изотопов накопление ^{244}Cm должно быть пропорционально шестой степени выгорания. Эксперимент подтверждает такую зависимость. Аналогичная зависимость величины нейтронного потока от выгорания получена и в работе [2].

На установке УНИТ-2 в "горячей" камере Нововоронежской АЭС было проведено нейтронное сканирование 11 твэлов кассеты реактора типа ВВЭР. Расчетная глубина выгорания топлива в кассете равна 18 000 МВт · сут/тU, время кампании – 320 эффективных суток и время выдержки после кампании – 760 суток.

Расположение измеренных твэлов в кассете показано на рис.3. Каждый твэл измерялся в десяти точках по длине через 25 см. Нейтронный фон в "горячей" камере во время измерений составил 0,5-3 импульсов в секунду.

В результате измерений для каждой измеряемой точки твэла было получены три величины N_1, N_2 и N_3 – скорости счета нейтронов с мишенями, содержащими соответственно обычную воду, тяжелую воду и бериллий. Эти данные обрабатывались по формулам:

$$C_{^{244}Cm} = \frac{T_{1/2}^{Cm}}{\bar{\nu} P \epsilon_1^1 \ln 2} N_1 = 5{,}92 \cdot 10^{14} \cdot N_1$$

$$C_{^{106}Ru} = \frac{T_{1/2}^{Ru}}{P \epsilon_2^2 \ln 2} \left[N_2 - \frac{\epsilon_2^1}{\epsilon_2^2} \cdot N_1 \right] = 1{,}12 \cdot 10^{15} [N_2 - 0{,}93 N_1]$$

$$C_{^{144}Ce} = \frac{T_{1/2}^{Ce}}{P \epsilon_3^3 \ln 2} \left[(N_3 - N_2 \frac{\epsilon_2^3}{\epsilon_2^2} - N_1 \left(\frac{\epsilon_2^1}{\epsilon_1^1} - \frac{\epsilon_2^1}{\epsilon_1^1} \cdot \frac{\epsilon_3^2}{\epsilon_2^2} \right) \right] =$$

$$= 0{,}6 \cdot 10^{15} [N_3 - 1{,}37 N_2 - 0{,}40 N_1]$$

где C – концентрация соответствующего изотопа в ат/г, P – вес на единицу длины твэла в гс/см, $\bar{\nu}$ – среднее число нейтронов на акт деления, $T_{1/2}$ – период полураспада соответствующего изотопа в секундах, ϵ_j^i – эффективность регистрации нейтронов и фотонейтронов для i-го калибровочного изотопа (1 – Cm; 2 – Ru; 3 – Ce) и j-ой мишени (1 - обычная вода; 2 – тяжелая вода; 3 – бериллий).

Результаты по распределению концентраций ^{244}Cm, ^{106}Ru и ^{144}Ce по длине одного из твэлов показаны на рис.4.

Распределение средних концентраций ^{244}Cm, ^{106}Ru и ^{144}Ce по диаметру кассеты показано на рис.5.

Полученные данные по содержанию ^{106}Ru хорошо согласуются с данными, измеренными γ-спектроскопическим методом, однако данные по концентрации ^{144}Ce приблизительно на 50 % превышают данные гамма-сканирования. Это превышение объясняется в основном влиянием тормозного излучения β-частиц ^{90}Sr на генерацию фотонейтронов на ядрах бериллия.

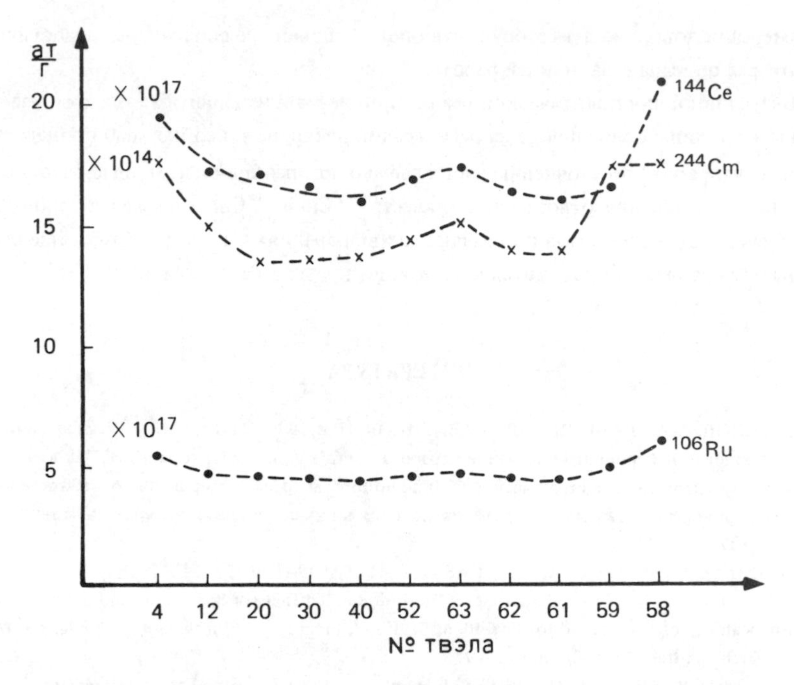

Рис.5. Распределение средних концентраций ^{244}Cm, ^{106}Ru и ^{144}Ce по диаметру кассеты.

Следует отметить, что приведенные данные по содержанию ^{244}Cm указаны без поправки на излучение нейтронов другими изотопами. По оценкам эта поправка для топлива с глубиной выгорания 18000 МВт · сут/тU не превышает 10%, а для топлива с выгоранием 30000 МВт · сут/тU – 15%. Погрешность в определении содержания ^{244}Cm без учета этой поправки составляет 5%. Погрешность в определении содержания ^{106}Ru – 7%. Определение содержания ^{144}Ce ввиду большой поправки на генерацию фотонейтронов за счет ^{90}Sr носит оценочный характер.

Исследование корреляции между величинами выхода собственных нейтронов и выгорания показывает, что значение выхода собственных нейтронов является очень чувствительным к величине выгорания. Формы кривых, характеризующих распределение интенсивности потока фотонейтронов по длине твэла и по диаметру кассеты хорошо согласуются с кривыми распределения ^{106}Ru и ^{144}Ce, полученными гамма-спектрометрическими методами. Наблюдаемая корреляция между величинами выхода собственных нейтронов и выгорания может служить основой создания аппаратуры для идентификации кассет с отработавшим топливом, для использования в системе гарантий. Накопление изотопов кюрия в легководных реакторах обуславливается цепочкой захватов нейтронов исходным ядром ^{238}U. Возможность подмены кассеты с указанным в паспорте начальным обогащением на кассету с меньшим начальным обогащением исключается совместным измерением интенсивности нейтронов и фотонейтронов. Такие измерения довольно легко осуществляются с помощью аппаратуры, аналогичной той, которая описана в настоящей работе.

Безусловно, для практической реализации неразрушающего метода анализа отработавшего топлива, основанного на регистрации нейтронов, необходимо проделать значительную работу по уточнению наблюдаемых корреляций с помощью разрушающих методов, по выяснению отношений количеств ^{242}Cm и ^{244}Cm, а также изотопов других трансурановых элементов при различных выгораниях и начальных обогащениях топлива. Такая работа проводится в настоящее время в Радиевом институте.

ЛИТЕРАТУРА

[1] ВОРОНОВ, А.А., ГАЛКИН, Б.Я., КАЗАРИНОВ, Н.М., ФЕДОТОВ, П.И., "Исследование характеристик нейтронного излучения отработавшего топлива реактора ВВЭР", Доклады совещания специалистов стран-членов СЭВ по вопросам транспортирования отработавших твэлов и неразрушающим методам определения в них содержания делящихся материалов, М., (1975).

[2] NATSUME, H., MATSUURA, Sh., OKASHITA, H., UMOSAVA, H., ESURE, H., "Collection of Gamma-Spectra Data of Irradiated Light Water Moderated Reactor Spent Fuel and Stady of the Applicability of the Method for Fuel Identification", Final Report on the Subject under the IAEA Research Contract 1119/El/RB, Oct. 1975.

[3] BARBERO, P., BIDOGLIO, G., BRSESTI, M., et al., "Post Irradiation Examination of the Fuel Discharged from the Trino Vercellese Reactor after the 2nd Irradiation Cycle", Joint Nuclear Research Centre (Ispra and Karlsruhe) Report EUR-5604 (1976).

IAEA-SM-231/109

A SAFEGUARDS SCHEME FOR 600-MW CANDU GENERATING STATIONS

D. TOLCHENKOV, M. HONAMI, D. JUNG
Department of Safeguards,
International Atomic Energy Agency, Vienna

R.M. SMITH, P. VODRAZKA
Mississauga, Ontario,
Atomic Energy of Canada Ltd.,

D.A. HEAD
Atomic Energy Control Board, Ottawa,
Canada

Abstract

A SAFEGUARDS SCHEME FOR 600-MW CANDU GENERATING STATIONS.

The on-power fuelling of CANDU reactors results in the continual discharge of spent fuel and input of fresh fuel, a feature which poses some unique safeguarding problems. This paper presents one possible safeguards approach for the CANDU 600-MW PHWR, developed by Atomic Energy of Canada Limited (AECL), Atomic Energy Control Board (AECB) and the International Atomic Energy Agency (IAEA). The safeguards approach suggested in this paper is based on the use of item accounting of fuel bundles and containment/surveillance measures. It enables the IAEA to meet its objective of the timely detection of the diversion of significant quantities of nuclear material. An inspection scenario, including the frequency of inspection and the inspection activities which could be used, is presented to indicate the function of various items of safeguards equipment in the scheme, although other frequencies and procedures can be equally valid.

1. INTRODUCTION

A growing number of countries are employing the CANDU nuclear reactor to satisfy their needs for electricity. The continuous fuel flow and impracticability of taking a core inventory at CANDU reactors require the development of a unique safeguards approach. As part of its programme to support international safeguards, Canada has worked in close co-operation with the IAEA to establish a safeguards scheme which is capable of giving adequate assurance that there would be timely detection of the diversion of significant quantities of nuclear material.

The first stage of this co-operative effort was to develop a general safeguards scheme for CANDU reactors (see paper IAEA-CN-36/185[1]). The second stage, completed in late 1977, concentrated on producing and demonstrating prototypes of the novel devices required to implement the concept. The third and final stage encompasses the engineering involved in adapting this general safeguards scheme to a particular power reactor. The development and installation of the necessary equipment to implement the scheme in this final stage is unique. For the first time a significant engineering effort has been expended to make safeguards an integral part of the reactor plant during construction rather than only an add-on after the reactor is operational.

This paper describes a safeguards approach taken for the CANDU 600-MW PHWR and the engineering involved in its realization. Although four of this type of reactor are now under construction, changes to plant are still possible before start-up, thus avoiding the difficulties of retrofit in a radioactive environment. Gentilly-2, which will commence operation early in 1980, will provide a working demonstration of the entire scheme.

2. METHOD OF THE SCHEME DEVELOPMENT

There are a number of separate steps that may be involved in establishing an acceptable safeguards scheme:

- (i) Define IAEA safeguards objectives and criteria
- (ii) Prepare diversion path analysis (DPA)
- (iii) Prepare design description of the proposed scheme
- (iv) Review the design and modify it if required for acceptance by the IAEA
- (v) Undertake detailed applications engineering needed for the installation of the equipment proposed in the scheme
- (vi) Acceptance of the equipment by IAEA
- (vii) Procure, install and commission the safeguards equipment
- (viii) Acceptance of the scheme as operational by IAEA
- (ix) Train IAEA staff in maintenance and operation of the equipment

The DPA identifies and characterizes the routes by which nuclear material may be diverted from the facility. The DPA for a 600-MW CANDU generating station was done in Canada by a consultant firm in 1977. A design description

---

[1] WALIGURA, A., et al., "Safeguarding on-power fuelled reactors – Instrumentation and techniques", Nuclear Power and its Fuel Cycle (Proc. Conf. Salzburg, 1977) **7**, IAEA, Vienna (1977) 587.

of an instrumented safeguards scheme to monitor the routes by which the irradiated fuel may be diverted was prepared in mid-1977 and critically evaluated as one of possible safeguards approaches by the System Studies Section in the IAEA to determine whether it met the IAEA objectives and criteria. The scheme, as described, with a few minor modifications was found to be acceptable, provided that:

Each reactor is constructed and operated in the way specified in the proposal;
The safeguards equipment described in the document is built into each of the reactor facilities and placed under the sole control of the IAEA; and
The described equipment operates as specified in the document.

Based on this design description and the results of the audit, detailed applications, engineering and procurement functions are now proceeding. A brief description of the 600-MW reactor followed by a description of the methods, instruments and inspection activity comprise the remainder of this report.

3. DESCRIPTION OF A 600-MW(e) CANDU GENERATING STATION

The reactor is fuelled with natural uranium in the form of small uranium dioxide pellets stacked end-to-end and sealed in a zirconium alloy tube. Thirty-seven of these tubes are jointed by two end-plates to make a cylindrical fuel bundle containing 18.7 kg U. There are 4560 of these bundles in the reactor core.

New fuel is received and stored in crates in the Service Building and, when required, is transferred to the Reactor Building (Fig.1) through the main airlock. The new fuel is manually loaded into the magazines of the new fuel ports which penetrate into the fuelling machine maintenance locks. The fuelling machines are charged from these new fuel ports in a remotely controlled operation.

The two identical remotely controlled fuelling machines are normally parked in the fuelling machine maintenance room suspended from fixed sets of tracks. They can travel along the tracks into the fuelling machine rooms and on to a bridge which can be raised to position the machine in front of the reactor. From this position they move forward and lock on to the fuel channel either to load new fuel or to accept spent fuel. Therefore, the travel is only in one fixed plane parallel to the face of the reactor.

The fuelling system is the on-power, bi-directional, push-through method and usually several refuelling operations are carried out daily. Normally eight fresh fuel bundles are added to one fuelling machine and eight spent fuel bundles are discharged into the other at each refuelling. Shield and closure plugs are removed before the refuelling operation and replaced after it by the fuelling machines. As well as the new fuel ports and the fuel channels the fuelling machines

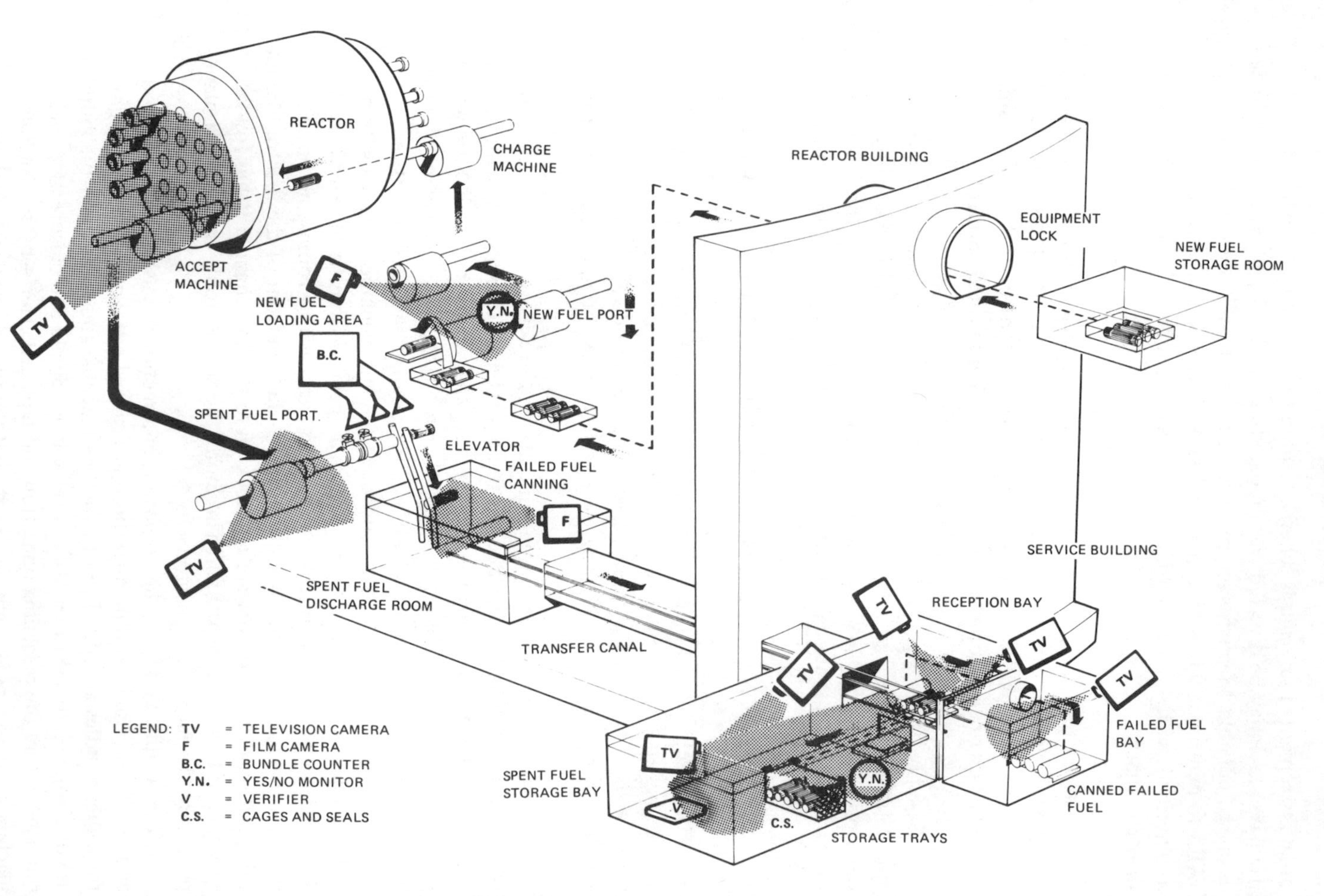

FIG.1. Location of safeguards instrumentation.

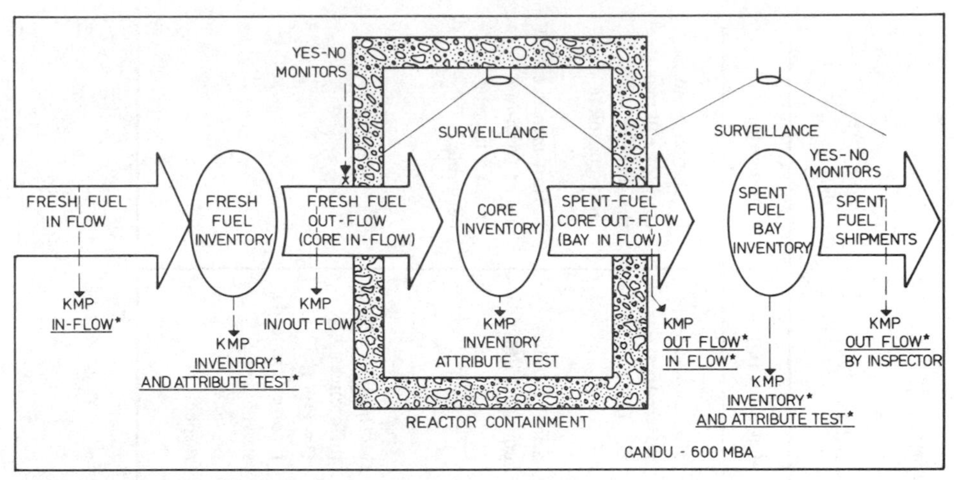

FIG.2. Nuclear material flow within CANDU-600 MBA.

can also lock on to spent-fuel ports to discharge fuel and on to the service port for service, tool exchange and calibration functions.

Shielding doors divide the maintenance locks from the fuelling machine rooms and, when closed, permit access to the fuelling machine maintenance locks even when the reactor is at power. Transfer doors separate the maintenance locks from the fuelling machine pit area, and there are personnel access doors connecting the main airlock area to the fuelling machine maintenance areas. The transfer doors are sized to allow the removal of the fuelling machine heads from the maintenance area to the service building via the main airlock in case of serious breakdown.

Spent fuel discharged from the fuelling machine through the spent-fuel port enters the discharge elevator in the spent-fuel discharge room (Fig.2) and is lowered on to an underwater cart, which carries it through the transfer canal to the reception bay of the spent-fuel storage area in the Service Building. Twenty-four bundles are loaded on to a storage tray which is transferred through an opening into the storage bay and placed on closely spaced stacks.

When defective fuel is discharged the bundle is inserted in a can in the spent-fuel transfer room at the head of the transfer canal. The can is transported through the reception bay into the defective fuel storage bay which is physically separate from the main storage bay.

TABLE I. SOURCE OF INDEPENDENT DATA FOR MATERIAL BALANCE CLOSING

Storage area	Inventory count (Q)	Inventory attribute	In-flow (I)	Out-flow (O)	Comments
Fresh fuel	Annual physical inventory-taking (counting) by inspection	Verification on sampling plan by method yet to be specified	Shipper's records	Method of differences O = Q – I	Fresh-fuel bundle counter will make the accounting more complete
Core area	Initial core inventory-taking coupled with strict surveillance measures	None. Possibly flux mapping information useful	Method of differences (I = O) workable with strict surveillance	Spent bundle counter data	In flow measurement use of fresh-fuel counter may be desirable
Spent-fuel bay area	Physical inventory-taking (counting) coupled with sealed bulk storage	Spent-fuel verifier in conjunction with bundle-counter data used at every inspection in a sampling plan	Spent-fuel bundle-counter data	Method of differences O = Q – I, receiver's records, surveillance; presence of inspectors	

4. SAFEGUARDS APPROACH

The safeguards approach for CANDU 600-MW reactors is based on a combination of item accounting and containment surveillance measures. The fuel inventory and fuel flow at designated key measurement points (KMPs) are established using independant data and thereby the reports and records supplied by the State concerning this nuclear material can be verified. Containment surveillance instruments will provide the assurance that any unusual fuel-handling activity which could be diversion significant would be detected. This approach takes into account two major diversion scenarios:

(i) The unreported removal of declared nuclear material.
(ii) The clandestine production and removal of plutonium using undeclared feed material.

Within these two scenarios the plutonium-related activities were considered to be more attractive to the divertor.

For the purposes of safeguards implementation the IAEA considers that approximately 112 irradiated fuel bundles will contain a significant quantity of plutonium (8 kg) and that about 560 fresh-fuel bundles contained a significant quantity of natural uranium (75 kg ^{235}U contained). These numbers will be used to establish the sampling plans employed in verification activities.

The nuclear material within the CANDU 600 MW resides in three main inventory areas:

(i) Fresh-fuel store
(ii) Reactor core
(iii) Spent-fuel storage bays

Each of these areas could be safeguarded using data on fuel attributes, the number of fuel bundles and incoming and outgoing flows. Table I indicates how independant data could be obtained for each of the inventory areas.

The safeguards approach outlined herein is considered able to give timely detection of the diversion of significant quantities. Specifically, the following will be detected:

(i) The unreported removal of fresh fuel from the facility
(ii) The unreported removal of irradiated fuel from the core by abnormal means
(iii) The unreported removal of irradiated fuel from the spent-fuel storage bays
(iv) The substitution of irradiated fuel by dummy or unirradiated fuel.

The remainder of this paper describes safeguards instruments and inspection activities which will provide this timely detection.

Examination of Table I (and Fig.2) shows, for the case of material balance around the core, that the IAEA is lacking a direct means of verifying core inventory. Surveillance equipment monitoring the spent-fuel transfer area will provide the verification that core inventory has not been changed without being reported and that irradiated fuel has not been transferred unreportedly by a route that bypasses the spent-fuel counter. With such assurance the material balance around the core can then be verified on the assumptions of constant core inventory, and that input flow is equal to verifiable output flow.

5. DESCRIPTION OF CONTAINMENT AND SURVEILLANCE EQUIPMENT

The safeguards scheme employs TV surveillance systems, photo-surveillance cameras, radiation yes-no monitors, bundle counters for spent fuel, an attribute verifier for trays of spent fuel, and security seals and cages. The equipment location is shown in Fig.3 and their functions are described below.

5.1. Spent-fuel transfer area surveillance equipment

The spent-fuel transfer surveillance area encompasses the route which the irradiated fuel takes during transfer from the reactor core to the bays. A television surveillance system is used to observe all movements of the fuelling machines in the fuelling machine rooms and fuelling machine maintenance locks. Two television cameras are on top of removable assemblies located in emergency access openings in the floor of the south and north fuelling machine rooms, and similar arrangements are provided for a second pair of cameras in the south and north fuelling machine maintenance locks. For maintenance or repair work the cameras can fairly readily be removed through the emergency access floor plugs from accessible rooms located in the basement of the reactor building.

The television control unit is located in the service building and consists of motion detection systems, floppy disc video recorders, video cassette recorders, system control microprocessor, date and time and camera identification generator, television monitor and an array of tamper-indicating and fault-indicating devices. This arrangement permits free access to the unit by the inspectors and also enables them to have uninterrupted observation of the recorded events. This control unit records pictures taken in the spent-fuel transfer area as well as pictures from cameras in the spent-fuel storage area.

Two film cameras in the new fuel area provide the necessary coverage to detect movement of irradiated fuel bundles through this route. Both fuelling machine transfer doors are viewed. A radiation sensor is used to trigger the

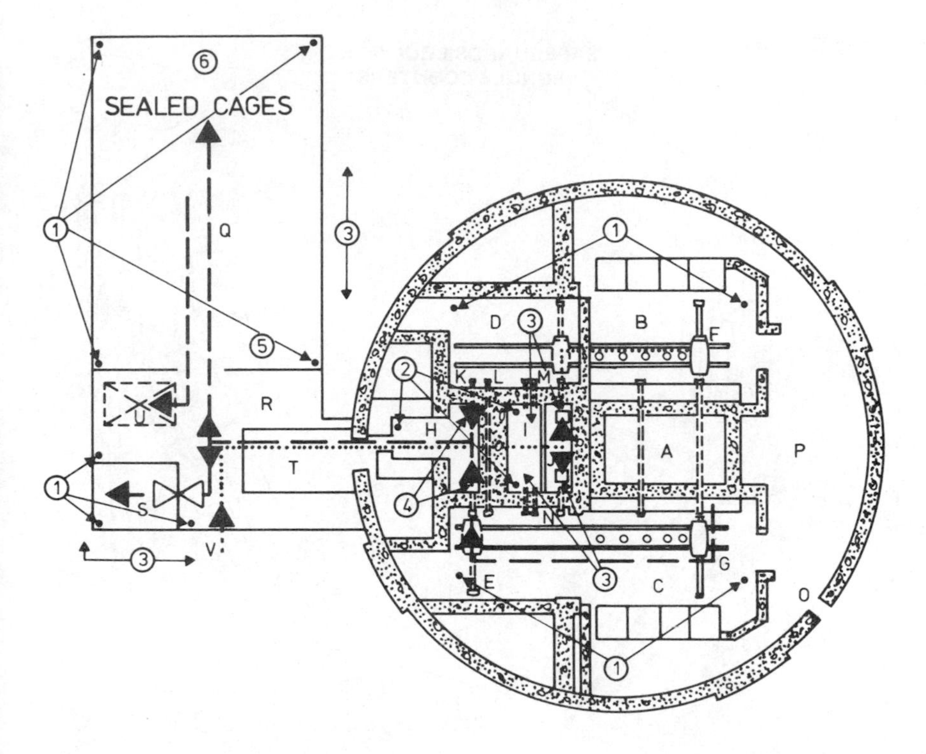

A REACTOR VAULT
B NORTH FUELLING MACHINE ROOM
C SOUTH FUELLING MACHINE ROOM
D NORTH FUELLING MACHINE MAINTENANCE LOCK
E SOUTH FUELLING MACHINE MAINTENANCE LOCK
F NORTH FUELLING MACHINE
G SOUTH FUELLING MACHINE
H SPENT FUEL TRANSFER ROOM
I FUELLING MACHINE PIT AREA
J NEW FUEL LOADING ROOM
K SPENT FUEL PORT
L REHEARSAL FACILITY PORT
M SERVICE PORT
N NEW FUEL PORT
O AUXILIARY EMERGENCY AIRLOCK
P MODERATOR ROOM
Q SPENT FUEL STORAGE BAY
R RECEPTION BAY
S DEFECTIVE FUEL BAY
T EQUIPMENT AIRLOCK
U CRANE ACCESS TO RECEPTION BAY FOR SHIPPING SPENT FUEL
V NEW FUEL STORE

1 TV CAMERAS
2 FILM CAMERAS
3 YES/NO MONITORS
4 SPENT FUEL BUNDLE COUNTERS
5 SPENT FUEL VERIFIER
6 SEALED CAGES

······ FRESH FUEL PATH
— — — SPENT FUEL PATH

FIG.3. Safeguards equipment location.

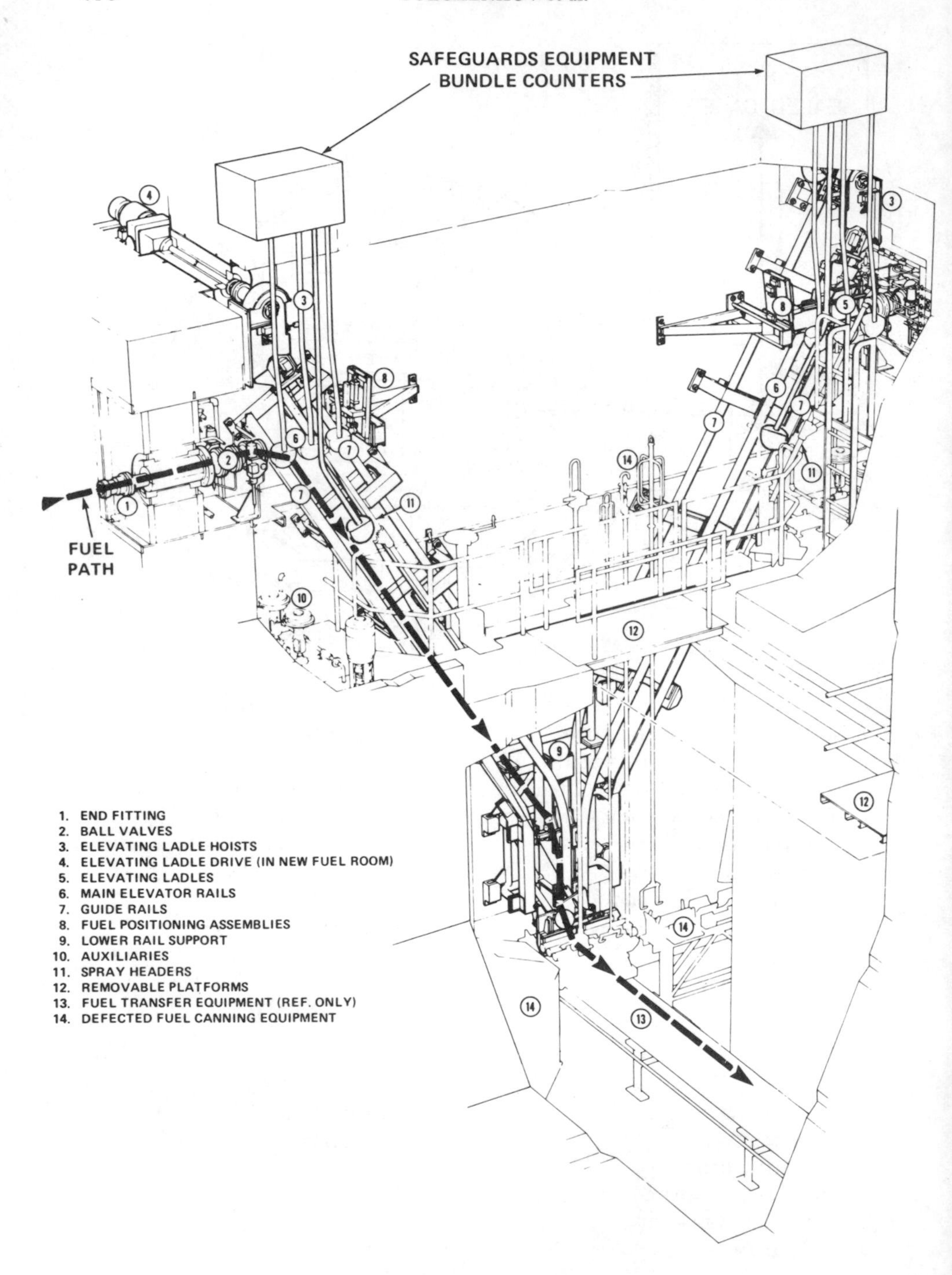

FIG.4. 600-MW(e) spent-fuel discharge equipment.

cameras to reduce to a minimum the number of pictures taken in order to simplify subsequent film review.

Another film camera equipped with a motion sensor is positioned above the exit from the spent-fuel discharge room to provide surveillance of a number of openings in the walls (such as ventilation vents and an access door). Although direct observation of the top portion of the fuel transport elevator is partially obscured, those operations connected with the manual canning of defective fuel will be viewed by the camera. A side-angle mirror attachment is used to ensure that a space directly adjacent to the exit door is included in the camera field of view.

Yes-no monitors placed on each of the new fuel and service ports will ensure that radioactive irradiated fuel does not pass through these ports (quite an abnormal procedure) without being detected. These monitors are small radio-photoluminescent glass dose meters inside standard IAEA seals.

Bundle counters near the top of each of the two discharge elevators (Fig.4) monitor the movement of irradiated fuel between the reactor and the storage bay. They consist of processing/logic electronics and power supply in a tamper-proof package and several Geiger tubes spaced to differentiate between single- and double-bundle transfers and positioned so as to be direction sensitive. Thus they are able to record the number of bundles passing the measurement point, the direction of the movement, and the time of the transfer. This flow information is used to verify records regarding core and spent-fuel bay inventory.

The bundle counter electronics packages will be located above the spent-fuel transfer room, which is an area of good accessibility. The eight penetrations of the four-foot-thick concrete ceiling, provided to permit positioning of the Geiger tubes close to the spent-fuel ports, and the necessary collimation and shielding will be designed to comply with reactor containment specifications and operational safety requirements. In addition, they will comply with environmental, seismic, and other criteria appropriate to nuclear power stations.

The IAEA has expressed considerable interest in determining the number of fresh fuel bundles loaded into the fuelling machine to provide an item count of the input to (and throughput from) the reactor core. Such a count would be useful in keeping a running core inventory and in providing a certain level of redundancy in case of the failure of some other instrumentation (such as a bundle counter). A study has just been completed to demonstrate in a laboratory mock-up of the fresh-fuel loading port the technical feasibility of interrogating fresh-fuel bundles to characterize their uranium content. Consideration is now under way to establish the requirements for fresh-fuel accounting, although there has not yet been a commitment to develop and test a prototype device. The requirement for this instrument had not originally been considered and it is not possible to have the design for this device ready in time to meet the construction schedule of the first four CANDU-600 reactors. The problem of retrofit has not been analysed.

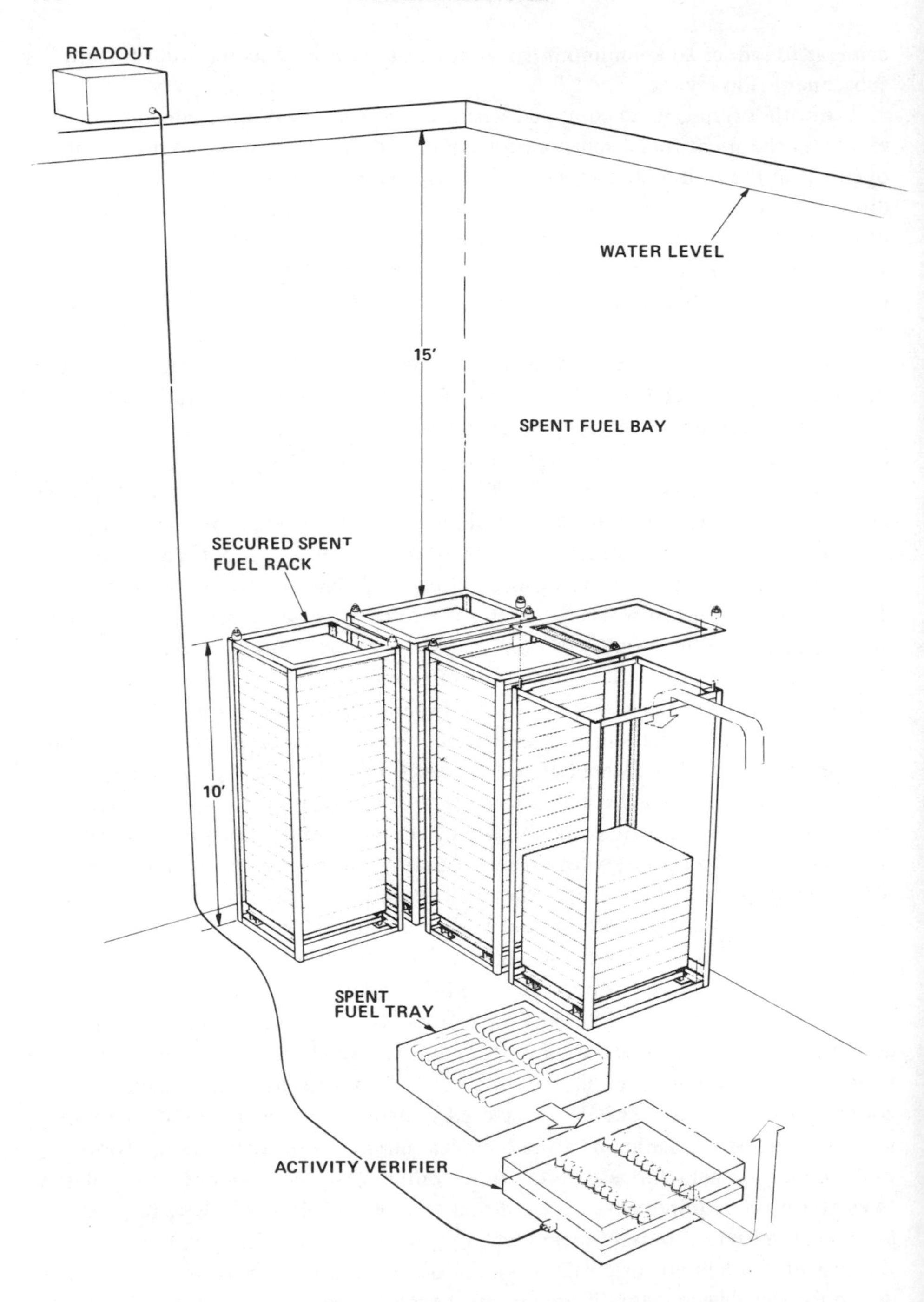

FIG. 5. 600-MW(e) spent-fuel verification and storage.

5.2. Spent-fuel storage surveillance and containment area equipment

Surveillance of all three spent-fuel bays is provided by a television system using seven cameras. The complicated topography of the bays dictates the location of two cameras in the receiving bay, four cameras in the main bay, and one in the defective fuel bay. These cameras are positioned to observe the water surface, access doors, and other above-surface wall openings.

Radiation yes-no monitors applied to the water purification piping system detect and record removal from the bay through this route of irradiated nuclear material (as fuel elements of pieces of broken bundles). Ready access to these monitors for checking and/or replacement is achieved by attaching them to the above-water portion of the system.

In the main storage bay the following equipment is available:

(i) A "bundle verifier" which verifies the gamma activity from each bundle in a loaded tray;
(ii) Tamper-indicating mesh cages, with covers, to enclose a stack of trays of spent fuel; and
(iii) Ultrasonic seals, verifiable in situ under water, to seal the cover on the cage.

The function of the bundle verifier (Fig.5) is to verify that the objects on a tray are highly radioactive, with radiation fields similar to that expected from spent fuel. The bundle verifier is an array of diodes sensitive to gamma radiation with a frame on which a loaded tray can be placed. Each detector is fitted with a collimator so that it sees the radiation from a single bundle, and there are sufficient detectors in the array to test a complete tray simultaneously.

The function of the cage, cover and ultrasonic seals is to contain a block of spent fuel in a secure way. These and the camera surveillance of the storage bays provide two separate methods of coverage of stored irradiated fuel. Because there are many fewer seals than fuel bundles, doing an annual physical inventory is simplified. The security cage is a light structural steel frame covered with flattened expanded stainless steel mesh, open at the top and attached to a stand at the bottom. The stand supports the spent fuel trays. A cover of similar construction can be fitted to the top of the cage to enclose the contents. The cover is about 4.5 m below the bay water surface.

An ultrasonically identifiable seal is employed to seal the cover to the cage. The seal consists of two parts, a cap and a stud to it by a collet which must be broken to remove the cap. Broken parts are removable with simple hand tools. The cap, stud and collet are ultrasonically interrogated to establish the seal signature, and they are fabricated from a low melting material which would change identity if thermal cutting were used to remove or replace parts of the sealing system.

6. INSPECTION ACTIVITIES

6.1. General

The IAEA implements inspection activities for the purpose of verifying design information, verifying initial inventory, verifying subsequent flow and inventory information, inspecting the shipment of fuel and auditing the records and reports provided by the State.

Although the functions of the safeguards equipment for this CANDU 600-MW scheme have been defined, the procedures for its use are still flexible. The procedures in use at any particular site at any particular time will depend upon a number of factors, including optimization of resources, the status of the reactor, and the safeguards agreement and facility attachment provisions. It is not possible, therefore, to lay down a single set of operation procedures which would necessarily be applicable to all 600-MW stations.

It would appear that the goal of timely detection of diversion of the nuclear material in a CANDU 600 MW could be satisfied with an inspection regime of about six routine inspections per annum, spaced at two-month intervals. At one of these inspections a physical inventory would be performed. The IAEA has estimated that the conversion time to remove plutonium from irradiated fuel is between one and three months and for natural uranium and fresh fuel approximately one year. A routine inspection frequency of every two months and an annual physical inventory have therefore been suggested to be in accord with these conversion times. It is possible that other inspection routines could be devised by which the IAEA safeguards goal could be met.

This instrumented safeguards scheme has no great impact on normal reactor operations except for the spent-fuel verification activities in the storage bay. Because these activities are manpower intensive for both the reactor operator and the IAEA inspectorate, variations of the procedures to be used must be investigated to achieve a balance between the depth of safeguards coverage and the inspection effort. For example, it may be possible to change the man-days expended per inspection and the interference with reactor operations if a different degree of common coverage is accepted.

The performance of safeguard equipment will be tested and the inspection procedures developed during the safeguard demonstration period that is to be conducted at the G-2 nuclear station in 1980. At that time, the inspector can establish the procedures that they will follow at this facility, and possibly at the other 600-MW nuclear stations. They can then ensure that the most efficient use is made of the inspector's and operator's time, taking into account the safeguard objectives and the agreements respecting the application of safeguards at that site. Saving in manpower can probably best be made in the fuel storage area. In the opinion of the authors the following procedure would appear to give assurance that diversion would be detected.

6.2. Initial inventory

Owing to the continuous operation of CANDU reactors the verification of initial inventory should be done prior to and during the charging of fresh fuel into the reactor. This will provide assurance as to the exact number of fuel bundles loaded into the core and from this moment those bundles are under surveillance. Verification may include NDA checks of the fuel bundles using a sampling plan.

6.3. Routine inspections

To meet the timely detection of the diversion of plutonium in irradiated fuel, six IAEA routine inspections are proposed per year, spaced at two-month intervals. At one of six inspections annual physical inventory-taking will be performed. Estimated routine inspection effort for a year is approximately 40–45 man-days. Transfers of spent fuel out of the spent-fuel bay would require additional effort.

6.3.1. Fresh-fuel inspection activities

The fresh fuel has low strategic significance (natural uranium) and one annual inspection for the purpose of physical inventory verification may suffice. This verification may consist of:

(a) The identification of bundles
(b) Physical counting of bundles
(c) Some form of NDA measurement using a sampling plan
(d) Auditing and comparison of records.

6.3.2. Spent-fuel routine inspection activities

The inspector may perform the following activities at every routine inspection:

(a) Updating inventory records for the period since the previous inspection;
(b) Determination from the spent-fuel bundle counter and, when included, from the fresh-fuel bundle counter, the number of bundles that have been counted by these instruments;
(c) Physical counting of unsealed fuel inventory;
(d) Verification of the spent fuel not in the sealed cages, by means of the spent-fuel verifier using sampling plan;
(e) Supervising the sealing of the fuel verified in (d) into cages and sealing these cages;

(f) Observation of information from TV or film cameras so as to note any significant unreported activities. Alternatively, observation may be done at a later date at IAEA Headquarters;
(g) Verification of the cage seals in the spent-fuel bay on a random basis;
(h) Examination of yes-no monitors and seals on safeguards equipment to ensure they are satisfactory;
(i) Partial inventory verification using above findings;
(j) Servicing the safeguards equipment;
(k) Verification of receipts and shipments if appropriate;
(l) Auditing station accounting and operating records to ensure that the material balance calculations are in accord with the inspector's findings.

6.3.3. Physical inventory inspection

Once a year, coinciding with the fresh-fuel inventory verification, a complete physical inventory inspection will be performed on all nuclear material. The annual physical inventory should be done by checking seals on all sealed cages of verified fuel and by checking on unsealed fuel. Sampling plans are to be developed to detect with adequate confidence diversion of significant quantities of nuclear material.

An annual physical inventory of the core content is impractical and here the book inventory must be used. Since it is possible to operate the reactor with many fewer fuel bundles than the maximum core loading, assurance is required that the core inventory has not been decreased significantly from that reported. Surveillance equipment in the area of the core will provide assurance that there has been no unreported removal of fuel bundles from the core. If it should fail, data gathered in other parts of the safeguards scheme may be used to provide a degree of confidence in the reported inventory.

6.4. Shipment of spent fuel

Detailed approaches for the safeguarding of the shipment of spent fuel will depend on the particular shipping procedures and only some general considerations are given. Cages may be moved as long as they remain under surveillance without the presence of an inspector. Sealed cages should be unsealed and fuel transferred to shipping flask in the presence of an inspector. When filled prior to shipment the cask should be sealed and until such time the cask should remain under surveillance. The content of any unsealed cage not opened in the presence of an inspector may require re-verification before being resealed.

7. CONCLUSION

The IAEA accepts the safeguards approach presented here comprising item accounting and containment/surveillance as one of the possible ways in which the CANDU 600-MW PHWR can be safeguarded. Before this scheme is considered to be operational Canadian safeguards development and IAEA safeguards staff will collaborate in its demonstration so as to have a high level of confidence in the reliability of the equipment and the assessment of the scheme effectiveness.

BIBLIOGRAPHY

ATOMIC ENERGY OF CANADA, The Development of Irradiated Fuel Bundle Counters for 600-MW CANDU Reactor Safeguards System, AECL-6209 (1978).

INTERNATIONAL ATOMIC ENERGY AGENCY, A Safeguards Approach for a CANDU-600 Reactor, IAEA-STR-72 (August 1978).

WALIGURA, A., et al., "Safeguarding on-power fuelled reactors – Instrumentation and techniques, Nuclear Power audits Fuel Cycle (Proc. Conf. Salzburg, 1977) **7**, IAEA, Vienna (1977) 587.

DISCUSSION

W.C. BARTELS: I recently heard of your experience with verification of spent fuel at an Ontario Hydro plant of the CANDU type in Canada. The spent fuel verifier referred to in your paper is presumably designed to reflect that experience? Can you describe the design principle?

D.A. HEAD: The principles of operation chosen for an attribute verification instrument depend upon the system in which it is used. The gamma-radiation-field monitoring device shown here is acceptable as it simply provides one more data input in a total system, and gives additional confidence that diversion has not occurred without detection. The IAEA has indicated that an attribute verification instrument used independently rather than as part of a total system must be capable of giving higher quality data.

G.B. MUMMERY: Will the field of view of your cameras include personnel? If so, do you expect any labour problems?

D.A. HEAD: Personnel might well be observed by the TV cameras inside the reactor containment during reactor shutdowns, and in the other areas at all times. On the basis of past experience with many IAEA cameras we do not expect any labour trouble in connection with the use of these devices.

EXPERIENCE IN THE IAEA SAFEGUARDING OF THE BOHUNICE A-1 ON-LOAD POWER REACTOR, CSSR, INCLUDING GAMMA-SPECTROMETRIC DETERMINATIONS OF BURNUP IN THE SPENT FUEL

V. PETENYI, S. ROHAR
NPP Research Institute, Bohunice,
Czechoslovakia

E. DERMENDJIEV, J. HORNSBY, V. POROYKOV
Department of Safeguards,
International Atomic Energy Agency, Vienna

Abstract

EXPERIENCE IN THE IAEA SAFEGUARDING OF THE BOHUNICE A-1 ON-LOAD POWER REACTOR, CSSR, INCLUDING GAMMA-SPECTROMETRIC DETERMINATIONS OF BURNUP IN THE SPENT FUEL.

The report describes the experience gained by the Central and Northern Europe Section of the Agency's Department of Safeguards in safeguarding the on-load reactor at Bohunice A-1 Nuclear Power Plant, Czechoslovakia. Measures to re-establish the inventory in the case of failure of surveillance devices are indicated. The correlation between the different fission product yields and burnup values have been established.

INTRODUCTION

Under the terms of the Agreement [1] between the Czechoslovak Socialist Republic and the Agency (IAEA) for the application of Safeguards in connection with the Treaty on the Non-Proliferation of Nuclear Weapons (NPT), which entered into force 3 March 1972, regular routine inspections have been performed at Bohunice A-1 "on-load" Power Plant, CSSR.

Nuclear material accountancy and control at the A-1 Power Plant is performed by the operator in accordance with the conditions of the general regulations [2] issued by the Czechoslovak Atomic Energy Commission (CAEC). This activity by the operator is periodically audited and verified by inspectors from both the CAEC and the IAEA.

This paper briefly describes IAEA inspection procedures and spent-fuel measurements which have been performed by both the IAEA and Bohunice research staff at A-1 Nuclear Power Plant.

DESCRIPTION OF FACILITY AND ASSOCIATED FUEL CYCLE

Bohunice A-1 Nuclear Power Plant is a continuous on-load refuelling reactor which went critical in late 1972 and commenced regular delivery of electricity to the national grid system from 1973. It is a natural-uranium metal-fuelled, heavy-water-moderated, carbon-dioxide (at 60 atm pressure) cooled power reactor (HWGCR) with a rated 600-MW thermal output. The reactor is refuelled on-load by two refuelling machines. The heterogeneous active core, in the form of a regular lattice of channels, is divided into concentric central periphery zones (Fig.1).

The fuel assemblies, which are assembled at the facility from natural-uranium metal rods clad in Mg/Be alloy and manufactured in the USSR, consist of two basic types – one contains 63 fuel rods and the other 75 fuel rods. Each rod is 6.3 mm in diameter and 4 m long (and contains approximately 2.3 kg natural uranium metal). The assembly is contained in an external zirconium alloy sheath of either 112 mm diameter (if of the 75-rod type) or 100 mm (if of the 63-rod type). The reactor core consists of 104 peripheral zone assemblies of the 63-rod type (designated 'O' type) and 44 central zone assemblies of the 75-rod type (designated '2' type). Assemblies are expected to remain in the core until an average burnup of 4600 MW·d/t is achieved (this is from 100 to 235 days in the core, with the reactor working at or near its stated power, and dependent upon the core position and reactor parameters).

Fuel assemblies are first discharged to a cooled short-term storage and later, when the short-term fission-product activity and associated heat have been sufficiently reduced, to the long-term storage – by the refuelling machines. The long-term storage has capacity for 900 irradiated fuel assemblies. When the maximum capacity is approached shipments of irradiated fuel to the USSR for reprocessing are planned.

DESCRIPTION OF NUCLEAR MATERIAL FLOW AND HANDLING

Bohunice A-1 Power Plant is divided into two Material Balance Areas (MBA). The first MBAconsists of receipt and storage of fresh fuel rods from the USSR and assembly area and storage for fresh fuel assemblies. The second MBA also includes a storage for fresh fuel assemblies as well as the reactor KS-150, short-term

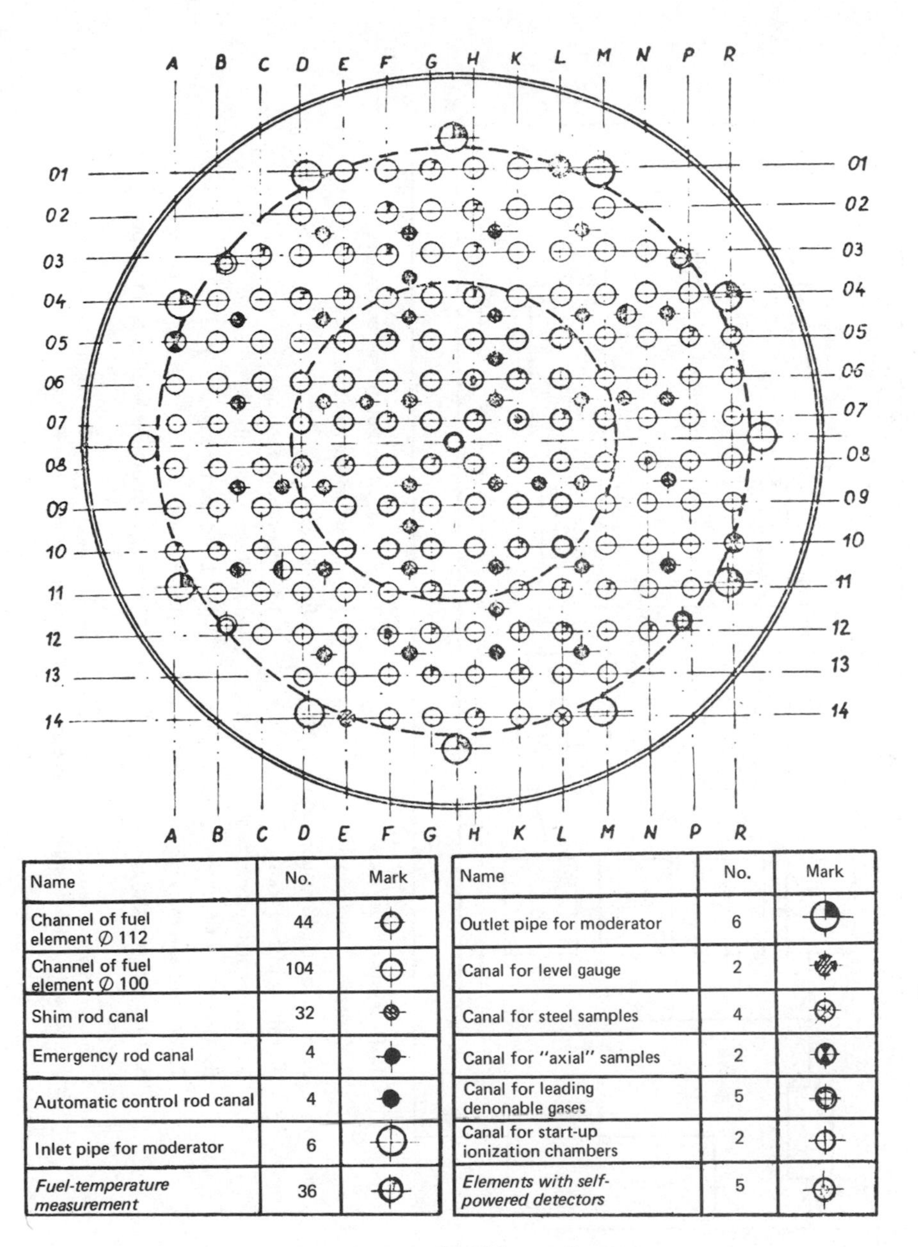

Name	No.	Mark
Channel of fuel element ∅ 112	44	
Channel of fuel element ∅ 100	104	
Shim rod canal	32	
Emergency rod canal	4	
Automatic control rod canal	4	
Inlet pipe for moderator	6	
Fuel-temperature measurement	36	

Name	No.	Mark
Outlet pipe for moderator	6	
Canal for level gauge	2	
Canal for steel samples	4	
Canal for "axial" samples	2	
Canal for leading denonable gases	5	
Canal for start-up ionization chambers	2	
Elements with self-powered detectors	5	

FIG.1. Reactor KS-150 core diagram.

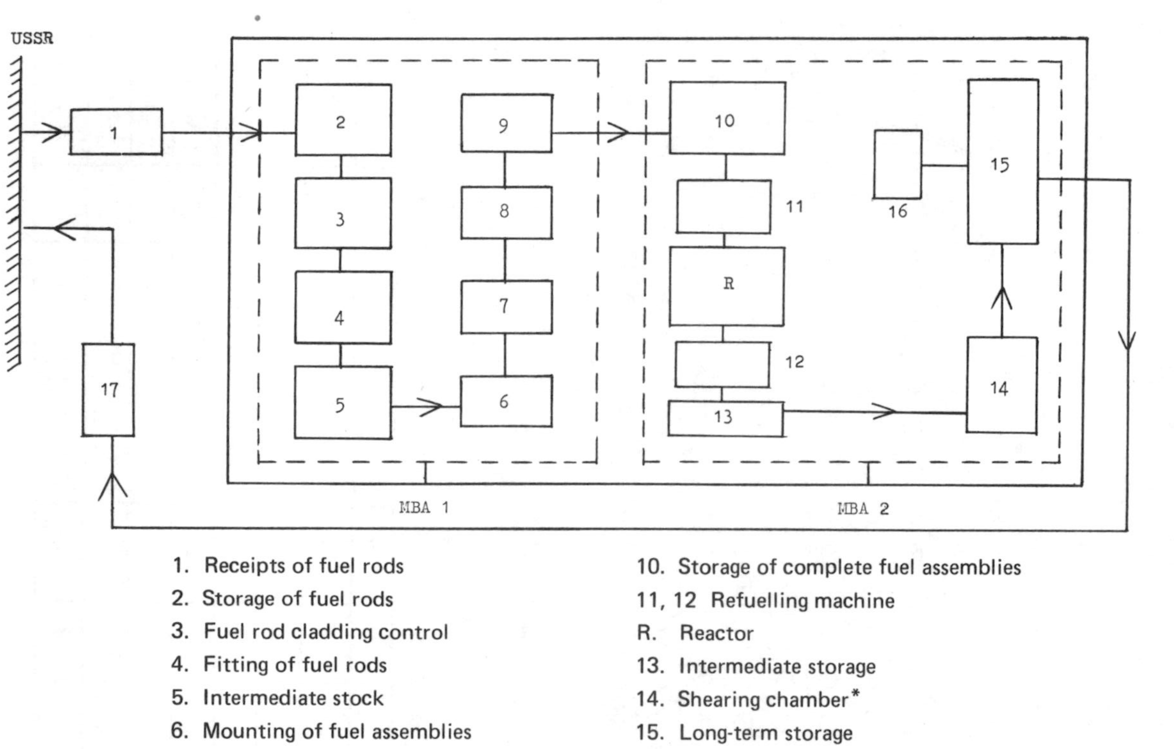

1. Receipts of fuel rods
2. Storage of fuel rods
3. Fuel rod cladding control
4. Fitting of fuel rods
5. Intermediate stock
6. Mounting of fuel assemblies
7. Control and marking
8. Intermediate storage
10. Storage of complete fuel assemblies
11, 12 Refuelling machine
R. Reactor
13. Intermediate storage
14. Shearing chamber*
15. Long-term storage
16. Hot cell**
17. Wagon-container loading and transport

* Shearing chamber is used to remove extraneous parts of the assembly (the fuel remains as an integral unit)
** In hot cell — one assembly at a time (can be as part assembly and cut rods)

FIG.2. Diagram of fuel movements in Bohunice nuclear power station.

storage for recently discharged irradiated fuel assemblies, two refuelling machines, a hot cell and long-term storage for spent-fuel assemblies.

The material flow (Fig.2) commences with the receipt of fuel rods from the USSR. After checking these rods are stored in storage racks in storage boxes. Records are kept of the number and type of rod in each storage box. When required the fuel rods are taken to the assembly area, physically examined, and then mounted into sheaths (casettes) to form the two types of fuel assembly. Each fuel assembly has a unique identification number stamped on to the sheath. Accounting and control data for each new assembly are maintained, and details such as irradiation history and burnup are completed as appropriate. Completed assemblies are then stored in the assembly storage area until they are transferred to the reactor hall storage area (and hence the second MBA) by means of a crane situated in the reactor hall. The new assemblies are then available to the refuelling machines for re-charging the core.

The irradiated fuel assemblies are initially discharged to a short-term storage for a short cooling period before being taken to the shearing chamber to have the fuel region separated from the rest of the assembly. This fuel region (still referred to as the assembly) is then transferred by the refuelling machines to the long-term storage. The refuelling machines will also be used to fill the transport cask with spent fuel when this occurs. When required irradiated fuel assemblies are taken from the long-term storage, by means of the refuelling machines, to the hot cell where the assembly can be subjected to post-irradiation examination, and samples taken for subsequent analysis or for other necessary examinations.

INTERNAL FUEL MANAGEMENT

An accounting office with an associated cross-reference recording system was established at the facility to account for and to control the nuclear material. External and inter-MBA movements are notified to the CAEC and hence by ICRs to the IAEA. Internal movements are recorded and controlled by the operator. Each assembly history is known and burnup calculations from reactor data are performed on a routine basis.

The operator takes his physical inventory once a year, on the basis of identification and counting of individual items, and prepares an itemized inventory listing, according to location and type of item — this listing is the basis for the verification activities performed by the CAEC and IAEA inspectors.

INSPECTION ACTIVITIES

The IAEA usually performs four inspections annually at this facility. Three of these consist of routine inspections to verify material flow, record audit and

FIG.3. View of corridor where movements of spent fuel casks are observed.

service and evaluation of surveillance records and service of surveillance equipment. The fourth inspection is to verify the operator's physical inventory taking plus evaluation of surveillance records and service of surveillance equipment.

(a) Fresh fuel rods

These are verified by visual identification and item counting, followed by qualitative NDA measurements (to establish the identity of the material and as a check of self-consistency) on rods randomly selected according to a prepared sampling plan. Storage boxes are also sealed, following verification, when permitted by the operational programme of the facility.

(b) Fresh fuel assemblies

These are verified by identification of serial numbers, recognition regarding type, counting and qualitative NDA measurements (to establish the identity of the material and as a check of self-consistency from assembly to assembly and from inspection to inspection) on assemblies randomly selected according to a prepared sampling plan.

(c) Assemblies within the core

A check of criticality is performed through the observation of the reactor control room "in-core" instrumentation plus visual checks of cooling-water flow temperature and the operation of the turbines. Track-etch monitors have also been installed in the core by the operator on behalf of the Agency to check upon the reactor operational programme (as part of an experimental programme).

(d) Spent fuel assemblies

Because of the design of the long-term storage it is not possible to identify and count spent-fuel assemblies. Identification of spent-fuel assemblies, by serial number and type, is possible in the shearing chamber and in the hot cell. But both these operations would require transportation by the refuelling machines, which is time-consuming and unacceptable to the facility. Indirect verification is achieved by auditing the operating records plus use of surveillance equipment installed to observe all movements to and from the reactor hall, fresh fuel storage and long-term storage, respectively (Fig.3).

In the case of failure of the surveillance system it will be necessary to re-establish the inventory. The spent-fuel inventory is re-established, using a prepared sampling plan, by randomly selecting irradiated fuel assemblies for check weighing in the long-term storage and then transporting some of these assemblies, by means of the refuelling machines, to the hot cell for visual identification and qualitative estimation of burnup and plutonium production using a high-resolution gamma-spectroscopic technique.

SURVEILLANCE OBJECTIVES

Surveillance is maintained on the exit through which the irradiated fuel assemblies must be moved (using the refuelling machines) from either the long-term storage or the reactor hall to the shipping cask. This surveillance also covers the access to the reactor hall and fresh fuel storage and assembly area (Fig.3).

In the past two twin-movie camera surveillance units, operating in the single-frame mode, were used to survey the access corridor in an attempt to minimize the failure of the surveillance equipment and to increase the surveillance of this strategic location. Problems with poor illumination and equipment failure have necessitated the re-establishment of the inventory. The operator has since agreed to maintain an acceptable level illumination at all times, including the installation of a "stand-by" lighting system.

One of the twin-movie camera units has now been replaced by a twin-camera closed-circuit TV system which has enabled the surveillance record to be evaluated at the facility during an inspection.

SPENT-FUEL MEASUREMENTS

The results of spent-fuel burnup measurements presented in this paper were obtained during the completion of IAEA-Bohunice research contracts Nos 1434/RB [3] and 2118/RS [4] and during routine IAEA inspections. All the NDA measurements on the irradiated assemblies were performed in the A-1 facility dry hot cell using a special "slit"-type collimator [3].

The selected irradiated fuel assemblies were measured using high-resolution gamma-ray spectroscopy (HRGRS) and later some were further investigated destructively. Burnup profiles of each selected assembly were obtained by scanning. Measurements were also carried out with the aim of determining the radial power distribution across the fuel assemblies.

The following fission products were used as burnup monitors: ^{137}Cs, ^{144}Ce and $^{134}Cs/^{137}Cs$ atomic ratio. The $^{106}Ru/^{137}Cs$ atomic ratio was used to determine the plutonium content in the spent fuel.

Destructive radiochemical analysis was used to establish the correlation between Pu content and $^{106}Ru/^{137}Cs$ ratio for the central type of fuel assembly. The correlation between burnup and uranium isotopic composition of the spent fuel was also established following the mass-spectroscopic analysis of nine selected samples.

The experimental equipment consisted of a planar-type Ge(Li) detector with a sensitive volume of 3.5 cm^3 (resolution of 2.7 kV at 662 kV) and a 4000-channel DIDAC multichannel analyser (MCA) with associated accessories.

The burnup measurements completed during the course of routine IAEA inspections were performed using coaxial-type Ge intrinsic detectors with sensitive volumes of 10 and 50 cm^3 and a 1018-channel SILENA MCA. The spectra obtained were recorded and later transferred to a ND 6620 Processing System memory for further treatment.

METHODS AND RELATIONSHIPS USED FOR BURNUP DETERMINATION

The initial measurements made on each assembly were to scan over the length of the assembly by taking approximately 15-minute measurements at 30-cm intervals. The measurement locations were carefully selected to be between the spacer grids of the assembly. Near the location of maximum gamma intensity two to three one-hour-long measurements were performed.

The evaluation of fuel burnup using the measured atomic ratios of $^{134}Cs/^{137}Cs$ requires the calculation of the relative fission product activities in the measured volume using the following relationship:

$$A_1 = P.\frac{I_{ij}}{J_{ij}.E_j} \qquad (1)$$

where

A is the activity of the i-th monitor;

J_{ij} is the corresponding yield of gamma quanta/β-decay for the E_j energy;

E is the relative efficiency for the energy E_j includes geometric factor, detector efficiency and self-absorption;

I_{ij} is the registered number of gamma quanta from the i-th product decay for energy E_j;

P is the normalization factor for the corresponding measurement, independent of energy and monitor.

NB To determine the absolute burnup only the one-hour measurement is used. For the other co-ordinates the burnup is assigned using $^{137}Ba^m$ gamma-radiation intensity ratio.

The gamma spectra obtained were evaluated using HOREC [5] computer code which generates the I_{ij} values (with corrresponding standard deviations). This program performs a Gaussian least-squares fit for the peaks of the spectrum with a linear background.

To obtain the relative efficiency of the detector the ^{134}Cs and ^{106}Ru peaks were calculated by the ENRI code using the method described by Harry et al. [6] and the corresponding J_{ij} values were taken from the paper by Erdtmann and Soyka [7]. Although the conditions required for the application of the above-mentioned method [6] are not met, the same distribution of ^{134}Cs and ^{106}Ru was assumed through the cross-section of a fuel assembly. In reality, the different radial distribution of these isotopes (see Fig.4) contributes a less than 2% error to the calibration curve. The measured activities were corrected for radioactive

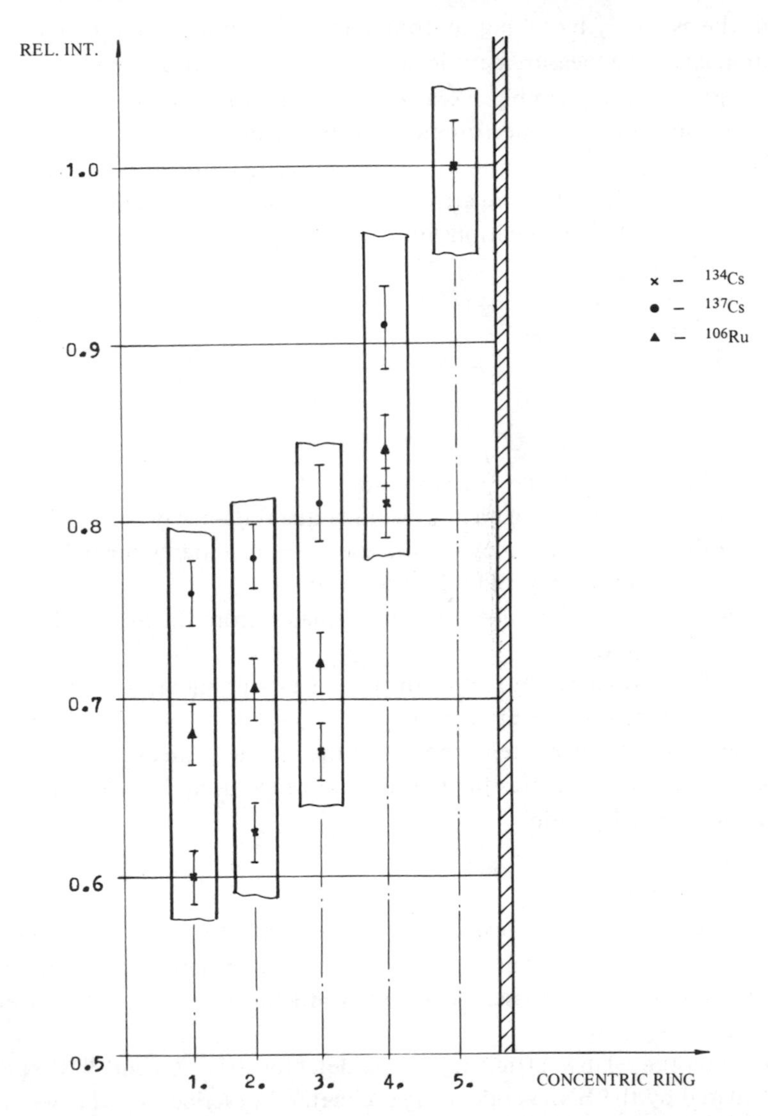

FIG.4. Fission-product distribution along the radius of the fuel assembly ZNOO53 for a burnup value of 6180 MW·d/t.

decay and radiation capture during the fuel irradiation and cooling periods by means of the FIPO code [8] using the following relationship:

$$B(MW \cdot d/t) = a + b \left(\frac{n_{134\text{-}Cs}}{n_{137\text{-}Cs}}\right)^{Corr} \quad (2)$$

where:

B = is the burnup value in MW·d/t U

a = = −340 ± 88

b = 171830 ± 2594

and $\left(\frac{n_{134\text{-}Cs}}{n_{137\text{-}Cs}}\right)^{Corr}$ is the corrected atomic density ratio of ^{134}Cs and ^{137}Cs for the given volume of fuel.

The average fuel assembly burnup values were obtained by corresponding integration of the axial burnup profiles. The measured fuel assemblies burnup values, as well as examples of axial profiles, are illustrated in Fig.3 and Table I.

In the case of low burnup, i.e. 1800 MW·d/t U the accuracy of formula (2) is reduced to 8% [3] but, for lower values, a greater deviation is to be expected.

ERRORS INTRODUCED IN THE SPECTROMETRIC BURNUP MEASUREMENTS

When measuring the burnup by a non-destructive method, it is practically impossible to determine the burnup profile through the fuel assembly cross-section. From Figs 4 and 5 it can be seen that the fission-product distributions used for burnup determination have various forms through the fuel assembly cross-section and change with the burnup. Evaluation of these effects has been done by the codes UNIK-HET and ENRI [9]. In the case of ^{106}Ru and ^{134}Cs the calculated change was less than 2%. The error introduced by the change of the fission-product distribution profiles due to burnup may be neglected.

RELATION BETWEEN THE PRODUCED PLUTONIUM AND ^{106}Ru CONTENT IN THE SPENT FUEL

As for the spent-fuel burnup determination, simple correlations were derived for plutonium produced in spent fuel. On the basis of calculations it has been shown by Petényi and Mikušová [10] that there is a linear relation between the generated atom number ratios of ^{106}Ru, ^{137}Cs and plutonium concentration ratio for a broad burnup interval. To get experimental dependencies of the above

TABLE I. COMPARISON OF MEASURED BURNUP VALUES WITH VALUES GIVEN BY THE OPERATOR

Fuel assembly	Burnup (MW·d/t U)		
	Reactor operator	Non-destructive gamma-spectrometric	Destructive and non-destructive
ON 0076	5175	4960	5020
2N 0005	590	530	–
2N 0053	4763	4530	4430
ON 0042	2047	1780	–
2N 0068	4728	4500	–
2T 0019	5090	4910	–
ON 0183	4857	4502	–

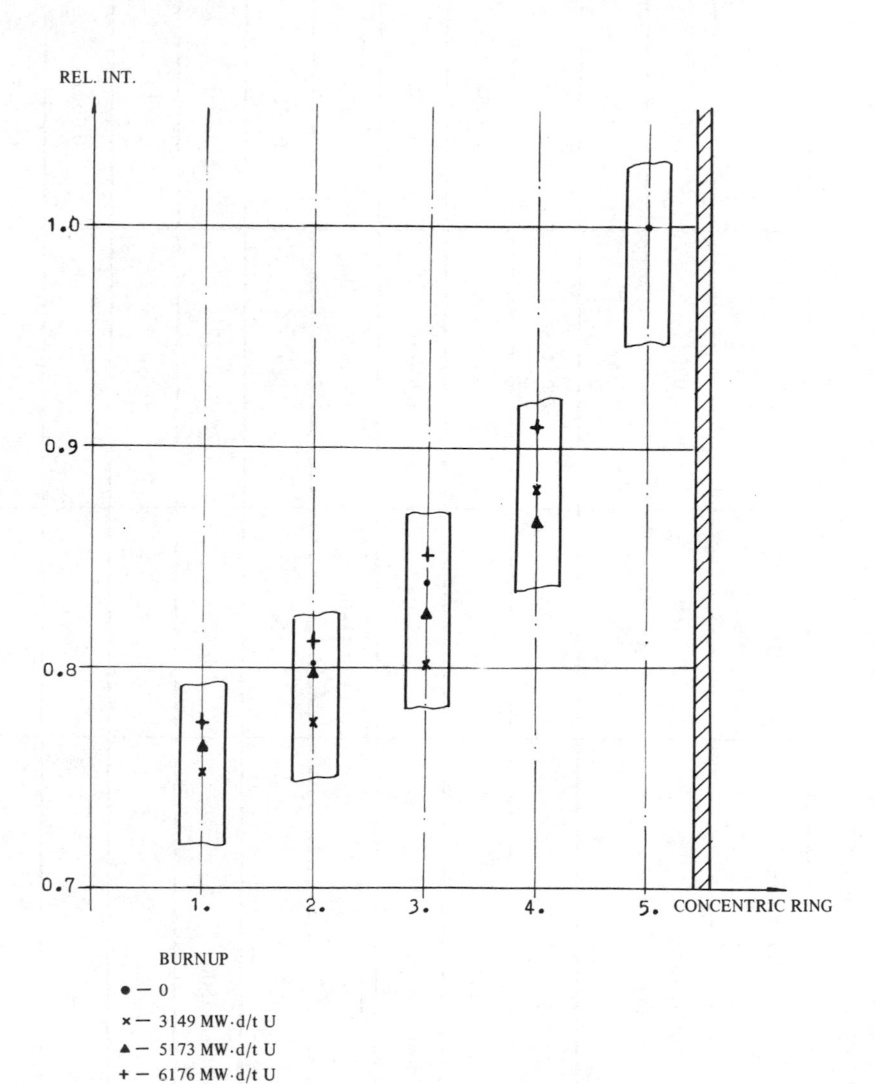

FIG.5. Change of relative profile of ^{137}Cs distribution versus burnup (during physical tests depicts relative activity profile).

TABLE II. MEASURED BURNUP VALUES AND ISOTOPIC COMPOSITION OF URANIUM IN SAMPLES OF SPENT FUEL FROM FUEL ASSEMBLY 2N0053

Sample	Burnup (MW·d/t U)	ΔU_5 wt%	ΔU_6 wt%	$\Delta U_6/\Delta U_5$
1	1330	0.1269	0.02011	0.1584
2	3590	0.2939	0.04613	0.1570
3	5810	0.4240	0.06486	0.1530
4	6850	0.4659	0.07165	0.1538
5	4450	0.3550	0.05562	0.1570
6	6200	0.4360	0.06676	0.153
7	5770	0.4147	0.06326	0.1525
8	5500	0.3995	0.06177	0.1546
Mean	4937	0.3645	0.05627	0.1549
Standard dev.	–	–	–	0.00225
Variation	–	–	–	0.0145

ratios, destructive radiochemical analyses were carried out. From the results of the radiochemical analysis [11] an experimental correlation was established:

$$Pu/U\ [kg/t\ U] = a + b \left(\frac{n_{106\text{-}Ru}}{n_{137\text{-}Cs}}\right)^{Corr} \tag{3}$$

where:

$\left(\frac{n_{106\text{-}Ru}}{n_{137\text{-}Cs}}\right)^{Corr}$ is the ratio between the ^{106}Ru and ^{137}Cs isotopic concentration values corrected for decay and radiation capture using FIPO code.

The a and b coefficients have the following values:

Central-type assemblies:	*Peripheral-type assemblies:*
a = −1.84 ± 0.39	a = −1.65 ± 0.145
b = 19.14 ± 1.48	b = 17.35 ± 0.49

These coefficients are valide for the burnup interval of 2000–8000 MW·d/t U. Correlation (3) may be advantageously used for inspection procedures within the IAEA Safeguards System.

CORRELATION BETWEEN THE CONCENTRATIONS OF ^{236}U AND ^{235}U IN THE SPENT FUEL

By means of mass-spectrometric measurements of the uranium isotopic composition in samples taken from the fuel element 2NOO53, simple correlations were established. From the results of the analysis the existence of a simple linear correlation between the isotopes ^{235}U and ^{236}was found. The mass-spectrometric measurement results [11] and the corresponding burnup values, obtained by destructive analysis of ^{137}Cs contents in the burned-up samples are given in Table II. The experimental value obtained for the ratio of $^{236}U/^{235}U$ = 0.1549 ± 0.0023 and this result is quite close to the value of 0.1523 obtained for the NPD reactor [12].

CONCLUSIONS

Routine Agency inspections (usually four to five annually) have been performed at Bohunice A-1 Nuclear Power Plant since the NPT agreement between

Czechoslovakia and the IAEA was concluded, and much experience in the safeguarding of this type of facility has been gained.

Owing to the facts that the reactor is refuelled "on load" and that the irradiated fuel in the long-term storage is not amenable to the normal identification and counting techniques, it has been necessary to re-establish the inventory by NDA techniques following surveillance failure. Improved surveillance of the fuel access points should ensure that the facility continues to be effectively safeguarded.

In the event of future surveillance failure it will be again necessary to re-establish the inventory but, nonetheless, it is the intention of the Agency to perform spent-fuel measurements (with the willing co-operation of the facility) as a matter of routine during inspections.

It should be emphasized that correlations between different fission-product yields and burnup values have been established. The correlation between Pu content and $^{106}Ru/^{137}Cs$ ratio is particularly important from an inspection viewpoint, as it can be used to verify the plutonium content of stored spent fuel.

The development in measurements on fuel and improvements in the surveillance at this facility have been greatly assisted by the ready and helpful co-operation provided by the facility personnel.

REFERENCES

[1] INTERNATIONAL ATOMIC ENERGY AGENCY, The Agreement between the Czechoslovak Socialist Republic and the Agency for the Application of Safeguards in Connection with the Treaty for the Non-Proliferation of Nuclear Weapons INFCIRC/173, 1972.

[2] CZECHOSLOVAKIAN ATOMIC ENERGY COMMISSION, Regulations Governing Nuclear Material Accountancy and Control (1977).

[3] INTERNATIONAL ATOMIC ENERGY AGENCY, IAEA Research Contract No.1443/RS.

[4] INTERNATIONAL ATOMIC ENERGY AGENCY, IAEA Research Contract No.2118/RB.

[5] PETENYI, V., KRAJČÍ, L., Code HOREC for automatic gamma spectrum analysis measured by Ge(Li) detectors, VÚJE (1978).

[6] HARRY, R.J.S., AALDIJK, J.K., BRAAK, J.P., "Gamma spectrometric determination of isotopic composition without use of standards" Safeguarding Nuclear Materials (Proc. Symp. Vienna 1975) 2, IAEA, Vienna (1976) 235.

[7] ERDTMANN, G., SOYKA, W., Die Gamma Linien der Radionuklide, Jül-1003-AC (Sep. 1973).

[8] PETENYI, V., Code FIPO, EBO (1975).

[9] KRAJČÍ, L., Code ENRI for relative peak efficiency versus the gamma energy calculation, VÚJE (1978).

[10] PETENYI, V., MIKUŠOVÁ, D., Non-destructive burn-up measurements at the A-1 Nuclear Power Plant, EBO 33/75.

[11] SEBESTA, F., Written communication concerning supporting results, letters dated 1978-02-08 and 1978-05-10, FJFI, Prague (1978).

[12] BEETS, C., Contribution to the Joint Safeguard Experiment at the Eurochemic Reprocessing Plant, Mol, Belgium, (Mol-IV), BLG486.

DISCUSSION

D. GUPTA: In your paper you have reported on the extensive use of cameras for safeguarding on-load refuelled reactors. Would you like to comment on your experience with the use of these cameras, particularly with regard to (a) the number of pictures to be evaluated within a given time interval, (b) the method of evaluation of these pictures, and (c) the frequency of false interpretation encountered so far.

V. POROYKOV: Up to 7000 frames of film are evaluated per camera within a surveillance period of up to four months. The method of evaluation is for the inspector to review the film frame by frame. If, during this evaluation, the inspector observes any movement of equipment or containers, he compares with the relevant facility operating records to ascertain the reason. If there are minor discrepancies, these are discussed with the operator. Practically no false interpretation has occurred.

IAEA NON-DESTRUCTIVE ANALYSIS MEASUREMENTS AS APPLIED TO IRRADIATED FUEL ASSEMBLIES

N. BEYER, M. DE CAROLIS, E. DERMENDJIEV,
A. KEDDAR, D. RUNDQUIST
Department of Safeguards,
International Atomic Energy Agency, Vienna

Abstract

IAEA NON-DESTRUCTIVE ANALYSIS MEASUREMENTS AS APPLIED TO IRRADIATED FUEL ASSEMBLIES.

Among the important measurement tasks for IAEA Safeguards is the verification of the declared plutonium content of irradiated fuel elements (nuclear production). Although no method exists to measure the plutonium content in spent-fuel elements directly, the production of plutonium can be calculated from burnup measurements. This paper describes the recent experiences of the IAEA in applying non-destructive assay (NDA) techniques to burnup determination. Most of the discussion concerns techniques based upon the measurement of gamma rays emitted from fission products accumulated in the fuel. High-resolution gamma-ray spectroscopy was used to determine ratios of characteristic gamma rays emitted from fission products such as ^{137}Cs, ^{134}Cs, ^{106}Ru, ^{144}Ce etc., and a correlation with burnup established. Measurements of irradiated fuel from PWRs, BWRs, HWRs, HTGRs and several research reactors are reported. For these reactors, the burnup values ranged from 1000 to 30 000 MW·d/t and in most cases the cooling time varied from several months to several years. The accuracy of the IAEA NDA burnup measurements, averaged over all cases, was about 15% compared with the operators' calculated values, which are generally reported with a stated accuracy of from 5 to 10%.

1. INTRODUCTION

The fact that the largest amount of plutonium in the world is in the form of spent fuel makes its safeguarding an important task of the IAEA, and because it is in the form of spent-fuel assemblies it is necessary to use NDA-type measurements. The goal of the IAEA NDA measurements is to:

(i) Establish an initial inventory verification of the material if it has not been safeguarded by containment and surveillance techniques, and to
(ii) Re-establish an inventory verification if containment and surveillance techniques fail.

To accomplish this goal two types of measurement are used:

(i) Qualitative verification (i.e. identification of material and detection of replacement by dummies).
(ii) Quantitative verification (i.e. determination of burnup and plutonium content).

Other authors [1–3] have shown that NDA measurements of spent fuel can be realized by analysing the gamma rays emitted from fission products or by detecting spontaneous fission neutrons emitted from some transuranium isotopes accumulated in the fuel. Recently some promising results were obtained by Dowdy et al. [4], who evaluated the use of different techniques, such as Cherenkov radiation detectors, thermoluminescent detectors and fission and ion chambers, for spent-fuel verification.

This paper discusses some of the IAEA safeguards NDA measurements made of different types of irradiated assembly, the different techniques employed, the results obtained for these spent-fuel measurements and possibilities for improving the measurements. Most of this paper is concerned with the application of NDA techniques based upon the use of high-resolution gamma-ray spectroscopy (HRGRS) for burnup measurements, since most of our work has involved these techniques. Determination of plutonium content is of the highest priority to the IAEA, but experience in this problem is much more limited and we therefore mention it only briefly in the paper.

2. MEASUREMENT PROCEDURE

2.1. Introduction

Some years ago it was found that the activities of some fission products (FP) such as ^{137}Cs, ^{144}Ce, ^{106}Ru, are proportional to the burnup values and can be used as burnup monitors [2, 3]. More convenient are the measurements of the activity of ratios of different FPs – ^{134}Cs/^{137}Cs, ^{154}Eu/^{137}Cs [2, 3] – which are the ones most used in the IAEA's spent-fuel measurements. This is accomplished by employing HRGRS.

2.2. Apparatus

The favourite geometry for spent-fuel measurements is to isolate an assembly under water and to use a long pipe as a collimator [2]. In addition, the assembly can usually be moved up and down for scanning. However, for some of our measurements the assemblies were placed in a dry hot cell [5], or another special type of container [6]. In either case, if the rotation of the assembly is impossible then a detector/collimator arrangement which "sees" a full cross-section of an assembly is desirable in order to obtain a gamma-ray spectral measurement averaged over all the rods, which contribute to the measured count-rate.

We used intrinsic germanium detectors with sensitive volume from 10 to 70 cm^3 in combination with a Silena portable multichannel analyser (MCA). After a preliminary analysis, the measured spectra can be recorded on magnetic tape and later transferred to our ND 6620 gamma-ray data-processing system memory for further evaluation. The energy region that is usually measured is between 0.5–1.5 MeV because it includes the most intensive gamma lines of ^{137}Cs, ^{134}Cs, ^{106}Ru, ^{95}Zr etc. Unfortunately, these measurements can only be realized if the irradiated assembly is isolated from neighbouring assemblies, either in the storage pond or in a dry hot cell, because of the strong gamma and neutron background radiation emitted by other nearby assemblies.

2.3. Measurement technique

It is important to obtain measurement values that are representative of the true burnup and these can be dependent upon the point on an assembly where a measurement is made.

It has been found that the burnup profile varies for different reactors, and for a given reactor depends on the core position of assembly, position of the control rods, etc. [5]. If HRGRS burnup measurements are realized only in a few points of the assembly, then the obtained values should be averaged over axial burnup distribution. The IAEA has made profile measurements using CdTe detectors and, recently, simple but sensitive fission and ion chambers have been used. The preliminary results show that the use of such chambers in combination with germanium detectors can significantly reduce the total measuring time per assembly, as compared with using germanium detectors alone.

In the case of multi-rod assemblies (such as PWR 15 × 15 rods or BWR 8 × 8 rods) a non-uniform transverse burnup distribution should be expected, because of the strong thermal neutron absorption in outer rows. In the measurements of Natsume et al. [7, 9], the horizontal burnup non-uniformity of an irradiated BWR assembly was about 20–30%. For an 8 × 8 rod assembly, the difference in $^{134}Cs/^{137}Cs$ ratios calculated for uniform and non-uniform burnup distribution (outer rods assumed to have 10–12% higher burnup) may reach ~ 16% [3]. This difference is due to the effect of the different source distribution and transmission of ^{137}Cs (E_γ = 662 keV) and ^{134}Cs (E_γ = 605 keV and 796 keV) gamma rays. Another significant effect observed for BWR assemblies is the axial dependence of the horizontal burnup distributions due to the different neutron spectra of the upper and lower parts of the assemblies [7, 8]. The

TABLE I. DETERMINATION OF IRRADIATION HISTORY AND COOLING-TIME CORRECTIONS

Case 1.
(1) Known irradiation history of spent fuel
(2) Known cooling time T_c
(3) Measured high-resolution gamma spectra are available; use of program SPFUEL for exact $^{134}Cs/^{137}Cs$ activity ratio calculations.

Case 2.
(1) Known irradiation history
(2) Unknown cooling time T_c, but $T_c \neq 1$ year
(3) Measured high-resolution gamma spectra are available; use of the following simple formulas [5]:

$$T_c = \frac{1}{\lambda_2 - \lambda_1} \cdot \ln \left[\frac{0.5359 \cdot f(t)}{1 - 0.4641 \cdot (A_{Nb}/A_{Zr})} \right]$$

where

$$f(t) = \frac{q_n \cdot (1 - e^{-0.0197 \cdot t_U}) + \sum_{l=1}^{n-1} q_i \cdot (1 - e^{-0.0197 \cdot t_i}) \cdot e^{-0.0197} \cdot \sum_{k=i+1}^{n} t_k}{q_n \cdot (1 - e^{-0.0106 \cdot t_U}) + \sum_{i=1}^{n-1} q_i \cdot (e - 1^{-0.0106 \cdot t_i}) \cdot e^{-0.0106} \cdot \sum_{k=i+1}^{n} t_k}$$

These formulas can be used for different fission product activity ratio (Nb/Zr, $Zr/^{137}Cs$, etc.) calculations.

Case 3.
(1) Unknown irradiation history
(2) Unknown cooling time
(3) Measured high-resolution gamma-spectra are available *only;* In this case some assumptions should be made –
(a) Constant reactor power;
(b) One or more irradiation cycles with equal duration (e.g. 100 days $\neq$ t $\neq$ 300 days).
Then, using activity ratios of different fission products, the cooling time T_c can be estimated.

effects mentioned above were found for a BWR assembly with a burnup value of 5570 MW·d/t. Unfortunately, it is difficult to obtain a correct estimate of the horizontal non-uniformity of burnup for PWR assemblies and for higher burnup values (about 25 000–35 000 MW·d/t) normally reached in modern BWR and PWR reactors. It might be assumed that the larger the burnup the smaller the burnup non-uniformity, because of the decrease of ^{235}U in outer rows. It is clear that the experimental data available are incomplete and might be used only in limited IAEA inspection of spent-fuel measurements.

This short discussion would be incomplete without mention of the migration of fission products (FP), their radial distribution in rods [5, 10] and the possible change of relative "yield velocities" of ^{137}Cs and ^{134}Cs at high burnup values [11]. These problems are still far from being completely solved.

3. DATA EVALUATION

3.1. Corrections

3.1.1. Relative efficiency

Corrections for the germanium detector efficiency and gamma-ray absorption in the material of the assembly have been calculated with our RELEFF program and our ND 6620. The relative efficiency functions have been calculated using ^{134}Cs gamma lines; in the case of poor statistics, both ^{134}Cs and ^{106}Ru lines have been used. We estimate the accuracy of calculated relative efficiency corrections as about 5–10%.

3.1.2. Irradiation history and cooling time

Interpretation of the gamma-ray FP ratio measurements must include corrections for the effects of irradiation history and cooling periods. The ways in which these corrections are calculated depends upon the amount of information made available to the inspector by the operator. These appear to fall into three cases, which are presented in Table I, and are briefly as follows:

(i) Complete information available (e.g. history, cooling, etc.)
(ii) Limited information (e.g. history only)
(iii) Little or no information (i.e. only measured gamma spectra are available).

In the first case, irradiation history corrections and cooling-time corrections can be calculated with reasonably good accuracy. We use our SPFUEL program, which calculates corrected $^{134}Cs/^{137}Cs$ ratios or other FP ratios for measured assemblies.

If only limited information is available, then (see Table I) the cooling time T_c can be calculated if $T_c < 1$ year. The $^{95}Zr/^{95}Nb$ activity ratio allows the determination of T_c with an accuracy better than 10% [2,5,12]. In the case of no available information, only estimates of T_c can be made, assuming constant reactor power and one or more irradiation cycles with equal duration (in our calculations $100\ d < t < 300\ d$).

3.2. Burnup determination

3.2.1. Introduction

We now briefly discuss some possible ways used in IAEA inspections to treat spent-fuel data and to verify burnup. We again emphasize the specific difficulties of inspection burnup measurements as contrasted with laboratory burnup measurements: limited time, non-optimal conditions, no burnup distribution measurements for rods or assays, no means of establishing any relationships between burnup and FP yields for given fuel, etc. On the other hand, correlations between burnup and FP yields are available only for limited types of fuel in any case. These circumstances force us to use the following approximate methods.

3.2.2. Consistency method (dependent on operator's burnup data)

Using the operator's burnup values for several well-measured "standard" assemblies a relationship between burnup and measured FP activity ratios can be established. It can then be used for checking the operator's burnup values for other measured assemblies [13]. This method is relatively simple and does not require any detailed investigations of a given type of spent fuel. The most important disadvantages of this method are that it is impossible to detect any systematic errors in declared burnup values, and that it is not independent of the operator's declaration.

3.2.3. Comparative method (dependent on burnup data from other sources)

This approach establishes a relationship between the measurements of the $^{134}Cs/^{137}Cs$ or other FP activity ratios and burnup values (and their associated FP activity ratios) obtained elsewhere for similar fuel. Unfortunately, there are only a few detailed non-destructive investigations of irradiated fuel [5, 7, 9, 11, 14, 15]. However, the relationships between burnup and the FP activity ratio $^{134}Cs/^{137}Cs$ for TRINO PWR [14], Novo-Voronezh PWR [11] and Bohunice HWR [5], shown on Fig.1, are close to each other to within a 10–15% error. This similarity was assumed to be more or less general, and it was used for several IAEA spent-fuel measurements with unknown relationship between burnup and Cs ratios.

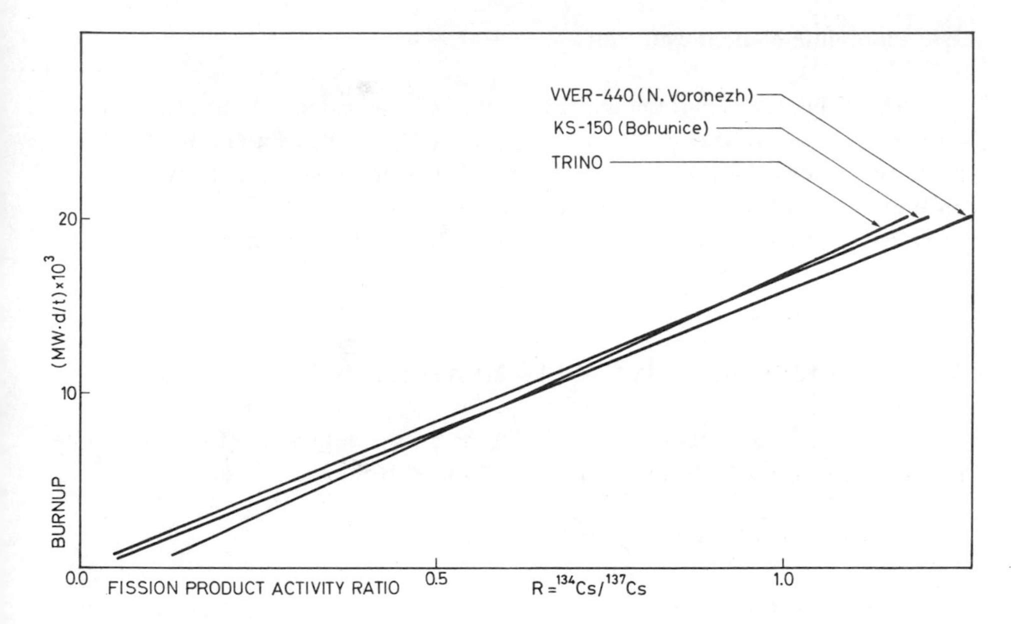

FIG.1. Relationships between burnup and fission-product ratio of $^{134}Cs/^{137}Cs$ for spent fuel from three reactors.

This method has some attractive features because sometimes destructive analysis data are available and an accuracy of 5% for burnup values can be reached [5, 14]. However, if destructive analysis data are extrapolated to other types of irradiated assembly, then the error might increase to 15–20%, both because the curves of Fig.1 are assumed to be applicable, and because of the unknown burnup profiles.

3.2.4. Absolute method (calculated from FP measurements)

For this method the burnup is calculated based on the measured FP activities, known detector efficiency, calculated assembly self-absorption corrections, known reactor neutron flux values, irradiation history, cooling time etc. [5]. The accuracy of burnup values for HWR assemblies calculated by the FIPO code [5] has been found to be about 5–7%. Our program SPFUEL, used for IAEA spent-fuel data treatment, is similar. These burnup calculations evidently require additional information, especially for PWRs and BWRs – core position of the measured assembly, core distribution and absolute values of neutron flux, neutron fission and capture cross-sections etc., to obtain good accuracy.

3.3. Plutonium content estimates

The estimates of plutonium content in irradiated assemblies can be made on the basis of isotope correlation techniques (ICT), such as Pu-burnup or Pu/U-burnup correlations established for different uranium enrichments and different types of fuel: PWR [14], BWR [7–9], HWR [5]. It was found that these relationships are practically insensitive to uranium enrichment [14].

4. RESULTS OF IAEA SPENT-FUEL MEASUREMENTS

This section includes discussions of some results obtained in IAEA spent-fuel measurements since 1976 and are presented by reactor type.

4.1. LWR (PWR and BWR)

Several Kozlodui PWR (WWER) irradiated assemblies were measured using the NDA HRGRS technique [16]. The relationship between $^{134}Cs/^{137}Cs$ ratios and burnup values was established on the basis of destructive analysis data of a WWER irradiated assembly. The accuracy of burnup values obtained in Kozlodui spent-fuel measurements was about 15%. Recent IAEA spent-fuel measurements of WWER assemblies gave burnup values with an accuracy of about 8%, which is better than in previous measurements, perhaps because an improved $^{134}Cs/^{137}Cs$ burnup correlation obtained for irradiated assemblies [11] was used:

$$B = (16.6 \pm 0.9)R \qquad (1)$$

where B is the burnup in kg/t U and R is the $^{134}Cs/^{137}Cs$ activity ratio.

In both measurements all assemblies were scanned and burnup values were averaged over burnup profile functions.

One of the authors participated in an investigation of PWR 15 × 15 rod assemblies. Fission and ion chambers were used for rapid burnup profile measurements of PWR 15 × 15 rod assemblies. The FP activities were measured by a germanium detector. Preliminary results show that the use of these chambers in combination with a germanium detector may significantly reduce the measuring time per assembly. In both cases plutonium content might be estimated using Pu-burnup correlations obtained for the TRINO PWR [14].

In one of the IAEA HRGRS measurements of BWR irradiated assemblies there was no operator's information about the measured assemblies available (see section 3.1.2). Assuming constant reactor power and irradiation cycle duration 100 days $< t <$ 300 days (see Table I, case 3), the burnup values were estimated. Later they were compared with operator's values and the standard deviation was found to be about 22%. Other detailed investigations of BWR spent fuel have been done under an IAEA research contract [7]. BWR 6 X 6 rod uranium oxide irradiated assemblies were analysed by HRGRS and destructive analyses were also made [7–9]. Some significant perturbing effects were found – non-uniform burnup cross-sections, spatial variation of neutron spectra, etc. The concentrations of ^{237}Np, ^{241}Am, $^{242}Am^m$, ^{242}Cm and ^{244}Cm were determined destructively and different heavy isotope correlations were investigated and established. This information seems to be very important for further improvement of NDA safeguards HRGRS spent-fuel measurements. However, the knowledge of ^{242}Cm, ^{244}Cm and plutonium isotope concentrations in correlation with burnup values might be useful for developing an independent passive neutron method of burnup determination [3, 17].

4.2. HWR

The safeguarding of HWR spent fuel is one of the most important IAEA safeguards problems, because plutonium accumulated in HWR irradiated assemblies is mostly "weapon grade" at the low burnup values which might be realized in such reactors. The data that were obtained in Bohunice HWR under an IAEA research contract [5] should be noted. Several irradiated assemblies were measured by the HRGRS technique and then analysed destructively. The radial distribution of the main FP in rods, burnup cross-section profiles, heavy isotopes correlations etc., were investigated. It was found that the accuracy of NDA determination of burnup values was about 7% [5]. Some other measurements of this HWR spent fuel were done using the $^{134}Cs/^{137}Cs$ vs burnup relationship:

$$B = -(339.7 \pm 88.2) + (171\,830 \pm 2594)R \quad (2)$$

established in previous measurements. Here B is in MW·d/t and R is $^{134}Cs/^{137}Cs$ concentration ratio. The standard deviation between measured and decleared burnup values was less than 7%. Plutonium content was estimated by a Pu/U vs $^{106}Ru/^{137}Cs$ correlation function [5] and by using the calculations of Zaritzkaya et al. [18]. Problems with safeguarding the Bohunice HWR are discussed in more detail in another report [19].

In HRGRS measurements of some other HWR irradiated assemblies, burnup values were calculated by using our SPFUEL code and were compared with operator's declared values. It was found that for six measured assemblies with 1.35% ^{235}U enrichment, for which irradiation history and cooling-time information was available, the standard deviation of measured burnup values was about 26%.

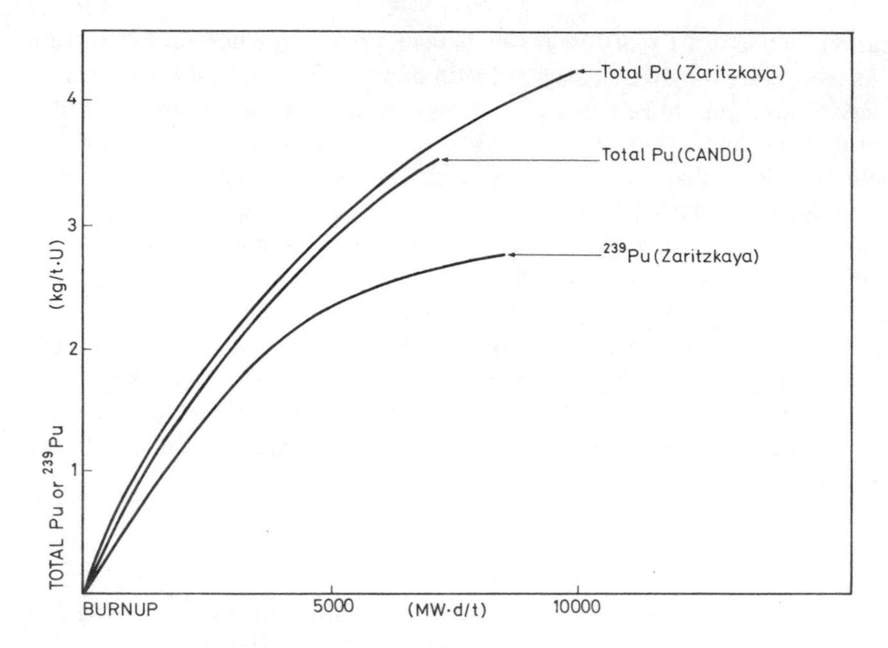

FIG. 2. Relationships between plutonium growth and burnup as calculated for CANDU spent fuel and the work of Zaritzkaya et al. [18].

Burnup values determined for four natural uranium irradiated assemblies have indicated a standard deviation of 17%. Using one of the measured assemblies as a standard, the standard deviation for the enriched uranium assemblies was reduced to 10%.

The same procedure described above for HWR spent fuel was used for data treatment of NDA burnup measurements of CANDU-type irradiated bundles. Six bundles were selected, isolated and measured by the HRGRS technique. A thin lead absorber between the germanium detector and the measured bundle was used to decrease low-energy background and lower the count-rate to a useable range. The declared burnup values of selected bundles ranged from 4500 to 8500 MW·d/t. The standard deviation between measured and declared burnup values was found to be approximately 15%. Because of the complicated irradiation history and long cooling time (T_c = 2 years) of selected bundles, constant reactor power was assumed.

Much more difficult were the burnup measurements of non-isolated CANDU bundles stored under water in special baskets. In this case, bundles were pushed out 12 cm from the basket, one after the other, and one-point HRGRS measurements were done. Using one bundle as a reference bundle, and assuming the same burnup profiles for all measured bundles, an accuracy of about 15% was obtained for burnup values. Unfortunately, there is no destructive analysis data of CANDU spent fuel and hence no direct plutonium content estimations can be made. Indirect estimations might be proposed. Indeed, the total Pu vs burnup curves shown in Fig.2 were taken from Halsall's calculations [20] for CANDU spent fuel and from the calculations of Zaritzkaya et al. [18]. Both curves are similar and are in fact nearly identical. On the other hand, experimentally defined plutonium content values of irradiated assemblies were close to the Zaritzkaya calculated values. Thus, Bohunice destructive data and particularly the Pu/U vs $^{106}Ru/^{137}Cs$ correlation function might be used for CANDU NDA spent-fuel measurements, at least until destructive analysis data for CANDU spent fuel become available.

4.3. HTGR

Several experimental NDA methods and techniques have been developed under an IAEA research contract for safeguarding of the HTGR uranium-thorium fission cycle [21]. Uranium-thorium irradiated pebbles have been measured using the HRGRS technique. The burnup values of these pebbles were determined by ^{137}Cs activities, which were measured with a Ge-Li detector, and calculated by the ORIGEN code. Agreement between measured and calculated ^{137}Cs activities was found. The calculations of FP activities accumulated in the same ^{233}U, ^{235}U or ^{239}Pu single-isotope irradiated samples show significant differences in total FP activities [22]. Thus, comparisons between measured and calculated gamma-ray spectra with different U/U or U/Pu isotopic ratios might allow uranium and plutonium isotopic fractions to be determined [21, 22].

4.4. Research reactors

Some NDA burnup measurements of swimming-pool-type research reactor spent fuel have been done by using HRGRS. An attempt to estimate burnup by detecting spontaneous fission neutrons emitted from WWR spent fuel has also been done under an IAEA research contract [23].

5. CONCLUSIONS

Table II summarizes most IAEA results of NDA spent-fuel measurements. As can be seen, the best accuracy (7–8%) was reached using the comparative method in PWR Kozlodui and HWR Bohunice burnup measurements, mostly because destructive analysis data were available and detailed NDA measurements of irradiated assemblies were done [5, 16]. At present, this method seems to be the most useful.

TABLE II. IAEA POWER-REACTOR SPENT-FUEL MEASUREMENTS

No.	Type of reactor and place	Type of measured fuel	Operator's available information	Range of burnup values (MW·d/t U)	Standard deviation of measured vs declared burnup values (%)[a]
1.	BWR – Kozlodui	WWER-type irradiated assemblies	CP; IH; CT; $^{134}Cs/^{137}Cs$-burnup correlation for WWER assemblies was used	15 000–25 000	15
2.	BWR – Kozlodui				8
3.	BWR	Assemblies	No information; $^{134}Cs/^{137}Cs$-burnup correlation for PWR was used	10 000–20 000	22
4.	HWR – Bohunice	Natural uranium; assemblies	CP; IH; CT; $^{134}Cs/^{137}Cs$-burnup correlation for Bohunice HWR was used	3 000–7 000	7
5.	HWR – Bohunice	Natural uranium; assemblies			7
6.	HWR –	1.35% enrichment; assemblies	CP; IH; CT; NF; FGS; SPFUEL code was used for burnup calculations	3 000–7 000	26
		Natural uranium; assemblies			17
			The same measurements; reference assembly used		10
7.	HWR – CANDU	Natural uranium; bundles	IH; CT; Bohunice $^{134}Cs/^{137}Cs$-burnup correlation was used	3 000–7 000	15
8	HWR – CANDU	Natural uranium;	Measurements of non-isolated bundles; reference bundle used	3 000–7 000	15

[a] These standard deviations include the errors of operator's declared burnup values. CP = core position; IH = irradiation history; CT = cooling time; NF = neutron flux; FCS = fission cross-sections.

There are not enough IAEA measurements of BWR spent fuel to give a real impression of the possible accuracy of NDA safeguards BWR burnup measurements. However, taking into account the data obtained for BWRs [7, 9], one might expect an accuracy for BWR burnup measurements at least comparable with that obtained for PWR spent fuel.

It appears that the use of reference assemblies improves the accuracy of burnup determinations, especially where several measured assemblies have a similar irradiation history and burnup profiles. In the case of HWR spent-fuel measurements, this accuracy was about 10% (see Table II). It can be concluded that the use of a reference assembly decreases the relative errors, but to minimize systematic errors the reference should be measured accurately and, if possible, the burnup value should be determined independently.

For HWR burnup measurements (see Table II, item 6) the accuracy of burnup values calculated by the SPFUEL code is relatively low. Further measurements are probably needed to explain these large differences.

It is important to emphasize that, at least at present, plutonium content in spent fuel can only be estimated because of the very limited ICT information for most types of irradiated assembly.

Finally, in our opinion, more destructive analysis data of different spent-fuel and ICT heavy-isotope correlations are urgently needed to strengthen the NDA measurements.

ACKNOWLEDGEMENTS

The authors appreciate the valuable safeguards spent-fuel data which the IAEA contractors provided. We thank Messrs A. von Baeckmann and A.J. Waligura for their support of this work, and R. Sher for his help in reviewing the manuscript.

REFERENCES

[1] EDWARDS, E.R., Nucl. Appl. **4** (1968) 245.
[2] DRAGNEV, T.N., IAEA/STR-48, IAEA, Vienna (1975).
[3] HSUE, S.T., CRANE, T.W., TALBERT, W.L., Jr., LEE, J.C., LA-6923 (ISPO-9); LASL (1978).
[4] DOWDY, E.J., CALDWELL, J.T., LASL Progr. Rep. (1978).
[5] VALOVIC, J., PETENYI, V., RANA, S.B., MIKUSOVA, D., KMOSENA, J., IAEA Research Contract No.1443, Final Rep. (1977).
[6] GRABER, H., HOFFMAN, G., NAGE, S., IAEA Technical Contract No.1956/TC, Progr. Rep. (1977).

[7] NATSUME, H., MATSUURA, S., OKASHITA, H., UMEZAVA, H., EZURE, H., IAEA Research Contract No.1119, Final Rep. (1975).
[8] MATSUURA, S., TSURUTA, H., SUZAKI, T., OKASHITA, H., UMEZAVA, H., NATSUME, H., J. Nucl. Sci. Technol. **12** (1975) 24.
[9] NATSUME, H., et al., J. Nucl. Sci. Technol. **14** (1975) 745.
[10] PHILLIPS, J.R., LA-5260-T, LASL (1973).
[11] GOLUBEV, L.I., GOROBTZOV, L.I., SIMONOV, V.D., SUNCHUGASHEV, M.A., At. Ehnerg. **41** (1976) 197.
[12] HICK, H., LAMMER, M., in Safeguards Techniques (Proc. Symp. Karlsruhe 1970) **1**, IAEA, Vienna (1970) 533.
[13] DRAGNEV, T.N., DE CAROLIS, M., KEDDAR, A., KONNOV, Y., MARTINEZ-GARCIA, G., WALIGURA, A.J., in Safeguarding Nuclear Materials (Proc. Symp. Vienna 1975) **2**, IAEA, Vienna (1976) 37.
[14] BRESESTI, A.M., et al., EUR-4909E (1972).
[15] EURATOM, Contribution to the Joint Safeguards Experiment MOL IV at the Eurochemic Reprocessing Plant (BEETS, C., Ed.), BLG-486 (1973).
[16] INTERNATIONAL ATOMIC ENERGY AGENCY, Final Rep. IAEA Research Contract No.1547 (1977).
[17] KAZARINOV, N.M., FEDOTOV, L.I., Rep. of Khlopin Radium Institute, Leningrad (1977).
[18] ZARITZKAYA, T.S., KRUGLOV, A.K., RUDIK, A.P., At. Ehnerg. **41** (1976) 321.
[19] PETENYI, V., DERMENDJIEV, E., HORNSBY, J., POROYKOV, V., IAEA-SM-231/106, these Proceedings, Vol.I.
[20] HALSALL, W.J., AECL-2631 (1967).
[21] HECKER, R., et al., IAEA Research Contract No.1559, Final Rep. (1977).
[22] SUSANTI, K., Diplomarbeit, Technische Hochschule Aachen (1977).
[23] INTERNATIONAL ATOMIC ENERGY AGENCY, IAEA Research Contract No.1575 (1977).

DISCUSSION

D. GUPTA: To be quite frank, I have been unable to understand the basic purpose of the investigation discussed in your paper. May I ask, for example, what would be the relation between verification of the operator's burnup data with very large uncertainties and detection of a diversion of a significant quantity? Also, the uncertainties in the assay of Pu content in the irradiated fuel elements using the NDA techniques discussed in the paper appear to be fairly high. They may be too high for the Agency to make a statement on the diversion of a significant quantity (e.g. 8 kg Pu), especially since containment and surveillance are used extensively in parallel, and in most cases the Pu content in irradiated fuels can be determined with a much higher accuracy at the input of a reprocessing plant. Would you like to comment on these points?

E. DERMENDJIEV: As can be seen from Table II of our paper, the uncertainties in burnup values determined by NDA spent-fuel measurements are about 7% for the Bohunice HWR, and less than 10% for the Kozlodui PWR.

This is mainly because destructive analysis data are available. In most cases the accuracies of the declared (i.e. calculated) burnup values are about 5–10%. This means that both the operator's and our measured burnup values are comparable and, hence, our measurements could be used for confirmation of the operator's statement.

What you say, however, could be correct when errors in measured values are much larger than errors in declared values. Of course, I am not discussing such extreme cases as our BWR measurements, for which there was no information available on the measured assemblies, or the HWR measurements (see Table II, item 6), for which there were no additional data (e.g. core positions, irradiation history, neutron fluxes, etc.).

At present, of course, nobody uses NDA HRGRS spent-fuel measurements to make quantitative statements relative to the material accountancy for irradiated fuel elements. The Agency currently uses HRGRS as a means of extending surveillance for positive detection of the presence of spent fuel and to confirm the identity of a specific, selected fuel assembly.

IAEA-SM-231/116

PRACTICAL EXPERIENCE OF NUCLEAR MATERIAL SAFEGUARDS AT LIGHT-WATER REACTORS

L. OUDEJANS, V. POROYKOV
Department of Safeguards,
International Atomic Energy Agency, Vienna

Abstract

PRACTICAL EXPERIENCE OF NUCLEAR MATERIAL SAFEGUARDS AT LIGHT-WATER REACTORS.

The Central and Northern Europe Section of the Agency's Department of Safeguards has been inspecting light-water reactors in the countries for which this Section has been responsible for many years. A description is given of the experience obtained. Mathematical formulae for verification purposes have been developed. Reference is also made to future burnup measurements on irradiated fuel assemblies.

INTRODUCTION

In many articles concerning implementation of safeguards at different facilities, authors have mentioned that the application of safeguards at LWRs is one of the easiest cases and that this problem has to a certain extent been settled. The basis for forming such a conclusion has normally been the fact that nuclear material at LWRs is in the form of assemblies. Up till now practically no simple methods and equipment exist which can permit the measurement of the amount of nuclear material in assemblies.

Some types of LWR have assemblies which cannot be disassembled and re-assembled at the facility without special equipment including welding. This is why methods of accounting for and identification, together with surveillance and containment, were chosen as the main means to verify nuclear material at LWRs. Counting and identification are the safeguards methods applied by the inspectorate. Some limitations have been mentioned to underline the difficulties of applying measurement methods for verifying material in assemblies.

However, there are no methods or recommendations to indicate the probability that any diversion occurred at the facility. At the same time the number of LWRs is rising steadily and a significant quantity of nuclear material is involved in this part of the fuel cycle. This is why the increase of the detection probability that no diversion took place at the LWRs throughout the world would be a valuable contribution to our general ideals.

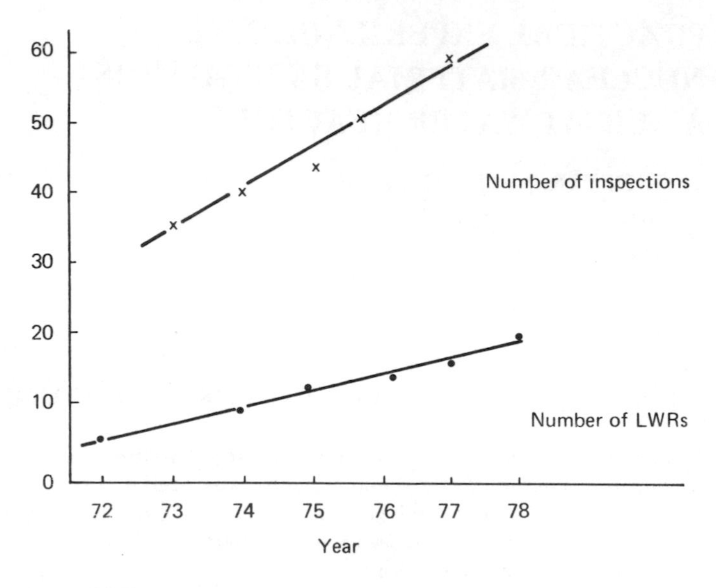

FIG. 1. Number of LWRs and inspections carried out from 1972–78 by the Central and Northern Europe Section, IAEA Department of Safeguards.

The Section for Central & Northern Europe of the Department of Safeguards, IAEA, has had long experience in carrying out safeguards at LWRs (Fig. 1).

This report discusses the existing safeguards approach as applied to LWRs at present, and analyses the weak points of this approach to suggest ways which could bring about the possibility of calculating the detection probability that no diversion occurred at a facility.

1. CHARACTERISTICS OF PLANTS

With a few exceptions, the off-load refuelled nuclear power plants are either boiling water reactors (BWR) or pressurized water reactors (PWR) both of which are widely used throughout the world today. Because both types of reactor use light water as coolant and moderator, they are also called light-water reactors. There are a significant number of light-water reactors of different sizes, constructed by different manufacturers under Agency safeguards. Although technical differences exist between the types of reactor, or differences in design due to different manufacturers, in general the diversion possibilities are the same, so that there is one safeguards approach. Some interesting features of light-water reactors from the safeguards aspect are:

(1) The reactor's core is contained in a steel pressure vessel which is closed during normal operation. The refuelling is done with the reactors shut down and the pressure vessel head removed.
(2) Refuelling is usually carried out once annually and lasts for about 60 days. There exist nuclear power plants with two reactor cores in one reactor hall. Both reactor cores are usually refuelled once annually but on different dates. The period in between is about three to four months. At each refuelling approximately one third of the spent-fuel assemblies in the core are replaced by fresh fuel assemblies.
(3) Light-water reactors use low enriched uranium as fuel. More than one enrichment can be used. The enrichments are normally between 1.5 and 4.0%. The fuel is in the form of UO_2 pellets which are contained in cylindrical rods. The rods are bundled together in fuel assemblies. There are nuclear power plants that receive ready fuel assemblies which are shipped back to the supplier after having been used in the reactor and after a certain cooling time, without any change to the assembly itself, while there are other reactors that have the facility to assemble the fuel rods or are able to perform rod exchange.

2. GENERAL PRINCIPLES OF SAFEGUARDS IMPLEMENTATION

At an early period in setting up the safeguarding of LWRs a safeguards approach was developed using strategic points. Two types of strategic points were established which are key measurement points (KMPs):

(a) To determine nuclear material flow, and
(b) To determine physical inventory.

This mechanism was used as a technical basis for safeguards. The reporting system of the State to the Agency makes use of the two different types of KMP. Facility Attachments, which are practical rules and guidelines for implementing safeguards at facilities, are based on the same principle. Experience has shown that up to now this principle is to a certain extent successful in all countries and at all types of facility where safeguards are applied.

In this report we try to give a simple mathematical description of the flow and inventory of nuclear material at typical light-water reactors (period of time is the period between two consecutive material balance periods of two inspections).

A typical light-water reactor consists of the following KMPs:

For flow:
KMP 1 – Receipt of nuclear material
KMP 2 – Nuclear loss and production in fuel discharged from the reactor
KMP 3 – Shipments

For material at facility: KMP A – Fresh fuel storage
KMP B – Fuel in the reactor core
KMP C – Spent-fuel storage
KMP D – Other location of nuclear material at facility
Stratification of nuclear material at NPS

Since LWRs represent facilities of the attribute type, where nuclear material normally is accounted for on the basis of item counting and identification using original shipper's data, it is not necessary to apply a difficult mathematical approach. Terms in the form of the sum ($\sum$) of items are enough for a determined period of time.

Let us denote items by i and categories of material by j. The flow of material between two consecutive inspections can be expressed using the following equation:

$$\Delta Q_j = \sum_i q_{R_{ij}} + \sum_i q_{S_{ij}} + \sum_i \xi(q)_{p_{ij}} + \sum_i q_{0_{ij}} \tag{1}$$

where j denotes the category of nuclear material, j = 1 is for ^{238}U, j = 2 for ^{235}U etc., and where

- Q_j – the algebraic sum of inventory changes occurred during verified periods
- $\sum_i q_{R_{ij}}$ – the sum of inventory changes as receipt, which corresponds to KMP 1
- $\sum_i q_{S_{ij}}$ – the sum of inventory changes as shipments, which corresponds to KMP 3
- $\sum_i \xi q_{p_{ij}}$ – the sum of nuclear loss and/or production which corresponds to KMP 2
- ξ – the function to describe loss and/or production of nuclear material, normally calculated upon discharge for each assembly
- $\sum_i q_{0_{ij}}$ – the sum of flow of nuclear material other than assemblies.

The material on inventory can be described using Eq.(2):

$$Q_j = \sum_i q_{F_{ij}} + \sum_i q_{C_{ij}} + \sum_i q_{S_{ij}} + \sum_i q_{0_{ij}} \tag{2}$$

where

- Q_j – the quantity of nuclear material; j – category at the facility on the day of verification
- $\sum_i q_{F_{ij}}$ – the sum of non-irradiated material at the facility (fresh fuel assemblies)

$\sum_i q_{C_{ij}}$ – the sum of nuclear material in the core

$\sum_i q_{S_{ij}}$ – the sum of nuclear material in spent-fuel assemblies in spent-fuel storage

$\sum_i q_{O_{ij}}$ – nuclear material in a form different from assemblies (for example, fission chambers etc.)

These equations to describe the material at a facility are commonly used in the Section's inspection activities.

2.1. Description of activities during inspections of nuclear power plants

(a) Verification of design information and nuclear material handling at the facility
(b) Auditing of records kept at the facility and comparison with reports submitted to the Agency
(c) Verification of nuclear material at the facility
(d) Evaluation of inspection results.

2.2. Verification of the construction of the facility and handling the nuclear material at the facility

This information is provided by the States to the Agency and is contained in the Design Information (DI). To verify the DI, the Section organizes a pre-operational visit together with the people concerned in the State. The purpose of this visit is:
(1) To check information in DI
(2) To establish key measurement points
(3) To consider with the operator the strategic points and to establish surveillance and containment devices.

Usually, after the pre-operational visit, the draft of the facility attachment is finally agreed upon.

2.3. Auditing of records at the facility and comparison with the reports to the Agency

Material accountancy is a safeguards measure of fundamental importance (INFCIRC/153, Art. 29).

Auditing of records at the facility has the purpose of ensuring that they are internally consistent and reflect all inventory changes to permit the book inventory to be determined at any time.

Verification of documents concerning accountability at the facility is normally limited to the following examination activities:

(a) Auditing of records kept at the facility concerning inventory changes which occurred during the inspection period (time between two consecutive inspections)
(b) Auditing of records concerning material at present at the facility
(c) Up-dating of nuclear material at the facility
(d) Comparison of records against reports to the Agency.

2.3.1. *Auditing of records kept at the facility concerning inventory changes*

The first term in Eq.(1), for the majority of LWRs can be reduced to:

for ^{238}U – sum of U_{tot} in all received assemblies
for ^{235}U – sum of ^{235}U in all received assemblies

It means that the auditing of records concerning receipt of fresh fuel assemblies consists of comparing data in the shipper's documents with the data included in the General Ledger and subsequently summing up these data in the General Ledger.

During the period since 1972 no significant discrepancies were noted in the shipment documents and records at facilities which were not resolved immediately.

Further steps are to compare the General Ledger data with the ICRs sent to the Agency and, finally, make a comparison between the relevant ICRs at the shipper's facility and at the receiver facility after any inconsistencies have been compared and resolved. We can then say that all relevant documents are internally consistent. This auditing is a very important step as it is the strategic point for calculating the total quantity of material at the facility.

2.3.2. *Shipments*

In the case of shipment, the shipping documents are compared with the General Ledger history cards and the ICRs (in the event that an ICR had already been officially received by the Agency). Checks are made on nuclear loss of total uranium, fissile uranium, on plutonium and on burnup values.

2.3.3. *Nuclear loss and production*

Documents concerning these inventory changes can be checked only by verifying the operating records on internal consistency. Normally they are the core loading diagram, information on thermal and electrical power produced at the facility, operator's calculation burnup for each assembly etc. Since nuclear loss and production are calculated and reported to the Agency upon discharge, this activity is connected with spent fuel verification. Data collected by examination of relevant operating records are compared with data included in history cards for fuel assemblies. These history cards give information on a fuel assembly during its stay in the facility and indicate the location of a fuel assembly at any time. In practice almost 100% of shipper's documents for fresh fuel and spent fuel are at present audited with follow-up comparison with the operator's reports to the Agency. As far as documents regarding nuclear loss and production are concerned, they are checked on a random basis only to check their internal sequence.

2.3.4. Up-dating of nuclear material

Up-dating is one of the most important steps of inspection. It gives the total quantity of nuclear material at the facility at any given time and the breakdown of locations and categories. This operation is normally done during each inspection. It is simply the summing-up of all inventory changes which occurred during the inspection period. To avoid possible falsification, which could occur after the inspection is completed, during each inspection the inspector completes a special working paper, in which all types of inventory change are summarized separately and the sum is given as book ending. This figure is the starting point for book up-dating during the next inspection. Only a few cases have occurred when, during the subsequent inspection, the inspector's figures received during the previous inspection differed from the operator's. In all cases these discrepancies occurred because some insignificant adjustments were made in the General Ledger but relevant ICRs had not been sent to the Agency. All inconsistencies were resolved immediately.

2.4. Verification of the material at the facility

The purpose of independent Agency inspection is to verify nuclear material including independent measurement of all nuclear material subject to safeguards at the facility (INFCIRC/153, Art.74).

The quantity of nuclear material at the facility is described by Eq.(2). For ^{238}U it will be $Q_1 = q_{Fi1} + q_{Ci1} + q_{Si1}$.

If we consider light-water nuclear power plants without exchangeable pins, the only item for accountability is the assembly. We can assume that nuclear material at the NPS in a form other than assemblies is negligible and the last term in Eq.(1) equals 0.

2.5. Fresh fuel storage

The number of items in the fresh fuel storage for the inspection of a reactor varies on an average between 5–10 to approximately 120–200. The normal approach is to account and identify by serial numbers. The requirements for increasing the efficiency of safeguards have made it necessary to develop more

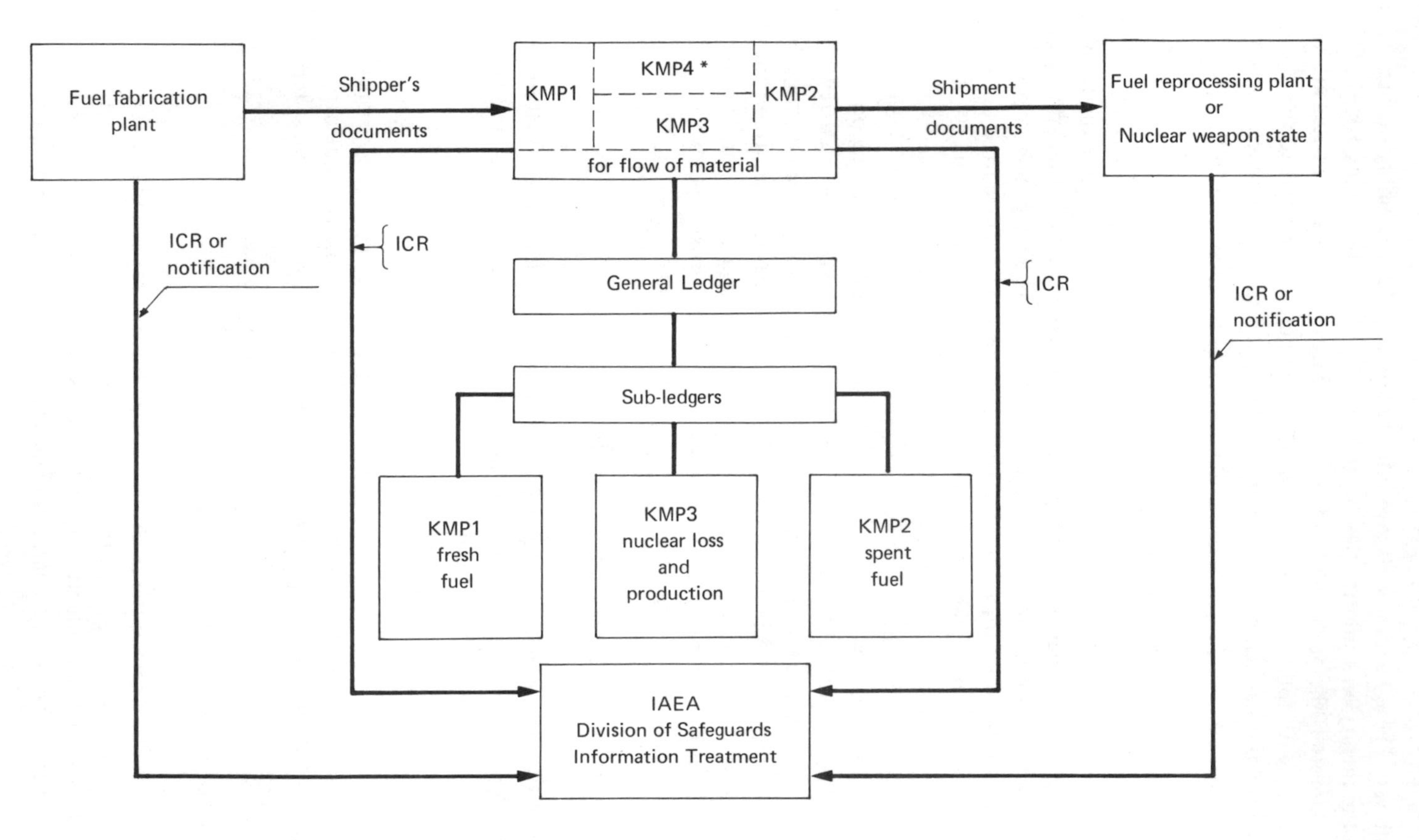

* KMP4 only in the case of pin exchange at the facility

FIG.2. Documents related to flow of nuclear material at LWRs.

accurate methods. Up till now no equipment is available in the Agency for the quantitative measurements of fuel. This is why the Section has reached the conclusion that use must be made of non-destructive assay qualitative methods for verifying fresh fuel (Fig.2). For a certain type of reactor a standard which represents a part of a real assembly was established.

For other types of reactor a method based on comparison of measurement results is applied. A number of fuel assemblies are randomly chosen and average counts for fixed positions are calculated. Using the same instrument at other facilities randomly selected assemblies are measured in the same way. All measurements performed at the same facility during several inspections are combined in one working paper and statistically evaluated. The results are compared with results received at other similar facilities. Taking into account that measurements made at different facilities were carried out at randomly selected assemblies and gave statistically comparable results, we can state, with a certain probability calculated in advance, that no diversion of fresh fuel had taken place.

The number of assemblies needed to be measured are calculated in accordance with attribute sampling plan[1].

2.6. Spent-fuel pond verification

The spent fuel at light water reactors has strategic value owing to the presence of Pu produced in the reactor. Spent fuel has a high level of radiation which causes difficulties in handling spent fuel assemblies. Therefore, for the time being the activity during inspection is limited to the counting and verification of assemblies. The use of surveillance devices, chiefly movie cameras, permits films to be seen and the conclusion to be made that the spent-fuel cask had not been used for illegal purposes. The examination of films goes together with the comparison of operating records such as a log book for operation in the reactor hall, work permission book, refuelling machine record book etc. It is clear that the absence of more precise methods for the verification of material in spent fuel is the weak part of the verification. At present the Section has begun to apply qualitative measurement methods at WWR-440-type reactors. For this a multichannel analyser, Silena, was used. Next year the optimization of conditions of this type of measurement will permit us to verify the operator's burnup calculation.

---

[1] JAECH, J.L., Statistical Methods in Nuclear Material Control, USAEC, TID-26298 (1973); and SHERR, T.S., Attribute Sampling Inspection Procedure based on the Hypergeometric Distribution, USAEC, WASH-1210 (1972).

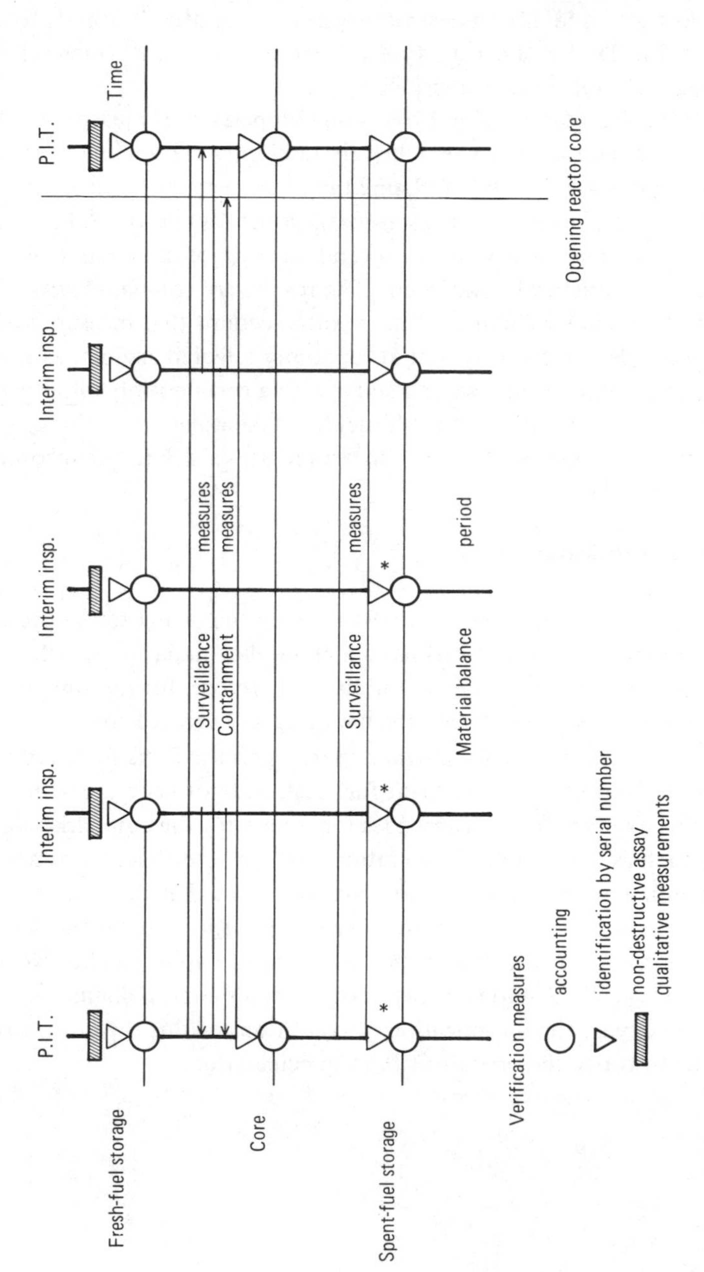

FIG.3. Periodical inspections at LWRs.

2.7. Surveillance and containment measures

Regarding assurances that no diversion occurred at the facility we can state that the level of probability is high that no diversion of fresh fuel has taken place.

At the same time the verification of fuel in the core and spent fuel storage is limited by the absence of convenient non-destructive measurement methods which can be applied to irradiated fuel.

The lack of measurement methods can to a certain extent be compensated for by the application of surveillance and containment methods, which it is possible to apply from the moment the reactor core is closed. Usually physical inventory-taking is scheduled after core refuelling is completed and just before the vessel is closed (Fig.3).

Inspectors verify the core by counting and identification and by cross-checking relevant documents. Simultaneously the fresh fuel and the fuel in the spent-fuel pond are verified. This gives the inspectors the opportunity to conclude that all items indicated on the inventory listing are indeed at the facility. After the reactor is closed, a seal should be applied to guarantee containment of the core. Verification of integrity of this seal during the material balance period ensures that no fuel assemblies were removed from the core.

Counting and identifying spent fuel cannot give full assurance that assemblies between two consecutive inspections were not substituted. Since, at the majority of facilities, spent fuel is kept for many years and since the inspectorate has no adequate equipment for quantitative non-destructive measurements of spent fuel, surveillance methods have special importance for the application of safeguards. Special movie surveillance cameras and/or TV surveillance units are used to observe the movement of the spent-fuel cask.

The frequency between frames is chosen on the basis of the analysis of the crane's speed for moving the cask between the pond and the cask's channel connecting the reactor hall with the outside area. In this event we do not consider moving separate fuel rods which can be taken from assemblies or bundles.

If these frames reflect the movement of the spent-fuel cask, or any unknown containers, the comparison with a log book might show a discrepancy.

SUMMARY

This report describes the activity of the Central and Northern Europe Section of the Department of Safeguards, IAEA in the implementation of safeguards at LWRs.

We have shown that safeguards are implemented in accordance with existing requirements. This task is not simple and many difficulties face inspectors in their practical work. We described mathematically the flow and inventory of nuclear material at the facility. It helps us to collect data concerning the verification activity for each group of material and to make analyses regarding which steps should be taken to increase the probability of timely detection of the diversion of significant quantities of nuclear material which could occur at the facility.

We believe that data collected by the Section can provide the necessary data to create a mathematical model to calculate the probability that, for the whole facility, with a certain level of confidence, no diversion of a significant quantity of nuclear material occurred at the facility during the period inspected.

INTERNATIONAL SAFEGUARDS FOR CRITICAL FACILITIES*

J.F. NEY, J.L. TODD
Sandia Laboratories, Albuquerque,
New Mexico, United States of America

Abstract

INTERNATIONAL SAFEGUARDS FOR CRITICAL FACILITIES.

A study was undertaken to investigate various approaches to provide international safeguards for critical facilities and to select an optimized system. Only high-inventory critical facilities were considered. The goal of the study was to detect and confirm the protracted or abrupt diversion of 8 kg of plutonium or 25 kg of the uranium isotope 235 within approximately a week of the diversion. The general safeguards alternatives considered were (1) continuous inspections by resident inspectors, with varying degrees of comprehensiveness, (2) periodic inspections by regional inspectors at varying time intervals, (3) unattended containment/surveillance measures, and (4) various combinations of the above. It was concluded that a practical and effective international safeguards system can be achieved by employing a method of continuously monitoring facility activities which could lead to diversion. This is in addition to the routine inspections typical of current international safeguards. Monitoring detects inventory discrepancies and violations of agreed-upon procedural restrictions, as well as unauthorized removal of Special Nuclear Materials (SNM). A special inventory is used following detection to confirm any suspected diversion. Comparison of 28 safeguards options led to the selection of a system for further development which uses a combination of surveillance and inspection by resident IAEA personnel, containment/surveillance by unattended equipment, and routine inventory sampling. A development programme is described which is intended to demonstrate the feasibility of several containment and surveillance measures proposed in the study. Included are a personnel portal and an instrument/material pass-through as well as associated recording and tamper-protection features.

1. INTRODUCTION

Nuclear critical facilities are used extensively in support of reactor research. These facilities play a major role in evaluating and characterizing new reactor designs and configurations. Studies on reactor safety, reactor kinetics, neutron spectra, and many other areas pertinent to nuclear research and development have been conducted using nuclear critical facilities. To carry out these activities, critical facilities may have large inventories (hundreds of kilograms) of plutonium and/or highly enriched uranium in the form of

* Work supported by the United States Department of Energy.

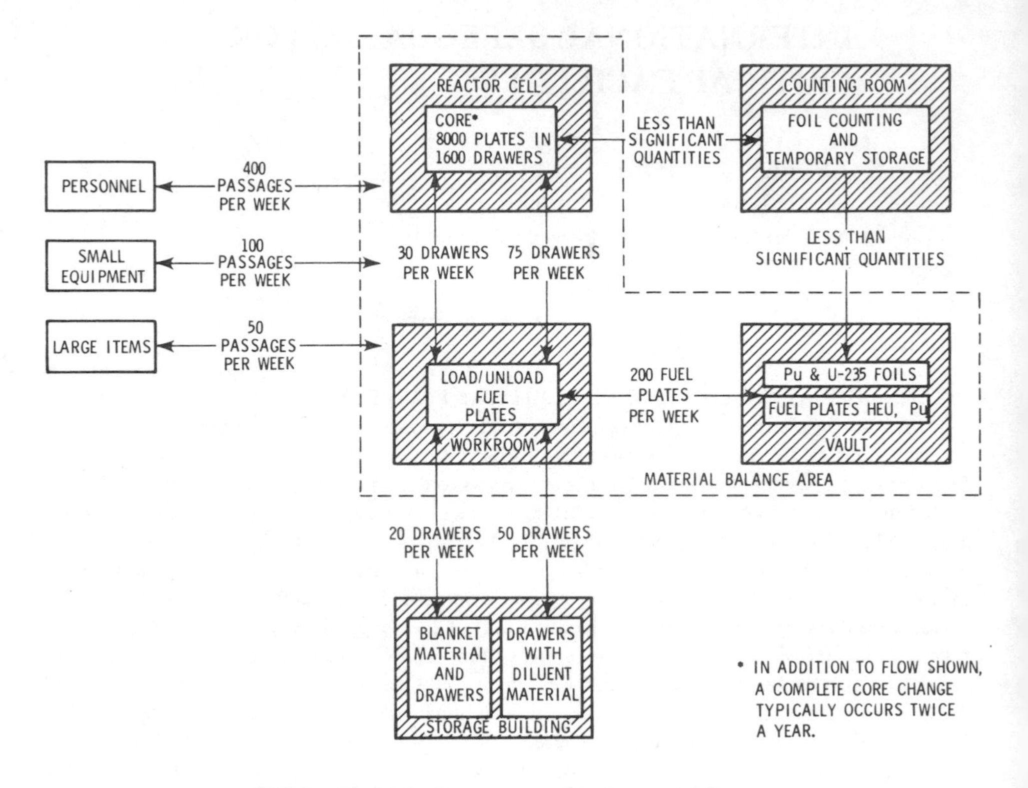

FIG.1. Model facility personnel and material flow.

small fuel elements or plates. Figure 1 depicts typical flow of material and personnel in a reference facility. The fissile material may be in either metallic or oxide form.

The stated IAEA objective of international safeguards in conjunction with the Treaty for Non-Proliferation of Nuclear Weapons is set forth as "The timely detection of diversion of significant quantities of nuclear material from peaceful activities...and the deterrence of such diversion by risk of early detection."[1] Significant quantities are defined as 8 kilograms of plutonium or ^{233}U and 25 kilograms of ^{235}U. The IAEA guidelines for timeliness are based on estimated times required to convert the diverted material to material suitable for the manufacture of nuclear explosive devices. IAEA guidelines [2] for timeliness are presented in Table 1. For purposes of this study, the goal established is the ability to detect and to verify diversion of 8 kilograms of plutonium or 25 kilograms of ^{235}U within approximately a week of the completion of the diversion. Both protracted diversion of small quantities of material in multiple attempts and abrupt diversion of large quantities of material in one attempt were considered.

An international safeguards system in addition to providing high confidence in providing timely detection of diversion, must have: (1) a low false alarm rate; (2) minimum interference with facility operations; and (3) acceptable cost to the IAEA, nation, and operator. The intent throughout the system study was to develop options that meet all objectives recognizing some compromise must be made since a number of these requirements tend to be mutually exclusive.

2. SAFEGUARDS APPROACH

International safeguards inspections conducted by the IAEA rely on routine reporting by the state and periodic inspection by IAEA personnel. To meet the timeliness criteria established for this study, these periodic inspections would have to occur at intervals of less than a week and could consist of a level of inventory verification that would be an extreme burden to both the IAEA and the facility operator. Figure 2 outlines the international safeguards system approach which is used as the basis for the options considered in this study.

"Continuous" monitoring bridges the time between the safeguards decision points now associated with periodic inspections. This monitoring may be accomplished by an inspector, by instrumentation, by routine inventory, or by a combination of these methods. Unauthorized actions which could indicate a diversion are sensed by the monitors and cause an alarm to be generated. The alarm in turn is received by the international authority. Response to an alarm is a special inventory verification procedure which assesses the alarm and could include a sampling of the inventory. This sequence provides the information necessary to reach a safeguards decision. The "timeliness" factor in this approach is the interval from the unauthorized action to the safeguards decision point at the end of the special inventory verification. The times between unauthorized action, alarm, receipt of alarm, and the beginning of the special inventory are system variables that depend upon inspection procedures, communication, and safeguards equipment.

3. OPTION DEVELOPMENT, EVALUATION AND SELECTION

Figure 3 illustrates the matrix used to generate 28 options selected for preliminary evaluation. For example, Option 16 assumes weekly inspection by one inspector with hand-held search instrumentation, unattended containment/surveillance, and remote reporting capability.

Comparison of these 28 options led to the selection of Option 8, which is believed to be a reasonable compromise between technology and number of inspectors.

TABLE I. IAEA GUIDELINES FOR TIMELINESS

Required conversion of nuclear material to the form suitable for the manufacture of nuclear explosive devices	Material Form	Approximate range of times required to convert nuclear materials to a form suitable for manufacture of nuclear explosive devices
Physical change or chemical and physical change but no purification	Plutonium and highly enriched uranium such as metal, oxide or solution	Days to weeks
Chemical and physical change with purification	Irradiated fuel, radioactive solution, and cold scrap	Weeks to months
Isotopic, chemical, and physical change	Natural and low enriched uranium	Less than one year

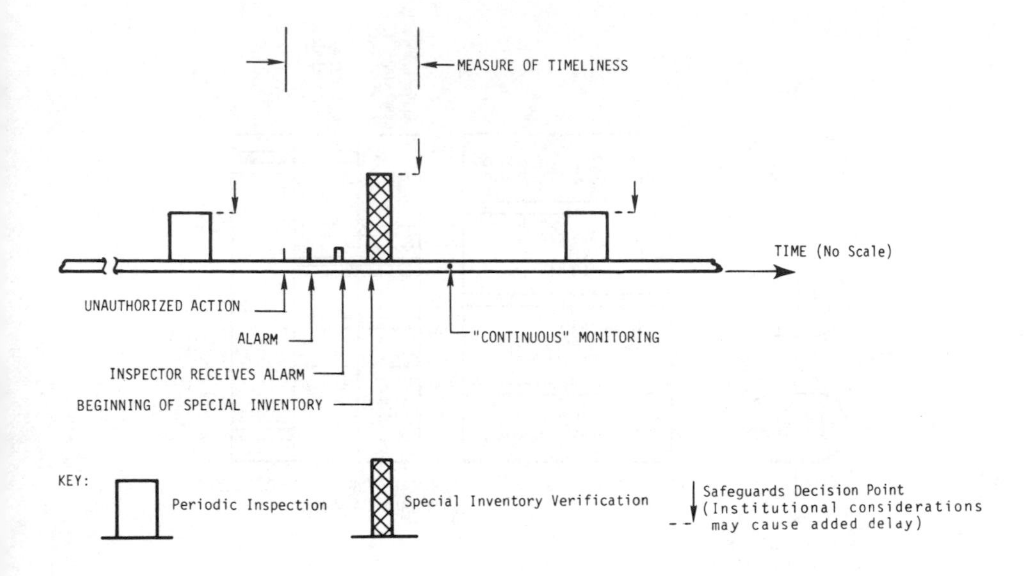

FIG.2. International safeguards.

INSPECTION OPTIONS (ALL INCLUDE MATERIAL ACCOUNTANCY)

EQUIPMENT OPTIONS	ON-SITE 24 HRS ONE INSPECTOR ON DUTY TOTAL OF 5-7 REQUIRED				ON-SITE 8 HRS ONE INSPECTOR WORKING HOURS DAILY VISITS NONWORKING HOURS						WEEKLY INSPECTION ONE INSPECTOR REGIONAL OFFICE ON CALL						MONTHLY INSPECTION SEVERAL INSPECTORS REGIONAL OFFICE ON CALL						ANNUAL INSPECTION INSPECTION TEAM HEADQUARTERS ON CALL					
NO EQUIPMENT OBSERVATION ONLY	X				X						X						X						X					
HAND-HELD SEARCH INSTRUMENTS RADIATION MONITORING EQUIPMENT		X		X		X		X		X		X		X		X		X		X		X		X		X		X
UNATTENDED CONTAINMENT/SURVEILLANCE PERSONNEL & MATERIAL MONITORS, CAMERAS, CCTV			X	X			X	X	X	X			X	X	X	X			X	X	X	X			X	X	X	X
REMOTE REPORTING REAL-TIME TRANSMISSION, MONITORING OF DATA & CCTV									X	X					X	X					X	X					X	X
OPTION NUMBER	1	2	3	4	5	6	7	8	9	10	11	12	13	14	15	16	17	18	19	20	21	22	23	24	25	26	27	28

FIG.3. Option identification.

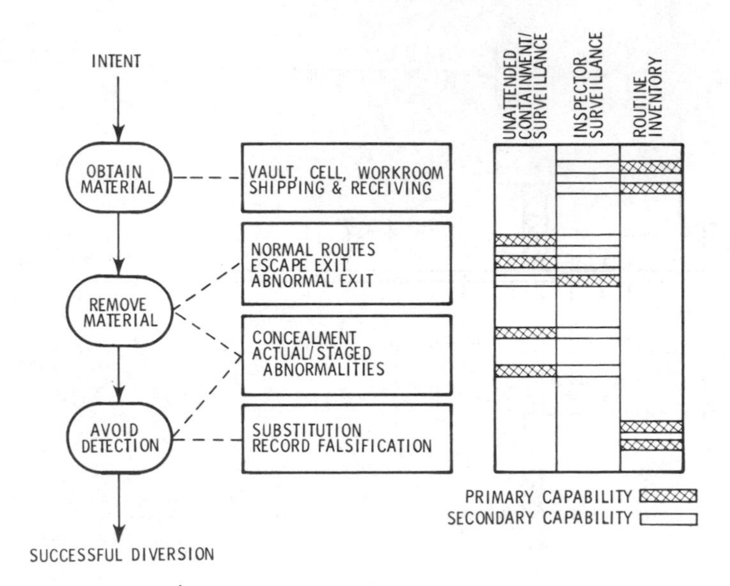

FIG.4. Detection capabilities of option 8.

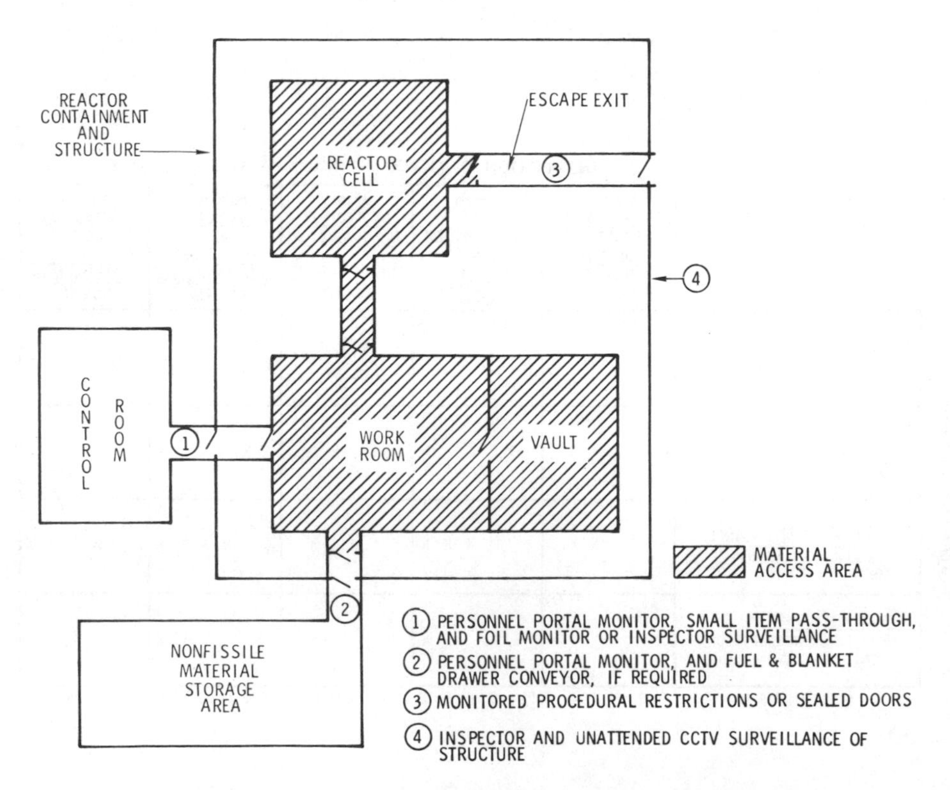

FIG.5. Conceptual design of model reference containment/surveillance system.

The key elements of the selected international safeguards system are:

1. Routine material accountancy and inventory verification on a sampling basis,
2. Containment of nuclear material by unattended SNM portal monitors,
3. Inspector presence during movement of any material that cannot pass through portal monitors,
4. Visual inspection of the facility to detect unusual activity or structural alterations,
5. Use of seals on containers of stored SNM to reduce inventory verification requirements, and
6. Special inventory procedures when unauthorized actions or indications of diversion occur.

The detection capabilities of Option 8 are reviewed in Figure 4, which presents a flowchart related to the generic sequence of actions required for successful diversion. Routine inventory provides the primary means for detecting inventory discrepancies both within the MAA (Material Access Area) and during shipping and receiving operations. A backup capability is provided by inspector surveillance. Diversion strategies involving a material substitution and/or record falsification can be detected only through routine inspection. Detection by routine inventory may not always be timely but it does provide high confidence in assuring that protracted diversion has not occurred.

On-site inspector surveillance provides the principal capabilities to detect unauthorized structural changes which could facilitate removal of material along abnormal exits such as holes in the reactor containment or modifications of the ventilation system. Inspector surveillance also provides valuable secondary capabilities for detecting abnormal activities which might occur during allegedly normal operations.

Unattended containment/surveillance measures are the primary means for detecting removal of SNM through facility exits either by concealment or during actual/staged emergencies. These measures effectively increase the capability of the inspector by providing unattended monitoring of exits at times when the inspector is not present or when he is engaged in other activities such as routine inventory or observation of large item movement.

A diagram of the conceptual containment and surveillance design is shown in Figure 5. Unattended personnel portals would be utilized at the two main entrances/exits into the MAA. From the control room provisions would also be made for allowing instrumentation and tools to pass. The use of the emergency exits would be checked by the use of tamper indicating seals. The integrity of the entire containment structure would be checked by the inspector backed up by unattended CCTV surveillance.

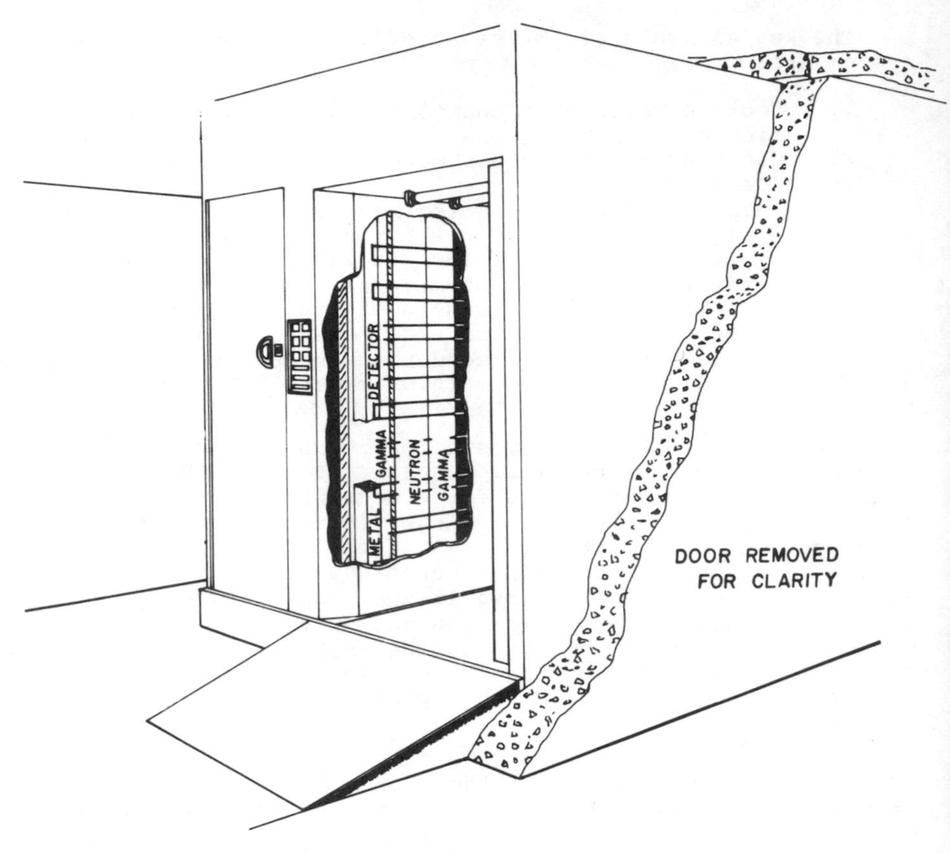

FIG.6. Personnel portal.

4. DEVELOPMENT PROGRAM

Development activities at Sandia Laboratories are primarily concerned with containment and surveillance measures. Other activities are addressing inventory verification techniques. The current effort to determine the feasibility of the containment and surveillance portions of the system consists of development to provide unattended portals for placement at each exit from the MAA, TV surveillance systems to verify the integrity of the facility structure and high security seals to be used in the storage vault to reduce the inventory verification effort. Seals and TV surveillance systems are the subject of a separate paper presented at this symposium [3].

The personnel portal (Figure 6) is being designed for unattended operation and will automatically interrogate personnel leaving the MAA for SNM and metal. SNM detection will be accomplished by measuring gamma radiation resulting from the natural radioactive decay of the SNM. Neutron detectors are also being investigated to determine the feasibility of using this type of instrument in the personnel portal to improve the

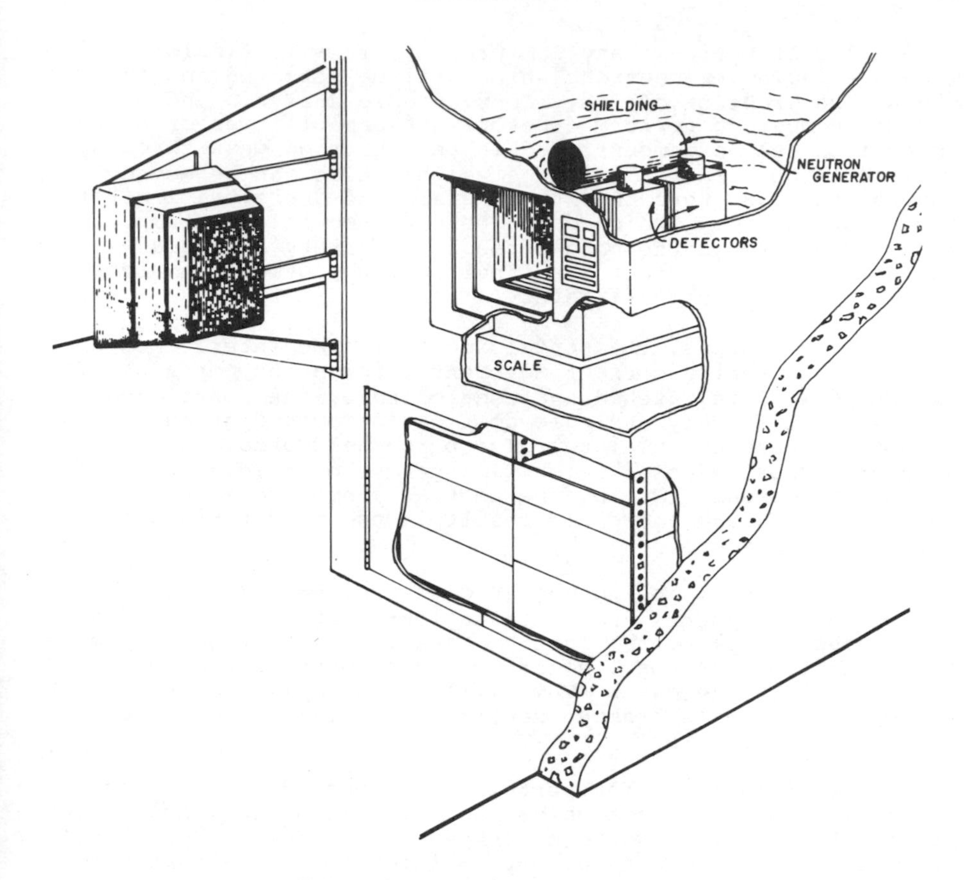

FIG.7. Instrument/material pass-through.

detection sensitivity for shielded plutonium. The metal detector will perform two primary functions: (1) detect the presence of metal containers which could be used to conceal and shield SNM; and (2) detect metallic SNM. Both functions require the capability to detect small amounts of metal within the volume enclosed by the portal. An active metal detector which generates a time varying magnetic field within a defined volume and responds to changes in the field caused by metal introduced into this volume is currently being evaluated for use in the portal.

The instrument/material pass-through (Figure 7) is also being designed for unattended operation and will automatically interrogate instruments, tools, and other material prior to removal from the MAA. To provide an acceptable level of detection sensitivity, a neutron activation technique using a pulsed neutron source is under development. Although this technique looks promising, it has not yet been employed for this type application. When material is to be interrogated, the neutron source will provide a burst of 14 MEV neutrons to bombard the material. If there is any SNM present, it will fission and generate secondary neutrons which will be detected and thus signal the presence of SNM. Considerable analysis and experimentation must be performed before the capabilities of neutron activation can be demonstrated in an operating pass-through. A scale will be used to limit the weight which can be moved through the pass-through without alarm and therefore limit the amount of shielding which could be employed in an attempt to divert SNM through the pass-through. A safety analysis will be conducted to verify that the final design provides an acceptable level of personnel safety.

In both designs, the portal doors will be interlocked and monitored to verify correct movement through the portals. The two doors will be latched and controlled by the portal controller so that only a single door can be opened at any time without generating an alarm. The personnel portal will contain emergency door releases which will permit overriding the door latches on demand. Use of the emergency door releases will result in an alarm and could result in the need for a special inventory procedure.

The portals are being designed to provide tamper protection for portal equipment. For international safeguards applications, the entire portal must be capable of unattended, tamper-safe operation for the longest period between inspections (approximately one day). This will be accomplished through the use of secure seals, tamper-sensing electronics and tamper-indicating materials.

A design goal for the portals is to provide a self-test capability for all electronics including an operational status monitor for the international inspector. A message panel will also be provided to display portal interaction instructions. At appropriate times, the panel will direct the user to close the doors and proceed through the portal. It will also display alarm indications. Recorders will be included in the portal to provide the capability for registering alarms and to store information, such as date, time, and type of portal detector alarms. The information will be stored until it is observed and acknowledged by an international inspector during his periodic equipment checks.

5. CONCLUSION

An international safeguards system which can effectively detect and confirm the abrupt or protracted diversion of a significant quantity of nuclear material within days to weeks of the diversion should have provision for continuously monitoring facility activities either with instrumentation, inspectors or a combination thereof. Twenty-eight system options were identified which could accomplish this monitoring. Evaluation of these options led to the selection of a conceptual system which is being used as a basis for further development. This option consists of a combination of surveillance and containment by unattended equipment, surveillance and inspection by resident IAEA personnel, routine inventory sampling and provision for a rapid special inventory to confirm or refute a suspected diversion. The developmental efforts are directed at determining the technical feasibility and operational acceptability of this type of safeguards system by a program of test and evaluation of equipment and procedures at an operating critical facility.

REFERENCES

[1] INTERNATIONAL ATOMIC ENERGY AGENCY, The Structure and Content of Agreements Between The Agency and States Required in Connection with the Treaty on the Non-Proliferation of Nuclear Weapons, INFCIRC/153, IAEA, Vienna (May 1971).

[2] INTERNATIONAL ATOMIC ENERGY AGENCY, IAEA Safeguards Technical Manual, IAEA-174, Part A, IAEA, Vienna (1976).

[3] CAMPBELL, J.W., JOHNSON, C.S., STIEFF, L.R., "Containment and surveillance devices", these Proceedings, Vol. I.

DISCUSSION

A.G. HAMLIN: Has any analysis been made of the secondary effects of such a system on the operator? A multiplicity of electromechanical checks and counter-checks will increase the possibility of false alarms and alarms resulting from innocent activities. The operator will therefore be exposed to re-inventory costs and the cost of investigations necessary to explain non-significant alarms. How much are these costs likely to amount to?

J.L. TODD: Experts familiar with operations at fast critical facilities have performed preliminary evaluations of the possible impact of the conceptual system on operations. These initial estimates, which are based upon evaluation of the concepts rather than on operational testing, indicate a possible 10 to 20% reduction in facility availability. We recognize the need for minimizing this impact and our development efforts are being directed to that goal.

W.C. BARTELS: I would like to add to Mr. Todd's answer to Mr. Hamlin. The model reference facility used in obtaining this estimate of 10 to 20% has a plutonium inventory several times greater than that of any fast critical facility in the western world. The physical inventory taking is correspondingly more difficult. Other things being equal, therefore, the reduction in facility output would be correspondingly less at the other facilities.

J.L. TODD: Mr. Bartel's point is entirely correct. The model reference facility used for this study has a large inventory consisting of tens of thousands of individual fuel items.

IAEA-SM-231/5

STATUS OF THE SAFEGUARDS SYSTEM DEVELOPED FOR THE LMFBR PROTOTYPE POWER PLANT SNR-300 (KKW KALKAR)

H. KRINNINGER*, Chr. BRÜCKNER*+, U. QUANDT+,
E. RUPPERT+, H. TSCHAMPEL**, P. VAN DER HULST++
* Internationale Natrium-Brutreaktor-Bau GmbH, Bensberg/Cologne
*+ Kernforschungszentrum Karlsruhe
+ INTERATOM, Bensberg/Cologne
** Dr. Seufert GmbH, Karlsruhe
++ Schnell-Brüter-Kernkraftwerks GmbH, Essen

Federal Republic of Germany

Abstract

STATUS OF THE SAFEGUARDS SYSTEM DEVELOPED FOR THE LMFBR PROTOTYPE POWER PLANT SNR-300 (KKW KALKAR).

In this report the features of the safeguards system developed for the LMFBR Prototype Power Plant SNR-300 (KKW Kalkar) are described. Due to the fact that LMFBR fuel assemblies are mostly handled in a sodium or inert gas environment, visual control and counting of the assemblies containing nuclear material, as routinely used in a LWR power plant, cannot be applied for LMFBRs. Consequently, for safeguarding the Kalkar Nuclear Power Station, an automatic and continuous monitoring of the fissile material inventory is the main objective of special instrumentation, the "Inaccessible Inventory Instrumentation System" (IIIS). For a qualitative distinction between the various types of assembly the detection of the emitted neutron and gamma radiation turned out to be most adequate. Based upon this conclusion and taking into account its tasks, the IIIS is composed of an activity measurement system, a position-sensing system and a micro-computer system (MCS). The high reliability, high fraud resistance and low maintenance requirements have had an important influence on the IIIS design. Access to IIIS data is protected by special hardware and software measures. To protect the system against the results of power failures the IIIS possesses an automatic restart facility and a battery-buffered power supply for the RAM-unit of the microprocessor system used for data acquisition and evaluation.

1. TASK DESCRIPTION AND PROBLEM IDENTIFICATION

Shortly after signature of the turn key-contracts between SBK (owner/operator of KKW Kalkar) and INB (constructor and manufacturer of KKW Kalkar) at the end of 1972 safeguards measures for surveillance of the special nuclear material (SNM) at the SNR-300 growed up to an important viewpoint. The motivation for entering into discussion with the safeguards authorities were mainly 4 points of consideration:

(1) The ratification of the verification agreement to the NPT treaty between IAEA, EURATOM and the member states of the European Communities (with the exception of France).
(2) The about 400 times higher physical inventory of special nuclear material expressed in kg eff of SNR-300 compared to modern commercial size LWR's (1300 MWe Power Plants).
(3) The facility attachments outlined up to that date didn't consider a reference scheme for safeguarding LMFBR Power Plants.
(4) The fact, that the KKW Kalkar is the first LMFBR Prototype Power Plant built in a non-nuclear weapon state.

During the meetings with the responsible safeguarding authorit it became evident that safeguarding a LMFBR power plant is much more complex than safeguarding a LWR power plant, due to the properties typical for LMFBR systems [1]:

(1) The reactor core is composed of a variety of assembly types corresponding to its different objectives (fuel-, blanket-, absorber- and steel reflector assemblies).
(2) Fresh fuel- and irradiated fuel- and blanket assemblies contain several kg of plutonium.
(3) Fuel and blanket assemblies are not designed for dismantling on site, therefore the assembly structure composed of wrapper tube, head and foot acts as second barrie preventing access to the fuel. The fuel pin cladding together with the welded-end-plugs forms the first barrier.
(4) Most of their residence time on site, the assemblies remain under sodium or inert gas environment and are hence invisible and inaccessible.
(5) The handling scheme of the assemblies on site is more complex, due to the various handling locations and handling machines in use. The leak-tight vessels and containe for the assembly-handling under sodium or inert gas environment are a third physical barrier for the access to the special nuclear material (SNM) contained in the relevant assemblies.

2. BRIEF DESCRIPTION OF THE KALKAR NUCLEAR POWER PLANT FROM A SAFEGUARDS POINT OF VIEW

The reactor building is subdivided into three main areas
- the primary heat transfer circuit area (including the reactor vessel)
- the fuel handling area and
- the auxiliary equipment area.

All handling operations with individual assemblies (fuel assem blies, blanket assemblies, absorber assemblies, reflector assemblies, B_4C- or sodium "diluent" assemblies) will occur o within that area. With the exception of receipt and shipment all assemblies remain in the fuel handling area during their residence time at the nuclear power plant.
Outside the handling area assemblies will be handled in shipping casks only.
The SNM available on site is only contained in either the fres and irradiated fuel assemblies or in the irradiated blanket assemblies. The further description of the plant can hence be

limited to those locations where these assemblies will be in use, in store or subject of different activities to be performed with themselves.

2.1 Nuclear material inventory of fuel and blanket assemblies

The typical properties of fast breeder reactor cores lead in all present power reactor designs to a "closed" assembly concept. Hence head and foot fitted to the wrapper tube form a barrier for the access to the nuclear material. The fuel assembly consists of 166 fuel pins, containing altogether about 7.7 kg Pu_{equiv} and 41.8 kg U in the outer core zone and 5.3 kg Pu_{equiv} and 44.7 kg U in the inner core zone. One kg of Pu_{equiv} corresponds to about 1.235 kg Pu of Magnox quality. The blanket assembly consists of 61 blanket pins containing altogether about 83 kg of depleted uranium. At end-of-life about 1.5 to 3.1 kg Pu will be built up. The most important and dominating part of the SNM available on site, is the plutonium in either the fresh and irradiated fuel assemblies or in the irradiated blanket assemblies. All types of assemblies are "closed" boxes from which pins cannot be extracted without destroying it.

2.2 Features of the assembly handling on site

The main features of the assembly handling on site are [2]:

(1) After arrival and passage of the acceptance inspection at the New Fuel Inspection Cell fresh fuel assemblies will be stored in closed cans in air atmosphere (New Fuel Store).
(2) Reloading of fuel and blanket assemblies occurs with the reactor shut-down. Spent fuel assemblies are unloaded from the reactor core and transferred to a Sodium-Cooled Store. This store can take over a complete core loading.
(3) During refuelling fuel and/or blanket assemblies may be transferred to the Irradiated Fuel Observation Cell for intermediate inspection and visual control.
(4) Assembly transfers between the Reactor Vessel and the different handling locations of the handling area occurs with leak-tight and shielded handling equipment (EX-vessel Handling Machine I).
(5) Handling operations inside the Reactor Vessel are carried out by means of a multiple rotating shield plug system and an In-vessel Handling Machine (IHM).
(6) All handling locations arranged underneath the handling floor are kept under an argon atmosphere, when they contain radioactive and sodium-wetted materials. They are equipped with locks closed by gas-tight and radiation-shielded blocks.
(7) After decay and sodium cleaning the assemblies will be temporarily stored in the Gas-Cooled Store which is cooled by forced nitrogen circulation.
(8) Before shipment to the reprocessing plant irradiated fuel and blanket assemblies must be cleaned from sodium. Hence they are transferred from the Sodium-Cooled Store to the Gas-Cooled Store, passing the Washing Cell. To avoid sodium contamination after cleaning they will be moved from the Washing Cell to the Gas-Cooled Store by means of a second EX-vessel Handling Machine (EXHM II).

TABLE 1. EVALUATION OF THE ACCESSIBILITY OF THE SNM AVAILABLE ON SITE

Location	Type of assembly	Environment of assemblies	Condition of assemblies	Valuation of accessibility
New Fuel Store	fresh fuel assembly	- under air atmosphere by ambient temperature - inside tubes which are blinded at the lower end and closed by removable plugs on topside - no significant radiation	- unirradiated - fissile material tight wrapped and easy to handle - weakly active	Relatively easy
Sodium-Cooled Store	fresh fuel assembly	- inside a tight and shielded vessel with a rotating plug - only accessible by a small channel in the rotating plug - under sodium with a temperature between 180 to 350 °C	- sodium-wetted - contaminated of a high degree - weakly active - unirradiated	Rather difficult
	irradiated fuel and blanket assembly		- highly active - contaminated of a high degree - sodium-wetted - heat producing	Extremely difficult
Gas-Cooled Store	irradiated fuel and blanket assembly	- inside a tight and shielded vessel with a rotating plug - only accessible by a small channel in the rotating plug - under radioactive argon atmosphere with approx. 100 °C	- highly active - heat producing - washed and inside gas-filled cans, closed by welded plugs - sodium-wetted inside sodium-filled open cans (only blanket assemblies)	Very difficult
Reactor vessel during operation	irradiated fuel and blanket assembly	- inside of a tight and shielded vessel with a rotating plug surrounded by inert gas - under sodium with approx. 560 °C - intense radiation - contamination of a high degree	- highly active - contaminated of a high degree - sodium-wetted - heat producing	Impossible (access only during refuelling)

2.3 Accessibility of the Special Nuclear Material (SNM) available on site

The fuel and blanket assemblies stored or in use in the different handling locations of the handling area are only accessible by special handling equipment (EXHM I and II, IHM, auxiliary lift). Due to health physics requirements the handling area is inaccessible for plant personnel during assembly handling operations. The handling area is surveilled by TV from the control room and only accessible via a guarded gate and air locks.
For the valuation of the accessibility of the SNM in the different assemblies, the assembly environment and the physical and chemical properties of the assemblies themselves were used for a qualitative analysis, summarized in table 1.

3. QUALITATIVE ASSESSMENT OF THE STRATEGIC VALUE OF THE SNM AVAILABLE ON SITE

The attractiveness of assemblies for diversion will depend mainly on the degree of processing necessary for transformation of the SNM into a form usable for weapons. The strategic value of the SNM is thereby assumed as inversely proportional to the difficulties involved in that processing. Other criteria influencing the strategic value of the SNM are:
- accessibility of the nuclear material
- isotopic composition of the nuclear material.

An evaluation of the strategic value based upon these criteria is given in table 2.

4. RESTRICTIONS FOR DEVELOPMENT AND REALIZATION OF THE KALKAR SAFEGUARDS SYSTEM

The engineering for the development and realization of the Kalkar safeguards system was and is subjected to restrictions resulting from
- NPT/Verification Agreement requirements
- plant specific requirements defined by the operator
- technical requirements specified by EURATOM
- limitations predetermined by the state of the art of technology achieved in the development of safeguards instruments
- limits in governmental funds for sponsorship of the development of the safeguards system
- the time schedule for plant construction and commissioning.

5. BRIEF DESCRIPTION OF THE KALKAR SAFEGUARDS SYSTEM / 3 /

The developed safeguards system takes into account the characteristics of the fuel handling schemes and the results of a qualitative assessment of the strategic value of the SNM available on site (§ 3). The previously described characteristics of the fuel and blanket assemblies (§ 2.1) and the features of the fuel handling system (§ 2.2) together with the safeguards measures form the basis for the safeguards system. Furthermore for its set-up the restrictions mentioned in § 4 were considered.

TABLE 2. EVALUATION OF THE STRATEGIC MATERIAL AVAILABLE ON

Item no.	Plutonium contained in	Isotopic composition	Accessibility of assemblies	Transportation difficulties	Type of pretreatment necessary before processing
1	Fresh fuel assemblies in New Fuel Store	typical MAGNOX or LWR-Pu-vector	relatively easy	normal care for mechanical damage	dismantling of assembly, dismantling of fuel pins
2	Fresh fuel assemblies in Sodium Cooled Store	typical MAGNOX or LWR-Pu-vector	rather difficult	handling of sodium contaminated assembly; normal care for mechanical damage	cleaning to remove the sodium, dismantling of assembly, dismantling of fuel pins
3	Irradiated blanket assemblies in Sodium Cooled Store	94 % Pu239 5 % Pu240 1 % Pu241	extremely difficult	special heavily shielded shipping casks; provisions for decay heat removal	remote handling for sodium cleaning and dismantling of assembly and blanket pins necessary
4	Irradiated fuel assemblies in Sodium Cooled Store	typical MAGNOX or LWR-Pu vector in core region, 97 % Pu239 3 % Pu240 axial blanket	extremely difficult	as for item 3	as for item 3

SITE FROM A SAFEGUARDS ASPECT

Processing	Inherent difficulties involved in pre-treatment and processing	Inherent difficulty in use for nuclear explosive devices	Number of assemblies needed to get a cirtical mass	Potential for diversion of the assembly
dissolution, chemical separation and reduction	toxity, neutron- and γ-emission, α-activity	extreme fast ignition technique necessary to cope with the problems caused by the high spontaneous neutron source strength	one fresh fuel assembly	highest of all assemblies
dissolution, chemical separation and reduction	handling rather difficult due to sodium, sodium contaminated with activated particles and fission products, toxity, neutron and γ emission, α activity	as for item 1	one fresh fuel assembly	smaller than that of item 1
dissolution, extraction, chemical separation (Uranium and fission products) and reduction	decay heat; highly radioactive; sodium contamination; toxity α-activity, γ emission	———	some (4 to 10) irradiated blanket assemblies	low
as for item 3	as for item 3 plus neutron and higher γ-emission	as for item 1	one irradiated fuel assembly	lowest of all assemblies

TABLE 3. RADIATION EMITTED BY FUEL AND BLANKET ASSEMBLIES UNDER DIFFERENT PHYSICAL CONDITIONS

Type of radiation	Fuel assembly			Blanket assembly		
	fresh	irradiated assembly (cooling time ~d; residence time 441 efpd)	decayed irradiated assembly (cooling time 240 d; residence time 441 efpd)	fresh	irradiated assembly (cooling time ~d; residence time 3 yr)	decayed irradiated assembly (cooling time 240 d; residence time 3 yr)
Neutron source strength[a] (spontaneous +(α,n) neutrons) [s^{-1}]	2.10^6 - 6.10^6	6.10^7 - 3.10^8	2.10^7-1.10^8	0	1.10^5-1.10^6	8.10^4-5.10^5
γ-dose rate (on assembly surface for core mid-plane) [R/h]	-	3.10^6 - 8.10^6	1.10^5-3.10^5	-	3.10^5-2.10^6	1.10^4-6.10^4

[a] neutron sources calculated for plutonium of MAGNOX-quality.

To avoid as much as possible the interference of safeguards activities with plant operations not subject to safeguards, the safeguards system restricts all safeguards measures only to that area of the plant, where SNM is in store, in process (energy production in reactor core) or is subject of handling operations.
Physical inventory taking, verification and maintenance is based upon the application of the principal safeguards measures
- accountancy
- surveillance and
- containment.

The environmental conditions existing on site for the assemblies containing SNM divide the handling locations of the fuel handling area into two groups. The first group involves these locations where the assemblies remain in air atmosphere under ambient temperatures, whereas the second group combines all these locations where the assemblies are in sodium or inert gas atmosphere. Assemblies in the first group of handling locations are accessible for inventory verification. Assemblies located in the second group of handling locations are inaccessible and hence can not be routinely subjected to direct safeguards measures.
The highest potential for a diversion, as outlined in § 3 possesses the fresh fuel assemblies. Single fresh fuel assemblies are for the first time accessible on site, after opening of the plugs on the shipping cask. Having passed the acceptance inspection the fresh fuel assemblies arrive at the New Fuel Store, in which the physical inventory taking by the inspector can be executed by item counting. Thereafter the occupied positions of this store may be sealed by the inspector so that inventory verification is replaced by a control of these seals.
The transfer of fresh fuel assemblies from the New Fuel Store to the Transfer Position which is the interface for the transition of the assembly environment from air atmosphere to sodium or inert gas atmosphere, is covered by a camera. This camera simultaneously watches the New Fuel Store to detect diversions from this area by non-operational handling equipment.
Physical inventory taking for the handling locations of the second group is the objective of a special instrumentation, called the "Inaccessible Inventory Instrumentation System" (IIIS) described in the following chapter.
Diversions with non-operational handling equipment from the region where the IIIS maintains the physical inventory are detected by cameras installed and operated by the safeguarding authorities.
The tasks of the inspector are therefore limited to the
- verification of the physical inventory for the New Fuel Store
- evaluation of the physical inventory provided for by the IIIS
- evaluation of the records of the camera surveillance system.

By a combination of the safeguards measures
- sealing of New Fuel Store (containment)
- Inaccessible Inventory Instrumentation System (accountancy)
- camera system (surveillance)
an efficient, tamper-resistant and complete supervision of the SNM at the Kalkar Nuclear Power Plant is assured.

6. THE "INACCESSIBLE INVENTORY INSTRUMENTATION SYSTEM (IIIS)" FOR AUTOMATIC AND CONTINUOUS FISSILE MATERIAL MONITORING

6.1 Objective of Inaccessible Inventory Instrumentation System

Handling operations on site with assemblies in sodium or inert gas environment are performed with the handling flasks of the EX-vessel Handling Machine's (EXHM I and II) only. Physical inventory verification can hence be performed by monitoring the flow of SNM utilizing an appropriate instrumentation.

The main objective of this special instrumentation (IIIS) are:

1. Monitoring of all handling operations with assemblies containing SNM.
2. Identification of the in and output handling locations.
3. Identification of the different types of assemblies, i.e. fresh fuel assemblies, irradiated fuel assemblies, irradiated blanket assemblies and other assemblies not subject to safeguards measures.
4. Updating of the actual physical inventory and recording of all significant data on mass memory.
5. Recording of all significant faults and errors in the IIIS

6.2 The IIIS design

The emitted passive neutron- and gamma radiation of assemblies was found to be a powerful tool for a qualitative distinction between the different types of assemblies. A comparison of the nuclear radiation emitted from the different assembly types containing SNM is given in table 3. Neutrons emitted from fresh fuel assemblies originate from spontaneous fissions of plutonium-isotopes (mainly Pu-240, Pu-242) and from $(\alpha,n)^{18}O$ reaction of α particles emitted mainly from Pu-239 and Pu-240. In fact the neutron source strength of the fresh fuel assemblies depends upon

- the Pu enrichment
- the Pu quality (isotopic composition, impurities)
- the multiplication of the fuel pin bundle
- the decay time of the Pu-241 since conversion, i.e. americium separation (Am-241 build-up by β^- decay of Pu-241

Concerning the objectives 1. to 5. of § 6.1 the IIIS is therefore composed of three sub-systems

- an activity measurement system
- a position sensing system
- a micro computer system.

Special features of the IIIS design are application of "redundancy" principle to achieve a high reliability and application of the black box concept (IIIS completely separated from plant operators instrumentation) to improve the fraud resistance. Access to any part of the IIIS by an unauthorized person is only possible by mechanical damage of housings or destruction of seals.

6.3 Subsystems of IIIS

6.3.1 Activity measurement system

An outline of the mechanical set-up of the detector assemblies and a block diagram of the electronic set-up is shown

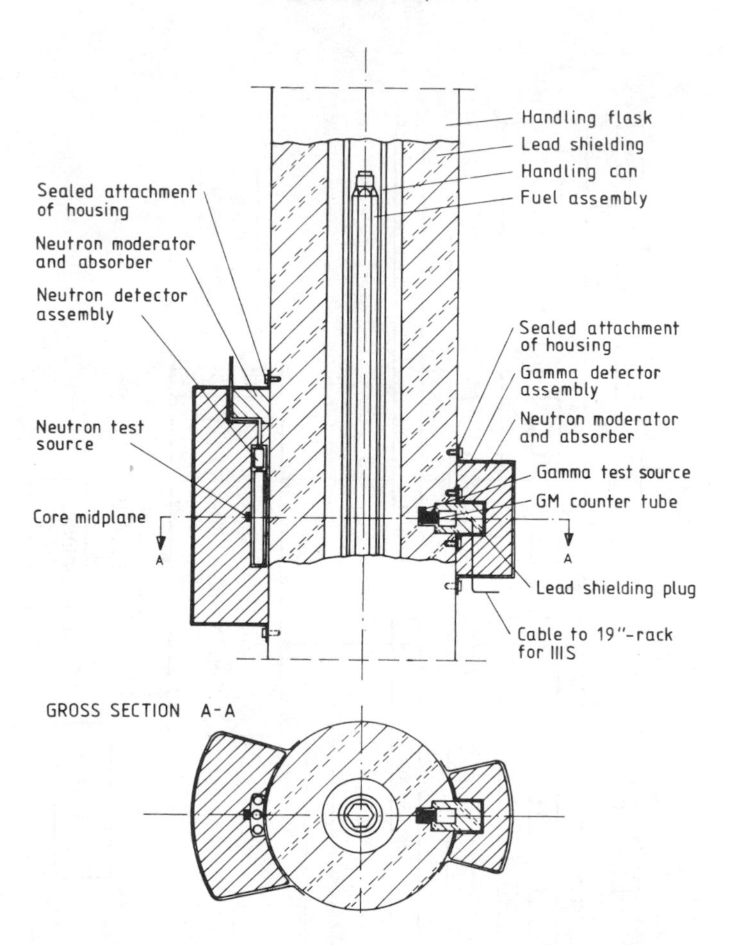

FIG.1. Scheme of mechanical set-up of the detectors for the activity measurement system on EXHM I and II.

in fig. 1 and 2. Standard AEC-NIM modules are used. Small neutron and gamma test sources serve for a continuous check of the detector system functioning.

The detectors are fastened on the handling flasks of the EXHM's at a location which is symmetric to the midplane of the fuel column. This location is insensitive against small variations in the final assembly positioning within the handling flask. The temperature conditions prevailing on this spot of the outer surface of the lead shielding under normal and accidental handling operations predetermine the design conditions for these detectors.

A comparison of He^4, CH_4 and BF_3 counters revealed that BF_3 counters offer some advantages:

(1) higher detector efficiency

(2) lower sensitivity to HV voltage fluctuations

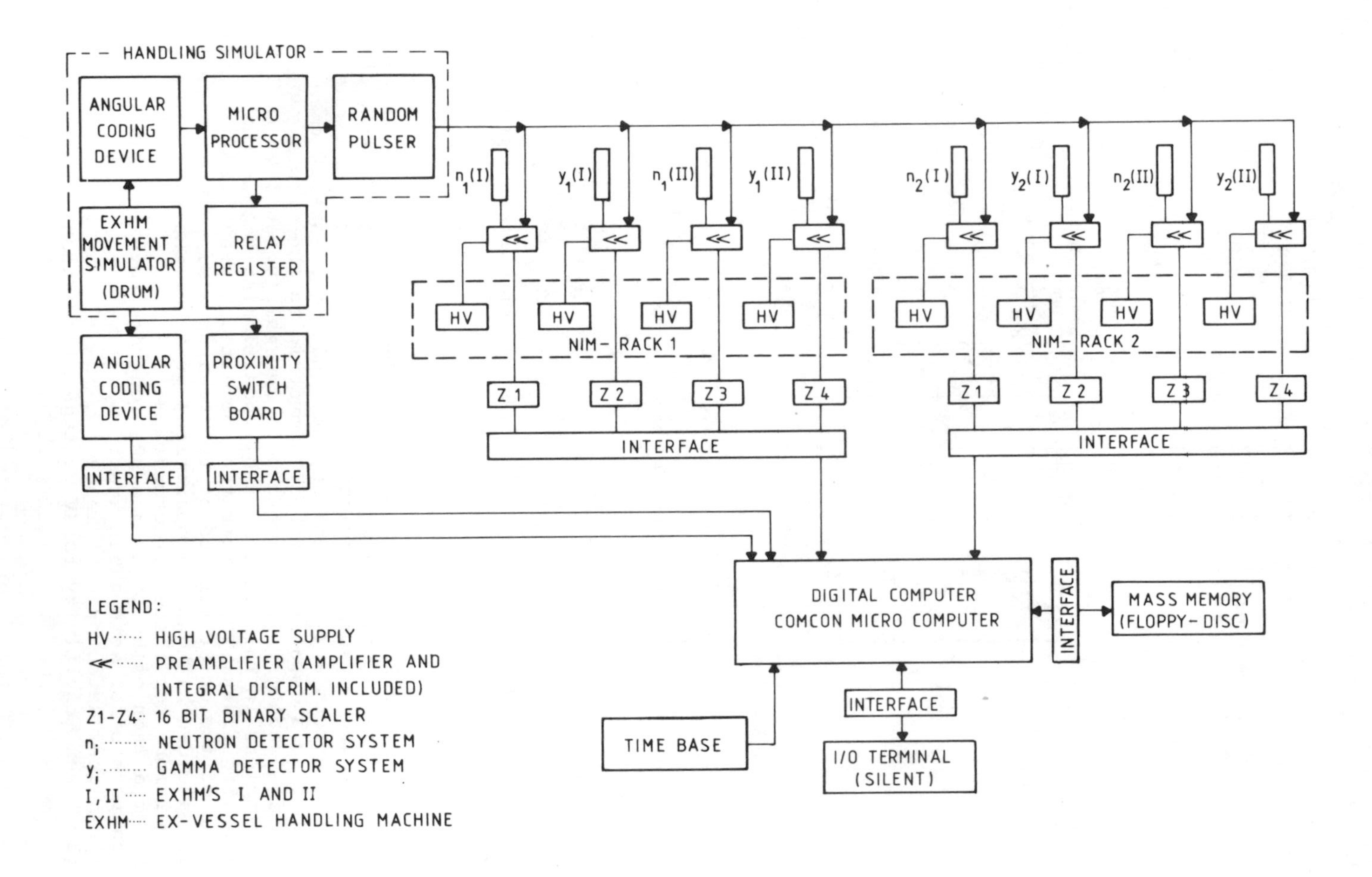

FIG.2. Block diagram of the electronic set-up (IIIS).

(3) lower gamma sensitivity
(4) lower operating voltage
(5) lower filling gas pressure.

The gamma detectors of the activity measurement system must only differentiate between dose rates. Properties typical for γ-spectrometry are not required. Because of their simplicity, reliability and stability Geiger-Müller counter tubes were selected. They are installed into drillings of the lead shielding of the EXHM's.

6.3.2 Position sensing system

To improve the tamper resistancy and the reliability of the position sensing system two different measurement techniques were applied: a proximity switchboard and an angular coding device.

The borders of each handling location are equipped with a binary-coded invisible array of actuators which will be read by the proximity switchboard fastened on the lower lorry of the EXHM's. Due to the geometric arrangement (see fig. 3) of the handling locations within the handling area the installation of actuators along the route of the lower lorry of the EXHM's is sufficient. For self-surveillance reasons negative logic is used. Using an appropriate code an one bit error can be corrected by the software of the MCS.

The angular coding device is coupled to an axis of the lower lorry. After calibration the position of the EXHM's can be determined (resolution power ~1 cm). Self-surveillance is realized by a parity check of the supplied signals.

6.3.3 Micro Computer System (MCS)

The block diagram of the MCS is shown in fig. 2. This system fulfils the following main tasks:

(1) remote operation and supervision of the IIIS
(2) execution of the tasks listed in § 6.1
(3) protection of IIIS against unauthorized intervention or access to data transmission, data processing and data recording.

The MCS is composed of

- COMCON micro processor system
- FLOPPY disc units
- I/O typewriter SILENT 700
- process interfaces for data transmission from peripheral devices.

The MCS is installed in the 19"-rack of the IIIS located at the platform on the upper lorry of the XY-coordinate drive machine.

6.4 Reliability analysis of the IIIS

To check whether the IIIS design fulfils the high standard of reliability requested by the safeguarding authority [4] a detailed reliability analysis with the aid of the INTERATOM computer codes IASAPL and IASAPD [5,6] was performed.

For that purpose a fault tree of the IIIS was prepared using

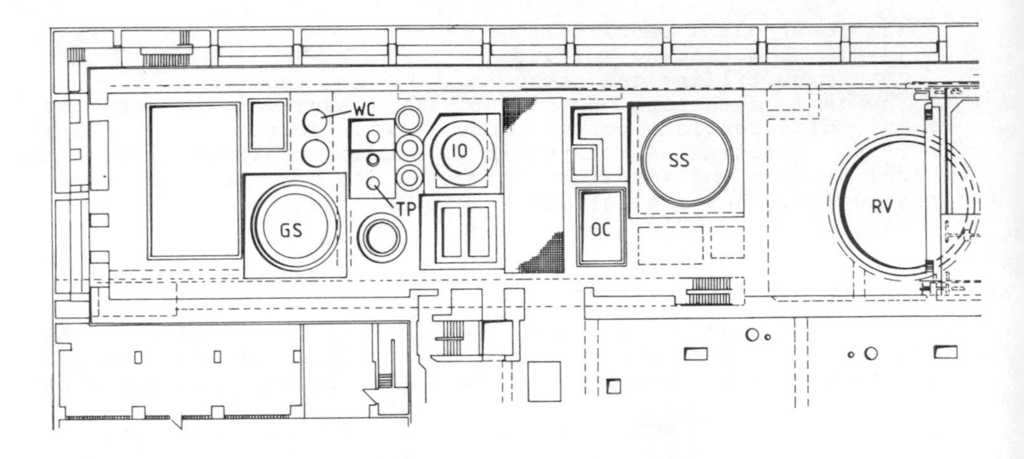

		OC	irradiated fuel observation cell
TP	transfer position		
RV	reactor vessel	GS	gas cooled store
SS	sodium cooled store	IO	shipping cask loading and unloading station
WC	washing cell		

FIG.3. Geometric arrangement of handling locations within the handling area.

the following events as serious failures liable to lead to the non-recording or false recording of handling operations:

(1) simultaneous fault of both neutron detection systems either on EXHM I or EXHM II
(2) simultaneous fault of both systems of the position sensing system
(3) simultaneous fault of both data recording units
(4) fault of COMCON system.

Because of the fact that the A.C. for the IIIS is supplied by the "protected" instrumentation grid, essential for the execution of handling operations, a fault of the A.C. power supply to the IIIS needs not to be considered. Since the assembly identification is primarily based on the readings of the neutron detection systems, a fault on both gamma detection systems either on EXHM I or EXHM II does not lead to a serious fault. One of the most important and most difficult tasks was the compilation of fault rates for the different modules setting up the IIIS. Using information gathered from manufacturers, from the research centers [7,8] and derived from own experience [9] the reference list of reliability data given in table 4 was prepared.

Assuming periodic preventive maintenance in quarterly intervals the predicted failure probability was $7.4 \cdot 10^{-2}$ for a quarter of a year. A maintenance strategy based on a repair of defective components within 48 hours after fault identification led to a predicted failure probability of $1.9 \cdot 10^{-2}$ per quarter of a year and to 3.6 ± 2 repairs a year. Therefore it was recommended to the safeguarding authorities to select this method as the reference maintenance strategy for

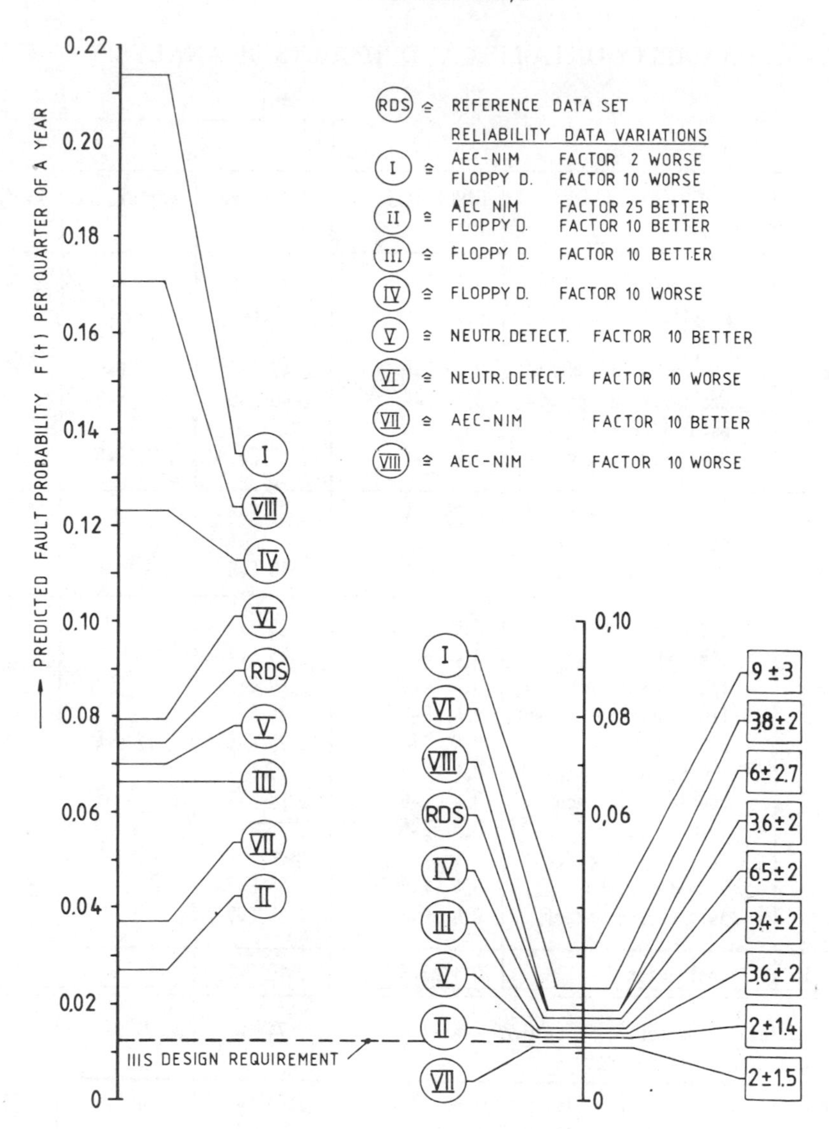

FIG.4. Comparison of predicted reliability of IIIS applying different maintenance strategies.

the IIIS. The results of the reliability analysis must be compared with the requirements imposed to IIIS design, requesting a predicted probability of an error in the local and total inventories less than $1.25 \cdot 10^{-2}$ per quarter of a year and which allows for 4 repairs a year.
In a sensitivity analysis MTTF data variations (see table 4) were performed for the neutron detector system (variations V, VI), the AEC-NIM modules (VII,VIII) and the FLOPPY disc (III, IV) to identify its influence on IIIS reliability.

TABLE 4. RELIABILITY DATA LIST AND RESULTS OF ANALYSIS

IIIS SUB-SYSTEM	COMPONENT	FAULT RATE (h^{-1})	Reference data	Variat. I	Variat. II
activity measurement system	n-detector assembly	0.5 E-5 [a]	. 20 E+6		. 10 E+7
	γ-detector assembly	0.5 E-5	. 20 E+6		
	HV-supply (AEC-NIM)	0.25 E-4	. 40 E+5	. 20 E+5	. 10 E+6
	amplifier, discriminator (AEC-NIM)	0.25 E-4	. 40 E+5	. 20 E+5	. 10 E+6
	NIM-power supply	0.25 E-4	. 40 E+5	. 20 E+5	. 10 E+6
position sensing system	proximity switchboard	0.33 E-4	. 33 E+5		
	angular coding device	0.2 E-4	. 50 E+5		
mini computer system	COMCON [b]	0.4 E-5	. 25 E+6		
	time base	0.4 E-6	. 25 E+7		
	I/O typewriter [b]	0.5 E-4 [c]	. 20 E+5		
	I/O typewriter interf.	0.4 E-5	. 25 E+7		
	FLOPPY disc [b]	0.4 E-4 [c]	. 25 E+5	. 25 E+4	. 25 E+6
	FLOPPY disc interf.	0.4 E-5	. 25 E+7		
	counter interface	0.4 E-6	. 25 E+7		
	16 bit counter	0.4 E-6	. 25 E+7		
	angular coding device interface	0.4 E-6	. 25 E+7		
	switchboard interf.	0.4 E-6	. 25 E+7		
19"-rack	ventilator	0.1 E-5	. 10 E+7		
predicted failure probability for 0,25 years/ periodic quaterly maintenance			. 74 E-1	. 21 E 0	. 27 E-1
repair of faulted component within 48 hours	failure probability for 0,25 years		.185 E-1	. 32 E-1	.135 E-1
	number of repairs per year		3,6 ± 1,9	9 ± 3	2 ± 1,4

[a] E-5 corresponds to 10^{-5}

[b] power supply fault rates for COMCON, FLOPPY and I/O typewriter are assumed to be included

[c] valid only for the expected typical operating conditions of the Inaccessible Inventory Instrumentation, e.g. 1000 data transfers a year.

mean time to failure (h)					
Variat. III	Variat. IV	Variat. V	Variat. VI	Variat. VII	Variat. VIII
		. 10 E+7	. 10 E+6	. 10 E+6 . 10 E+6 . 10 E+6	. 20 E+5 . 20 E+5 . 20E+5
. 25 E+6	. 25 E+4				
. 67 E-1	. 12 E 0	. 70 E-1	. 80 E-1	. 37 E-1	. 17 E 0
.155 E-1	.175 E-1	.145 E-1	.235 E-1	.125 E-1	. 19 E-1
3,4 ± 2,0	6,5 ± 2,4	3,6 ± 1,8	3,8 ± 1,9	2,4 ± 1,5	5,9 ± 2

It was intended to deduce therefrom guide lines for further IIIS design improvements. In the scope of variations I and II best and worst MTTF data were attributed to different components at the same time. A comparison of the predicted failure probability for both maintenance strategies is given in table 4 and in fig. 4 including all abovementioned variations. The variations II and VII yielded reliability data very close to or better than the design requirements. The results for the recommended maintenance strategy have been separately drawn although their deviations from the reference data set (RDS) partially represent more the standard deviation (10%-20% of the Monte Carlo simulation than the significance of the data variation itself. E.g. to get for the variations II and VII a better precision, the number of Monte Carlo games had to be increased by about two orders of magnitude (from 500 to 50,000).
Again as a conclusion one must state that the recommended maintenance strategy leads to results which are significantly better than those of a quarterly maintenance procedure.

6.5 Fraud resistance of the IIIS

A high level of fraud resistance - i.e. insensitivity to external measures acting upon the final accountancy result - was demanded. In that context two general considerations should be kept in mind:

- Practice fraud to the instrumented safeguards system requires a detailed knowledge of the design and lay-out of the system. Classification by the safeguarding authorities of the most sensitive design details and parameters can already do a lot.
- The level of fraud resistance should be consistent to the attractiveness to divert a particular type of fuel from the power plant. In a previous section (§ 3) it has been shown that fresh fuel assemblies are most attractive from a diversion point of view.

6.5.1 Recording of fresh fuel assembly handlings

With the envisaged instrumentation it could be tried to simulate a fuel assembly in EXHM I by introduction of a neutron source, external to this handling flask. The generation of a detector signal is prevented in this case by a combination of moderator and absorber material surrounding the counter tubes.
Another possibility of falsification is introduced by internal neutron sources in a dummy assembly. Detection of this kind of falsification is possible by controlling the time dependence of the neutron detector signal during loading and unloading of fuel assemblies from the handling flasks.
The opposite way of practice fraud is the handling of a fresh fuel assembly without detecting it. This can only be arranged if a sufficient amount of shielding material can be brought between assembly and detector.
The available space in the fuel handling flask however is far insufficient for this purpose.

6.5.2 Recording of irradiated fuel and blanket assembly handlings

The diversion of irradiated fuel assemblies is very unattractive because of the extreme high neutron and gamma radiation, which is to be simulated in a dummy assembly on the one hand and which rendered the diversion of the fuel assembly itself on the other hand.

6.5.3 Fraud resistance of the position sensing system

In order to falsify the data coming from the position sensing system:
- the position code of the proximity switchboard has to be decoded
- the rotation of the angular coding device has to be simulated
- the peripheral sensors have to be removed.

When the fuel handling machine arrives at a fuel handling location, the position code of this location must be consistent with the previous code and the length of the pathway between the two locations, indicated by the angular coding device. Deviation from this consistency will indicate irregularities.
Unauthorized removal of the peripheral sensors will be detected by seals on the corresponding fastening screws.

6.5.4 Fraud resistance of the Micro Computer System

The following measures are foreseen to prevent falsification of the MCS:
- sealing of the housing of the electronic equipment
- safety-key controlled access to the electronic equipment
- use of "read-only" memory for MCS software
- access to the MCS software by the use of a changeable password only known to the safeguarding authority
- automatic control of the proper functioning of the activity measurement system
- automatic control of the data transmission lines between peripheral sensors and MCS.

6.6 Assembly identification technique

For the assembly identification the emitted neutron dose rate is used. The gamma dose rate is applied for a consistency check and as back-up information to tackle unexpected problems in assembly classification. The predicted neutron- and gamma source strengths for fuel and blanket assemblies are given in table 3 and fig. 5. The full straight lines in fig. 5 cover the influence of the irradiation history on neutron- and gamma source strength for all fuel and blanket assemblies in the reactor core. The influence of the cooling time spent in the assembly storages on the emitted dose rates is also shown in fig. 5. A cooling time of 240 days is representative for the first shipments of fuel and blanket assemblies to the reprocessing plant. The range of uncertainties in the prediction of the dose rates is indicated by the dotted lines (hatched area).
Assembly identification will be done by a comparison of the measured count rate with stored reference count rates of the different assembly types.

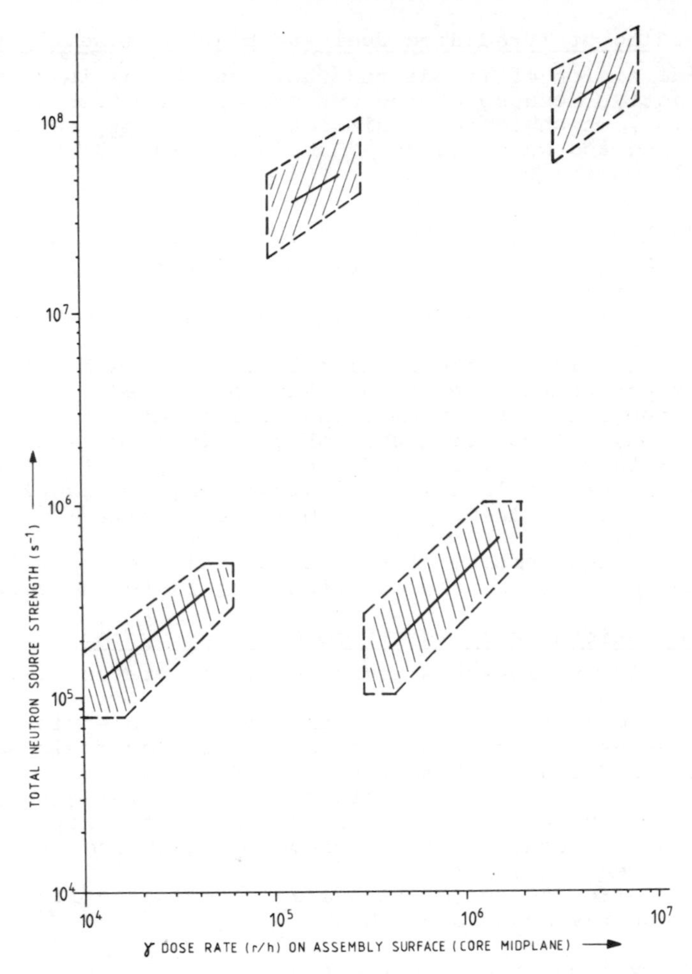

FIG.5. Neutron and gamma source strengths emitted from spent fuel and blanket assemblies.

6.7 Software of the IIIS

The software of the IIIS covers all functions for defining assembly classes, assembly handlings and supervision of all components of the IIIS.

6.7.1 User interface

Users of the Micro Computer System (MCS) - inspector and/or maintenance personnel - are able to communicate with the system in a dialogue session. First they have to identify themselves by typing in a given password. If the password is accepted by the system they may continue the session with commands which define or update data of the system.
Other important functions for the inspector are:

- listing a part of or all assembly handlings since last inspection

- listing a part of or all error messages occurred since last inspection
- changing a FLOPPY disc (may be there is no more space available on it).

When the inspector ends the dialogue session the system checks whether it is completely able to work. If not,it requests for the removal of the disturbance.
Different from this routine dialogue during IIIS operating time is the procedure requested for the first dialogue after installation of the system. In this case all system relevant data and parameters as

- identification of assemblies of different classes with corresponding sets of neutron and gamma count rates
- definition of handling locations with corresponding values of proximity switches and angular coding device
- definition of the actual assembly inventories (local and total)
- initiation of date and time
- formatting of FLOPPY discs

have to be defined.
The MCS proves correctness and completeness of all data and checks the functions of all subsystems when the dialogue is terminated. If all data are well defined the system begins the surveillance of the EX-vessel Handling Machines (EXHM)s).

6.7.2 Structure of the software

The software of the IIIS is divided into two levels:

- a so-called micro-level, which contains the operating system, the interpreter for the macro-assembler and the drivers for I/O-devices
- and a macro-level containing all programs for implementation of the features of the IIIS.

All programs on the micro-level are written in micro-assembler. Time critical problems and interrupt handling are executed on this level. The used operating system provides multiprogramming: different programs of the macro-level can be suspended under time conditions or while execution of I/O-functions proceed. This allows permanent supervision of the EX-vessel Handling Machines, even if the inspector is communicating with the system in a dialogue session. The major advantage of the use of such an operating system is the aid for modular structuring of a software system (on macro-level).

6.7.3 Components of the program system

The classification of the tasks is obtained from its different requirements:

- permanent cyclic execution
- timed cyclic execution
- demanded execution.

The first category contains all test functions like time-base test, power supply test and scaler test. The second category is necessary for cyclic supervision of the position of the EX-vessel Handling Machines (position sensing system) and the count rates produced by the activity measurement system. It is also used for updating the actual date and time.

The third category mentioned above contains all I/O requests. One program identifies the request of the inspector to start a dialogue session. It provides also print-out of actual count rates, stops the continuous print-out of error- and assembly-handling listings and prepares the system for hardware or software diagnosis in case of a system malfunction. A second program executes all commands of a dialogue session. The last program of this category provides the print-out of error messages and assembly handling listing on teletype.

6.7.4 Data protection

The data protection is assured in several ways:

- All programs are protected against power failure by a special save-area restart mechanism. After a power failure the system is restarted automatically. No data will be lost nor will the system enter in an undefined state.
- All errors of special importance are recorded too, even if they have been eliminated.
- The disc-unit is redundant. If one FLOPPY disc is full of data or the unit is inoperable then the other disc is used automatically. Moreover, it is not possible to overwrite information which is stored on FLOPPY disc.

6.8 Recovery of physical inventory in case of serious IIIS failur

The reliability analysis (§ 6.4) predicts a probability of once in a period of ten years, that a serious failure on IIIS may occur.
Due to the application of the "black box" concept the inoperability of the IIIS is unknown to the plant operator. Therefore the use of the plant operator records should enable a correction of the IIIS data. The possibility of verification of operators records with the aid of the records supplied by the camera surveillance system should also be taken into account for this very special situation.
After IIIS fault elimination a random-sampling plan can additionally be used to check the correctness of the operator dat by means of a very limited number of assembly identifications

6.9 Importance of instrumented safeguards system versus an inspector-based safeguards system

Radiological protection requirements applied on site prohibits the access of the handling area during handling operations. Therefore a resident inspector on site can only watch via a TV system the movements of the EXHM's.
The use of leak-tight handling equipment prevents that assemblies can be physically verified by an inspector. Hence the information gained from an instrumented system like the IIIS can not be substituted by simply delegating safeguards inspectors assigned to take over this duty.

6.10 Status of the IIIS

All components have been already delivered to INTERATOM laboratories.
The software package of the Micro Computer System was at first tested with a simulator program. Thereafter the peri-

pheral devices have been connected to the MCS in order to check the performance of the complete III system. Thereafter the MCS was exposed to extreme environmental conditions in a climate test chamber. During these tests the MCS was operated for a period of seven days with abnormal high temperature and humidity conditions. Simultaneously it was intended to "artificially age" the MCS hardware and to possibly detect weak components of the hardware.

For the purpose of the acceptance inspection of the MCS and the IIIS long endurance test a micro processor-controlled fuel handling simulator was assembled.

The fuel handling simulator (see fig. 2) is composed of

- a mechanical equipment for simulation of movements of the EXHM's during handling operations
- a random pulser for the simulation of handling operations with different assemblies
- a micro processor system including a tape reader for programming of assembly handlings.

With the aid of the abovementioned fuel handling simulator, the MCS software was successfully tested. A sequence of 60 different assembly handling operations containing normal and abnormal handling operations was executed in periodical runs.

All detectable IIIS errors were simulated and the corresponding error messages checked. Special tests were conducted to demonstrate the IIIS operation with one of the redundant systems switched off.

7. OUTLOOK ON FUTURE ACTIVITIES

In the period covering the fourth quarter of 1978 and the first quarter of 1979 the long endurance test of the IIIS will be performed, simulating the assembly handlings of several years of reactor operation. The main objective of this test is to gain information on IIIS reliability and maintenance.

Based on this experience

- the operating manual
- the service instruction manual and
- the instructions for the installation, check-out and commissioning of IIIS at the Kalkar Nuclear Power Plant

will be prepared.

The cables for the IIIS will be installed at the EXHM's according to the IIIS wiring schemes at the shops of the EXHM manufacturer. After shipment of the EXHM's to Kalkar site, the installation of the peripheral sensors and of the electronic equipment will occur.

The engineering activities will concentrate mainly on

- the recalculation of neutron source strength for fresh and irradiated fuel assemblies taking into account as parameter
  - . Pu-quality
  - . time between reprocessing (conversion of plutonium) and in-pile use of fuel assemblies
- the reevaluation of the IIIS reliability based upon the experience gained with the IIIS components during the long endurance test

- the information exchange with IAEA in the framework of the safeguards cooperation agreement between IAEA and the Federal Republic of Germany
- an assessment of the tamper and fraud resistancy of the IIIS including the identification of possible counter measures
- the continuation of the information exchange with EURATOM.

After arrival of the first fuel assemblies at Kalkar the IIIS will be calibrated and thereafter handed over to the safeguarding authority.

REFERENCES

[1] BÖDEGE, R., BRAATZ, U., HEGER, H., Spaltstoffüberwachung in Kernkraftwerken, Atomwirtsch. **21** (1976) 135–42.

[2] SNR-300 Liquid Metal-Cooled Fast Breeder Reactor Prototype Plant, Nucl. Eng. Int. (July 1976).

[3] BRÜCKNER, Chr., VAN DER HULST, P., KRINNINGER, H., "Safeguards system for the LMFBR prototype power plant SNR-300 (KKW Kalkar)", Safeguarding Nuclear Materials (Proc. Symp. Vienna, 1975) **2**, IAEA, Vienna (1976) 581–87.

[4] EURATOM, Frame Specification for the Inaccessible Inventory Instrumentation System, Document No. 52.1583.4 (1976).

[5] SCHÜLLER, F., Reliability of the Safeguards System, Interatom Note 32.2314.8.

[6] SCHÜLLER, F., ROSENHAUER, W., Zuverlässigkeitsanalyse der Spaltstofffloßkontrolle, Interatom Note 32.2210.3.

[7] BÖHNEL, private communication, INR Karlsruhe Nuclear Research Centre.

[8] MENLOVE, H.O., private communication, LASL.

[9] KRINNINGER, H., RUPPERT, E., "Operational experience with the automatic lead spectrometer facility for nuclear safeguards", Am. Nucl. Soc. Winter Meeting, Wash. D.C., Nov. 1972.

DISCUSSION

A.G. HAMLIN: There is a golden rule which should be borne in mind in regard to plant instrumentation, namely "If it can go wrong, it will". The complex system described seems to me to present two risks which I should like you to comment on. First, the system may malfunction and accuse the operator of making a move when he has not done so. Second, a diverter may attempt to discredit the system by making real moves, which are logged by the system but which he subsequently denies. What protection is there against these risks?

H. KRINNINGER: The IIIS design has taken account of such events. A reliability analysis of IIIS performed with standard computer codes widely used in LMFBR safety system analysis predicted an IIIS breakdown once in a period of ten years. Such a breakdown would always result in non-recording of fuel-handling operations. A fault which could lead to the recording of a non-existent handling operation will be identified by the IIIS thanks to its self-checking capabilities and the event will be recorded as erroneous. The recording of handling operations subsequently denied by the operator can be cross-checked with the aid of the records supplied by the camera surveillance system (independent redundant information).

S.J. CRUTZEN: In a paper you presented at a previous symposium[1], you described the TUID (Tamper-resistant Unique Identity Device) which could be used for safeguarding the SNR-300 fuel bundles. Are there any technical reasons why you do not mention this device in your new safeguards system for the SNR-300 Kalkar reactor?

H. KRINNINGER: The TUID can only be attached to the foot of the fuel assembly and re-identification of the TUID fingerprint at the plant can be performed only during acceptance inspection of fresh fuel or before shipment of irradiated fuel to the reprocessing plant (it is not at present certain that TUID re-identification will be feasible on fuel assemblies being prepared for shipment in the irradiated fuel observation cell), so that the timely detection of diversion is not possible with the TUID. The Kalkar safeguards project team recognizes the importance of the TUID from the safeguards authorities' point of view as a tool for covering the closed part of the LMFBR fuel cycle. This is the reason why the offer was made to Euratom to attach TUIDs to a limited number of fuel assemblies on an experimental basis.

S. ERMAKOV: I think you have worked out a very comprehensive safeguards system for SNR-300. I have only one question: Is it possible to distinguish an experimental fuel sub-assembly from a fuel sub-assembly for normal operation?

H. KRINNINGER: If the fissile material content in the experimental fuel assemblies differs from that of the regular SNR-300 fuel assemblies, the IIIS can identify and differentiate between these different fuel assemblies. An experimental fuel assembly irradiation programme has not yet been finally drawn up. Such a programme will probably be directed more towards the development of advanced structural materials than fuels. Hence the fissile material contents will not differ significantly between the two assembly types.

---

[1] Brückner, Chr., et al., "Safeguards system for the LMFBR prototype power plant SNR-300 (KKW Kalkar)", Safeguarding Nuclear Materials (Proc. Symp. Vienna, 1975) **2**, IAEA, Vienna (1976) 581.

IAEA-SM-231/139

МЕТОД КОЛИЧЕСТВЕННОГО ОПРЕДЕЛЕНИЯ НЕСАНКЦИОНИРОВАННОГО ИЗЪЯТИЯ ЯДЕРНЫХ МАТЕРИАЛОВ ПО ИЗМЕНЕНИЯМ СПЕКТРАЛЬНЫХ ИНДЕКСОВ

В.М.ГРЯЗЕВ, Г.И.ГАДЖИЕВ, Н.Р.НИГМАТУЛЛИН,
Л.И.ДЕМИДОВ, И.В.ЯКОВЛЕВА
Научно-исследовательский институт
атомных реакторов им.В.И.Ленина
Государственного комитета по использованию
атомной энергии СССР,
Димитровград,
Союз Советских Социалистических Республик

Abstract–Аннотация

A METHOD FOR QUANTITATIVE DETERMINATION OF UNAUTHORIZED REMOVAL OF NUCLEAR MATERIALS BASED ON CHANGES IN THE SPECTRAL INDICES.

The paper describes a study of the effect of changes in the quantity of nuclear materials on the spectral indices measured in the core of the SPEKTR critical assembly. The authors determined the volume from which the removal of a given amount of ^{235}U can be detected using a single pair of indicators. This made it possible to fix the control points in such a way that any diversion exceeding a certain level can be reliably detected. To determine the magnitude of the diversion an algorithm was developed, based on the assumption of a linear relationship between the variation of the spectral indices and the magnitudes of local diversions. Using this algorithm it is possible to calculate the magnitude of diversions of not more than 10.5 kg with an accuracy of 30–40%.

МЕТОД КОЛИЧЕСТВЕННОГО ОПРЕДЕЛЕНИЯ НЕСАНКЦИОНИРОВАННОГО ИЗЪЯТИЯ ЯДЕРНЫХ МАТЕРИАЛОВ ПО ИЗМЕНЕНИЯМ СПЕКТРАЛЬНЫХ ИНДЕКСОВ.

В работе исследовалось влияние изменений количества ядерных материалов на спектральные индексы, измеряемые в активной зоне критической сборки СПЕКТР. Определены размеры объема, изъятие из которого фиксированного количества урана-235 контролируется одной парой индикаторов. Это позволило выбрать такое размещение контрольных точек, что любое несанкционированное изъятие ядерных материалов, превышающее некоторый порог,будет надежно обнаружено. Для определения количества изъятых материалов разработан алгоритм, основанный на предположении о существовании линейной связи изменений спектральных индексов с размерами локальных изъятий. Расчет по этому алгоритму позволяет вычислить величину несанкционированного изъятия ядерных материалов в размере, не превышающем 10,5 кг с точностью 30-40%.

1. ВВЕДЕНИЕ

Затраты времени и усилий на инспекцию быстрой критической сборки могут быть значительно сокращены применением так называемых "коллективных" методов контроля, позволяющих измерить количество ядерных материалов в активной зоне в целом.

Первая попытка реализации таких методов была выполнена в Научно-исследовательском институте атомных реакторов (НИИАР) им.В.И.Ленина во время исследований по контракту с Международным агентством по атомной энергии №1435/RB в 1973-75 гг., когда был разработан метод, основанный на измерении декремента затухания мгновенных нейтронов в зависимости от изменения загрузки активной зоны [1]. Экспериментальная градуировка метода показала, что наряду с высокой чувствительностью к малым отклонениям от конфигурации активной зоны, взятой как базис для дальнейших исследований, наблюдается значительная неопределенность при изъятии ядерных материалов.

Это объясняется тем, что декремент затухания является глобальным параметром активной зоны и зависит не только от количества ядерных материалов, но и от их расположения в активной зоне.

Тогда же была оценена чувствительность некоторых локальных нейтронно-физических функционалов реактора к изменению количества делящихся материалов и получены обнадеживающие результаты в отношении метода спектральных индексов.

В настоящей работе излагаются некоторые результаты исследований метода спектральных индексов, выполняемых в связи с контрактом №2038/RB с МАГАТЭ и являющихся продолжением упомянутых экспериментов.

2. НЕСАНКЦИОНИРОВАННОЕ ИЗЪЯТИЕ ЯДЕРНЫХ МАТЕРИАЛОВ НА КРИТИЧЕСКОЙ СБОРКЕ

Рассматривалось только изъятие делящихся материалов из ТВС активной зоны с одновременным введением необходимого количества полиэтилена для компенсации потери реактивности, т.е. предполагалось, что изъятие выполняется так, чтобы запас реактивности критической сборки был сохранен. Такое изъятие может быть осуществлено различными способами, в том смысле, что конечное распределение по ак-

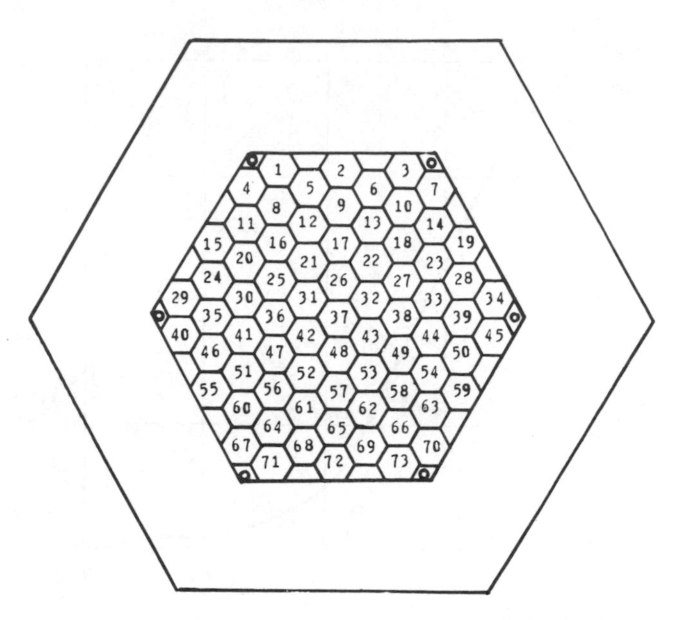

Рис.1. Картограмма активной зоны критической сборки СПЕКТР: 1-73 – номера ячеек с ТВС; 37, 48, 52, 57, 61, 64, 65, 68, 71, 72 – номера ячеек, в которых последовательно проводились калибровочные изъятия ядерного материала.

тивной зоне делящихся и замедляющих материалов может быть произвольным при условии сохранения запаса реактивности.

Частным случаем таких действий является локальное изъятие, состоящее в изъятии делящегося материала из достаточно малого компактного объема в активной зоне и введении замедлителя тоже в малый объем. Возможные размеры таких объемов позже будут определены.

3. ЧУВСТВИТЕЛЬНОСТЬ МЕТОДА СПЕКТРАЛЬНЫХ ИНДЕКСОВ К НЕСАНКЦИОНИРОВАННОМУ ИЗЪЯТИЮ ЯДЕРНЫХ МАТЕРИАЛОВ

С помощью моделирования локальных изъятий была исследована чувствительность спектрального индекса $I = \bar{\sigma}_{In}/\bar{\sigma}_{Au}$ к величине изъятия и к расстоянию от контрольной точки до места, где оно было произведено.

Под контрольной точкой понимается любая точка активной зоны, выбранная для установки индикаторов.

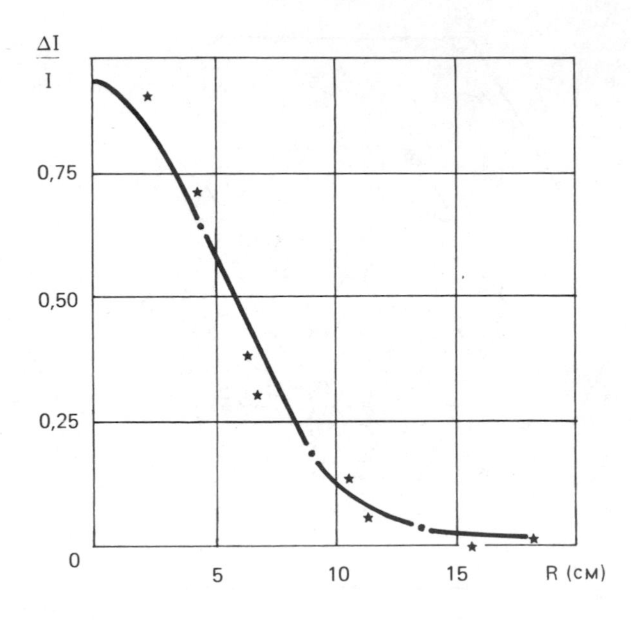

Рис.2. Относительное изменение спектрального индекса по радиусу критической сборки для изъятия двух килограммов урана-235 из ячейки 37: — • — — расчет; * — эксперимент.

На рис.1 показана картограмма активной зоны критической сборки СПЕКТР и используемая в дальнейшем нумерация ее ячеек. В экспериментах по определению чувствительности контрольные точки размещались между чехлами соседних ТВС.

Локальные изъятия ядерных материалов состояли в поочередном изъятии из ячеек, указанных на рис.1, штатных ТВС с ураном 90% обогащения и в замене их либо на макетные пакеты без урана (при этом из активной зоны изымается 2 кг урана-235), либо на ТВС с ураном 21% обогащения, что обеспечивало удаление 1,5 кг урана-235. Компенсация потери реактивности выполнялась введением необходимого количества полиэтилена.

В каждом эксперименте измерялись спектральные индексы и контролировалось их изменение относительно их значений в зоне базисной конфигурации в точках на различных расстояниях от места изъятия ядерного материала. Это позволило получить зависимость типа, показанного на рис.2, относительного изменения спектрального индекса от расстояния до центра области, где было произведено изъятие при его фиксированной величине. Рис.2 соответствует локальному изъятию размером в 2 кг урана-235.

По известным экспериментальным ошибкам в определении спектрального индекса с помощью этих кривых можно оценить расстояние, на котором изъятие указанного объема может быть надежно обнаружено. Это расстояние в дальнейшем называется радиусом контроля R и для ошибки в 5% при доверительном интервале 2σ составляет около 10 см. Радиус контроля позволяет определить объем части активной зоны, контролируемый одной парой индикаторов $V = \frac{4}{3}\pi R^3$.

Естественно стремление разместить контрольные точки по активной зоне таким образом, чтобы любая ее область попадала хотя бы в один контролируемый объем. В этом случае любое локальное изъятие ядерного материала будет надежно зарегистировано.

Количество контрольных точек зависит от типа регулярной решетки, по которой они располагаются, и для активной зоны критической сборки СПЕКТР их число может колебаться от 24 до 34. Причем, при любом распределении контрольных точек, удовлетворяющем приведенному требованию, некоторая часть зоны входит в контрольные объемы сразу нескольких контрольных точек, что повышает вероятность обнаружения локального изъятия ядерных материалов.

4. МЕТОД КОНТРОЛЯ

Применялась следующая процедура обнаружения изъятия. В активной зоне базисной конфигурации в выбранных контрольных точках производилось измерение спектральных индексов для определения опорных значений, которые сохраняются все время, пока существует соответствующая базисная конфигурация. Во время инспекции измерения выполняются в тех же точках и вновь полученные спектральные индексы используются для проверки величины их относительного изменения по сравнению с опорными значениями. Если эти изменения не превышают допустимых пределов (2σ), то можно гарантировать, что с точностью лучшей, чем два килограмма (это количество использовалось при определении радиуса контроля), изъятие ядерных материалов в активной зоне не имело места.

Выход хотя бы одного значения за пределы ошибок сигнализирует о возможности изъятия (или ошибки в измерении). В этом случае требуется оценить его размер.

5. МЕТОД РАСЧЕТА РАЗМЕРОВ НЕСАНКЦИОНИРОВАННОГО ИЗЪЯТИЯ ЯДЕРНЫХ МАТЕРИАЛОВ

Как известно, величины спектральных индексов определяются спектром нейтронного потока в окрестности той точки, где они измеряются. В то же время сам спектр формируется за счет состава и геометрии активной зоны, причем существенное значение имеет распределение по объему активной зоны делящихся материалов.

Попытка теоретического расчета зависимости между изменениями спектральных индексов в некоторых точках и изменениями количества делящихся материалов, например, с помощью теории возмущений, приводит к нелинейной задаче, решить которую пока не удается.

Тем не менее, существование этой связи очевидно и поэтому имеет смысл попытаться определить ее экспериментально в рамках какой-либо гипотетической модели.

Воспользуемся предположением о линейной связи между изменениями количества ядерных материалов в контролируемых объемах и изменениями спектральных индексов, выражаемой соотношением:

$$\Delta F_i = \sum A_{ij} (\Delta I_j / I_j^0) \qquad (1)$$

где $\Delta F_i = F_i^0 - F_i$ – изменение количества делящихся материалов в окрестности контрольной точки с номером i; F_i^0 и F_i – количество ядерных материалов для базисной и инспектируемой конфигураций, соответственно; $\Delta I_j / I_j^0 = (I_j^0 - I_j)/I_j^0$ – относительное изменение спектрального индекса; $I_j^0 = \bar{\sigma}_{In} / \bar{\sigma}_{Au}$ – спектральный индекс в базисной конфигурации активной зоны; I_j – то же для инспектируемой конфигурации; A_{ij} – матрица оператора, отражающего линейную связь (матрица влияния).

Практическое использование соотношения (1) возможно, если известны элементы матрицы A_{ij}. Их можно определить из эксперимента или расчетным путем.

Все экспериментальные модели изъятия ядерных материалов, организованные на критической сборке СПЕКТР, сопровождались расчетными исследованиями, которые показали возможность достаточно верного предсказания спектральных индексов в различных конфигурациях активной зоны почти во всех точках, за исключением точек, близких к границе активной зоны и графитового отражателя. Согласие расчета и эксперимента хорошо видно на рис.2.

Поэтому проверка методики вычисления величины изъятия выполнялась не только на экспериментально смоделированных изъятиях, но и на расчетных. В последнем случае моделировались ситуации, которые огранизовать на сборке практически невозможно. Такие, например, когда делящиеся материалы извлекаются равномерно из всех ТВС активной зоны.

Матрицы влияния строились в основном по расчетным калибровочным актам изъятия ядерных материалов.

Для нахождения элементов матрицы A_{ij} можно использовать два алгоритма. Если моделируются локальные изъятия, то удобно искать элементы матрицы, обратной к A_{ij}, по формулам

$$A_{ij}^{-1} = \Delta I_i / (I_i \Delta F_j)$$

так как каждое такое калибровочное изъятие определяет один столбец обратной матрицы. Матрица A_{ij} при этом получается обращением. Возможно также отыскание непосредственно элементов матрицы A_{ij} по методу наименьших квадратов при достаточном числе калибровочных изъятий. В этом случае надо позаботиться, чтобы число линейно независимых экспериментальных векторов было не меньше размерности матрицы (числа контрольных точек).

Число действительно выполняемых калибровочных экспериментов или расчетов может быть резко сокращено при наличии свойств симметрии у инспектируемой критической сборки.

Результаты вычисления величины изъятия для экспериментально и расчетно смоделированных случаев приведены в табл.I. Расчеты в этой таблице соответствуют использованию двух матриц, построенных по калибровочным изъятиям различной величины: первая – для двухкилограммовой калибровки; вторая – для одного килограмма.

Характер зависимости элементов обратной матрицы от величины калибровочного изъятия можно понять из рис.3. Так как матричный элемент есть коэффициент наклона хорды соответствующей кривой, проведенной из начала координат в точку ее пересечения с вертикальной прямой, проходящей через калибровочный вес, то ясно, что матрица, вычисленная по большим калибровкам, будет занижать действительный размер изъятия, тогда как матрица, построенная по калибровочным эксперимен-

ТАБЛИЦА I. РЕЗУЛЬТАТЫ РАСЧЕТА ВЕЛИЧИН ИЗЪЯТИЯ ЯДЕРНЫХ МАТЕРИАЛОВ

Варианты изъятия	Место		Фактически изъято (кг)	Матрица (кг)	
	изъятия урана-235.	введения замедлителя		первая	вторая
1	16, 18, 57	16, 18, 57	4,5	4,4	19,3
2	37, 42, 48	на границе активной зоны	4,5	4,7	8,7
3	46, 51, 55	на границе активной зоны	4,5	4,9	9,2
4	37, 42, 48	37, 42, 48	6,0	5,3	14,5
5	46, 51, 55	46, 51, 55	6,0	3,2	10,7
6	16, 18, 57	16, 18, 57	6,0	5,6	22,6
7	26, 31, 32, 37, 42, 43, 48	на границе активной зоны	10,5	7,7	15,8
8	равномерно	равномерно	6,0	4,6	14,5
9	равномерно	равномерно	27,6	14,1	39,0
10	равномерно	равномерно	50,1	18,3	48,6
11	равномерно	равномерно	78,0	19,8	51,2

ПРИМЕЧАНИЕ. Указанные места изъятия урана-235 и введения замедлителя соответствуют рис.1.

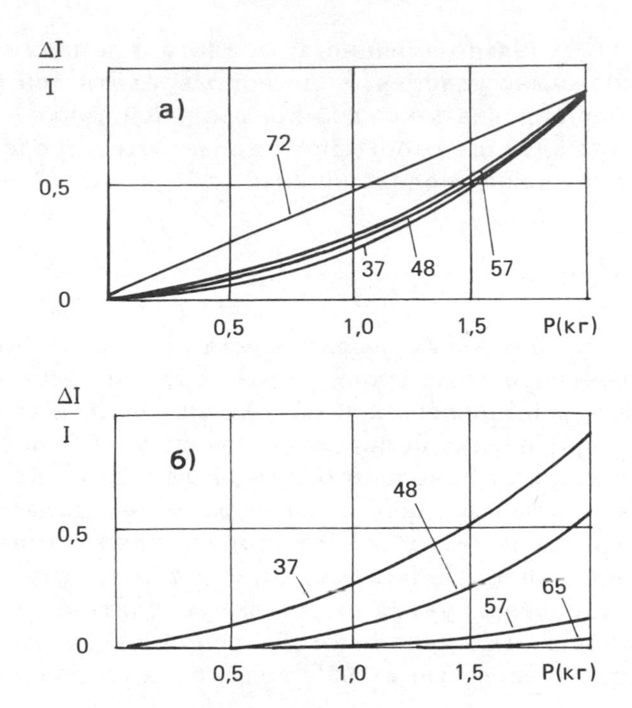

Рис.3. Зависимости относительных изменений спектральных индексов от величины элементарного изъятия: а – изъятие и измерение в позициях 37, 48, 57, 72; б – изъятие в позиции 37, а измерения в позициях 37, 48, 57, 65.

там с одним килограммом, будет давать правильные результаты для изъятий среднего размера, завышая маленькие.

Все это можно видеть в табл.I, где под номерами 2, 3 и 7 приведены экспериментально смоделированные изъятия.

6. ВОЗМОЖНОСТИ ДЛЯ УСОВЕРШЕНСТВОВАНИЯ МЕТОДИКИ

Трудность точного вычисления величины изъятия обусловлена грубостью приближения (1), которое приемлемо только для малых величин изъятия ядерного материала.

Кроме того, на точность влияет то, что метод учитывает только влияние изменения количества горючего на спектральные индексы, хотя очевидно, что введение замедлителя играет существенную роль.

Учет нелинейности возможен, если получено несколько матриц с различными калибровками.

Учесть отдельно влияние замедлителя и делящихся материалов возможно удастся, если использовать два различных спектральных индекса в каждой контрольной точке.

В обоих случаях необходимо дальнейшее экспериментальное и теоретическое исследования.

7. ЗАКЛЮЧЕНИЕ

Как показала экспериментальная проверка, описанный метод коллективного контроля ядерных материалов в активной зоне быстрой критической сборки еще нельзя считать готовым для определения количества изъятых ядерных материалов. Присущая ему довольно большая ошибка в определении изъятия материалов в значительных размерах делает его схожим с импульсным методом, у которого также имеется большая неопределенность при измерении больших изъятий и который также характеризуется высокой чувствительностью к начальным малым отклонениям от базисной конфигурации. Это явление, замеченное еще в 1975 году, пока не получило должной оценки.

Международное агентство по атомной энергии контролирует не одну, а целый ряд критических сборок и, несомненно, с течением времени их число будет возрастать. Поскольку маловероятно, что на всех критических сборках одновременно и постоянно будет происходить изъятие ядерных материалов, надо полагать, что в большинстве случаев инспекции будут проводиться на сборках, где изъятие не имело места. Тем не менее инспектор должен убедиться, что его действительно нет. Таким образом, усилия Агентства, в основном, будут затрачиваться на инспекцию установок, которые не используются в целях несанкционированного изъятия ядерных материалов. Экономия трудозатрат и времени на такие инспекции позволит сосредоточить их там, где они требуются.

Характерные свойства обоих упомянутых методов — высокая чувствительность к малым изменениям и высокая точность измерений, обеспечат уверенность в отсутствии изъятия там, где его нет и обнаружить изъятие там, где оно имеет место.

Это замечание, конечно, не снимает вопроса о дальнейшем развитии коллективных методов, расширении их возможностей в отношении количественного определения изъятия ядерных материалов и упрощении процедур, связанных с их использованием. Практическая проверка сигнализирующих свойств коллективных методов на быстрых критических сборках, более распространенной конструкции,чем СПЕКТР, была бы полезна и своевременна [2].

ЛИТЕРАТУРА

[1] ГРЯЗЕВ, В.М. и др., "Испытание системы учета и контроля ядерных материалов на быстротепловой критической сборке СПЕКТР", Отчет НИИАР по контракту с МАГАТЭ №1435/RB, Димитровград (1975).

[2] COBB, D.D., SAPIR, J.L., "Preliminary Concepts for Materials Measurement and Accaunting in Critical Fasilities", Draft report, Los Alamos (1977).

КОНТРОЛЬ НЕОБЛУЧЕННОГО ТОПЛИВА В ТЕПЛОВЫДЕЛЯЮЩИХ СБОРКАХ БЫСТРОГО РЕАКТОРА БОР-60 ПО НЕЙТРОННОМУ ИЗЛУЧЕНИЮ

В.М.ГРЯЗЕВ, Г.И.ГАДЖИЕВ, Ю.И.ЛЕШЕНКО,
А.К.ГОРОБЕЦ, А.Л.СЕМЕНОВ
Научно-исследовательский институт атомных реакторов
им. В.И.Ленина
Государственного комитета по использованию
атомной энергии СССР,
Димитровград,
Союз Советских Социалистических Республик

Abstract–Аннотация

CONTROL OF UNIRRADIATED FUEL IN BOR-60 FAST REACTOR FUEL ASSEMBLIES BY NEUTRON RADIATION.

Experimental studies have been made of the possibility of monitoring the characteristics of unirradiated fuel assemblies for the BOR-60 reactor by means of the natural neutron radiation and the delayed-neutron emission initiated by a ^{252}Cf source. The experimental apparatus uses a neutron detector with a high efficiency (~ 13% for the neutrons from the californium source) and a pneumatic transport system with a ~ 5×10^7 n/s californium source. A coincidence technique is employed to separate the spontaneous-fission and the (α,n)-reaction neutrons in the natural neutron radiation of the fuel. The possibility of monitoring the fuel characteristics by means of delayed neutron emission has been tested by studying the sensitivity of the measurement results to the removal of fuel from the assemblies. For a fuel with a known enrichment, the measured values seem to be linearly related to the amount of ^{235}U over a limited range of fuel contents. An analysis of the results has shown that an estimate of the amount of ^{235}U in a fuel assembly can be obtained to an accuracy of better than 2% (2b) in a measurement time of 30 minutes. For fuel where the enrichment is unknown, use is made of the dependence of the measured values on the spectrum of the neutrons irradiating the assembly. The results for unirradiated fuel in BOR-60 assemblies show good agreement with the rated ^{235}U content. In connection with the use of this apparatus in safeguards systems, attention has been given to the possibility of falsification of the results and to ways of detecting this.

КОНТРОЛЬ НЕОБЛУЧЕННОГО ТОПЛИВА В ТЕПЛОВЫДЕЛЯЮЩИХ СБОРКАХ БЫСТРОГО РЕАКТОРА БОР-60 ПО НЕЙТРОННОМУ ИЗЛУЧЕНИЮ.

На экспериментальной установке исследована возможность контроля характеристик необлученных топливных сборок реактора БОР-60 по естественному нейтронному излучению и по излучению запаздывающих нейтронов, инициированному с помощью изотопного ^{252}Cf-источника. В установке использован эффективный (~13 % для нейтронов калифорниевого источника) детектор нейтронов и пневмотранспортная система с калифорниевым источником интенсивностью

~5 · 10^7 н/с. Для разделения нейтронов спонтанного деления и нейтронов (α, n)-реакции естественного нейтронного излучения топлива применяется техника совпадений. Для обоснования возможности контроля содержания топлива по излучению запаздывающих нейтронов исследована чувствительность результата измерения к изъятию топлива из сборок. В ограниченном диапазоне содержания топлива с известным обогащением результат измерения можно считать линейно связанным с количеством урана-235. Анализ результатов контроля показал, что оценка количества урана-235 в топливной сборке может быть получена за 30 мин измерения с погрешностью менее 2 % (2σ). Для контроля топлива неизвестного обогащения использована зависимость результатов измерения от спектра облучающих сборку нейтронов. Результаты контроля содержания необлученного топлива в сборках реактора БОР-60 хорошо согласуются с паспортными данными о содержании в них урана-235. С точки зрения использования установки в системе гарантий рассмотрена возможность фальсификации результатов контроля и пути ее обнаружения.

1. ВВЕДЕНИЕ

В настоящее время атомные промышленные и исследовательские установки отличаются большим разнообразием форм и состояний используемых на них делящихся материалов. Проведение эффективного контроля таких установок и количественный анализ делящихся веществ требуют использования широкого арсенала средств и методик, позволяющих контролировать содержание делящихся материалов в топливных единицах без их разрушения [1].

В настоящей работе рассматриваются возможности контроля содержания высокообогащенного урана в необлученных топливных сборках (ТВС) реактора БОР-60. Разработанный способ контроля, который основан на регистрации запаздывающих нейтронов, инициированных с помощью изотопного ^{252}Cf-источника [2, 3], предполагается в дальнейшем использовать для входного контроля ТВС в зоне топливного баланса реактора БОР-60 в рамках выполняемых работ по исследовательскому контракту с МАГАТЭ 2039/RB.

2. УСТАНОВКА

Схема установки контроля содержания топлива в ТВС изображена на рис.1. В установке использованы детектор нейтронов и пневмотранспортная система с калифорниевым источником интенсивностью ~ 5 · 10^7 н/с.

2.1. Детектор

Детектор нейтронов представляет собой блок полиэтилена размерами 370 × 370 × 370 мм, имеющий центральную полость для помещения ТВС и концентрически расположенные полости для размещения счетчиков нейтронов. Всего в детекторе используются 12 счетчиков с ^{3}He, разбитых на четыре группы по три счетчика в каждой. Импульсы от каждой группы счетчиков усиливаются, дискриминируются, формируют-

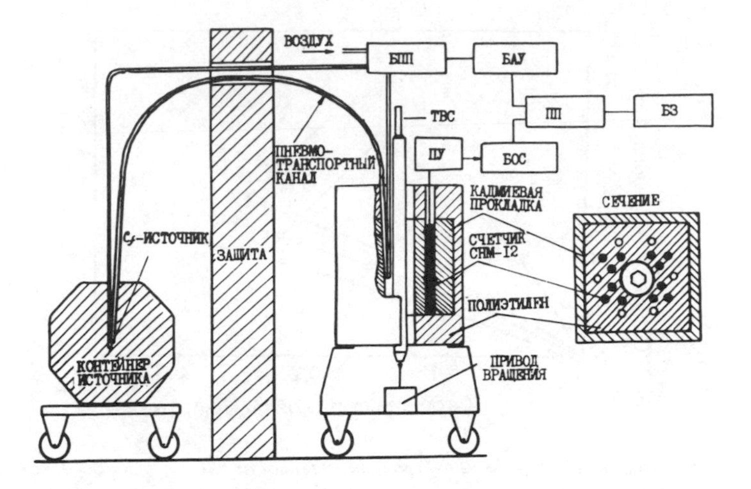

Рис.1. Схема установки: ПУ – предусилитель; БОС – блок обработки сигналов; ПП – пересчетный прибор; БАУ – система автоматического управления измерением; БЗ – блок записи; БПП – блок питания пневмосистемы.

ся и регистрируются пересчетным прибором. Разрешающее время тракта регистрации нейтронов детектором составляет ~ 6 мкс. Эффективность регистрации нейтронов калифорниевого источника, помещенного в центр измерительной полости, составляет ~ 13 %.

В установке предусмотрена возможность использования техники совпадений для решения задач контроля пассивным методом. Блок совпадений регистрирует импульсы со счетчиков, поступающие в течение 100 мкс после поступления первого импульса. Фон случайных совпадений определяется путем подсчета импульсов в течение интервала аналогичной длительности, начало которого задержано по отношению к началу первого на время 500 мкс.

2.2. Режим измерений

В процессе измерений установка работает в двух режимах:

1) измерение фона,

2) измерение излучения запаздывающих нейтронов.

В режиме измерения запаздывающих нейтронов калифорниевый источник с помощью пневмотранспортной системы периодически засылается из защитного контейнера к исследуемой ТВС. В промежутке между облучениями детектором, в измерительной полости которого помещена ТВС, регистрируются испускаемые последней запаздывающие нейтроны. Интервалы времени облучения ($t_{обл}$ = 1 мин), выдержки после облучения до начала счета ($t_{выд}$ = 3 с) и времени счета ($t_{сч}$ = 1 мин) поддерживаются постоянными системой автоматического управления процессом измерения.

Рис.2. Результат измерения собственной нейтронной активности ТВС: ●, □, Δ, ○ – двуокись естественного урана, 21 %, 36 %, 90 % обогащения, соответственно; 1, 2, 3, 4 – содержание в активной зоне 1280 г, 1630 г, 2020 г, 2210 г двуокиси урана, соответственно.

Так как облучение ТВС проводится в полости детектора, происходит искажение (смягчение) спектра нейтронов источника за счет рассеянных и замедлившихся в блоке замедлителя нейтронов. Поэтому, чтобы снизить эффект экранирования внешними слоями топлива внутренних, ТВС окружена кадмиевым экраном.

В качестве величины, характеризующей излучение запаздывающих нейтронов, используется среднее число нейтронов, регистрируемых за один цикл измерения. Эта величина подразумевается под термином "счет запаздывающих нейтронов" ("счет з. н.").

3. ИЗМЕРЕНИЕ СОБСТВЕННОЙ НЕЙТРОННОЙ АКТИВНОСТИ

Собственная нейтронная активность ТВС обусловлена спонтанным делением тяжелых ядер и (α, n)-реакцией на ядерах кислорода. На рис.2 приведены результаты измерений собственного нейтронного излучения необлученных ТВС реактора БОР-60 с двуокисью урана различного обогащения. Взаимно-однозначное соответствие между обогащением и количеством урана, с одной стороны, и общей скоростью счета и скоростью совпадений, с другой стороны, показывает принципиальную возможность определять количество ^{235}U. Однако низкая интенсивность собственного нейтронного излучения делает такие измерения неоперативными. Возможность пассивного контроля основана на предположении о неизменности изотопного состава топлива. Поэтому применение такого контроля во многом определяется постоянством процентного содержания в топливе ^{234}U, который является основным альфа-излучателем.

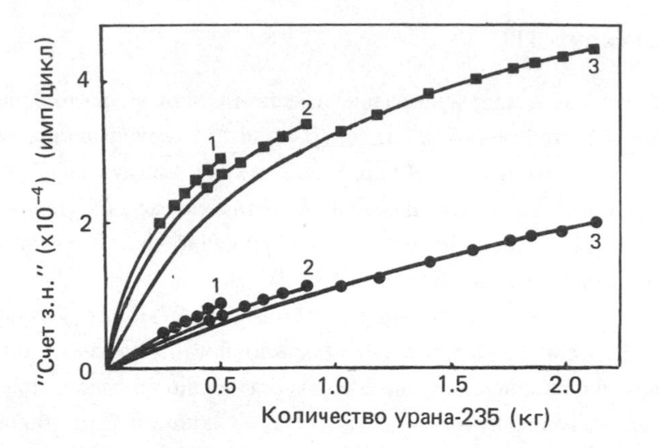

Рис. 3. Зависимость "счета з.н." от содержания в ТВС урана-235: 1, 2, 3 – двуокись урана 21%, 36%, 90% обогащения, соответственно; ■ *– измерения проводились без кадмиевого экрана.*

С точки зрения оперативности пассивный контроль представляется более перспективным для топлива, содержащего плутоний. Однако при этом сохраняются ограничения применимости данного метода, связанные с требованиями постоянства изотопного состава контролируемых топливных материалов. Интерпретация результатов измерения собственной нейтронной активности представляет большую трудность в случае флуктуаций содержания примесей высокоактивных изотопов (^{238}Pu, ^{241}Am и др.) и легких элементов-мишеней (α, n)-реакции. На результат измерения может влиять и качество (гомогенность) смеси легких элементов и альфа-активных изотопов. Измерения топливных сборок с известным содержанием двуокиси плутония показали, что в случае практической реализации метода вышеперечисленные причины могут приводить более чем к 10% погрешности контроля.

Трудности и дополнительные ограничения, связанные с гамма-активностью ТВС, приводят к переоценке применимости пассивного анализа по нейтронному излучению [1, 4].

4. ИЗМЕРЕНИЕ ИЗЛУЧЕНИЯ ЗАПАЗДЫВАЮЩИХ НЕЙТРОНОВ

На рис. 3 представлена зависимость результатов измерения излучения запаздывающих нейтронов от содержания ^{235}U в ТВС с двуокисью урана различного обогащения. Удалением кадмиевого экрана из измерительной полости детектора имитировалось изменение спектра нейтронного облучения. Нелинейный вид зависимостей отражает влияние процесса экранирования внешними рядами твэлов внутренних в ТВС и это влияние тем сильнее, чем мягче спектр нейтронов.

4.1. Контроль штатных ТВС

При контроле ТВС в малом диапазоне изменения количества топлива одинакового обогащения зависимости "счета з.н." от содержания ^{235}U можно аппроксимировать прямой с коэффициентом наклона $W = \Delta N/\Delta X$, где ΔN – изменение "счета з.н." при изменении количества ^{235}U на ΔX. Величина W в этом случае характеризует абсолютную чувствительность контроля и определяется при калибровке с помощью образцовых ТВС с известным содержанием топлива.

Следует отметить, что с течением времени значение W может меняться из-за дрейфа аппаратуры. Поэтому необходимо периодически повторять измерения эталонной ТВС с известным содержанием топлива, а в качестве калибровочного коэффициента удобно использовать величину $K = N_{эк}/W$, где $N_{эк}$ – "счет з.н." эталонной ТВС в момент калибровки.

Содержание ^{235}U в ТВС определяется путем сравнения результатов, полученных для исследуемой и эталонной ТВС в процессе контроля согласно соотношению:

$$X = X_э + K\,(N/N_э - 1) \qquad (1)$$

где $X_э$, $N_э$ – количество ^{235}U и "счет з.н." для эталонной ТВС, N – "счет з.н." исследуемой ТВС.

Измерение на установке в течение ~ 30 мин штатной ТВС реактора БОР-60 с двуокисью урана 90 % обогащения позволяет оценить количество ^{235}U в ней с ошибкой менее 2 % при доверительной вероятности 0,95.

4.2. Контроль топлива неизвестного обогащения

Деление ^{238}U быстрыми нейтронами обусловливает зависимость кривых, представленных на рис. 3, от обогащения топлива. Предположение, что содержание ^{238}U (Y) в ТВС увеличивает "счет з.н." на величину, соответствующую изменению количества ^{235}U на $\Delta X = \delta \cdot \Delta Y$, позволяет аналитически описать семейство этих кривых функцией вида:

$$N = \{1 - \exp[-a_1 (X + \delta Y)]\}\,[a_2 + a_3 (X + \delta Y)] \qquad (2)$$

где a_1, a_2, a_3 и δ – константы, определяемые при калибровке. Величина δ, отражающая эффект деления ^{238}U, зависит от спектра нейтронов, инициирующих деление в ТВС. Это открывает возможность оценивать количество ^{235}U в ТВС с топливом неизвестного обогащения, используя результаты измерения в различных спектрах. Для каждого спектра параметры a_{ij} и δ_j ($j = 1, 2;\ i = 1, 2, 3$) определяются аппроксимацией функцией (2) результатов калибровочных измерений в этом спектре образцовых ТВС с известным содержанием топлива различного обогащения. Измерение "счета з.н."

ТАБЛИЦА I. РЕЗУЛЬТАТЫ КОНТРОЛЯ ТВС С ТОПЛИВОМ РАЗЛИЧНОГО ОБОГАЩЕНИЯ

№ п/п	Паспортные значения количества топлива в ТВС (г)		"Измеренные" значения количества топлива в ТВС (г)	
	^{235}U	^{238}U	^{235}U	^{238}U
1	365	1371	369 ± 34	1350 ± 370
2	419	1574	446 ± 28	1340 ± 330
3	473	1777	499 ± 25	1480 ± 310
4	495	1254	483 ± 26	1510 ± 310
5	569	1440	566 ± 24	1520 ± 290
6	606	1532	610 ± 23	1510 ± 280
7	643	1625	658 ± 23	1440 ± 270
8	500	1878	525 ± 25	1540 ± 290
9	589	1046	576 ± 24	1190 ± 290
10	636	1130	628 ± 23	1290 ± 280
11	683	1214	634 ± 23	1670 ± 280
12	730	1298	695 ± 23	1590 ± 270
13	777	1381	750 ± 23	1570 ± 260
14	824	1465	845 ± 23	1250 ± 260
15	858	734	848 ± 23	650 ± 260
16	999	848	972 ± 22	1230 ± 260
17	1070	905	1094 ± 22	830 ± 240
18	1183	679	1146 ± 22	860 ± 240
19	1424	823	1464 ± 22	550 ± 250
20	1505	871	1561 ± 22	580 ± 250
21	1992	217	1970 ± 29	440 ± 310
22	2105	229	2077 ± 33	530 ± 360

от исследуемой ТВС и решение уравнений (2) относительно двучленов

$$X + \delta_j Y \equiv Z_j \qquad (3)$$

позволяют получить систему линейных уравнений вида (3). Из этой системы легко определяются X и Y.

На установке выполнены измерения ТВС с известным содержанием топлива различного обогащения, которые проводились с кадмиевым экраном в измерительной полости детектора и без него. Результаты контроля представлены в табл. I.

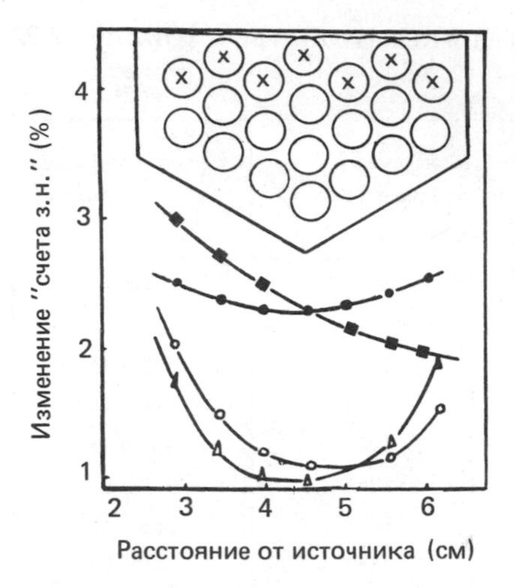

Рис. 4. Изменение счета при извлечении одного твэла в зависимости от места извлечения: ○ – *ТВС не вращается, нет кадмиевого экрана;* Δ – *нет экрана, ТВС вращается;* ■ – *на ТВС экран, нет вращения;* ● – *ТВС вращается в кадмиевом экране. Извлекаемые твэлы помечены знаком* ×.

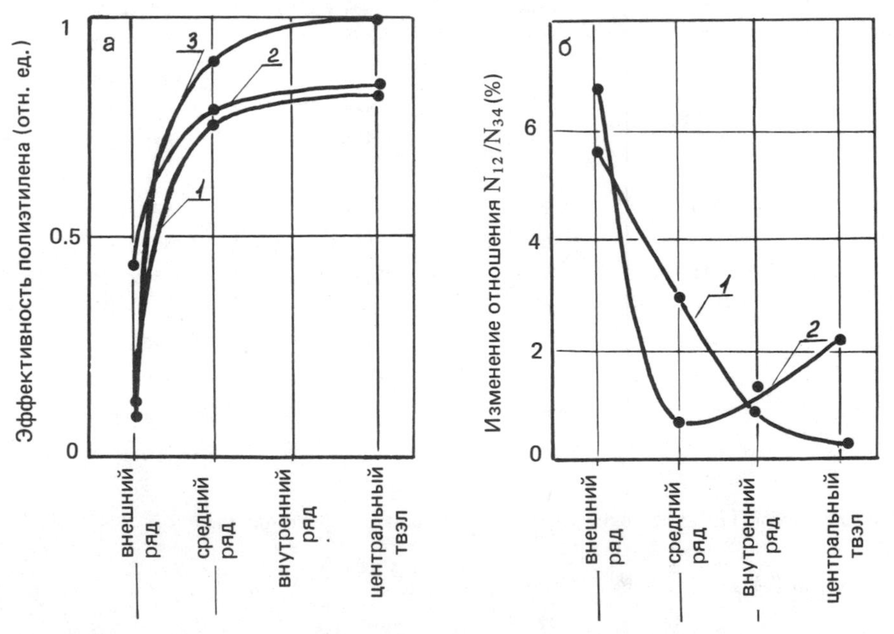

Рис. 5. Влияние помещенного в ТВС полиэтилена на результаты измерения: а) относительная эффективность компенсации полиэтиленом извлекаемых из различных рядов твэлов: 1 – измерения без кадмиевого экрана, 2 – измерения в кадмиевом экране, 3 – ТВС заполнена водой, кадмиевый экран отсутствует; б) изменение отношения результатов измерения различными группами счетчиков при изъятии твэлов и помещении полиэтилена в ТВС: 1 – изъяты твэлы, 2 – вместо них помещен полиэтилен.

5. ЧУВСТВИТЕЛЬНОСТЬ К ИЗЪЯТИЮ ТОПЛИВА

Количественный контроль, проводящийся для целей гарантий, предполагает возможность хищения топлива и умышленной фальсификации результатов измерений. Поэтому особый интерес представляет чувствительность метода контроля к изъятию топлива из ТВС и мероприятия, гарантирующие надежность результатов.

Для оценки возможностей контроля на установке выполнены эксперименты, которые показали, что изменение "счета з.н." при извлечении отдельных твэлов из ТВС существенно зависит от их положения в ТВС (рис.4). Вращение исследуемой ТВС и экранирование ее кадмием уменьшает эту зависимость и диапазон неопределенности контроля до 3 ÷ 5 %. Очевидно, что удаление замедляющих материалов и источника от ТВС позволит свести радиальную неравномерность контроля до минимума.

Однако жесткий спектр нейтронов открывает другой вариант для маскировки изъятия топлива: введение в ТВС замедляющих материалов, присутствие которых приводит к более интенсивному делению в ТВС. Для оценки возможности обнаружения факта фальсификации, не прибегая к помощи дополнительных методов и аппаратуры контроля, выполнялись эксперименты, в которых изъятие твэлов компенсировалось помещением в ТВС равных им по объему полиэтиленовых стержней. Эксперименты показали, что наибольший эффект достигается при помещении полиэтилена в центр исследуемой ТВС (рис.5а). Присутствие полиэтилена может быть обнаружено, если заполнить ТВС жидким замедлителем или поглотителем нейтронов, что изменит эффективность компенсации полиэтиленом изъятого топлива.

Представляет интерес возможность обнаружить факт хищения топлива путем сравнения результатов измерений отдельных групп счетчиков без вращения ТВС. Эта возможность основана на том, что изъятие твэлов и помещение вместо них полиэтилена влияет на распределение интенсивности источников деления в ТВС, что отражается на "счете з.н." отдельными группами счетчиков в различной степени. Изменение отношения N_{12}/N_{34} "счета з.н.", регистрируемого ближними (N_{12}) и дальними (N_{34}) от источника группами счетчиков при изъятии твэлов и помещении полиэтилена в ТВС, показано на рис.5б.

Не исключено, что рациональный (с точки зрения похитителя) выбор замедлителя и способа его размещения в ТВС позволит значительно снизить эффективность предложенных процедур. Однако в этом случае задача похитителя представляется не менее сложной, чем задача инспектора МАГАТЭ.

6. ЗАКЛЮЧЕНИЕ

На этапе внедрения неразрушающих методов контроля необходимым требованием является простота и доступность их практической реализации. Как показал опыт, этому требованию удовлетворяет описанный в докладе способ контроля ТВС по запаздывающим нейтронам. Практическое применение такого способа анализа топлива, очевидно, не ограничивается использованием его только для входного контроля необлученных ТВС быстрого реактора БОР-60. Аналогичные установки могут быть применены для экспериментальной оценки выгорания топлива неразрушающим способом, для контроля содержания малообогащенного топлива в ТВС тепловых реакторов и для других физических и технологических исследований. Эта возможность отвечает естественному стремлению сделать аппаратуру, предназначенную для контроля в системе гарантий, полезной и желательной для оператора ядерно-физической установки.

ЛИТЕРАТУРА

[1] ФРОЛОВ, В.В., Ядерно-физические Методы Контроля Делящихся Веществ, М., Атомиздат, 1976.

[2] BREMNER, W., J.Brit. Nucl. Soc. **13** 1 (1974) 79.

[3] STEPHENS, M., System Control for the Modulated ^{252}Cf Sourse "Shuffler", LA-6007-MS, Los Alamos Scientific Laboratory (1975).

[4] AUGUSTSON, M., REILLY, T., Fundamentals of Passive Nondestructive Assay of Fissionable Material, LA-5651-M, Los Alamos Scientific Laboratory (1974).

ОСНОВНЫЕ ПРИНЦИПЫ ОРГАНИЗАЦИИ СИСТЕМЫ УЧЕТА И КОНТРОЛЯ ЯДЕРНЫХ МАТЕРИАЛОВ НА ОПЫТНОМ РЕАКТОРЕ НА БЫСТРЫХ НЕЙТРОНАХ БОР-60

В.М.ГРЯЗЕВ, Г.И.ГАДЖИЕВ, И.Н.АЛЕКСЕЕВ,
А.К.ГОРОБЕЦ, Л.И.ДЕМИДОВ, Н.Р.НИГМАТУЛЛИН,
Ю.И.ЛЕШЕНКО, А.Л.СЕМЕНОВ, С.Н.ХИЛЕНКО,
И.В.ЯКОВЛЕВА
Научно-исследовательский институт
атомных реакторов им. В.И.Ленина
Государственного комитета по
использованию атомной энергии СССР,
Димитровград,
Союз Советских Социалистических Республик

Abstract–Аннотация

BASIC PRINCIPLES OF ACCOUNTING AND CONTROL OF NUCLEAR MATERIALS IN THE BOR-60 EXPERIMENTAL FAST REACTOR.

Under a contract with the International Atomic Energy Agency, the V.I. Lenin Atomic Reactor Research Institute is currently carrying out a study of ways of organizing a nuclear materials accounting and control system for the BOR-60 fast reactor. Some results of this study are presented in the paper. The special physical and technological features of fast reactors create additional difficulties in safeguards systems and give rise to a number of new possibilities for the illicit removal of nuclear materials. These questions are discussed with reference to the BOR-60 reactor but the conclusions are probably applicable to all fast reactors. The proposed accounting and control system is based on non-destructive measurements of the amount of fissile materials in the operating fuel assemblies and screened bundles of the reactor, on the independent control of the principal facility parameters (a list of which is given) and on an automated information collection and evaluation system. Visual means of inspection can be very effective in fast reactor safeguards systems, especially for controlling storage, but they are not used with the BOR-60 reactor.

ОСНОВНЫЕ ПРИНЦИПЫ ОРГАНИЗАЦИИ СИСТЕМЫ УЧЕТА И КОНТРОЛЯ ЯДЕРНЫХ МАТЕРИАЛОВ НА ОПЫТНОМ РЕАКТОРЕ НА БЫСТРЫХ НЕЙТРОНАХ БОР-60.

По контракту с Международным агентством по атомной энергии в Научно-исследовательском институте атомных реакторов (НИИАР) им. В.И. Ленина в настоящее время проводится исследование путей организации системы учета и контроля ядерных материалов на реакторе на быстрых нейтронах БОР-60. Некоторые результаты этого исследования приведены в настоящей работе. Физические и технологические особенности быстрых реакторов с одной стороны порождают дополнительные трудности в системе гарантий, а с другой – предоставляют некоторые новые возможности для скрытого хищения ядерных материалов. Эти вопросы обсуждаются на примере реактора БОР-60, но

выводы, по-видимому, применимы ко всем быстрым реакторам. Предлагаемая система учета и контроля основана на использовании неразрушающих методов измерения количества делящихся материалов в рабочих тепловыделяющих сборках (ТВС) и в экранных пакетах (ЭП) реактора, на независимом контроле основных параметров установки, перечень которых приводится, и на автоматизированной системе сбора и обработки учетной информации. Средства наблюдения могут быть очень эффективны в системе гарантий на быстром реакторе, особенно для контроля хранилищ, но на реакторе БОР-60 они не используются.

1. ВВЕДЕНИЕ

Быстрые ядерные реакторы-размножители представляют собой особую проблему с точки зрения применения к ним системы гарантий Международного агентства по атомной энергии. Это объясняется следующими причинами:

- высокое обогащение топлива позволяет использовать его для изготовления взрывных устройств с минимальной переработкой и в короткое время;
- большая по сравнению с тепловыми реакторами загрузка делящихся материалов расширяет возможности скрытого хищения значительных количеств ^{235}U или ^{239}Pu;
- жесткий спектр нейтронного потока создает условия для маскировки несанкционированного изъятия ядерных материалов;
- воспроизводство ^{239}Pu в экранах представляет собой дополнительную проблему для инспектора и дополнительную возможность для хищения;
- обычно используемый в реакторе, а иногда и в хранилищах, натриевый теплоноситель затрудняет доступ к отработанным ТВС и ЭП, выгружаемым из реактора.

Система контроля ядерных материалов должна учитывать как перечисленные общие особенности быстных реакторов, так и специфику каждой конкретной установки.

Нельзя рассчитывать, что простой поштучный учет и проверка отчетных документов могут обеспечить уверенность в отсутствии несанкционированного изъятия материалов. Необходима разработка методов и приборов для прямого или косвенного измерения количества ядерных материалов в ТВС и ЭП, так же, как способов контроля топлива в активной зоне. Существенную помощь при осуществлении гарантий могут оказать системы контроля и регистрации положения и перемещений перегрузочной машины, подобные системе, предлагаемой для быстрого реактора SNR-300. Разработка и испытание таких систем [1, 2] позволяет выработать рекомендации по их внедрению на строящихся быстрых реакторах.

Организация системы учета и контроля ядерных материалов на реакторе БОР-60 основывается на следующих предпосылках:

- система должна обеспечивать непрерывный контроль за положением и состоянием каждой ТВС и каждого ЭП, находящихся в зоне баланса материалов;
- система должна позволять независимо от оператора установки контролировать ряд физических и технологических параметров реактора;

- система должна позволять дублирование измерений наиболее важных характеристик ТВС и реактора;
- система не отказывается от помощи оператора, но использует ее в минимальной степени;
- система должна быть защищена от вмешательства оператора и иметь возможность нормально функционировать даже в условиях скрытого сопротивления оператора;
- система должна снабжать оператора установки полезной для него технологической информацией.

Это означает, что основная задача системы состоит в обнаружении несанкционированного изъятия ядерных материалов, осуществляемого скрытно на уровне государства, т.е. с использованием всех доступных для государства средств изъятия ядерных материалов и способов фальсификации результатов измерений. Не утверждается, что эта задача может быть полностью решена в настоящее время, но для создания действительно эффективной системы гарантий необходима работа именно в этом направлении.

2. КРАТКОЕ ОПИСАНИЕ РЕАКТОРА БОР-60

Быстрый опытный реактор БОР-60 построен для проведения исследований по физике быстрых реакторов и для накопления технологического опыта работы на таких установках. Тепловая мощность реактора – 60 МВт, электрическая – около 12 МВт. Реактор использует топливо, состоящие из двуокиси урана, обогащенного по ^{235}U до 90%. Топливо находится в твэлах с оболочкой из нержавеющей стали, которые в свою очередь упакованы в ТВС. Каждая ТВС содержит до 2200 г ^{235}U. В активной зоне реактора размещено обычно около 95 ТВС. Боковой экран состоит в основном из стальных пакетов, но в настоящее время часть этих пакетов заменена на ЭП с двуокисью обедненного урана. Всего в реакторе может быть размещено 265 тепловыделяющих сборок и экранных пакетов. Высота активной зоны составляет 400 мм. Сверху и снизу активная зона окружена торцевыми экранами высотой 100 мм из двуокиси обедненного урана.

На установке имеются следующие места размещения ядерных материалов:

- хранилище свежего топлива;
- активная зона и экран реактора;
- хранилище отработавших ТВС и ЭП;
- промежуточное хранилище;
- обмывка;
- разогрев.

Все места размещения ядерных материалов, кроме хранилища свежего топлива, обслуживаются одной перегрузочной машиной.

Общий вид установки БОР-60 показан на рис.1. Более подробное описание можно найти в работах [3, 4].

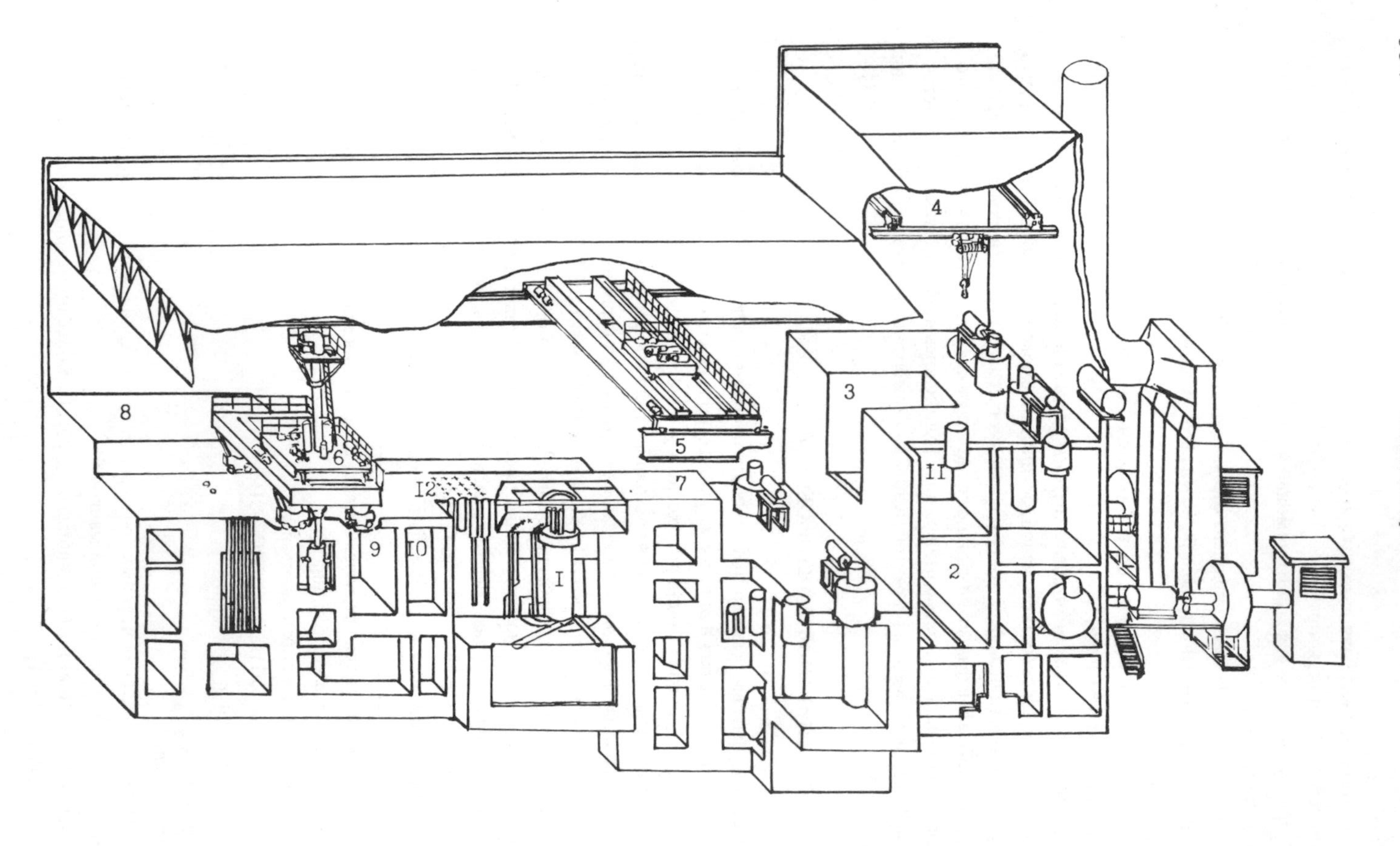

Рис.1. Разрез здания реактора БОР-60: 1 – корпус реактора; 2 – транспортный проезд; 3 – транспортный проем; 4 – кран-балка; 5 – мостовой кран; 6 – разгрузочно-загрузочная машина; 7 – участок подготовки ТВС; 8 – хранилище облученного топлива; 9 – установка для неразрушающих измерений; 10 – пульт установки неразрушающих измерений; 11 – хранилище свежих ТВС.

2.1. Транспортная технология

Ядерные материалы поступают на установку непосредственно перед перегрузкой реактора в металлических контейнерах, вмещающих по три ТВС. Через транспортный проем (рис.1) контейнеры с помощью мостового крана переносятся в центральный зал и размещаются для временного хранения в хранилище свежего топлива. Разгрузка свежих ТВС и контейнеров выполняется вручную.

Перед загрузкой в реактор каждая ТВС и ЭП осматривается на участке подготовки, проходит контроль размеров и передается в устройство подогрева, откуда она уже может быть захвачена разгрузочно-загрузочной машиной. Все дальнейшие операции выполняются с помощью этой машины.

После наведения на соответствующую ячейку активной зоны или экрана разгрузочно-загрузочная машина (РЗМ) переносит ТВС или ЭП к реактору и устанавливает его в ячейку. Наведение перегрузочного канала и загрузочного устройства РЗМ на ячейку реактора контролируется автоматической координатной системой и отображается на табло пульта оператора установки.

Выгружаемые из реактора отработавшие ТВС и ЭП прежде всего проходят отмывку от остатков натрия и затем помещаются на временное хранение в промежуточное хранилище, где может находиться одновременно не более двадцати четырех ТВС.

Из промежуточного хранилища ТВС и ЭП переносятся к хранилищу облученного топлива. Доставка ТВС и ЭП в хранилище выполняется наклонной тележкой. Для их перемещения внутри хранилища используется кран-балка.

Из хранилища облученного топлива ТВС и ЭП могут быть извлечены в обратном порядке: сначала с помощью кран-балки на наклонную тележку, затем с наклонной тележки РЗМ может перенести их в подготовленный для отправки транспортный контейнер. Транспортный контейнер отправляется через транспортный проем мостовым краном.

По различным причинам описанный путь движения ядерных материалов иногда может быть изменен.

3. ВОЗМОЖНЫЕ СПОСОБЫ НЕСАНКЦИОНИРОВАННОГО ИЗЪЯТИЯ ЯДЕРНЫХ МАТЕРИАЛОВ

Как это неоднократно отмечалось, наибольшую привлекательность с точки зрения осуществления несанкционированного изъятия ядерных материалов на реакторе представляет свежее необлученное топливо. В не меньшей степени это справедливо для быстрых реакторов и, особенно, для реактора БОР-60.

Изъятие ^{235}U из отработавших ТВС или ^{239}Pu из ЭП представляет собой гораздо более сложную задачу, но пренебрегать учетом этих возможностей тоже не следует.

Имеется также некоторая возможность для накопления плутония в экранных пакетах или специальных устройствах, которая может быть реализована путем скрытия их от инспекторов Агентства и не отражения в отчетах.

Подчеркнем, что все эти возможности могут быть реально осуществлены лишь в том случае, если несанкционированное изъятие ядерных материалов выполняется на уровне государства. Необходимость создания установок (или использования имеющихся) для переработки изъятого топлива, необходимость использования громоздких защитных устройств – все это затрудняет осуществление несанкционированного изъятия ядерных материалов, но не делает его невозможным.

Изъятие ядерных материалов может происходить как на самой установке, так и вне ее. В первом случае на установке должно иметься соответствующее оборудование, во втором – появляется необходимость скрытой отправки и получения ядерных материалов. И то и другое может быть обнаружено хорошо организованной системой контроля.

Главное внимание в исследованиях по системе гарантий должно быть уделено следующим вопросам:

- как повлияет на показания приборов, использующихся в системе контроля, изъятие делящихся материалов из всех или некоторых ТВС и ЭП;
- как отразится такое изъятие на режиме работы реактора;
- какие возможности имеются у оператора установки для маскировки этих влияний.

3.1. Несанкционированное изъятие ядерных материалов в виде свежего топлива

Для осуществления несанкционированного изъятия ядерных материалов в виде свежего топлива на быстрых реакторах, также как и на тепловых, имеется две возможности [5]. Первая – это изъятие части ядерных материалов из тепловыделяющих сборок при объявлении в отчетах номинального количества. Она может быть реализована как на самой установке, так и вне ее. Вторая возможность – это завышение в отчетных документах действительного количества ядерных материалов, полученных на установке. Последнее имеет смысл только в случае, если горючее поступает на установку с завода, ноходящегося под гарантиями в этом же государстве.

Во влиянии изъятия ядерных материалов первого типа (называемого обычно – "недогрузка") на режим работы быстрого и теплового реактора имеются характерные особенности. Выполненное в значительных масштабах на тепловом реакторе, оно может привести вообще к невозможности выхода его на номинальные параметры. При этом не существует никаких средств для компенсации его влияния. В быстрых реакторах дело обстоит иначе. При помощи введения вместо изъятого ядерного материала подходящего количества замедляющих материалов имеется возможность его компенсации при сохранении общего запаса реактивности. Например, изъятие топлива из одной ТВС реактора БОР-60 может быть полностью скомпенсировано введением меньшего (0,7-0,8) объема бериллия, который зависит от места последующей установки ТВС. Таким путем можно изъять очень значительные количества делящихся материалов.

Влияние изъятия ядерных материалов второго типа на быстрый реактор аналогично его влиянию на тепловые реакторы. Естественно, оно не влияет на режим реактора и может быть обнаружено только применением методов неразрушающего контроля ядерных материалов к ТВС.

Изъятие ядерных материалов "недогрузкой" может выполняться различными способами, наиболее эффективным из которых является равномерное изъятие небольших количеств ядерных материалов из каждого изделия и соответствующая их компенсация замедлителем. При этом обеспечивается спокойный режим работы ТВС в активной зоне при номинальных нагрузках. Однако, при этом фальсифицируются все ТВС и первая же проверка на установках неразрушающего контроля может обнаружить наличие несанкционированного изъятия ядерных материалов.

Изъятие топлива в больших количествах из отдельных ТВС приведет к значительным перекосам нейтронного потока в активной зоне реактора и к недопустимым перегрузкам ТВС, стоящих вблизи от фальсифицированных. Это может создать серьезные трудности при работе реактора на номинальной мощности.

3.2. Несанкционированное изъятие ядерных материалов в виде облученного топлива

Осуществление изъятия облученного топлива, если пренебречь связанными с ним техническими трудностями, содержит в себе большие возможности для маскировки. Это объясняется прежде всего слабой разработанностью методов контроля облученного топлива.

В отработавших ТВС находится еще очень много делящихся материалов и хищение их может представлять интерес с точки зрения осуществления несанкционированного изъятия ядерных материалов. Однако, изъятие облученного топлива практически неосуществимо на реакторе – слишком сложное и громоздкое оборудование требуется для этого и слишком мала вероятность, что инспектор не обнаружит это оборудование. В то же время оператор установки имеет большие возможности для обеспечения такого изъятия и для создания препятствий к его обнаружению. Он может дать завышенные данные о достигнутом выгорании, т.е. объявить в отчете меньшее содержание делящихся материалов, чем есть на самом деле. Лишнее фактическое количество может быть изъято на одной из установок государства, а заявленное количество возвращено. У него имеется также возможность скрытно подменять ТВС, т.е. оставлять на более длительный срок в активной зоне часть выгоревших ТВС (конечно, если они не разрушились), отправляя вместо них сборки с низкой степенью выгорания или даже необлученные сборки.

3.3. Несанкционированное изъятие вторичных делящихся материалов

Осуществление несанкционированного изъятия вторичных делящихся материалов связано с теми же трудностями, что и в предыдущем случае, и практически не может быть выполнено на реакторе. Оператор может способствовать его проведению или путем занижения количества загруженного в реактор сырьевого материала в отчетных документах, или занижая действительные темпы накопления, или, наконец, занижая действительную тепловую мощность реактора.

4. ВОЗМОЖНЫЕ СПОСОБЫ КОНТРОЛЯ

Любой из перечисленных выше типов несанкционированного изъятия ядерных материалов приводит в конце концов к расхождению величины количества ядерного материала, заявленного оператором в отчетах, с его фактическим количеством во всех или в части изделий, содержащих ядерные материалы. Поэтому самым надежным способом обнаружения несанкционированного изъятия ядерных материалов является измерение этого фактического количества.

В настоящее время имеется несколько достаточно хорошо разработанных методов неразрушающего измерения характеристик ядерных материалов в ТВС, и они должны применяться так часто, как это только необходимо.

В связи с тем, что сохранность необлученного топлива является самым слабым местом в системе контроля, следует признать необходимым измерение загрузки необлученных ТВС перед установкой их в активную зону. Такие измерения, если они охватывают все без исключения сборки, значительно снижают возможность изъятия ядерного материала "недогрузкой", оставляя для нее довольно узкий диапазон, лежащий в пределах ошибок измерения. Кроме того, снижается и диапазон возможного изъятия ядерных материалов путем завышения в отчетных документах действительного их количества. Следует отметить, однако, что эти измерения имеют смысл только в том случае, когда они выполняются в присутствии инспектора. Если инспекции приурочиваются к моменту перегрузки, то такие измерения могут быть выполнены в течение двух-трех дней при односменной работе инспектора.

Измерение характеристик облученных сборок представляет собой гораздо более сложную проблему. Это связано прежде всего с необходимостью использования транспортно-технологического оборудования установки, что будет создавать нежелательные помехи оператору при выполнении перегрузочных работ. Даже если оператор предоставит в распоряжение инспектора РЗМ и соответствующий персонал, каждое измерение облученной сборки потребует не менее двух часов, а возможно, и больше. Учитывая, что на реакторе БОР-60 одновременно может находиться до 500 различных сборок, а инспектор при круглосуточной работе не сможет измерить более 12 сборок, проведение полной инспекции потребует около двух месяцев непрерывной работы. Такой объем инспекционных работ не может устроить Агентство и, тем более, оператора.

Таким образом, использование только средств неразрушающего контроля в настоящее время не может обеспечить надежного и своевременного обнаружения несанкционированного изъятия ядерных материалов. Поэтому в системе контроля ядерных материалов следует использовать и другие методы контроля, т.е. система должна быть комплексной.

Представляется привлекательным в качестве дополнительного метода измерения использовать контроль реактивности или подкритичности при перегрузочных работах [6, 7]. На реакторе БОР-60 как извлечение облученного пакета, так и установка в активную зону необлученной ТВС заметно влияют на показания штатных приборов контроля мощности в подкритическом состоянии. Скачок нейтронной мощности при этом достигает 7-10% от ее среднего уровня. Установка специальных датчиков и аппаратуры для контроля мощности или подкритичности позволит получить еще одну возможность измерения загрузки ТВС. Этот прибор может быть проградуирован, хотя и довольно грубо, в килограммах и использоваться для непрерывного контроля перегрузочных операций. Более того, он может быть полезен и при работе на номинальном уровне мощности для контроля запаса реактивности.

Еще одну возможность для контроля загрузки реактора можно получить при расчетном моделировании состояния реактора и физических процессов в нем на основе объявленной оператором суммы делящихся материалов в активной зоне и экранах. Известно, что расчет критической массы и запаса реактивности может быть выполнен достаточно хорошо. Соответствие расчетов величинам, измеренным на реакторе, может служить подтверждением факта отсутствия изъятия ядерных материалов в активной зоне. И наоборот, значительная рассогласованность их сигнализирует о возможном изъятии и о необходимости более тщательной проверки.

5. СОСТАВ СИСТЕМЫ УЧЕТА И КОНТРОЛЯ

При проектировании системы учета и контроля ядерных материалов на реакторе БОР-60 была сделана попытка одновременного и согласованного использования различных методов измерений.

5.1. Установка неразрушающего контроля

На реакторе БОР-60 спроектирована и изготовлена многоцелевая установка для неразрушающих измерений количества ядерных материалов в облученных и необлученных ТВС и ЭП реактора. Она обеспечивает возможность активными и пассивными нейтронными методами контролировать загрузку сборок и имеет аппаратуру для гамма-сканирования с детектором высокого разрешения. Для активных методов применяются интенсивные нейтронные источники трех типов: а) нейтронный генератор НГ-150М с выходом, равным 10^{11} н/с; б) калифорниевый источник; в) сурьмяно-бериллиевый источник.

Установка будет использоваться для контроля содержания ядерных материалов как на входе, так и на выходе. Кроме того, на ней будут проводиться исследования различных способов фальсификации результатов измерений и методов их обнаружения.

Для градуировки установки применяются специально изготовленные экспериментальные ТВС с различным содержанием делящихся материалов и облученные в реакторе БОР-60 до различных величин выгорания.

5.2. Система контроля физических и технологических параметров

Эта система должна обеспечивать независимые от оператора установки измерения следующих параметров, необходимых для контроля за состоянием реактора:

- нейтронная мощность;
- спектр нейтронного потока;
- тепловая мощность;
- запас реактивности;
- эффективность стержней регулирования и их положение;
- координаты РЗМ и загрузочного канала;
- подкритичность заглушенного реактора.

Не все перечисленные возможности реализованы. В частности, измерения спектра нейтронов выполняются активационным методом и пока нет никаких предпосылок для полной автоматизации таких измерений.

В идеальном случае все параметры должны измеряться или вычисляться на основе измерений автоматически и непрерывно.

5.3. Автоматизированная система учета

Автоматизированная система учета представляет собой комплекс вычислительных программ, обеспечивающих автоматическое выполнение следующих функций:

- сбор, кодирование и хранение входной учетной информации;
- сбор, кодирование и хранение параметров реактора;
- физический и тепловой расчеты реактора;
- сбор и переработка информации о перемещениях топлива;
- обработка учетной информации на основе результатов физического расчета и имеющихся экспериментов;
- подведение баланса материалов и выдача отчетов.

Система может быть использована для введения информации о поступлениях и отправках материалов и получения необходимых форм отчетов как оператором установки, так и инспектором.

В реально действующей системе гарантий подобный комплекс программ должен базироваться на специальной вычислительной машине, защищенной от вмешательства оператора. В нашем случае он разрабатывается на ЭВМ общего пользования БЭСМ-6 и некоторые его функции моделируются через обычный ввод с перфокарт и вспомогательными программами.

5.4. Обеспечение сохранности

На реакторе БОР-60 в качестве мер по обеспечению сохранности можно применять опечатывание реактора и хранилищ. Эти меры позволят зачастую значительно экономить усилия, затрачиваемые на инспекцию, особенно в том случае, когда инспекция совпадает по времени с перегрузкой реактора.

5.5 Наблюдение

Не используется из-за отсутствия приборов.

6. РАБОТА СИСТЕМЫ УЧЕТА И КОНТРОЛЯ

Ядром системы учета и контроля является автоматизированная система учета. Она выполняет многие функции оператора установки и инспектора, освобождая их от рутинной работы. Основные массивы исходных данных система получает автоматически, но некоторую информацию она получает от оператора и от инспектора.

Запуск системы, который должен быть выполнен инспектором на основе имеющейся у него проверенной информации, начинается с ввода информации о конструкции установки. Эти данные используются в системе для организации расчетных схем, идентификации хранилищ, подготовки бланков ТВС в архивах системы и т.д.

Вводится информация об имеющихся на установке ТВС согласно данным поставщика (паспорта ТВС) и о местах их хранения. Ввод при первоначальном запуске системы выполняется инспектором, а в дальнейшем при каждом поступлении ядерных материалов на установку – оператором или инспектором, если поступление было в его присутствии.

При обычной работе вручную требуется вводить только информацию о поступлении ядерных материалов по данным поставщика, координаты размещения ТВС в хранилищах и в реакторе и данные об измерениях каждой ТВС на установке неразрушающего контроля. Этот ввод может выполняться и оператором и инспектором. Кроме того, оператор должен ввести в систему номера всех отправляемых с установки ТВС и ЭП. Система имеет возможность различать информацию оператора и инспектора по их шифрам.

На основании полученной информации формируются задания на расчет физических характеристик активной зоны по реальной загрузке реактора и определяется запас реактивности, распределение и спектральные характеристики потока нейтронов, распределение реактивностей ТВС и стержней регулирования и готовится расчет выгорания и изменения изотопного состава. Эти расчеты выполняются в нашем случае один раз в неделю. При этом выполняется корректировка изотопного состава в каждой расчетной зоне каждой ТВС и ЭП согласно их расположению в активной зоне и в экране. Данные поставщика, однако, сохраняются в течение всего времени пребывания изделия на установке.

Расчетные характеристики ТВС система может корректировать на основании результатов экспериментальных измерений.

По требованию оператора или инспектора можно получить любой из учетных или отчетных документов, соответствующий текущему моменту времени. Кроме того, предусмотрена возможность выполнения прогнозирующих расчетов, не изменяющих учетных данных в архивах системы, но позволяющих оператору получить данные об ожидаемом выгорании, о запасе реактивности и о других характеристиках реактора на любой момент.

6.1. Инспекция

Для инспектора представляется необходимым выполнять следующие действия:

- по прибытии на установку убедиться в исправности вычислительной машины и измерительных приборов;
- запросить у системы информацию о расположении ТВС и ЭП;
- проверить соответствие этих данных фактическому состоянию;
- получить от системы информацию о количестве ядерных материалов по каждой ТВС и ЭП и общий баланс материалов и убедиться в соответствии этих данных отчетам оператора;
- измерить содержание делящихся материалов во всех ТВС, предназначенных к очередной загрузке в реактор, и ввести эти данные в систему;
- выполнить выборочные измерения некоторых отработавших ТВС, убедиться в их соответствии или расхождении с расчетными результатами и в последнем случае ввести экспериментальные данные в систему;
- если применяются меры обеспечения сохранности, проверка печатей при прибытии и установка их перед отъездом – очевидные действия инспектора.

Таким образом, если инспекция совпадает по времени с выполнением перегрузочных работ, инспектор и система будут иметь объективные, подтвержденные экспериментом, данные о содержании ядерных материалов в каждом загружаемом в реактор изделии и об их расположении в реакторе. Это обеспечит возможность обоснованной работы системы в промежутке между инспекциями. Содержание изотопов в выгоревших ТВС и ЭП будет определяться из расчетов, что в настоящее время применяется почти повсеместно, но эти данные могут сравниваться с выборочными измерениями и корректироваться. При автоматическом контроле подкритичности с помощью системы можно было бы оценивать содержание ядерных материалов в извлекаемых изделиях, но, к сожалению, сейчас пока не ясно, как это делать.

Учитывая, что изъятие ядерных материалов из облученных изделий на реакторе невозможно, необходима проверка хранилищ выгоревшего топлива, которая должна состоять из визуального, дистанционного контроля номеров ТВС и ЭП и их соответствия номерам, объявленным оператором, ячеек хранилища. Кроме того, периодически следует применять выборочный контроль на установке для неразрушающих испытаний, чтобы исключить возможность подмены ТВС какими-либо макетами с теми же номерами. Использование системы наблюдения может серьезно помочь при организации контроля содержания ядерного материала в хранилище.

Незагружаемое необлученное топливо обычно имеется на реакторе в очень небольших количествах и может быть вполне проверено инспектором неразрушающими методами и опечатано до следующей инспекции. При нарушении печатей это топливо перед следующей загрузкой должно быть обязательно проверено.

7. ЗАКЛЮЧЕНИЕ

Описанная система контроля и учета ядерных материалов на реакторе БОР-60 является попыткой реализации комплексного подхода к применению гарантий на быстрых реакторах. Несмотря на то, что некоторые из ее существенных компонентов только моделируются, она позволит получить опыт контроля характеристик ядерных материалов в условиях, близких к реальным, и выявить те трудности, которые ожидают инспектора на таких установках.

Оценка эффективности и работоспособности системы, оценка экономических затрат на ее внедрение будут полезны при разработке требований к конструкции быстрых реакторов, подпадающих под гарантии, и к системам контроля и учета ядерных материалов на них.

ЛИТЕРАТУРА

[1] BRÜCHNER, Chr., VAN DER HULTS, P., KRINNINGER, H., "Safegurads System for the LMFBR Prototype Power Plant SNR-300 (KKW KALKAR)", Safeguarding Nuclear Materials, v. II (Proc. Symp. Vienna, 1975) IAEA, Vienna (1976) 581.

[2] THALGOTT, F.W., Evaluation of the Safeguards System Proposed for Fast Breeder Reactor, Draft report (1977).

[3] ЛЕЙПУНСКИЙ, А.И. и др., Ат. Энерг. 30 2 (1971).

[4] ГРЯЗЕВ, В.М. и др., Исследование условий применения системы гарантий на реакторе БОР-60 и разработка системы контроля ядерных материалов, Отчет НИИАР по контракту № 2039/RB, Димитровград (1978).

[5] LOVETT, J.E., Safeguarding Nuclear Power Reactors and Nuclear Power Reactor Fuel, Draft report (1976).

[6] TOFFER, H., NIELSEN, L.A., Trans. Am. Nucl. Soc. 22 1 (1975) 292.

[7] GREEN, L., Trans. Am. Nucl. Soc. 22 1 (1975) 141.

DISCUSSION

D. GUPTA: I should like to express my appreciation of the excellent work reported in the papers summarized. In particular, the possibility of detecting a diversion through criticality measurements and the possibility of assaying the plutonium and uranium content in complete fuel assemblies are of great interest. Now I should like to put the following questions:

(a) Have you considered the possibility of using sealing and camera techniques to supplement your measurement techniques?

(b) Are you proposing to measure the assemblies only once during their lifetime?

(c) I understand that you propose to apply the safeguards system that you are developing to the BOR-60 reactor. Do you plan to use this system in other critical assembly systems as well?

V.M. GRYAZEV: I will answer your questions in the order in which you put them:

(a) We consider it extremeley useful to employ different means of surveillance and containment on the reactor as additional safeguards measures.

(b) The fresh fuel assemblies are in the material balance area of the reactor a short time before they are loaded into the reactor. We propose to measure them on a special facility immediately before loading, and to measure them again to determine the amount and quality of fissile materials present after a scheduled period of operation in the reactor, i.e. once a given burnup is reached.

(c) We do not intend to put the BOR-60 under safeguards. The safeguards system we are developing under contract will be tried out on the BOR-60 and its operation will be demonstrated to Agency officials using special fuel assemblies having different concentrations of uranium and plutonium.

W.C. BARTELS: I did not hear the details you gave regarding the neutron spectrum of the SPEKTR critical experiment facility. Could you repeat what you said?

V.M. GRYAZEV: The "SPEKTR" critical assembly has a high concentration of fissile materials (90% enriched uranium-235) in the core. The median neutron energy is 470 keV.

H. KRINNINGER: You mentioned that the assay of SNM in fuel assemblies on LMFBR sites is foreseen as a routine safeguards measure. We think it more appropriate to assay the fissile material content of fuel assemblies at the fuel fabrication site and thereafter to apply containment and surveillance measures during transfer from the fuel fabrication plant to the reactor site to ensure that the SNM content remains unchanged.

V.M. GRYAZEV: In the fuel assemblies of a fast reactor a considerable change takes place in the number of parent fissile nuclei (in our case ^{235}U) and there is an accumulation of secondary plutonium. We consider that measurements should be made at the reactor both on fresh fuel assemblies and on used ones; only in this way is it possible to verify the material balance with any confidence. In my opinion it is not enough merely to assay the fresh fuel assemblies at the factory, especially if the factory and the reactor are in different countries, as there may be differences in the accuracy of the assay systems employed.

Session V (Part 1)

CONTAINMENT AND SURVEILLANCE

Chairman: A.G. HAMLIN (United Kingdom)

Rapporteur summary: *Ultrasonic sealing techniques*

Papers IAEA-SM-231/21, 22 and 124 were presented by
S.J. CRUTZEN as Rapporteur

Rapporteur summary: *Containment and surveillance equipment and techniques*

Papers IAEA-SM-231/38, 39, 63 and 119 were presented by
G. STEIN as Rapporteur

ROLE OF CONTAINMENT AND SURVEILLANCE IN IAEA SAFEGUARDS

T. SHEA, D. TOLCHENKOV
Department of Safeguards,
International Atomic Energy Agency, Vienna

Abstract

ROLE OF CONTAINMENT AND SURVEILLANCE IN IAEA SAFEGUARDS.

Nuclear materials accountancy is the measure of fundamental importance in IAEA safeguards. Containment and surveillance (C/S) have been described as important complementary measures. In this paper, the types of application of C/S measures are discussed, leading to consideration of the logic associated with the implementation of specific C/S devices. Examples of current practice are given. Containment and surveillance contribute to the effectiveness of safeguards in a manner directly related to the mode of application. Preliminary thoughts on the quantification of these contributions are described in terms of two performance indices: cost-effectiveness and systems reliability.

I. INTRODUCTION

IAEA safeguards are built around nuclear materials accountancy as the major safeguards measure, in combination with containment and surveillance (C/S) as importance complementary measures. In those States where the use of containment and surveillance is permitted, combinations of these measures permit the Agency to implement effective, safe and efficient approaches for safeguarding different facilities, with acceptable levels of intrusion into normal plant operations.

As described in a separate paper [1], technical criteria for the Agency's safeguards system translate into specific design and evaluation requirements for each facility to be adequately safeguarded. Combinations of nuclear materials accountancy, containment and surveillance are important and often are essential for the Agency to meet its safeguards objective : timely detection of diversion of significant quantities of nuclear material and the deterrence of such diversion through the risk of early detection.

Containment and surveillance functions may be accomplished by inspectors or by various items of equipment, including instruments and devices. In essence, C/S devices are designed to detect and provide a record of the occurrence of specific situations or activities. This generally means either access to material within a container, storage cubicle or room or a reactor vessel, or the movement of nuclear material containers. Containment and surveillance go hand-in-hand to provide assurance that undeclared access or material movement has not occurred. It should be noted that, as in the case of accountancy such mechanisms do not in themselves detect diversion but provide indications of anomalous material control which require resolution.

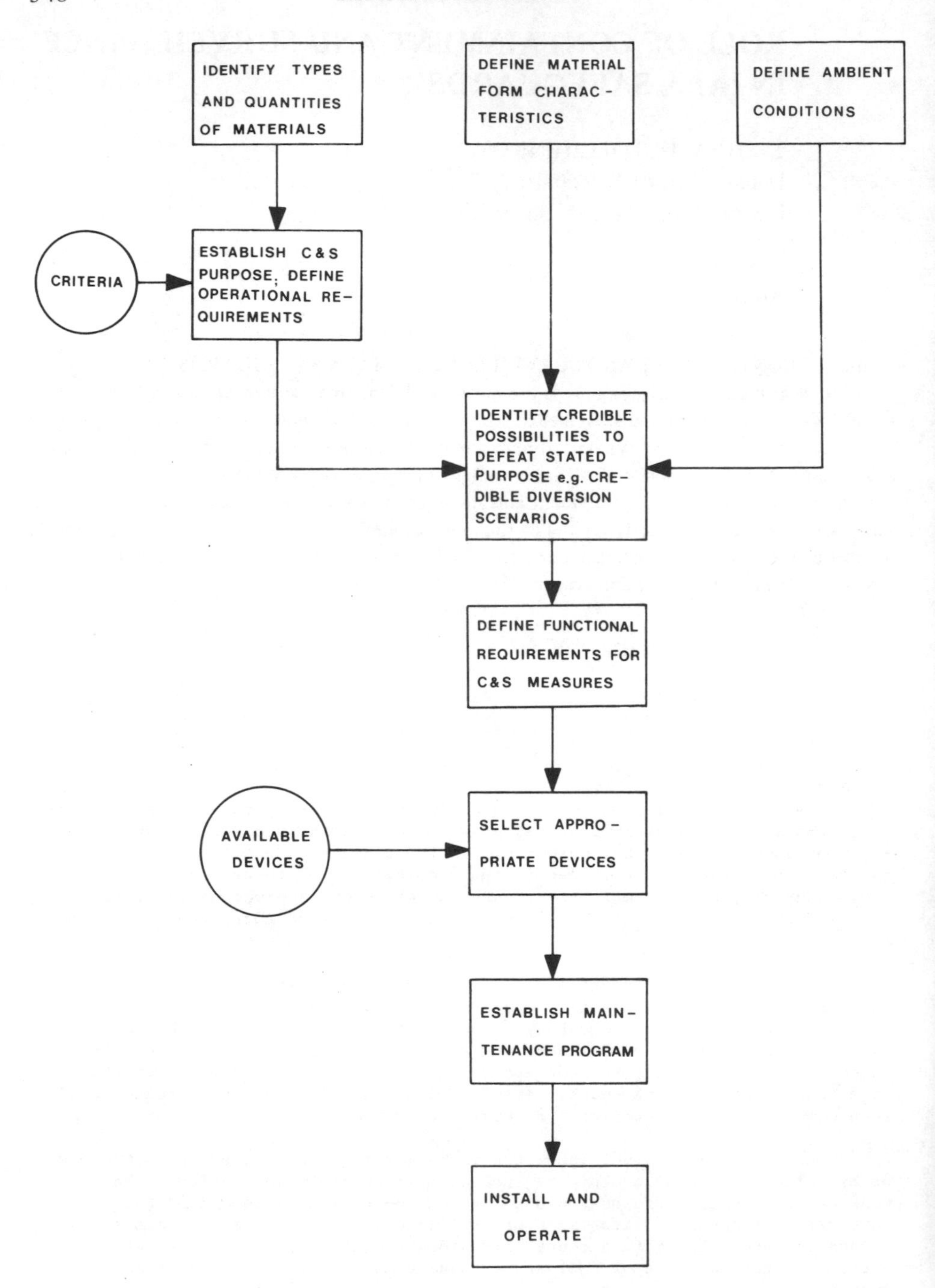

FIG.1. Containment and surveillance application sequence.

II. TYPES OF APPLICATION

Three types of applications of C/S mechanisms may be considered as distinct. The first two directly support nuclear materials accountancy, facilitating the conclusions to be derived through this fundamental safeguards measure. The third type will be necessary in large scale bulk processing facilities of the future when irreducible measurement uncertainties prevent the Agency's safeguards objectives to be realized using accountancy.

In the first type of application, C/S measures are applied during flow or inventory verification inspections to ensure that : each item is inventoried without duplication; sample integrity is preserved; and that Agency instruments, devices, working papers and supplies are not tampered with in the course of the inspection. C/S measures used in this type of application make it unnecessary to have a large number of inspectors on site to accomplish these functions.

In the second type of application, C/S measures provide a means to extend the validity of verified physical inventories, especially at facilities handling only discrete fuel items [e.g. fuel assemblies]. In such cases, it is generally possible to reduce the frequency with which the Agency must re-measure an inventory in order to meet its objectives. In this type of application, C/S measures are applied in a manner to provide an indication of any change in the content or location of material verified in conjunction with previous inspection activities. The assumption is made that as long as all changes are detected and satisfactorily explained, the last physical inventory verification remains valid. This approach may be used to achieve a short detection time capability where frequent inventories are impractical. Note that in any case, physical inventories are required as noted in a separate paper [1] at a frequency of one-to-four times per year. In the event that a C/S mechanism fails or an observed inventory change is not satisfactorily explained, a special inventory is generally required to re-establish effective safeguards.

In the third type of application, containment and surveillance measures are applied in a systematic manner to detect all diversion possibilities considered credible at a facility. This application will be increasingly important in large future facilities where Agency safeguards objectives cannot be realized exclusively through nuclear material accountancy measures. It is important now in facilities where the safeguarded materials could be converted to nuclear explosives in a short time period, or when the inventory of a facility is difficult to access for verification.

III. DEVICE APPLICATION SEQUENCE

The logic sequence of events leading to the application of C/S measures is illustrated in figure 1. As illustrated, the specific operational objective is determined in consideration of the types and quantities of nuclear material and the technical criteria relevant to the facility.

These operational objectives, together with consideration of the material forms and ambient conditions at the facility, translate into functional objectives.

For example, in conjunction with an LWR spent fuel storage pond, the second type of application is relevant, the material (plutonium and low enrichment uranium) is contained in highly radioactive LWR fuel assemblies, and the ambient conditions are underwater storage in a vertical storage rack. It is assumed that only access through the pond storage is credible, material movements outside the pond can only be accomplished by crane and more than 20 minutes is required to move one assembly. The functional objective for

TABLE I. CHARACTERISTICS OF CONTAINMENT/SURVEILLANCE MECHANIS

Mechanism	Principles of Operation	Typical Examples	Typical/Potential Applications
Seals (tamper indicating devices)	Movable segments of containment structure joined by seal which must be violated for access through sealed segments	Fiber-optic; metal; weld; defect/inclusion; paper tape	Containers for storage, shipment; reactor vessels; sample containers; process or storage areas during inspection; C/S devices
Optical Surveillance Cameras	Provide visual record of activities in defined field of view for later review to prove that material has not been removed, sequence for making and storing frames determined by required time to move objects being safeguarded (may include motion sensor to simplify analysis)	CCTV; Film	Spent fuel storage ponds, storage or process areas during inspection
Passage of Flow Monitors	Observe, record (and possibly alarm) on passage of nuclear material past a point in defined corridor, channel, pipe, etc.	yes/no monitors; Bundle counters; solution flow/level monitors; portal monitors	Personnel/equipment passages near on-load power reactors which should not be used for fuel transport; flow indicators for on-load reactors, reprocessing plants
Integral Presence Indicators	Observe (and possibly record) various phenomena arising from presence of specific nuclear material quantities in specific configurations	Neutron physics parameters e.g. reactor power level, flux distribution, energy spectra, reactivity etc.	Power and research reactors, critical facilities; stores of nuclear material in defined arrays; hold-up monitors, attribute check instruments
Human Surveillance	Observe, record, assess and act on various information related to safeguards		All safeguards activities

Containment Function	Surveillance Function
Seals applied at all points of access to containment structure; containment assumed valid while seals intact	Identity and integrity of seals checked at appropriate intervals; total number of seals (issued and unused) accounted for
Field of view considered intact unless visual change noted. Often change to be detected in specific to movement of safeguarded items, not personnel (example : LWR spent fuel pond-pond surface viewed to detect insertion/removal of assemblies or shipping cask)	Review of visual record to identify actions potentially related to diversion; resolution of such actions with operator. Review of device to assure normal operation. Service.
Material entering containment assumed to remain unless transfer indicated through observed passage; indicated transfers serve as input to next containment area or as update for source data	Periodic check of indicators on transit passages for any abnormal indication; update of information base to track changes in inventory through allowed transit passages at appropriate time intervals.
Containment assumption is that material must be present in proper quantity and configuration to produce observed phenomena	Parameters observed at appropriate intervals to assure continual presence of material; system observed to ensure phenomena are real, not fabricated
Establish and periodically verify containment of safeguarded nuclear material within stipulated boundaries	Observe and reconcile anomalous material control activities

application of C/S in this example is : provide an indication of possible assembly movement out of the pond for any assembly in the pond at a maximum interval of 15 minutes.

Given this requirement, a device or combination of devices is selected from available stores, a maintenance programme is devised, and the devices are installed and operated. The service period selected is chosen on the basis of detection time criteria (2 months for the example) or by the recording period associated with the device if that is less than the desired detection period.

IV. BACK-UP MEASURES

Questions frequently arise related to re-establishing effective safeguards under those circumstances when containment or surveillance devices fail. In most instances, a failed device is repaired or replaced and the safeguarded material is re-inventoried. In the example of the LWR spent fuel pond, the assemblies in the racks are counted, samples may be identified by serial number and compared to records, and a second set of samples may be examined non--destructively to assure that the assemblies viewed are, in fact, spent fuel.

To minimize the rate at which the protection afforded by C/S fails, it is essential that the devices are designed for high reliability. Redundancy may be important when device failures occur with unacceptable frequency. In the third type of applications discussed, redundancy is essential for the system to provide the high confidence required.

V. CHARACTERISTICS OF C/S DEVICES

Table 1 presents a summary description of five different categories of C/S mechanisms, including principles of operation, typical examples, applications and a description of the containment and surveillance functions provided by the devices.

VI. REQUIREMENTS FOR USE

A summary of the C/S devices currently in use at the Agency and under development is given in a separate paper [2].

VI.1. Seals

Seals are devices installed to provide assurance that containment access is impossible with violating the integrity of the seal. Requirements for acceptable Agency seals are as follows :

1. Tamper-Resistance - it must be essentially impossible to open and reclose a seal without providing an abnormal indication [this is generally provided by use of single wires, fiber-optic bundles or electrical conductors attached to the seal head; intrinsic feature, weld and adhesive seals are under investigation].

2. Counterfeit Manufacture - seals must be sufficiently unique as to be impossible to counterfeit [for example, in a random uniqueness scheme, the possibility of random reproduction should be less than $1:10^6$].

3. Installation - when installed, it must be impossible to gain access to the sealed volume without violating the seal or creating an additional opening in the containment.

4. Verification - seals should be readily verified (identity and integrity), using simple, rapid methods, as installed.

5. Lifetime - seals must survive until broken or until a convenient opportunity arises for the replacement. This requirement covers normal failures from material fatigue or aging, including degradation of power sources in electronic seals. It also covers requirements for use in radiation environments, underwater, adverse weather, and requirements for adequate ruggedness in rough-handling applications.

6. Cost - seals must be affordable.

7. Acceptability - seals must be accepted for use by the facility operator; e.g. they must allow access in emergency situations.

Seals are extremely important. Safeguards technology would be enhanced greatly with better seals - seals used routinely on fuel assemblies over their entire life cycle, which when verified could provide unambiguous assurance that the assemblies verified are the correct assemblies and that the assembly has remained intact since manufacture. Or adhesive based seals which could be used on thousands of fuel items at bulk processing plants, facilitating frequent re-verification of static items required especially in short detection time applications.

VI.2. Surveillance monitors

Surveillance monitors are devices installed to provide an indication of access or material movement, which is then resolved through follow-up inquiries at the plant. Two types of surveillance monitors have been used : optical and radiometric.

Optical surveillance monitors include both photographic and television systems. These devices are installed to view an area in which nuclear material is contained. They provide a visual record which is periodically reviewed by an inspector.

Radiometric surveillance monitors are installed in accessways or material passageways to provide an indication of the passage of any nuclear material, to count the number of fuel items or measure the flow in solution passing a point, as the basis for later verification.

Requirements for acceptable Agency surveillance monitors are :

1. Observational Sensitivity - the surveillance monitors must be capable of providing an unambiguous indication of the situation or activity under surveillance, related to inventory changes.

2. Recording Requirement - the monitors must produce an easily interpreted record of surveillance for the period between inspector service visits. Preferably, items of interest should be denoted by the system, rather than requiring extensive review of monitor data.

3. Ambient Operational Requirement - the monitors must be capable of operating in the intended environment, through the anticipated range of ambient conditions, including temperature, humidity, radiation, light, vibration and power fluctuations, as appropriate.

4. Tamper-Resistance - surveillance monitors must be designed to resist or to indicate modifications during use which would obscure their performance.

5. Counter-Deception Requirement - Surveillance monitors must be designed and operated so as to provide protection against credible efforts to circumvent the devices. This may involve random interval data collection, or complementary motion detection capabilities, for example.

6. Installation/Maintenance Requirement - Surveillance monitors must be designed for ease of installation, service and for low maintenance.

7. Cost Requirement - surveillance monitors must be affordable.

8. Safety Requirement - surveillance monitors must not impose a safety risk to the facility operator or to the inspector.

9. Acceptability Requirement - the surveillance monitors must be accepted for use at the facility by the facility operator.

VI.3. Human surveillance

At sensitive facilities, Agency criteria call for a safeguards capability designed to counter both abrupt and protracted diversion strategies [1]. The short detection time capability (1-3 weeks) required in bulk processing facilities may be achieved through a mode of inspection activities referred to as "observation of ongoing operational activities" related to nuclear material accountancy and control. In essence, inspectors are in the facility on a continuous basis during routine plant operating periods, and on call to observe operations which impact the Agency's ability to maintain continuity of knowledge of the flow and inventory of nuclear material within the plant.

The basis for diversion detection in this case is pattern recognition. Inspectors observe plant operations over a period of time, accumulating experience in the normal operating practices and process control patterns, including operator instrument data used for process control purposes.

These patterns become increasingly rigid with experience, and quantitative when supported by inventory data derived at a later time.

At reprocessing plants, as described in a separate paper [3], Agency inspectors observe all main process operations, including assembly chopping, dissolution, sampling from the accountability vessel, volume determinations and calibrations, sample analysis, transfers from process to storage and measurements of hulls.

Through careful analysis of the operator's data on input, output in-process and waste quantities, of the consistency of findings for each period compared with previous findings, with input/output measurements, records, operator process data, shipment verification and the inspector's direct observation of on-going operational activities, the inspectors can conclude that no evidence exists for diversion and thus the Agency can conclude that for the period under consideration, no diversion has in fact occurred.

It is important to note that human surveillance as described is intended to verify information provided by facility operators, related to the flow and inventory of material within a facility.

VII. QUANTIFICATION

As described in reference 1, the performance objectives for the implementation of IAEA safeguards at a specific facility are determined by the types and quantities of nuclear material to be encountered, while the approach to be pursued depends on the requirements, the material form, and ambient characteristics at the facility.

The manner and extent to which containment and surveillance contribute to safeguards depends on the purpose to be achieved, as denoted by the type of application and the frame of reference. There are three relevant frames of reference: one for expected or desired effectiveness over a time period in the future (design effectiveness), one for effectiveness in use (operational effectiveness) and one for actual effectiveness achieved over a period in the past (performance effectiveness).

VII.1. Design effectiveness

Will the system satisfy its objectives over the next period of interest? This measure of effectiveness must be gauged in a different manner for each type of application. All applications have in common the underlying assumption that the devices are acceptable for use - acceptable in the sense of satisfying all requirements identified in Section VI, and available.

For the first type of application, containment and surveillance is essential for nuclear materials accountancy to be valid. The design effectiveness of containment and surveillance is the degree to which a clear "yes" or "no" can be expected in this regard. The question is not whether to apply containment and surveillance, but how either a device (or combination of devices) fulfills a need satisfactorily or an inspector must be assigned the task. Clearly the decision to use C/S devices here is judged by the relative cost-effectiveness of the alternative. Even assuming that the alternatives are equally effective, the costs will most often favour C/S devices over human surveillance.

In the second type of application, assuming that accountancy could be employed with material balance closings at a frequency which is adequate to meet the desired detection time objectives, C/S contributes as a cost-effective alternative to high frequency physical inventory measurements. For containment and surveillance, design effectiveness is a measure of the degree to which a clear "yes" or "no" can be expected to the question: "Is the last verified physical inventory still valid?" This alternative is only available when containment and surveillance can be applied in such a manner as to assure that any inventory change will be detected, in time to permit appropriate inquiries.

The anticipated costs associated with closing a material balance, to the Agency, the State and to the facility operator, including manpower, logistical support, lost operating time, etc., constitute a base line. Avoiding the need for high frequency physical inventory re-measurement through the use of containment and surveillance is generally a much more attractive alternative. Thus, in those facilities where Agency objectives can be met using accountancy, containment and surveillance safeguards are used to complement that practice.

In the third type of applications, where Agency safeguards objectives cannot be met through exclusive reliance on accountancy , effective Agency safeguards require shifting emphasis from accountancy to a comprehensive system comprised of combinations of C/S mechanisms with accountancy implemented to the extent possible.

A preliminary examination of how and to what extent it would be possible to develop procedures for the quantification of safeguards effectiveness provided by C/S measures in this application has been undertaken. The basic scheme for such quantification involves :

1. Identifying all credible acts which could be undertaken in conjunction with removing material from a facility, (i.e., a diversion path analysis);

2. Devising C/S countermeasures to each sequence of acts leading to diversion; and

3. Estimating the likelihood that a diversion act would be detected, if attempted.

For this scheme to work, it is necessary to identify all credible diversion possibilities. There is no theoretical basis for determining when the set of possibilities identified is complete, but analysts can agree on reasonable boundaries and through comparisons of investigations at similar facilities, arrive at a set of possibilities which require consideration. Design effectiveness for containment and surveillance for this type of application is the estimated likelihood that a diversion act would be detected, if attempted. This requires consideration of the following for each C/S mechanism:

a) Normal performance capability - the sensitivity of each component to detecting the phenomenon indicating anomalous activity along one or more diversion paths.

b) Normal failure rate, expressed as the probability that each component may fail from natural causes between successive checks; and

c) The susceptibility of each component to sabotage or misinterpretation.

Once the set of alternative possibilities is defined, and the degree of redundancy required is stipulated, the safeguards system can be designed. Often there will exist no acceptable means to restore confidende in the safeguards systems in the event of all mechanisms countering a diversion possibility malfunction. Thus, it is essential that the likelihood that this can occur, be kept very low.

To estimate the design effectiveness of this type of safeguards system, the likelihood that diversion along all paths judged credible must be assessed. In effect the question is how reliable is the system expected to be in detecting diversion through any path in the set?

No scientifically rigorous methods exist to estimate e.g. the overall effectiveness of a facility safeguarded in this manner. However, heuristic rules can be created which can, at least, make the analysis transparent and transferable.

It should be noted that this approach is being considered in the absence of more attractive alternates. In developing the safeguards approach at large scale facilities, a number of options will be explored.

1. Continuous inspections are necessary for all such plants;

2. The inspector should maintain continuous knowledge of inventories and flow;

3. Special design features should be foreseen to enhance effective safeguards;

4. Proper positioning, stratification, accessibility of nuclear material, minimizing of unknown inventories in the plant, and itemizing of nuclear material;

5. Automated real-time control to improve the availability and the transparency of nuclear material information system to provide running inventory;

6. Continuous accountancy activities carried out in combination with C/S measures;

7. Continuous random verification of some portion of the inventories;

8. The increasing accuracies on nuclear material measurements; and

9. Development and use of more sophisticated C/S devices which can provide direct information on the quantities of nuclear material under surveillance or in containment.

Using these measures, the authors are of the opinion that effective safeguards approaches can be found for future large scale facilities.

VII.2. Operational effectiveness

The questions are relevant to operational effectiveness: (1) Is the system functioning as designed? and (2) Are anomalies being resolved in a satisfactory manner?

Once designed and implemented, the system can be maintained at the design effectiveness level by performing preventative maintenance to reduce the probability of device failure, by promptly repairing or replacing failed devices and restoring confidence through appropriate back-up procedures, and by periodically reviewing the design assumptions to ensure that conditions have not changed in a manner which might effect the system's performance capability.

In operation, the containment and surveillance system operates in a "cleared" or "not cleared" status. Any anomalous indication noted by an inspector is brought to the attention of the designated contact at the facility for prompt resolution. When the inspector is satisfied with the explanation supported by corroborative observations or measurements if necessary, the system is again considered in a "cleared" status.

VII.3. Performance effectiveness

In an after-the-fact review, it becomes useful to view the system operating over a period of time past. Such a review should cover a variety of concerns including:

(1) Questions related to difficulties or interference with implementation need to be posed and answered.

(2) Has the device failure rate exceeded the anticipated rate?

(3) Is there any reason to be suspicious if not?

(4) Is the failure rate consistent with that observed in similar installations?

(5) Have observed anomalies been resolved in a satisfactory manner? In a timely manner?

(6) Is the rate of observed anomalies consistent with similar applications? and

(7) In effect, does past performance warrant continued use?

VIII. SUMMARY AND CONCLUSIONS

The role of containment and surveillance has long since passed the point of importance of being an interesting alternative. The Agency could not meet its objectives today without such mechanisms and the clear trend is for more extensive use of C/S in future safeguards. As an example, note the approach described in the paper related to safeguarding CANDU 600 reactors [4]. In current applications, such measures make it possible to derive meaningful conclusions based on mass balance accounting, and to dramatically reduce intrusiveness into plant operations by reducing the frequency of physical inventory measurements. In future large scale facilities, expanded usage of C/S is essential if the Agency is to maintain its ability to "detect the diversion of significant quantities of nuclear material".

Quantifying the effectiveness of Agency safeguards incorporating C/S measures for present modes of application requires resolution of any and all discrepancies detected and resorting to back-up measures when the devices fail in service. In such cases, nuclear materials accountancy is the mechanism of fundamental safeguards importance and the index of effectiveness is, to the extent practicable, based on verification of the material balance.

The outlook for quantifiable, effective safeguards for large future plants is not hopeless. Much additional work needs to be done to extend the range of mass balance accounting and to define an acceptable composite index of safeguards effectiveness for such applications. It is unquestionable that the role of containment and surveillance in such facilities will increase in such facilities, with attention devoted to predicting effectiveness in the design phase, and evaluating effectiveness during operations and for periodic summaries.

REFERENCES

[1] HOUGH, G., SHEA, T., TOLCHENKOV, D., "Technical criteria for the application of IAEA safeguards", IAEA-SM-231/112, these Proceedings, Vol.I.

[2] KONNOV, Y., SANATANI, S., "Development of containment and surveillance measures for IAEA safeguards", IAEA-SM-231/119, these Proceedings, Vol.I.

[3] SUKHORUCHKIN, V., et al., "Development of the safeguards approach for reprocessing plants", IAEA-SM-231/113, these Proceedings, Vol.II.

[4] TOLCHENKOV, D., et al., "A safeguards scheme for 600-MW CANDU generating stations", IAEA-SM-231/109, these Proceedings, Vol.I.

DISCUSSION

D.A. HEAD: In cases where material accountancy is difficult or impossible, can containment/surveillance measures be used as the sole method of safeguards? Is it possible that cost/benefit considerations will make C/S measures the prime safeguards technique to be used?

D. TOLCHENKOV: Under certain conditions (large throughput or inventory), material accountancy might not be fully adequate for safeguards purposes. In these circumstances, material accountancy should be applied to the extent possible in combination with continuous C/S measures in order to provide assurance that no undeclared removal of nuclear material has occurred. I think it is meaningless to apply C/S measures to unknown quantities of nuclear material. The quantities should be known even if with a large uncertainty.

The cost/benefit factor is an important feature of Agency safeguards and the application of C/S measures is helping in this respect, but is not replacing material accountancy.

D. GUPTA: Would you like to comment on (a) the responsibilities and obligations of the facility operators in ensuring that the Agency obtains adequate assurance from the use of C/S measures that no diversion of nuclear material has taken place, and (b) the consequences for the operation of the facility should such C/S measures lead to controversial interpretations? For example, could they involve establishment of a new inventory?

D. TOLCHENKOV: Any event detected by C/S measures which might indicate a possible diversion should be carefully investigated by the Agency before a statement is made on diversion. It is in the interest of the operator to provide the necessary proof that no diversion has taken place. In this sense, the effectiveness and functioning of a State system of accounting and control and of a facility accountancy system are very important for the Agency verification, along with co-operation between the operator and the inspector. Of course, the establishment of a new inventory is the ultimate step and would be necessary in the case of clear indication of diversion.

IAEA-SM-231/21

REMOTE-CONTROLLED AND LONG-DISTANCE UNIQUE IDENTIFICATION OF REACTOR FUEL ELEMENTS OR ASSEMBLIES

S.J. CRUTZEN, C.J. VINCHE,
W.H. BÜRGERS, M.R. COMBET
Joint Research Centre,
Materials Science Division,
Ispra Establishment,
Ispra, Italy

Abstract

REMOTE-CONTROLLED AND LONG-DISTANCE UNIQUE IDENTIFICATION OF REACTOR FUEL ELEMENTS OR ASSEMBLIES.

The unique identification method which has been developed at Ispra J.R.C. since 1970, and presented previously as a possible method for surveillance by the inspectorate of containments such as fuel elements, fuel assemblies or fuel storage, was greatly improved during 1977 and 1978. Such improvements aimed at automatization; simplification of the procedure (few or no handlings of fuel or equipment); possibility of identification by the plant operator (at the request of the inspector); identification at distance by the inspector (radiotelephone); and continuous interrogation. These goals are now achievable and various work is in progress in collaboration with many European bodies. The method, based on the ultrasonic signature principle, now uses a simplified procedure based on a portable device developed at Ispra and produced by Nukem in the Federal Republic of Germany. This device has automatic calibration; digital output; uses contact transducers; electronic scanning (scanning mechanism is no longer needed); and key identification number (making it possible for the inspector to give numerous different unique identities to a same seal), which can be easily integrated in a Remote Continual Verification System (RECOVER).

INTRODUCTION

The Remote Continual Verification System (RECOVER) is an automated data acquisition system for providing remote, secure and timely detection of the integrity and/or operational status of containment, surveillance, and measurement devices used for international safeguards [1].

The ultrasonic identification technique [2, 3], which has been under development since 1970 and applied experimentally, can be a part of the RECOVER system because the seals, or identified units and the ultrasonic transducer, can in fact be the monitor unit, and the portable digital identification device (Nukem-J.R.C. licence) [4] is part of the on-site multiplexer, or the interface between both.

If the remote verification unit of the RECOVER system is compatible with the JRC portable identification device, the ultrasonic signature method can easily be used for long-distance remote control and timely detection as part of an international safeguards procedure such as RECOVER.

This paper describes the latest developments of the ultrasonic signature technique with the aim of incorporating this method in a general system using different monitor units such as fibre optic seals, TV networks and ultrasonic seals.

1. ULTRASONIC SIGNATURE

The identification, whether of a seal or of a fundamental unit requiring safeguarding, depends on the presence of natural or artificial external or internal marks. The methods used to detect and identify these marks depend on optical, radiographic, electronic or acoustic techniques.

In the present case, the ultrasonic technique is used with strict reproduceability criteria [2]. The method is based on:

The natural or artificial marking of the piece to be identified; this marking is done by inclusions or defects randomly dispersed in a matrix; and

The identification is carried out by ultrasonics. The output signals are electric at analog signals.

1.1. Item to be uniquely identified

1.1.1. Inherent marks

Natural features such as welds can give good identification when the image of the weld is obtained by ultrasonics. Such a method was developed at Ispra while testing 1000 welds of cladding tubes [5].

1.1.2. Seals

Ispra developed seals containing internal marks (inclusions or voids) which were identified by using ultrasonics. The principle was applied to several types of seal (see Section 3). In principle, these seals can be identified under water or in contact with the transducers, in radiation fields and under temperature, using the same portable device.

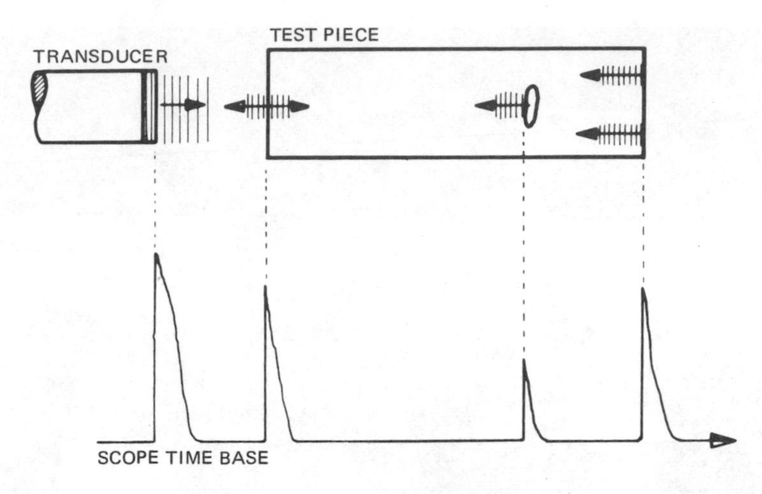

FIG.1. Ultrasonic identification principle.

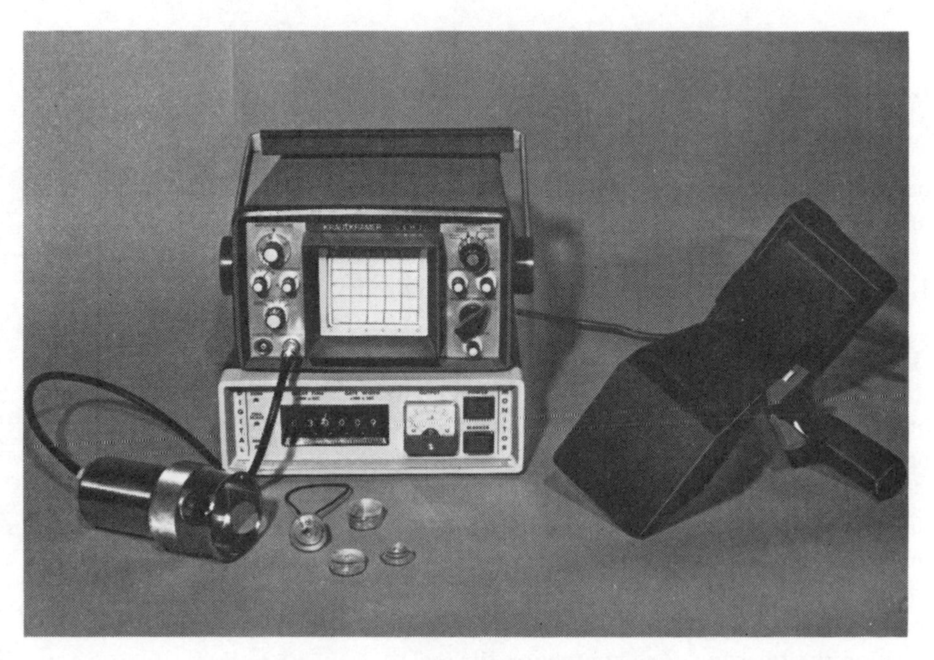

FIG.2. Steady identification equipment.

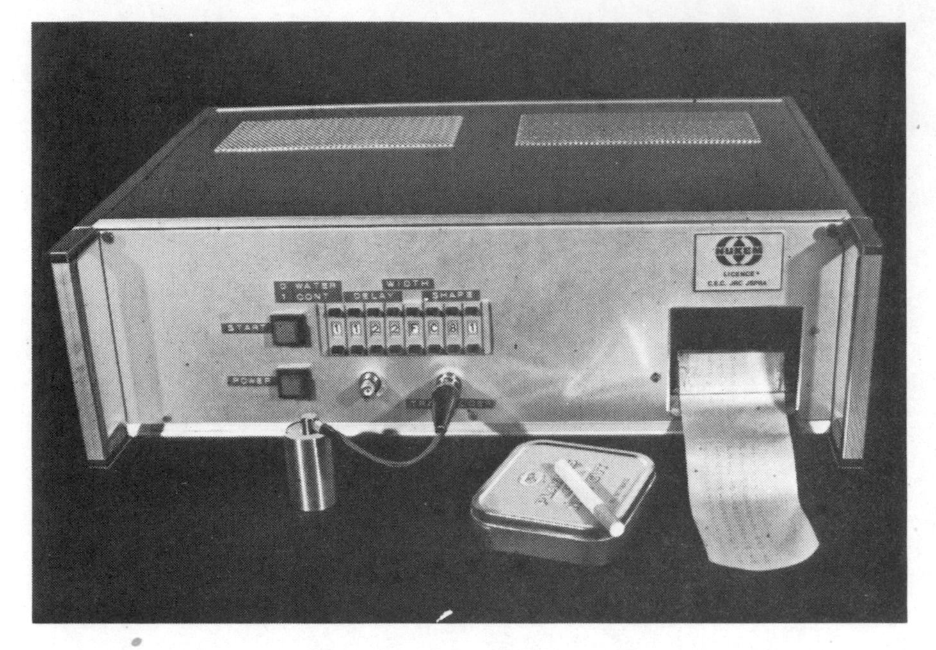

FIG.3. Portable identification device.

1.1.3. Fabrication of seals [2]

Two fabrication methods are used to obtain material with randomly dispersed inclusions or voids: power metallurgy techniques (for plastic, aluminium or steel); and brazing techniques (for steel).

1.2. System of identity measurements by ultrasonics

Using industrial ultrasonic equipment, the signals obtained by sonifying the seals are very satisfactory. The general method is based on reflection because of the great difference in acoustical impedance of the matrix and the inclusions or voids (Fig.1).

The transducers used are standard but carefully characterized and corrected in order to achieve sufficient reproduceability. The characteristics of the transducers, even of the same type and from the same fabrication batch, have to be evaluated and the specification of the essential parameters must be defined and verified for each application. The mechanical part of the installation has to position the transducers accurately on the seal.

The effect of variation in many important characteristics can be suppressed by calibrating the installation with a standard which is an artificial defect or an inclusion in the seal. The principle of using commercially available equipment and of scanning of the seal (rotation or translation of the transducer) is now discontinued, even if successfully applied in different cases; it introduces rotating mechanisms under water for remote-controlled seals on irradiated fuel bundles. Electronic scanning is more suitable for automatic identification and for portable devices.

2. PORTABLE IDENTIFICATION DEVICE

For measurement on the spot, although the chain shown on Fig.2 is easily transportable, a more portable device was developed by our laboratories and taken over, under licence by Nukem, Federal Republic of Germany, for production [4] (Fig.3).

2.1. Principle of the device

The ultrasonic receiver gives an electric signal according to the inclusion shape and position and on the acoustic impedance characteristics of the matrix and inclusions; digital switches select the signal in various zones of the seal; in each zone, the selected signal is digitally measured; and an automatic gain control makes the signal measurement reproducible.

2.2. Main characteristics of the device

The main characteristics of this device are: Digital input and digital output; automatic measurement – no operator intervention during measurement; and output independence fron transducer-seal distance.

In addition,

(i) It is easy to use. When the key number is put (on the spot or remotely) on the digit switches (number to be defined by the inspector, taking into account the type of seal; this number can be changed as desired) the identity is printed on request.

(ii) It can be used either by the inspector or by the plant operator at the request of the inspector – provided that the transducer is positioned on the seal, the input is a suite of characters which constitutes the identification key number; the output is the identity of the seal or item in the form of a suite of characters.

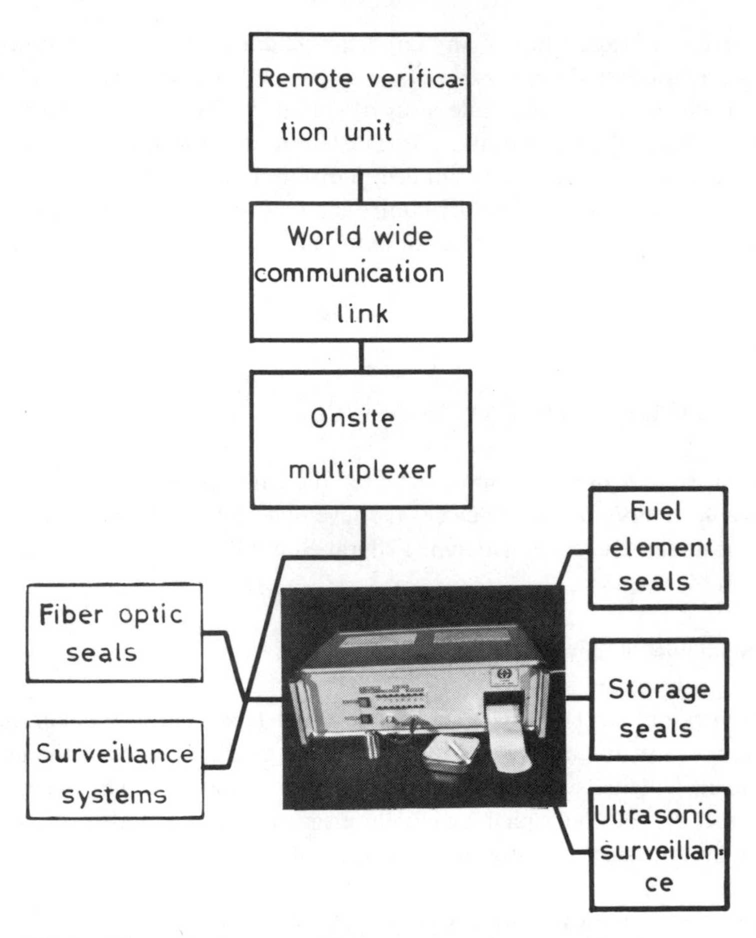

FIG.4. Ultrasonic signature with a remote continuous verification system.

(iii) It can be used for interrogation at a distance – if the seal bears its own transducer, or if a transducer is positioned by the plant operator on a seal, to introduce the output (identification key number) and to receive the output (identity) is very easy by telephone, telex or radio.

(iv) It is suitable for continuous interrogation of the identity and integrity of a container, either from a central point in the plant or at distance. This is useful for safeguards and physical protection purposes.

(v) Since the device is fully automatic and as input and output are digital, its insertion in a remote continual verification system is easy. The illustration of the RECOVER system as IAEA Safeguards Equipment could thus include ultrasonic seals (Fig.4).

3. SPECIFIC APPLICATIONS

3.1. Cap seals

This type of seal was designed for identifying light-water reactor fuel elements. The seal is a stainless-steel cap which, following the general principle, contains randomly distributed inclusions. The sealing is done by applying a marked "cap seal" on one or more tie-rods, in order to lock the screws.

The seal is designed to enable these long-life (five years) fuel elements to be dismantled if necessary (e.g. to change failed fuel pins); although it can be removed without destroying the marks, it is made useless in the process and a *new seal would have to be attached.* These seals are already used for experimental sealing of BWR fuel bundles in the Lingen KWL Gundremmingen KRB reactors. It is intended that the same principle be applied on larger bases.

For re-identifying the seals, two schemes can be used, *with or without* remote-controlled equipment:

(a) The seal, irradiated or not, can be taken off the fuel bundle and identified later. This scheme can introduce some difficulties and dangers such as:

  (i) Handling of fuel bundles;
  (ii) For irradiated seals, the necessity of shielding for transportation and for the identification apparatus; and
  (iii) Delay in giving inspection results.

(b) The seal, irradiated or not, can be identified on the fuel bundle (in the cooling pool) using the portable identification device (Fig.5) whether or not it is inserted in a RECOVER-type system.

3.2. Rivet seals

This type of seal was designed and fabricated for use on the box of MTR fuel elements [6]; the seal rivets the nozzle on to the edge plate of the fuel so that the element cannot be dismantled and plates cannot be pulled off. Only the disc of the seal is marked with inclusions; the cylindrical foot is destroyed by removal of the seal which is then unusable for a new sealing.

This type of seal has already been used for identifying the fuel elements of the HFR Petten, BR2-Mol, CAMEN Pisa, MERLIN Jülich, and will soon be applied to other European MTRs.

Re-identification can be made either after extraction of the seal, or on the spot using the portable identification equipment together with the positioning device of the transducer on the fuel-element seal (Fig.6).

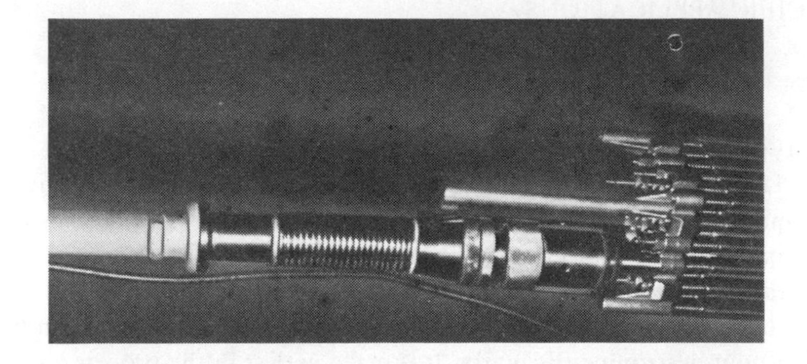

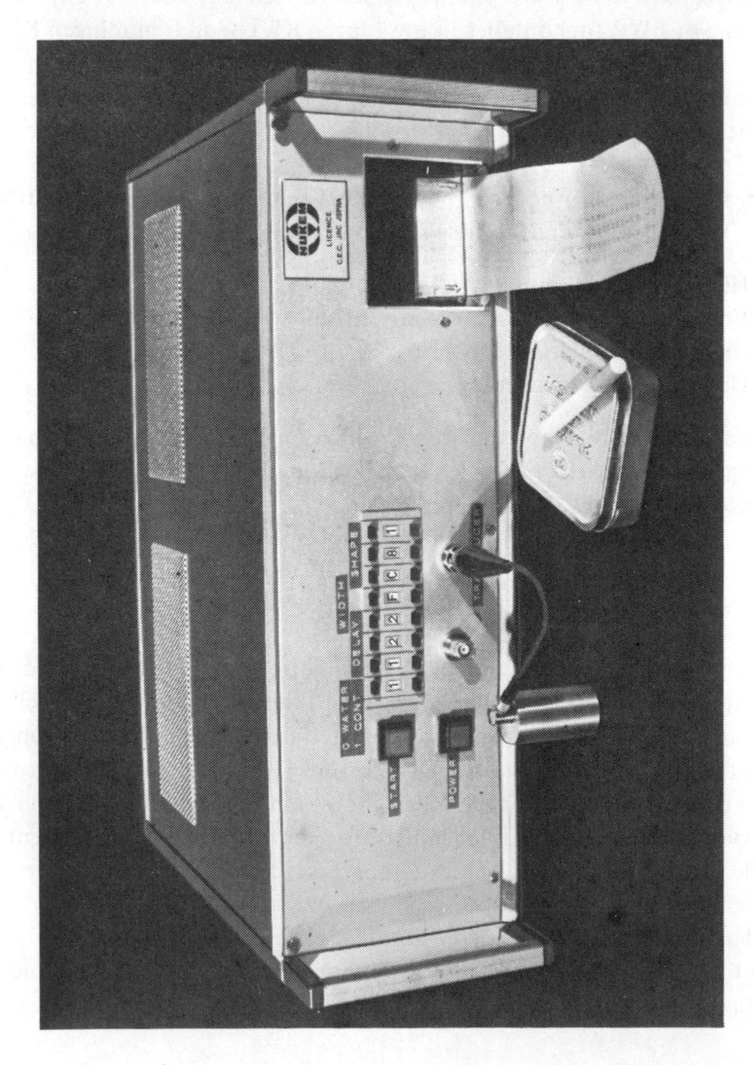

FIG.5. Cap seal identification on a BWR fuel bundle.

FIG.6. Rivet seal identification on a MTR fuel box.

3.3. FBR fuel-bundle seal

When the fuel bundle can be assumed to be non-dismountable, an identification piece can be welded to a suitable part of its structure to confer a unique identity to this "container". Development work has been done at Ispra in collaboration with INTERATOM, SBK and Belgonucléaire to define and propose a tamper-resistant unique identification device (TUID) [7] for the SNR 300 fuel bundles. A stainless-steel marked piece is welded in the foot of the seal bundle and can be identified by using the portable device. In principle identification under sodium is possible (Fig.7).

3.4. Storage pool for irradiated fuel bundles [8]

For the special case of CANDU fuel-bundle storage, a seal has been designed in collaboration with AECL (Canada) for underwater sealing of large quantities of fuel bundles. The portable identification equipment has to be used for re-identification. If a transducer is continuously connected to the seal, or constitutes part of the seal, the continuous interrogation or identification and structure integrity control, at distance, are simple.

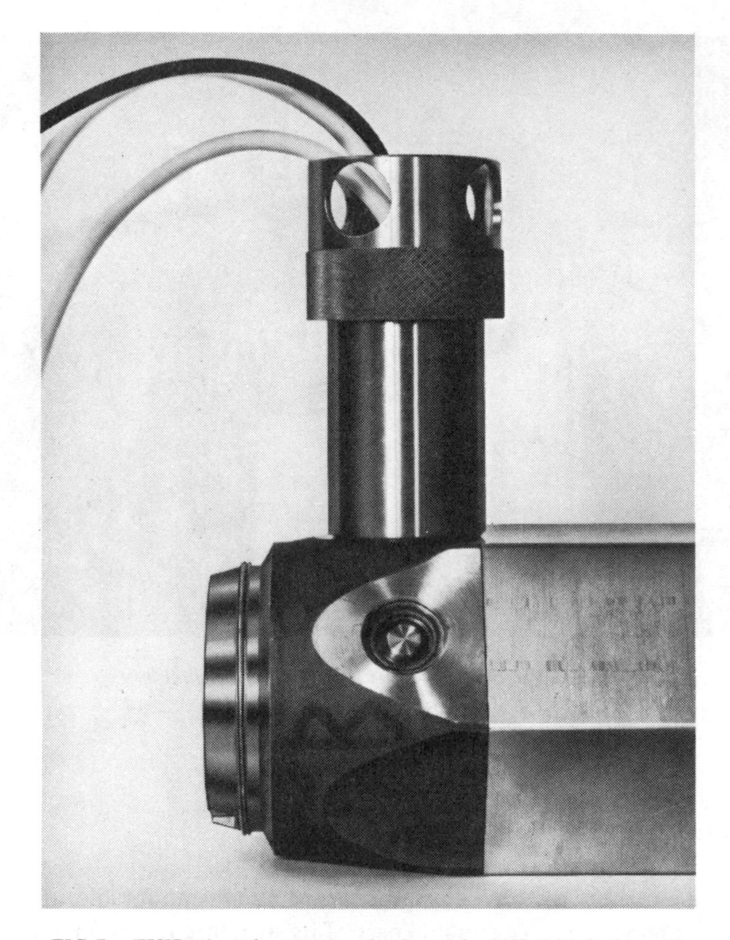

FIG. 7. TUID identification on the foot of a SNR 300 fuel bundle.

3.5. Container identification without seals

In the case of the pins of standard and of welded containers, the welds or plugs could be identified using the portable device. The welds or structures generally contain enough information to be taken as marking.

3.6. Ultrasonic surveillance

Fuel elements stored under water, or high strategic material containers (Pu storages), can be surveyed by ultrasonics without using any seals. Items stored under water constitute inclusions in the matrix "water". Ultrasonic transducers (of commercial quality) positioned under water continuously take the identity of the system "fuel elements-water-transducer", within a precise zone of the pool defined by the digital gating of the portable identification device.

This method is perfectly suitable for a remote continual verification system such as the RECOVER system.

4. CONCLUSIONS

The principle of the ultrasonic signature, together with the ultrasonic identification equipment (fabricated under Euratom licence by Nukem) can solve many surveillance problems connected with safeguards and physical protection:

Tamper-resistant unique identification of seals or items;
Tamper-resistant control of the integrity of a container or structure;
Identification at Safeguards Authority headquarters, or on the spot;
Identity check by the plant operator at the request of the safeguards inspector;
Remote-controlled (even long-distance) identification and integrity checking; and
Continuous interrogation.

Its insertion in the RECOVER system seems to be easy; TV camera, fibre optic electronic seals and ultrasonic seals could thus be part of a global IAEA international safeguards system.

REFERENCES

[1] RECOVER: Remote continual verification of surveillance/containment systems, USA Program for Technical Assistance to IAEA Safeguards, Atlantic Research Corp. Alexandria, Virginia, under contract to the US Arms Control and Disarmament Agency.
[2] CRUTZEN, S.J., BORLOO, E., Ultrasonic signature, EUR 5108e (1974).
[3] CRUTZEN, S.J., et al., "Application of tamper-resistant identification and sealing techniques for safeguards", Safeguarding Nuclear Materials (Proc. Symp. Vienna, 1975) 2, IAEA, Vienna (1976) 305.
[4] CRUTZEN, S.J., FAURE, J., WARNERY, M., System for Marking of Objects, in Particular Fuel Elements in Nuclear Reactors, Euratom Patent UK No.1241.287, 10.7.1969; and Euratom/Nukem, Licence Contract No.169/LDB, Luxembourg, 22.12.1972.
[5] BORLOO, E., An ultrasonic technique for the inspection of magnetic and explosive welds, using a facsimile recording system, Non-Destr. Test. **6** 1 (Feb.1973) 25–28.
[6] CRUTZEN, S.J., et al., A Rivet Seal for Safeguarding the MTR Fuel Elements, EUR 5110e (1974).
[7] BRUCKNER, Chr., et al., in Safeguarding Nuclear Materials (Proc. Symp. Vienna, 1975) 2, IAEA, Vienna (1976) 581.
[8] CRUTZEN, S.J., et al., Ultrasonic Uniquely Identified Seal for CANDU Spent Fuel Bundles Surveillance, AECL-Rep. 17/30085 (Canada) (1977).

IAEA-SM-231/22

STATISTICAL ASPECTS OF ULTRASONIC SIGNATURES

W.L. ZIJP
Netherlands Energy Research Foundation ECN,
Petten, The Netherlands

S.J. CRUTZEN
Joint Research Centre,
Ispra Establishment,
Ispra, Italy

Abstract

STATISTICAL ASPECTS OF ULTRASONIC SIGNATURES.

The unique identification method, developed at JRC Ispra, and based on an ultrasonic non-destructive technique, is in principle tamper-resistant. The method uses the responses of random properties of materials (e.g. random inclusions in a matrix material). Application of this technique requires high reproducibility of standard apparatus and transducers. The identification method is based on: The natural or artificial marking of the piece of materials to be identified. This marking is done by inclusions or defects randomly dispersed in a matrix; and the identity reading, which is carried out by an ultrasonic scanning method. From rather simple statistical considerations as presented here one may conclude that an identity pattern should include at least eight elementary informations (amplitude peaks). With this approach one can have at least a million seals with random inclusions, which will be distinguished as different seals. From a statistical aspect one may conclude that, with the applied seal fabrication technique and the ultrasonic signature detecting device, a unique identification of seals can be realized.

INTRODUCTION

The unique identification method, developed at JRC Ispra, and based on an ultrasonic non-destructive technique, is in principle tamper-resistant [1–3]. The method uses the responses of random properties of materials (e.g. random inclusions in a matrix material). Application of this technique requires high reproducibility of standard apparatus and transducers. In this paper attention is given to the main reliability parameters of the ultrasonic signature. The identification method is based on the natural or artificial marking of the piece of material to be identified. This marking is done by inclusions or defects randomly dispersed in a matrix; and the identity reading, which is carried out by an ultrasonic scanning method.

The main question of interest is: Which is a good number of inclusions to be included in a matrix material with random dispersion, so that the seal can properly function? Proper functioning implies unique identification of the seal; and tamper-proofness of the seal.

1. INSTRUMENTATION

The characteristics of the ultrasonic testing equipment have been described elsewhere [4]. The response pattern of a seal scanned by this type of equipment may be influenced by a series of circumstances, e.g. the presence of air bubbles attached to the seal by immersion in water for the measurement. The following influences have been studied:

The effects of temperature;
The stability of the electronic devices; and
The mechanical positioning of the seal, relative to the scanning device.

The present considerations are based on the availability of good ultrasonic scanning equipment, which functions in an easy, stable, reproducible and reliable way. The main interest here concerns the statistical aspects of the application of seals to be identified by the ultrasonic signatures.

2. RESPONSE PATTERN (Fig.1)

The ultrasonic response pattern (also called here the identity pattern) is a record of the analog signal as function of rotation angle, with a periodically repeating structure. The pattern shows some extremely sharp peaks, which correspond to the presence of inclusions in the focal area within the seal. The abscissa of a peak position is a measure of the position of the inclusion scanned by the equipment. The ordinate of the peak amplitude is a measure of the reflecting characteristics of the inclusion (determined by material and geometry). Our main interest here is the peak amplitude, not the peak position.

3. PARAMETERS OF INTEREST

I = number of inclusions in that region of the seal, which is seen by the ultrasonic scanning device;

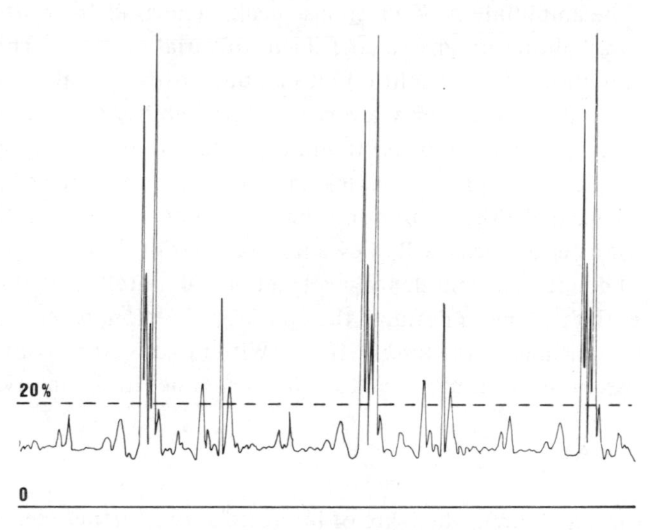

FIG.1. Example of an identity pattern.

A = amplitude of the response peak under consideration, sometimes expressed as a fraction or percentage of the maximum value. Low signals (e.g. signals smaller than a certain threshold, say 10 or 20%) may be discarded for the identification procedure;

$s(\bar{A})$ = standard deviation of the mean value for the amplitude, based on a long series of measurement. It is a measure for the small and random variations in the measurements, due to fluctuations in the experimental conditions. The parameter $s(\bar{A})$ may be considered as a measure for the reproducibility.

4. PRECISION OF MEASUREMENTS

A good estimate for $s(\bar{A})$ is obtained from a series of measurements on 200 freshly prepared seals before their use, and also on those seals which have been returned after two years after their application. Let $v(\bar{A})$ denote the coefficient of variation, i.e. the standard deviation relative to the average value

$$v(\bar{A}) = s(\bar{A})/\bar{A}$$

Extensive measurements in the Ispra Laboratory yield the following value

$$v(\bar{A}) = 0.0225$$

5. AMPLITUDE INTERVALS

Consider the amplitude A of a response peak. The available routine equipment in the Ispra Laboratory gives a coefficient of variation $v(\bar{A})$. This amplitude is now thought to be split up into a number of intervals with width ΔA, in such a way that there is only a very small probability that amplitudes are classified in an incorrect interval. It will be evident that, for increasing interval widths, the probability of making an incorrect amplitude assignment decreases. Now we make the assumption that the observed values in the amplitude measurement of a certain peak will obey a normal distribution.

Let $(1-\alpha)$ denote the confidence level and $u(\alpha)$ denote the confidence coefficient, i.e. the parameter defining the confidence region, when a normal distribution is applicable to the observations. With $(1-\alpha) \cdot 100\%$ confidence one can then assign to the amplitude A a value which lies in the interval

$$\bar{A} \pm u(\alpha) \cdot s(\bar{A})$$

Some values for α and $u(\alpha)$, which are of interest for our further considerations, are as follows. The parameter α describes the probability for occurrence of errors larger than the confidence coefficient times the standard deviation.

α	$u(\alpha)$
10^{-1}	1.64
10^{-2}	2.58
10^{-3}	3.29
10^{-4}	3.89
10^{-5}	4.42
10^{-6}	4.89

This means, for instance (and popularly speaking) that, with a confidence of 99.99%, one may state that the true value for A lies within the interval $\bar{A} \pm 3.89 \cdot s(\bar{A})$ or, more precisely, if the amplitude measurements were to be repeated over and over again, then 99.99% of the intervals defined by $\bar{A} \pm 3.89 \cdot s(\bar{A})$ will be expected to include the true value for the amplitude A.

The level width which we introduce is thus a function of the level of confidence. So ΔA is in fact a function of α. Since ΔA comprises the full width of the interval, one defines

$$\Delta A(\alpha) = 2u(\alpha) \cdot s(\bar{A})$$

The number of levels comprised in an amplitude is equal to

$$Q = A/\Delta A$$

This leads to

$$Q = \frac{A}{2 \cdot u(\alpha) \cdot s(\bar{A})} = \frac{1}{2 \cdot u(\alpha) \cdot v(\bar{A})}$$

or, remembering that $v(\bar{A}) = 2.25\%$,

$$Q = \frac{1}{(0.045) \cdot u(\alpha)}$$

6. CODIFICATION OF IDENTITY PATTERN (Fig.2)

The identity pattern can be codified without losing its character of unique identity. Each amplitude is split up into a certain number of intervals, where the interval width is chosen to be ΔA, as defined above. A practical limit for the number of intervals is $Q = A/\Delta A$. In practice one can take the largest integer value, starting from the maximum amplitude. The number K of different identity patterns, which can be obtained using a number of I inclusions in the seal, is given by:

$$K = [Q]^I$$

The assumption behind this formula is that all inclusions are randomly dispersed, and do not influence each other. More precisely – the amplitudes of the I inclusions are independent of each other.

Example

Let $Q = 4$ and $I = 5$.

For the first amplitude one has four choices; for the second amplitude one has again four choices; the same for the third, fourth and fifth amplitude. The number of possibilities is therefore $4 \times 4 \times 4 \times 4 \times 4 = 4^5$ or 1024.

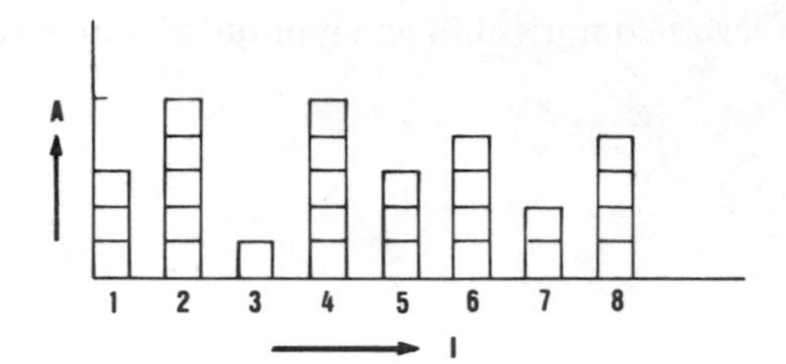

FIG.2. Codification scheme.

Note: In information theory, dealing with codification languages, Q corresponds to the possible alphabet, I to the format of the words, and K to the power of the language. By substitution in the expression above the previously found relation for Q, one obtains

$$K = \left[\frac{1}{2 \cdot u(\alpha) \cdot v(\bar{A})}\right]^{I}$$

We repeat that this formula is based on the assumption that a normal distribution applies for describing the variations in the amplitude measurements.

7. OCCURRENCE OF FAILURES

We will now consider the possibility for non-detection of false information in the signature, i.e. misinterpretation of a falsified seal.

Let β denote the probability for non-recognition of any discrepancy in the identity pattern after use of the seal.

The β parameter is the probability for the occurrence of a failure in the scanning and data treatment system (c.f. the probability for a failure in an alarm mechanism).

Or, popularly speaking, β is the probability for non-detection of a dicrepancy.

Let E_i denote the event that item i of elementary information (i.e. the i-th amplitude in the pattern) is false.

Let $P(E_i)$ denote the probability that item i of information which is false is not recognized as such – i.e. $P(E_i)$ is a probability that i-th amplitude fails to give a good signal.

Let $P(E_i + E_j)$ denote the probability that either the i-th amplitude or the j-th amplitude fail to give a good signal.

Let $P(E_i \cdot E_j)$ denote the probability that both the i-th and the j-th amplitudes fail to give a good signal.

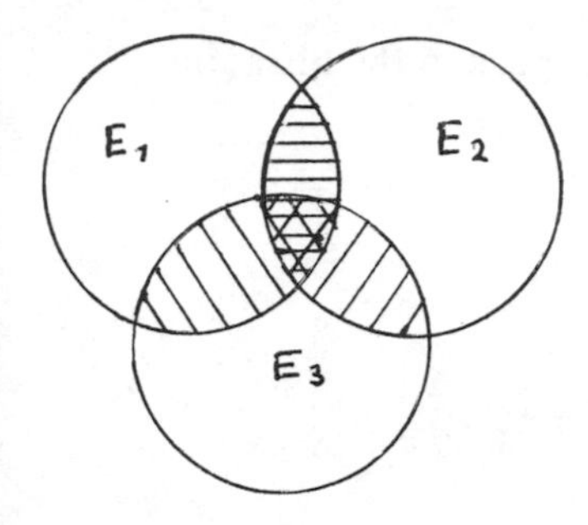

For illustration we assume for a moment that the identity pattern is composed of 3 items (3 amplitudes).
One then has

$$\beta = P(E_1 + E_2 + E_3)$$

$$\beta = P(E_1) + P(E_2) + P(E_3) - P(E_1 \cdot E_2) - P(E_2 \cdot E_3) - P(E_3 \cdot E_2) + P(E_1 \cdot E_2 \cdot E_3)$$

A similar expression can be written down when the number of items is larger than 3. Terms like $P(E_1 \cdot E_2)$ appear when some events are not mutually exclusive, i.e. when they may in principle occur in combination. We now can make the simplifying assumption that the events E_1, E_2, ..., and also the probabilities for their occurrence are the same. The $P(E_i)$ is the probability that the amplitude i is not correctly determined. This is related to the confidence level of the amplitude measurement. In fact, one may assume:

$$P(E_1) = P(E_2) = P(E_3) = \alpha.$$

Another simplification is now made – we assume that events E_i are mutually exclusive, i.e. that the number of incorrect amplitude measurements is not larger than 1. Thus $P(E_i, E_j) = 0$.

This leads to

$$\beta = I\alpha$$

where I is the number of amplitudes under consideration.

If the last assumption is not made one has

$$\beta \leqslant I\alpha$$

For the following considerations it is sufficient to realize that in good approximation one has

$$\beta \approx I \cdot \alpha$$

Whatever the situation, the maximum value of β can be estimated by $I \cdot \alpha$

$$\beta_{max} = I \cdot \alpha$$

Since one may assume that one needs at least one clear peak in the ultrasonic identity pattern (i.e. $I \geqslant 1$) one has

$$\beta \geqslant \alpha$$

Thus, finally

$$\alpha \leqslant \beta \leqslant I \cdot \alpha$$

8. CALCULATION OF K

K is the number of possible identity patterns, which should be distinguished as clearly different patterns.

Using the formulae

$$K = \left| \frac{1}{0.045\ u(\alpha)} \right|^{I}$$

and

$$\beta \approx I \cdot \alpha$$

one can calculate corresponding values for K, I, α and β.

For a good reliability of the seals, one is interested in very small values for α and β, and in very large values for K.

Let us assume that the safeguards inspectorate wishes to have available at least 1000 seals a year. Therefore, we will consider the cases $K = 10^3$, $K = 10^4$, $K = 10^5$ and $K = 10^6$. What is then the (minimum) number of elementary informations?

$$I = \frac{\log K}{-\log\{2\ u(\alpha).v(A)\}} = \frac{\log K}{-\log\{0.045\ u(\alpha)\}}$$

K=1000 $\quad \alpha=10^{-2};\ u_\alpha=2.58 \rightarrow I=3/0.94 \rightarrow I=4;\ \beta<4\times10^{-2}$

$\alpha=10^{-3};\ u_\alpha=3.29 \rightarrow I=3/0.83 \rightarrow I=4;\ \beta<4\times10^{-3}$

	$\alpha=10^{-4}$; $u_\alpha=3.89$ →	$I=3/0.76$	→	$I=4$; $\beta<4\times10^{-4}$
	$\alpha=10^{-5}$; $u_\alpha=4.42$ →	$I=3/0.70$	→	$I=5$; $\beta<4\times10^{-5}$
	$\alpha=10^{-6}$; $u_\alpha=4.89$ →	$I=3/0.66$	→	$I=5$; $\beta<4\times10^{-6}$
K=10000	$\alpha=10^{-2}$ →	$I=4/0.94$	→	$I=5$; $\beta<4\times10^{-2}$
	$\alpha=10^{-3}$ →	$I=4/0.83$	→	$I=5$; $\beta<4\times10^{-3}$
	$\alpha=10^{-4}$ →	$I=4/0.76$	→	$I=6$; $\beta<4\times10^{-4}$
	$\alpha=10^{-5}$ →	$I=4/0.70$	→	$I=6$; $\beta<4\times10^{-5}$
	$\alpha=10^{-6}$ →	$I=4/0.66$	→	$I=7$; $\beta<4\times10^{-6}$
K=100000	$\alpha=10^{-2}$ →	$I=5/0.94$	→	$I=6$; $\beta<4\times10^{-2}$
	$\alpha=10^{-3}$ →	$I=5/0.83$	→	$I=7$; $\beta<4\times10^{-3}$
	$\alpha=10^{-4}$ →	$I=5/0.76$	→	$I=7$; $\beta<4\times10^{-4}$
	$\alpha=10^{-5}$ →	$I=5/0.70$	→	$I=8$; $\beta<4\times10^{-5}$
	$\alpha=10^{-6}$ →	$I=5/0.66$	→	$I=8$; $\beta<4\times10^{-6}$
$K=10^6$	$\alpha=10^{-4}$ →	$I=6/0.76$	→	$I=8$; $\beta<4\times10^{-4}$

9. ROLE OF PEAK LOCATIONS

So far we have used only part of the information contained in the identity pattern. Only the amplitudes of the response peaks were taken into account. The response peak obtained for a single inclusion is characterized not only by its amplitude, but also by the peak location (determined by the co-ordinates of the inclusion).

This location is in fact a time interval on the oscilloscope screen or an abscissa on the analog record. This abscissa corresponds with the depth of the inclusion inside the matrix. This abscissa value is also a real random information, and can be used in a way similar to that described above for the peak amplitude.

Combining information on abscissae and amplitudes one can imagine that one will arrive at an equation of the type

$$K = \left\{\frac{1}{2\,u(\alpha)\cdot v(A)}\right\}^{I} \cdot \left\{\frac{1}{2\,u(\alpha)\cdot v(x)}\right\}^{I} = \left\{\frac{1}{2\cdot u(\alpha)}\right\}^{I} \left\{\frac{1}{v(A)\cdot v(x)}\right\}^{I}$$

The reproducibility of the measurements of the abscissae values is rather large for all types of seal, so that the reliability for the unique identification is increased.

However, according to the present opinion of the investigators at Ispra, a falsification, comprising the preparation of seals giving responses at the same abscissae as the ones of a real identity pattern, can be imagined under special circumstances when the confidence interval around these abscissae is about $2\frac{1}{2}\%$.

The determination of the abscissae can therefore only be used as a verification of the integrity of the seal. The real identity check must be mainly based on the determination of the amplitudes.

It has been shown that laboratory trials to copy a seal had no success at all. From a mechanical aspect one may also conclude that it is (practically) impossible to prepare a second seal with the same response pattern.

10. CONCLUSION

From rather simple considerations as presented here one may conclude that an identity pattern should include at least eight elementary informations (amplitude peaks). With this approach one can have at least a million seals with random inclusions, which will be distinguished as different seals.

The probability of an incorrect identification can be made extremely small by taking a conservative value for the width of the amplitude intervals introduced.

From a statistical aspect one may conclude that, with the applied seal fabrication technique and with the ultrasonic signature detecting device, a unique identification of seals can be realized.

REFERENCES

[1] CRUTZEN, S.J., FAURE, J., WARNERY, M., System for Marking of Objects, in Particular Fuel Elements in Nuclear Reactors, Euratom Patent UK No.1241.287, 10.7.1969.

[2] CRUTZEN, S.J., DENNYS, R., IAEA-SM-231/124, these Proceedings, Vol.I.

[3] CRUTZEN, S.J., JEHENSON, P.S., HAAS, R., LAMOUROUX, A., "Application of tamper-resistant identification sealing techniques for safeguards", Safeguarding Nuclear Materials (Proc. Symp. Vienna, 1975) 2, IAEA, Vienna (1976) 305.

[4] BORLOO, E., CRUTZEN, S.J., Ultrasonic signature, Rep. EUR-5108-e, CEC, Luxembourg (1974).

USE OF ULTRASONICALLY IDENTIFIED SECURITY SEALS IN THE 600-MW CANDU SAFEGUARDS SYSTEM

S.J. CRUTZEN
Joint Research Centre, Euratom,
Ispra, Italy

R.G. DENNYS
Atomic Energy of Canada Ltd,
Power Projects, Mississauga,
Ontario, Canada

Abstract

USE OF ULTRASONICALLY IDENTIFIED SECURITY SEALS IN THE 600-MW CANDU SAFEGUARDS SYSTEM.

The CANDU 600-MW reactors require, as part of the proposed safeguards system, tamper-resistant containers. Verified quantities of irradiated fuel would be sealed into these containers located in the spent-fuel storage bays. This requires the use of a security seal which could be installed, verified and re-verified in situ, under water. Ultrasonically identified seals, because of the probability of their meeting all the requirements, were selected for application. A cable, or type of customs seal, which was initially developed by JRC, Euratom, Ispra, Italy, proved the concept. Difficulties in installation led to the development of a "cap" seal. This has been tested and will be manufactured for use with the 600-MW reactors.

1.0 INTRODUCTION

The safeguards system which is being applied to 600 MW CANDU reactors, uses a number of specialized items of equipment developed for this specific use. One of these items is a special security seal used in the irradiated fuel storage bay. These seals are used to secure a specific quantity of irradiated fuel bundles within a tamper-resistant container. After an inspector has counted and verified the quantity of fuel, a seal is applied and henceforth only the container and seal need be inspected to verify the fuel contained within. This paper describes the evaluation and development of the ultrasonically verified security seal used with the CANDU reactor safeguards system.

1.1 General Requirements

The concept of verifying and then sealing fuel into underwater tamper-resistant containers established the necessity for a container seal. At the initial stage, the only practical type of seal appeared to be an underwater customs type. This consists of a cable and sealing device for the cable ends. The possibility of installing or "making" and verifying the seal above water was considered, but did not appear practical. The general requirements for a seal were, therefore, established as follows:

- It must be possible to install the seal from above the water surface but the seal itself would be underwater.
- The seal must withstand the consequence of water immersion as well as radiation to which it would be subjected, since it would be very close to irradiated fuel.
- It must be possible to verify, from above the water surface, whether or not tampering with the seal had taken place.
- The seal would be tested at the Douglas Point Nuclear Generating Station (operated for AECL by Ontario Hydro). It must, therefore, be compatible with the requirements of that fuel storage system as well as the 600 MW system.

1.2 Types of Seals Considered

The only category of seal which seemed to be available (when the field was surveyed during 1975) was the general type of wire or customs seal. This is a type in which a wire or cable of some kind is joined in a tamper resistant closure. Various versions of this type of seal were in use or being developed.

1.2.1 General Use Seal (IAEA)

The IAEA common seal uses a wire, of varying length, which is inserted through one part of the seal closure. The ends are joined by mechanical means such as twisting or crimping. The other part of the seal is then permanently installed completely enclosing the wire joint. The seal must be opened to determine if tampering or replacement has taken place. Tamper resistance was not considered to be particularly good and the requirement to destroy the seal each time it was checked was considered unacceptable. Equipment could be designed for underwater installation.

1.2.2 Fibre Optic Seal

The physical application of the fibre optic seal was similar to that of the general purpose cable and closure seal. The cable, consisting of optical fibres, had its ends permanently joined by a special closure. The special lighting and camera system photographs the closure pattern produced when light is passed through the cable. The random pattern of the optical fibres is impossible to reproduce and is changed by tampering. There was, however, no current method of identifying the seal underwater. The life of this type of seal in a high radiation environment is also unproven.

1.2.3 Ultrasonically Identified Seal

The ultrasonically identified seal considered was similar to the general use seal. Two parts are used to permanently close the mechanically joined ends of the wire. The closure could, however, be identified or "fingerprinted" ultrasonically. This was done by manufacturing the parts from material containing random inclusions. The seal parts once closed, could not be opened without changing the ultrasonic identity. This identity was obtained by scanning specific parts of the seal. The seal is removed by cutting the wire, identified and inspected for tampering. This seal was developed by the Euratom Joint Research Centre (JRC), Ispra, Italy.

1.3 Seal Choice for CANDU System

The seal chosen for development for the irradiated fuel storage bay was the ultrasonically identified seal. This general type of seal met, or was expected to meet, all of the requirements established. It was necessary to design a version of the seal which could be identified and verified underwater without removal.

2.0 SEAL DEVELOPMENT - CABLE SEALS

At the initiation of the irradiated fuel seal development program, the Douglas Point Nuclear Generating Station was to be utilized in the development of CANDU safeguards equipment. The seal, sealing equipment and security covers were, therefore, designed to suit the irradiated fuel storage trays and tray handling equipment in use at Douglas Point.

2.1 Seal and Cover Design

The Douglas Point fuel trays are handled in the bay with a tool which slides from the side over a lip at each end of a tray. Pickup of a tray requires a clear area equal to the tray area beside the tray, for attaching and removing the tool. The trays are stacked 30 high and in reasonably close proximity, progressing from one side of the bay toward the other. This will provide the required tool clearance. This situation made it impractical to lower trays into a security cover. It was decided, therefore, to design a cover which could be lowered over, and sealed to, a completed stack of trays.

Many materials and designs of security covers were assessed, including ceramic, glass and metal. The type finally chosen was considered to have a good combination of tamper resistance, low cost and ease of handling. The cover was a structure made from stainless steel light rolled sections. It was covered with a flattened expanded stainless steel mesh, with openings about 30 mm x 15 mm and a thickness of 1.6 mm. The Douglas Point fuel trays have a 19 mm hole on the upper surface, near one end. This would be utilized to seal the cover, by means of the seal cable, to the upper tray. Access to the tray through the cover was too limited to allow lifting of the tray, even with a special tool projecting through the mesh. It was felt that this

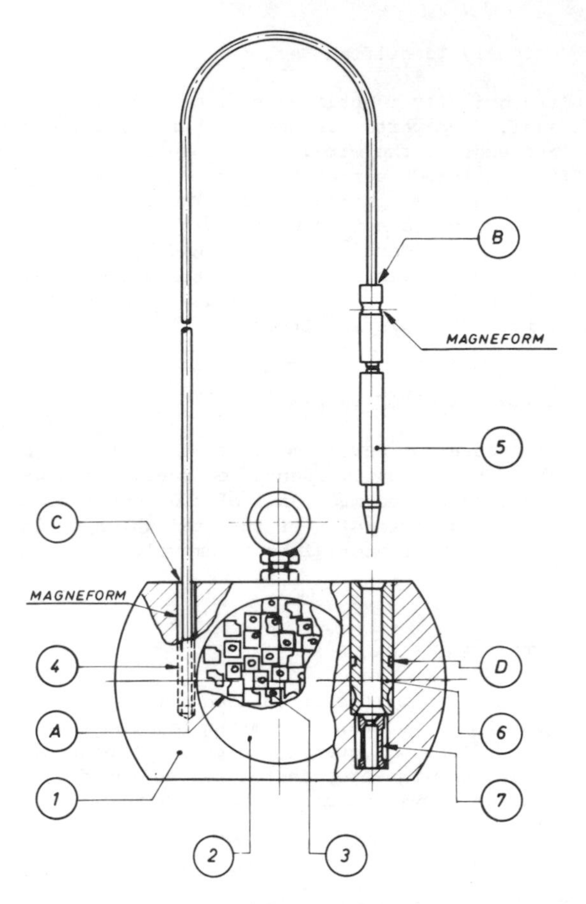

FIG.1. Scheme of the seal.

combination of seals and covers would be reasonably tamper resistant. Also it would determine the suitability of the underwater sealing concept.

The concept was based on and in fact defined the seal, as a type of wire customs seal. The application placed no further limitation on seal size, except that the wire or cable should be of sufficient length to thread through the cage and fuel tray.

2.2 Euratom Program

The Ispra Joint Research Centre (JRC) of the European Community had developed a sealing technique using ultrasonics for unique identification of the seal. Identification can be performed underwater and under irradiation via remote control. In order to adapt the ultrasonic identification technique to the CANDU fuel sealing requirements, a collaboration agreement was signed between Atomic Energy of Canada Ltd (AECL) and JRC, Ispra. Seals would be designed and developed at Ispra to meet the specific requirements of the Douglas Point sealing system as defined by AECL. A handling tool was to be designed and fabricated at Ispra for the installation of seals and for remote identification. Commercially available electronics were to be chosen by JRC to fit the identification method.

2.2.1 Seal Development at Euratom

The scope of the agreement, seen from the JRC side, was the demonstration of the validity of the ultrasonic technique for unique identification of underwater seals. This first approach using wire seals provided AECL with a seal which it was possible to place on trays without substantial modification to the storage system. But a better design was possible (δ 3.0) to combine easy underwater sealing with the reliability of an ultrasonic signature.

Figures 1 and 2 show the principle of the seal made from stainless steel pieces brazed together with a nicrobraze ($1000^{\circ}C$ - 10^{-3} torr). Identification marks are scanned by the ultrasonic transducer (see Figure 3) so the system is tamper resistant.

Scanning is performed by translation of an ultrasonic transducer. This transducer is commercially available, but fully characterized and corrected to enable a good degree of reproducibility of ultrasonic response. This is necessary to obtain tamper-resistant identities.

Marks are obtained by a random fabrication process. One scan line is sufficient to provide information leading to about 10^{8} different identities (4). This total takes into account uncertainties arising from aging or replacement of transducers subjected to humidity and irradiation.

2.2.2 Installation/Identification Tool

Installation/identification equipment with such a non-optimized design of seal appears to be rather complicated (see Figure 4). The operations required to close the seal are:

- threading the cable through the cover and fuel tray,
- introducing the cable end (needle) into the closure system of the installation tool,
- lifting the wire to lock the needle in the sliding part of the installation tool,
- lowering the sliding part to engage the needle into the seal and press to obtain non-return closure of the seal,
- connecting cable to ultrasonic equipment,
- recording the identity (see Figure 5) and
- taking the seal out of the holder of the installation tool.

The feasibility of such a complicated scheme was demonstrated under water, with remote controlled identification of the seal.

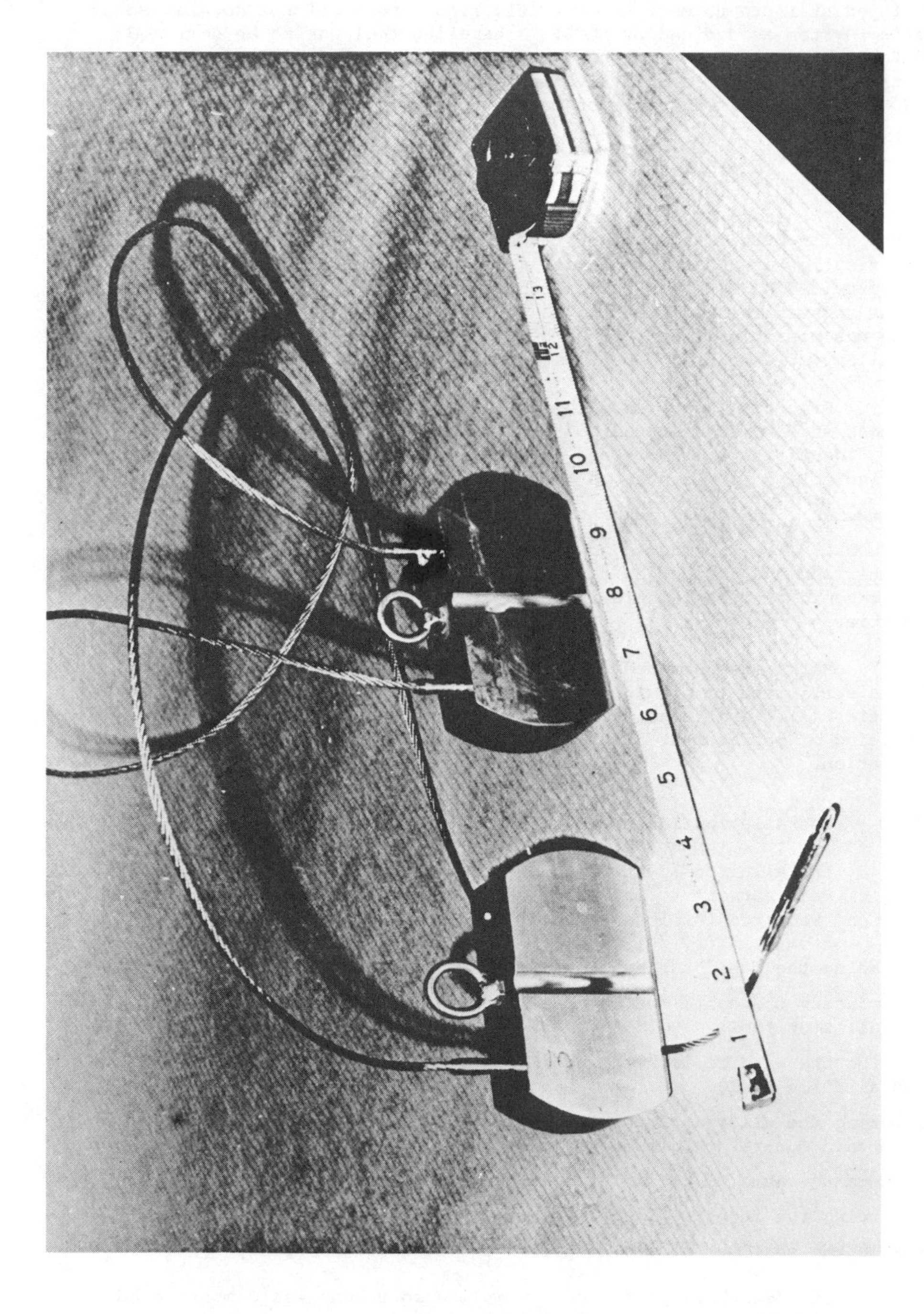

FIG.2. CANDU spent-fuel cover seals.

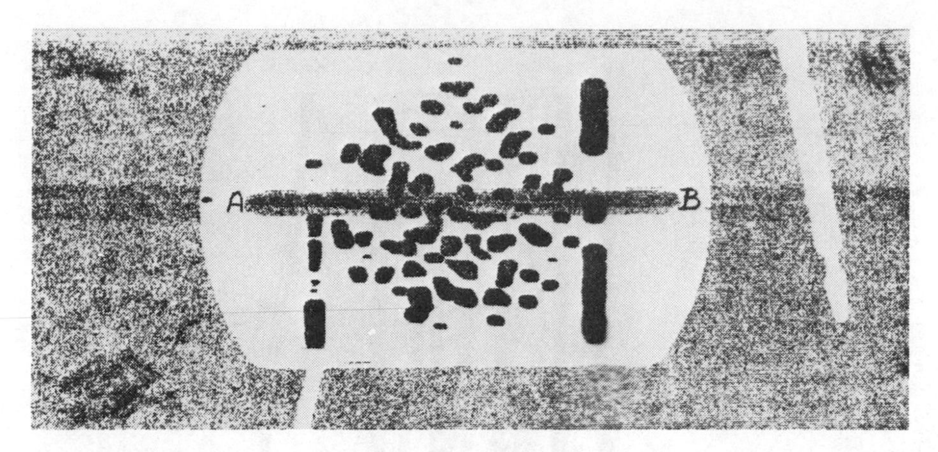

FIG.3. Scanned zone of the seal: A, B.

2.3 Douglas Point Testing

2.3.1 Hand Tools

To utilize the installation/verification tool developed at Ispra, hand tools are required to lift the seal body into and out of the verification tool, as well as to thread the cable through the cover and fuel tray.

These hand tools were developed at AECL's Sheridan Park Laboratory, near Toronto, prior to delivery of the verification tool from Italy. Mock-ups of an upper end security cover and the lower part of the verification tool were fabricated and used to develop three hand tools (see Figure 6). The hand tools consist of a hook for lifting the seal in and out of the verification tool and two pincher type tools, one with vertical and one with horizontal jaws. These tools and the verification tool would be operated from the storage bay gantry, positioned over the security cover to be sealed. The verification tool would be supported by an existing electric hoist on the gantry. Hand tools could be hooked onto the gantry railing when not being manipulated.

Prior to delivery to Douglas Point for on-site testing, the equipment designed and manufactured at Ispra was shipped to AECL's Sheridan Park Laboratory. The same hand tool development set-up was used to check the operation of the complete seal system equipment. A wooden tank allowed all equipment, including the ultrasonics, to be checked out underwater (see Figure 7). This exercise showed that, while fairly straightforward in theory, the actual installation of the seal, particularly the threading of the cable, was a rather difficult task underwater.

Ultrasonic identification of seals was made using the test set-up, the electronic instrument and recorder purchased Identifications were also made with the portable ultrasonic equipment built by Ispra for identifying general purpose seals. Both recorders gave good identities which could be readily cross-referenced (see Figure 8).

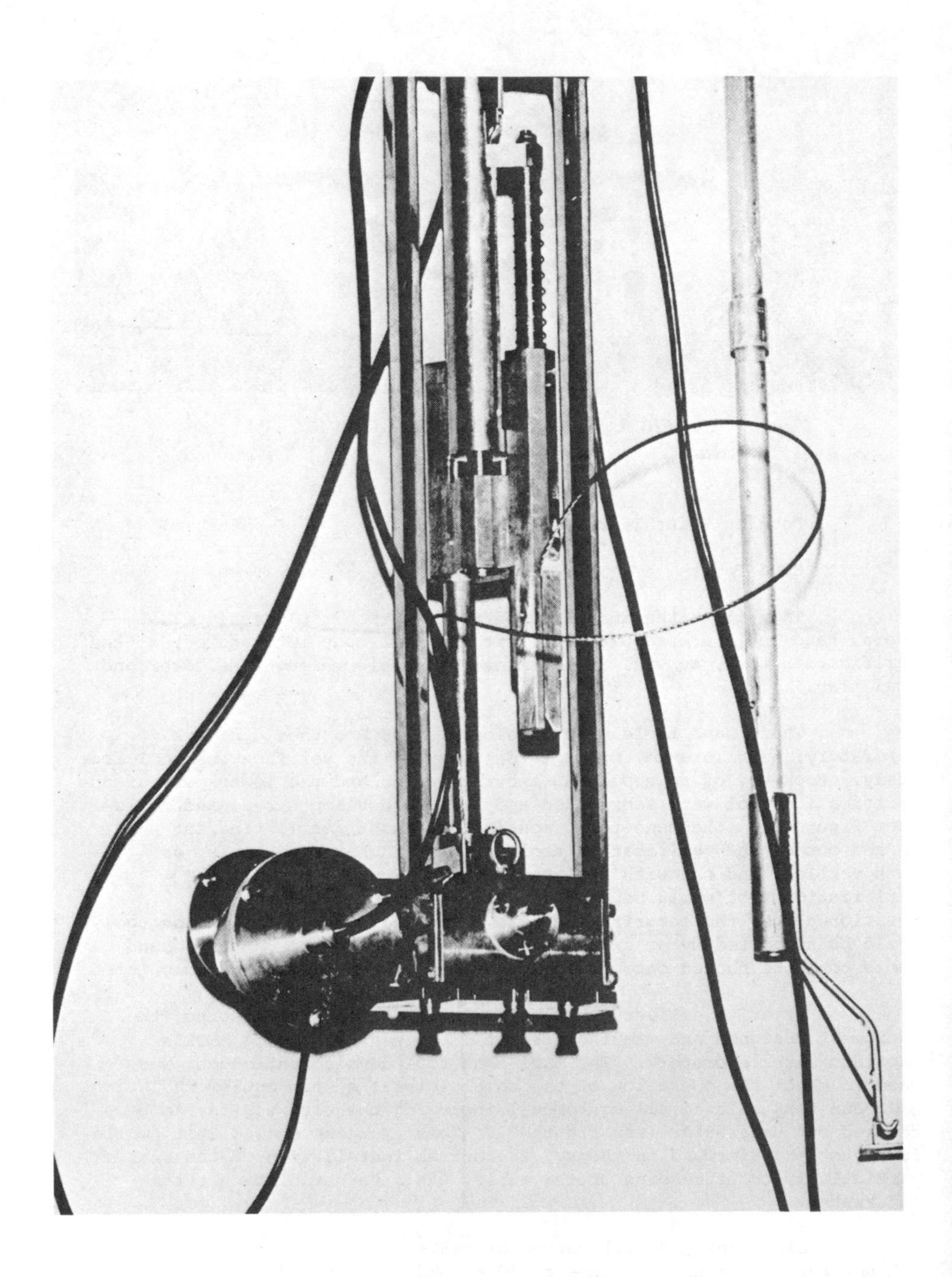

FIG.4. Seal installation and verification apparatus.

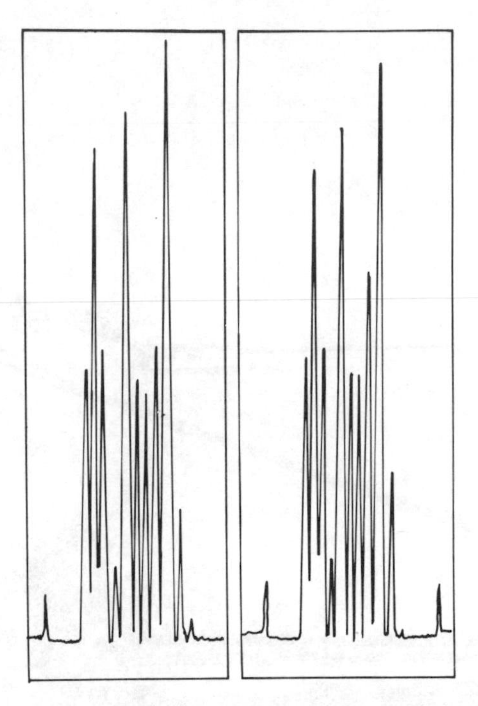

FIG.5. Identities of the same seal taken with two different transducers.

2.3.2 Site Testing

Upon completion of check-outs and testing, the equipment was taken to the Douglas Point Nuclear Power Station for tests in the irradiated fuel storage bay. Two stacks of irradiated fuel trays (30 trays high) has been covered with security covers in readiness for seals to be applied. The seal equipment was installed in the bay and set up for seal installation and verification.

2.3.2.1 Seal Application

A dummy seal (without locking collets) was used for the first test. The installation of the seal and cable had the same degree of difficulty previously encountered in threading the cable. The initial installation of the dummy seal took about an hour. Subsequently an identification of the seal was made ultrasonically in a matter of minutes. The dummy seal was removed and a proper seal was installed.

Installation and verification of this seal took approximately one half hour (see Figure 9).

It was apparent during the tests that good lighting was required. A portable underwater light proved very effective. A floating doughnut-shaped viewing glass, through which the hand tools could be operated, was of considerable assistance during seal installation.

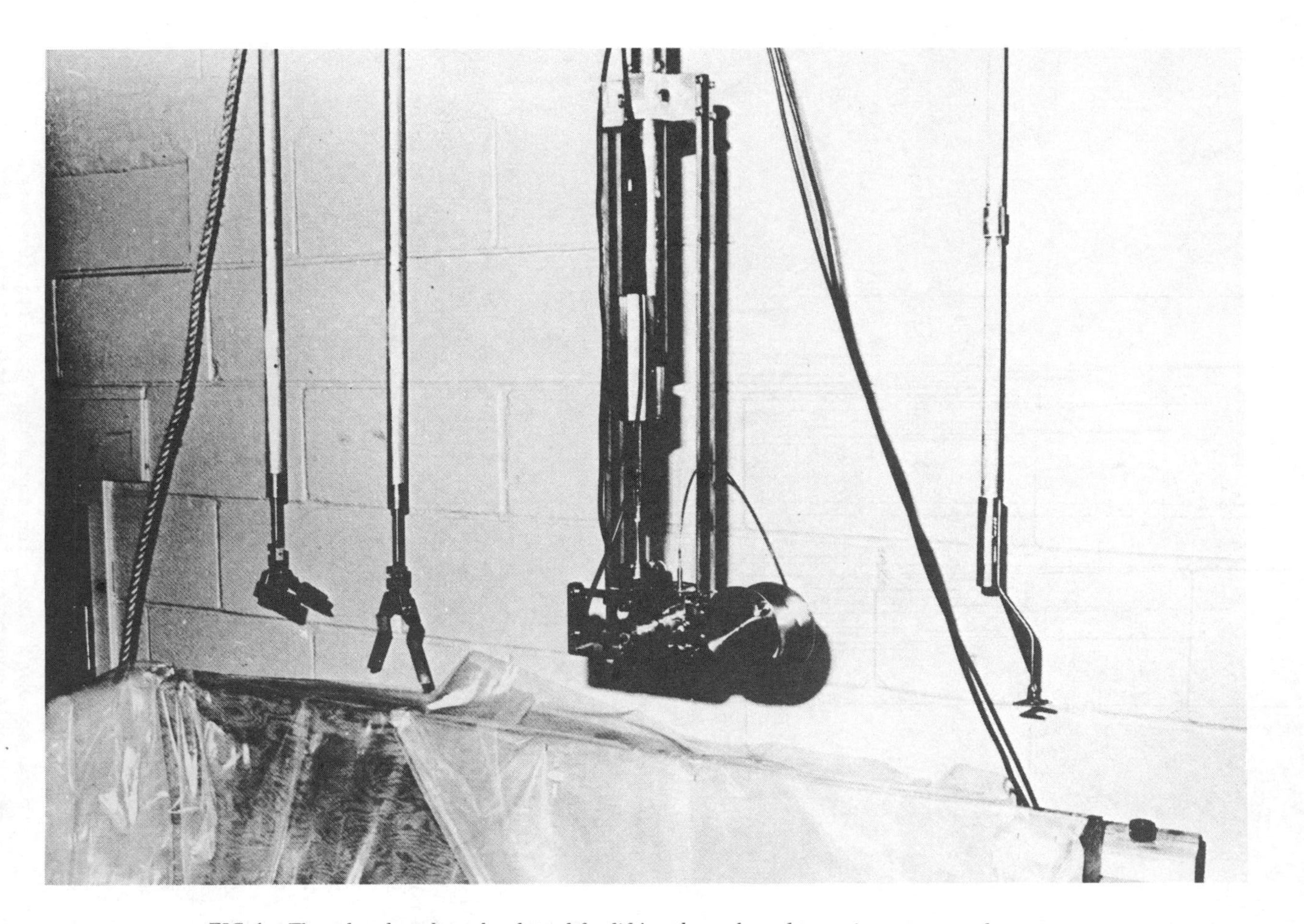

FIG.6. Three hand tools: a hook tool for lifting the seal; and two pincer-type tools, one with vertical and one with horizontal jaws.

FIG. 7. Installation and identification apparatus and handling tool check at Sheridan Park Laboratory.

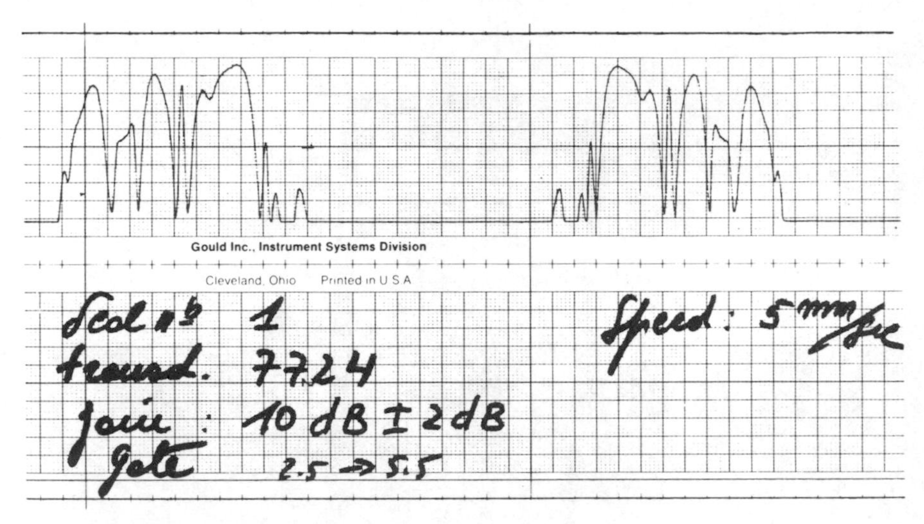

FIG.8. Identification of seal No.1 installed at Douglas Point Nuclear Generating Station, 15 July 1976.

FIG.9. Seal installed on a security cover in the spent-fuel bay.

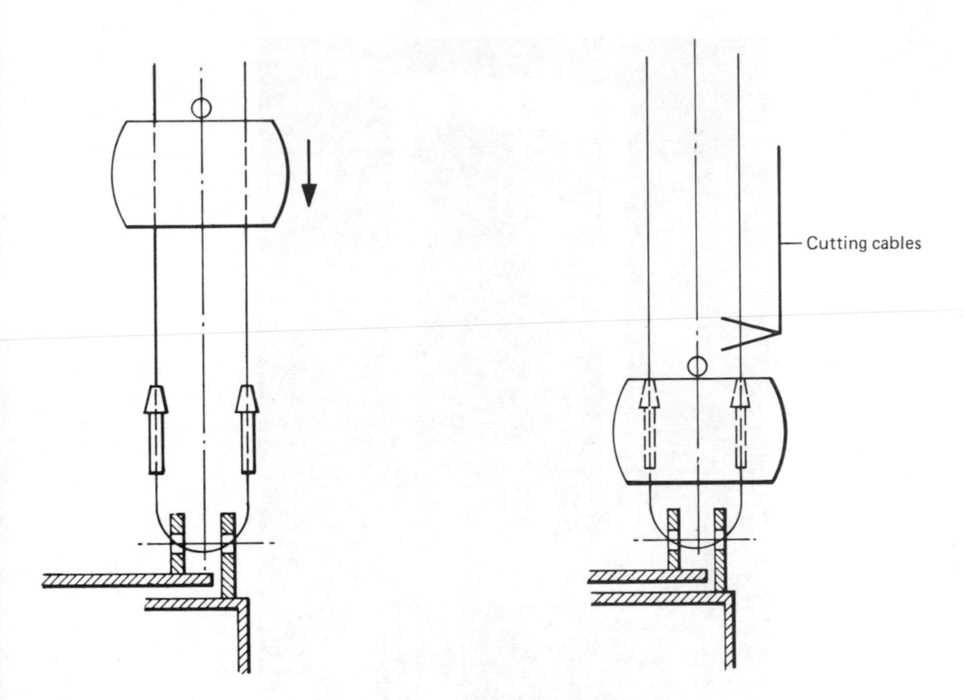

FIG.10. Sealing with a long cable seal.

2.3.2.2 Results and Observations

The seal identification equipment functioned well. There were no problems with operation of the mechanical equipment and adequate identities were obtained in each case. The major problem with this seal was the practical difficulty concerning the installation of the cable, both through the cover and fuel tray and into the installation mechanism. This task proved very tedious and time consuming. Considerable practice and a good deal of luck were required to install the cable in any reasonable time. Even with practice, the time required for installation of a seal varied from ten minutes to two hours. Visibility and depth perception caused the major difficulties associated with the cable installation.

The degree of tamper resistance provided by the cable has been questioned, but it is considered as tamper-resistant as the security cover. It would be very difficult to remove and repair without visible evidence. It does, however, rely on visual checking to determine if the cable was tampered with. Ultrasonics identify only the seal.

One of the major difficulties encountered in installing the seal cable was threading it through the cover openings and through the hole in the fuel tray. This was a problem inherent in the use of the existing fuel trays. A simpler method of applying the seals would be used if the fuel trays could be lowered into a cage with a cover sealed on top. In this case, horizontal threading of the cable could be used in conjunction with studs projecting through the cover.

FIG.11. BWR fuel-element upper part sealed with a cap seal (KRB and KWL reactors, Federal Republic of Germany).

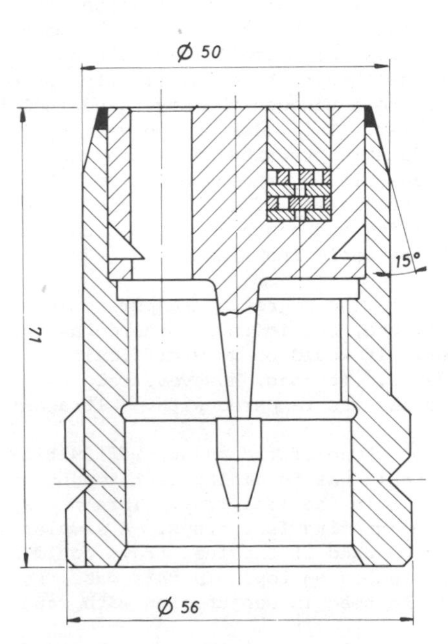

FIG.12. Cap design.

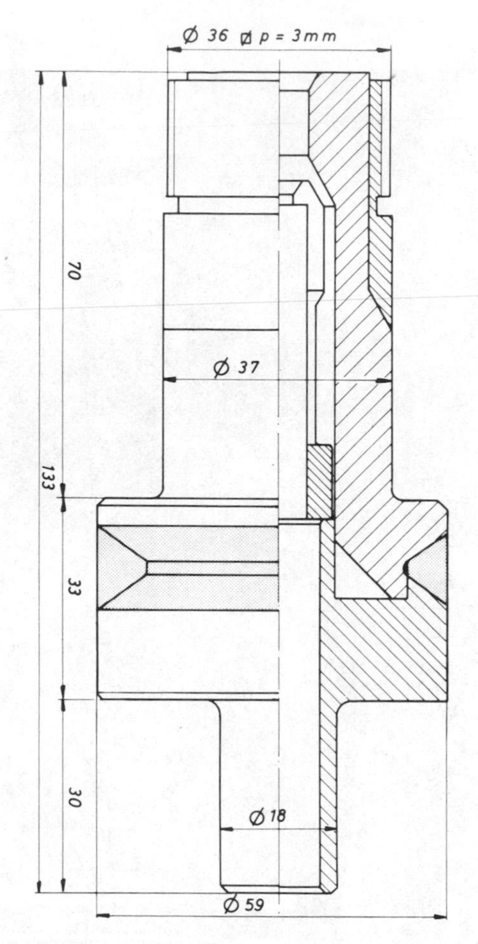

FIG.13. Stud design.

2.3.2.3 Improved Cable Seals

The use of sealing studs with cross holes would allow horizontal threading of the seal cable. This use would reduce the practical problems of cable seal installation. To overcome the problem of installing the cable end into the insertion mechanism, a modified seal has been proposed. This seal (see Figure 10) would have an extension on the end of the locking "needle" of sufficient length to reach above the bay water surface. One end would be threaded through the stud to be sealed and then raised to the surface. The seal, which would have holes completely through the body, would have both cable ends inserted together. Then the cable would slide down the extension wires until the needle entered and locked into the body. The extension wires would be cut off and the seal would be treated as before. The installation mechanism would require a holder only for the seal block and the transducer assembly.

FIG.14. Photo of cap and stud in frame.

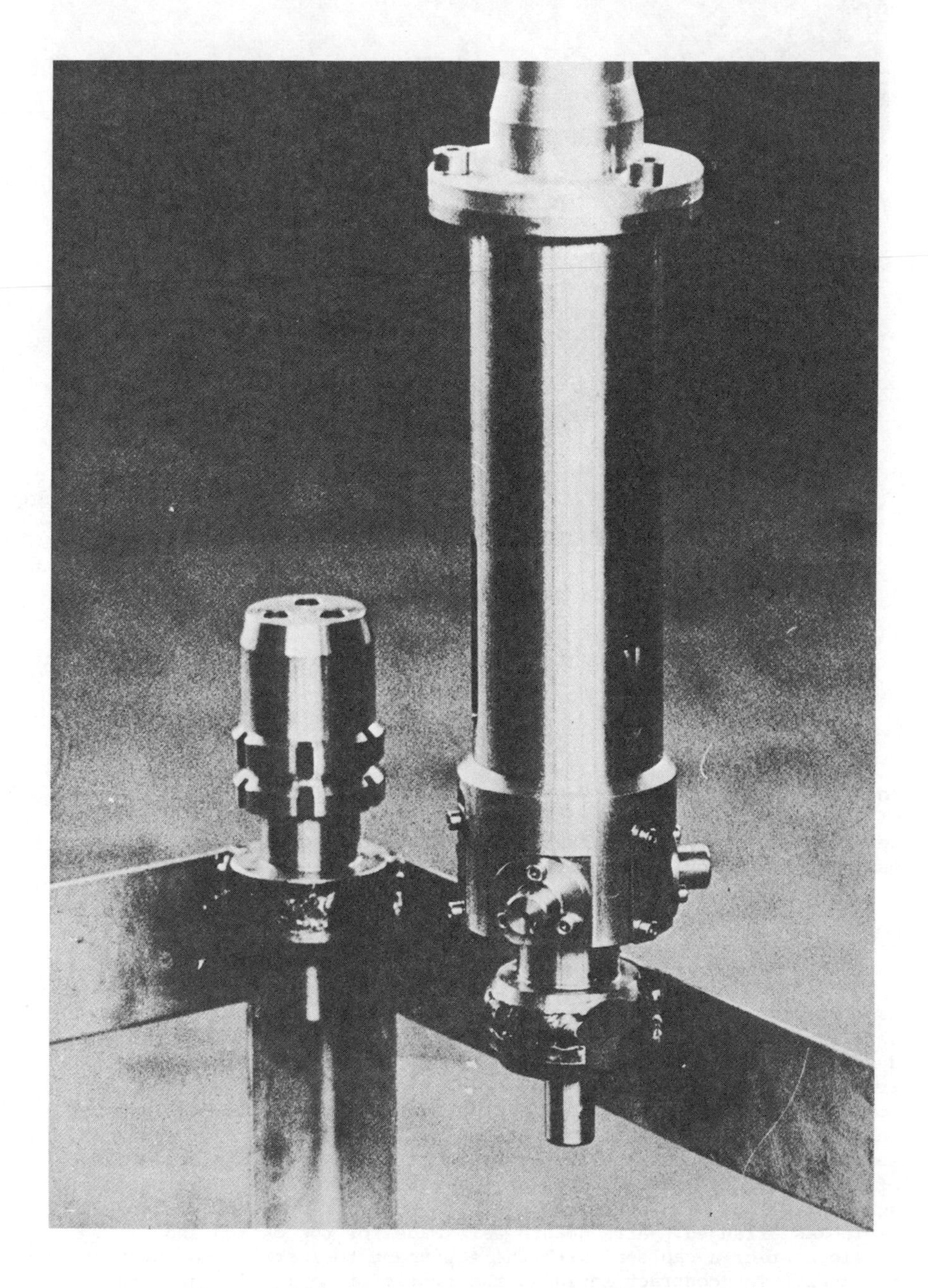

FIG.15. Installation-identification tool. Installation of cap.

FIG.16. Foot of the installation-identification tool containing transducer.

This type of seal has not yet been tested and may not be used because of the evolution of a cap seal for this type of application. The cap seal approach, similar to that used on BWR fuel bundles, would have many advantages over the cable seal.

3.0 CAP SEAL

JRC Ispra, developed a cap seal for use on BWR fuel bundles (see Figures 3 and 11). Because of the inherent advantages of such a seal in installation and verification, it was decided to proceed with design and development of a seal suitable for application to the CANDU irradiated fuel storage bay system. The cap seal would be used in conjunction with a stud welded onto the security cage to hold a cover on top of the cage. This type of cap seal would have to be designed so that the stud could be re-usable and the cap replaceable. Because of the experience of JRC Ispra in the ultrasonically identified cable seal as well as with other cap seals, a contract between AECL and JRC Ispra was arranged. Arrangments were made for the design and development of a cap seal with the equipment to install and interrogate the seal. The contract included the supply of an initial quantity of seals and tools to install and verify the seals, as well as definition of the electronic equipment required. The seals would utilize an electronic scanning system and would produce a printed digital identity for the seal.

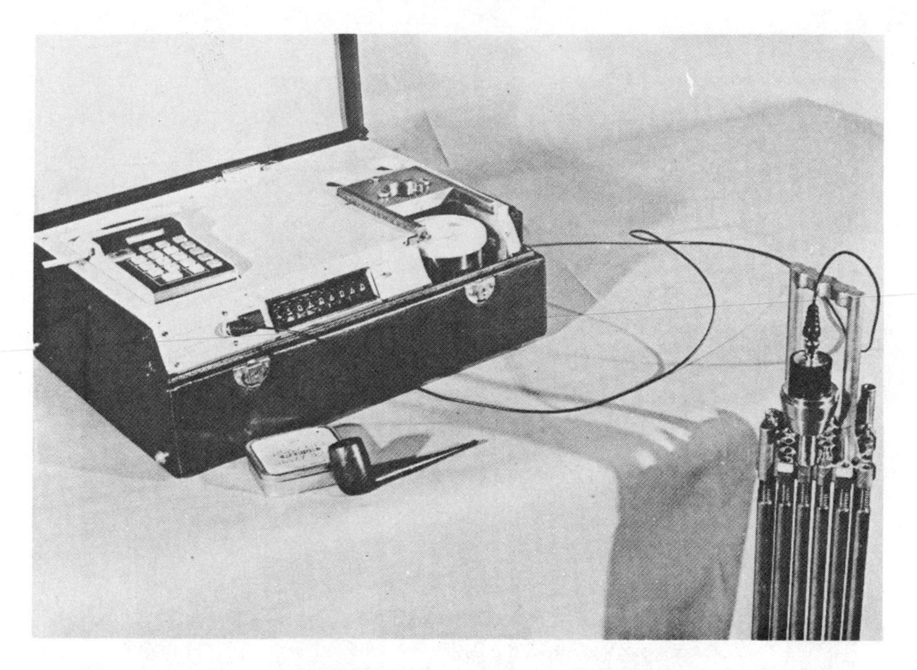

FIG.17. Portable digital identification equipment.

3.1 Development of Cap Seal at JRC Ispra

CANDU cap seals include a cap and stud, both items being ultrasonically identifiable. The cap is composed of three main parts (see Figure 12) which are:

- cap housing with thread for screwing the seal into the stud,
- cap body with unique identification marks,
- cap pin, part of the non return system, controlled with ultrasonics and breaking when extracting the seal.

The stud is a tube welded onto the structure of an enclosure (see Figure 13). It contains a collet, the second part of the non-return closure system, which retains the pin of the cap when the cap is screwed on. The weld is part of the identity of the stud. Figure 14 details a cap placed on a stud, welded onto a test structure.

3.2 Installation/Identification Tool

Installation of the cap seal is made by a simple keyed socket at the extremity of a 6 metre long tube. This allows easy positioning of the cap above the stud. Then it is screwed onto the stud until the pin

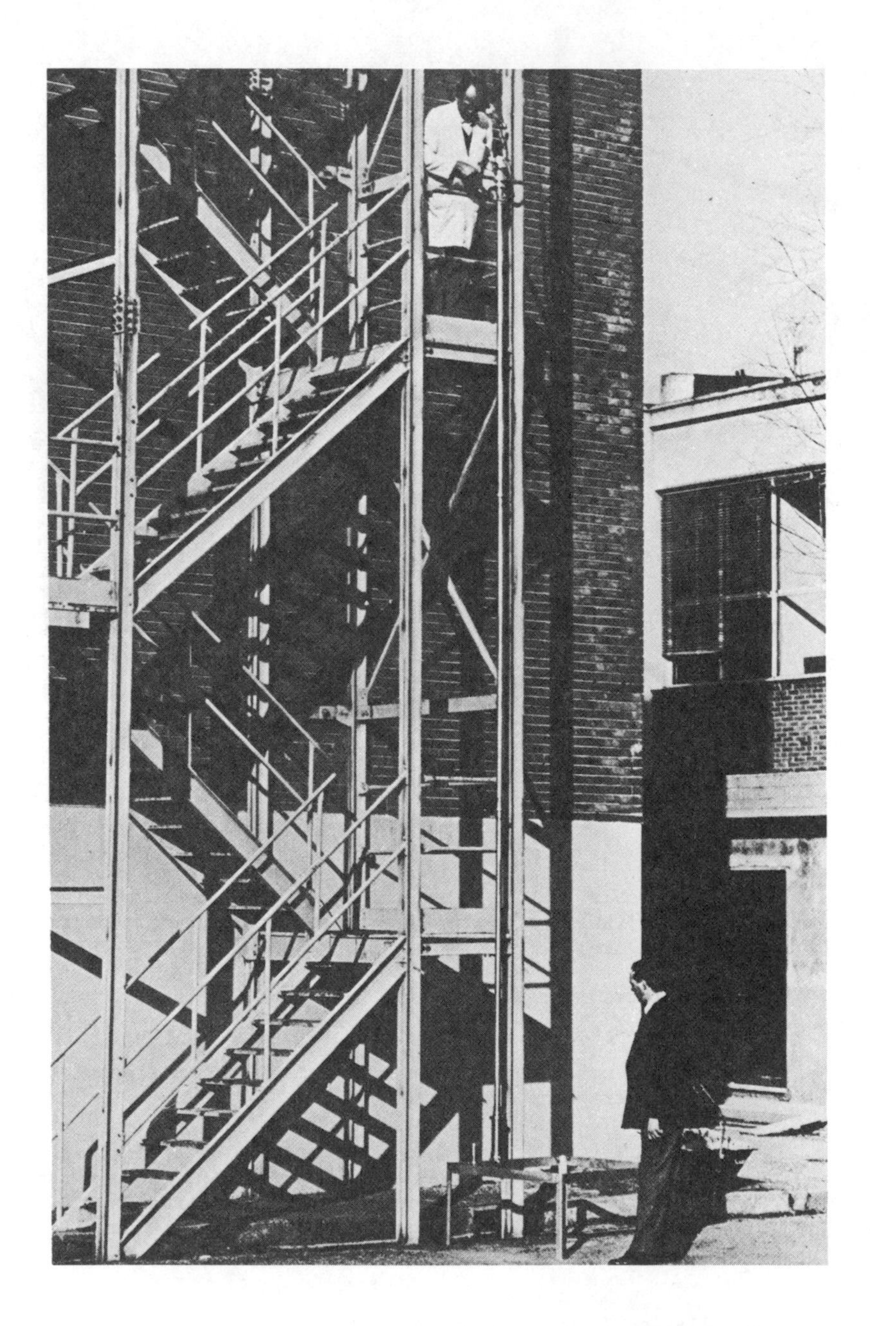

FIG.18. Tests of the identification-installation tool at Ispra.

is gripped by the collet on the stud (see Figure 15). Identification of seal and stud is performed using the same installation tool with a transducer contained in the socket (see Figure 16) , which can easily be lowered onto the cap to scan the cap identity zone, the cap body, the cap pin, the stud body and the stud identity zone. This scanning provides cap identification, cap integrity check, stud identification and stud integrity check which can be performed in one operation.

A portable identification unit (5) is connected to the installation/identification tool and presents the digital result in less than 10 seconds (see Figure 17). Portable equipment will be commercially available from Nukem by the end of 1978.

3.3 Results of Cap Seals Testing at Ispra (see Figure 18)

Little time was given to JRC Ispra, to extrapolate their knowhow about cap seals for BWR to large cap seals to be used in CANDU fuel storage. Well adapted transducers were not available at the time, but positive conclusions can be drawn from these laboratory tests:

- installation/identification tool is simple,
- installation and extraction of seal is easy and requires only a few minutes,
- identification and integrity check of cap is easy and reproducibility is consistent,
- identification and integrity check of stud could not be demonstrated because of the non-appropriate transducer,
- test of tempering with seals demonstrated variations in identity or integrity patterns.

Final results will be available when the three following elements of the project will be put together:

- industrially fabricated seals (series of 50)
- adapted transducer (Aerotech supply)
- Nukem portable equipment (Euratom Licence).

3.4 Further Development of Cap Seal

To reduce the cost and shipping expense of the security cage, a cover and two seals are required to seal a stack of fuel. An alternative approach, using the cap seal, has been considered. This sealing method would use a long rod inserted down between fuel bundles through the stack of trays. In conjunction with a cover and keying system, this would allow sealing of a stack of trays with a relatively simple cover, a long rod with a stud end and a single seal. There are problems to overcome in obtaining sufficient tamper resistance, but the advantages over the cage system would be considerable.

4.0 CONCLUSIONS

The use of ultrasonic interrogation techniques in sealing systems underwater, has been shown to be an acceptable approach to the problem. Seals have been developed to provide excellent tamper resistance. They are fast, simple to install and interrogate. This type of seal and this approach in general, will probably be further improved to reduce cost. It will most certainly have other applications in addition to those described herein.

REFERENCES

[1] CRUTZEN, S.J., FAURE, J., WARNERY, M., System for Marking of Objects, in Particular Fuel Elements in Nuclear Reactors, Euratom Patent UK No. 1241.287, 10.7.1969.

[2] CRUTZEN, S.J., DENNYS, R., Ultrasonic Uniquely Identified Seal for CANDU Spent Fuel Bundles Surveillance, AECL Rep. 17-30085 (Canada) (1977).

[3] CRUTZEN, S.J., JEHENSON, P.S., HAAS, R., LAMOUROUX, A., "Application of tamper-resistant identification sealing techniques for safeguards", Safeguarding Nuclear Materials (Proc. Symp. Vienna, 1976) **5**, IAEA Vienna (1976).

[4] ZIJP, W.L., CRUTZEN, S.J., "Statistical aspects of ultrasonic signatures", IAEA-SM-231/22, these Proceedings, Vol.I.

[5] CRUTZEN, S.J., VINCHE, C., BURGERS, W., COMBET, M., "Remote-controlled and long-distance unique identification of reactor fuel elements or assemblies", IAEA-SM-231/21, these Proceedings, Vol.I.

DISCUSSION

R.W. FOULKES: If an ultrasonic seal is used under water, how long will its integrity be maintained? Will it last several years or will the contents of the bundle have to be re-verified after a much shorter period?

S.J. CRUTZEN: Our experience is based on three cases: (1) Rivet seals. Several hundred of these have been used since 1973 (HFR at Petten in the Netherlands). On the seals recovered at Marcoule and, last summer, at the Savannah River plant no deterioration was observed; (2) LWR cap seals. Sixteen of these were tested for five years on fuel bundles in a reactor core and no damage was observed; (3) CANDU cap seals. These have been used for about two years and no deterioration of the seal integrity has been reported.

S. SANATANI: You have described several types of ultrasonic seal. Which of these are commercially available now and how much do they cost?

S.J. CRUTZEN: General-use seals (plexiglass and aluminium) are made by different firms. The cost is about US $2. Rivet seals (for MTR fuel boxes) are made by the firm Nukem and cost about US $10. CANDU cap and stud seals are made by an Italian firm and cost US $250. The Nukem portable seal identification unit mentioned in the paper is expected to cost around DM 37 000.

W.C. BARTELS: A vital aspect in the development of any sealing system is the establishment of its capability to resist a sophisticated effort to defeat the seal. For instance, the type of metal seal now used by the IAEA has been subjected to careful study to determine the minimum effort needed to defeat the seal and hence to establish its level of credibility. To what extent has this been done for these ultrasonic seals?

S.J. CRUTZEN: The ultrasonic seals were studied for tampering in two cases – the plexiglass general-use seal, and the CANDU wire seal. In the case of plexiglass seals it was impossible to reproduce the pattern even with seals having only two inclusions. In the case of the CANDU wire seal tampering with the seal wire was nearly successful, but this was a prototype seal made to demonstrate the principle.

Tests for tamper-resistance of the seals should be performed for each type of application by some group not involved in the manufacture of the seals. No real data on the effective reliability of the different applications of the "ultrasonic signature" are available, but the results presented in paper IAEA-SM-231/22[1] give figures of the order of 10^{-6}.

---

[1] ZIJP, W.L., CRUTZEN, S.J., these Proceedings, Vol.I.

APPLICATION OF SAFEGUARDS DESIGN PRINCIPLES TO THE SPENT-FUEL BUNDLE COUNTERS FOR 600-MW CANDU REACTORS

A.J. STIRLING, V.H. ALLEN
Atomic Energy of Canada Limited,
Chalk River Nuclear Laboratories,
Chalk River, Ontario, Canada

Abstract

APPLICATION OF SAFEGUARDS DESIGN PRINCIPLES TO THE SPENT-FUEL BUNDLE COUNTERS FOR 600-MW CANDU REACTORS.

The irradiated fuel bundle counters for CANDU 600-MW reactors provide the IAEA with a secure and independent means of estimating the inventory of the spent-fuel storage bay at each inspection. Their function is straightforward – to count the bundles entering the storage area through the normal transfer ports. However, location, reliability, security and operating requirements make them highly "intelligent" instruments which have required a major development programme. Moreover, the bundle counters incorporate principles which apply to many unattended safeguards instruments. For example, concealing the operating status from potential diverters eases reliability specifications, continuous self-checking gives the inspector confidence in the readout, independence from continuous station services improves tamper-resistance, and the detailed data display provides tamper indication and a high level of credibility. Each irradiated fuel-bundle counter uses four Geiger counters to detect the passage of fuel bundles as they pass sequentially through the field-of-view. A microprocessor analyses the sequence of the Geiger counter signals and determines the number and direction of bundles transferred. The readout for IAEA inspectors includes both a tally and a printed log. The printer is also used to alert the inspector to abnormal fuel movements, tampering, Geiger counter failures and contamination of the fuel transfer mechanism.

1. *CANDU 600-MW REACTOR SAFEGUARDS*

The IAEA Safeguards System for 600-MW CANDU[1] reactors combines inspection with unattended surveillance and unit bundle accounting to detect and deter spent fuel diversion [2]. The irradiated fuel bundle counters provide the Agency with a secure and independent means of estimating the inventory of the spent fuel bay at each inspection. Checking the bundle counter tallies with the contents of the spent fuel storage bay gives the first indication of diversion. Once alerted, the inspectorate has film and videotape records to check for unreported activities.

[1] CANDU: **CAN**ada **D**euterium **U**ranium.

[2] TOLCHENKOV, D., HONAMI, M., JUNG, D., SMITH, R.M., VODRAZKA, P., HEAD, D., IAEA-SM-231/109, these Proceedings, Vol.I.

The instruments described here are suitable for the 600-MW CANDU reactors at Gentilly and Point Lepreau in Canada, at Cordoba in Argentina and at Wolsung in Korea (assuming appropriate agreements with the IAEA are in force). Though the design of an instrument to count fuel bundles is relatively straightforward, unique constraints are imposed when it is to be used as an unattended safeguards tool. Many of the design approaches taken here may therefore be applicable to other unattended safeguards instruments.

2. *SPENT FUEL BUNDLE COUNTER: CRITERIA FOR DESIGN*

For the 600-MW CANDU reactor, the unit of accounting is the fuel bundle [1]. Each fully irradiated spent fuel bundle contains an average of 45 g of fissile plutonium. Thus, approximately 168 bundles would be required to assemble an amount of plutonium equal to the IAEA Threshold Diversion Amount.

The principle function of the spent fuel bundle counters is to determine the total number of irradiated fuel bundles transported into (or out of) the spent fuel bay through the normal entry ports. However, the instrument design is strongly influenced by the fuel handling system, and the security, tamper-indicating, credibility and maintainability criteria, which are discussed to explain how the specifications were determined (see Table 1).

Spent Fuel Handling System

Fuel removed from the 600-MW CANDU reactor will be discharged into one of the two fuelling machines (see Figure 1). The fuelling machine will then be aligned with the spent fuel port, and either one or two bundles will be pushed onto the discharge elevator in the spent fuel discharge room (see Figure 2). Figure 2 shows two fuel bundles on the elevator ladle in position opposite the spent fuel port.

During normal operation, bundles will be discharged in pairs at the rate of approximately 100 bundles per week. However, it is possible for bundles to be unloaded singly, and in this case the bundle may be pushed into any position along the length of the ladle. After the bundles have been loaded onto the ladle, it is lowered into the water-filled bay on guide rails at an angle of about 30 degrees to the vertical. While on the elevator, the bundles are cooled by a water spray that covers the whole of the ladle.

With the use of grapples, a diverter could attempt to transfer bundles in reverse, from the spent fuel area into the reactor vault. Thus the spent fuel counter must be capable of counting bundles passing in either direction.

Since there are two points of entry to the spent fuel bay (see Figure 1) there must be two irradiated fuel bundle counters per reactor. Each must be located close enough to the spent fuel port that bundles must pass the counters on entering or leaving the spent fuel storage area.

Detection and Deterrence

The objectives of an instrumented safeguards system are to *detect*, and by threat of detection, to *deter* the diversion of irradiated fuel

TABLE I. BUNDLE-COUNTER SPECIFICATIONS

FUNCTIONAL:	Count the number of fuel bundles transferred into and out of the spent fuel area through a spent fuel port.
AVAILABILITY:	MTBF greater than three years.
ACCURACY:	Exceeding 98%
CREDIBILITY:	Provide time and date of each irradiated fuel transfer. Provide automatic self-checking during periods of unattended operation. Provide test feature for Agency inspectors prior to leaving instrument unattended.
SECURITY:	Provide no means whereby the potential diverter can determine whether the counter is operating. Provide data to inspectors only. Resist tampering and accidental damage. Indicate tampering. Operate from internal battery power supply. Employ IAEA seals. Indicate irregular bundle transfer operations.
MAINTAINABILITY:	First-line maintenance by unit or module replacement. Second-line repair by Agency or Agency contractor.
ENVIRONMENTAL:	Meet environmental specifications applied to control and safety instrumentation in CANDU 600 MW reactors.
DOCUMENTATION:	Operate without the need for coded or secret documentation.

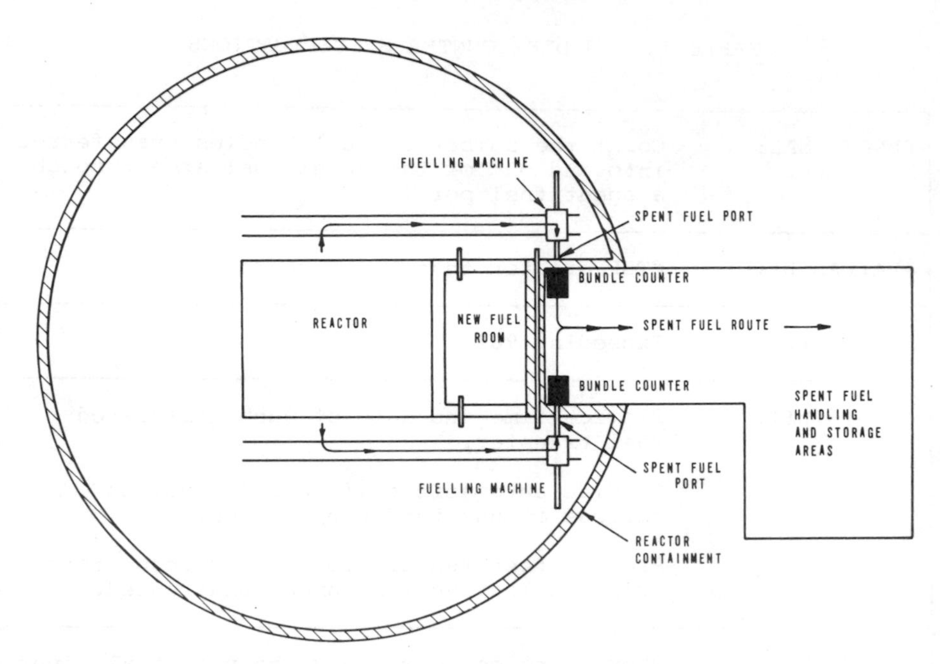

FIG. 1. Spent-fuel route and bundle-counter locations for CANDU 600-MW reactors (schematic).

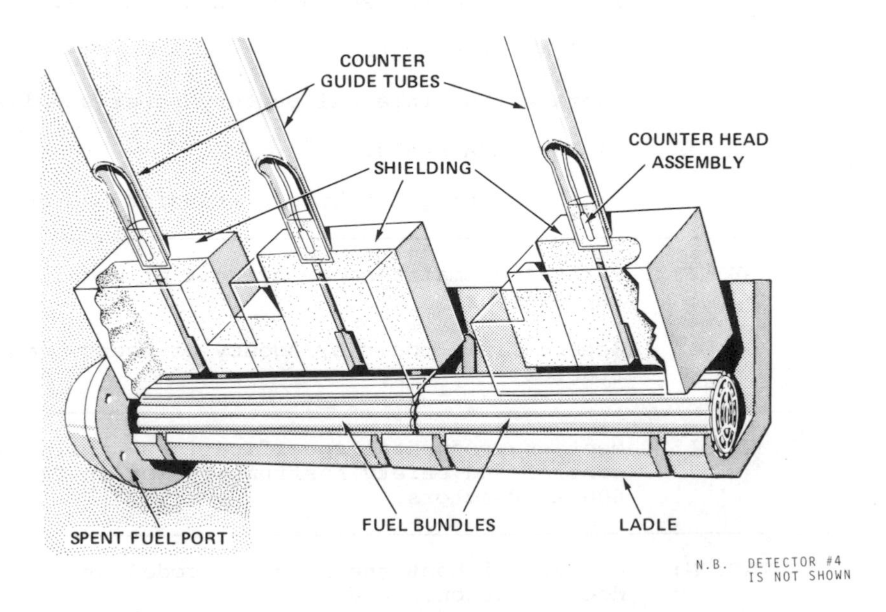

FIG. 2. Spent-fuel elevator ladle with collimators for safeguards counter, shown schematically.

bundles. There exists no accepted method for establishing the criteria which may be used to write precise specifications for these objectives. However, it is clear that there are two sets of design criteria. One set relates to the reliability as a counter, i.e. to detect, and the other to its ability to act as a deterrent.

The reliability of the counter to detect the passage of irradiated fuel can be quantified by its availability, derived from expected failure rates and inspection periods. The availability must be high enough that the data provided to the inspector is credible.

Availability must also be high enough that the maintenance load is not excessive. For the purpose of the present design, a target for Mean Time Between Failure (MTBF) of three years has been chosen. This corresponds to an availability of 97%, assuming an inspection period of two months. This level of reliability can be achieved using the design practices normal for safety-grade instrumentation.

The bundle counter must not only *be* reliable, but must be *known* to be reliable. It must therefore establish its own credibility which is done by automatically checking itself each day, and by recording times and dates of bundle transfers.

The accuracy of the bundle counter must also be high enough that the inspector will believe that the tally is credible. It is believed that an accuracy of 98% will be acceptable, because such discrepancies between counter tally, inventory and station records will be less than 5% of the Threshold Diversion Amount.

As yet there is no quantitative measure of the effectiveness with which a safeguards system deters diversion. The motivation of the diverter, his access to technology and the risks he is prepared to take, can never be assessed.

One way to make an instrument effective in deterring diversion is to provide no clue of its operating status to potential diverters. The operating status of the counter must be known only to the inspector, and data must be sufficiently credible that the inspector does not need to give any clue of the operating status through excessive questioning of station staff. When a diverter is unaware that a bundle counter has failed, he is unlikely to risk being exposed.

A diverter could use the strategy of sabotage, but to be effective in the long term, he must do this in such a way that

(i) the Agency is unaware of the sabotage, or
(ii) the Agency can be convinced that a malfunction resulted from a component failure or accident, and
(iii) the diverter is sure that the instrument has been rendered inoperative.

A level of protection against sabotage is achieved by

(i) providing no means to determine whether the instrument is operating,
(ii) incorporating tamper indicating devices, and
(iii) establishing high levels of confidence in the instrument reliability, so that simultaneous failures themselves are indication of sabotage.

IAEA Inspection Procedure

It is anticipated that the Agency inspectors will visit CANDU 600 MW reactors approximately six times per year, for a few days on each visit. Since inspectors are not necessarily skilled in the maintenance of instrumentation, the inspection procedure imposes the following constraints on the irradiated fuel bundle counter:

(i) operate unattended for at least 100 days;
(ii) provide simple-to-understand data display;
(iii) require a minimum of operating materials; and
(iv) be maintainable by inspector.

3. DESIGN PRINCIPLES AND IMPLEMENTATION

Detection and Analysis

The passage of spent fuel bundles is established by measuring emitted gamma radiation with four Geiger counters. The Geiger counters are located in massive collimators set close to the fuel transfer ladle, and adjacent to the opening of the spent fuel port. They are spaced such that at least one counter receives radiation when a single bundle is transferred, irrespective of its position in the ladle. The collimators prevent single bundles appearing as a pair.

Each Geiger counter assembly consists of a replaceable Geiger counter and a small gamma ray source. The source is used in a watchdog circuit to detect tampering. The whole assembly is housed in a metal guide tube embedded in the collimator block.

The mean direct current flowing in a counter increases when a bundle passes, and triggers a threshold circuit. The logic levels from the threshold circuits are processed by electronic signal sequence analysis circuits to determine the number and direction of bundles. Each category of bundle motion generates a unique sequence of signals. The electronic circuits examine the order of occurrence of the signals and thereby determine the type of bundle motion. The normal categories of bundle motion which the electronic circuits are designed to recognize are

- pair of bundles to the bay, and
- single bundle to the bay (in any position on the ladle).

The logic circuits can also recognize a limited number of categories of bundle motion which may occur during abnormal operation of the equipment, including

- transfers in the reverse direction, and
- transfers without the use of the ladle.

In the event that a signal sequence is unrecognizable by the logic, a warning message is displayed.

When no bundles are present, a low range counting ratemeter responds to Geiger counter pulses produced by the watchdog source. Absence of this signal indicates failure of Geiger counter, failure of high voltage bias, or disconnection of the Geiger counter from the circuit. This tamper-indicating feature is important since the electronic circuitry is located in a cabinet remote from the Geiger counters. Should pulses from the watchdog source cease, the date and time are recorded.

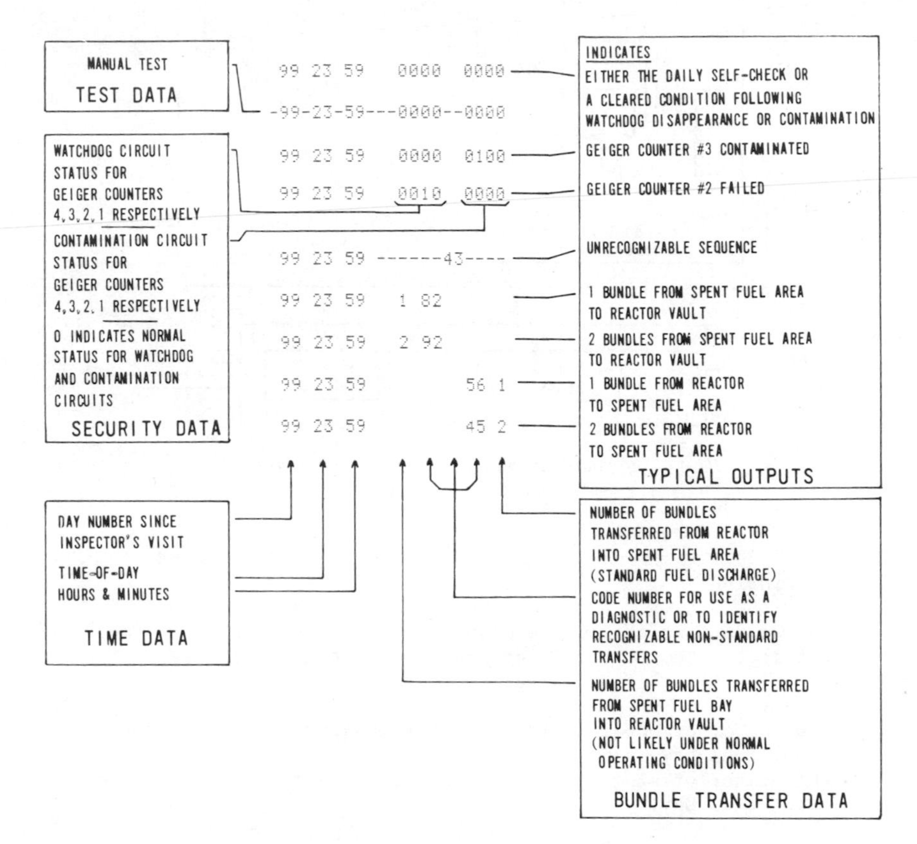

FIG.3. Examples of safeguards bundle-counter log available to inspector.

The analysis of signal sequences is made by a stored program microprocessor (RCA Model 1802), with memories and peripheral elements selected from the CMOS (Complementary Metal Oxide Silicon) range of semi-conductor elements. The microprocessor program continuously interrogates the Geiger counter logic outputs. When any counter detects a bundle, the program seeks to recognize a valid sequence and print the result.

In addition to analyzing the sequence of the logic signals, the microprocessor

- reads date and time information from an in-circuit clock,
- performs the self-checking function once daily, and
- controls the data displays.

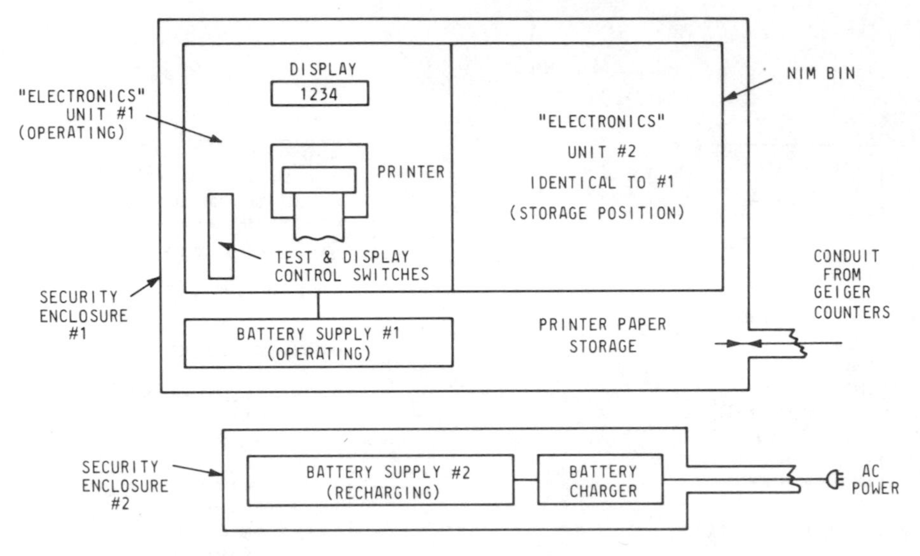

FIG.4. Safeguards bundle-counter physical schematic.

Data Display

Information from the bundle counter is displayed in two forms, tallies and logs. The four tallies are read on a digital display which provides

- the number of bundle pairs transferred into and out of the spent fuel bay, and
- the number of single bundles transferred into and out of the spent fuel bay.

The tallies are stored in four groups of latches which are updated as transfers occur.

The log printed by an electrostatic printer lists the same information, plus the times and dates of each transfer. In addition, abnormal or irrational transfers and supervisory information are printed (see Figure 3 for typical examples).

The printer display provides sufficient data to establish credibility while the digital readout provides an easy-to-use display when cross checking with station records is not needed. The digital display also provides redundancy against printer failure and plays a role in preventing a diverter from knowing whether the bundle counter is operating. (Since the printer generates acoustic noise, its failure would be relatively easy to detect. However, a diverter could not readily determine whether the digital display had failed too.)

A series of push buttons enables the inspector to test the bundle counter. The buttons simulate Geiger counter signals and the inspector can test any given sequence of bundle motion.

Power Supply

To maximize tamper resistance and to maintain independence from station sources, power is provided by interchangeable batteries. Battery packs will be changed by the inspector at scheduled visits and recharged at the station between inspections. The battery packs are light enough for the inspector to handle. The recharger units will be located near, but not within, the sequence analysis and display circuits cabinet. The recharger units include monitors to show that the batteries are fully charged, and the units will be sealed by the Agency to indicate tampering. The battery pack drives the printer directly at 24 V and supplies required for the electronics and Geiger counters are derived from dc/dc converters. Battery life of 100 days between charges is provided using a battery with a capacity of 12 A·h at 24 V.

Physical

The irradiated fuel bundle counters will be located in steel boxes on the walls near the Geiger counter locations and connected to them by buried conduits. The instrument enclosures and the battery chargers will be covered by hinged steel doors under Agency seal (see Figure 4 for physical arrangement).

Operating Status Concealment

As discussed, concealing the operating status of the bundle counter means that the instrument deters diversion, even if it has failed. The specific design features which conceal the operating status from the potential diverter are:

- located within steel box;
- cable connections embedded in concrete;
- battery supply is independent of continuous operation of the station supply;
- low power circuits generate no detectable heat at points accessible to the diverter; and
- no acoustic noise (except from printer which has back-up silent display).

Tamper Resistance and Indication

Tamper resistance is inherent in the design of the instrument housing. Located in a steel box and operating from its own power, there is little likelihood that sabotage could be attributed to accidental damage or station power failure.

Other tamper-indicating features include

(i) the watchdog gamma ray source;
(ii) a maximum-minimum temperature indicator within the circuit enclosure;
(iii) a radiation dosimeter; and
(iv) Agency seals.

Maintenance

The design provides for a replacement bundle counter within the sealed enclosure housing the operating counter. Interconnection to the Geiger counters is simple, so that the inspector can switch to the spare circuit modules. With the reliability expected, this should be required only once every few years. Housing a spare electronic unit with the operating unit provides, in effect, a "bonded store" for the Agency. Under normal operating conditions Geiger counter tubes should not need replacing due to wear-out failures. If deliberately overexposed, the printer will indicate tube failure as a "loss of watchdog" signal.

4. SUMMARY

Designing a spent fuel bundle counter as part of the IAEA safeguards system presents unique challenges. The designers for the 600 MW CANDU irradiated fuel bundle counters have incorporated features which should be applicable to the design of other unattended safeguards instruments. The most important of these features are operating status concealment, continuous self-testing, provision of sufficient data to maintain instrument credibility, and independence from continous station services.

Prototype spent fuel bundle counters have been constructed and will be undergoing Agency acceptance tests in late 1978 for installation as the CANDU 600-MW reactors are completed.

ACKNOWLEDGEMENT

This work is being undertaken as part of a co-operative programme with the IAEA and the Atomic Energy Control Board of Canada.

THE ELECTRONIC SEALING SYSTEM, VACOSS, AS A CONTROL MEASURE FOR INTERNATIONAL SAFEGUARDS

K. KENNEPOHL, D. MAECKELBURG, G. STEIN
Kernforschungsanlage Jülich GmbH,
Jülich, Federal Republic of Germany

Abstract

THE ELECTRONIC SEALING SYSTEM, VACOSS, AS A CONTROL MEASURE FOR INTERNATIONAL SAFEGUARDS.

The electronic sealing system VACOSS (Variable Coding Sealing System) was developed at KFA Jülich within the framework of an IAEA research contract. The seal is equipped with a light guide which acts as a lock. Statistical infra-red wavelength light pulses are transmitted via this light guide. The electronic components of the seal can store the opening and closing events of the light guide. In addition, the quasi-statistic treatment of code data fed into the seal by means of an adapter box guarantees an extremely high protection against unauthorized access. The portable seal is easy to handle, can be operated with a battery and can be re-used immediately after replacement of this battery. To make the seal operational code data are fed in, and the information stored in the seal can be read via the adapter box. Two versions of VACOSS were developed. VACOSS I permits the state of the seal in the facility to be checked on site only. VACOSS II can be monitored remotely by an operator passing on coded information.

1. INTRODUCTION

Recent discussions showed that in order to increase the effectiveness of the safeguards system and to realize the control objectives the IAEA must attach special attention to the introduction of containment and surveillance measures. In this connection it seems to be inevitable that these containment and surveillance measures no longer are to be considered supplementary but rather complementary measures of the established nuclear material accounting system. This fact becomes evident particularly when considering facilities with a large annual throughput of nuclear material where a considerable diversion potential is produced due to the measurement errors in connection with the determination of nuclear material content. For this reason, the development of containment and surveillance instruments is urgently required that will facilitate nuclear material accounting during physical inventory taking on the one hand and at the same time for the early detection of the diversion of significant quantities permit remote interrogation of the system to be employed. Systems that serve both purposes mainly are based on electronic components such as electronic surveillance cameras and electronic seal systems.

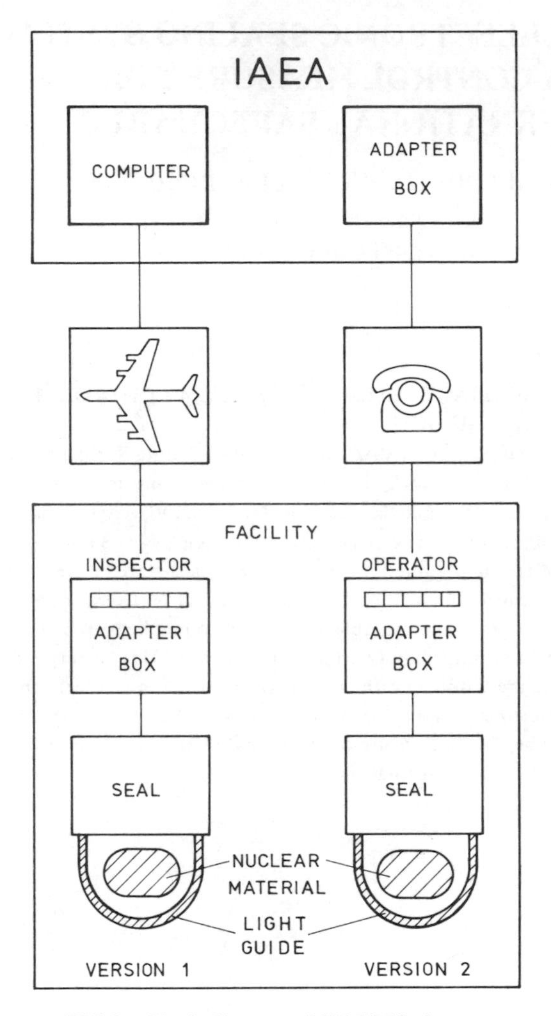

FIG.1. Block diagram of VACOSS elements.

Within the framework of an IAEA research contract in the Juelich nuclear research center the electronic seal system VACOSS (Variable Coding Seal System) [1,2] has been developed. During the development work prototypes for two different VACOSS versions were built. VACOSS I is exclusively used for verification of the seal integrity on site and may be employed for facilitation of physical inventory taking. VACOSS II permits a transmission of information to the control center by telephone regarding the state of the seal. In addition, the electronic equipment of both systems is capable of storing information, which, if needed, can remotely be interrogated with the aid of suitable peripheral systems.

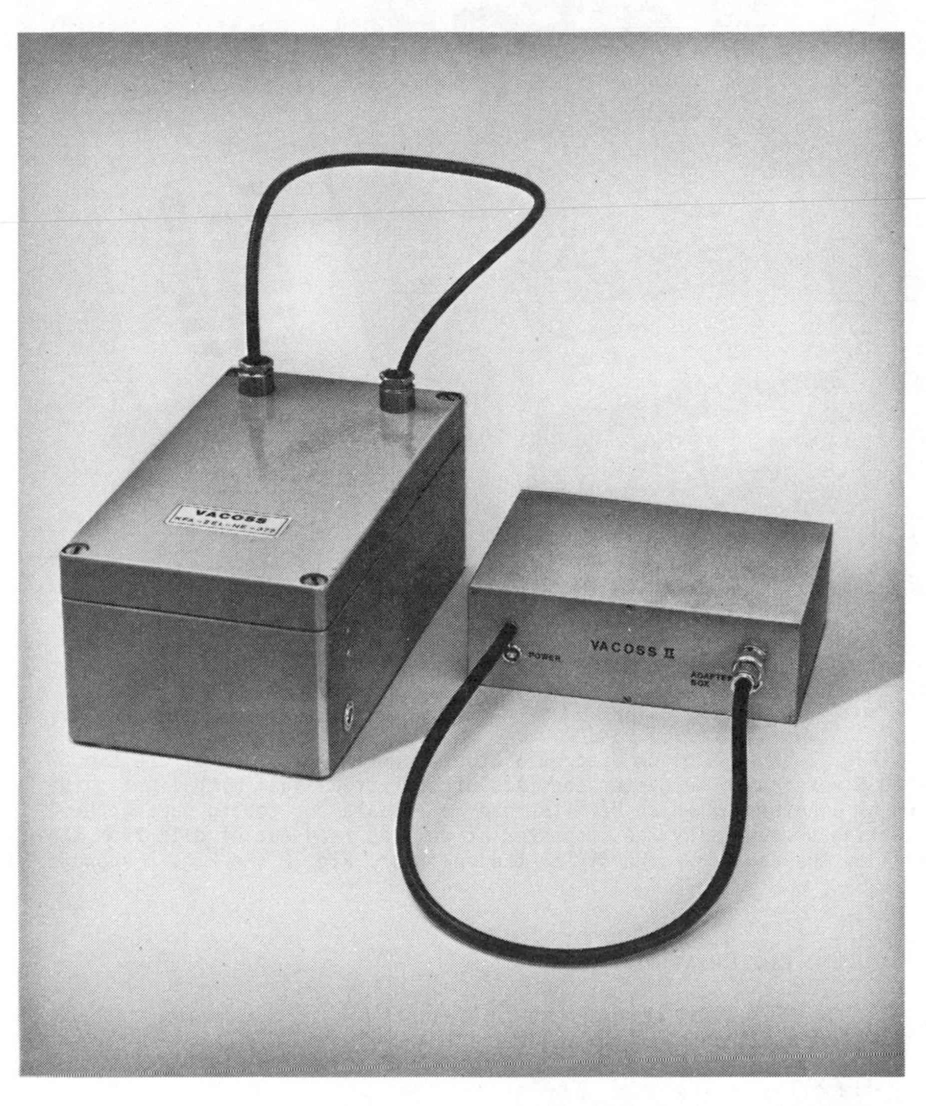

FIG. 2. VACOSS seal.

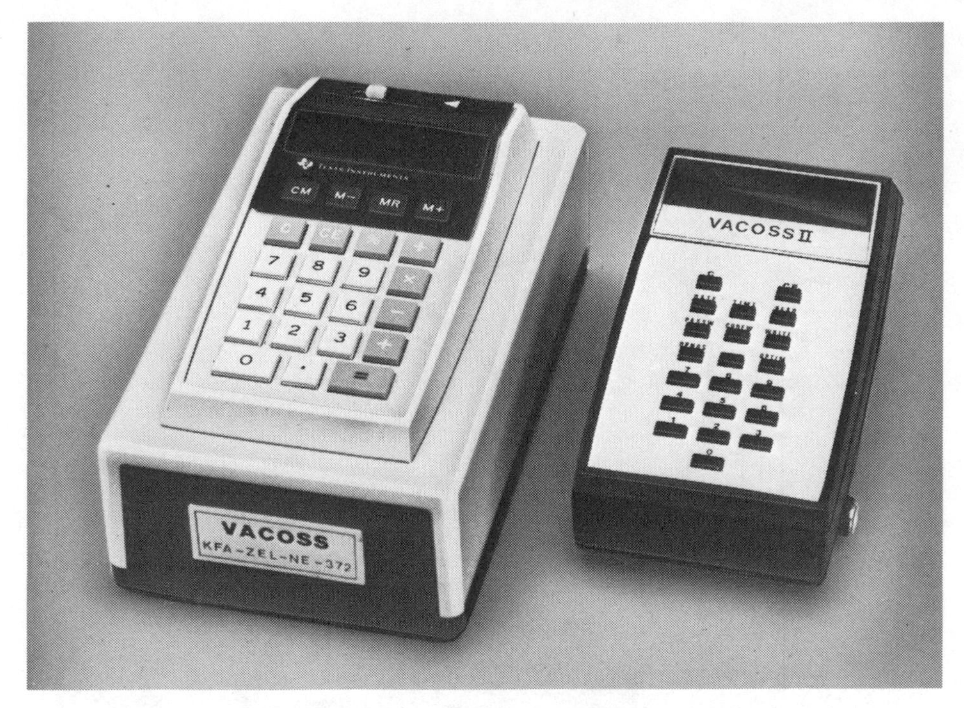

FIG.3. VACOSS adapter box.

Fig. 1 shows a block diagram of the essential elements of the two VACOSS versions. The system consists of the actual seal with light guide and an adapter box which permits read-in of data for coding during the starting procedure by the inspector as well as read-out of data from the seal by the control authority or the operator. Fig. 2 and Fig. 3 show for

2. DESIGN CRITERIA

The essential criteria that played a role in the development and design of the electronic components are:
- access security
- handling
- reliability
- acceptance by the operator
- potential of applicability

Access security

The security of the system against manipulation from outside and against tampering of the state of the system enjoy maximum priority in the application of the seal system for nuclear material safeguards. The VACOSS seal system meets these requirements by the light guide technology and the statistical treatment of data that are read-in the seal as well as by the special seal electronics. The light guide consists of a fiber guide and serves as a lock. By means of a light emitter - with a frequency of 100 Hz - light impulses in the infrared wave range are transmitted over the light guide which are detected in a appropriate manner. The opening and closing times are coded and stored. For VACOSS I the decoding procedure will be done in the center of the supervisory authority with the aid of a simulation program. In the case of VACOSS II the decoding will be performed by suitable electronics in the adapter box itself.

Handling

The seal has small dimensions and may be carried in the baggage of an inspector. The seal operates on mains power supply and on batteries. Following exchange of the discharged batteries, the seal can be used again. Control of the state of the seal is carried out with the aid of the adapter box by reading out the stored contents of the seal.

Dimensioning of the seal mainly depends on the capacity of the batteries which provide the seal with an operating period of about 1 year.

Reliability

The electronic components of the seal system consisting of C-MOS, IC's as well as ROM's and RAM's for VACOSS II provide a high degree of reliability. The temperature range for an admissible operation is between -30^{0} C and $+70^{0}$ C. The seal may be exposed to a radiation of up to 10^{4} R.

Acceptance by the operator

The adapter box with which information can be read out from the seal may be designed in different versions permitting also the operator to control the state of a seal system, if this is required.

Potential of application

Version I of VACOSS exclusively permits control of the seal system by an inspector. In the seal of version II a coded figure is generated which can be read out by the operator with an adapter box and transmitted by telephone to the supervisory authority. The supervisory authority is able to decode this information with a special adapter box and ascertain whether the seal was opened during their absence.

3. DESCRIPTION OF THE SEAL

3.1 VACOSS I

VACOSS I consists if three components: the seal, an adapter box and a simulation program which remains exclusively in the hands of the supervisory authority.

The seal:

Apart from a memory and a statistic generator further essential components of the seal are a light guide with light emitter and receiver in the infrared light range. The light guide assumes the function of a seal during the sealing procedure. A maximum of two opening and closing events can be stored. The information on the temporal association of these events are coded with the aid of a statistic generator. This statistic generator varies pseudostatistically two code words read in by the inspector with a preselectable time interval of 2-2o minutes. These statistically changed code words are stored during the opening and closing procedures. The seal is equipped with lithium batteries having an operating period of 2 years. The dimensions are 22 x 12 x 9 cm^3. Access from outside by opening of the case is indicated by appropriate electronic measures. The seal consists of the following electronic components:

- light guide with light emitter and receiver
- memory
- data-coding logic
- statistic generator
- clock generator
- data input and output logic
- control logic
- crystal oscillator
- battery
- metal case

Adapter box:

The adapter box is designated for reading in data in the seal and/or reading out data from the seal and may be designed in two versions. One for exclusive use by the supervisory authority, permitting the inspector to read in data for turning on the seal and to interrogate the coded information on opening and closing events stored in the seal, and one to be used, for instance, by operators, permitting the interrogation of information on the integrity of the seal. The integrity control of the seal may be carried out very simply and quickly, namely by connection the adapter box with the seal and by successively indicating the 4 stored figures on the display of the adapter box. If these are set at zero no access took place during the inspection period. The adapter box has the size of a pocket calculator and consists of the following electronic components:

- calculator
- display
- keyboard
- BCD/octal binary decoder
- instruction logic
- crystal oscillator
- case

The simulation program is located in the office of the supervising office, permitting the decoding of coded data read out from the seal and, thus, determinating of the opening and closing events in terms of time.

3.2 VACOSS II

VACOSS II consists of only two components, the seal and the adapter box. The simulation program is no longer required here as the required calculations and decoding of data can be performed with the adapter box.

The seal:

The light guide and light control are designed in the same way as in version I. By installation of these electronic components it was possible to reduce considerably the size of the seal which measures 15.0 x 9.0 x

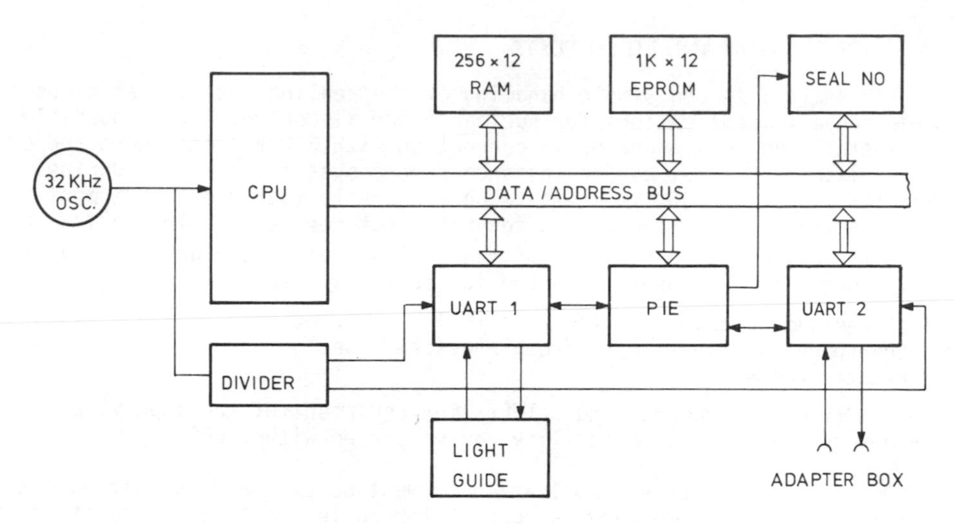

FIG.4. Electronic components of VACOSS 2.

5.0 cm^3. The lithium batteries guarantee an operating period of approx. 1 year. The electronic system of the seal is equipped with programmable ROM's and RAM's. Fig. 4 shows a block diagram of the seal's electronic components. In detail, the seal is suited to carry out the following functions:

- up to 2o opening and closing events can be stored in the seal
- it is possible to store a code-number containing the information on the integrity state of the seal. For this reason a variable number is changed pseudo-statistically at fixed time intervals. This pseudo-statistical series is modified in a reproducible manner on each opening and closing event of the light guide. This code-number can be read out and transmitted by the operator to the supervisory authority without risk of being falsified, so that the said authority is in a position to decide whether or not and how often the seal might have been opened in the course of the supervision period.
- an information about the operating conditions can be stored in the seal.

Adapter box:

The electronic components of the adapter box are housed in a pocket calculator box whose keyboard and display are utilized. The information stored in the seal can basically be read out with the adapter box. In this connection, the adapter box may be designed in different versions, depending on which information is to be read out and made available to the operator. The supervisory authority possesses an adapter box, permitting both read-in and read-out and, consequently, enabling the seal to be turned on. Contrary to VACOSS I, the closing and opening times are calculated directly in the adapter box. The variable code-number transmitted by phone can be decoded by the adapter box itself too.

5. APPLICATION AND FIELD TESTS

The small size and simple handling of the sealing system guarantees a wide range of applications for nuclear material control. It is possible to install the seal alone or in connection with other containment and/or surveillance systems, as for instance camera systems. However, during application of VACOSS, it seems to be reasonable that the containment to be sealed has similar access security features as the electronic seal itself. The following sections of the nuclear fuel cycle are particularly well suited for the use of electronic sealing systems:

- stores and containers with fresh fuel
- transport containers with fresh and spent fuel
- reactor vessels.

In addition, instruments and calibration equipment of the supervisory authority stationed in a facility can be sealed with VACOSS.

During the respective application it must be examined to what extent version I is to be employed as support for nuclear material accounting or version II for control by telephone. At the moment, VACOSS I is being used for the nuclear material safeguards system of the Juelich nuclear research center. Successful field tests with VACOSS I have been carried out particularly in connection with the transport of fissionable and/or radioactive material. It is planned to employ VACOSS within the frame of the control system for the HTR-reactor facility THTR-300.

6. CONCLUSION

The tests so far made with the VACOSS sealing system which was developed at the Juelich nuclear research center showed that the system can successfully be employed for effective nuclear material safeguards. The simple handling of VACOSS I can facilitate nuclear material accountancy. In this connection, VACOSS II may be employed as complementary measure especially for early detection of a diversion. The remote interrogation of both sealing systems is basically possible with additional peripheral systems.

REFERENCES

[1] KENNEPOHL, K., STEIN, G., VACOSS, Variable coding seal system for nuclear material control, Jül-1472 (1977).

[2] STEIN, G., KENNEPOHL, K., BÜKER, H., "The application of electronic sealing syster for nuclear material control", INMM-Meeting, Cincinnati, 1978, to be published.

CONTAINMENT AND SURVEILLANCE DEVICES

J.W. CAMPBELL, C.S. JOHNSON
Sandia Laboratories,
Albuquerque, New Mexico

L.R. STIEFF
Fiber Lock Corp.,
Kensington, Maryland,
United States of America

Abstract

CONTAINMENT AND SURVEILLANCE DEVICES.

The growing acceptance of containment and surveillance as a means to increase safeguards effectiveness has given an impetus to the development of improved surveillance and containment devices. Five recently developed devices are described. They include one photographic and two television surveillance systems and two high security seals that can be verified while installed.

The growing acceptance of containment and surveillance as a means to increase safeguards effectiveness and to optimize IAEA inspection efforts, has provided impetus to the development of improved containment and surveillance devices. Emphasis is now being placed on programs to demonstrate reliable equipment which can be conveniently used in the field by Agency inspectors. Several efforts in the United States to achieve this goal include the development of unattended photographic and television systems for surveillance, and high security seals for containment.

OPTICAL SURVEILLANCE SYSTEMS

Surveillance cameras are used by IAEA to provide information about plant operations and the handling of special nuclear material during an inspector's absence. Applications include the surveillance of spent fuel storage bays, access points to storage bays or controlled areas, enrichment plant feed and takeoff points and product loadout points of reprocessing plants.

Photographic Surveillance Camera

A prototype photographic camera system has been developed to provide the IAEA with several capabilities in an instrument of small size. The major subsystems of the secure surveillance camera are a single frame camera, timer, day and time display, tamper indicating features, and power supply. The camera photographs a scene and the day and time display when it is actuated by the programmable timer. The tamper-indicating features irrevocably record attempts at altering the operation of the camera or the recorded data.

FIG.1. Secure surveillance camera.

The system incorporates a Minolta XL-400 Super 8 movie camera, Figure 1, which is used in the single-frame mode. This camera has a through-the-lens automatic exposure control which enables it to be used in any reasonable existing light. The only modification made to the camera is the addition of a battery-charging circuit which keeps the two camera batteries charged.

The time base for the clock, calendar, and periodic and random interval timers is a crystal-controlled oscillator. The periodic timer circuitry triggers the camera at selectable intervals (10 to 90 minutes in 10-minute steps). The pseudo-random timer produces a pseudo-random time interval sequence that will not repeat for over a year. The minimum time interval is selectable (1, 2, 4, 8, or 16 minutes), and the maximum interval is 16 times the selected minimum.

The day and time are displayed on a 7-digit, light-emitting-diode display. The display is superimposed on the picture frame through a series of two mirrors and a lens. A 4-digit mechanical counter is used to indicate the number of frames exposed. The counter is visible through the front window of the tamper-indicating enclosure.

FIG.2. Advanced unattended television surveillance system.

For any unattended instrument, the validity of its recorded data must be assured. To provide this assurance, the camera and its associated electronics are housed within a tamper-indicating enclosure consisting of a mirrored glass cylinder, and an anodized aluminum end cover which are joined by a seal. The surveillance camera system includes an electrical power monitor which actuates an irreversible electromechanical counter also contained within the enclosure. Any attempt to remove or alter the electrical input voltage to the camera causes the counter to increment.

The power supply is the only part of the surveillance camera system which is outside of the tamper-indicating enclosure. In addition to converting 110 Vac to 16 Vdc, it includes batteries to provide power for 24 hours if the main power fails.

Advanced Television Surveillance System

An advanced television surveillance and recording system has been developed to provide long-term unattended surveillance of activities at nuclear facilities. The TV system for the long-term surveillance has some special features such as dual CCD cameras in tamper indicating/environmental housings;

tamper-detecting transmission lines; slave video recording unit; and video cassettes for the ease of tape threading. Access points to the cameras and recording console are secured by tamper-indicating seals. The television system is controlled by a microprocessor which permits various unique operational features to be incorporated into the system. The microprocessor and the remainder of the system utilize CMOS logic to reduce power and to extend system operational life when loss of main power requires operation in the battery-backup mode.

The mechanical configuration of the unattended television system is modular (Figure 2) and permits adapting the basic system to different surveillance requirements. The base for the system can be used as a support base or it can be filled with batteries to supply power for longer periods of operation without AC mains power. A master console contains all the electronic assemblies and also supplies a limited battery backup capability with rechargeable cells. The console contains two video recorders which can be operated individually or with one serving as a backup for the other. Both recorders can also be operated simultaneously to provide reliability through redundant recording. The slave recording unit contains a third recorder that utilizes a special locking video cassette. The video cassette can be removed from the slave unit without disturbing the system operation. Operation of the locking mechanism in the video cassette is controlled by the microprocessor when it receives a request from the inspector for access to the slave unit. The microprocessor will command the recorder to fast forward the cassette thereby locking up the cassette before access is permitted to the slave unit. Playback of the video information contained in the slave recorder cassette requires the drilling of special holes to remove locking pins in the cassette. The slave recorder and its cabinet are also optional to the system and may or may not be installed, as desired. The unattended television system uses two CCD cameras which are designed to operate at low voltages from the battery supply. The cameras will receive from 12 to 35 volts depending upon the length of the coaxial cable between the camera and the master console. This single cable is all that is required for installation of the CCD camera. A special multiplexing line supervision system protects the single coax against line tampering. The TV cameras are driven by special dual sync generators inside the master console that automati- cally detect failure in either of the generators and switchover to the operating generator if a failure occurs. An auto iris on the camera will adjust it for varying light levels. The camera which utilizes a unique design whereby the printed circuit boards are fitted around the lens assemblies can accept a wide variety of focal length lenses varying from 12.5 milli- meters to 75 millimeters.

The master console receives the incoming video and demultiplexes it to feed it to a video mixer where the two cameras are combined. Date, time and status information concerning tampers, battery operation or motion detection is added in the character generator. The output of the video mixer is sent to

FIG.3. Portable battery-operated television surveillance system.

the three recorders, the video analysis circuit, and the motion detect circuit. The motion detect circuit samples the video at designated points and determines that motion is occurring in the combined picture.

A video analysis circuit is used to determine if the recorders are recording the video signal and will detect a failure if either of the video signals or sync is absent. This circuitry and the other features provided by the CMOS 1802 microprocessor enables the system to develop self-diagnostics information. The results of the self-diagnostics are available to the Agency via an interface which will permit connecting the system to the remote monitoring units being developed for the Agency.

The microprocessor system also enables a number of the inspector's operational functions to be simplified. All functions are initiated via a keyboard next to the video monitor contained in the master console. The potential for human error has been reduced by appropriate software programming of the microprocessor to provide automatic operation, test sequences and fail-safe start of the systems.

Battery Operated Television System

A battery-operated portable television surveillance system has been developed to provide a surveillance capability to monitor special activities such as refueling of a LWR reactor. The system is packaged in two aluminum cases (Figure 3). One case consists of a video cassette tape recorder, a small television monitor, a battery charger, a sealed lead acid battery and the control circuitry for the system. The second case houses a tripod, a Charge Coupled Device (CCD) camera, associated cable, and extra cassettes.

The battery-operated system is designed to be set up and placed into operation in a minimum of time. The system provides an inspector with the capability of monitoring an activity at selectable time periods between one and 15 minutes (in one minute increments) for a surveillance duration of up to 24 hours. At each interval, the controller for the battery-operated system turns on the tape recorder and records approximately one second of video. The controller also places the time of occurrence of the recording and the day number into the video. The case containing the controller and recorder receives the video signal from the sealed tamper-indicating camera housing via a fiber optic cable The use of this type of cable makes attempts to tamper with the video quite difficult. If the power line to the camera housing is interrupted or shorted, the system will detect the loss of power and automatically record a tamper indication in the video. If the video cable is broken, then a second tamper-indication will appear in the video recording. Both tamper indications can be observed by the inspector during playback.

A special tamper-indicating circuit permits an inspector to detect if the recording case has been opened in his absence. The inspector, before closing the case, enters a code, selectable from one million different numbers into the controller. When he reopens the case, he enters the same number for comparison with the number stored when the case was last closed. If the numbers match, a green light appears. If he does not see a green light, he can be reasonably assured that someone has opened the case in his absence.

SEALS

Seals are often used to provide containment of special nuclear material or security for unattended instrumentation. The integrity and identity of the seals currently being used cannot be determined while the seal is installed. Two new seals, one passive and one active, have been developed to provide the Agency with this capability. Both seals utilize fiber optic bundles as the sealing "wire." Fiber optic seals are a new class of high security, tamper-resistant/indicating seals whose integrity and unique identity can be established in the field without removal or disassembly. In addition, this type of seal has the capability of being both continuously and remotely monitored.

Passive Fiber Optic Seal

The unique identity or fingerprint of the passive seal is established at the time the seal is assembled in the field by

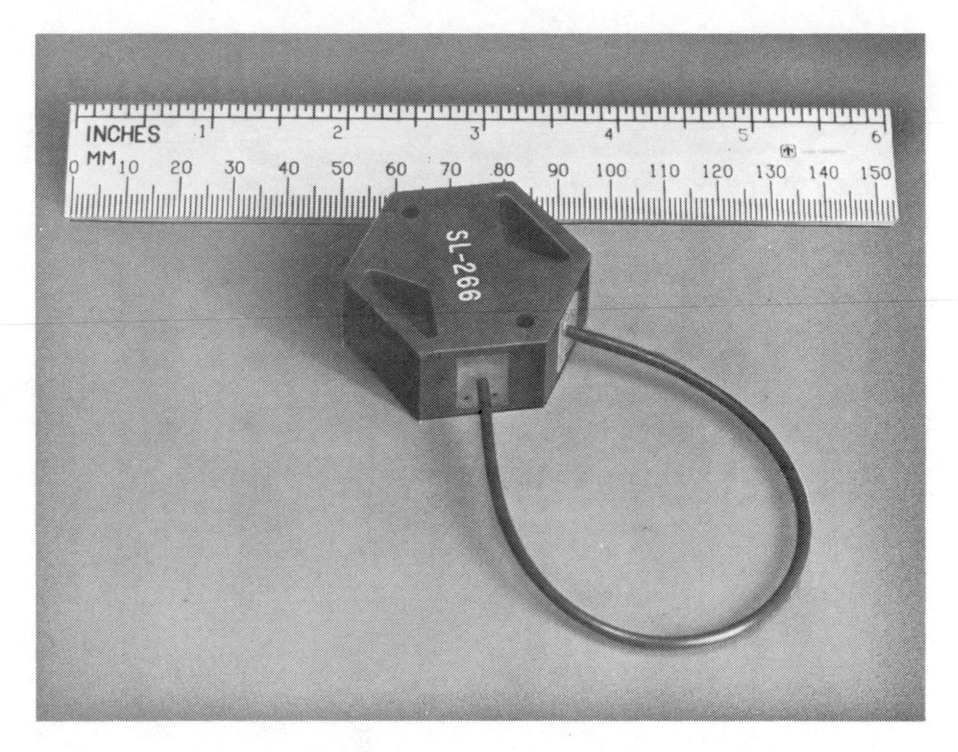

FIG.4. Passive fiber optic seal.

recording the random positions of the ends of glass optical fibers which make up the sealing loop. This recording can be accomplished either photographically or by noting the coordinates of a small subset of fibers. A direct comparison of a negative taken at the time of the seal assembly with a positive print taken at a later date (when the integrity of the seal is being checked) provides the highest level of confidence that a seal has not been compromised. For less demanding situations, a comparison of the coordinates or relative positions of a small number of fibers should be satisfactory.

In principle, the high level of security which is offered by fiber optic sealing devices depends upon the unique fingerprint which is generated during assembly by the completely random pattern of the 225 ends of the 0.06mm diameter optical fibers in the bundle. The uniqueness of this fingerprint is further enhanced by the imperfections in shape and optical characteristics of the individual fibers in the bundle. This fingerprint will be destroyed during any seal disassembly, complete withdrawal of one end of the fiber bundle from the seal assembly block or severance of the bundle. Duplicating this unique fingerprint would be a formidable task should the original seal be reassembled or replaced with a substitute.

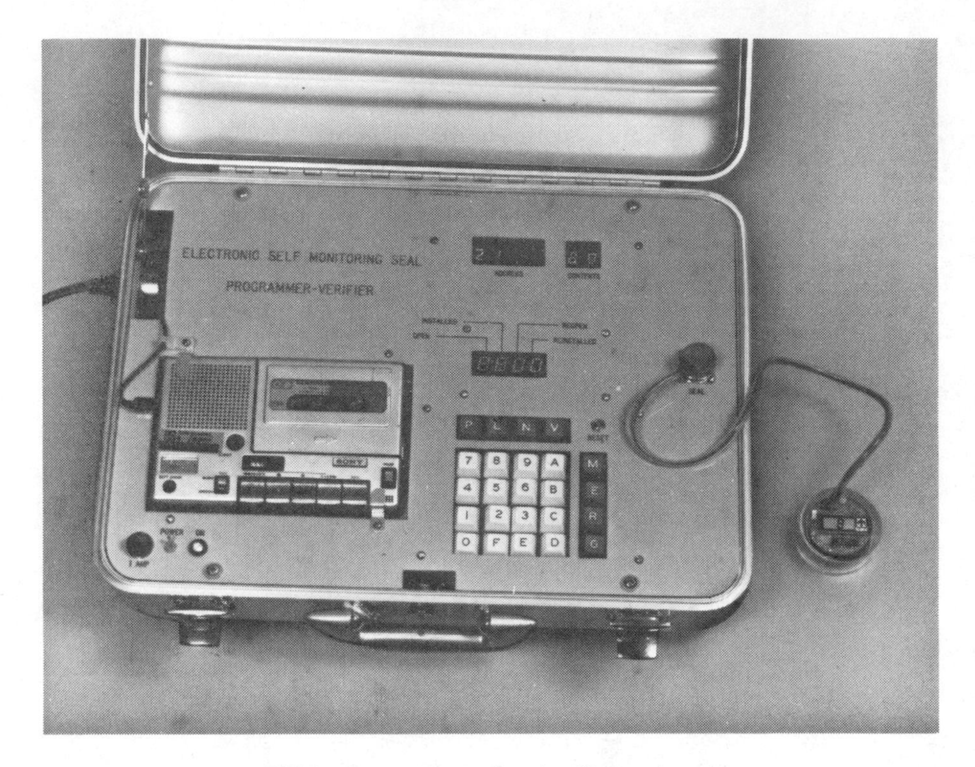

FIG.5. Programmer and active fiber optic seal.

The passive fiber optic sealing system consists of four elements; the seal (Figure 4), assembly tool, hand-held microscope and instant print camera. The fiber optic seal employs a plastic, hexagonal shape assembly block and a polyvinyl-jacketed bundle of glass fibers. The block holds the two ends of the bundle firmly in place and insures the complete mutual interpenetration of the fibers during the assembly procedure. During manufacture, one end of the fiber bundle is permanently emplaced in the assembly block. At the time the seal is completed in the field, the free end of the fiber optic bundle is inserted into the assembly block and secured with the aid of an assembly tool. The unique fingerprint of the seal is then immediately documented by either visual observation or by photographic means using the hand-held microscope and special instant print camera. During subsequent inspections, verification of the fingerprint is accomplished by comparing the seal fingerprint with either the visual observation or the original photomicrograph.

Active Fiber Optic Seal

The active fiber optic seal is a security seal that continuously monitors the integrity of a fiber optics loop and displays the status, opened or closed, in a simple manner. The status of the seal can be identified by observing the seal's optical display, as shown in Figure 5. The observation can be made by a representative of the inspectorate that installed the seal or by a representative of the operator within whose facility the seal is installed (with the observation reported to the inspectorate).

As a security seal, the active (or self-monitoring) seal provides several unique capabilities: High security, field verification while installed, remote verification, time resolution of integrity, and reusability. This seal is intended for use in applications that require one or preferably more of these features. The sealing of containers for large quantities of strategically or economically valuable materials is one potential application. Unattended instrumentation used to monitor such material may also require the use of a seal with these features to assure the validity of the data collected.

Each seal is programmed by a special piece of equipment to display unique sequences of different numbers and letters. These sequences of displays provide the identity of the seal. For each seal, the display will change at preset intervals, once every 1, 2, 4, 8, 16, or 32 hours. Unlike other seals that provide their complete identity at all points in time, the information that identifies this seal is distributed through time.

Each seal can be reused by reprogramming the display generator with one of more than two million unique sequences and by installing fresh batteries, if necessary. The batteries contained within the seal are sufficient to operate the seal for six months over a temperature range of 0°C to 50°C.

This seal consists of two major parts: a fiber optics loop and the electronic monitor module that verifies the loop's integrity. When a seal is in use, the loop integrity sensor transmits light pulses into one end of the fiber optics loop. If the pulses reentering the electronics package from the other end of the loop do not correspond to the pulses transmitted, the display generator indicates a violation by changing the sequence of displays produced after that time. The module utilizes a custom designed large-scale integrated circuit to control the generation of display sequences. The integrated circuit and batteries are enclosed in a tamper-responding container. Any attempt to gain access through the container to the integrated circuit results in the interruption of electrical power to the circuit. Since the programming information is stored in a volatile form, loss of electrical energy causes loss of this information. Correct display sequences cannot be reported after this has occurred.

The normal operational cycle for the active fiber optics seal is as follows: (1) the display generator is programmed and started at the Agency's headquarters just prior to the module's deployment to the host's facility; (2) an inspector attaches the module to the fiber optic loop on the item to be sealed; (3) the facility host reads and records the display at the intervals requested by the Agency and reports the information at times selected by the Agency; (4) during each visit by an inspector to the host's facility, the seal's point of application and the seal's integrity (correct display value) are verified. (5) during these visits, the electronic modules approaching the end of their operational phase, either due to battery life or number of display changes, are replaced with modules that have been reprogrammed for future use.

The five devices described above are part of a larger program to develop and make prototype surveillance and containment equipment available for evaluation by the Agency. The results of the Agency's evaluation will determine what additional development or design changes may be necessary before initiating production or developing a production capability by commercial suppliers.

DEVELOPMENT OF CONTAINMENT AND SURVEILLANCE MEASURES FOR IAEA SAFEGUARDS

Y. KONNOV, S. SANATANI
Department of Safeguards,
International Atomic Energy Agency, Vienna

Abstract

DEVELOPMENT OF CONTAINMENT AND SURVEILLANCE MEASURES FOR IAEA SAFEGUARDS.

Containment and surveillance are important measures which complement material accountancy in serving the main safeguards objective: timely detection of diversion of significant quantities of nuclear material. Surveillance, in the context of IAEA Safeguards, is defined as instrumental or human observation to indicate or detect the movement of nuclear material. The paper describes surveillance instruments and devices which are intended to indicate that either no nuclear material has left a certain location or that it has left only via legitimate routes. Surveillance instruments and devices can be classified into three categories: sealing systems; camera and television systems; and monitors. Taking advantage of the development and improvement of instruments and techniques that are a continuous process in the IAEA supported by Member States, the IAEA has accumulated considerable field experience in the routine use of surveillance equipment. The paper briefly reviews the present status of instrumental surveillance techniques, which are being used by the IAEA or are under development and seem promising for safeguards use. The results of the first Advisory Group Meeting, held in Vienna in June 1978, on Development of Containment/Surveillance Measures and Surveillance Instruments and Techniques for IAEA Safeguards, are briefly discussed. Some of the recommendations of the Advisory Group and Agency's future plans conclude the paper.

INTRODUCTION

As used in international safeguards, surveillance means observation and use of instruments to indicate or detect the movement of nuclear materials. Surveillance instruments and devices as used for international safeguards are intended to provide assurance that either no nuclear material has left a certain location or that it has left only via legitimate routes.

Containment measures are used by plant operators for a number of reasons, e.g. physical protection of material, safety of personnel, convenience of operational procedures. In general, such measures are not provided specifically for international safeguards purposes, but their existence in a facility often simplifies the application of surveillance devices by the IAEA.

Surveillance and containment are important complementary measures for material accountancy, which provide information on the movement of, or access to, nuclear material in the absence of inspectors, in order to preserve the integrity of prior measurement of nuclear material by the IAEA and to provide knowledge of material flow at important points in the fuel cycle. The objectives of containment/surveillance (C/S) measures with respect to IAEA safeguards are to provide information concerning nuclear material movement and integrity of items, to monitor activities independently that could have relevance to safeguards, to improve IAEA inspectors' efficiency without increasing their personal involvement and to complement measures of material accountancy. In this respect it is useful to bear in mind the difference in purposes for which C/S measures are used by an international safeguarding authority on the one hand, and by a Member State or plant operator on the other. An international organization like the IAEA is primarily interested in preventing diversion of nuclear material for production of explosive devices or for unknown purposes, "by the risk of early detection",[1] whereas the plant operator or Member State may use the same or similar techniques as those used by the IAEA for reasons besides international safeguards, e.g. for physical protection of material and facilities, safety of personnel and processing control.

During the past few years, the increasing number of nuclear facilities coming under IAEA safeguards has led to an increasing deployment of instruments and devices which were found to be routinely usable by the IAEA. This situation of rapid growth has enhanced the pressure on the development of new systems which are reliable, effective and economic in cost. The IAEA, with the collaboration of Member States, and on the basis of experience gained by its own Inspectorate, is actively engaged in the development, procurement and field testing of various surveillance systems and devices to meet its immediate and growing future needs.

A recent activity of the IAEA's Division of Development and Technical Support in this area was the convening of an Advisory Group Meeting on Containment Surveillance Measures and Surveillance Instrumentation and Techniques for IAEA Safeguards, in Vienna from 26 to 30 June 1978. Seventeen experts from twelve Member States and Euratom attended the meeting and made detailed proposals on the development and application of practical containment and surveillance measures which will be useful for safeguarding different types of facility.

---

[1] INTERNATIONAL ATOMIC ENERGY AGENCY, INFCIRC/153 (Corrected), IAEA, Vienna (1972) para. 28.

1. CLASSIFICATION OF SURVEILLANCE TECHNIQUES

For the purpose of a systematic discussion we can classify surveillance instruments and devices into three categories:

Sealing systems,
Camera and television systems, and
Monitors.

Seals are devices used to verify the integrity of a containment. For a seal to be effective, the containment must be chosen so that access, other than through the sealed opening, is difficult and can readily be detected. Also the sealing wire or cable should be immune against surreptitious cutting and rejoining. A seal can also be used to identify uniquely an item such as a fuel element. If a seal can be checked on site, without opening it or, even better, if it can be remotely monitored for its status, a great saving in inspection time and effort can be achieved. For this reason current effort is directed towards the development of such sealing systems. Metallic seals, now routinely used by the IAEA, do not have these features; they have to be individually checked at Headquarters to rule out substitution or tampering.

Various types of still- and single-frame movie cameras are routinely used by the IAEA for the surveillance of strategic areas in a nuclear facility. Among the optical surveillance systems now in use, Super 8-mm movie cameras with specially fitted timers have proved to be most suitable in many cases. The IAEA possesses a number of video recording TV systems with remotely controlled cameras. These have distinct advantages over other optical surveillance systems (e.g. the possibility of taking pictures in poor light and in radiation fields, large picture capacity, the possibility to replay tape and to check surveillance records immediately on site, etc.) and are in use in many countries. The IAEA is vigorously pursuing the further improvement and development of such systems.

Various types of monitor are now in limited use and are being developed for the IAEA. These include instruments which observe or keep track of processes or operations involving nuclear materials, monitors which operate unattended for long periods and indicate the movement or lack of movement of nuclear material, and radiation detectors which measure roughly the radiation level in the area.

In Section 2 a brief discussion of a selection of the above types of surveillance instrument and device is presented [1]. These are either in use by the IAEA or are being developed and tested and look promising for practical use.

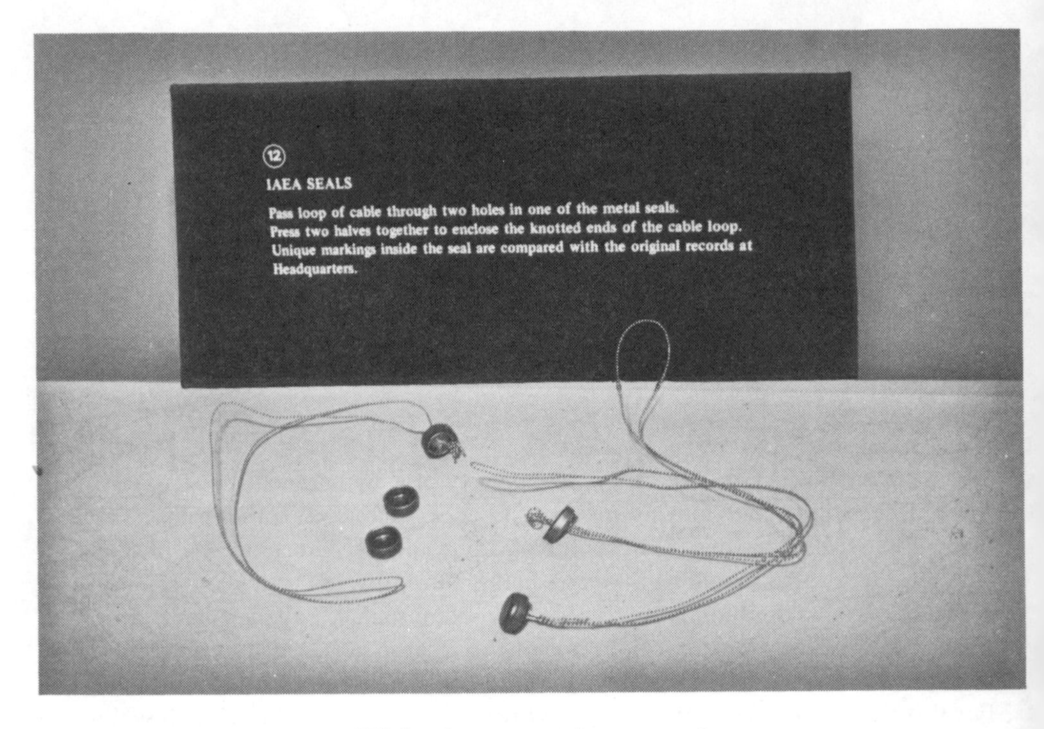

FIG.1. Agency general-purpose seal.

2. REVIEW OF SURVEILLANCE INSTRUMENTS AND DEVICES

2.1. Seals

The most commonly used seals by the IAEA today are the IAEA general-purpose seals (Fig. 1), which have now been in regular use for a number of years and have served their purpose well. Because of the extreme simplicity in their installation and the well-developed and reliable method of their re-identification at Headquarters, these seals are not likely to be dropped from IAEA use in the very near future. However, an improved version, which has increased resistance to tampering as well as corrosion, but which is equally easy to apply, is currently being developed in a Member State. Lack of in-situ verification possibility is the major limitation of these seals.

A type of pressure-sensitive paper seal label has recently been developed. These adhesive strips [3], once attached, are almost impossible to remove without tearing; and the use of solvents for removal and re-attachment, without leaving marks, can also be ruled out. However, substitution of a similar paper seal with

FIG.2. Fiber optic sealing system.

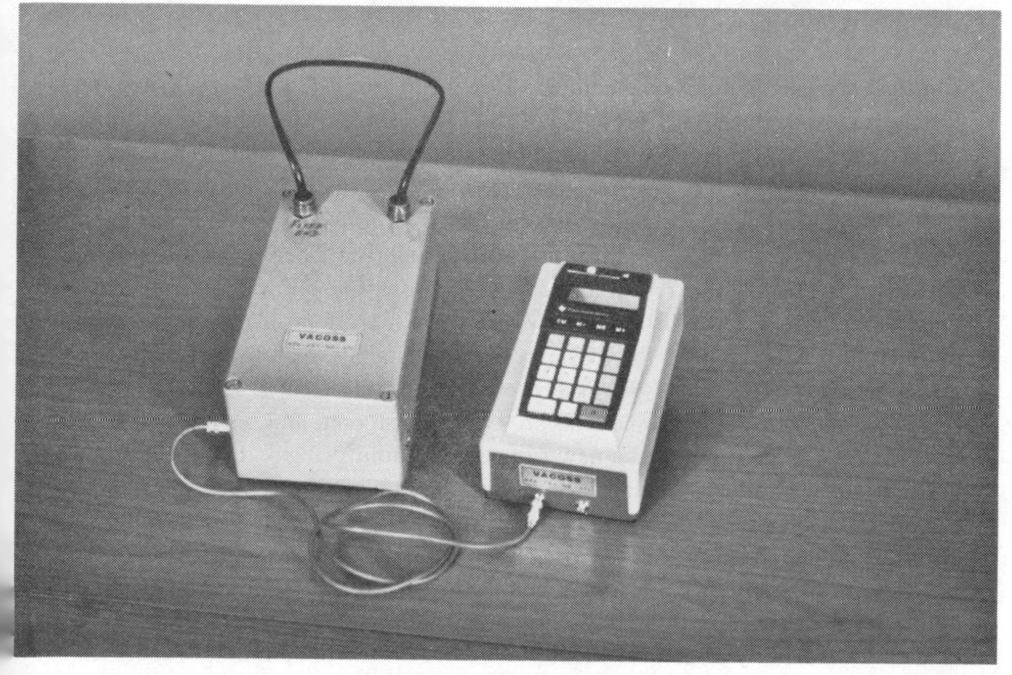

FIG.3. VACOSS electronic seal.

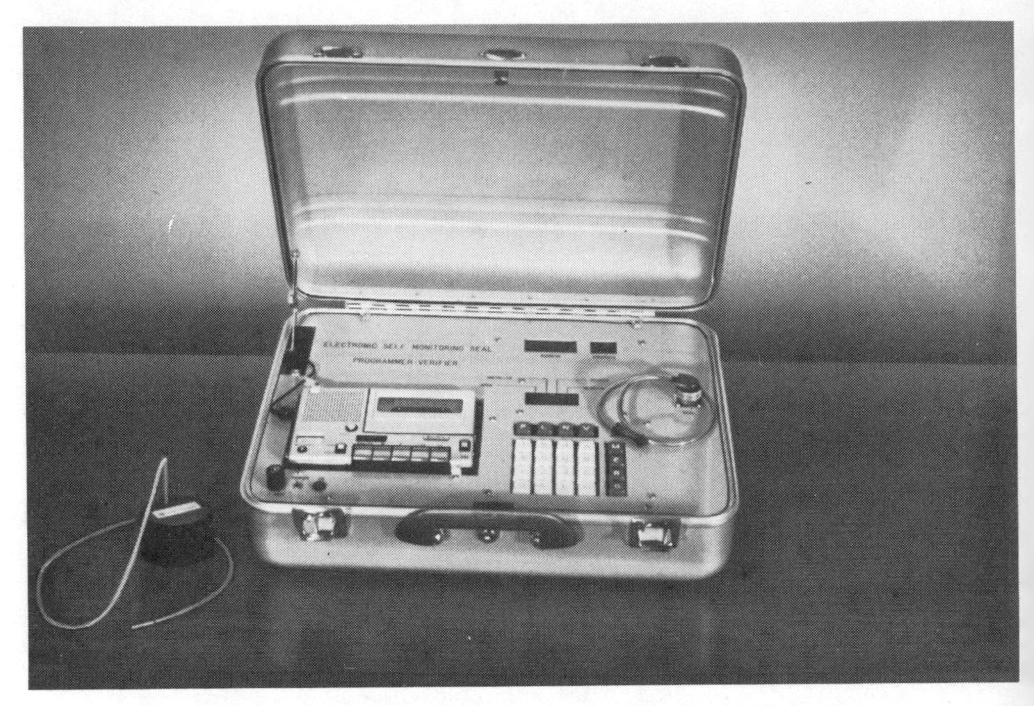

FIG.4. Sandia self-monitoring seal and programmer/verifier unit.

identical markings cannot be excluded. For this reason the IAEA uses these seals for short periods during the presence of the inspector on site, when conditions for attaching the seal labels obtain and when the use of these seals facilitates work of the inspector.

After a prolonged evolution through many versions, we now have a new model of a fiber-optic seal (Fig. 2), which looks promising [4]. The unique fiber-optic pattern in the seal can be repeatedly checked, visually or photographically, in situ. Work is under way to build a remote verification feature into these seals. The main limitation is the need for special tools for installation and verification which are somewhat bulky. The IAEA is at present field-testing these seals.

Two versions of electronic seals are being considered and tested by the IAEA – the VACOSS (Fig. 3) [5], and a new type of self-monitoring seal (Fig. 4) [6]. Both use fiber-optic bundle as sealing wire, but different electronics packages to produce the digital display which is an indication of the status (opened, closed, re-opened, etc.) of the seal. Because of the completely unpredictable (except by one in possession of the initially injected coded message into the seal and a microprocessor/computer with appropriate soft-ware) sequence of displays, the seals can be remotely verified by asking the facility operator on the telephone to report the displays at specific times and checking them against results obtained in Headquarters

FIG.5. Two Super 8-mm camera surveillance systems.

The IAEA has limited experience of seals based on ultrasonics. Various versions have been developed and successfully tested by Member States. Ultrasonic seals may also be used for item identification and their use for sealing under water trays of spent fuel looks promising [7].

2.2. Electro-optical surveillance systems

Among various photographic camera systems, the Super 8-mm movie camera with a single-frame picture-taking feature has found widespread use by the IAEA Inspectorate. The most satisfactory system are the two independently working cameras, each with a special timer, mounted in a single enclosure (Fig. 5).

Both movie and still photographic camera systems have certain limitations for prolonged use in nuclear facilities and these are fully discussed in Ref. [2]. For example, some of these limitations are

Limited picture capacity (Minolta camera 7200 frames)
Necessity to process the film to see the record
Inability to take good pictures in poor light
Fogging of pictures owing to overexposure to radiation
No recording of date and time on picture frames in most models
Occasional failure under extreme environmental conditions.

FIG.6. Closed-circuit television surveillance system.

Work is under way to develop camera surveillance systems that seem best for meeting the IAEA's needs.

To overcome the limitations of film cameras, television cameras with video-magnetic recording were developed (Fig. 6) [8]. In general a TV surveillance system consists of a camera remotely operated by a control unit. The components of the control unit (video recorder, triggering device, video calendar/clock, emergency power supply, etc.) are built into a sealed housing. The principal advantages of the TV surveillance system are

It has very large frame capacity
No film processing is required
The recorded information can be obtained without entering a contaminated area
The recorded information is always ready for evaluation on the spot or at Headquarters
It incorporates on each picture the date and time when it was recorded
It is more resistant to radiation than film cameras
It has the ability to record events under a wide range of lighting conditions and the use of infra-red illumination promises to provide the capability of recording events when there is no visible light.

The limitations of the TV systems are that they are costly, bulky and sophisticated, needing maintenance by experts.

An interesting new development project started by the IAEA in connection with the TV surveillance system is based on the sophisticated technology of freeze-picture transmission over long distances using commercial telephone lines. The technique would enable one to check remotely the operational status of a TV surveillance unit installed many miles away, and thus considerably save the time spent by an inspector for travelling. For this reason, the importance of developing this capability was stressed by the Advisory Group on Containment/Surveillance at its recent meeting.

2.3. Monitors

Efforts have been made in the past to develop suitable monitors to count, in a tamper-proof way, the number of irradiated fuel elements discharged from the core of a reactor (usually on-load refuelled type) during a given period of time. Some demonstrations of such monitors [9] have been reported.

No fuel counters are yet used routinely by the IAEA, but development and testing of such monitors, now under way in Member States, are considered to be an important activity for improving IAEA safeguards of on-load fuelled reactors.

FIG.7. Track-etch reactor power monitor.

With the prospect of safeguarding large stores of spent fuel under water, the need arises of developing a technique of rapidly and reliably verifying whether the submerged items are indeed spent fuel elements or dummies. Some equipment has already been developed for this purpose and is undergoing tests in a Member State [10].

Monitoring the thermal or electrical power of a reactor over a period of time independently would provide the IAEA a means to check operator's records and estimate roughly the plutonium produced in the discharged fuel. A device based on track-etch technique was developed some years ago [11] and the IAEA has several monitors of this type in its possession (Fig. 7). The IAEA has also now acquired the equipment and capability to process and read out exposed tapes in its Seibersdorf laboratories. A re-evaluation of this technique and careful field testing of the monitors are under way. An alternative system, based on sensitive neutron detectors (e.g. ^{3}He) and conventional electronics, has been considered and a prototype instrument has been built and successfully tested in a Member State [12].

Passive gamma/neutron detectors are used to detect the removal of nuclear material through small openings or ports. The IAEA has gathered satisfactory experience in the use of radio photoluminescent fluoroglass dose meters as yes/no monitors in on-load fuelled reactor facilities [13].

3. FUTURE PLANS AND ADVISORY GROUP RECOMMENDATIONS

The IAEA has, over the past few years, gained considerable experience in the use of surveillance equipment in the field. It has been found that different systems are suitable in different situations. Above all, reliable systems are sought because failure of an instrument might in some cases imply the need to take the inventory again. In the special circumstances of international safeguards, where the surveillance equipment is supposed to work unattended for long periods in the sites of installation, the need of instrument reliability cannot be over-emphasized. In addition, we have to consider problems of acceptance by the operator, the cost of equipment, ease of servicing, interpretation of surveillance records, tamper-resistance, safety features and the inherent limitation of each instrument or device.

The Advisory Group on Development of Containment/Surveillance Measures which met in June 1978 [2], was the first meeting of this kind on containment and surveillance. The Group noted that out of a long list of instruments discussed in the meeting, only a small number was in routine use by the IAEA. There is need to seek improvement of these and also to complete the development and testing of other promising techniques so that these could also be incorporated into the list of routinely usable techniques in the near future.

The Advisory Group also noted that successful application of C/S measures requires close co-operation between the facility operator and the IAEA. Goodwill on the part of the operator would facilitate an efficient and effective surveillance system. Notwithstanding this fact, the Group recommended that legal implications of using C/S measures should be adequately considered in addition to technical questions, e.g. reliability, tamper-proofness, safety.

The Advisory Group pointed out that broadly based system studies are needed to develop and evaluate new concepts and approaches to plant layout and to improve existing designs of facilities for safeguards applications. It also recommended that the IAEA should substantially strengthen its efforts in developing surveillance and containment measures in order to cope with the growing needs.

Various other points of interest were discussed by the Group and numerous recommendations were formulated. Because of limitations in manpower and resources that can be allotted by the IAEA for developing containment/surveillance measures, it may not be possible to fulfil many of the recommendations of the Advisory Group, unless substantial co-operation is received from Member States.

4. CONCLUSION

The importance of containment and surveillance measures in the application of IAEA safeguards is growing continually. In some situations, the role of C/S may exceed their originally prescribed complementary role to material accountancy. There is a need to improve cost-effectiveness and reliability of currently used instruments and devices and to complete the development and testing of promising sealing systems, camera and TV surveillance units. Similarly, monitors beneficial for safeguards application should be carefully tested under field conditions so that these can also be routinely used by the IAEA if found suitable. The IAEA appreciates the support it has received so far and continues to receive from Member States in its efforts towards developing better surveillance instrumentation in order to strengthen IAEA safeguards.

REFERENCES

[1] DRAGNEV T.N., et al., "Some Agency contributions to the development of instrumental techniques in safeguards", Safeguarding Nuclear Materials (Proc. Symp. Vienna 1975) **2**, IAEA, Vienna (1976) 37; IAEA Safeguards Technical Manual, Part E (1975) Chap. 10.

[2] INTERNATIONAL ATOMIC ENERGY AGENCY, Report of Advisory Group Meeting on Development of Containment/Surveillance Measures and Surveillance Instruments and Techniques for IAEA Safeguards, AG-190 (1978).

[3] INTERNATIONAL ATOMIC ENERGY AGENCY, Instructions for the Use of Pressure Sensitive Seal Labels, IMI No. 27, IAEA, Vienna (1977).

[4] Mark 1 Fiber Optic Sealing System, Operations Manual, Fiber-Lock Corporation (1978).

[5] KENNEPOHL, K., STEIN, G., VACOSS variable coding seal system for nuclear material control, Jül-1472 (1977).

[6] Self-monitoring seal, Operations Manual, Sandia Laboratories (1978).

[7] CRUTZEN, S., JEHENSON, P., "Ultrasonically identified seals for safeguards and physical protection purposes", INMM Meeting, Cincinnati, June 1978. To be published.

[8] INTERNATIONAL ATOMIC ENERGY AGENCY, IAEA Bull. **19** 5 (1977).

[9] SINDEN, D.B., et al., "Testing of techniques for the surveillance of spent fuel flow and reactor power at Pickering Generating Station", Safeguarding Nuclear Materials (Proc. Symp. Vienna 1975) **2**, IAEA, Vienna (1976) 279.

[10] STIRLING, A.J., ALLEN, V.H., "Application of safeguards design principles to the spent-fuel bundle counters for 600-MW CANDU reactors", IAEA-SM-231/38, these Proceedings, Vol. I.

[11] GINGRICH, J.E., et al., Development of nuclear reactor and fuel monitors for unattended safeguards applications, NE DG-12463, General Electric Co. (1973).

[12] DOWDY, E., et al., An operator independent reactor power monitor, Los Alamos Rep. LA-UR-78 1727.

[13] SCHAER, R., "Health dosimeter plays an unusual role as a safeguards device", IAEA Bull. **19** 2 (1977).

DISCUSSION

D.F. RAWSON: My question concerns surveillance devices installed by plant management. In a plutonium or highly enriched uranium fabrication facility there are times when maintenance operations may allow real access to material. If, at such a time, a scram alarm is sounded (for contamination or criticality) all of the staff will evacuate immediately. On return to the plant, the resident inspector may either demand a physical inventory, which could be expensive and take several days (even when such an unplanned inventory could be effective) or accept the evidence of operator-installed devices to show that no diversion has taken place. I should therefore like to know whether the inspector would accept evidence from such operator-installed devices or only that from IAEA-installed equipment.

S. SANATANI: Situations arising out of unusual occurrences such as false alarms, emergencies or instrument failures are being actively considered by the IAEA. The need for clarification with the facility operator as well as with the State concerned about what the inspectors should do in such situations was pointed out at the recent meeting of the Advisory Group. Depending on the particular situation, if containment surveillance measures fail, or accidents occur, the physical inventory may have to be re-established.

Session V (Part 2)

DESTRUCTIVE AND NON-DESTRUCTIVE MEASUREMENT TECHNOLOGY

Chairman: A.G. HAMLIN (United Kingdom)

Rapporteur summary: *Mass-spectrometric analytical techniques*

Papers IAEA-SM-231/6, 108 and 122 were presented by
J.G. VAN RAAPHORST as Rapporteur

PLUTONIUM ACCOUNTANCY IN REPROCESSING PLANTS BY CERIC OXIDATION, FERROUS REDUCTION AND DICHROMATE TITRATION
A Novel Method

A. MACDONALD, D.J. SAVAGE
UKAEA, Dounreay Nuclear Establishment,
Thurso, Caithness, Scotland
United Kingdom

Abstract

PLUTONIUM ACCOUNTANCY IN REPROCESSING PLANTS BY CERIC OXIDATION, FERROUS REDUCTION AND DICHROMATE TITRATION – A NOVEL METHOD.

The performance of some existing titrimetric methods for plutonium estimation is reviewed in the context of plutonium accountancy in nitric acid solutions of irradiated mixed plutonium-uranium oxide fast reactor fuels. A novel titrimetric procedure is described in which plutonium is oxidized to plutonium VI by cerium IV in nitric acid solution, the excess oxidant destroyed chemically, and plutonium VI reduced by a measured excess of iron II which is back-titrated with potassium dichromate. The procedure is suitable for use in both glove-boxes and remote-handling facilities. The precision and accuracy of the procedure are described, both for pure plutonium solutions, and for plutonium in the presence of simulated fission products representing highly burned-up fuels.

1. INTRODUCTION

Reprocessing of irradiated plutonium based fuels is an integral part of the Fast Breeder Reactor concept. During reprocessing it is essential to obtain fissile material mass balance data on the plant throughput, both for criticality control purposes and for International Safeguards Control. In such cases, main plutonium streams must be analysed for plutonium content with a precision of about 0.1% and a bias of less than 0.1%. Several plutonium analysis methods can meet these requirements for pure, relatively concentrated plutonium solutions; the problem is more severe when the plutonium is not pure.

One of the principal accountancy sample points is the irradiated fuel solution fed to the reprocessing plant. This presents additional analysis problems from two sources.

a. The high $\beta\gamma$ radiation levels associated with the fission products necessitate remote manipulation in a shielded cell during the plutonium analysis.

b. Fuel constituentsother than plutonium, including uranium and fission products are capable of introducing bias into otherwise acceptable techniques.

In cell manipulation must be kept as simple as possible if adequate precision is to be achieved. Additionally the equipment used must be readily maintained or replaced.

A chemical separation is normally used to overcome interference problems, using ion exchange or solvent extraction (1)(2)(3) to remove the interfering species or to separate the plutonium. However both techniques increase the amount of in-cell manipulation and introduce the risks of plutonium loss and of contamination pick up, with consequent loss of precision and introduction of bias.

The ideal solution would be a technique which can be applied directly to irradiated fuel solutions without separative pretreatment. This paper considers the interference limitations of some typical plutonium analysis procedures, and shows that by tighter selection and control of the redox conditions used it is possible to develop a precise method that retains its accuracy when applied to solutions of irradiated fuel.

2. SELECTION OF PLUTONIUM TITRATION PROCEDURE

Practical limitations on sample size and the precision requirement of accountancy analysis rule out most techniques other than weight titration or coulometry, using weight aliquots of sample. These methods rely on chemical or electrochemical redox reactions, using either the conversion of plutonium III to plutonium IV, or of plutonium VI to plutonium IV.

Reduction of plutonium to plutonium III with titanium III, and subsequent titration of plutonium III to IV with dichromate (2)(4) is subject to interference from uranium, copper, iron, nitrate and several fission products. Modifications using copper I in chloride solution as reductant, and a phosphate/sulphate complexing medium for the titration (5) adjust the redox condition sufficiently to remove most of these interferences. The method is still subject to interference from technetium and iodine when applied directly to solutions of irradiated fuel. The related coulometric technique relying on the conversion of plutonium III to IV (1) (4) is subject to interference from either iron or the presence of Plutonium VI, and probably some fission products.

Methods based on the reduction of plutonium VI to IV are generally subject to interference from ruthenium, cerium, chromium and manganese (1)(2)(4)(6) (7) but tolerate iron and uranium.

The interference is caused if foreign ions are oxidised during the formation of plutonium VI, and subsequently reduced by the iron II. The usual oxidant employed is silver II oxide, a less vigorous oxidant would be adequate for plutonium oxidation, and would be less liable to oxidise other species. Preliminary trials demonstrated that cerium IV is a potentially suitable oxidant in dilute nitric acid medium and approximately quantitative oxidation took place in a reasonable time.

The excess oxidant must then be destroyed before subsequent titration of the plutonium. Three possible procedures have been investigated.

a. Titration with iron II. This entails locating an additional end point, with a possible loss of precision. This procedure is successful with reagent blanks but fails with plutonium containing solutions due to premature reduction of some plutonium VI.

b. Reduction with oxalate: The reaction between cerium IV and oxalate is known to be slow (8). However, some excess oxalate can be added as the reduction of plutonium VI by oxalate is very slow even with heating (9). Initial trials of this technique were not successful.

c. Reduction with arsenite. This reaction was used by Hedrick (10) to overcome the interference from up to 0.2 mg of cerium in a version of the silver II plutonium method. In the present work, initial trials suggested that arsenite can be used to destroy up to 200 mg of cerium IV and this approach was selected for further study.

After removal of the excess oxidant, standard techniques for the titration of plutonium can be selected to give maximum precision. We have added excess iron II, and back titrated with standard potassium dichromate using amperometry for end point location.

3. EXPERIMENTAL

3.1 Recommended Procedure

1. Add an aliquot of sample containg 5 - 60 mg plutonium and up to 15 milli-equivalents of free acidity to a 100 ml beaker.

2. Add 1 ml of 1M ammonium hexanitratocerate solution and wash the beaker walls with 50 ml of 1M nitric acid/0.1M sulphamic acid. Stir gently for 5 minutes.

3. Add 1 ml of 0.24M ferric nitrate in 1M nitric acid solution and 0.1 ml of 0.25% osmium tetroxide in 10% sulphuric acid solution.

4. Add 1M sodium arsenite solution dropwise until the cerium IV colour disappears then add 1 - 2 drops in excess.

5. Add 0.2N potassium permanganate solution dropwise until a stable pink colour is formed.

6. Add 0.1M oxalic acid solution until the permanganate colour is destroyed then add 1 - 2 drops in excess.

7. Add 15 ml 2N H_2SO_4 and sufficient ammonium ferrous sulphate solution to reduce the plutonium VI to plutonium IV and leave a suitable excess for back-titration. Wait 2 - 3 minutes.

8. Titrate the excess iron II with standard potassium dichromate solution.

3.2 Experimental Details

To obtain maximum precision in this work, the plutonium solutions were aliquoted by weight, using polythene ampoules. The titration was also performed by weight, using both 0.05N and 0.005N titrant for convenience.

The end point of the iron II, chromium VI titration was located amperometrically, using a 200 mV potential between a pair of 1 cm^2 gold electrodes about 1 cm apart. This end point may also be detected by measuring the potential across the gold electrodes when polarised by a heavily stabilised 1 μA current.

TABLE I

CERIUM IV OXIDATION OF PLUTONIUM

Molar Ratio Cerium IV / Plutonium	Acidity	Oxidation Time (minutes)	Plutonium measured (%)
2.3	1.0	5	97.7
2.3	1.0	15	99.42
2.5	1.0	3	98.3
2.5	1.0	5	99.82
2.8	1.0	5	99.95
5.5	1.0	5	99.95
3.6	1.3	5	99.27
4.2	1.3	5	99.70
2.6	1.5	5	94.8
4.0	1.5	5	99.36
4.0	1.5	15	100.43
4.9	1.5	5	100.04

The reduction of the cerium IV excess with arsenic III, manganese VII and oxalic acid may be followed visually or amperometrically.

The chromium VI solutions used in this work were prepared by weight from potassium dichromate as primary standard. For the performance trials NBS standard potassium dichromate was used.

The iron II solution was prepared from ammonium ferrous sulphate and was standardised twice daily by titration with chromium VI.

Most of the development work was carried out using plutonium control solutions containing 10 to 12 mg/ml of (mainly) plutonium IV in about 2N nitric acid. Their plutonium contents were cross checked using a plutonium III/IV titrimetric procedure (5) and agreed within the experimental errors. Plutonium recoveries of optimisation experiments have been expressed relative to the mean concentration found by repeat determinations using the recommended procedure.

The overall performance of the recommended procedure was checked by using a solution of NBS standard plutonium metal in dilute HCl, after removal of the chloride by repeated evaporation with nitric acid.

The fission product simulate solution used contained all the fission products expected from a fast reactor irradiation. Most were added as nitrates in nitric acid solution. Exceptions to this included tin, antimony and rhodium (added as chlorides), ruthenium (nitrosyl - nitrate complexes) tellurium (tellurate) iodine (iodide) and technetium (pertechnetate).

TABLE II

EFFECT OF CATAYSTS ON REAGENT BLANK

Catalyst added (in addition to Osmium VIII)	% error in iron factor
No catalyst	-1.9
0.5 mls $K1O_3$/K1	-0.25
2 mls $K1O_3$/K1	-0.62
0.5 mls Fe(III)	0.01
1 ml Fe(III)	-0.05 (mean of 3 pairs)
$K1O_3$/K1 plus Fe(III)	-1.53

TABLE III

EFFECT OF CATALYSTS ON PLUTONIUM ESTIMATION

Solution Used	Catalyst Used	Mean Results	Coefficient of Variation (from 8 results)
Plutonium Control A	$K1O_3$/K1	12.247	0.17
Plutonium Control A plus fission product simulate	$K1O_3$/K1	12.268	0.29
Plutonium Control B	Fe(III)	11.120	0.11
Plutonium Control B plus fission product simulate	Fe(III)	11.136	0.06

4. OPTIMISATION OF PROCEDURE

The overall requirement for precision and bias will only be achievable if (a) the oxidation to plutonium VI is complete (b) the excess oxidant is destroyed completely without reducing any plutonium and (c) fission products and other chemicals do not interfere.

These 3 stages are considered in detail in the following sections. The experimental examination of the separate areas is to a large extent dependend upon the other conditions being fulfilled. It follows that performance improvements in one area permit a more acurate examination of the others. Results obtained with the recommended procedure are used in this paper where ever possible.

TABLE IV

ASSESSMENT OF PUBLISHED DATA ON REACTIONS USED IN CERIUM IV REDUCTION AND POSSIBLE SIDE REACTIONS

Possible Reaction	Summary of literature data	Reference	Interference Status
CeIV+AsIII $\xrightarrow{OsVIII}$ CeIII+AsV	Kinetic data for 2N H_2SO_4 shows that initial reaction rate is 50% per second, and at CeIV conc equivalent to 0.01% bias, reaction rate is 5% per second.	16	Rapid and complete
AsIII+MnVII $\rightarrow$ AsV+MnII	Direct titrations reported Catalysed by OsVIII.	17, 18 17	expected to be rapid and complete
MnVII+Ox $\rightarrow$ MnII	Slow at room temp. iodide or iodate as catalyst or iron III catalyst, increase reaction rates.	12 13, 14	completeness uncertain
PuVI+Ox $\rightarrow$ PuIV	Kinetic data extrapolated to R/T shows 0.001% PuVI reduction per minute.	9	too slow to cause bias
CrVI+Ox $\rightarrow$ CrIII	Reaction catalysed by Mn^{2+} under markedly different conditions.	19	no relevant literature data
PuVI+CeIII $\rightarrow$ PuIV+CeIV	Redox potential unsuitable.	Table V	no reaction expected
PuVI+MnII $\rightarrow$ PuIV+MnIII	Redox potential unsuitable.	Table V	no reaction expected
AsV+FeII $\rightarrow$ AsIII+FeIII	Redox potentials unsuitable unless iron II complexed.	Table V	no reaction expected
PuVI+AsIII PuIV+AsV	Redox potentials suitable. No other data.	Table V	no relevant literature data

TABLE V

RELEVANT REDOX POTENTIALS (20)

Redox Change	Actual System	Potential (Volts)
Ce IV ⇌ Ce III	$Ce^{4+} + e^- \rightleftharpoons Ce^{3+}$	1.61 (HNO_3)(21) 1.44 (H_2SO_4)(21)
Pu VI ⇌ Pu IV	$PuO_2^{2+} + 4H^+ + 2e^- \rightleftharpoons Pu^{4+} + 2H_2O$	1.05
Pu IV ⇌ Pu III	$Pu^{4+} + e^- \rightleftharpoons Pu^{3+}$	0.98
Cr VI ⇌ Cr III	$Cr_2O_7^{2-} + 14H^+ + 6e^- \rightleftharpoons 2Cr^{3+} + 7H_2O$	1.33
Fe III ⇌ Fe II	$Fe^{3+} + e^- \rightleftharpoons Fe^{2+}$	0.77
As V ⇌ As III	$H_3AsO_4 + 2H^+ + 2e^- \rightleftharpoons HAsO_2 + 2H_2O$	0.58
Mn VII ⇌ Mn II	$MnO_4^- + 8H^+ + 5e^- \rightleftharpoons Mn^{2+} + 4H_2O$	1.49
Mn IV ⇌ Mn II	$MnO_2 + 4H^+ + 2e^- \rightleftharpoons Mn^{2+} + 2H_2O$	1.21
Mn III ⇌ Mn II	$Mn^{3+} + e \rightleftharpoons Mn^{2+}$	1.51
Oxalic Acid	$2CO_{2(g)} + 2H^+ + 2e^- \rightleftharpoons H_2C_2O_4$	0.49 (22)

4.1 Oxidation of Plutonium with Cerium IV

Initial work showed that in 3M nitric acid a molar ratio of cerium IV to plutonium of about 30 was required to give complete plutonium oxidation within 30 minutes. It was found that plutonium oxidation is much quicker in 1.5M or 1.0M nitric acid, and requires a lower cerium IV excess, as would be expected from the acid dependency of the plutonium IV/plutonium VI reaction. Some typical results in Table 1 show that a cerium IV to plutonium ratio of less than 3 is adequate for complete oxidation within 5 minutes in 1N nitric acid. The recommended procedure uses a 1N nitric acid medium, which is similar to conditions selected independently by Philips (11). Sulphamic acid is added to prevent nitrite induced side reactions. Additions of aliquots of strongly acid sample solutions will increase the acidity at the oxidation stage. In the recommended procedure addition of 15 milli-equivalents of acid with the sample aliquot will result in oxidation taking place in 1.3N acid. At this acidity, with the maximum recommended plutonium present, the recommended cerium IV addition will give complete oxidation within the recommended 5 minutes. Higher acid addition with the sample aliquot can be tolerated by reducing the acidity of the nitric acid diluent. However, the acidity at the end of the oxidation stage must not be below 1N, or the consumption of acid in the arsenic III/cerium IV reaction can reduce the acidity to below a level at which precipitation occurs (probably of cerium).

4.2 Cerium IV Reduction

Hedrick (10) reduced 0.2 mg of cerium IV by reaction with arsenic III catalysed by osmium VIII. Excess arsenic III was then removed by potassium

TABLE VI

EXAMINATION OF POSSIBLE SIDE REACTIONS

Reaction Examined	Conditions	Result	Conclusion
PuVI+AsIII ⇌ PuIV+AsV	Normal Plutonium Control procedure ie 1 to 2 drops excess arsenite, less than 1 minute reaction time.	Plutonium concentration 10.088 mg/g	
	Plutonium control procedure but with 10 drops excess arsenite 5 minutes reaction time.	Plutonium concentration 10.080 mg/g	No significant reaction
CrVII+Ox ⇌ CrIII	1 ml excess 0.005N dichromate and 1 ml excess oxalate added to simulated titration mixture, dichromate concentration followed spectro photometrically.	1.5% of dichromate reduced in 1 hour	Reaction too slow to cause bias
	Direct iron standardisation.	Iron factor: 9.779	
FeII+AsV ⇌ FeIII+AsIII	Iron standardisation in presence of 0.5 milli moles of arsenic V, and 30 minute reaction time.	iron factor: 9.777	No significant reaction

permanganate, followed by a slight excess of oxalate. Molybdate was used to catalyse the oxalate-permanganate reaction. Initial trials of a scaled up version of this procedure failed due to precipitate formation on addition of the molybdate catalyst. Using potassium iodate or iodide as an alternative catalyst was more successful, but was eventually shown to leave a small reagent blank as shown by the results in Table II. Use of iron III as a further alternative catalyst eliminated this problem (Table II) and also improved the attainable precisions (see Table III).

The complicated series of reactions in this procedure for cerium IV reduction provides several possible sources of bias from incomplete reactions and unwanted side reactions. Examination of the published data on these possible reactions, summarised in Table IV and of the relevant redox potentials in Table V confirms that the most likely source of bias is from incomplete reaction between permanganate and oxalate. Both potassium iodide and iron III are reported to act as catalysts in this reaction, with iodide catalysing the initial reduction of permanganate by oxalate (12) while iron III speeds up the final stage of reaction between manganese III and oxalate (13)(14). This is consistent with the experimental results (Table II).

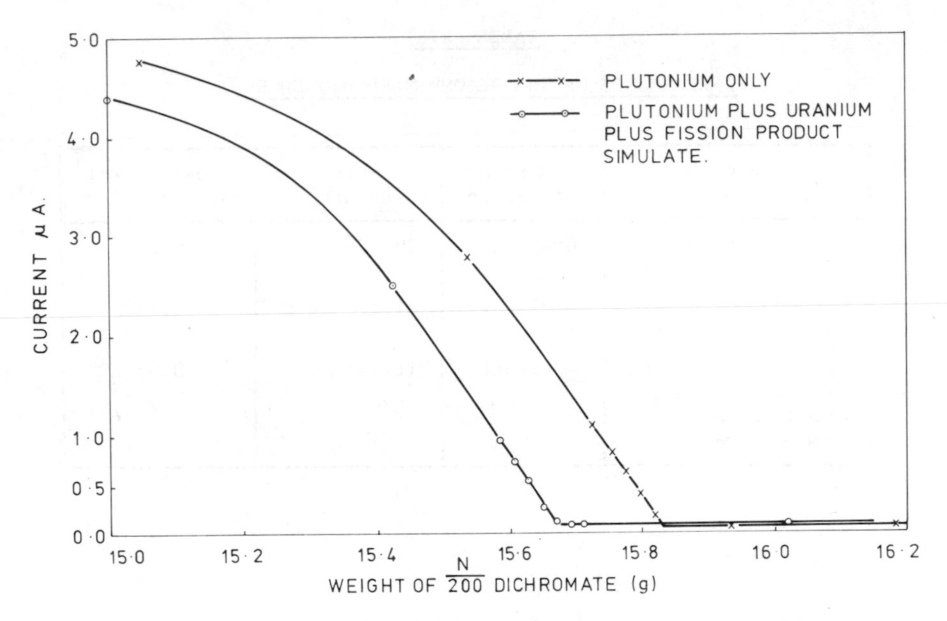

FIG.1. Amperometric titration curves.

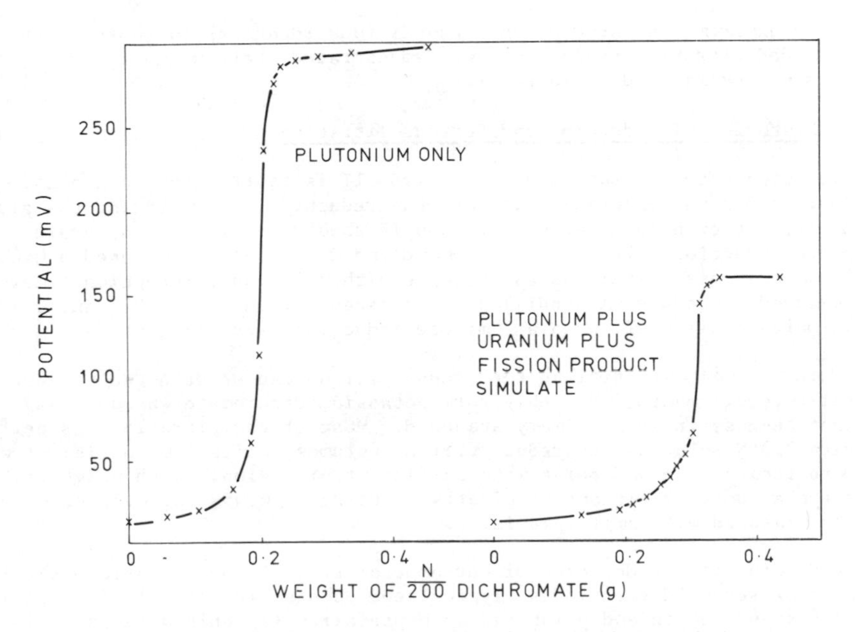

FIG.2. Potentiometric end-point (polarized gold electrodes).

TABLE VII

PERFORMANCE OF THE RECOMMENDED PROCEDURE

Plutonium solution used	End Point location	Mean Result	Coefficient of Variation
NBS standard 50-60 mg aliquots	Graphical	99.92%	0.03
Pu Control B 50-60 mg aliquots	visual	11.120 mg/g	0.11
Pu control B plus fission product simulate plus 150 mg. uranium	visual	11.136 mg/g	0.06

Three possible side reactions cannot be ruled out by the available literature data,

a. reduction of plutonium VI by the arsenic III added to reduce cerium IV,
b. reaction between dichromate and oxalate and
c. reaction between iron II and arsenic V.

Specific experimental studies under conditions identical to those in the recommended titration procedure have demonstrated their unimportance. These results are summarised in Table VI.

4.3 Plutonium VI Reduction and Ferrous Titration

The reaction between plutonium VI and iron II is fairly slow in a nitric acid medium. Addition of sulphate assists the reduction by complexing the plutonium IV, but even so an excess of iron II should be present to ensure complete reduction (2)(7)(15). Drummond and Grant (6) recommended a 0.001 milli-mole excess of ferrous as adequate with a 5 minute reduction time. The recommended procedure (Appendix) uses an excess of between 0.025 and 0.125 milli-moles or iron II, when a 2 minute reduction is sufficient.

Standard procedures to titrate the excess ferrous can be selected to meet specific requirements. For this work potassium dichromate was used as titrant because it is a primary standard. Most of the titration was performed with 0.05<u>N</u> solution, to reduce titrant volumes, while 0.005<u>N</u> titrant was used to located the end point with sufficient precision. With weight titrations the amounts of titrant (typically about 2g of 0.05<u>N</u> and 0.5g of 0.005<u>N</u>) can be measured with ample precision.

The end point can be detected either amperometrically or potentiometrically with polarised gold electrodes. Typical end points are shown in figures 1 and 2. The amperometric end point has been preferred for this work as it is unaffected by the presence of fission product simulates.

TABLE VIII

EFFECT OF ANIONIC IMPURITIES

Plutonium added (mg)	Anion added (mg)	Other impurities present (mg)	% Plutonium recovered
26.07	Sulphate 200 mg	75 mg Uranium	100.04
24.48	Chloride 70 mg	None	100.19
37.35	Phosphate 2 mg	74 mg Uranium	99.95

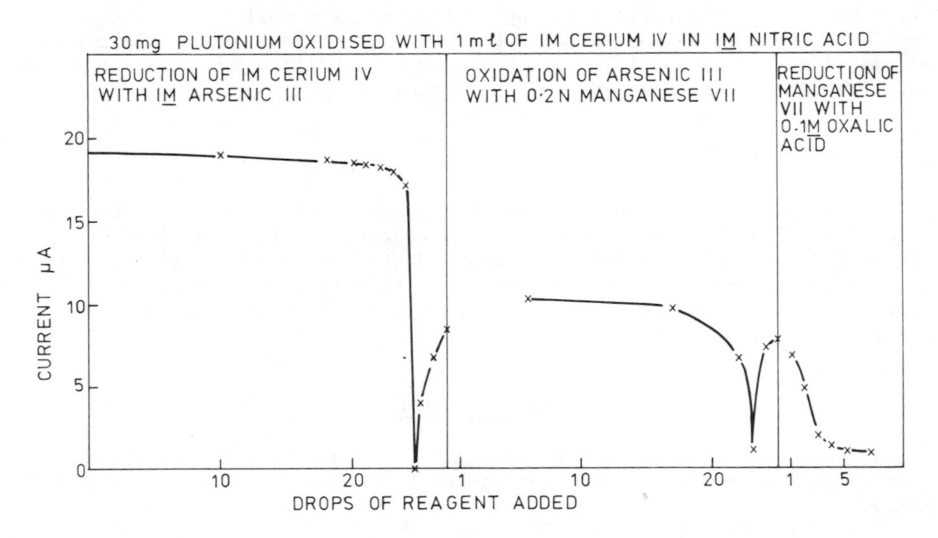

FIG.3. Amperometric detection of intermediate reactions.

5. PERFORMANCE OF THE RECOMMENDED TECHNIQUE

The precision of the method was measured using a plutonium control solution and was found to be about 0.1% coefficient of variation for a single result. The accuracy of the method with pure plutonium standard solutions was estimated using a solution prepared from NBS standard plutonium metal. A plutonium control was also used to observe the effect of uranium, and fission product simulates equivalent to 7% burn up of a fast reactor fuel. These results are shown in Table VII. Work is continuing to clarify the status of the marginal bias effects suggested by these results.

Quantities of common foreign anious greatly in excess of the levels expected in the intended application have minimal effect as shown by the results in Table VIII.

5.1 Remote Application

The basic operations of weight aliquoting and titration can be performed remotely using manipulators, although some loss of precision is possible. The amperometric end point is suitable for in-cell use, and the same equipment may be used to follow the course of the cerium IV, arsenic III, manganese manganese VII, oxalic acid reaction sequence, rather than visually observing the colour changes (see Fig 3 for typical current/reaction curves). The addition of excess manganese VII and its subsequent destruction with oxalate, can also be readily followed visually through a lead glass window provided some white lighting is available in the cell.

5.2 Other Possible Applications

This procedure is being developed specifically for precise plutonium estimation in solutions of irradiated fast reactor fuel. The large range of ions which must be tolerated in this application suggest that the method may be applicable to a wide range of sample types. The method is capable of considerable procedural modification to accommodate differing requirements such as plutonium aliquot size and titrant strength.

6. CONCLUSIONS

Plutonium can be estimated precisely by oxidation with cerium IV, chemical reduction of the excess oxidant, and subsequent iron II reduction/chromium VI titration. Bias effects from reagents and fission products are low, permitting direct application of the procedure to solutions of irradiated plutonium based fuels. The manipulations involved can all be performed in a remote handling facility.

ACKNOWLEDGEMENT

The authors wish to thank their colleagues at Dounreay for assistance, in particular the practical contributions of C Digby-Grant, J K Smith, and E M McKay.

REFERENCES

1. GUTMACHER, R G, STEPHENS, F, ERNST, K, HARRAR, J E, MAGISTAD, J. USAEC Rep WASH 1282 (1973).

2. VENKATASUBRAMANIAN, V, DURHAM, R W, CORRIVEAU, V. AECL Rep AECL-3206 (1968).

3. SHULTS, W D. Talanta 10 (1963) 833.

4. American Society for Testing and Materials Standard No ASTM C 697-72 (1972).

5. DAVIES, W, TOWNSEND, M. UKAEA Rep TRG R 2463(D) (1974).

6. DRUMMOND, J L, GRANT, R A. Talanta 13 (1966) 477.

7. CORPEL, J, REGNAUD, F. Anal Chim Acta 35 (1966) 508.

8. DODSON, V H, BLACK, A H. J. Am. Chem. Soc. 79 (1957) 3657.

9. ZAKHAROVA F A, ORLAVA, M M. Sov Radiochem 15 (1973) 796.

10. HEDRICK, C E, PIETRI, C E, WENZEL, A W, LERNER, M W. Analyt chem 44 (1972) 377.

11. PHILLIPS, G, CROSSLEY, D, VENKATARAMANA. UKAEA Rep AERE R8885 (1977).

12. MISRA, D D, GUPTA, Y K. Bull Acad Polon Sci Ser Sci Chem 9 (1961) 379.

13. MAPSTONE, G E, SMITH, J W. Chemistry and Industry (1952) 238.

14. MAPSTONE, G E. J Appl Chem Biotechnol 21 (1971) 238.

15. SEILS, C A, MEYER, R J, LARSEN, R P. Analyt Chem 35 (1963) 1673.

16. HABIG, R L, PARDUE, H L, WORTHINGTON, J B. Analyt Chem 39 (1967) 600.

17. CRIMMINS, E, POUND, J R. Chem Eng Mining Rev 31 (1939) 457.

18. SANDHU, S S. Indian J Chem 5 (1967) 455.

19. CHAKRAVARTY, D N, GHOSH, S. Proc Nat Acad Sci India. Sect A 29 (1960) 199.

20. WEAST, R C. Handbook of Chemistry and Physics 52nd ed. The Chemical Rubber Co, Cleveland (1971) D111.

21. WADSWORTH, E, DUKE, F R, GOETZ, C A. Analyt Chem 29 (1957) 1824.

22. LATIMER, W M. The Oxidation States of the Elements and their Potentials in Aqueous Solution 2nd Ed. Prentice-Hall (1952) 131.

DISCUSSION

S.V. KUMAR: The U/Pu ratio reported in the paper is about 3. What will be the effect of higher U/Pu ratios, such as will normally be encountered? In other words, what is the influence of uranium on the method?

W. DAVIES: There is no gross interference by uranium. Work done very recently indicates that large amounts of uranium may produce a positive bias in the plutonium measurement procedure, though this has still to be confirmed. The magnitude of the apparent effect appears too small to cause any problem for fuel having a U/Pu ratio of 3, but for fuels having very much higher U/Pu ratios the effect might conceivably be greater.

J. HURE: What is the effect of traces of solvent (tributylphosphate) dissolved in the solutions analysed by the two methods proposed?

W. DAVIES: The effect of TBP/OK (tributylphosphate/odourless kerosene) on the method is being studied by the authors, one of the objects being to develop a procedure for the determination of plutonium in TBP/OK solutions. As far as I am aware, small amounts of TBP dissolved in the sample would not seriously affect the procedure.

RECENT DEVELOPMENTS IN THE APPLICATION OF CONTROLLED-POTENTIAL COULOMETRY TO THE DETERMINATION OF PLUTONIUM AND URANIUM

G. PHILLIPS, D. CROSSLEY
Chemistry Division,
AERE Harwell,
Harwell, United Kingdom

Abstract

RECENT DEVELOPMENTS IN THE APPLICATION OF CONTROLLED-POTENTIAL COULOMETRY TO THE DETERMINATION OF PLUTONIUM AND URANIUM.

Controlled-potential coulometry is a widely accepted technique for the determination of plutonium and uranium in solution at the milligram level. The major difference between controlled-potential coulometry and other destructive analytical methods is the avoidance of complicated chemical manipulation. This constitutes an advantage from the health and safety aspect and also for the recovery of plutonium and uranium from analytical residues. The benefit of this technique for fissile material accountancy lies in the fact that the coulometer can be calibrated against physical standards of current and time, and hence absolutely in terms of the Faraday. This double check on the accuracy of the instrument is of considerable importance in satisfying the requirements of national and international organizations whose purpose is to safeguard fissile material. Recent developments in controlled-potential coulometry at AERE Harwell have included three procedures: One to enable plutonium and uranium to be determined sequentially on the same sample aliquot using a solid electrode; another to enable uranium to be determined reversibly at a solid electrode as an alternative to the irreversible determination at a mercury pool electrode. These two procedures depend upon a preliminary reduction of UO_2^{2+} to U^{4+} by electrogenerated hydrogen at a platinum or gold mesh electrode. The uranium is then oxidized back to UO_2^{2+} with either (1) electrogenerated PU^{4+} from plutonium already present in the solution, or (2) electrogenerated Fe^{3+} from iron added to the solution. This oxidation step gives the coulomb equivalent of Pu + U or Fe + U. The plutonium or iron is then determined coulometrically by reduction back to the lower valency state leaving the uranium unchanged. A third procedure to oxidize plutonium quantitatively to the hexavalent state with ceric ion in the working compartment of a controlled-potential coulometric cell: The excess ceric ion is reduced in situ and the plutonium is then determined coulometrically by reduction to Pu^{3+} followed by oxidation to Pu^{4+}. This procedure avoids interference from iron present in the sample and is an alternative to procedures using argentic oxide or fuming perchloric acid as oxidants for plutonium. These latter reagents are undesirable constituents of plutonium residue solutions. The precisions of all the procedures employed are $\leqslant \pm 0.2\%$ and the accuracies obtained are consistent within those limits. The behaviour of some expected impurity elements has been investigated.

1. INTRODUCTION

The use of controlled-potential coulometry as an analytical technique for the determination of plutonium or uranium in solution offers advantages which are becoming increasingly important from several points of view. Firstly, the technique requires only mg amounts of fissile material which minimises the holdings in the analytical laboratory and the hazards to the operators. Secondly no reagent additions are required for the analytical determination which facilitates the recovery of analytical residues. Thirdly the technique is amenable to both physical and chemical calibration which is an important point from the accountancy point of view. Many years experience of controlled-potential coulometry at AERE Harwell have emphasised the above advantages and have at the same time drawn attention to several shortcomings to the application of the technique in practical situations. As an example, it is a common request for both plutonium and uranium to be determined on the same sample and separate coulometric determinations on the same solution have always been possible. A number of procedures [1] [2] [3] have been described to enable plutonium and uranium to be determined sequentially by controlled-potential coulometry on a single aliquot of solution. This paper describes a procedure enabling this to be accomplished without any chemical adjustment of the valency states or the addition of further reagents other than those required to dissolve the sample. The procedure described for the sequential determination of plutonium and uranium changes the uranium determination from being an irreversible controlled-potential reduction of uranium at the mercury pool electrode to a reversible reaction at the platinum electrode. This change in the nature of the controlled-potential coulometric determination of uranium has proved to have some advantages in cases where the sample is limited or the valency state of the uranium is in doubt or in avoiding some interferences. In the absence of plutonium it has been shown that iron can be used to effect the reversible reduction and oxidation of uranium. The major disadvantage to the controlled-potential coulometric determination of plutonium based upon the reversible Pu^{3+}/Pu^{4+} change has always been the interfering effect of iron impurity. Although this effect can be minimised by choosing an appropriate base medium or corrected for by means of an independent iron determination, the only satisfactory method of avoiding iron interference completely is by basing the controlled-potential coulometric determination on the irreversible PuO_2^{2+}/Pu^{4+} change [4]. This necessitates a chemical oxidation of plutonium to the hexavalent state. This is commonly achieved in other analytical methods either with argentic oxide or by fuming with perchloric acid. Argentic oxide is very effective in oxidising plutonium to the hexavalent state but interference can arise during a coulometric measurement due to the reduction of the Ag^+ ion [5]. Fuming perchloric acid is equally effective as an oxidant for the preparation of hexavalent plutonium but acid decomposition products have been found to interfere with the subsequent coulometric determination. The use of ceric ion as an alternative oxidant has been investigated and a procedure developed. Although this involves the addition of a reagent to the solutions, cerium is a less objectionable constituent of plutonium residue solutions than silver salts or perchloric acid.

2. EXPERIMENTAL

The controlled-potential coulometer used for this work has been described elsewhere [6]. The instrument was employed in the semi-automatic mode with control of the various steps by time rather than by terminal current. The electrolysis cell was of the three compartment type [7] with a working compartment of about 7 ml capacity, and separated from the counter electrode and control electrode compartments by means of discs of anion exchange membrane. Working electrodes of either platinum or gold gauze were used.

2.1 The determination of plutonium and uranium by sequential controlled-potential coulometric titration using a solid electrode

The direct controlled-potential coulometric oxidation of U^{4+} to UO_2^{2+} has been attempted in complexing media [8] [9] but in dilute mineral acid media the reaction is too slow to be analytically useful at potentials lower than the decomposition potential of the solvent. In the presence of plutonium however the reaction is observed to be fast [3] [10]. It is feasible therefore to postulate a controlled-potential coulometric determination of uranium plus plutonium based upon the two reactions:-

$$U^{4+} + Pu^{4+} \longrightarrow UO_2^{2+} + Pu^{3+}$$

and

$$Pu^{3+} \longrightarrow Pu^{4+}$$

proceeding simultaneously at a potential governed by the E_0^1 of the plutonium reaction. This is immediately followed by a controlled-potential coulometric reduction of plutonium:-

$$Pu^{4+} \longrightarrow Pu^{3+}$$

leaving the uranium in the hexavalent state. This approach is dependent upon the uranium and plutonium in the solution being conditioned quantitatively to U^{4+} and Pu^{3+} respectively at the commencement of the electrolysis. A number of reagents, notably Cr^{2+}, Ti^{3+} [3] and Bi^{3+}, are capable of effecting these reduction steps, but these have the disadvantage of introducing an additional substance into a waste stream and also adding to the difficulties of automatic analysis. These objections can be overcome by using hydrogen as the reductant, and this can be conveniently accomplished by electrogenerating hydrogen in situ at the working electrode at a potential of about -0.3 volts vs the S.C.E. This potential is chosen to be negative with respect to the H^+/H_0 electrode, but not so negative that excessive gassing of the solution occurs during the latter stages of a reduction.

The preferred medium for the controlled-potential coulometric determination of plutonium and uranium is molar sulphuric acid. In the case of oxide and carbide fuels this is normally achieved by igniting to a mixture of $PuO_2.U_3O_8$ and then dissolving in nitric acid with the addition of a trace of hydrofluoric acid. The solution is then conditioned by fuming with sulphuric acid to remove hydrofluoric and nitric acids followed by diluting to molar sulphuric acid to give a solute concentration of about 5 mg/ml. In such a medium the procedure envisaged for the controlled-potential coulometric determination of plutonium and uranium at a solid electrode would contain the following steps:-

(a) Electrolysis of the solution at -0.30 volts vs the S.C.E. to reduce UO_2^{2+} to U^{4+} and PuO_2^{2+} + Pu^{4+} to Pu^{3+}
(b) Electrolysis of the solution at +0.20 volts vs the S.C.E. to remove adsorbed hydrogen from the working electrode and allow degassing of the solution.
(c) Electrolysis of the solution at +0.70 volts vs the S.C.E. to oxidise Pu^{3+} to Pu^{4+} and U^{4+} to UO_2^{2+}. The integral of the current passed in this step corresponds to the sum of the two reactions.

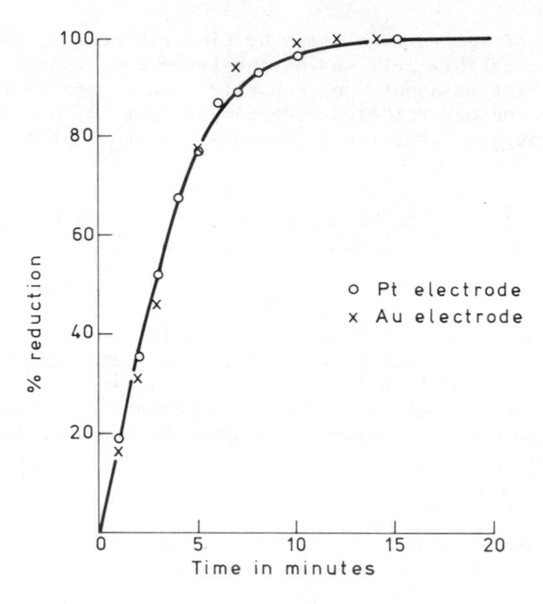

FIG.1. Efficiency of reduction of UO_2^{2+} at the Pt or the Au electrode in MH_2SO_4.

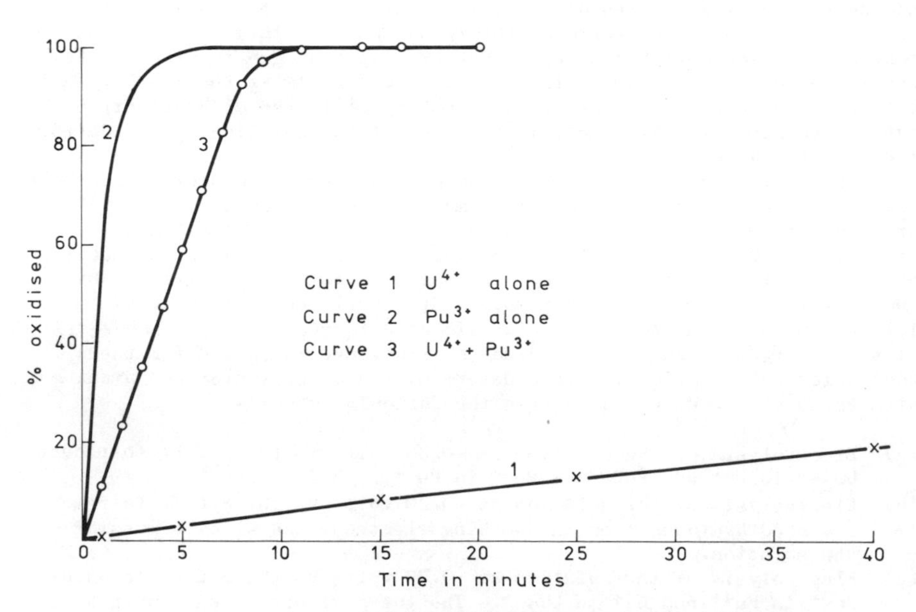

FIG.2. Efficiency of oxidation of U^{4+} at the Au mesh electrode in MH_2SO_4.

(d) Electrolysis of the solution at +0.30 volts vs the S.C.E. to reduce Pu^{4+} to Pu^{3+}. The integral of the current passed in this step corresponds to plutonium only.
(e) Electrolysis of the solution at +0.70 volts vs the S.C.E. to oxidise Pu^{3+} to Pu^{4+}. The integral of the current passed in this step corresponds to plutonium only.

The critical aspect of step (a) is the time required to effect quantitative reduction of the uranium and the plutonium. The progress of the reduction cannot be followed in the usual way for controlled-potential coulometric reactions, i.e. observation of the current until a low background value is reached, since there will be a high current at the completion of the reaction due to reduction of the hydrogen ion. It is necessary therefore to establish experimentally for each working electrode area, electrolyte concentration and volume, that sufficient time of electrolysis is allowed to achieve quantitative reduction of the uranium and the plutonium. The results of a series of experiments to establish the efficiency of reduction of UO_2^{2+} at the platinum or gold electrode at a potential of -0.30 volts vs the S.C.E. in the presence of plutonium are summarised in FIG 1. It can be seen that complete reduction of the uranium and plutonium can be achieved in 15 minutes. It was also shown that the rate of reduction of the uranyl ion at -0.30 volts vs the S.C.E. was independent of the presence of the plutonium or of its valency state.

If the reduction step (a) is followed immediately by the oxidation step (c) then the initial current surge contains a component corresponding to the oxidation of hydrogen adsorbed on the electrode. This 'hydrogen' current is high relative to that required to charge the electrical double layer, but unlike the latter it cannot be evaluated by means of a 'blank' determination. A brief electrolysis (3-5 mins) at an intermediate potential eliminates the hydrogen adsorbed on the electrode, and also allows time for dissolved hydrogen in the solution to be displaced by the stream of inert gas normally used for stirring and removing dissolved air.

The efficiency of oxidation of U^{4+} at +0.70 volts vs the S.C.E. (step c) at the platinum electrode in MH_2SO_4 is shown in FIG 2. Quite clearly the rate of oxidation of U^{4+} in the absence of plutonium is too slow to be analytically usable whereas in the presence of plutonium both reactions are complete in 15 minutes or less. The ratio of Pu:U corresponds to a typical fuel element composition of about 21.5 weight % plutonium. The efficiency of oxidation of U^{4+} by the above technique should be unaffected at higher Pu:U ratios since the generation of Pu^{4+} is clearly the rate controlling step. At lower Pu:U ratios the general shape of curve 3 is unaffected even with a tenfold reduction of the plutonium concentration. At this lower level however the titration time is prolonged due to the slowness with which an acceptable background current of less than 20 μA is approached. It can be concluded that a reasonable operational range for the technique lies between a Pu:U ratio of 1:2 and 1:10.

The electrochemical reduction of Pu^{4+}, step (d) or the oxidation of Pu^{3+}, step (e) in the presence of hexavalent uranium does not call for any special comment. The integral of either of these steps corresponds to plutonium only, and in the interests of a shorter overall analytical procedure, step (d) can be used as the plutonium determining step. It is necessary, however, to establish for a particular design of coulometer that the integrator is exactly reversible and that step (d) is equivalent to step (e). It should also be noted that the residual current at +0.70 volts vs the S.C.E. will differ from that at +0.30 volts and hence the integrated background correction values will not be equivalent.

TABLE I.

The sequential determination of plutonium and uranium by controlled-potential coulometry at the platinum electrode in M H_2SO_4

Uranium taken 10.973 mg) Corresponding to
) approximately 25%
Plutonium taken 3.946 mg) PuO_2/UO_2 sample

Uranium found		Plutonium found	
mg	% recovery	mg	% recovery
10.984	100.10	3.924	99.44
11.031	100.53	3.939	99.82
10.925	99.56	3.948	100.05
10.949	99.78	3.965	100.48
11.009	100.33	3.910	99.10
11.025	100.47	3.930	99.59
Σ/n 10.987	100.13	3.936	99.75
SD 0.043	CV 0.39%	SD 0.019	CV 0.48%

TABLE II.

The effect of U:Fe ratio on the uranium recovery

Weight of uranium taken, mg	U : Fe ratio	Mean uranium recovery, %
5.431	2.7	100.75
5.431	3.4	100.40
7.253	4.0	100.05
9.067	7.6	100.09
9.067	11.4	100.07
5.431	13.6	100.01
9.067	22.7	99.65

TABLE III.

The determination of uranium by electrogenerated ferric ion in molar sulphuric acid

Weight of uranium taken	U : Fe ratio	Mean uranium recovery	Number of determinations	Coefficient of variation
9.785 mg	4.7	100.08%	12	0.16%
5.431 mg	4.5	99.95%	24	0.20%

2.2 Results for the determination of plutonium and uranium by sequential controlled-potential coulometric titration using a solid electrode

Two solutions were prepared in molar sulphuric acid, one containing plutonium at 1.973 mg Pu/ml, the other containing uranium at 2.656 mg U/ml. A 2.0 ml aliquot from the plutonium solution together with a 3.0 ml aliquot from the uranium solution were placed in the working compartment of a controlled-potential coulometric cell designed for use with a platinum gauze working electrode. After degassing with nitrogen the solution was put through the following sequence of controlled-potential coulometric steps:-

(a)	-0.30 volts vs S.C.E.	20	minutes
(b)	+0.20 "	5	"
(c)	+0.70 "	20	"
(d)	+0.30 "	10	"
(e)	+0.70 "	15	"

Repetitive determinations of plutonium and uranium, based upon the above sequence of operations, were made in order to establish the accuracy and precision of the measurements. The results are shown in Table I.

2.3 The reversible determination of uranium by controlled-potential coulometry at a solid electrode using electrogenerated ferric ion

The use of ferric ion, either as an added reagent or in the electrogenerated form, has been previously recommended in reversible controlled-potential coulometric procedures for the determination of uranium [2] [3]. This work describes the use of electrogenerated ferric ion as an oxidant for U^{4+} following reduction with electrogenerated hydrogen. The procedure employed was identical to that described for plutonium plus uranium mixture in 2.2 but using potentials in steps (c), (d) and (e) appropriate to the Fe^{2+}/Fe^{3+} couple. In molar sulphuric acid these are +0.62 volts vs the S.C.E. for step (c) and (e) and +0.22 volts vs the S.C.E. for step (d). The effect of varying the U:Fe ratio on the uranium recovery was investigated and the results are given in Table II. Satisfactory results were obtained over a range of U:Fe ratios from about 4:1 to 14:1 with a tendency to obtain high recoveries at low ratios and low recoveries at higher ratios. In the latter case the time required to complete step (c) was unacceptably long. A further series of recovery experiments were carried out at the optimum U:Fe ratio of about 5:1 in order to estimate the accuracy and precision of the method. The results are shown in Table III. These results compare favourably with those normally obtained by controlled-potential coulometry using a mercury pool electrode.

2.4 The controlled-potential coulometric determination of plutonium based upon cerium oxidation and the PuO_2^{2+}/Pu^{4+} valency change

The standard potentials (Table IV) of the cerium III/IV, plutonium VI/V, plutonium VI/IV and plutonium VI/III systems are such that there should be no difficulty in oxidising plutonium to the hexavalent state with ceric ion but consideration has to be given to the effect of the acid medium on the formal potentials. In molar nitric acid solution the formal potentials do not change significantly from the standard values and this would be the preferred medium for the first oxidation step. In molar sulphuric acid however, the cerium III/IV formal potential is much lower (circa 1.44 volts vs the N.H.E.) and the plutonium potentials more positive leading to greater difficulty in effecting complete oxidation.

TABLE IV

Standard potentials of the redox systems involved in the proposed procedure

Element	Oxidation numbers	Reaction	Standard potential E_0 volts vs the NHE
Pu	VI-V	$PuO_2^{2+} + e^- = PuO_2^{1+}$	0.933
	VI-IV	$PuO_2^{2+} + 4H^+ + 2e^- = Pu^{4+} + 2H_2O$	1.024
	VI-III	$PuO_2^{2+} + 4H^+ + 3e^- = Pu^{3+} + 2H_2O$	1.022
Ce	IV-III	$Ce^{4+} + e^- = Ce^{3+}$	1.62
Fe	III-II	$Fe^{3+} + e^- = Fe^{2+}$	0.77

The selective controlled potential reduction of excess ceric ion in the presence of hexavalent plutonium is again more favourably accomplished in nitric acid rather than sulphuric acid. Having achieved the quantitative oxidation of the plutonium to the hexavalent state and the selective reduction of the excess cerium, the plutonium may then be determined coulometrically in either nitric acid or sulphuric acid medium. In the case of nitric acid medium it is essential to add iron and carry out a secondary reduction of the hexavalent plutonium with electrogenerated ferrous ion followed by oxidation of ferrous to ferric and trivalent plutonium to quadrivalent plutonium. In the case of sulphuric acid medium the addition of iron is not essential since hexavalent plutonium can be electrolytically reduced to the trivalent state in that medium at an acceptable rate. In either case the plutonium may be determined without iron interference by calculation from the coulombs recorded for the nett reaction $PuO_2^{2+} \rightarrow Pu^{4+}$.

Preliminary experimental work established several important aspects of the proposed method. Firstly, the oxidation of 3 mg of plutonium could be readily accomplished in about 5 minutes using electrogenerated ceric ion or chemically added ceric ion in molar nitric acid containing 0.2 molar sulphamic acid. Similar attempts to oxidise plutonium in molar sulphuric acid were completely unsuccessful. Secondly the excess ceric ion could be selectively reduced at +1.17 volts vs the S.C.E. without reduction of the hexavalent plutonium. The overall results obtained in attempting to carry out the proposed determination in nitric acid medium throughout were disappointing in terms of accuracy and precision particularly when using electrogenerated ceric ion. This was attributed to a surface effect on the platinum gauze electrode which manifested itself as a high and variable 'blank' corresponding to about 8.8% of the coulombs measured. The results showed slight improvement on using added ceric ion at a ratio of 2.3 with respect to the plutonium. The 'blank' value in this case corresponded to about 2.7% of the coulombs measured with a mean recovery of 99.83% with a precision of 0.72% (coefficient of variation). It was concluded therefore that although the oxidation of the plutonium could only be achieved in nitric acid, or possibly perchloric acid, and that chemically added ceric ion avoided the worst of the surface effects on the working electrode there might still be further advantage to be gained in changing from nitric acid to sulphuric acid medium immediately after the reduction of the excess ceric ion. This could be readily achieved by adding a small quantity of strong sulphuric acid to the working compartment of the electrolysis cell.

TABLE V.

The determination of plutonium by controlled-potential coulometry based upon PuVI/IV change following ceric ion oxidation

Standard solution A		Standard solution B	
Pu taken mg	Pu recovered mg	Pu taken mg	Pu recovered mg
5.400	5.4038	6.500	6.4954
	5.3789		6.5130
	5.3946		6.5039
	5.3762		6.5065
	5.3962		6.5169
	5.3876		6.5022
	5.4135		6.5130
	5.4086		6.4967
	5.4059		6.4954
	5.4135		6.5039
	5.4059		6.4967

mean 5.3986 ≡ 99.97% 6.5040 ≡ 100.06%

σ 0.0131 0.24% σ 0.0077 0.12%

2.5 Results for the controlled-potential coulometric determination of plutonium based upon cerium oxidation and the PuO_2^{2+}/Pu^{4+} valency change in MH_2SO_4

The secondary controlled-potential coulometric determination of plutonium has been fully described by Shults [4] who showed that accurate plutonium determinations could be obtained even when the Fe:Pu ratio was as high as 10:1. At this ratio the reduction step must be terminated by the operator when all the plutonium has been reduced to the trivalent state and slight excess of ferrous ion has been generated. If this reduction step is allowed to go to completion at such high ratios of Fe:Pu then the coulombs passed in the back titration of plutonium to the tetravalent state and iron to the ferric state are equivalent to >95% of the coulombs passed during the reduction step and the determination of the plutonium being based on the difference becomes inherently less accurate. To perform the coulometric determination in this manner demands a fairly accurate knowledge of the quantity of plutonium present and the active co-operation of the operator. This is inconvenient in practice and impossible when the coulometric steps are performed automatically under time or current control [11]. Under circumstances where the reduction step has to be allowed to run to completion it is better to minimise the quantity of added iron. Using the type of electrolysis cell described and a controlled-potential coulometer with a limiting current of 23 mA, the advantage obtained from the addition of iron is slight and limited to a reduction time of 15 minutes rather than 20 minutes in the absence of iron. It was concluded therefore to eliminate the addition of iron from the final procedure.

The procedure was finally tested using two plutonium standard solutions prepared from selected plutonium metal. Pieces of the metal were cleaned by anodic polishing in 10% potassium carbonate solution, rinsed and dried in an argon atmosphere glove box. The metal was weighed, dissolved in the minimum quantity of 5.5M hydrochloric acid, evaporated repeatedly with nitric acid and finally made to volume in molar nitric acid. Aliquots were taken from these solutions using a calibrated pipette and analysed using the following procedure:

(a) Oxidise the plutonium solution in $MHNO_3$ + 0.2M sulphamic acid with added ceric nitrate such that $Ce^{4+}:Pu \geqslant 2.0$ allowing 5 minutes for the oxidation to take place.
(b) Reduce the excess Ce^{4+} electrochemically at a potential of 1.17 volts vs the S.C.E.
(c) Adjust the solution to molar sulphuric acid and reduce electrochemically at a potential of +0.25 volts vs the S.C.E. Record the coulombs.
(d) Oxidise the solution electrochemically at a potential of +0.72 volts vs the S.C.E. Record the coulombs.
(e) Repeat in the absence of plutonium to assess the blank for each step. The plutonium content is calculated on the difference between the coulombs recorded for steps (c) and (d).
The results are given in Table V.

3. DISCUSSION

The approach to the determination of plutonium and uranium on a single aliquot described in 2.1 has a number of advantages including the possibility of automating the determination and avoidance of the use of reagents which might prove an embarrassment to recovery processes. It is worthy of note that in cases where a Pu/U or a Pu/Pu+U ratio is required this can be obtained without the necessity to measure the aliquot used accurately. It is necessary however, to evaluate the various steps in the procedure in relation to the electrolysis cell and coulometer design as outlined in 2.1. The coulometer used for this work had a limiting reducing current of 23 mA, and a working electrode area large enough to sustain a current of that magnitude but not as large as that employed in high-speed controlled-potential coulometric techniques [12]. It is necessary therefore to evaluate carefully the time of application of the reducing current as against the approximate quantity of uranium in the cell. This may not be the same as described in this paper for a coulometer and electrolysis cell having different design parameters. The use of iron to effect a reversible determination of uranium at the solid electrode does not call for any special comment. It is a logical extension of the work described in 2.1. In practice the most useful application of this technique has been in cases where the available sample has been limited to 1-2 mg U. In these cases the results obtained have shown a higher precision than similar determinations obtained by controlled-potential coulometry using a mercury pool. The controlled-potential coulometric determination of plutonium based upon cerium oxidation and the PuO_2^{2+}/Pu^{4+} valency change is a useful technique in the presence of iron impurity but suffers from the disadvantages that the ceric addition has to be added chemically and that a change of acid medium is necessary during the course of the determination.

REFERENCES

[1] ANGELETTI, L.M., BARTSCHER, W.J., MAURICE, M.J., Z. Anal. Chem. 246 (1969) 297.
[2] DAVIES, W., GRAY, W., McLEOD, K.C., TALANTA 17 (1970) 937
[3] FARDON, J.B., McGOWAN, I.R., TALANTA 19 (1972) 1321
[4] SHULTS, W.D., TALANTA 10 (1963) 833
[5] PHILLIPS, G., MILNER, G.W.C., Proceedings SAC Conference Nottingham 1965, W. Heffer & Sons Ltd. Cambridge U.K.
[6] PHILLIPS, G., MILNER, G.W.C., The Analyst 94 (1969) 833
[7] MILNER, G.W.C., EDWARDS, J.W., UKAEA Rep. AERE-R3772 (1961)
[8] ZITTEL, H.E., DUNLAP, L.B., Anal. Chem. 35 (1963) 125
[9] BOYD, C.M., MENIS, O., Anal. Chem. 33 (1961) 1016
[10] UKAEA Unpublished Information
[11] PHILLIPS, G., NEWTON D.A., WILSON, J.D., J. Electroanal. Chem. 75 (1977) 77.
[12] GOODE, G.C., HERRINGTON, J., Analytica Chimica Acta 33 (1965) 413

DISCUSSION

J. HURE: What is the effect of traces of solvent (tributylphosphate) dissolved in the solutions analysed by the two methods proposed?

G. PHILLIPS: The effect of TBP/OK in trace quantities is not known but the effects of phosphate from TBP breakdown are not harmful and are almost beneficial owing to the 'cleaning up' effect of the steps taken to break down the TBP.

S.V. KUMAR: In the first part of your paper, the success of the estimation depends on the conversion of U(VI) to U(IV). How do you ensure that all the uranium is converted to U(IV)?

G. PHILLIPS: As stated in the paper, this cannot be followed in the usual way for controlled-potential coulometric reactions. A plot such as that shown in Fig. 1 must be prepared for each design of controlled-potential coulometer, electrolysis cell and quantity of uranium present.

INSTRUMENTAL REQUIREMENTS FOR EFFECTIVE NUCLEAR MATERIALS VERIFICATION AND CONTROL IN STATIC LOCATIONS

A.S. ADAMSON
NMACT Harwell

J.C. CLEGG
BNFL Risley

A.G. HAMLIN
NMACT Harwell,
United Kingdom

Abstract

INSTRUMENTAL REQUIREMENTS FOR EFFECTIVE NUCLEAR MATERIALS VERIFICATION AND CONTROL IN STATIC LOCATIONS.

Much non-destructive instrumentation so far applied or developed for nuclear materials verification and control has been derived from process control or laboratory applications. As such, it suffers from the inherent assumptions, among others, that (a) the sample offered to it, e.g. the thin "infinite depth" skin in the case of γ-spectrometry, is representative of the whole; and (b) the reading is not falsified by random or perhaps deliberate changes in parameters other than sample composition, such as background or sample position. In addition it is frequently bulky and expensive since the financial value of process control or laboratory measurement is often large enough to justify substantial expenditure at one site. All these are disadvantageous when the instrumentation is applied to nuclear materials verification and control away from the process situation. An analysis of the total safeguards system shows rapidly that the amount of nuclear material in process at any one time is a small fraction of the total. The bulk of the material to be safeguarded is in store or in static locations in a variety of isotopic and chemical forms, packages and backgrounds. This material may be held in such conditions for long periods and its continuing presence requires instrumental verification from time to time, even if it is to some extent protected by sealing. This paper discusses the instrumental and operational problems that are presented by the verification of this static material and assesses the contribution that can be made by currently developed techniques. It shows that these by no means solve the problem of frequent rapid and economical verification of the presence of nuclear material. The areas of uncertainty are discussed and the properties of materials that might contribute to the provision of effective instrumental techniques considered. Finally an attempt is made to define the broad specification for an acceptable instrument for verification and control of nuclear material away from the process area.

1. INTRODUCTION

1.1 The non-destructive instrumental methods currently being applied to, and developed for, nuclear materials verification and control, are based essentially on nuclear properties - γ and neutron radiation, and heat of decay.

1.2 Historically, radioactive material was first detected, then measured by its own radiations and on this basis many laboratory methods were eventually developed for process control applications. However, for many measurements the best technique remained by destructive analysis, which for many purposes provided more accurate values within the timescale required by the operator. It followed, therefore, that the non-destructive techniques which were applied were considered to be economically justified by plant management, bearing in mind their responsibilities for accountancy, safety or internal auditing of stocks, and therefore represented a technological elite among the choice available. It has seemed logical that the non-destructive methods needed for safeguards purposes should develop from these successful applications, particularly in view of the concentration of safeguards thinking upon the plant, and especially the process area situation, for which the instrumentation had been designed. Such apparent logicality can be questioned by reviewing first the distribution of the material in the total safeguards system and then the limitations of nuclear instrumentation as regards its control and verification. It is then possible to define more analytically the type of instrumentation that is required to maintain an efficient control and verification system.

2. THE DISTRIBUTION OF MATERIAL TO BE CONTROLLED

2.1 Nuclear materials are chemical products produced and processed in plants designed and operated for the most part on a basis of chemical industry practice. A specific feature of nuclear material, however, is that the properties which give rise to the need for safeguarding are scarcely affected by subsequent use. The bulk of the material is always recoverable with very small change in isotopic composition and it can be, and is, recycled in an almost closed cycle.

2.2 When the cycle is closed, the residence time outside the processing stages of the recovery plant or fuel fabrication plant will be at least two, and possibly three, orders of magnitude greater than that in the process areas. The amount of product held in static or slow moving stores or reactor cores will be greater, by the same factor, than the hold-up of the process plant. Where open cycles are used for economic or political reasons, the factor has no upper limit. The system is illustrated in Table I.

2.3 Therefore, consideration of the quantities of material involved would suggest that the major effort of verification should be applied to storage areas. It may be argued that, in terms of accessibility, the material in the process plant justifies more attention than the comparatively small total quantity would otherwise merit, but this increased accessibility is balanced by frequent quantitative measurements, plant security controls, and continuous observation.

2.4 In the storage areas, where most of the nuclear material is located, security measures and seals are applied, but the only assurances that these have not been circumvented are by verification measurements which prove that the correct material is still there in the correct quantity.

TABLE I. DISTRIBUTION OF MATERIAL IN THE NUCLEAR FUEL CYCLE

Stage	State of Material	Length of Stay
Ores, Residues	Stored	Months, years
Purification	In process	Days
Pure Materials	Stored	Months
Fabrication	In process	Days
Components	Stored	Days - months
Assembly	In process	Days
Fresh Fuel	Stored	Months
Reactor Operation	(Stored)	Years
Irradiated Fuel	Stored	Years
Reprocessing	In process	Days

The major problems of effective nuclear materials verification and control lie in the need to obtain this assurance in the static locations and the conditions found in such locations themselves constrain the design of effective instruments.

2.5 The four major factors affecting instrument design are as follows:-

(i) The unirradiated material is in a great variety of forms and in containers ranging from simple cans through drums of fresh fuel material to completed assemblies.

(ii) The items may be difficult to move from the location either because of size and weight or because of the necessary accounting and criticality control.

(iii) It is often necessary for economic reasons to pack material as closely as criticality considerations will allow. This automically increases the radiation background in the vicinity of the material to be verified and may also result in limitations on the working time permitted in the store.

(iv) Irradiated material comprises a very large part of the inventory of nuclear material, even in a closed cycle. This poses special problems both in technique and operation of equipment.

2.6 A brief specification of desirable instrumentation would therefore, at this point, indicate that it should be insensitive to the presentation of the item to be verified (2.5(i)), capable of verifying the contents of a fuel assembly or a large container in its storage location (2.5(ii)), be insensitive to background radiation (2.5(iii) and 2.5(iv)), and possibly be capable of remote operation. It would also be desirable that the complete equipment be portable and fast in operation. No instrumentation currently available even approaches such a specification.

3. LIMITATIONS OF THE PROCESS INSTRUMENTATION APPROACH

3.1 In making the apparently logical step from process instrumentation to verification equipment inadequate attention appears to have been paid to the following differences between management's routine measurements and measurements carried out for the purpose of verification.

(a) Confidence in the general nature of the item to be measured.

(b) Background radiation effects.

(c) Acceptable time scale for measurement.

(d) Accuracy required.

(e) Number of items to be measured.

3.2 Difference in confidence in the general nature of the item to be measured

3.2.1 A plant operator can justifiably make assumptions regarding the nature of material and measurement conditions which may be unacceptable to an independent inspector. He may assume, for example, that an effluent contains plutonium and a certain proportion of other gamma or neutron emitters only, that the background is stable or controlled, and that no attempt has been made to falsify the measurement by altering the background or the composition of the effluent. In general an inspector knows much less about a material and the circumstances under which he attempts to make a verification than the operator who handles the material. The process of confirming the nature of stored material, even when sound analytical data for it exist, is not easy. Material of hazardous nature may, prior to long term storage, be subjected to multiple containment - in extreme cases this could result in double polythene containment within double steel cans. The geometric arrangement of powder and containment, and the density of the powder are difficult to confirm and hence lower the confidence in a result when the gross item is examined by non-destructive techniques involving γ- or neutron-emission.

3.3 Differences in Background Radiation

3.3.1 On-line techniques such as effluent monitors, instruments to measure profiles in mixer settlers, or even low resolution γ-spectrometry to measure the uranium content of uranium/aluminium fuel plates, assume a background within known limits and of certain characteristics at an appropriate location with regard to the production line. Verification measurements

TABLE II. SOME TYPICAL ANALYSIS TIMES FOR NON-DESTRUCTIVE TECHNIQUE

Technique	Time for a single determination[a]
Monitoring instrument (either gamma, neutron or ultrasonic)	Instantaneous output
Determination of U-235 or Pu, using NaI	1 - 5 min
Segmented gamma scanner measurement of Pu	20 - 30 min
Neutron coincidence measurements	1 - 10 min
Active interrogation of soft waste (Cf-252 source)	10 min
Active interrogation using 14 MeV neutrons	30 - 60 min
Gamma spectrometric determination of plutonium isotopic composition	Overnight
Calorimetric measurement of plutonium in slags	3 days

[a] This does not include the time taken to present the sample to the measurement equipment, which is frequently more limiting on throughput than the actual analysis time.

frequently require to be made at less advantageous positions or in a storage location with high interfering background or with instruments which are not able to take advantage of the heavy shielding which may have been incorporated in the plant instrument.

3.4 Differences in Acceptable Timescale of Measurement

3.4.1 Plant control techniques are devised with a known requirement in terms of speed of measurement varying from very fast semi-quantitative on-line monitoring systems to methods, such as the measurement of residual fuel in hulls, where the time taken from the decision to make the measurement to receipt of the result could be several days for a single item, largely due to the transport and manipulation involved in transferring the item to and from the measurement station. It is also possible that long counting times may be required by plant operators to achieve the accuracy which they regard as necessary. Safeguarding authorities would certainly agree that accountancy must be based on such measurements, but they may not be compatible with the needs of verification when a large number of items may have to be measured in a short space of time. Some representative times for methods currently in use are shown in Table II.

3.5 Differences in Accuracy Required

3.5.1 At present, the major applications of non-destructive techniques to accountancy are to the assay of uranium/aluminium alloy and the measurement of waste [1][2]. The accuracy achieved in normal operation is as good or better than could be achieved by any available alternative techniques but if used for verification it might be necessary to sacrifice some accuracy to achieve adequate throughput.

3.5.2 Other non-destructive technique applications in the U.K. for verification purposes such as gamma spectrometry measurements and neutron coincidence counting, cannot match the accuracies obtained by the best destructive analytical techniques, but when applied to a range of similar items of well controlled isotopic composition, chemical composition and geometry, can provide a result which is no worse than $\pm$ 5% (1σ) on significant quantities of enriched uranium or plutonium. This is inadequate to give in itself sufficient assurance of the contents of a large store; one can be reasonably confident by other means that no more than a few percent of the contents of a large store have been diverted.

3.6 Differences in number of items to be measured

3.6.1 The speed of measurement required in order to verify an accumulation of items has an adverse effect on the accuracy of the determination and may reduce it to a point where measurements add nothing to existing assurances. This indicates a significant difference between the criteria of verification and process control.

3.6.2 The number of items available for verification in a large storage MBA may be of the order of 5,000. At a determination time of one minute per sample, the actual machine time taken to verify the stock would therefore be about 2½ working weeks. If the items were all similar and the random error of the short reading was $\pm$ 25% the total in the store would be known to $\pm$ 0.4% or the presence or absence of the equivalent of the content of 20 items in the store would be uncertain. In practice the items would not be similar and the error would be worse. On the other hand a similar instrument for process control could more realistically make 5,000 one-minute readings on a stable process stream and if it reported the integrated throughput to $\pm$ 0.4% it would be considered to have an excellent performance, and it would consume no operator time.

3.7 It will be seen that the foregoing confirm closely the need for a verification instrument to conform to the brief specification listed above (2.6), and it is reasonable to examine existing systems against these criteria.

4. PRESENT CAPABILITY AND FUTURE POTENTIAL OF ESTABLISHED NON-DESTRUCTIVE TECHNIQUES

4.1 The status of non-destructive techniques currently in use in the U.K. and certain states of the E.E.C. have been described previously [3] [4]. The main techniques are based upon γ- and neutron-radiation, and it is difficult to see that α- and β-techniques, which have even more severe transmission problems, can offer any potential for development for verification of material in static locations. Discussion is therefore confined to the two former and will be further limited to the discussion of the main fuel cycle elements, uranium and plutonium.

4.2 Gamma Spectrometry

4.2.1.1 Gamma spectrometry has a number of serious limitations in verification activities which it will be difficult to overcome by any development.

4.2.1.2 The γ-energies of interest range from roughly 120-420 keV, being limited at either margin by interfering emissions [5] [6]. They are therefore of relatively low penetration, and in the high density forms in which the materials measured occur, they have infinite depths of a few millimetres. Thus from a verification point of view, provided the container has a surface skin of approximately this thickness of the correct material, the remainder may be substituted by inactive material without detection.

4.2.1.3 Similar falsification is possible in n+γ-coincidence measurements for the determination of fission events e.g. Random Driver type instruments.

4.2.1.4 The small infinite depth for the radiations of interest means that the use of γ-spectrometry for quantitative measurements is limited. It gives useful information on the ratio of the isotopes of plutonium, but with uranium, since uranium 238 has no useful emission, the enrichment can be assessed only if the geometry of sources and detector used for calibration can be precisely reproduced with the samples to be measured. Where the sample is a rod or pellet in a larger container, a powder of unknown bulk density, in a non-standard container, or in multiple containment, this is not possible. Such cases are frequent in verification.

4.2.1.5 Where the activity of the sample is high, as with irradiated specimens or plutonium derived from high burn-up fuel, much of the time of the spectrometer is absorbed in sizing and rejecting unwanted pulses, while problems of peak overlap and pulse pile-up increase. These difficulties can be overcome in laboratory practice where counting times can be long and computer techniques used to unravel the spectrum, but in verification in the field with necessarily short counting times, the information ultimately available is rapidly reduced to a level where it is not possible to do more than approximately verify the label on the container.

4.2.1.6 Finally the physics of absorption in the detectors and in the case of high resolution detectors the necessity for intense cooling, renders current detectors bulky and suitable for use only externally to the sample.

4.2.2 The above problems relate to the generalised use of γ-spectrometry in verification. The technique is utilised successfully where they can be overcome, as in the measurement of MTR fuel elements or the enrichment of uranium hexafluoride in cylinders, and other such applications will no doubt be devised. More general applications to cases where the infinite depth limitation is not dominant will depend upon electronic developments to eliminate faster from the analyser the unwanted pulses, and reduction of the detector to a size where it can be used in the form of a probe in bulk material or to verify the inner pins of a fuel assembly.

4.2.3 In the U.K. there are a number of programmes aimed at developing improved high resolution detectors. Particular promise has been shown in the development of mercuric iodide and cadmium telluride crystals at the universities of Wales and Hull respectively in collaboration with the UKAEA. The former have not yet reached the stage of testing as detectors but encouraging results are being obtained from the cadmium telluride. A further development which seems likely to facilitate work in the field is the production of high purity germanium detectors cooled by the Stirling effect. The advantages are not limited to the elimination of the need for a liquid nitrogen supply, but also includes the increased ability to control the geometric arrangement of sample and detector.

4.2.4 It is perhaps a useful time to assess whether, in terms of Safeguards, it would be profitable to pursue the development of γ-spectrometry or to devote the available resources to other techniques.

4.3 Neutron methods

4.3.1 While neutron methods have the theoretical advantage that the sample under test is much more transparent than to γ-radiation, it rapidly becomes apparent in applying passive neutron techniques to large samples, that the behaviour of neutrons in such samples or in the counting head is not currently understood sufficiently for the measurements to form a precise accountancy or verification method. The response is not a linear function of mass, and varies with such factors as the chemical composition, spatial configuration and density of the material. If the behaviour of passive neutrons is in doubt the same limitations must apply to active interrogation measurements [7].

4.3.2 The response of materials to neutron methods is a function of the isotopic composition which cannot be inferred from the results. Thus the use of neutron methods in verification must be supported by auxilliary methods, e.g. γ-spectrometry for isotopic analysis and possibly by methods for non-destructive verification of density, spatial distribution and chemical composition. Combinations of determinations of this type must degrade the precision ultimately attainable.

4.3.3 As with γ-spectrometry the above difficulties arise with the general application of neutron methods to verification in static locations. The methods give valuable service in applications where the restrictions mentioned do not apply such as the rapid checking of numbers of nearly identical samples against a similar standard, or the measurement of small quantities of material in waste. [4] [8]

4.3.4 In the U.K. development is continuing on the construction and understanding of coincidence neutron detection systems. This has involved much computer simulation and it seems likely that quantities of plutonium containing up to 500g Pu-240 will be able to be assayed to within $\pm$ 3% (1σ) and it should also be possible to combine this with an isotopic composition determined by gamma spectrometry to give an assay of plutonium in a container to $\pm$ 5% (1σ).

4.4 Other Methods

4.4.1 Ultrasonic methods show promise of wide use in concentration measurements and volume measurements. At present the only proven application is to the measurement of heavy metal concentration in solvent where it is possible to guarantee the composition of the matrix. [9]

4.4.2 Thermo-luminescent dosimetry has been shown capable of giving useful results in the verification of inner fuel rods in fuel bundles. The technique is not rapid but the equipment is inexpensive and, since a large number of dosimeters can be employed, a high throughput can be achieved. The method should be capable of application to distributed inventories.

TABLE III. SPECIFIC GRAVITIES OF POSSIBLE SUBSTITUTE OXIDES

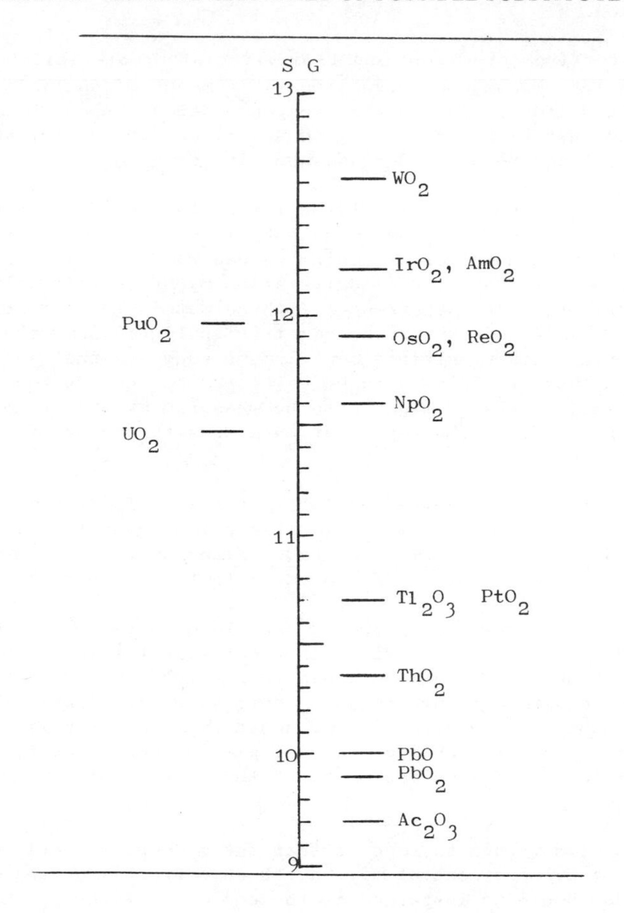

4.4.3 Calorimetry may be employed in verification measurements to the extent that it will provide a signature for a particular item [10]. It does not, at present, seem likely that the technique will be applied directly to the assay of any material for the purpose of verification because of the long equilibration times that are involved.

5. ULTIMATE POSSIBILITIES

5.1 Given the requirements of verification in static locations discussed in earlier sections, it seems unlikely that any of the above methods can be developed to give assurance in a reasonable time that not more than a significant quantity may be missing from a large quantity of material in a static location, bearing in mind the possibility of substitution anywhere in the location. One may therefore question whether the resources available for such development should not be switched to other systems.

5.2 The list of properties that may be used for non-destructive identification is not long.

5.3 Thermal output from plutonium and from irradiated material, but not from uranium, is large enough to give a possible means of analysis, but as with neutron methods for plutonium, the response depends upon the isotopic composition which must be determined by an auxilliary procedure, while with irradiated material the response depends upon its decay history.

5.4 The only other unique property of the materials considered appears to be density. Table III indicates that substantial substitution of material of similar density to uranium or plutonium oxides would be difficult on grounds of availability, but lesser adulteration might be difficult to detect with certainty. The occurrence of these materials in particle sizes ranging from a few microns to centimetres (pellets), and the remoteness of the material within multiple containment suggests that sonic rather than ultrasonic methods would need to be developed for non-destructive density measurement. Density might also be measured by absorption of radiation, but the result would represent bulk density and be of limited value in verification.

5.5 The possibility of development of instrumentation for rapid absolute verification of large numbers of items in static locations therefore appears slim, and it is necessary to consider alternative systems by which assurance can be obtained that the declared contents are correct.

5.6 Non-destructive methods are probably capable of development to a sufficient degree of precision provided that sufficient time can be allowed for the determination. This can be achieved for material in static locations if the requirement for frequent continuing verification can be avoided. This suggests a destructive or non-destructive verification when the material is first contained in can, pin or fuel element, followed by frequent verification of the integrity of the containment rather than of the contents.

5.7 This statement may seem to suggest that the concept of verification as an extension of process control may have been correct i.e. that the verification step should be installed at the end of whatever process results in containment, and that thereafter instrumentation should verify only the containment. However, unless containers and verification equipment can be absolutely standardised, which seems unlikely, it would be difficult to transfer assurance of containment from one MBA to another. Furthermore containment of an item at a static location provides opportunities for the containment assurance to be overcome. Whoever is responsible for the material must therefore assure himself that it is present when he receives the containers and that it remains in them until it is used or disposed of. He therefore still needs verification instruments.

5.8 Because of their technical limitations however these instruments can only be used effectively if they are part of a system which minimises the effect of these limitations. Such a system might involve three sub-systems:-

(1) A rapid system for proving that items had not been removed from the inventory.

(2) A slower system for checking that containment had not been compromised.

(3) A slow system of non-destructive analysis of high accuracy and precision for confirming receipts and shipments and for rechecking at intervals the items remaining on inventory.

5.9. For the first sub-system, rapid commercial systems, e.g. bar-code readers, are available and the data output can be made compatible with computerised accountancy systems so that this sub-system should present no difficulty. If highly secure labels, e.g. magnetic rather than optical bar-code, can be used, the system can contribute largely to the second sub-system.

5.10. For the third sub-system, some form of neutron measurement appears to be the most probable candidate with development of appropriate auxilliary measurements, such as isotopic and chemical composition and density, as may be necessary to uprate its present performance to the level of true accountancy. The methods available need to be developed as an integrated sub-system for general verification purposes and not as alternatives, since each has weaknesses, discussed earlier, which will prevent its providing alone the properties required.

5.11. Any weaknesses remaining in the sub-system should be as few as possible and should be protected by the other sub-systems. In the proposed three-part system the criterion that has been relaxed for the third sub-system is the time per determination. The maximum acceptable relaxation can only be determined by a careful analysis of the entire system but will obviously be a function of the number of receipts and despatches to be checked, the capital and operating costs that can be tolerated, and the intervals at which material retained in the inventory must be rechecked, taking into account the assurance provided by the second sub-system on integrity of containment. This frequency should be established on some logical and quantifiable procedure such as risk analysis.

5.12. The second sub-system is the one to which least effort has so far been dedicated.

5.13. Single-journey containers which can only be opened by destruction and which range from food cans to fuel pins are a possible element, but for economic reasons have somewhere to be in quantity production so that substitution is always possible.

5.14. Seals are another possible element but again many are objects of commerce, and so could be substituted. Increasing sophistication of seals, as with many aspects of physical security is likely to direct attack to other parts of the containment so that the second sub-system really needs to contain some element of overall surveillance.

5.15. The provision of this element allows a wider field of instrumentation to be developed than for the quantitative verification of the material since the measured factor is checked only for stability.

5.16. Thus possible lines of development are:-

- ultrasonic or other interrogation of the closures of single journey containers

- External interrogation of internal identifiers of sealed or single journey containers that would be difficult to duplicate

- Integration of radiation output from containers over consecutive periods or continuously
- Fingerprinting of radiation from containers by integration of two or more radiation outputs simultaneously and possibly continuously
- Continuous monitoring of weight
- Continuous monitoring of heat output
- Continuous check for absence of movement
- Continuous check for absence of approach

5.17. A number of these systems are under investigation in different laboratories. This will establish some elements of feasibility and choice but it seems unlikely that laboratory work will define the ultimate optimum application. This can only be determined by operational needs and practices, although these themselves may need modifying for safeguards purposes.

5.18. It is possible on general lines and by reference to the conditions in which they will be used, to define the general specification that such instrumentation will need to meet. This can be divided into portable and installed instrumentation.

5.19. Portable instrumentation must be:-

- Truly hand portable[1]
- Preferably battery operated
- Easily set up for operation with short stabilisation time
- Highly reliable, since false alarms can be expensive
- Rapid in measurement to reduce operator exposure
- Unaffected by background radiation
- Preferably capable of internally recording both coded label information from the item and the required data to cut down operator time, eliminate transcription errors, and permit rapid processing of data by computer
- Preferably capable of adaptation to remote operation for monitoring irradiated material

5.20. Installed instrumentation must be:-

- Cheap per unit if large numbers are required
- Highly reliable
- Capable of remote interrogation by data processing equipment either over consecutive periods or continuously

5.21. There is a distinct difference in these requirements. Portable instrumentation may prove to be a development from conventional nucleonic instruments although there is a clear trend towards non-nuclear aspects of non-destructive testing and electronic data processing. Installed instrumentation is almost entirely in the latter fields and may retain only a vestige of nucleonic instrumentation.

5.22. There is a clear need to develop these non-nucleonic, possibly less obvious lines of approach to the non-destructive verification of nuclear material if adequate control systems are to be achieved.

---

[1] Transportable equipment would be an inferior alternative where large numbers of observations distributed in a static location are concerned.

6. CONCLUSIONS

6.1 Instrumentation required for the verification of nuclear material held under safeguards has tended to evolve from laboratory and process control methods based on nucleonic principles, aided by an undue concentration of attention on process plants. Success has been achieved in particular applications but no generally applicable instrument has been developed.

6.2 An analysis of the situation shows that the bulk of material to be accounted for and controlled exists in static locations, and for the verification of this material different criteria apply.

6.3 Recognising this situation, it is possible to derive broad specifications of the type of instrument that is required to control and verify the material under operational conditions.

6.4 It is clear that no one nucleonic system will meet this specification and that rather than continuing the development of individual systems, the available choice should be combined to develop a system having the required precision and accuracy and capable of general application.

6.5 This system will itself have deficiencies with respect to the draft specification - the most probable being the time needed for a measurement - and will itself need to become a sub-system in a total system designed to overcome the deficiencies.

6.6 The remaining sub-systems offer the widest field for possible development effort at the present time and will probably involve areas of instrumental expertise very different from the nucleonic areas so far exploited for nuclear material control.

REFERENCES

[1] SINCLAIR, V.M. and ADAM, W.B. "Procedures used for the accountancy of Uranium-235 in the fabrication of highly enriched uranium fuels" Safeguarding Nuclear Materials (Proc. Symp. Vienna 1975) 1, IAEA, Vienna (1976) 501.

[2] McDONALD, B.J., FOX, G.H. and BREMNER, W.B. "Non-destructive measurement of plutonium and uranium in process wastes and residues" Ibid., 2, p. 589.

[3] ADAMSON, A.S. and HAMLIN, A.G. "Present Status and Future Outlook for Non-Destructive Analysis in Nuclear materials accountancy in the United Kingdom" American Society for Metals Symposium, Salt Lake City, 1978 In course of publication.

[4] ADAMSON, A.S. "Established application of non-destructive techniques for nuclear materials measurements, control or verification; reported to ESARDA" ESARDA-6 or AERE R-9167.

[5] GUNNINK, R. and EVANS, J.F. "In-line measurement of total and isotopic plutonium concentrations by gamma-ray spectrometry" UCRL-52220 (1977).

[6] BANHAM, M.F. "The Determination of the Isotopic composition of plutonium by gamma-ray spectrometry" AERE R-8737 (1977).

[7] LEES, E.W. and ROGERS, F.J.G. "Experimental and theoretical observations on the Euratom Variable Dead Time Neutron Counter for the passive assay of plutonium" IAEA-SM-231/51 (1978).

[8] TERREY, D.R., HORNSBY, J.B., BROWN, F. and ROSS, A.E. "Verification of the plutonium content of irradiated fresh elements from a zero energy fast reactor" UKAEA, COS 25 (1973).

[9] ASHER, R.C. et al "Ultrasonic technique for on-line surveillance and monitoring of process plant; some applications in nuclear fuel reprocessing" Proceedings of Soc.Chem.Industry Conference, London 1977.

[10] Mound Laboratory Progress Reports for the Division of Safeguards and Security, 1976 and 1977.

DISCUSSION

P. FILSS: Could you give some futher details regarding the measuring times indicated in Table II of your paper? What are the assumed contents of fissile material per sample and what is the accuracy required which will lead to the measuring times quoted?

A.G. HAMLIN: The times shown are indicative only, but in general the longer times relate to cases where information has to be obtained on the isotopic composition of the material. This is because certain count-rates are low. The same principle of long counts applies also if the actual quantity of an otherwise prolific emitter is low.

M. CUYPERS: The development and use of NDA equipment for wide application has the disadvantage of loss of accuracy and precision of the assay; great caution is therefore required. Specific developments generally have much higher reliability, and automation of operation and data treatment is simplified. Euratom inspection instruments tend to be specifically oriented for this reason. In regard to the definition of specifications, the final use of instruments for attribute, or attribute-in-variable-mode, verification is of fundamental importance. NDA is particularly suitable for the latter.

A.G. HAMLIN: In general I do not disagree with your comments. In fact, it was considerations such as these, together with the realization that no general instrument was ever likely to be developed that would meet all the requirements of verification, that prompted the development of the outline specifications for instrument systems described in my paper.

G. Robert KEEPIN: The major thrust of your paper is certainly well-taken, namely that the bulk of nuclear material in many fuel-cycle facilities resides in static inventory, so that there is a corresponding need to give appropriate attention to material in static locations and to utilize containment and surveillance technology to the fullest extent practicable.

The traditional development of NDA instrumentation has indeed focused on in-process material because the measurement and accountancy of nuclear materials under dynamic flow conditions in a process plant have generally posed the more difficult problems, thus requiring a greater sustained level of research and development activity. Also, the variety of physical and chemical forms found in process lines has required a correspondingly wide range of instruments for accurate measurement and accountancy of these in-process materials.

With reference to the contrast between the large van and the small car, I would comment that a larger vehicle is required to house the more complicated and wider range of instrumentation needed for in-process measurements, while the C/S equipment is generally more compact, more portable and more universally applicable in various types of facilities. Both the measurement/accountancy and the C/S functions, and both types of equipment, are clearly needed for effective overall safeguards and security of nuclear materials. I therefore certainly agree on the importance of achieving a balanced overall system incorporating modern measurement technology (both NDA and destructive assay), C/S technology and physical security in a total *integrated* safeguards and security system that is appropriately optimized for each type of fuel cycle facility.

J. BOUCHARD: Mr. Hamlin, there is a fundamental difficulty in reconciling the accuracy requirements you mentioned with standardization of combined physical measurements in more or less portable equipment. You have not considered special cases such as irradiated or non-irradiated materials or concentrated or diluted materials. High accuracy is of course essential but it can be obtained only by determining the most suitable method or combination of methods for each individual category of materials. Only then is it possible to consider the development of portable equipment based on these methods.

A.G. HAMLIN: I do not think that we are in disagreement. Accurate measurement of the contents of a container of nuclear material requires the best available method, which may be quite incapable of being made portable or of application to other forms of material. We are proposing that the integrity of this measurement should be assured by instruments which could be portable and standardized rather than that the original measurement should be repeated at each verification.

BEHAVIOUR OF THE QUADRUPOLE MASS-SPECTROMETER TOWARDS VARIOUS NOBLE GASES

H. ULLAH
Pakistan Atomic Energy Commission,
Islamabad, Pakistan

Abstract

BEHAVIOUR OF THE QUADRUPOLE MASS-SPECTROMETER TOWARDS VARIOUS NOBLE GASES.

This paper describes an experimental set-up for the analysis of noble gases with a view to subsequently extending its scope for the isotopic measurements on krypton and xenon, which evolve from the dissolver tank of a reprocessing facility. For safeguards purposes this analysis, in conjunction with isotopic correlation techniques, gives valuable aid in material accountancy of the fissile material present in the accountability vessels. General behaviour of the instrument towards helium, argon, krypton and xenon compared with air and nitrogen shows that satisfactory performance can be achieved by using leak valves capable of admitting large pressure variations of gases into the ion source and by frequent calibration of the instrument.

INTRODUCTION

Safeguards verifications of the contents of fissile material at the input accountability tank, at waste points and at the tail-end of a reprocessing facility, mainly involve volume-concentration measurements. These measurements could be numerous during a regular reprocessing campaign including those carried out by the inspectors themselves. The methods involved in these types of measurement are highly specialized and time-consuming. For routine purposes a back-up method is thus highly desirable. Isotopic correlation techniques have recently been evolved to serve this purpose. Correlations between fission isotopes (stable), especially those of fission noble gases such as krypton and xenon and plutonium build-up, have been found to be very useful as they involve fairly precise [1], tamper-proof and more rapid analysis procedures. However, very little experimental literature is available [2] in this field because of the recent evolution of the technique, and also because satisfactory interlaboratory comparison systems (especially those involving data-handling) are still in the course of final settlement. This paper is an attempt to present data on the mass-spectrometric behaviour towards various noble gases with a view to extending the experience thus gained for a regular and fool-proof

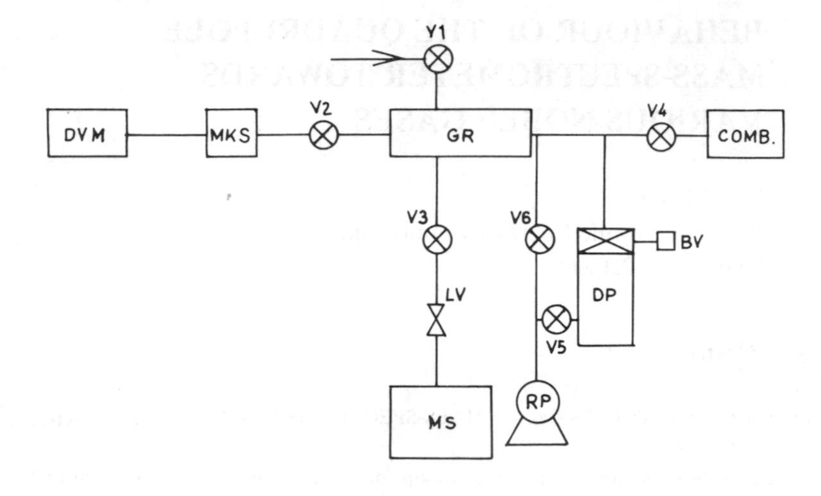

GR:	GAS RESERVOIR (4 LITRE)
DP:	OIL DIFFUSION PUMP
V_1 - V_6:	RIGHT-ANGLED MAKE-BREAK VALVE
BV:	BUTTERFLY VALVE
LV:	LEAK VALVE (VARIAN MODEL 951-5100 OR GRANVILLE PHILLIPS c-202)
COMB:	COMBITRON ION GAUGE (10^{-5}-10^{-2} TORR)
MKS:	CAPACITANCE MANOMETER (BARATRON - 210)
RP:	ROTARY PUMP
DVM:	DIGITAL VOLTMETER (DM - 2020)
MS:	QUADRUPOLE QMG - 311

FIG.1. Schematic diagram of apparatus set-up.

analytical procedure applicable to the determination of plutonium in the irradiated fuel. A quadrupole-type mass-spectrometer was used for all the analyses as this instrument is very sensitive for the gases studied.

1. APPARATUS

A schematic diagram of the set-up is shown in Fig.1. Apart from sample introduction and mass analysis, this arrangement can also be used for sample transfer from gas reservoirs to sampling bottles. Vacuum measurements on various points of the inlet system are determined by a membrane manometer (MKS-Baratron), whose signal is also fed to the digital voltmeter (DVM) for

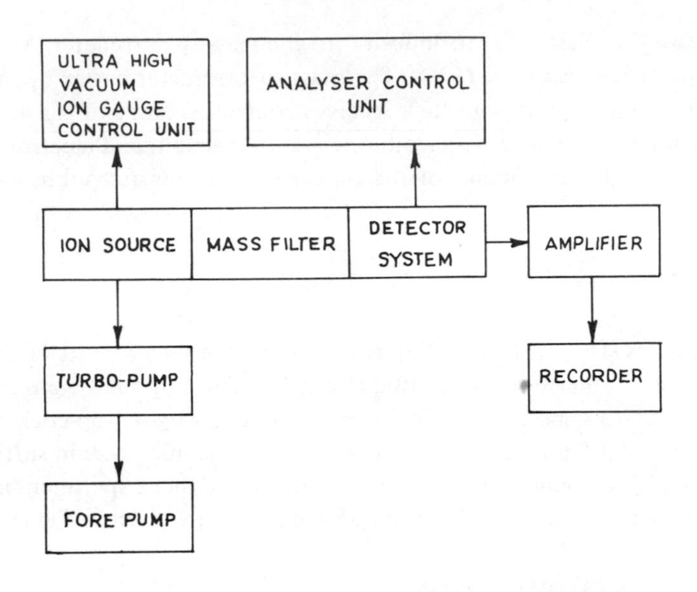

FIG. 2. Schematic diagram of the quadrupole mass-spectrometer.

TABLE I. WORKING PARAMETERS

1.	Warm-up time of the instrument	2 h
2.	Filament current	1 mA (ca. 95 eV)
3.	Resolution M/ΔM	260 (19% valley)
4.	Ion source pressure	Less than 10^{-6} torr
5.	Amplifier damping	40–50% of full scale
6.	Detectors	Faraday cup and SEM
7.	Peak heights	Accepted after normalization and correction. Average of 3 readings not differing from each other by more than 1 part in 1000.
8.	Focusing	Focusing of ion beams was performed by appropriately varying ion-optical voltages and optimizing the ion deflection in the analyser.

accurate readings. Gases are introduced into the mass-spectrometer (MS) through a precision leak valve (LV). The mass-spectrometer has an open ion source (selection impact type) which receives controlled entry of gas directly on to the filament. The mass-spectrometric set-up (vacuum and electronic parts) is given in Fig.2. Details of the mass-spectrometer are published elsewhere [3].

2. EXPERIMENTAL

Noble gases (He, Ar, Kr and Xe) were obtained in 99.9% purity from sealed ampoules supplied by Air Liquid (France). The ampoules were converted by glass-blowing to sample reservoirs by appropriate air-tight stop-cock system. Vacuum ($10^{-5}-10^{-6}$ torr range) was maintained in the inlet system sufficiently long so as to give the minimum background spectra. Other experimental conditions generally maintained for the present work are given in Table I.

3. RESULTS AND DISCUSSION

For most of the experimental runs a Granville Phillip's leak valve was preferred to a Varian valve as the former allowed a reasonably large admittance of samples over ion gauge readings which were practicable to measure. The linear relationship between the ion gauge (IG) reading (representation of ion source pressure) and the membrane manometer (MKS) was checked using air, N_2, He, Ar, Kr and Xe. The results are shown in Figs 3 to 8. (Ion gauge readings are given as a function of inlet pressure indicated by the MKS reading.) Relevant data are given on each figure itself including equations of linear regression and goodness of fit. MKS sensors used were 1000-mm head (1–1000 torr) and 1-mm head (0.1 to 1 torr). Pressure readings below 0.1 torr were uncorrected owing to the non-availability of calibration data from the manufacturer. However, a reliability test on this equipment was published by Lorist and Moran [4] using the volumetric pressure division method. This test showed linearity between 2×10^{-4} and 5×10^{-6} torr (1.006 ± 0.004 slope and $3.5 \pm 3.5 \times 10^{-7}$ intercept) with an accuracy of 0.6% ($+ 4 \times 10^{-7}$ torr). Thus, for the present work, the pressure correction at the experimental readings was not important. Nevertheless, non-linearity was observed (Fig.4) for air between 0.1 and 0.044 torr (0.044 torr was the minimum obtainable pressure in the inlet system under the experimental conditions as indicated by MKS). Linearity was observed for all gases studied under a wide range of inlet pressure variations.

An indirect test for checking the calibrations of both 1-mm and 1000-mm MKS heads was also provided by the above experiments, especially between 100 torr and 1 torr for a 1000-mm sensor and between 0.1 torr and the

Text continued on p. 703

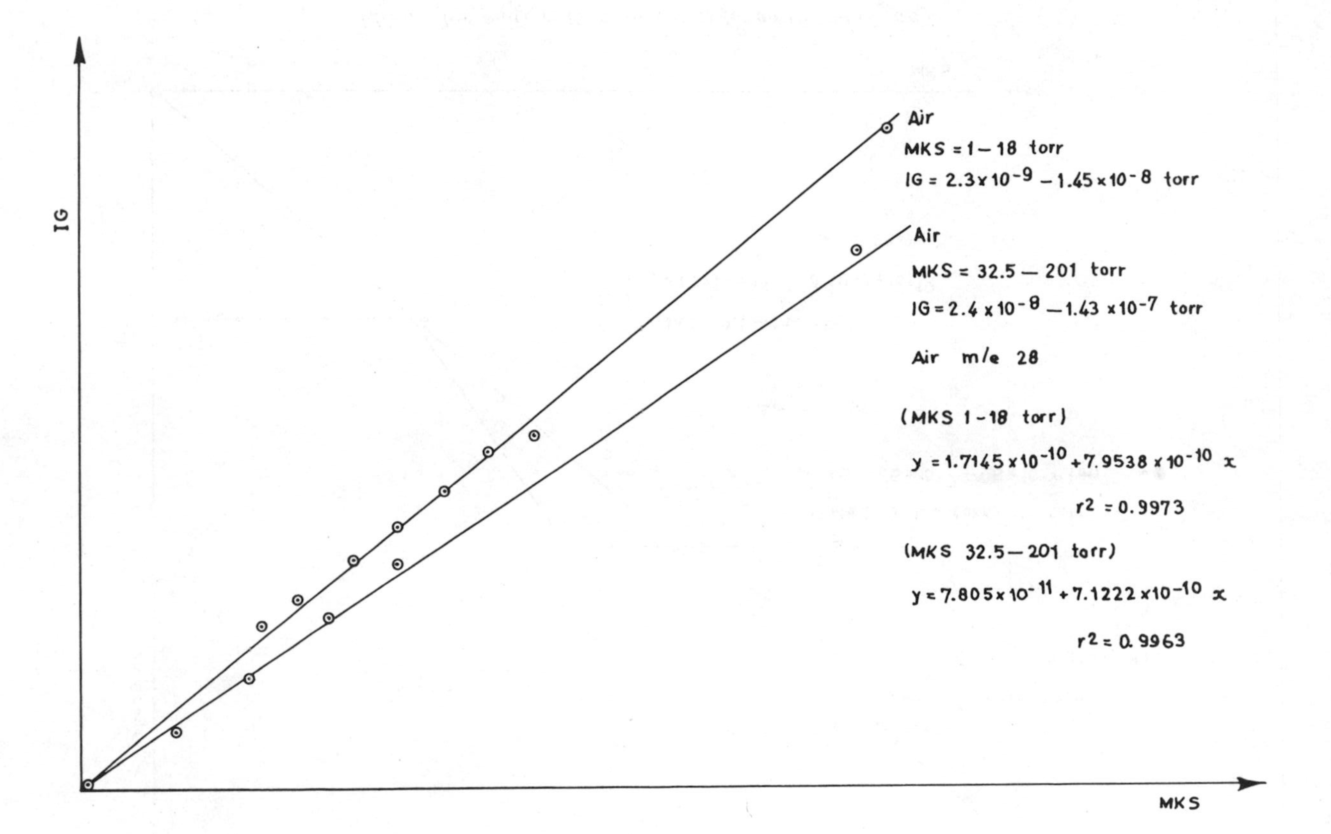

FIG.3. Ion gauge vs Baratron curves for air (1–18 torr).

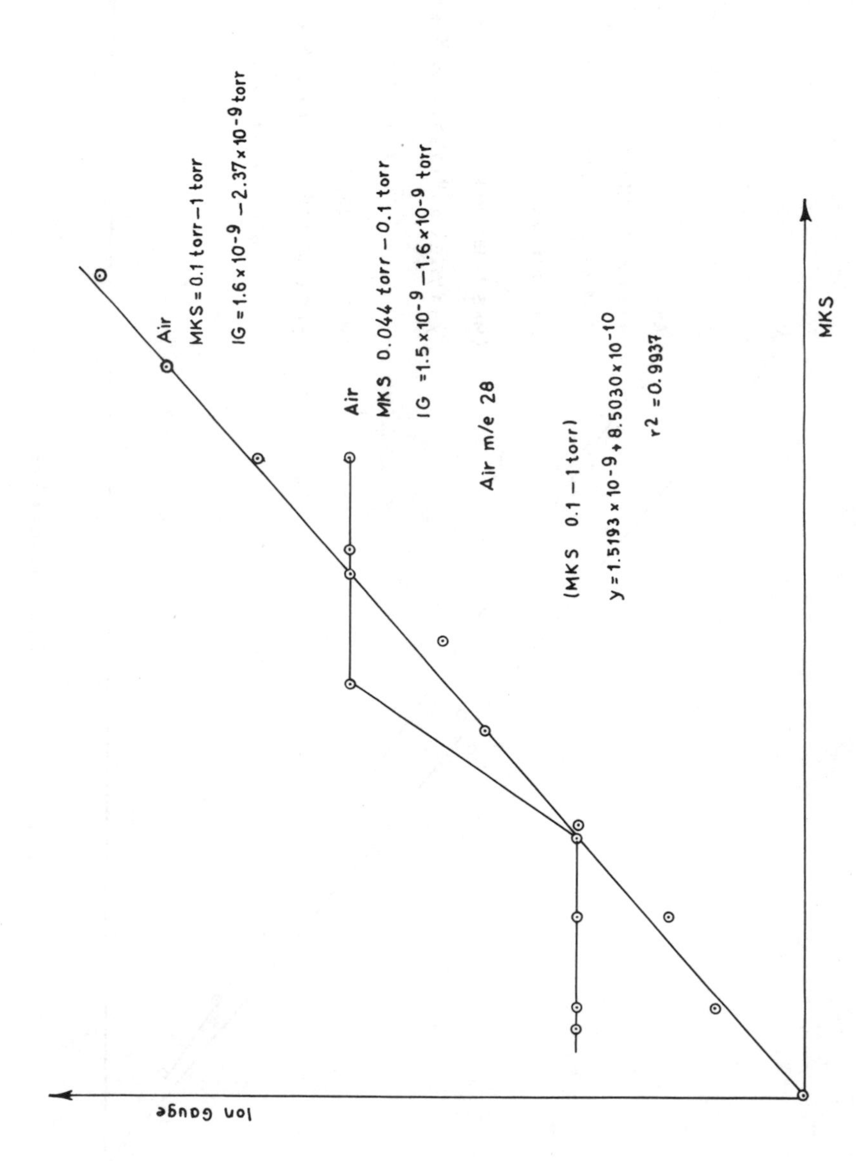

FIG.4. Ion gauge vs Baratron curves for air (0.044–1 torr).

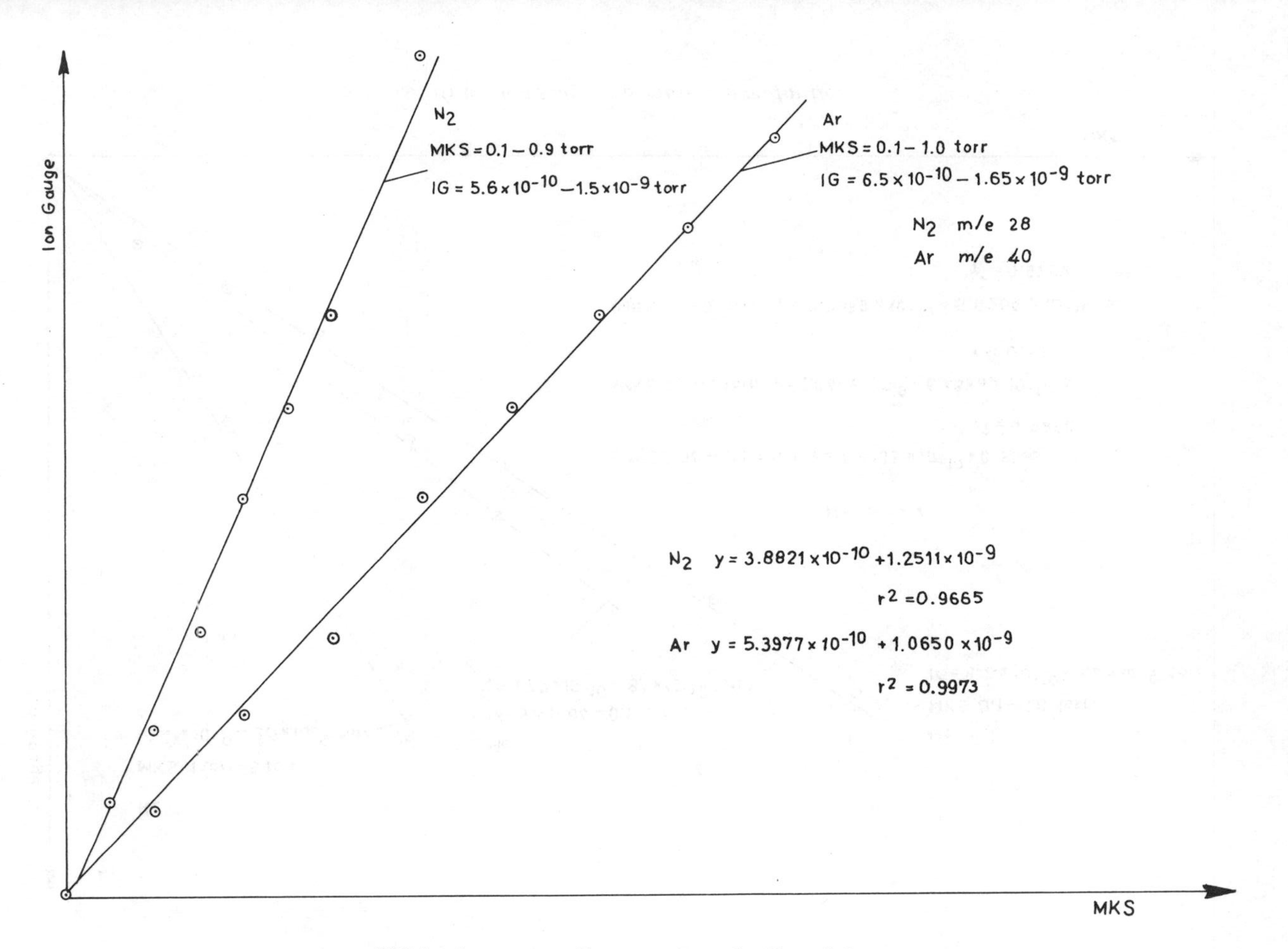

FIG.5. Ion gauge vs Baratron curves for N_2 and Ar.

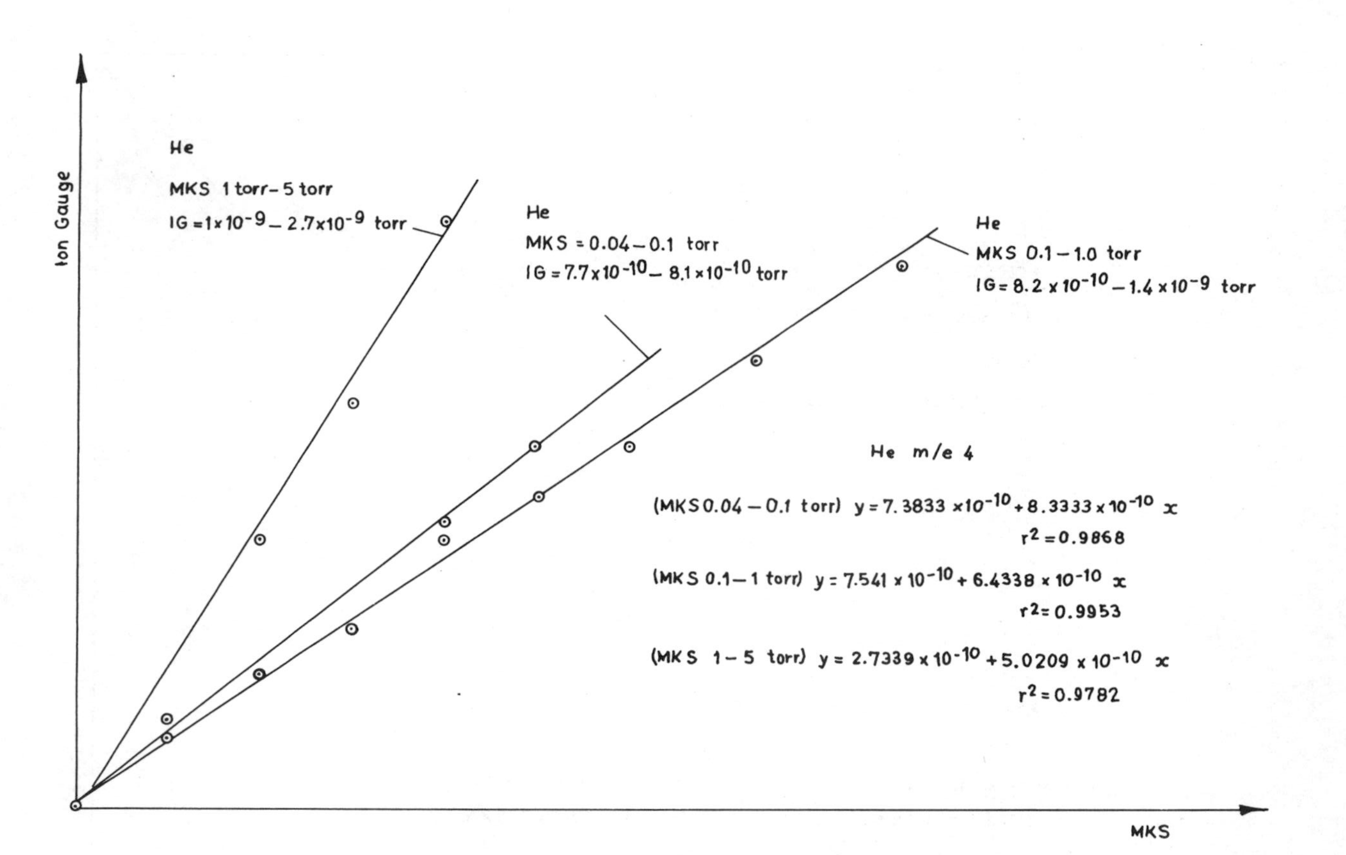

FIG.6. Ion gauge vs Baratron curves for He.

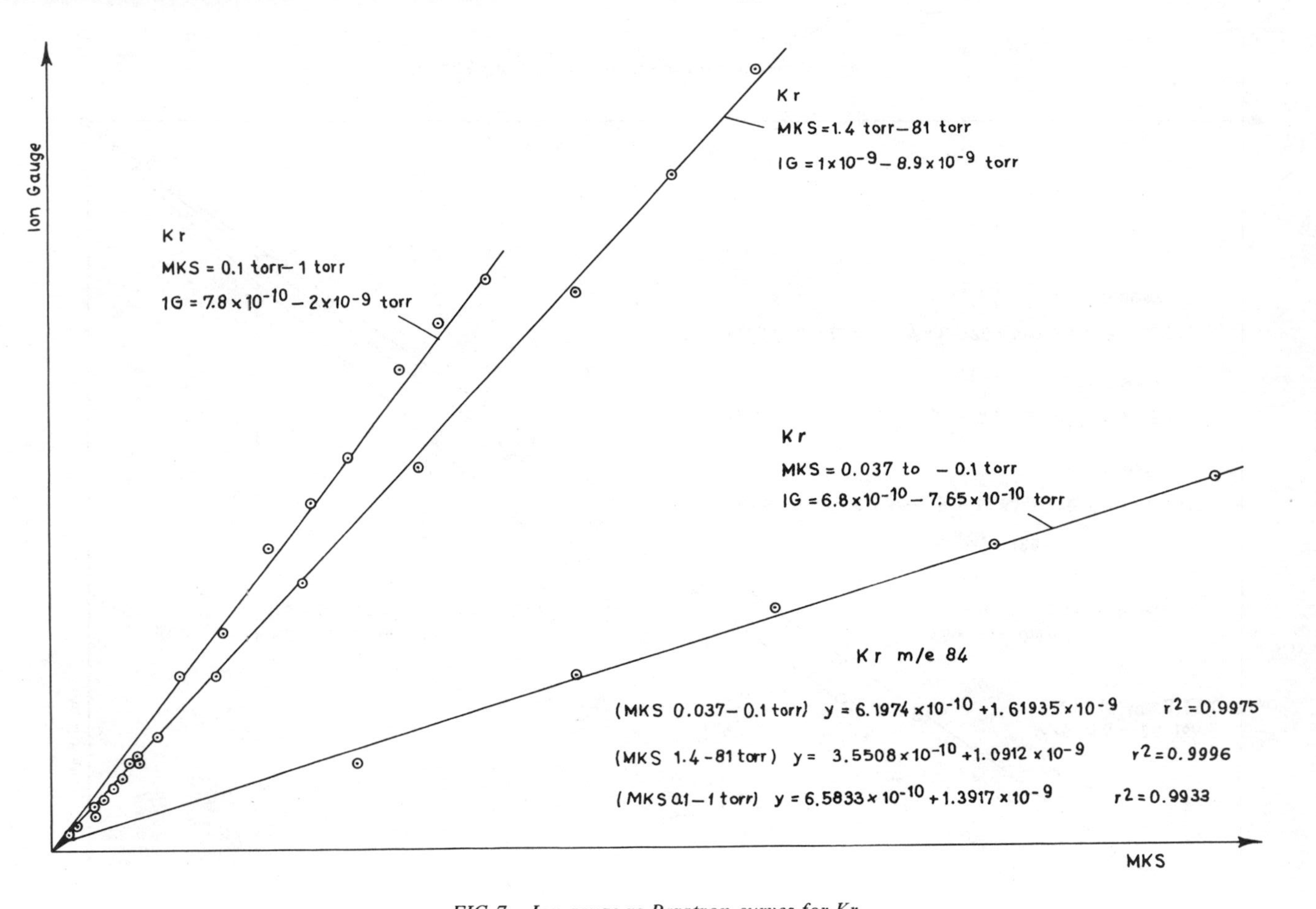

FIG.7. Ion gauge vs Baratron curves for Kr.

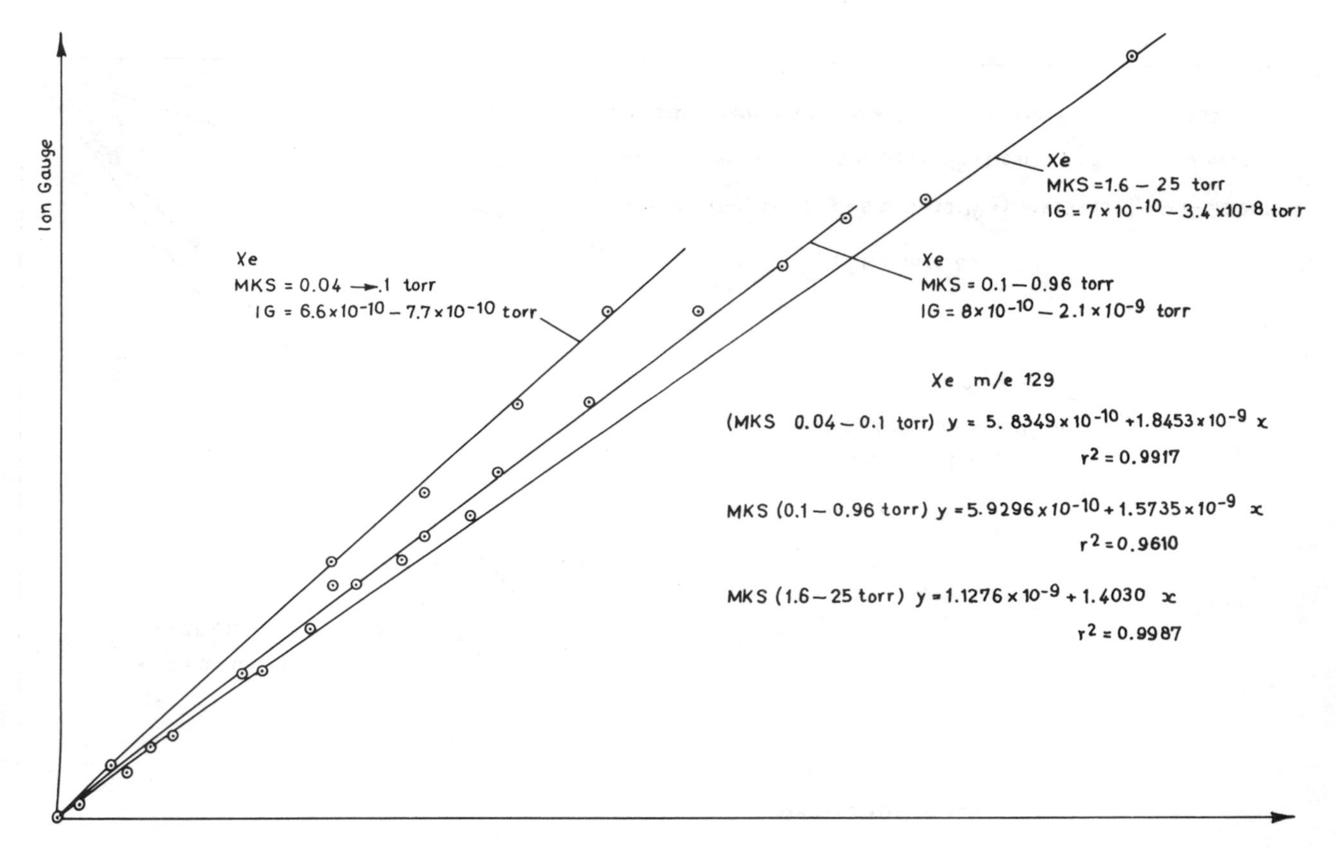

FIG.8. Ion gauge vs Baratron curves for Xe.

TABLE II. SENSITIVITY VARIATIONS

Gas	Date	Sensitivity (peak height/inlet pressure)	Variations (%)
Ar	16.12.76	1.25	4
	16.12.76	1.20	
	17.12.76	6.4	1
	17.12.76	6.3	
	20.12.76	5.2	
	22.12.76	28.0	1
	22.12.76	28.5	
	5.1.77	96.0	2
	5.1.77	98.0	2
	7.1.77	97.0	1
	7.1.77	98.0	
	12.1.77	60.0	1
	12.1.77	61.0	
	1.2.77	53.3	0.6
	1.2.77	52.0	
	7.2.77	51.6	7.7
	8.2.77	56.0	
He	17.1.77	63	0
	18.1.77	63	
Kr	12.1.77	20.5	0
	12.1.77	20.5	0.5
	13.1.77	20.6	
Xe	13.1.77	0.005	0
	14.1.77	0.005	

minimum possible torr value for the 1-mm sensor. This was important because the calibration data provided by the manufacturer consisted only of the 1000–100 torr ranges and 1–0.1 torr for a 1000-mm and a 1-mm head, respectively.

Consistency checks on the sensitivities of Ar, He, Kr and Xe indicated that short time consistency (1–2 days) was fairly good. Several experiments were carried out with Ar to study both short-term and long-term variations. Table II gives the typical results.

Short-term (1-day variations) results for He, Kr and Xe are also included in Table II. Argon shows a range of 1–4% variations in the sensitivities obtained at an interval of about eight hours on the same day. Sensitivity changes were inconsistent for more than a 24-hour-interval experiments. Thus, in one case a variation of 7.7% for a 32-hour interval, and in another a variation of 2% for 56 hours was obtained. For He, Kr and Xe the 32-hour-interval experiment showed no sensitivity change. All these sensitivity experiments were performed using the inlet pressure range of 0.1–1 torr. To check the validity of the experiments vis-à-vis the leak-valve performance over a wide range of inlet pressure, the air sensitivity was determined over an inlet pressure range of 1–100 torr at a 32-hour-interval period. A percentage difference of 1.8 (2.12 vs 2.16) was observed, which indicates a long-range inlet pressure admittance capability of the leak valve under the existing experimental conditions.

A careful scrutiny of the background contributions during the sensitivity determinations of all gases under study revealed that no significant memory effect existed. Adequate pumping between analysis intervals gave normal background spectra.

4. CONCLUSIONS

(1) The criteria of better performance of an entire set-up rests largely on the capability of the leak valve to allow a large pressure variation to be admitted into the ion source with fixed leak-rate and within the safe ion-gauge measurement capability.

(2) Short time calibrations (ca. 24-hour interval) should be performed to obtain consistency in sensitivity valves. This has direct bearing on the accuracy of the results. Ideally, each measurement run should be performed after calibrating the mass-spectrometer.

(3) Under the present experimental conditions, noble gases do not show memory effects.

REFERENCES

[1] BERG, R., FOGGI, C., KOCH, L., KRAEMER, R., WOODMAN, F.J., in Practical Aspects of R and D in the Field of Safeguards, Proc. ESARDA Symp., Rome (1974) 183–211.

[2] (a) FOGGI, C., FENQUELLUCCI, F., in Safeguarding Nuclear Materials (Proc. Symp. Vienna, 1975) 2, IAEA, Vienna (1976) 425–38.
(b) KOCH, L., GERIM, F., DE MEESTER, R., in Contributions to the Joint Safeguards Experiment MOL IV at the Eurochemic Reprocessing Plant (BEETS, Ed.), BLG-486 (SCK-CEN), pp KO 1–4 and Appendices.

[3] Quadrupole Mass Spectrometer QMS-311, Balzer's Aktiengesellschaft für Hochvakumtechnik und Donne Schichten, Jan 1974/DN 5480.

[4] LORIST, G., MORAN, T., Rev. Sci. Instrum. **46** (1975) 140.

DISCUSSION

L. KOCH: I find this application of a quadrupole very interesting, as we are also working in this field. For comparison with our own work I should like to know something about the reproducibility of the isotopic abundance measurement with constant inlet pressure and the change in mass discrimination as a function of inlet pressure.

H. ULLAH: I am afraid I cannot give you any information, as we have not yet performed any investigations in this direction. However, isotopic abundance measurements and mass discrimination will constitute the next step in our studies.

IAEA-SM-231/108

EXPERIENCE WITH THE AUTOMATIC EVALUATION OF MEASUREMENT DATA FROM A THERMAL IONIZATION MASS-SPECTROMETRY SYSTEM

J.G. VAN RAAPHORST, B. BEEMSTERBOER,
P.A. DEURLOO, J.E. ORDELMAN
Netherlands Energy Research Foundation,
Petten (N.H.),
The Netherlands

Abstract

EXPERIENCE WITH THE AUTOMATIC EVALUATION OF MEASUREMENT DATA FROM A THERMAL IONIZATION MASS-SPECTROMETRY SYSTEM.

A thermal ionization mass-spectrometry system with automatic data evaluation is described. The experimental procedure for determining the isotopic composition of uranium and plutonium is given. The experimental results obtained with standard materials, uranium as well as plutonium, are discussed. The stability of the system in terms of bias correction factor is described as a function of time. The influence of some experimental parameters on this stability is shown.

INTRODUCTION

The use of computers is often opposed by the statement that obtaining optimal results requires human interference. Also in mass spectrometry the role of the on-line computer is often a non-decisive one. The mass spectrum is measured in one or another way and the decision which data are used for the final calculation is made by the operator. In our system the computer is equipped with a rather complete statistical system which gives the opportunity to evaluate the data in a sophisticated way without interference of the operator. A description of the system is given and results obtained with standard materials are discussed.

INSTRUMENTATION

The mass spectrometer system used is a V.G. Micromass 30 B system, initially developed at AWRE, Aldermaston. It is a 12 inch radius 90° sector single focussing mass spectrometer. The ion source accepts a filament-barrel loaded with up to six filament beads. For detection a Faraday cup as well as a Daly detector are available. The data system is based on a DEC PDP 8 computer with 16 K of store, dual floppy disk, teletype and visual display.

The so-called Field Memory Unit provides a possibility to apply peak jumping by a stepwise change of the magnetic field controlled by a Hall probe. The basic unit has a program for sequential peak-jumping through up to sixteen selected channels (field values). The Digital Field Programmer supplies a digital signal to the electro-magnet supply via a digital to analog convertor. After changing the magnetic field there is a delay time for settling. Thereafter the computer reads signals from the digital voltmeter till the end of the prese t read time, in our case 4 seconds.

MODE OF OPERATION

Uranium

In the case of uranium the isotopes 238, 236, 235, 234 and sometimes, when dealing with spiked material also 233, are measured. As already mentioned the Field Memory Unit can be programmed up to sixteen steps.

For unspiked uranium the program is
238, 236, 235, 234
238, 236, 235, 234
238, 236, 235, 234
238, Background, Bg., Bg.
This cycle is carried out five times. In the case of spiked uranium only 14 positions from the 16 are used.
238, 236, 235, 234, 233
238, 236, 235, 234, 233
238, Bg., Bg., Bg.   This cycle is performed 8 times.

Plutonium

For plutonium again a sixteen step cycle is used.
239, 240, 241, 242, 238, 235
239, 240, 241, 242, 238, 235
239, B.g., B.g., B.g.
This cycle is carried out 8 times.

The intensity measured at the 235 position is used to correct the 238 intensity, in the case of uranium contamination. However, the value of the 238 intensity is usually obtained by alfa-spectrometry.

EVALUATION OF DATA (1)

a. Background correction

At the end of each cycle the background is measured three times. The first value is rejected to avoid memory effects in the background. The other two data are used for the correction of the other intensity data by subtraction.

b. Drift correction

Usually the intensity of the signal is decreasing or increasing. The change in the value of the intensity of the main isotope (238-U, 239-Pu) is used for the correction, by linear interpolation, of the other intensities.

c. Calculation of ratios

From the corrected intensities the ratios are calculated, all against the main isotope. For unspiked uranium three sets of 15 ratio values are obtained 236-U/238-U, 235-U/238-U and 234-U/238-U. In the case of spiked uranium 4 sets of 16 ratio values: the already mentioned ones and 233-U/238-U. A plutonium measurement delivers 4 sets of 16 ratio values 242-Pu/239-Pu, 241-Pu/239-Pu, 240-Pu/239-Pu and 238-Pu/239-Pu. For the application of the student-t-test, later in the program, also the ratios of the corrected background values to the reference peak are calculated. Of course the mean of these values is zero.

d. Application of Dixon criterion (2,3)

The sets of ratio values are tested on the presence of outliers. The test applied is the one of Dixon (2,3), with a 95 % confidence interval.

e. Student-t-test

This test is performed to check if there is statistical evidence for a difference between the concerning series of intensity ratios and the background, the so-called null hypothesis. In our case the null hypothesis is that the difference is zero. The confidence interval is 95 %.

f. The computational program ends with the calculation of the mean of all ratios and the coefficient of variation.

Experimental Procedure

In thermal ionisation mass spectrometry the reproducibility of the heating and stabilising procedure is of great importance. No high quality results can be obtained without a great attention for this part of the measurement. For that reason the procedures applied for obtaining the reported results are given.

Sample Preparation

Uranium

The uranium is applied on the rhenium side filaments in a 1 molar nitric acid solution. 1 microliter solution containing 2.5 microgram uranium is put on the filament and evaporated till dryness by passing a current of 1 A during 40 seconds. This procedure is repeated so, 5 microgram of uranium is applied totally. After the evaporation currents of 1.2, 1.4, 1.6, 1.8 and 2.0 A are passed successively through the filament during 5 seconds until the orange oxide is visible.

Plutonium

Total amount applied 0.1 microgram. Application of the same procedure as with uranium with only difference that always the whole procedure is followed as there is no visible oxide present.

Heating Procedure

The heating and stabilising procedure is the same for uranium and plutonium

t (minutes)	Centre Filament	Side Filaments
0 - 5	1 A	0.2 A
5 - 10	2 A	0.4 A
10 - 15	3 A	0.6 A
15 - 20	4 A	0.8 A
20 - 25	4.5 A	1.0 A
25 - 30	5 A	1.2 A
30 - 35	5.2 A	1.4 A
35 - 40	5.2 A	1.6 A
40 - 45	5.2 A	1.8 A
45 - 50	5.2 A	2.0 A
50 - 55	5.2 A	2.2 A
55 - 60	5.2 A	2.2 A
60	start measurement	

The aim is to create an ion current of 5.10^{-12} A for the main isotope (238-U, 239 - Pu). In here lies the first difficulty concerning reproducibility. In the case there is not a sufficient current at 2.2 A the filament current is increased in steps of 0.2 A with stabilising times of again 5 minutes. In the past the maximum current has been 2.8 - 2.9 A. When this happens the measurement will be delayed for at least 15 minutes which has consequences for the accuracy.

RESULTS

Mass discrimination

The term mass discrimination is misused in this context as it means the bias in the measurement caused by the difference in mass of two isotopes. In this article the term correction factor will be used for the ratio of the result obtained with a reference material and its certified value.

The most frequently used reference material for uranium is the NBS 500. The results with this material in a rather short period are shown in figure 1. It is obvious that in weeks 11 and 12 there was something wrong with the machine. Furthermore, in the period from week 13 - 26 the ratio varied between 1.0070 - 1.0021 which is quite acceptable. The results where obtained without human interference for determining which set of ratios should be used.

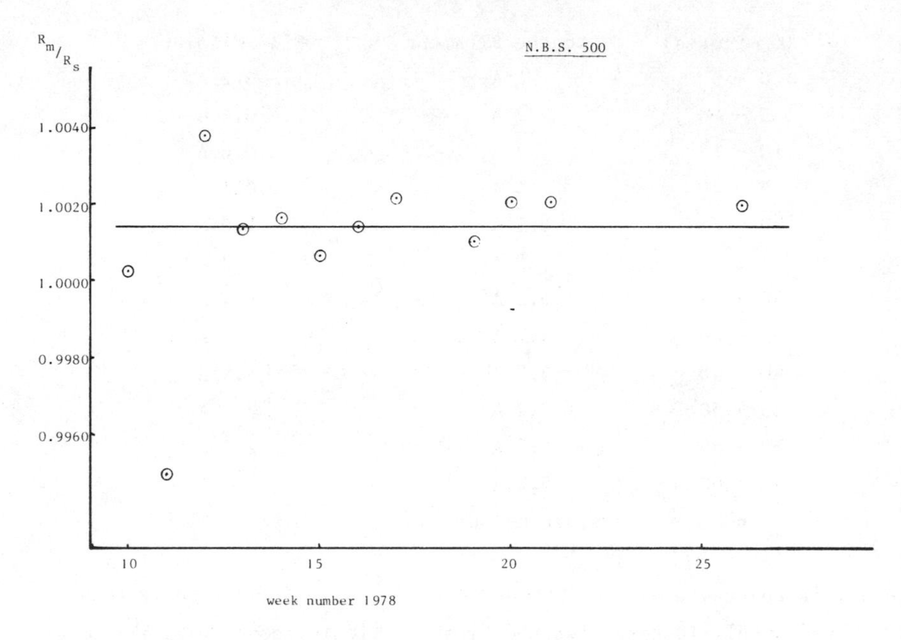

FIG.1. The variation of the correction factor for $^{235}U/^{238}U$.

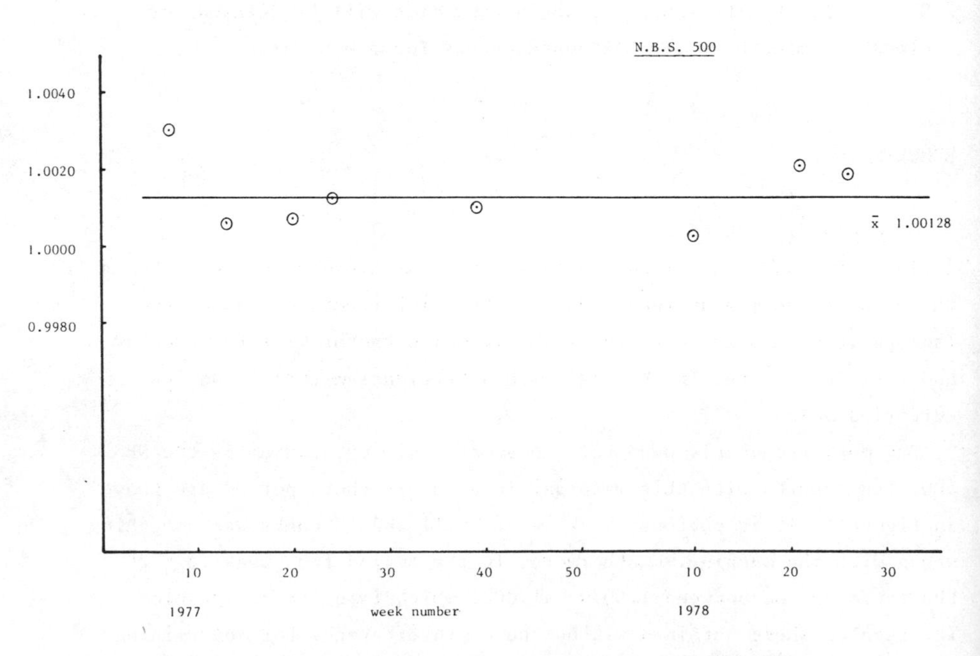

FIG.2. Long-term variation in the correction factor for $^{235}U/^{238}U$.

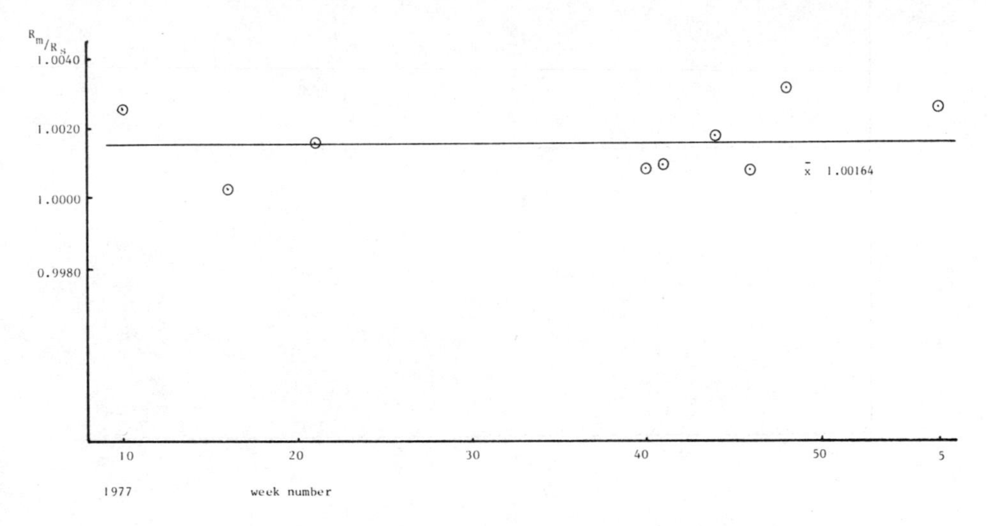

FIG.3. Variation in the correction factor for $^{235}U/^{238}U$.

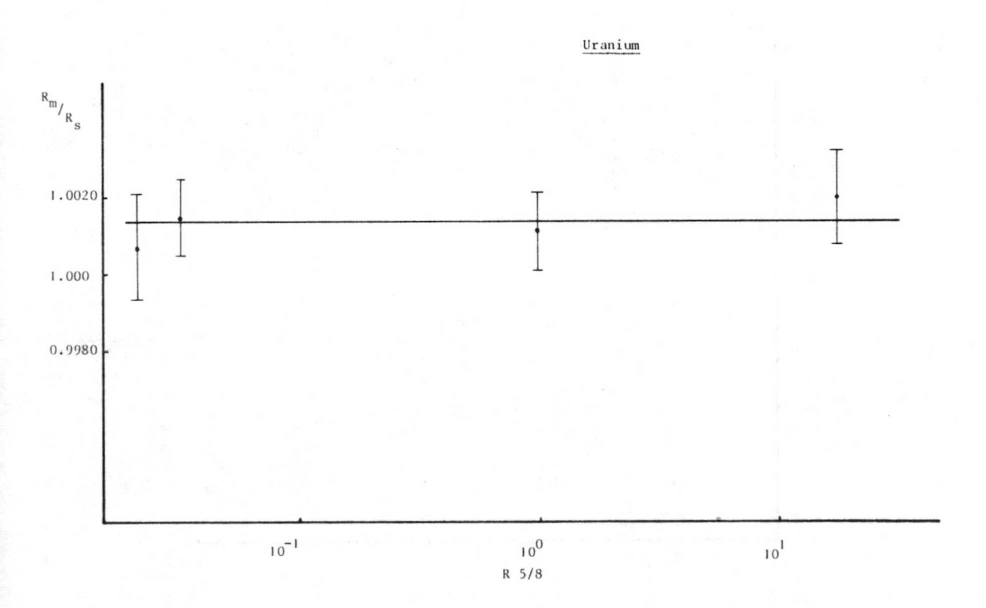

FIG.4. Variation of the correction factor with the enrichment.

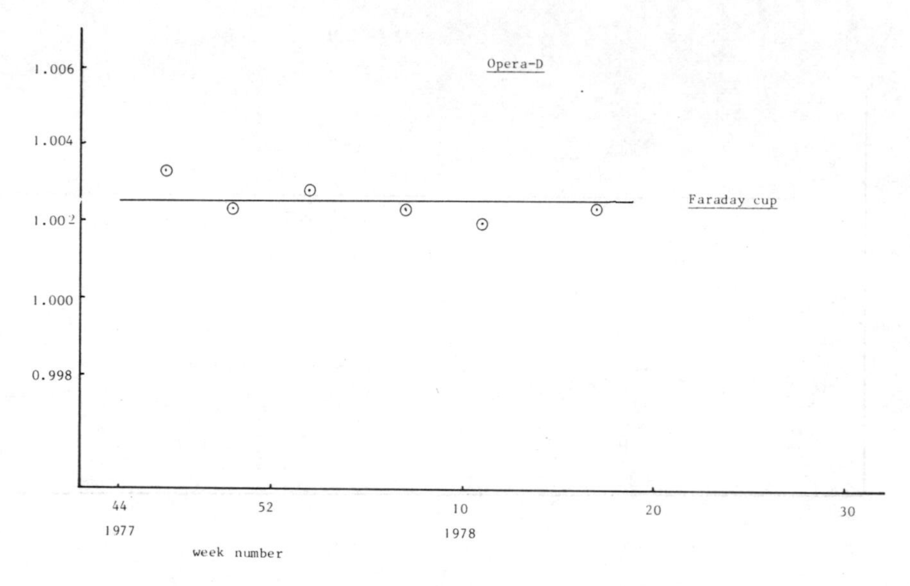

FIG.5. Variation of correction factor of $^{240}Pu/^{239}Pu$.

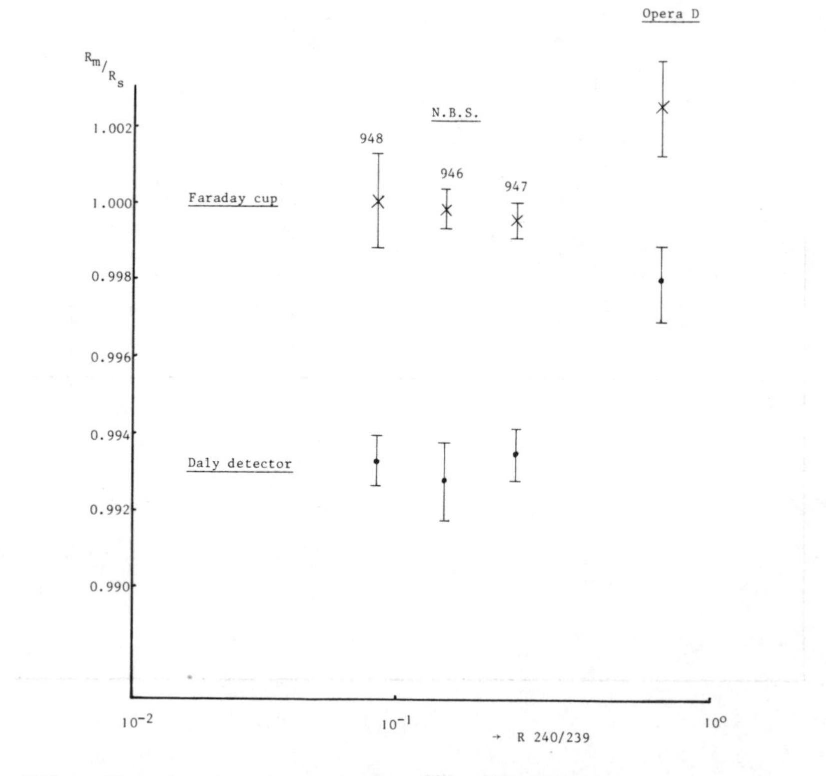

FIG.6. Variation of correction factor of $^{240}Pu/^{239}Pu$ with the value of the ratio.

In a longer period the variation showed to be somewhat larger (figure 2), the range is from 1.0030 (March 1977) to 1.0003 (March 1978).

Another uranium reference material frequently used is the NBS 030. The variation of the ratio is from 1.0031 (December 1977) - 1.0003 (April 77) (figure 3).

The next item is the correction factor as a function of the enrichment in 235-U (figure 4). The values of the main ratios are obtained with the help of NBS 020, 030, 500 and 930.

In the case of plutonium there are available two sets of reference materials, the ones from NBS, USA and those from the CEA, France. The reference materials used in this study are NBS 946, 947, 948 and OPERA-D. The spread in the results for the Faraday cup is comparable with the one obtained for uranium (figure 5).

With plutonium, next to the Faraday cup, also the Daly detector is used especially when dealing with small amounts. The correction factors are quite different for both (figure 6). Another remarkable fact is the discrepancy between the NBS standards and OPERA-D. A close comparison of NBS 947 and OPERA-D resulted in the following correction factors.

	Faraday cup		Daly detector	
	NBS 947	Opera-D	NBS 947	Opera-D
	0.9988	1.00270	0.9893	0.9994
	0.9987	1.00263	0.9890	0.9997
	0.9989	1.00242	0.9914	1.0007
	0.9991	1.00220	0.9916	0.9995
	0.9986	1.00218	0.9922	1.0008
	0.9988	1.00255	0.9911	0.9999
$\bar{X}$	0.9988	1.00245	0.9908	1.0000
	± 0.0001	± 0.00022	± 0.0013	± 0.0006

Operating Parameters

The results from a thermal ionisation mass spectrometer are in terms of precision and accuracy strongly affected by experimental circumstances. In a few publications (4, 5) evidence is given about those influences. With this computer program our results of the last one

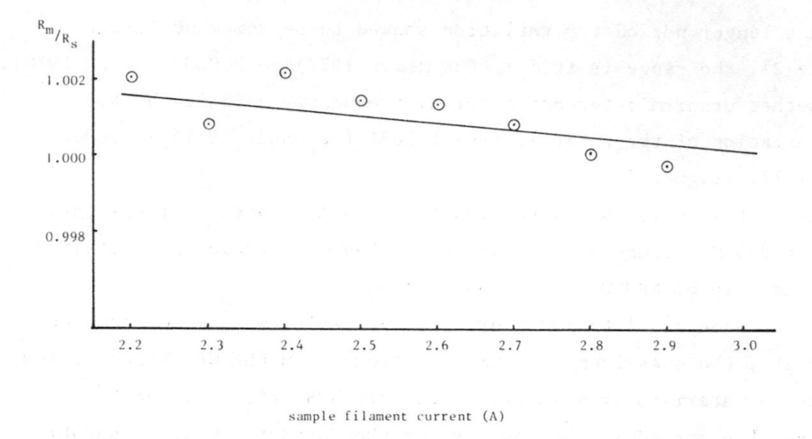

FIG.7. Influence of sample filament current.

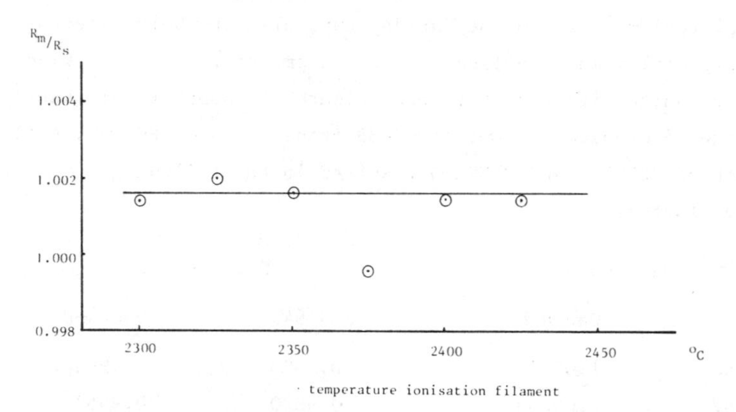

FIG.8. Effect of temperature of the ionization filament.

and a half year are evaluated to look for influences by some experimental parameters on results.

Influence of sample filament current

When the mean of the correction factors obtained for the individual sample filament currents is plotted against the current a declining curve is obtained. This is in so far not amazing as a larger filament current means more fractionating due to a longer time period between start and measurement or probably a higher temperature. As these results are from routine operation the spread in the results is large but a certain tendency is visible (figure 7).

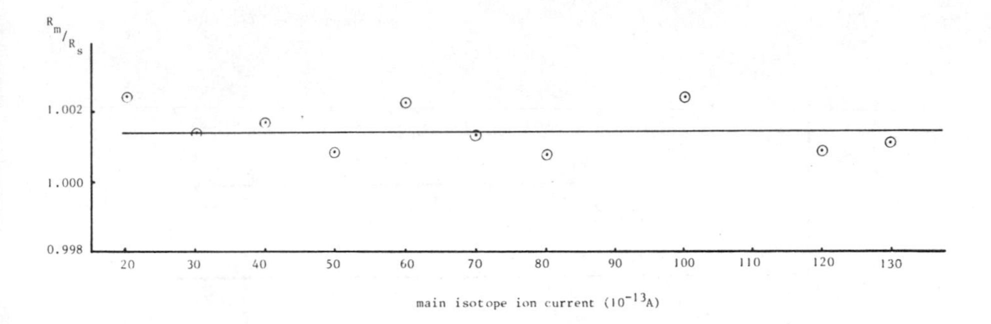

FIG.9. Correction factor against main isotope ion current.

Temperature of centre filament

From the theoretical picture of fractionating no influence is to be expected from the temperature of the centre filament. With our system no effect can be detected (figure 8).

Influence of the ion current

The obtained ion current during the measurement has been different To see if a depency existed between the main ion current and the correction factor both were plotted against each other (figure 9). The spread in the results does not allow any conclusion about a dependency if one exists.

Fractionation under routine operation

How large is the effect of fractionation in this system? In the outlined procedure a measurement is performed after reaching an ion current of 5.10^{-12} A. In the case of uranium a measurement takes 9 minutes followed by a second one. For 25 measurements of NBS 500 reference materials the correction factor of both measurements was calculated. The ratio of both correction factors has to be one, when no fractionating occurs, or at least there has to be a normal distribution around the value of one. From the 25 experiments 18 proved to be larger than one and 6 smaller. Application of a student-t-test on this population showed that there was a significant difference from one at a 5 % significance level. The mean ratio between the two measurements proved to be 1.00038 ± 0.00018.

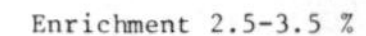

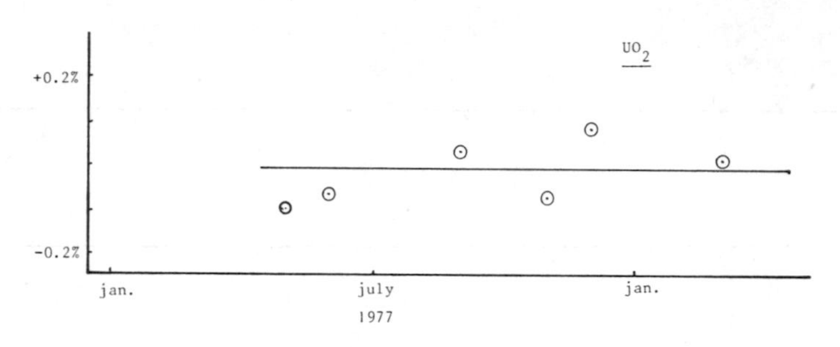

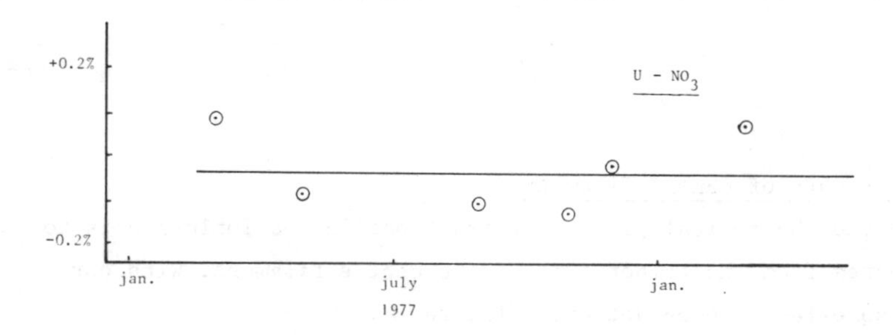

FIG.10. Results SALE program.

SALE Program

To give an indication what has been actually performed with this machine and this computer program results of the SALE Program from 1977 are shown (figure 10).For as well uraniumdioxyde as uranyl-nitrate results were within 0.1 % though the spread for uranylnitrate was somewhat larger and there also existed a small bias in the mean.

CONCLUSION

The purpose of this article was to show some results obtained with an automatic on-line statistical program. This program performs its job without human interference. Results are not better but also not worse. There are however a few advantages in applying a fully computer controlled data evaluation system. An operator may have the tendency to improve the results, yes or no on purpose. This influences the results especially the value of the standard deviation.

As a second point, the operator-machine combination has a certain bias. In the bias obtained with an automatic data evaluation system the bias will be less dependent of the operator.

REFERENCES

[1] VAN RAAPHORST, J.G., ORDELMAN, J., in press.

[2] DIXON, W.J., MASSEY, F.J. Jr., Introduction to Statistical Analysis, McGraw Hill (1957) 275–78.

[3] GARDINER, D.A., BOMBAY, F., Technometrics **7** (1965) 71.

[4] De BIEVRE, P., Advances in Mass Spectrometry **7A**, Heyden and Son Ltd, London (1978) 395–447.

[5] MOORE, L.J., HEALD, E.F., FILLIBEN, J.J., Advances in Mass Spectrometry **7A**, Heyden and Son, London (1978) 448–75.

URANIUM ISOTOPIC RATIOS BY THERMAL IONIZATION MASS SPECTROMETRY

N.M.P. MORAES, S.S.S. IYER, C. RODRIGUES
Institute of Atomic Energy,
São Paulo, Brazil

Abstract

URANIUM ISOTOPIC RATIOS BY THERMAL IONIZATION MASS SPECTROMETRY.

This paper describes the thermal ionization mass-spectrometric procedures used to determine the isotopic composition of uranium in a micro- and submicrogram range over a range of U isotopic composition. The results of a series of measurements carried out in this laboratory on NBS uranium isotopic standards using a thermionic mass spectrometer with on-line computer for data acquisition and processing are discussed. The mean measured isotopic ratios agree with the NBS certified values within the limits specified. The evaluation of accuracy and precision in the measurements of the NBS U-500 standard showed a typical internal precision of ±0.01% for $^{238}U/^{235}U$ ratios.

INTRODUCTION

In the nuclear fuel cycle an accurate knowledge of isotopic distribution of fissile materials is of great importance. The importance of the isotopic analysis is emphasized by their use in such areas as accountability, assessment of monetary values, criticaly consideration, materials safeguards, etc.

The isotopic compositions of nuclear materials are most conveniently determined by mass spectrometry techniques. The development of high resolution mass spectrometer with surface ionization sources has led to their almost exclusive use for isotope abundance measurements of uranium samples in the nuclear fuel cycle. To ensure the level of accuracy and precision that is necessary for these isotope abundance measurements a continous surveillance of the calibration factors for the mass spectrometer is required. For uranium isotopic abundance determinations the level of the instrument bias is generally determined by comparing the measured isotopic ratios with known and certified values of uranium isotopic standards issued from the National Bureau of Standards [1].

The present work focuses the results obtained for isotopic analysis of NBS uranium standards as a part of our analytical control program in the mass spectrometry laboratory. For the calculation of the mass discrimination bias factor isotopic measurements on NBS uranium standards, mainly the NBS U-500, were carried out.

A standardized procedure for chemical processing of the samples, obtaining adequate ion intensities and evaluating the internal and external precisions and accuracies was adopted.

URANIUM ISOTOPIC ANALYSIS

The uranium samples for isotopic analysis are prepared from U_3O_8 which is dissolved in 0.5N HNO_3. A few μg of uranium of this solutions is loaded on the side filament of a double rhenium filament assembly. The filament is heated under standardized conditions to dry the sample and to obtain the UO_3 form. The ionization cartridge with the double filament assembly is introduced into the mass spectrometer and the filaments are preheated by slowly increasing the filament currents. The gradual increase in the temperature is carried out under specified conditions yielding stable evaporation and ionization rates.

The mass spectrometer is a VARIAN – MAT, type TH-5 equiped with Spectrosystem 106TH programmed to perform control functions as well data acquisition and processing.

Data acquisition is controled by a list containing mass numbers to be measured, number of measurements and integration time for each isotope. Correction for background and drift of the ion current are applied to the data.

First the amplifier zero with and integration time T_{ST1} and number of measurements N_{ST1} is measured. Following the measurement of the amplifier zero the magnetic is automatically set to a value corresponding to a mass M_{ST2} and N_{ST2} measurements with an integration time T_{ST2} are carried out. Similar procedure is repeated for all other masses to be measured. After all the N_{ISO} isotopes have been measured the cycle is started again. The cycle is repeated N_{MES} times.

After the last cycle is completed the computer gives the proportions of the isotopes in weight and atom percentages, the isotope ratio relative to a reference mass and the relative standard deviation for the isotope ratios. Following the print of these data the system start automatically a new set of cycles with the same filament conditioning.

For the NBS U-500 the following procedure was adopted: three solutions of this standard are prepared and from each solution three separate filament assemblies are loaded; ten cycles (N_{MES} = 10) are carried out, where each cycle consists of five measurements of each isotope (N_{ST} = 5) with an integration time of 0.32 s (T_{ST} = 0.32 s). For each filament five sets of ten cycles are carried out yielding five sets of data.

From the data of each set of ten cycle one calculates the mean isotopic ratio value and the internal variance for each filament. The internal variance is calculated by

$$S_{IN} = \frac{\sum \sigma}{100(N_{MES}{}^{-1}) \cdot K}$$

where:

σ is the relative standard deviation in percent for each set of cycles, N_{MES} is the number of cycles – we use ($N_{MES}{}^{-1}$) instead of N_{MES} because the last cycle is omitted in the processing of the data – and K is the number of sets for each filament.

TABLE I – CALIBRATION FOR MASS DISCRIMINATION AND ACCURACY IN DOUBLE FILAMENT SURFACE IONISATION ISOTOPIC ANALYSIS OF $^{235}U/^{238}U$ FOR NBSU500 STANDARD

FILAMENT	$\overline{R}$	$\overline{R}_{corr.}$	S_{in}	$\frac{\overline{R}_{corr.} - R_T}{R_T}$
1 A	1.00054	0.99934	0.00011	−0.00035
2 A	1.00058	0.99938	0.00009	−0.00031
3 A	1.00064	0.99944	0.00010	−0.00025
1 B	1.00161	1.00041	0.00014	0.00071
2 B	1.00166	1.00046	0.00017	0.00076
3 B	1.00148	1.00028	0.00015	0.00058
1 C	1.00047	0.99927	0.00008	0.00043
2 C	1.00057	0.99937	0.00009	0.00033
3 C	1.00050	0.99930	0.00009	0.00039

Mass Discrimination

$$\frac{\Sigma \frac{\overline{R}}{R_T}}{9} = 1.0012 \qquad (R_T = 0,99970)$$

$$\overline{S}_{in} = \left(\frac{\Sigma S_{in}^2}{9}\right)^{½} = 0.00012 \qquad \sigma_t = \left(\frac{\sum^{n} S_{in}^2}{n} + S_{ex}^2\right)^{½} = 0.00053$$

$$S_{ex} = \left[\frac{\Sigma \left(\frac{\overline{R}_{corr.} - R_T}{R_T}\right)^2}{8}\right]^{½} = 0.00052$$

TABLE II – CALIBRATION FOR MASS DISCRIMINATION AND ACCURACY IN DOUBLE FILAMENT SURFACE IONISATION ISOTOPIC ANALYSIS OF $^{235}U/^{238}U$ USING SIX NBS STANDARDS

STANDARD	$\overline{R}_T$	$\overline{R}$	$\overline{R}_{corr.}$	$S_{in} \times 10^{-3}$	$\frac{\overline{R}_{corr.} - R_T}{R_T}$
NBS – U010	0.01014	0.01016	0.01005	0.00278	−0.00815
NBS – U020	0.020810	0.02081	0.02059	0.00135	−0.01019
NBS – U030	0.03143	0.03185	0.03153	0.00149	+0.00319
NBS – U050	0.052784	0.05407	0.05352	0.00237	+0.01415
NBS – U100	0.11360	0.11507	0.11389	0.00201	+0.00263
NBS – U200	0.25126	0.25346	0.25087	0.00104	−0.00151

Mass Discrimination

$$\frac{\Sigma \frac{\overline{R}}{R_T}}{6} = 1.0103$$

$$\overline{S}_{in} = \left(\frac{\Sigma S_{in}^2}{6}\right)^{1/2} = 0.00196$$

$$\sigma_t = \left(\frac{\sum^{n} S_{in}^2}{n} + S_{ex}^2\right)^{1/2} = 0.00904$$

$$S_{ex} = \left(\frac{\Sigma \left(\frac{R_{corr.} - R_T}{R_T}\right)^2}{5}\right)^{1/2} = 0.00883$$

From the results for all of the mean isotopic ratios, knowing the certified value for the NBS U-500, the mass discrimination factor is calculated, ratio values for each filament are corrected for this factor and we have then the final values for the isotopic ratios for each filament.

In the Table I we show our results. The table includes the measured isotopic ratio values, the corrected isotopic ratio for mass discrimination factor, the relative internal variance and the relative difference between the measured ratio corrected for mass discrimination and the certified ratio. Also shown in the Table I are the calculated mass discrimination factor, the relative internal variance for all of the filaments calculated from the individual internal variances, the relative external variance derived from the variation of mean values corresponding to different filaments and samples solutions of same standard and the overall error of the isotopic ratio determination derived from both internal and external variances.

A similar procedure was used to determine the overall precision on the isotope ratio measurements of other six NBS standards with different isotopic distribution. For these measurements, only one solution was prepared for each standard. Only one filament was loaded and six sets of then cycles were carried out for each filament.

The results for these measurements are shown in the Table II. They include the measured isotope ratio for each of the NBS standard, the corrected value of this ratio, the relative internal variance for each isotopic ratio determination, the mean internal variance, theexternal variance and the total standard deviation.

CONCLUSIONS

From the results obtained it is seen that for the NBS U-500 standard a typical internal precision within a single run is ± 0.01% whereas the between-filament reproducibility as measured by the external variance is ± 0.05%. Further it is seen that for this standard measurement our obtained isotopic ratio value compares well with the NBS certified value within the error limits. The NBS U-500 has certified atomic abundance ratio of 235/238 = 0.99970 ± 0.00142 [2]

For the other standards our measured ratios agree well with the certified values within the internal precision of a single run. For these measurements the variable fractional between runs is ± 0.09%. Since the standardization procedure for filament temperatures was not closely controlled it is possible, that a lesser spreed could be achieved by doing this.

REFERENCES

[1] GARNER, E. L. et alii. Standard reference materials: uranium isotopic standard reference materials (certification of uranium isotopic standard reference materials). Washington, D. C., National Bureau of Standards, April 1971. (NBS, Special publ. 260 27).

[2] ARDEN, J. W., GALE, U. H., Separation of trace Amounts of Uranium and thorium and Their Determination by Mass Spectrometric Isotope Dilution Analytical Chemistry. 46, (1974) 687.

GAMMA-RAY SPECTROMETRIC DETERMINATION OF PLUTONIUM ISOTOPIC COMPOSITION

T.D. REILLY
Los Alamos Scientific Laboratory,
Los Alamos, New Mexico,
United States of America

D. D'ADAMO, I. NESSLER, M. CUYPERS
CEC Joint Research Centre,
Ispra, Italy

Abstract

GAMMA-RAY SPECTROMETRIC DETERMINATION OF PLUTONIUM ISOTOPIC COMPOSITION.

A method for determining isotopic ratios of plutonium by high-resolution gamma-ray spectrometry is described. The performance of the method on arbitrary samples, independent of material type, geometry and containment, has been tested. No standards were used, and the absolute method, establishing experimentally a relative total efficiency curve in the gamma-ray energy regions of interest, has been applied. Well characterized Pu powder, mixed-oxide pellets and pins of different types were measured and the obtained accuracies for different isotopic ratios are discussed. The results of in-field measurements on PuO_2 cans during a physical inventory-taking are also reported.

INTRODUCTION

The potential application of gamma-ray spectrometry to the measurement of plutonium isotopic composition has long been recognized and studied. In special cases it is already used as a routine accountability measurement technique.
Pu-238-241 and Am-241 all have gamma-ray signatures, only Pu-242 is not directly measurable by gamma counting. Pu-239, Pu-241, and Am-241 have strong, readily measurable radiations. At higher concentrations (≥ 0.2%) Pu-238 becomes accurately measurable within reasonable counting times. Pu-240 is the most difficult isotope having only weak, highly-interfered gamma radiations. This paper will briefly discuss different types of measurement problems and then describe work performed at the Joint Research Centre (JRC) of the Commission of the European Communities (CEC) in Ispra, Italy. A technique is described and results presented for the measurement of an arbitrary sample independent of material type, geometry and containment.

Various physical properties of the sample can profoundly affect the applicable measurement technique. Several of these are physical state (liquid-solid), Pu concentration, age from chemical processing, homogeneity, density, containment geometry, and interfering radiations (e.g. fission products). How these can affect the applicability of a given measurement technique is disussed in ref. 1 and 2. These articles give a rather detailed classification of the different measurement problems.

A different classification is possible based on the degree of sample variability found in the measurement situation. Arbitrarily we call these three situations: quality control, small analytical sample and arbitrary sample. The quality control situation is the simplest case characterized by the relative measurement of small deviations from the counting behaviour of an identical standard. This applies to finished fuel material (rods, pellets) where rapid gamma-ray measurement against a standard may provide a useful check of product quality.
Minimal data interpretation is required as the finished item is expected to be identical to the standard (if not, it is rejected or subjected to further analysis).

The small analytical sample case may exibit a range of isotopic composition but other physical characteristics of the samples are well controlled. Furthermore, the sample is packaged (possibly even prepared - e.g. Am stripped) in such a way as to facilitate the gamma-ray measurement. This is basically the situation where mass spectrometry is applied today and is also where gamma-ray techniques have been used as routine accountability measurements /3/. Here sample attenuation is well-defined by calibration standards or easily and accurately calculated due to the fixed simple sample geometry. In a given plant situation sample variability is sufficiently limited to be adequately covered by a reasonable set of calibration standards. These techniques may be adapted to continuous flow (in-line) measurements /4/.

Wide variability is characteristic of the arbitrary sample situation, i.e. the need to measure very different materials including items which in other circumstances would fall into the previous situations. Sufficient calibration standards can not be reasonably constructed to cover this situation. Sample attenuation is extremely variable and unknown a priori. It can not be easily represented in standards. This is often the measurement situation facing the safeguards inspector. The technique described below pertains mostly to this case.

MEASUREMENT TECHNIQUE DESCRIPTION

The procedure described herein determines the ratios Pu-239/Pu-241 and Am-241/Pu-239 by analysing the 332-345 keV complex from three isotopes. The ratio Pu-238/Pu-241 is determined from the gamma-ray ratio 152.8/148.6, and the Pu-240/Pu-241 ratio from the 160 - 164.6 keV complex. All interferences are taken into account and corrections made for differences in total efficiency (relative) at different energies. These corrections are made by determining segments of the total relative efficiency curve using intense gamma rays from a single isotope. This curve is essentially the convolution of the detector efficiency and the absorption of the sample and any intervening filters. Published nuclear data for half-lives and gamma-ray branching intensities are used to convert activity ratios to isotope ratios.

Spectra are taken with high resolution Ge detectors. The most accurate results have been obtained with small (1 cm^3) hpGe detectors, but larger, high efficiency (5-15%) detectors are now available with nearly equal resolving power which should give similar results in less time. Results are reported here using a 1 cm^3 hpGe (FWHM = 560 eV at 122 keV, 920 eV at 414 keV) and a 75 cm^3 GeLi (FWHM = 1150 eV at 122 keV, 1390 eV at 414 keV). A judicious choice of filter materials (between sample and detector) can minimize counting time and rate related problems such as pulse pileup. If only the high energy radiation (>300 keV) is of interest, a 1-3 mm Pb filter will significantly reduce the intense gamma x-ray radiation below 200 keV while permiting a higher counting rate in the interesting energy region. In some cases this can reduce counting times a factor of two or better.

TABLE I. COEFFICIENTS FOR ANALYSIS OF 332 - 345 keV COMPLEX

	Coefficient	Error(%)
A	3.9824	0.48
B	4.6063×10^{-2}	0.43
C	1.3838×10^{-2}	0.43
D	-3.8875×10^{-2}	0.47
E	6.7037×10^{-3}	1.06
F	8.3624×10^{-2}	0.43
G	-1.0892×10^{-2}	1.05

If only the low energy region (100 - 200 keV) is of interest the best choice is a small detector with, at most, a 0.5-1 mm Cd sheet to reduce the intense 59.5 keV gamma from Am-241.

The results reported here were obtained using a channel-by-channel summation procedure and subtracting a smooth, step-like background determined from stationary regions on either side of the full energy peak or peak group /5/. This simple procedure appears to yield satisfactory results for most of the isotopic ratios and requires only a modest minicomputer-based MCA system. System energy resolution is measured for the energies of interest and 2.5x or 3.0x FWHM peak window limits are chosen to insure that the determined areas include the same fraction of the total full-energy events.

The crucial part of the procedure is the analysis of the 332-345 keV complex which is resolved as three composite peaks at approximately 332, 336, and 345 keV. The first two contain contributions from all three istopes (Pu-239, Pu-241 and Am-241) and the third (345.0 keV) is a clean, single peak from Pu-239. The relative efficiency (RE) curve is determined using the 345.0, 375.0, 392.8, and 413.7 keV gamma rays from Pu-239. The RE is calculated dividing each peak area by the appropriate gamma branching intensity and normalizing the resulting values to 1.0 at 413.7 keV. These four values are then least-squares fit to either a linear or power-law curve (E^{-a}). The latter form usually gives slightly better results.
With a large detector and a highly absorbing sample (or a thick Pb filter), the curve may have a maximum in this region. This requires a different fitting expression but has not been well studied as yet.

Next the fitted curve is used to predict the RE at 332 and 336 keV. An advantage of the region is that the efficiency curve is usually changing relatively slowly with energy and as such can be evaluated accurately. Finally the two isotopic ratios are calculated using the efficiency-corrected peak areas and the following expressions:

$$\text{Pu-239/Pu-241} = 345'A \,/\, (332'B - 336'C + 345'D)$$

$$\text{Am-241/Pu-239} = (-332'E + 336'F + 345'G) \,/\, 345'A$$

where :

$$345' = \text{area}(345)/\text{RE}(345)$$

A - G are coefficients listed in table 1.

The analysis may stop here for certain control measurements where these ratios are sufficient to verify declared isotopic data.

The other two ratios are determined using the series of gamma rays between 148.6 and 164.6 keV. The relative efficiency curve in this region is interpolated (E^{-a}) between the 148.6 and 164.6 keV gamma rays of Pu-241. As there are only two points here this section of the curve is usually not as well known as the section above 300 keV. The Pu-238/Pu-241 ratio is computed using the expression:

$$\text{Pu-238/Pu-241} = \frac{153}{148} \quad \frac{RE(148)}{RE\ (153)} \quad K_{38}$$

where:

153, 148 = peak areas 152.8, 148.6 keV

RE(148, 153) = relative efficiencies

K_{38} = coefficient given in table 2.

These are both quite interference free. The 152.8 gamma from Pu-238 is, however, fairly weak especially for the trace levels ($\ll$0.1%) found in low-burnup Pu. For reactor grade plutonium this level is 0.2% - 1.0% or higher and the measurement becomes faster and more accurate.

The Pu-240/Pu-241 ratio is determined by analysing the 160-164 keV complex and is the weakest point of this procedure. With a high resolution detector (FWHM $\leq$600 eV at 122 keV) the 161.45 gamma of Pu-239 is resolved from the 160 composite peak which includes contributions from Pu-240, Pu-241, and possibly Pu-239[1]. If the weak Pu-239 activity is ignored the ratio is given by :

$$\text{Pu-240/Pu-241} = K_1\left(\frac{160}{(164)_{41}} \quad \frac{RE(164)}{RE(160)} \quad - K_2\right)$$

where:

160 = area of composite peak at 160 keV

$164)_{41}$ = area of 164.6 peak corrected for small Am contribution

K_1, K_2 = coefficients given in table 2.

As mentioned above this is the most difficult ratio to determine. It works reasonably well for low burnup plutonium but becomes less accurate for reactor grade material because Pu-241 grows in relatively more rapidly than Pu-240 and the latter becomes swamped by the larger Pu-241 contribution to the composite 160 keV peak.

---

[1] R. Gunnink's latest listing UCRL-52139 gives a weak gamma at 160.19 keV from Pu-239. This was not given in the original study and is quoted with a 20% precision indicating a very weak (or even uncertain) transition. If this is included, the above formula becomes:

$$\text{Pu-240/Pu-241} = K_1\left(\frac{160}{(164)_{41}} \quad \frac{RE(164)}{RE(160)} - K_2 - K_3 \quad \frac{^{239}Pu}{^{241}Pu}\right)$$

TABLE II. COEFFICIENTS FOR ANALYSIS OF 148 - 164 keV REGION

	Coefficient	Error(%)
K_{38}	1.1818	0.61
K_1	51.08	0.98
K_2	0.1489	1.19
K_3	8.23×10^{-5}	20.0

All of the coefficients given in table 1 and 2 are derived using the half-lives suggested in American National Standard N15.22 /6/ and the most recent gamma branching intensities published by R. Gunnink /7/. These values should be adjusted to reflect newer measurements as they become available especially as regards half-lives (several half-life studies are presently being completed). One can see that the uncertainty in these coefficients is rather large (0.5-1.0%) reflecting the uncertainties in the fundamental data. This can be reduced experimentally by counting a sufficient number of well-characterized reference materials using the identical equipment and analysis procedures and then calculating the coefficients directly from the experimental data. This has the advantage of reducing biases which may result from the use of different peak area determination procedures for the fundamental nuclear data and the isotopic composition measurements.

The procedure described above has been adapted to an Intertechnique Multi-20 computer-based multichannel analyser system. The system controls data acquisition (spectra are stored on magnetic tape), performs all the necessary analysis, and outputs all the relative efficiency points and the four independent isotopic ratios. The program is written in the simple interpretive language of Intertechnique, LEM.

MEASUREMENT RESULTS

Three sets of measurement results will be presented. First, well-characterized materials (NBS Standard Reference Materials, fuel rods and pellets of several types) were measured with the 1 cm^3 hpGe detector. Next, these same samples were measured with the large GeLi (16%) to yield a quick analysis of the 330-414 keV region.
Lastly, large cans of PuO_2 were measured, again with the small detector, in a fuel fabrication facility in support of a safeguards physical inventory inspection.

The three Standard Reference Materials certified by the National Bureau of Standards of plutonium isotopic composition were measured as unknowns to compare results with the declared results of the SRM's. This was carried out in the framework of an interlaboratory comparison sponsored by the Nondestructive Analysis Working Group of the European Safeguards Research and Development Association (ESARDA) - /8/. The result of this study are summarized in table 3. Long, overnight spectra were used for these measurements; count rates were kept under 2000 cps to maintain optimum resolution. The Am-241/Pu-239 data is not included because the NBS materials are not certified for americium content. Included in table 3. are values for Pu-239/Pu-241 obtained from two different spectra, one filtered with 2mm Pb + 1mm Cd + 1mm Cu and the second (unfiltered) using only the cadmium and copper. Considering that this is an absolute measurement using published nuclear data, the agreement between measured and declared values is very satisfying.

TABLE III. MEASUREMENT OF NBS - SRM WITH SMALL DETECTOR

Sample	Ratio of Atom Percent 238/241	239/241	240/241
SRM - 946			
filtered		27.93 ± .16	
unfiltered	0.0769 ± .0008	27.55 ± .38	3.97 ± 0.15
declared	0.0795 ± .0023	28.07 ± .09	4.07 ± 0.015
SRM - 947			
filtered		22.24 ± .12	
unfiltered	0.0816 ± .0008	22.18 ± .29	5.38 ± 0.15
declared	0.0837 ± .0018	22.49 ± .08	5.43 ± 0.02
SRM - 948			
filtered		258.0 ± 4.7	
unfiltered	0.0298 ± .0010	253.4 ± 7.6	22.18 ± 0.35
declared	0.0294 ± .0028	254.0 ± 0.9	21.94 ± 0.08

TABLE IV. FUEL ROD MEASUREMENT RESULTS (ISOTOPIC RATIOS)

MEASUREMENTS - 9/1976

Type	238/241	239/241	240/241	Am/239
A (17)[a] meas.	0.0346 ± .0007	23.88 ± .15	6.22 ± .16	0.0038 ± .0001
decl.	0.022	24.04	6.24	0.0041
B (60) meas.	0.1202 ± .0007	11.91 ± .09	2.15 ± .13	0.0130 ± .0002
decl.	0.1249	11.89	2.23	0.0143
C (60) meas.	0.035 ± .002	151.6 ± 2.4	16.1 ± .5	0.0025 ± .0001
decl.	0.026	151.5	15.4	≥ 0.0025
MEASUREMENTS - 3/1978				
A (13) meas.	0.0376 ± .0007	25.71 ± .16	6.68 ± .13	0.0076 ± .0001
(3) meas.	0.0364 ± .0015	25.20 ± .33	6.60 ± .28	0.0077 ± .0002
decl.	0.023	25.90	6.73	0.0071
B (17) meas.	0.1275 ± .0006	12.92 ± .11	-[b]	0.0197 ± .0002
(2) meas.	0.1266 ± .0014	13.17 ± .33	-	0.0192 ± .0007
decl.	0.1329	12.81	2.40	0.0203

[a] The spectrum count time in hours is given in parenthesis

[b] The 240/241 ratio for this measurement was grossly in error, possibly due to a small gain shift which can critically effect the window selection in this complex region. The spectrum must be reanalysed to investigate this further.

TABLE V. DIFFERENCE (MEAS - DECLARED)/DECLARED IN PERCENT

Sample	RATIO OF wt. %			
	238/241	239/241	240/241	Am/239
SRM-946	-3.27 ± 3.07	-0.50 ± .66	-2.53 ± 3.80	
SRM-947	-2.51 ± 2.36	-1.11 ± .65	-0.92 ± 2.81	
SRM-948	1.36 ± 10.1	1.58 ± 1.85	1.09 ± 1.62	
A	57 ± 2	-0.67 ± .63	-0.32 ± 2.6	-7.3 ± 2.6
	63 ± 2	-0.73 ± .62	-0.74 ± 1.9	7.0 ± 1.3
B	-3.76 ± .58	0.17 ± .76	-3.59 ± 6.05	-9.1 ± 1.5
	-4.06 ± .47	0.86 ± .85	-	-3.0 ± 1.0
C	35 ± 6	0.07 ± 1.58	4.55 ± 3.1	-
	mean	-0.045	-1.65	
	sigma	0.91	1.98	

TABLE VI. COMPARISON FUEL ROD AND PELLET MEASUREMENT

Type	(PELLET - ROD)/ROD (%)			
	238/241	239/241	240/241	Am/239
A	-4.8 ± 4.6	-1.7 ± 2.4	-14.0 ± 5.2	0 ± 3.1
B	-2.4 ± 1.1	-0.1 ± 1.2	-5.0 ± 9.4	-3.4 ± 1.8
C	0.6 ± 8.4	0.7 ± 1.1	-5.7 ± 2.2	-1.3 ± 2.6

The stated uncertainties in the measured ratios include only the propagated counting errors. Uncertainties in the equation coefficients have not been included.

The same equipment was used to measure several mixed-oxide fuel pins and pellets. These come from three different reactor types and have been specially characterized for the JRC, Ispra. Three fuel rods were available for measurement along with three unencapsulated pellets from each fuel type. One pellet from each type has been selected for further destructive analysis to improve our knowledge of the isotopic composition, especially the values for Pu-238 and Am-241. Results of the fuel rod measurements are presented in table 4. All measurements were made with a gross count rate of 2000 cps or less to maintain optimum resolution. No attempt was made at this stage to answer the important practical question of how rapidly these measurements could be made. Shorter spectra (2-3 hours) were taken during the second measurement period and show the precision available under these conditions. The agreement between measured and declared values is quite similar to that shown for the NBS-SRM's. Table 5 shows the per cent difference between measured and declared values for these six samples. The 239/241 and 240/241 ratios agree well within statistics, however, some of the 238/241 and the Am measurements are quite far outside statistics. We believe this reflects inaccuracies in the declared data for the fuel rods.

TABLE VII. $^{239}Pu/^{241}Pu$ ISOTOPIC RATIO USING LARGE GELI AND 1000 sec COUNT TIME

Sample	Measured	Declared	(M - D)/D %
SRM-946	28.27 ± .37	28.42	-0.53 ± 1.31
SRM-947	22.03 ± .28	22.77	-3.25 ± 1.27
SRM-948	274.2 ± 12.5	256.0	7.11 ± 4.56
A - pellet	25.56 ± .40	25.90	-1.31 ± 1.56
A - rod	25.73 ± .41	25.90	-0.66 ± 1.59
B - pellet	12.98 ± .20	12.81	1.33 ± 1.54
B - rod	12.86 ± .17	12.81	0.39 ± 1.32
C - pellet	162.0 ± 2.3	162.8	-0.49 ± 1.41

TABLE VIII. MEASURED ISOTOPIC RATIOS FROM 1000-sec SPECTRA

	$^{239}Pu/^{241}Pu$		$^{241}Am/^{239}Pu$	
Sample	Measured	Declared	Measured	Declared
A1	23.85 ± 1.03	23.67	6.2 ± .7 x 10^{-2}	5.22 x 10^{-2}
A2	23.78 .98	"	5.3 .7 "	"
A3	22.78 .80	"	6.2 .7 "	"
A4	24.01 1.03	"	5.4 .7 "	"
A5	23.11 .30	"	6.5 .3 "	"
A6	22.83 .98	"	5.2 .6 "	"
A7	25.41 1.12	"	6.3 .7 "	"
A8	23.16 .25	"	5.9 .2 "	"
mean	23.62		5.88 x 10^{-2}	
sigma	0.87 (3.7%)		0.53 x 10^{-2} (9.0%)	
B1	55.0 ± 3.2	55.76		-
C1	8.3 ± 0.6	8.97		-

The individual fuel pellets were also measured to test their uniformity one to another and their agreement with ratios determined from the assembled rods /9/. The results of this comparison are presented in table 6 where it is seen that within statistics the pellets yield the same isotopic ratios as were measured from the fuel rods.

The second group of measurements involves all the samples previously described but measured this time with a large GeLi detector (16%). It was filtered with ½ mm Cu + 1mm Cd + 3mm Pb to optimize the counting chain operation for the 332 - 345 keV complex. Counting rates were still held to around 2000 cps as the resolution was just marginal to separate the 332 and 336 keV peaks. A 1000 sec spectrum was made of the NBS-SRM's and all the well-characterized fuel rods and pellets available. The results of these measurements, presented in table 7, show that a 1000 sec count is sufficient to yield a precision of 1 - 2% for the Pu-239/Pu-241 ratio. With one of the higher resolution detectors available today, it would be possible to use even higher

TABLE IX. ISOTOPIC RATIOS FROM 6-HOUR SPECTRUM OF SAMPLE A8

	238/241	239/241	240/241	Am/239
measured	0.0419 ± .0015	23.16 ± .25	6.61± .26	0.0059 ± .00014
declared	0.0435	23.67	6.58	0.00522

count rates and still resolve these peaks better than in the present detector. This simple measurement could be a useful technique for safeguards inspectors because Pu-239/Pu-241 and Am-241/Pu-239 should often be sufficient to verify a stated isotopic composition. The measurement is rapid and the analysis simple enough to be programmed into a small microprocessor-based instrument.

The last measurements to be described were performed in a mixed-oxide fuel fabrication plant in support of a safeguards physical inventory inspection. For this exercise the small detector and Multi-20 system were trucked to the facility and set up to measure small cans containing 1-2 kg of plutonium as PuO_2. Ten containers were measured (basically during one day), eight from one batch and one each from two different batches. Short measurements (1000 sec) were made of each item to measure the Pu-239/Pu-241 ratio using the 332-345 keV complex thereby verifying the declared composition and checking within batch uniformity. The results of this test are presented in table 8. They confirm, within counting statistics (~4%), that the eight samples of batch A have the identical plutonium isotopic composition and Am content. The measured Pu-239/Pu-241 ratios agree with those stated by the operator, in fact, the mean for batch A of 23.62 ± .31 agrees very well with the declared value of 23.67. In addition to these short spectra a six hour spectrum was made of one can selected from batch A, and this spectrum was analysed for all four isotopic ratios. These results, presented in table 9 again show excellent agreement with the operator declared values. Though only a few samples were measured, these results were encouraging and served as a useful demonstration to the Euratom inspectorate personnel. To be sure, the equipment used was of a type suitable for fixed laboratory operation and not optimized for "in field" use, but a simpler and more compact instrument can be designed to perform this simple analysis.

CONCLUSION

The measurement technique described in this paper is independent of sample type, geometry, uranium admixture, and even small levels of fission product contamination. Its only limitation is that U-237 must be in equilibrium with Pu-241 as many of the important gamma rays come from the daughter isotope. Practically this means the material must be at least one month from chemical processing. As indicated above the technique could be of use to safeguards inspectors especially the rapid determination of Pu-239/Pu-241. Future work will involve the study of larger, high resolution detectors, simple spectrum fitting procedures, and improved relative efficiency curve expressions (including the theoretical calculation of the efficiency curve). Other fuel materials are presently under study, and ultimately Euratom inspectors will be trained to apply these techniques during physical inventory takings.

REFERENCES

1. J.L. Parker and T.D. Reilly, "Plutonium Isotopic Determination by Gamma-Ray Spectroscopy", Los Alamos Scientific Laboratory report LA-5675-PR (August 1974) pp. 11-13. See also in same report "Gamma Spectroscopic Measurement of Plutonium Isotopic Concentration in the Arbitrary Sample", pp. 13-19.

2. R. Gunnink, "Status of Plutonium Isotopic Measurements by Gamma-Ray Spectroscopy", Lawrence Livermore Laboratory report UCRL-76418 (June 1975).

3. R. Gunnink, "A System for Plutonium Analysis by Gamma-Ray Spectrometry. Part 1: Description of the Techniques for Analysis of Solutions". Also "Part 2: Computer Programs for Data Resu and Interpretation", Lawrence Livermore Laboratory report UCRL-51577 (April 1974).

4. R. Gunnink and J.E. Evans, "In-Line Measurement of Total and Isotopic Plutonium Concentrations by Gamma-Ray Spectrometry", Lawrence Livermore Laboratory report UCRL-52220 (February 1977).

5. T.D. Reilly and J.L. Parker, "A Guide to Gamma-Ray Assay for Nuclear Material Accountability", Los Alamos Scientific Laboratory report LA-5794-M (March 1975) pp. 8-9.

6. "Calibration Techniques for the Calorimetric Assay of Plutonium-Bearing Solids Applied to Nuclear Materials Control", American National Standard ANSI-N15.22 - 1975, pp. 25-30.

7. R. Gunnink, J.E. Evans, and A.L. Prindle, "A Reevaluation of the Gamma-Ray Energies and Absolute Branching Intensities of U-237, 238, 239, 240, Pu-241, and Am-241, "Lawrence Livermore Laboratory report UCRL-52139 (October 1976).

8. R.J.S. Harry, IAEA-SM-231/23, these Proceedings, Vol. II.

9. D. D'Adamo and I. Nessler, "Misura dei Rapporti Isotopici del Pu su Barre e su Pastiglie di Combustibile ad Ossidi Misti (U - Pu) O_2 mediante Spettrometria Gamma", Joint Research Centre, Ispra, Italy, informal report FMM/N°34 NDA (February 1978).

DISCUSSION

M.R. IYER: In Table VII, showing the $^{239}Pu/^{241}Pu$ ratios, the difference seems to increase with increasing ^{239}Pu content. This is contrary to what one would expect. Could you comment on the large difference in the case of the NBS 948 sample?

M. CUYPERS: The difference was observed for the measurements with the 10% efficiency detector but not for the high-resolution detector. Further investigations are being carried out to try to explain this difference.

P. FILSS: I should like to ask a question on partially decontaminated nuclear material. What would you estimate to be the highest acceptable level of radioactivity which could be handled without overloading the system?

M. CUYPERS: I can give you no exact figures for this.

G. MALET: Perhaps I can comment on this point. Our experiments have shown that fission-product activities of around several μCi in a solution containing several grams of plutonium per litre are no problem. Actually, these fission products are mainly ^{95}Zr, ^{95}Nb, ^{106}Ru, with traces of ^{137}Cs, the various gamma lines of which do not interfere with those of the plutonium isotopes.

T.N. DRAGNEV: When I was working with the Agency, over two years ago, we performed Pu–Am–U isotopic ratio measurements several times under field conditions on mixed UO_2-PuO_2 and pure PuO_2 samples. The method of data processing developed and used was more or less the same as in Mr. Cuypers' paper. The results were very encouraging. The method and results were published as an Agency report (IAEA STR-60) and in the Journal of Radioanalytical Chemistry **36** (1977) 401.

A.G. HAMLIN: In paper (IAEA-SM-231/54)[1] we suggested that a combination of gamma spectrometry and nuclear and thermal measurements be used for verification, particularly of plutonium. To calculate total plutonium from the "plutonium-240 equivalent" response obtained, accurate knowledge of the isotopic composition of plutonium is needed. Would Mr. Cuypers care to predict how well and how quickly this might be obtained by future development of gamma spectrometry?

M. CUYPERS: With the recent availability of better quality detectors (higher efficiency without loss of resolution), it is expected that a substantial reduction in measurement time will be obtained. So far we have concentrated our efforts on ^{239}Pu/^{241}Pu ratio measurements.

H. OTTMAR: I should like to add a few remarks in this connection. In our work on plutonium isotope analysis using gamma spectrometry we investigated the precision attainable in a reasonable counting time and some data are presented in our paper[2]. Taking the ^{239}Pu/^{240}Pu ratio determination, for example, a precision as good as 1% can be obtained in an assay time of not more than 30 minutes when the experimental conditions are optimized for this particular measurement exercise.

---

[1] ADAMSON, A.S. et al., IAEA-SM-231/54, these Proceedings, Vol. I.

[2] EBERLE, H., et al., IAEA-SM-231/12, these Proceedings, Vol. II.

METHODS FOR PRECISE ABSOLUTE GAMMA-SPECTROMETRIC MEASUREMENTS OF URANIUM AND PLUTONIUM ISOTOPIC RATIOS

T.N. DRAGNEV, B.P. DAMJANOV
Institute of Nuclear Research and
Nuclear Energy,
Bulgarian Academy of Sciences,
Sofia, Bulgaria

Abstract

METHODS FOR PRECISE ABSOLUTE GAMMA-SPECTROMETRIC MEASUREMENTS OF URANIUM AND PLUTONIUM ISOTOPIC RATIOS.

A new gamma-spectrometric method for precise absolute (i.e. without use of standards) measurements of the uranium isotopic ratio is proposed. The method is based on intrinsic self-calibration of gamma-spectrometric measurements and on a new approach to the data treatment which allows even the most advanced pocket programmable calculators (e.g. TI-59) to be used to resolve a multiline 92.8 keV peak, unresolvable by gamma spectrometers. The narrow energy range of the uranium gamma spectrum (92–99 keV) is only used for uranium isotopic ratio measurements, which makes the results reliable and precise, and they do not depend on the size, shape and type of the material. Two new approaches to Pu-Am absolute gamma-spectrometric isotopic ratio measurements are reported. In the first the energy ranges, 250–414 and 148–208 keV, are used. The first is for ^{241}Pu/^{239}Pu and ^{241}Am/^{241}Pu isotopic ratio determination on the basis of gamma rays in the narrow range, 332.354 – 345.014 keV; the second is for ^{238}Pu/^{241}Pu and ^{240}Pu/^{241}Pu determinations using neighbouring gamma lines of corresponding isotopes, which is suitable for large samples when the thickness of the container wall is large. The precision of the ^{238}Pu determination is not very good. The second approach mainly uses the low-energy range, 94 – 105 keV, where practically all isotopes emit strong gamma or X-ray lines. The precision of the isotopic ratio determination in this case is significantly higher, particularly for ^{238}Pu/^{241}Pu and ^{240}Pu/^{241}Pu. A high-resolution gamma-spectrometric system in the 90–210 keV range is required for these measurements.

INTRODUCTION

At present the most severe restriction for the widespread application of non-destructive quantitative measurements (NDQM) of special nuclear materials (SNM) is the lack of suitable standards or methods for accurate absolute measurements. The preparation and accepted certification of standards suitable for NDQM and their use by IAEA nuclear material safeguards (NMS) are extremely difficult and time-consuming because of the large variety in size, shape, concentrations, cladding etc. of SNM items met in different countries. In addition, their transport from site to site in one or various countries is quite difficult because of existing regulations. Therefore, the development of absolute NDQM methods and corresponding procedures, which means methods without the use of standards, suitable for IAEA NMS applications, is of paramount importance.

1. $^{235}U/^{238}U$ AND ^{235}U ENRICHMENT MEASUREMENTS

Two main methods of gamma-spectrometric enrichment measurements are:

(i) The use of infinitely thick homogeneous samples — enrichment measurement method [1, 2].

(ii) The method proposed by Harry et al. [3] based on the comparison of the measured intensities of several high-energy gamma lines of $^{234}Pa^m$ (^{238}U daughter product) as a measure of ^{238}U concentration of the sample, and intensities of several relatively low-energy gamma lines of ^{235}U as a measure of its concentration.

When standards of the same material, size and shape are available, the enrichment measurement method is very convenient and precise, but is quite difficult to use as an absolute method.

Method (ii) is not very convenient and generally not very accurate because the $^{234}Pa^m$ and ^{235}U gamma lines, used for $^{235}U/^{238}U$ isotopic ratio measurements, are in widely separated energy regions.

The method proposed in this paper is based on the use of gamma and X-rays from the narrow 92–99 keV range. The lines used for isotopic ratio measurements are situated in the even more narrow 92.4–93.4 keV range. The intensities of the two – 92.367 and 92.792 keV – lines of ^{234}Th [4] are used as a measure of ^{238}U concentration of the sample, and the 93.35 keV Th-K_{α_1} line is used as a measure of its ^{235}U concentration. The ^{238}U contribution at the 93.35 keV intensity is negligible and is automatically taken into account by the calibration used. The same is valid for normal ^{234}U concentration, at least for low enriched materials. If, for some measured samples, the ^{234}U concentration is much higher, this will be easily detected in the same measurements through its most intensive 121.0 keV line.

1.1. The background of the method

The instrumental low-energy U spectra of several UO_2 samples with different enrichments, measured with commercially available instrumentation (small planar Ge detector and Silena multichannel analyser system) are shown in Fig.1. It is proposed that the well-resolved and easily recognizable U-K_α 94.660 and

98.441-keV peaks be used for energy and efficiency calibration of the narrow energy range near 92–99 keV. As can be seen from the spectra, the 92.367 and 92.792 keV lines of the ^{234}Th proposed for use as a measure of ^{238}U concentration, and the 93.35 keV line for use as ^{235}U concentration measure are so near each other that they are not resolved, which makes their use difficult. The case is even more complicated since 92.367 and 92.792 keV rays are gamma lines and 93.35 keV rays are X-ray lines, which means that they have a different shape [5–7] and line width. Still we have found an algorithm and have developed programs suitable for the most advanced programmable calculators (e.g. TI-59), which resolve these lines precisely and use them for the $^{235}U/^{238}U$ isotopic ratio and ^{235}U enrichment measurements.

1.2. Measurements and their analysis

Three groups of pellet and four powder samples with different enrichment were measured many times by the above-mentioned measurement system. The results of these measurements and their analysis when using the TI-59 calculator and our programs are shown in Table I.

There are three groups of measured ratios: $I_{K_{\alpha_2}}/I_{K_{\alpha_1}}$; $I_3/I_1 + I_2$ ratios and the corrected ratios of

$$\frac{I_3 \cdot \epsilon_1 \cdot \epsilon_2}{(I_1 \cdot \epsilon_2 + I_2 \cdot \epsilon_1) \cdot \epsilon_3}$$

where $I_{K_{\alpha_1}}$ and $I_{K_{\alpha_2}}$ are the intensities of U $K\alpha_1$ and $K\alpha_2$ lines; I_1, I_2 and I_3 are the corresponding intensities (in relative units) of the 92.367, 92.792 and 93.35 keV lines; ϵ_1, ϵ_2 and ϵ_3 are overall relative efficiencies for detecting corresponding gamma and X-rays [8] determined with respect to the efficiency of detecting 98.441-keV X-rays, supposing a linear dependance on their energy of the overall efficiency of detecting gamma and X-rays in this narrow region.

The last group of ratios was used as a measure of $^{235}U/^{238}U$ isotopic ratio and of $^{235}U/U_{tot}$ enrichment ratio. Then the dependence between these and the corresponding ratios from mass-spectrometric measurements was obtained using the least-squares method. The following equations were established:

$$E = 0.0147 + 14.3726\ R \text{ (linear fit)} \quad (1)$$

$$E = 0.1 + 13.398\ R + 2.513\ R^2 \text{ (parabolic fit)} \quad (2)$$

Some of the data were considered as standard data – the rest as measured sample data.

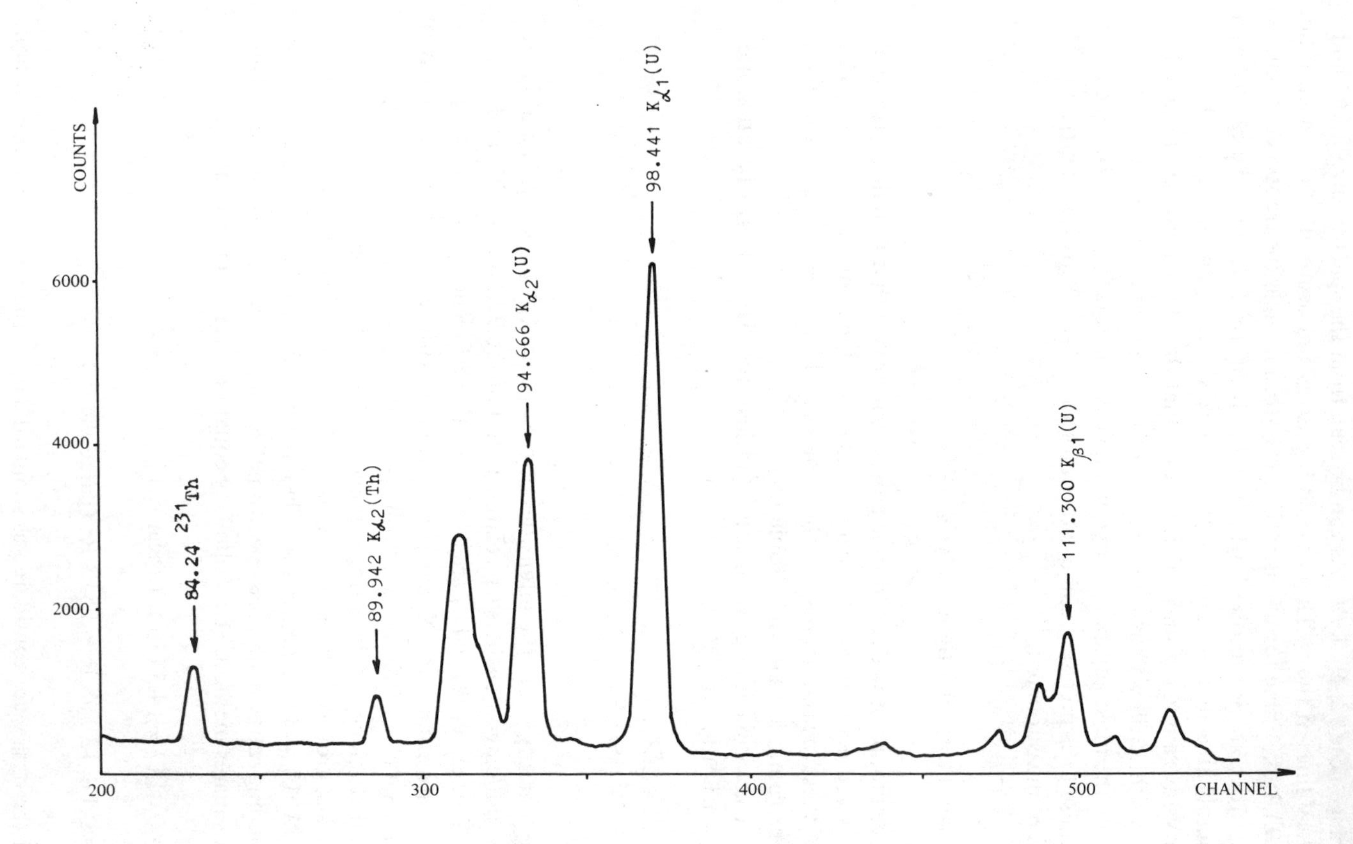

FIG.1. Instrumental low-energy spectra of UO_2 pellets.

TABLE I. RESULTS OF THE INTENSITY RATIOS MEASUREMENTS IN THE 94–98 keV RANGE

E enrichment (%)	$\frac{I_{K_{\alpha_2}}(U)}{I_{K_{\alpha_1}}(U)}$		$R = \frac{I_{93.35}}{I_{92.367} + I_{92.792}}$		$R_\epsilon = \frac{(I/\epsilon)_{93.35}}{(I/\epsilon)_{92.367} + (I/\epsilon)_{92.792}}$	
	Value	± Stand. dev.	Value	± Stand. dev.	Value	± Stand. dev.
1.80	0.594	0.005	0.1253	0.0015	0.1239	0.0013
2.27	0.583	0.006	0.1603	0.0043	0.1584	0.0051
2.38	0.596	0.003	0.1674	0.0029	0.1657	0.0028
2.55	0.588	0.004	0.1794	0.0058	0.1771	0.0058
3.04	0.586	0.007	0.2135	0.0045	0.2104	0.0040
3.19	0.594	0.004	0.2236	0.0017	0.2212	0.0019
3.96	0.597	0.003	0.2768	0.0015	0.2741	0.0018

TABLE II. COMPARISON OF MASS-SPECTROMETRIC AND GAMMA-SPECTROMETRIC DETERMINATIONS OF THE ENRICHMENT OF DIFFERENT U-CONTAINING SAMPLES

E (%)	Experiment		Parabolic fit: $Y = a_0 + a_1X + a_2X^2$				Linear fit: $Y = a_0 + a_1X$			
	R_ϵ	$\pm \Delta R_\epsilon$	R_ϵ	$\pm \Delta R_\epsilon$	E	$\pm \Delta E$	R_ϵ^a	$\pm \Delta R_\epsilon$	E^b	$\pm \Delta E$
1.80	0.1239	0.0013	0.1240	0.0013	1.80	0.02	0.1242	0.0012	1.80	0.01
2.27	0.1584	0.0051	0.1573	0.0012	2.29	0.02	0.1569	0.0009	2.29	0.01
2.38	0.1657	0.0028	0.1650	0.0013	2.39	0.02	0.1646	0.0009	2.40	0.01
2.55	0.1771	0.0058	0.1770	0.0015	2.55	0.02	0.1764	0.0008	2.56	0.01
3.04	0.2104	0.0040	0.2111	0.0015	3.03	0.02	0.2105	0.0009	3.04	0.01
3.19	0.2212	0.0019	0.2214	0.0015	3.19	0.02	0.2209	0.0010	3.19	0.01
3.96	0.2741	0.0018	0.2740	0.0018	3.96	0.02	0.2745	0.0015	3.95	0.02

$R_\epsilon = -00699 + 0.07428\,E - 0.00084\,E^2$

$E = 0.100 + 13.398\,R + 2.513\,R_\epsilon^2$

[a] $R_\epsilon = -0.00097 + 0.06955\,E$

[b] $E = 0.0147 + 14.3726\,R$

TABLE III. RESULTS OF THE MEASUREMENTS OF A SAMPLE WHILE DIFFERENT ABSORBERS ARE INSERTED BETWEEN THE SAMPLE AND THE DETECTOR

D thickness X 0.65 mm	$\frac{I_{K_{\alpha_1}}(U)}{I_{K_{\alpha_2}}(U)}$	$\frac{I_{K_{\alpha_1}}(U)}{I_{92.367}+I_{92.792}}$	$\frac{I_{93.35}}{I_{92.367}+I_{92.792}}$
0	1.674	1.765	0.2150
1	1.720	1.832	0.2165
2	1.747	1.928	0.2108
3	1.793	2.015	0.2164
4	1.819	2.070	0.2150
6	1.914	2.233	0.2173
7	1.935	2.286	0.2181
			Mean 0.2141
			Stand. dev. ± 0.0034

$$\frac{I_{K_{\alpha_1}}(U)}{I_{K_{\alpha_2}}(U)} = 1.6761 + 0.0378\,D + 0.0024\,D^2$$

$$\frac{I_{K_{\alpha_1}}(U)}{I_{92.367}+I_{92.792}} = 1.7590 + 0.0864\,D - 0.0015\,D^2$$

The results of the isotopic ratios determined through gamma-spectrometric measurements are shown in Table II. As can be seen from the Table the agreement between mass-spectrometric and gamma-spectrometric measurements is quite good. Differences with respect to the value are less than 1%, averaging 0.5%.

To investigate the influence of the container wall, special measurements were carried out. One of the samples was measured several times with different absorbers between it and the detector. The results of the measurements and the analysis are given in Table III. As was expected with the increase of the thickness there was a small increase of the ratios

$$\frac{I_3 \cdot \epsilon_1 \cdot \epsilon_2}{(I_1 \cdot \epsilon_2 + I_2 \cdot \epsilon_1) \cdot \epsilon_3}$$

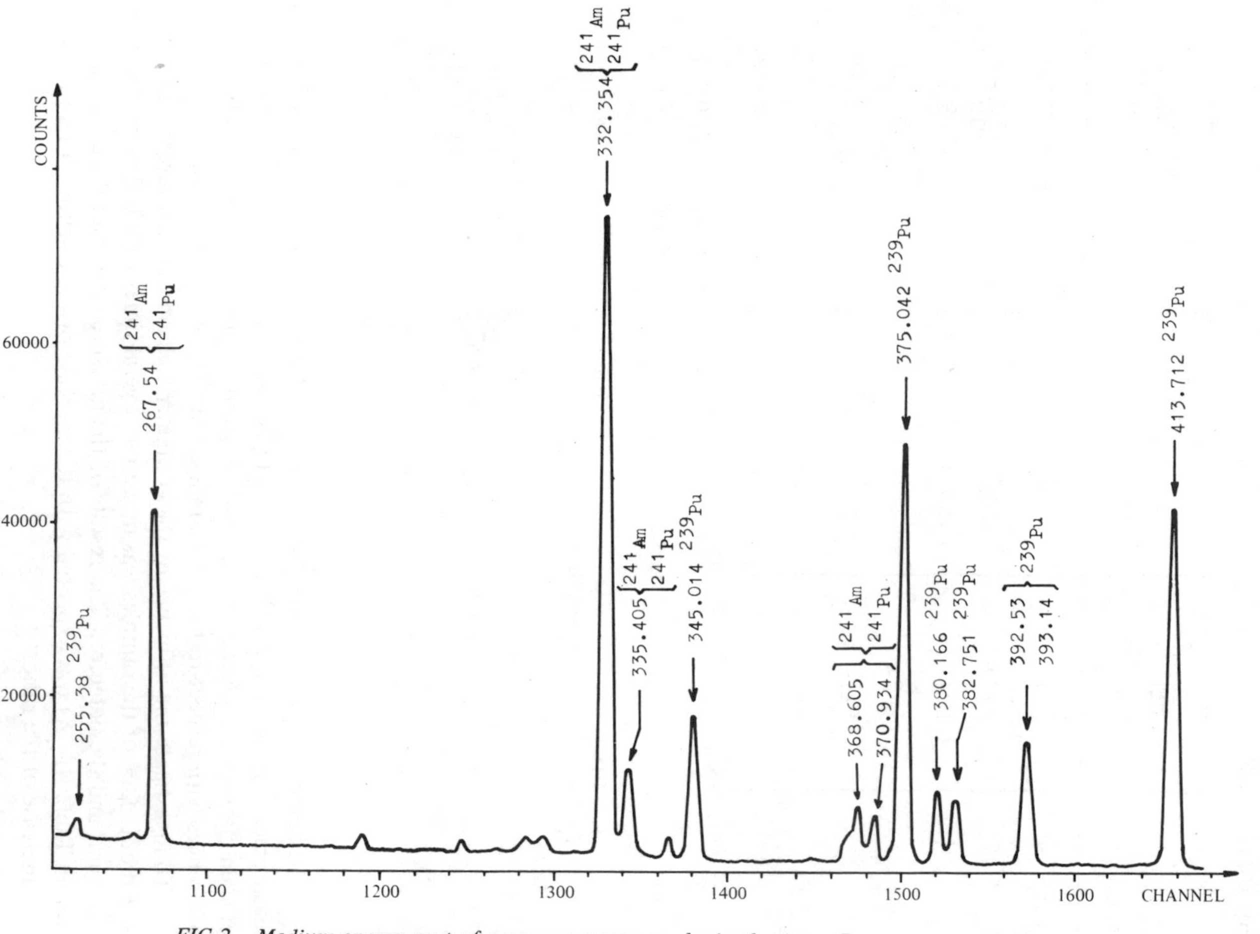

FIG.2. Medium-energy part of a gamma spectrum obtained using a Ge spectrometric system from a mixed (UO_2-PuO_2) fuel element.

TABLE IV. MASS-SPECTROMETRIC AND GAMMA-SPECTROMETRIC RESULTS ON Pu-Am ISOTOPIC RATIOS DETERMINED USING THE 145–208 AND 250–420 keV RANGES

Isotopic ratios	Sample No. 515		Sample No. 525		Sample No. 549	
	Known	Measured	Known	Measured	Known	Measured
$^{238}Pu/^{239}Pu$	0.00201	0.00174 0.00194[a]	0.00190	0.00178 0.00199[a]	0.00151	0.00134 0.00149[a]
$^{240}Pu/^{239}Pu$	0.2814	0.2804	0.2952	0.3019	0.2490	0.2498
$^{241}Pu/^{239}Pu$	0.0452	0.0456	0.0507	0.0508	0.0406	0.0415
$^{241}Am/^{241}Pu$		0.115		0.0491		0.115

[a] The values obtained on the basis of an "empirical" calibration.

but up to 4.5 mm thick it is possible to neglect them. Then, using the correlated changes of $U_{K_{\alpha_1}}/U_{K_{\alpha_2}}$ it is possible to introduce a quantitative correction as was done for the concentration measurements [9]. However, as most of the measured samples will in general have cladding or absorber thinner than 4.5 mm, this correction will not so often be used.

2. MEASUREMENTS OF Pu AND ^{241}Am ISOTOPIC RATIOS AND Pu ISOTOPIC COMPOSITION

There are several gamma-spectrometric methods for measuring Pu and ^{241}Am isotopic ratios [8, 10–13]. As Pu usually has five isotopes with significant concentrations and ^{241}Am is generally present, these measurements are usually more difficult than U isotopic ratio measurements. However, as all these isotopes, with the exception of ^{242}Pu, emit several intensive gamma lines, an accurate gamma-spectrometric determination of the isotopic ratio can be done.

TABLE V. COMPARISON OF THE YIELDS OF DIFFERENT ISOTOPES IN THE 94.105 AND 140–210 keV RANGES

Isotope	Energy (keV)	Yields ($(\gamma/\text{dis}) \times 10^6$)
^{238}Pu	99.864 152.68	72.4 9.56
^{239}Pu	98.441 203.537	67.6 5.60
^{240}Pu	104.244 160.28	69.8 4.02
^{241}Pu	101.066 148.567	6.20 1.87
^{241}Am	102.966 146.557	195.0 4.61

The methods proposed in Refs [10–12] are mainly for laboratory applications for small samples and, in some cases [11, 12], it is essential to separate Am chemically. In addition, the methods require quite a large minicomputer, which is not convenient for travelling inspectors. A method particularly suitable for IAEA application was proposed in Ref. [8]. To use it in the field, however, automated data-processing is required, which is quite possible, even with small programmable desk calculators such as the HP9815A/S.

The aim of this work is that, by using the significantly greater computing power of the present handy programmable calculators (e.g. TI 59), and presenting a new approach to the gamma-spectrometric data processing, which we recently have developed [14], the possibilities of the NDQM for large Pu samples will be further improved. Two new approaches were tested.

2.1. Measurements using the medium energy part of the Pu spectrum

The basic ideas used in Ref. [8] were further developed. The instrumental Pu gamma spectrum of a medium energy range is shown in Fig.2. Using the new approach to the gamma-spectrometric data treatment it was possible to resolve the two groups of lines in the 332.567 and 335.405 keV range, which are the most suitable for ^{241}Pu/^{239}Pu and ^{241}Am/^{239}Pu isotopic ratio measurements. Then,

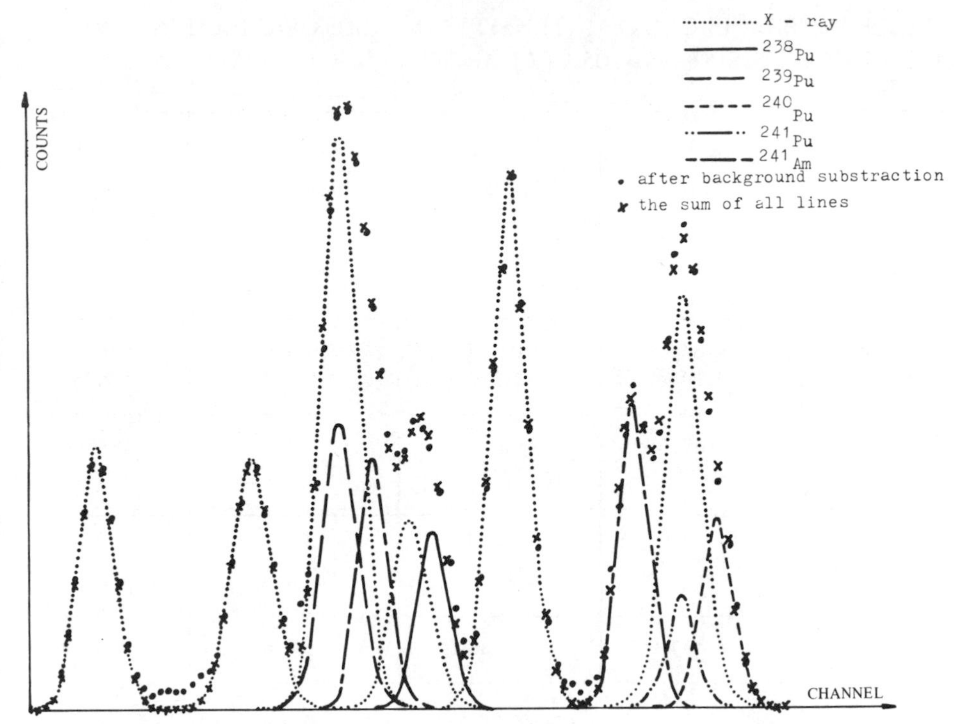

FIG.3. 94–105 keV part of a Pu spectrum, after background subtraction, and the lines of different isotopes.

using the ratio of ^{241}Am/^{241}Pu thus determined, the overall relative efficiency curve in the 148–208 keV region was determined. The nearest 152.68 keV ^{238}Pu line and 148.567 keV ^{241}Pu line were used to determine the ^{238}Pu/^{241}Pu isotopic ratio, and the 160.28 keV ^{240}Pu line and the 164.58 keV ^{241}Pu line to determine the ^{240}Pu/^{241}Pu isotopic ratio. The results from the measurements of three different samples are shown in Table IV. The main advantage of this approach is that the penetrability of the gamma rays used is significantly higher than the penetrability of the low-energy gamma rays in the 98–105 keV region. The main problem is the relatively low yields of 152.68 keV ^{238}Pu gamma rays and 160.28 keV ^{240}Pu gamma rays compared with the 98–105 keV region, which makes it rather difficult to obtain precise results, particularly for samples with low ^{238}Pu and ^{240}Pu concentrations.

2.2. Measurements of Pu-Am isotopic ratios using the 94.105 keV and 148–208 keV parts of the spectrum

The 94–105 keV range of the Pu-Am spectrum is in many cases the optimal one for precise ^{238}Pu, ^{240}Pu and ^{241}Am determinations with respect to ^{241}Pu

TABLE VI. PRECISIONS OF INTENSITY AND ISOTOPIC RATIOS DETERMINED USING 94–105 keV RANGE

Intensity ratio	$\frac{I_{102.966}}{I_{101.066}}$	$\frac{I_{98.441}}{I_{101.066}}$	$\frac{I_{99.864}}{I_{101.066}}$	$\frac{I_{104.244}}{I_{101.066}}$
Average value	0.554	1.214	0.322	0.327
Standard deviation	± 0.006 1.10%	± 0.017 1.39%	± 0.009 2.85%	± 0.016 4.78%
Standard error	± 0.0015 0.28%	± 0.009 0.35%	± 0.002 0.71%	± 0.004 1.20%
Isotopic ratios	$\frac{^{241}Am}{^{241}Pu}$	$\frac{^{238}Pu}{^{241}Pu}$	$\frac{^{239}Pu}{^{241}Pu}$	$\frac{^{240}Pu}{^{241}Pu}$
Value	0.497	0.186	108.4	11.77

because in this range all these isotopes emit intensive gamma and X-ray lines. ^{239}Pu also emits gamma and X-rays but, for normal reactor-grade Pu, its lines are weaker and its precise measurements are possible mainly for weapon-grade materials. As the penetrability of gamma rays with such energies is sufficient for non-destructive measurements, this energy range should always be considered for ^{238}Pu and ^{240}Pu determinations (see Table V and Fig.3). The main problem in using the 94–105 keV range is how to resolve the groups of many closed lines of different isotopes which are not resolved by the best semiconductor spectrometers. These lines are grouped in two five-line peaks, 98.441 – 99.864 keV and 102.966–104.244 keV. An additional difficulty in resolving these lines arises from the fact that several of these lines are those of X-rays and consequently have a different shape and width compared with gamma-ray lines of the same energy region. As our efforts were directed towards solving this problem by using small powerful programmable calculators we are now assuming X-ray lines also as Gaussian, but with a larger width. Later it will probably be possible to use the shape of the X-ray lines [7]. However, it should be indicated that if suitable samples with different isotopic compositions are used for calibrating the measurements then this will partly correct for assuming the X-ray lines to be Gaussian. Using only this simplification, an algorithm and programs for the TI-59 programmabl calculator were developed to resolve and determine the areas of all gamma and

X-ray lines in both the unresolved peaks. An artificial test was designed and performed in order to investigate the program performance. It has been demonstrated that the areas of lines can be determined with high precision when the distance between the neighbouring lines is at least equal to 1σ, and σ is equal or larger than two channels. When the distance between two lines is smaller than two channels the areas of the two lines have to be considered together for better precision. Individual line areas in this case may not be quite precise. Using this possibility to resolve the lines of the unresolved peaks the following procedure was developed for isotopic ratio measurements. In addition to the two multiline peaks there are also five well-resolved X-ray lines: 94.660, 110.421, 111.300 keV lines of U K X-rays, and 97.071 and 101.066 keV lines of Np K X-rays. The 94.660, 97.071 and 101.066 keV lines are used to determine the lines' parameters, energy self-calibration of the measurements and, together with the two other lines, for overall relative detecting efficiency calibration.

Using this overall relative detecting efficiency calibration, the activity and isotopic ratios of the following isotopes were determined with respect to ^{241}Pu through its dominating activity in the 101.066 keV line:

(i) ^{241}Am/^{241}Pu, using 102.966 keV line of ^{241}Am

(ii) ^{238}Pu/^{241}Pu, using 99.864 keV line of ^{238}Pu

(iii) ^{240}Pu/^{241}Pu, using 104.244 keV line of ^{240}Pu

(iv) ^{239}Pu/^{241}Pu, using 98.441 keV U K_{α_1} line with a significant contribution of ^{239}Pu activity.

When the ^{239}Pu participation in the 98.441 keV line is small (less than 30%), then the couple of 203.534–208 keV lines has to be used. In this case the overall relative detecting efficiency curve has to be determined in the 148–208 keV range. The ^{241}Am/^{241}Pu isotopic ratio already determined in the 94–105 keV range is used to determine ^{241}Pu participation in the 164.58 and 208 keV lines and, on this basis, the ORDE[1] curve in the 148–208 keV range is determined. Finally the ^{239}Pu/^{241}Pu ratio is determined by using the 203.534 keV line of ^{239}Pu and the ^{241}Pu-dominated portion of the 208 keV line. To investigate the precision of the isotopic ratio determinations, by using this procedure and these programs, one of the samples was measured sixteen times in the course of several days and spectra processed by a TI-59 programmable calculator were obtained. The results of these measurements and data-processing are shown in Table VI.

It is necessary to note that the spectrometric system used was drifting during the long time measurements. As the measured sample was small and in a heavy container, at a distance from the detector the measurement time was rather long

---

[1] ORDE = Overall relative detecting efficiency.

which had a somewhat adverse effect on the results. In addition, the procedure should still be optimized. Nevertheless, the precision of the measurements, as can be seen from Table VI, is quite good. It is also necessary to indicate that, for this type of measurement, a gamma-spectrometric system with high resolving power in the 90–210 keV energy range is required.

With all these isotopic ratios and by using the isotopic correlation technique, it is also possible to determine the ^{242}Pu concentration and finally to determine the isotopic composition of Pu. Even if the ^{242}Pu determination is not very accurate this will not significantly influence the concentration of other isotopes because ^{242}Pu concentration is always small [8].

REFERENCES

[1] REILLY, T.D., WALTON, R.B., PARKER, J.L., Los Alamos Rep. LA-4605-MS (1970).
[2] BEETS, C., COENS, J., DRAGNEV, T., GOOSENS, H., MOSTIN, N., in Peaceful Uses of Atomic Energy (Proc. 4th Int. Conf. Geneva, 1971) **9**, IAEA, Vienna (1972) 449.
[3] HARRY, R., AALDIJK, J., BRAAK, J., in The Safeguarding of Nuclear Materials (Proc. Symp. Vienna, 1975) **2**, IAEA, Vienna (1976) 235.
[4] SAMPSON, T.E., Nucl. Instrum. Methods **111** (1973) 209.
[5] WILKINSON, D.L., Nucl. Instrum. Methods **95** (1971) 259.
[6] ROBERTS, B.L., RIDDLE, R.A.J., SQUIER, G.T., Nucl. Instrum. Methods **130** (1975) 559.
[7] GUNNINK, R., Rep. UCRL 78707 (1975).
[8] DRAGNEV, T.N., J. Radioanal. Chem. **36** (1977) 401.
[9] DRAGNEV, T.N., DAMJANOV, B.P., KARAMANOVA, J.S., IAEA-SM-231/131, these Proceedings, Vol. II.
[10] GUNNINK, R., TINNEY, J.F., Rep. UCRL-73274 (1971).
[11] GUNNINK, R., Rep. UCRL-75 (1973) 105.
[12] UMEZAWA, H., SUZUKI, T., ICHIKAWA, S., J. Nucl. Sci. Technol. **13** 6 (1976) 327–32.
[13] DRAGNEV, T., SCHARF, K., Int. J. Appl. Radiat. Isot. **26** (1975) 125.
[14] DAMJANOV, B., et al., to be published.

DISCUSSION

H. OTTMAR: Your approach for measuring ^{235}U enrichment using gamma rays in the 92 keV region is interesting in that it permits enrichment analysis without reference to physical standards. However, since this approach relies on the assumption that ^{238}U is in equilibrium with its daughter product ^{234}Th, it is only applicable to those samples where this equilibrium has been established. Otherwise your ^{235}U enrichment analysis might be quite erroneous.

T.N. DRAGNEV: I agree. More details about the cases where the equilibrium is disturbed are given in paper IAEA-SM-231/131.[1]

W.C. BARTELS: You mentioned the work of your colleague in this field, Mr. Gunnink of the Lawrence Livermore Laboratory in the United States of America. At a topical meeting of the American Nuclear Society last spring, Gunnink reported that he had obtained a combined precision and accuracy of two or three per cent after long counting times on the ^{238}Pu content of samples containing 0.01 wt%. What figure did you obtain?

T.N. DRAGNEV: As indicated in our paper, it is difficult to obtain high precision in the gamma-spectrometric determination of $^{238}Pu/^{241}Pu$ ratios using 152.68 keV gamma rays of ^{238}Pu, because the branching ratio of this line is low, particularly for solid samples with low ^{238}Pu concentration. The branching ratio of the 99.864 keV line is about 7.6 times greater and the precision attainable using this energy range should be better than 1%. However, we do not have enough data at present, particularly for samples with ^{238}Pu concentrations as low as 0.01 wt%.

H. OTTMAR: I would like to comment on this problem of ^{238}Pu abundance measurements. In reactor-grade materials, the abundance of this isotope is quite easily measured by gamma spectrometry with a precision of 1% or even better. As regards the accuracy of the measurements, the major problem at present is the lack of suitable plutonium-isotope standard reference materials, which are necessary for testing the gamma measurements in terms of absolute values.

[1] DRAGHEV, T.N. et al., these Proceedings, Vol. II.

CHAIRMEN OF SESSIONS

Session I	H.W. SCHLEICHER	Commission of the European Communities (CEC)
Session II	A. BURTSCHER	Austria
Session III	D. GUPTA	Federal Republic of Germany
Session IV	C. CASTILLO-CRUZ	Mexico
Session V	A.G. HAMLIN	United Kingdom
Session VI	V.M. GRYAZEV	Union of Soviet Socialist Republics
Session VII	H. KRINNINGER	Federal Republic of Germany
Session VIII	G. Robert KEEPIN	United States of America
Session IX	A. PETIT	France
Session X	A. VON BAECKMANN	International Atomic Energy Agency (IAEA)

SECRETARIAT

Scientific Secretary	J.E. LOVETT	Department of Safeguards, IAEA
Administrative Secretary	Edith PILLER	Division of External Relations, IAEA
Editor	Monica KRIPPNER	Division of Publications, IAEA
Records Officer	D.J. MITCHELL	Division of Languages, IAEA

The following conversion table is provided for the convenience of readers and to encourage the use of SI units.

FACTORS FOR CONVERTING SOME OT THE MORE COMMON UNITS TO INTERNATIONAL SYSTEM OF UNITS (SI) EQUIVALENTS

NOTES:

(1) SI base units are the metre (m), kilogram (kg), second (s), ampere (A), kelvin (K), candela (cd) and mole (mol).

(2) ▶ indicates SI derived units and those accepted for use with SI;
▷ indicates additional units accepted for use with SI for a limited time.
[*For further information see The International System of Units (SI), 1977 ed., published in English by HMSO, London, and National Bureau of Standards, Washington, DC, and International Standards ISO-1000 and the several parts of ISO-31 published by ISO, Geneva.*]

(3) The correct abbreviation for the unit in column 1 is given in column 2.

(4) * indicates conversion factors given exactly; other factors are given rounded, mostly to 4 significant figures.
≡ indicates a definition of an SI derived unit: [ ] in column 3+4 enclose factors given for the sake of completeness.

Column 1 *Multiply data given in:*	Column 2	Column 3 *by:*	Column 4 *to obtain data in:*	
Radiation units				
▶ becquerel	1 Bq	(has dimensions of s^{-1})		
disintegrations per second (= dis/s)	1 s^{-1}	≡ 1.00×10^{0}	Bq	*
▷ curie	1 Ci	= 3.70×10^{10}	Bq	*
▷ roentgen	1 R	[= 2.58×10^{-4}	C/kg]	*
▶ gray	1 Gy	[≡ 1.00×10^{0}	J/kg]	*
▷ rad	1 rad	= 1.00×10^{-2}	Gy	*
sievert *(radiation protection only)*	1 Sv	[= 1.00×10^{0}	J/kg]	*
rem *(radiation protection only)*	1 rem	[= 1.00×10^{-2}	J/kg]	*
Mass				
▶ unified atomic mass unit ($\frac{1}{12}$ of the mass of ^{12}C)	1 u	[= $1.660\,57 \times 10^{-27}$	kg, approx.]	
▶ tonne (= metric ton)	1 t	[= 1.00×10^{3}	kg]	*
pound mass (avoirdupois)	1 lbm	= 4.536×10^{-1}	kg	
ounce mass (avoirdupois)	1 ozm	= 2.835×10^{1}	g	
ton (long) (= 2240 lbm)	1 ton	= 1.016×10^{3}	kg	
ton (short) (= 2000 lbm)	1 short ton	= 9.072×10^{2}	kg	
Length				
statute mile	1 mile	= 1.609×10^{0}	km	
nautical mile (international)	1 n mile	= 1.852×10^{0}	km	*
yard	1 yd	= 9.144×10^{-1}	m	*
foot	1 ft	= 3.048×10^{-1}	m	*
inch	1 in	= 2.54×10^{1}	mm	*
mil (= 10^{-3} in)	1 mil	= 2.54×10^{-2}	mm	*
Area				
▷ hectare	1 ha	[= 1.00×10^{4}	m^2]	*
▷ barn *(effective cross-section, nuclear physics)*	1 b	[= 1.00×10^{-28}	m^2]	*
square mile, (statute mile)^2	1 $mile^2$	= 2.590×10^{0}	km^2	
acre	1 acre	= 4.047×10^{3}	m^2	
square yard	1 yd^2	= 8.361×10^{-1}	m^2	
square foot	1 ft^2	= 9.290×10^{-2}	m^2	
square inch	1 in^2	= 6.452×10^{2}	mm^2	
Volume				
▶ litre	1 l *or* 1 ltr	[= 1.00×10^{-3}	m^3]	*
cubic yard	1 yd^3	= 7.646×10^{-1}	m^3	
cubic foot	1 ft^3	= 2.832×10^{-2}	m^3	
cubic inch	1 in^3	= 1.639×10^{4}	mm^3	
gallon (imperial)	1 gal (UK)	= 4.546×10^{-3}	m^3	
gallon (US liquid)	1 gal (US)	= 3.785×10^{-3}	m^3	
Velocity, acceleration				
foot per second (= fps)	1 ft/s	= 3.048×10^{-1}	m/s	*
foot per minute	1 ft/min	= 5.08×10^{-3}	m/s	*
mile per hour (= mph)	1 mile/h	= { 4.470×10^{-1} 1.609×10^{0}	m/s km/h	
▷ knot (international)	1 knot	= 1.852×10^{0}	km/h	*
free fall, standard, g		= 9.807×10^{0}	m/s^2	
foot per second squared	1 ft/s^2	= 3.048×10^{-1}	m/s^2	*

This table has been prepared by E.R.A. Beck for use by the Division of Publications of the IAEA. While every effort has been made to ensure accuracy, the Agency cannot be held responsible for errors arising from the use of this table.

Column 1 *Multiply data given in:*	Column 2	Column 3 *by:*	Column 4 *to obtain data in:*
Density, volumetric rate			
pound mass per cubic inch	1 lbm/in^3	= 2.768×10^4	kg/m^3
pound mass per cubic foot	1 lbm/ft^3	= 1.602×10^1	kg/m^3
cubic feet per second	1 ft^3/s	= 2.832×10^{-2}	m^3/s
cubic feet per minute	1 ft^3/min	= 4.719×10^{-4}	m^3/s
Force			
▶ newton	1 N	[≡ 1.00×10^0	$m \cdot kg \cdot s^{-2}$] *
dyne	1 dyn	= 1.00×10^{-5}	N *
kilogram force (= kilopond (kp))	1 kgf	= 9.807×10^0	N
poundal	1 pdl	= 1.383×10^{-1}	N
pound force (avoirdupois)	1 lbf	= 4.448×10^0	N
ounce force (avoirdupois)	1 ozf	= 2.780×10^{-1}	N
Pressure, stress			
▶ pascal	1 Pa	[≡ 1.00×10^0	N/m^2] *
▷ atmosphere[a], standard	1 atm	= $1.013\,25 \times 10^5$	Pa *
▷ bar	1 bar	= 1.00×10^5	Pa *
centimetres of mercury (0°C)	1 cmHg	= 1.333×10^3	Pa
dyne per square centimetre	1 dyn/cm^2	= 1.00×10^{-1}	Pa *
feet of water (4°C)	1 ftH_2O	= 2.989×10^3	Pa
inches of mercury (0°C)	1 inHg	= 3.386×10^3	Pa
inches of water (4°C)	1 inH_2O	= 2.491×10^2	Pa
kilogram force per square centimetre	1 kgf/cm^2	= 9.807×10^4	Pa
pound force per square foot	1 lbf/ft^2	= 4.788×10^1	Pa
pound force per square inch (= psi)[b]	1 lbf/in^2	= 6.895×10^3	Pa
torr (0°C) (= mmHg)	1 torr	= 1.333×10^2	Pa
Energy, work, quantity of heat			
▶ joule (≡ W·s)	1 J	[≡ 1.00×10^0	N·m] *
▶ electronvolt	1 eV	[= $1.602\,19 \times 10^{-19}$	J, approx.]
British thermal unit (International Table)	1 Btu	= 1.055×10^3	J
calorie (thermochemical)	1 cal	= 4.184×10^0	J *
calorie (International Table)	1 cal_{IT}	= 4.187×10^0	J
erg	1 erg	= 1.00×10^{-7}	J *
foot-pound force	1 ft·lbf	= 1.356×10^0	J
kilowatt-hour	1 kW·h	= 3.60×10^6	J *
kiloton explosive yield (PNE) (≡ 10^{12} g-cal)	1 kt yield	≃ 4.2×10^{12}	J
Power, radiant flux			
▶ watt	1 W	[≡ 1.00×10^0	J/s] *
British thermal unit (International Table) per second	1 Btu/s	= 1.055×10^3	W
calorie (International Table) per second	1 cal_{IT}/s	= 4.187×10^0	W
foot-pound force/second	1 ft·lbf/s	= 1.356×10^0	W
horsepower (electric)	1 hp	= 7.46×10^2	W *
horsepower (metric) (= ps)	1 ps	= 7.355×10^2	W
horsepower (550 ft·lbf/s)	1 hp	= 7.457×10^2	W
Temperature			
▶ temperature in degrees Celsius, t where T is the thermodynamic temperature in kelvin and T_0 is defined as 273.15 K	$t = T - T_0$		
degree Fahrenheit	$t_{°F} - 32$	} $\times \left(\frac{5}{9}\right)$ gives	t *(in degrees Celsius)* *
degree Rankine	$T_{°R}$		T *(in kelvin)* *
degrees of temperature difference[c]	$\Delta T_{°R}$ (= $\Delta t_{°F}$)		ΔT (= Δt) *
Thermal conductivity[c]			
1 $Btu \cdot in/(ft^2 \cdot s \cdot °F)$ *(International Table Btu)*		= 5.192×10^2	$W \cdot m^{-1} \cdot K^{-1}$
1 $Btu/(ft \cdot s \cdot °F)$ *(International Table Btu)*		= 6.231×10^3	$W \cdot m^{-1} \cdot K^{-1}$
1 $cal_{IT}/(cm \cdot s \cdot °C)$		= 4.187×10^2	$W \cdot m^{-1} \cdot K^{-1}$

[a] atm abs, ata: atmospheres absolute;
atm (g), atü: atmospheres gauge.

[b] lbf/in^2 (g) (= psig): gauge pressure;
lbf/in^2 abs (= psia): absolute pressure.

[c] The abbreviation for temperature difference, deg (= degK = degC), is no longer acceptable as an SI unit.

HOW TO ORDER IAEA PUBLICATIONS

An exclusive sales agent for IAEA publications, to whom all orders and inquiries should be addressed, has been appointed in the following country:

UNITED STATES OF AMERICA UNIPUB, 345 Park Avenue South, New York, NY 10010

In the following countries IAEA publications may be purchased from the sales agents or booksellers listed or through your major local booksellers. Payment can be made in local currency or with UNESCO coupons.

ARGENTINA	Comisión Nacional de Energía Atomica, Avenida del Libertador 8250, RA-1429 Buenos Aires
AUSTRALIA	Hunter Publications, 58 A Gipps Street, Collingwood, Victoria 3066
BELGIUM	Service du Courrier de l'UNESCO, 202, Avenue du Roi, B-1060 Brussels
CZECHOSLOVAKIA	S.N.T.L., Spálená 51, CS-113 02 Prague 1 Alfa, Publishers, Hurbanovo námestie 6, CS-893 31 Bratislava
FRANCE	Office International de Documentation et Librairie, 48, rue Gay-Lussac, F-75240 Paris Cedex 05
HUNGARY	Kultura, Hungarian Trading Company for Books and Newspapers, P.O. Box 149, H-1389 Budapest 62
INDIA	Oxford Book and Stationery Co., 17, Park Street, Calcutta-700 016 Oxford Book and Stationery Co., Scindia House, New Delhi-110 001
ISRAEL	Heiliger and Co., Ltd., Scientific and Medical Books, 3, Nathan Strauss Street, Jerusalem
ITALY	Libreria Scientifica, Dott. Lucio de Biasio "aeiou", Via Meravigli 16, I-20123 Milan
JAPAN	Maruzen Company, Ltd., P.O. Box 5050, 100-31 Tokyo International
NETHERLANDS	Martinus Nijhoff B.V., Booksellers, Lange Voorhout 9-11, P.O. Box 269, NL-2501 The Hague
PAKISTAN	Mirza Book Agency, 65, Shahrah Quaid-e-Azam, P.O. Box 729, Lahore 3
POLAND	Ars Polona-Ruch, Centrala Handlu Zagranicznego, Krakowskie Przedmiescie 7, PL-00-068 Warsaw
ROMANIA	Ilexim, P.O. Box 136-137, Bucarest
SOUTH AFRICA	Van Schaik's Bookstore (Pty) Ltd., Libri Building, Church Street, P.O. Box 724, Pretoria 0001
SPAIN	Diaz de Santos, Lagasca 95, Madrid-6 Diaz de Santos, Balmes 417, Barcelona-6
SWEDEN	AB C.E. Fritzes Kungl. Hovbokhandel, Fredsgatan 2, P.O. Box 16356, S-103 27 Stockholm
UNITED KINGDOM	Her Majesty's Stationery Office, P.O. Box 569, London SE 1 9NH
U.S.S.R.	Mezhdunarodnaya Kniga, Smolenskaya-Sennaya 32-34, Moscow G-200
YUGOSLAVIA	Jugoslovenska Knjiga, Terazije 27, P.O. Box 36, YU-11001 Belgrade

---

Orders from countries where sales agents have not yet been appointed and requests for information should be addressed directly to:

Division of Publications
International Atomic Energy Agency
Kärntner Ring 11, P.O. Box 590, A-1011 Vienna, Austria

HOW TO ORDER IAEA PUBLICATIONS